Introduction to Linear Algebra for Science and Engineering

Daniel Norman **Dan Wolczuk**
University of Waterloo

Illustrated by
Michael A. LaCroix
University of Waterloo

Second Edition

Toronto

Vice-President, Editorial Director: Gary Bennett
Executive Acquisitions Editor: Cathleen Sullivan
Senior Marketing Manager: Michelle Bish
Developmental Editor: John Lewis
Project Manager: Sarah Lukaweski
Manufacturing Manager: Susan Johnston
Production Editor: Denise Botelho, Element LLC
Copy Editor: Ashley Thomason
Proofreader: Denne Wesolowski
Compositor: Thomson Digital India
Art Director: Julia Hall
Cover and Interior Designer: Anthony Leung
Cover Image: Shutterstock

Credits and acknowledgments for material borrowed from other sources and reproduced, with permission, in this textbook appear on the appropriate page within text.

10 9 8 7 6 5 4 3 2 1 EB

Library and Archives Canada Cataloguing in Publication

Norman, Daniel, 1938–
Introduction to linear algebra for science and engineering / Daniel Norman, Dan Wolczuk.—2nd ed.

Includes index.
ISBN 978-0-321-74896-6

1. Algebras, Linear. I. Wolczuk, Dan. II. Title.

QA184.N67 2012 512'.5 C2011-905480-9

PEARSON

ISBN 978-0-321-74896-6

Contents

A Note to Students

Linear Algebra—What Is It?

Linear algebra is essentially the study of vectors, matrices, and linear mappings. Although many pieces of linear algebra have been studied for many centuries, it did not take its current form until the mid-twentieth century. It is now an extremely important topic in mathematics because of its application to many different areas.

Most people who have learned linear algebra and calculus believe that the ideas of elementary calculus (such as limit and integral) are more difficult than those of introductory linear algebra and that most problems in calculus courses are harder than those in linear algebra courses. So, at least by this comparison, linear algebra is not hard. Still, some students find learning linear algebra difficult. I think two factors contribute to the difficulty students have.

First, students do not see what linear algebra is good for. This is why it is important to read the applications in the text; even if you do not understand them completely, they will give you some sense of where linear algebra fits into the broader picture.

Second, some students mistakenly see mathematics as a collection of recipes for solving standard problems and are uncomfortable with the fact that linear algebra is "abstract" and includes a lot of "theory." There will be no long-term payoff in simply memorizing these recipes, however; computers carry them out far faster and more accurately than any human can. That being said, practising the procedures on specific examples is often an important step toward much more important goals: understanding the *concepts* used in linear algebra to formulate and solve problems and learning to interpret the results of calculations. Such understanding requires us to come to terms with some theory. In this text, many of our examples will be small. However, as you work through these examples, keep in mind that when you apply these ideas later, you may very well have a million variables and a million equations. For instance, Google's PageRank system uses a matrix that has 25 billion columns and 25 billion rows; you don't want to do that by hand! **When you are solving computational problems, always try to observe how your work relates to the theory you have learned.**

Mathematics is useful in so many areas because it is *abstract*: the same good idea can unlock the problems of control engineers, civil engineers, physicists, social scientists, and mathematicians only because the idea has been abstracted from a particular setting. One technique solves many problems only because someone has established a *theory* of how to deal with these kinds of problems. We use *definitions* to try to capture important ideas, and we use *theorems* to summarize useful general facts about the kind of problems we are studying. *Proofs* not only show us that a statement is true; they can help us understand the statement, give us practice using important ideas, and make it easier to learn a given subject. In particular, proofs show us how ideas are tied together so we do not have to memorize too many disconnected facts.

Many of the concepts introduced in linear algebra are natural and easy, but some may seem unnatural and "technical" to beginners. Do not avoid these apparently more difficult ideas; use examples and theorems to see how these ideas are an essential part of the story of linear algebra. By learning the "vocabulary" and "grammar" of linear algebra, you will be equipping yourself with concepts and techniques that mathematicians, engineers, and scientists find invaluable for tackling an extraordinarily rich variety of problems.

Linear Algebra—Who Needs It?

Mathematicians

Linear algebra and its applications are a subject of continuing research. Linear algebra is vital to mathematics because it provides essential ideas and tools in areas as diverse as abstract algebra, differential equations, calculus of functions of several variables, differential geometry, functional analysis, and numerical analysis.

Engineers

Suppose you become a control engineer and have to design or upgrade an automatic control system. The system may be controlling a manufacturing process or perhaps an airplane landing system. You will probably start with a linear model of the system, requiring linear algebra for its solution. To include feedback control, your system must take account of many measurements (for the example of the airplane, position, velocity, pitch, etc.), and it will have to assess this information very rapidly in order to determine the correct control responses. A standard part of such a control system is a Kalman-Bucy filter, which is not so much a piece of hardware as a piece of mathematical machinery for doing the required calculations. Linear algebra is an essential part of the Kalman-Bucy filter.

If you become a structural engineer or a mechanical engineer, you may be concerned with the problem of vibrations in structures or machinery. To understand the problem, you will have to know about eigenvalues and eigenvectors and how they determine the normal modes of oscillation. Eigenvalues and eigenvectors are some of the central topics in linear algebra.

An electrical engineer will need linear algebra to analyze circuits and systems; a civil engineer will need linear algebra to determine internal forces in static structures and to understand principal axes of strain.

In addition to these fairly specific uses, engineers will also find that they need to know linear algebra to understand systems of differential equations and some aspects of the calculus of functions of two or more variables. Moreover, the ideas and techniques of linear algebra are central to numerical techniques for solving problems of heat and fluid flow, which are major concerns in mechanical engineering. And the ideas of linear algebra underlie advanced techniques such as Laplace transforms and Fourier analysis.

Physicists

Linear algebra is important in physics, partly for the reasons described above. In addition, it is essential in applications such as the inertia tensor in general rotating motion. Linear algebra is an absolutely essential tool in quantum physics (where, for example, energy levels may be determined as eigenvalues of linear operators) and relativity (where understanding change of coordinates is one of the central issues).

Life and Social Scientists

Input/output models, described by matrices, are often used in economics, and similar ideas can be used in modelling populations where one needs to keep track of subpopulations (generations, for example, or genotypes). In all sciences, statistical analysis of data is of great importance, and much of this analysis uses linear algebra; for example, the method of least squares (for regression) can be understood in terms of projections in linear algebra.

Managers

A manager in industry will have to make decisions about the best allocation of resources: enormous amounts of computer time around the world are devoted to linear programming algorithms that solve such allocation problems. The same sorts of techniques used in these algorithms play a role in some areas of mine management. Linear algebra is essential here as well.

So who needs linear algebra? Almost every mathematician, engineer, or scientist will find linear algebra an important and useful tool.

Will these applications be explained in this book?

Unfortunately, most of these applications require too much specialized background to be included in a first-year linear algebra book. To give you an idea of how some of these concepts are applied, a few interesting applications are briefly covered in sections 1.4, 1.5, 2.4, 5.4, 6.3, 6.4, 7.3, 7.5, 8.3, 8.4, and 9.2. You will get to see many more applications of linear algebra in your future courses. In addition, an essay on linearity and superposition in physics, provided on the Companion Website, describes how linear algebra can be used in a variety of physical problems.

A Note to Instructors

Welcome to the second edition of *Introduction to Linear Algebra for Science and Engineering*. It has been a pleasure to revise Daniel Norman's first edition for a new generation of students and teachers. Over the past several years, I have read many articles and spoken to many colleagues and students about the difficulties faced by teachers and learners of linear algebra. In particular, it is well known that students typically find the computational problems easy but have great difficulty in understanding the abstract concepts and the theory. Inspired by this research, I developed a pedagogical approach that addresses the most common problems encountered when teaching and learning linear algebra. I hope that you will find this approach to teaching linear algebra as successful as I have.

Changes to the Second Edition

- Several worked-out examples have been added, as well as a variety of mid-section exercises (discussed below).
- Vectors in $\mathbb{R}^n$ are now always represented as column vectors and are denoted with the normal vector symbol $\vec{x}$. Vectors in general vector spaces are still denoted in boldface.
- Some material has been reorganized to allow students to see important concepts early and often, while also giving greater flexibility to instructors. For example, the concepts of linear independence, spanning, and bases are now introduced in Chapter 1 in $\mathbb{R}^n$, and students use these concepts in Chapters 2 and 3 so that they are very comfortable with them before being taught general vector spaces.

- The material on complex numbers has been collected and placed in Chapter 9, at the end of the text. However, if one desires, it can be distributed throughout the text appropriately.
- There is a greater emphasis on teaching the mathematical language and using mathematical notation.
- All-new figures clearly illustrate important concepts, examples, and applications.
- The text has been redesigned to improve readability.

Approach and Organization

Students typically have little trouble with computational questions, but they often struggle with abstract concepts and proofs. This is problematic because computers perform the computations in the vast majority of real-world applications of linear algebra. Human users, meanwhile, must apply the theory to transform a given problem into a linear algebra context, input the data properly, and interpret the result correctly.

The main goal of this book is to mix theory and computations throughout the course. The benefits of this approach are as follows:

- It prevents students from mistaking linear algebra as very easy and very computational early in the course and then becoming overwhelmed by abstract concepts and theories later.
- It allows important linear algebra concepts to be developed and extended more slowly.
- It encourages students to use computational problems to help understand the theory of linear algebra rather than blindly memorize algorithms.

One example of this approach is our treatment of the concepts of spanning and linear independence. They are both introduced in Section 1.2 in $\mathbb{R}^n$, where they can be motivated in a geometrical context. They are then used again for matrices in Section 3.1 and polynomials in Section 4.1, before they are finally extended to general vector spaces in Section 4.2.

The following are some other features of the text's organization:

- The idea of linear mappings is introduced early in a geometrical context and is used to explain aspects of matrix multiplication, matrix inversion, and features of systems of linear equations. Geometrical transformations provide intuitively satisfying illustrations of important concepts.
- Topics are ordered to give students a chance to work with concepts in a simpler setting before using them in a much more involved or abstract setting. For example, before reaching the definition of a vector space in Section 4.2, students will have seen the 10 vector space axioms and the concepts of linear independence and spanning for three different vector spaces, and they will have had some experience in working with bases and dimensions. Thus, instead of being bombarded with new concepts at the introduction of general vector spaces, students will just be generalizing concepts with which they are already familiar.

Pedagogical Features

Since mathematics is best learned by doing, the following pedagogical elements are included in the book.

- A selection of routine mid-section exercises is provided, with solutions included in the back of the text. These allow students to use and test their understanding of one concept before moving on to other concepts in the section.
- Practice problems are provided for students at the end of each section. See "A Note on the Exercises and Problems" below.
- Examples, theorems, and definitions are called out in the margins for easy reference.

Applications

One of the difficulties in any linear algebra course is that the applications of linear algebra are not so immediate or so intuitively appealing as those of elementary calculus. Most convincing applications of linear algebra require a fairly lengthy buildup of background that would be inappropriate in a linear algebra text. However, without some of these applications, many students would find it difficult to remain motivated to learn linear algebra. An additional difficulty is that the applications of linear algebra are so varied that there is very little agreement on which applications should be covered.

In this text we briefly discuss a few applications to give students some easy samples. Additional applications are provided on the Companion Website so that instructors who wish to cover some of them can pick and choose at their leisure without increasing the size (and hence the cost) of the book.

List of Applications

- Minimum distance from a point to a plane (Section 1.4)
- Area and volume (Section 1.5, Section 5.4)
- Electrical circuits (Section 2.4, Section 9.2)
- Planar trusses (Section 2.4)
- Linear programming (Section 2.4)
- Magic squares (Chapter 4 Review)
- Markov processes (Section 6.3)
- Differential equations (Section 6.4)
- Curve of best fit (Section 7.3)
- Overdetermined systems (Section 7.3)
- Graphing quadratic forms (Section 8.3)
- Small deformations (Section 8.4)
- The inertia tensor (Section 8.4)

Computers

As explained in "A Note on the Exercises and Problems," which follows, some problems in the book require access to appropriate computer software. Students should realize that the theory of linear algebra does not apply only to matrices of small size with integer entries. However, since there are many ideas to be learned in linear algebra, numerical methods are not discussed. Some numerical issues, such as accuracy and efficiency, are addressed in notes and problems.

A Note on the Exercises and Problems

Most sections contain mid-section exercises. These mid-section exercises have been created to allow students to check their understanding of key concepts before continuing on to new concepts in the section. Thus, when reading through a chapter, a student should always complete each exercise before continuing to read the rest of the chapter.

At the end of each section, problems are divided into A, B, C, and D problems.

The A Problems are practice problems and are intended to provide a sufficient variety and number of standard computational problems, as well as the odd theoretical problem for students to master the techniques of the course; answers are provided at the back of the text. Full solutions are available in the Student Solutions Manual (sold separately).

The B Problems are homework problems and essentially duplicates of the A problems with no answers provided, for instructors who want such exercises for homework. In a few cases, the B problems are not exactly parallel to the A problems.

The C Problems require the use of a suitable computer program. These problems are designed not only to help students familiarize themselves with using computer software to solve linear algebra problems, but also to remind students that linear algebra uses real numbers, not only integers or simple fractions.

The D Problems usually require students to work with general cases, to write simple arguments, or to invent examples. These are important aspects of mastering mathematical ideas, and all students should attempt at least some of these—and not get discouraged if they make slow progress. With effort, most students will be able to solve many of these problems and will benefit greatly in the understanding of the concepts and connections in doing so.

In addition to the mid-section exercises and end-of-section problems, there is a sample Chapter Quiz in the Chapter Review at the end of each chapter. Students should be aware that their instructors may have a different idea of what constitutes an appropriate test on this material.

At the end of each chapter, there are some Further Problems; these are similar to the D Problems and provide an extended investigation of certain ideas or applications of linear algebra. Further Problems are intended for advanced students who wish to challenge themselves and explore additional concepts.

Using This Text to Teach Linear Algebra

There are many different approaches to teaching linear algebra. Although we suggest covering the chapters in order, the text has been written to try to accommodate two main strategies.

Early Vector Spaces

We believe that it is very beneficial to introduce general vector spaces immediately after students have gained some experience in working with a few specific examples of vector spaces. Students find it easier to generalize the concepts of spanning, linear independence, bases, dimension, and linear mappings while the earlier specific cases are still fresh in their minds. In addition, we feel that it can be unhelpful to students to have determinants available too soon. Some students are far too eager to latch onto mindless algorithms involving determinants (for example, to check linear independence of three vectors in three-dimensional space) rather than actually come to terms with the defining ideas. Finally, this allows eigenvalues, eigenvectors, and diagonalization to be highlighted near the end of the first course. If diagonalization is taught too soon, its importance can be lost on students.

Early Determinants and Diagonalization

Some reviewers have commented that they want to be able to cover determinants and diagonalization before abstract vector spaces and that in some introductory courses, abstract vector spaces may not be covered at all. Thus, this text has been written so that Chapters 5 and 6 may be taught prior to Chapter 4. (Note that all required information about subspaces, bases, and dimension for diagonalization of matrices over $\mathbb{R}$ is covered in Chapters 1, 2, and 3.) Moreover, there is a natural flow from matrix inverses and elementary matrices at the end of Chapter 3 to determinants in Chapter 5.

A Course Outline

The following table indicates the sections in each chapter that we consider to be "central material":

Chapter	Central Material	Optional Material
1	1, 2, 3, 4, 5	
2	1, 2, 3	4
3	1, 2, 3, 4, 5, 6	7
4	1, 2, 3, 4, 5, 6, 7	
5	1, 2, 3	4
6	1, 2	3, 4
7	1, 2	3, 4, 5
8	1, 2	3, 4
9	1, 2, 3, 4, 5, 6	

Supplements

We are pleased to offer a variety of excellent supplements to students and instructors using the Second Edition.

The new **Student Solutions Manual** (ISBN: 978-0-321-80762-5), prepared by the author of the second edition, contains full solutions to the Practice Problems and Chapter Quizzes. It is available to students at low cost.

The new **Companion Website** (www.pearsoncanada.ca/norman) includes practice quizzes, additional applications, and the first edition's "Essay on Linearity and Superposition in Physics."

The new **Instructor's Resource CD-ROM** (ISBN: 978-0-321-80759-5) includes the following valuable teaching tools:

- An **Instructor's Solutions Manual** for all exercises in the text: Practice Problems, Homework Problems, Computer Problems, Conceptual Problems, Chapter Quizzes, and Further Problems.
- An **Instructor's Resource Manual** featuring additional Problems and teaching notes.
- A **Test Bank** with a large selection of questions for every chapter of the text.
- Customizable **Beamer Presentations** for each chapter.
- An **Image Library** that includes high-quality versions of the Figures, Theorems, Corollaries, Lemmas, and Algorithms in the text.

Finally, the second edition is available as a **CourseSmart** eTextbook (ISBN: 978-0-321-75005-1). CourseSmart goes beyond traditional expectations—providing instant, online access to the textbook and course materials at a lower cost for students (average savings of 60%). With instant access from any computer and the ability to search the text, students will find the content they need quickly, no matter where they are. And with online tools like highlighting and note taking, students can save time and study efficiently.

Instructors can save time and hassle with a digital eTextbook that allows them to search for the most relevant content at the very moment they need it. Whether it's evaluating textbooks or creating lecture notes to help students with difficult concepts, CourseSmart can make life a little easier. See all the benefits at www.coursesmart.com/instructors or www.coursesmart.com/students.

Pearson's **technology specialists** work with faculty and campus course designers to ensure that Pearson technology products, assessment tools, and online course materials are tailored to meet your specific needs. This highly qualified team is dedicated to helping schools take full advantage of a wide range of educational resources by assisting in the integration of a variety of instructional materials and media formats. Your local Pearson Canada sales representative can provide you with more details about this service program.

Acknowledgments

Thanks are expressed to:

Agnieszka Wolczuk: for her support, encouragement, help with editing, and tasty snacks.

Mike La Croix: for all of the amazing figures in the text and for his assistance on editing, formatting, and LaTeX'ing.

Stephen New, Martin Pei, Barbara Csima, Emilio Paredes: for proofreading and their many valuable comments and suggestions.

Conrad Hewitt, Robert Andre, Uldis Celmins, C. T. Ng, and many other of my colleagues who have taught me things about linear algebra and how to teach it as well as providing many helpful suggestions for the text.

To all of the reviewers of the text, whose comments, corrections, and recommendations have resulted in many positive improvements:

Robert André
University of Waterloo

Luigi Bilotto
Vanier College

Dietrich Burbulla
University of Toronto

Dr. Alistair Carr
School of Applied Sciences and Engineering, Monash University

Gerald Cliff
University of Alberta

Antoine Khalil
CEGEP Vanier

Hadi Kharaghani
University of Lethbridge

Greg Lewis
Faculty of Science
University of Ontario Institute of Technology

Eduardo Martinez-Pedroza
Department of Mathematics and Statistics, McMaster University

Dorette Pronk
Department of Mathematics and Statistics, Dalhousie University

Dr. Alyssa Sankey
University of New Brunswick

Manuele Santoprete
Wilfrid Laurier University

Alistair Savage
University of Ottawa

Denis Sevee
John Abbott College

Mark Solomonovich
Grant MacEwan University

Dr. Pamini Thangarajah
Department of Mathematics, Physics and Engineering
Mount Royal University

Dr. Chris Tisdell
School of Mathematics and Statistics
The University of New South Wales

Murat Tuncali
Nipissing University

Brian Wetton
University of British Columbia

Thanks also to the many anonymous reviewers of the manuscript.

Cathleen Sullivan, John Lewis, Patricia Ciardullo, and Sarah Lukaweski: For all of their hard work in making the second edition of this text possible and for their suggestions and editing.

In addition, I thank the team at Pearson Canada for their support during the writing and production of this text.

Finally, a very special thank you to Daniel Norman and all those who contributed to the first edition.

Dan Wolczuk
University of Waterloo

CHAPTER 1

Euclidean Vector Spaces

CHAPTER OUTLINE

Some of the material in this chapter will be familiar to many students, but some ideas that are introduced here will be new to most. In this chapter we will look at operations on and important concepts related to vectors. We will also look at some applications of vectors in the familiar setting of Euclidean space. Most of these concepts will later be extended to more general settings. A firm understanding of the material from this chapter will help greatly in understanding the topics in the rest of this book.

1.1 Vectors in $\mathbb{R}^2$ and $\mathbb{R}^3$

We begin by considering the two-dimensional plane in Cartesian coordinates. Choose an origin O and two mutually perpendicular axes, called the x_1-axis and the x_2-axis, as shown in Figure 1.1.1. Then a point P in the plane is identified by the 2-tuple (p_1, p_2), called coordinates of P, where p_1 is the distance from P to the x_2-axis, with p_1 positive if P is to the right of this axis and negative if P is to the left. Similarly, p_2 is the distance from P to the x_1-axis, with p_2 positive if P is above this axis and negative if P is below. You have already learned how to plot graphs of equations in this plane.

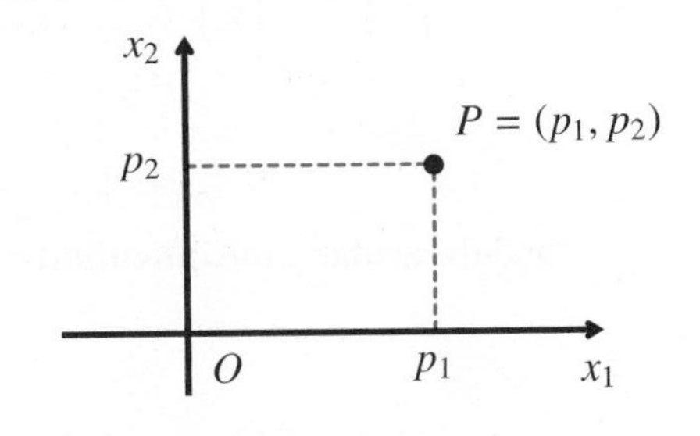

Figure 1.1.1 Coordinates in the plane.

For applications in many areas of mathematics, and in many subjects such as physics and economics, it is useful to view points more abstractly. In particular, we will view them as **vectors** and provide rules for adding them and multiplying them by constants.

Definition
$\mathbb{R}^2$

$\mathbb{R}^2$ is the set of all vectors of the form $\begin{bmatrix} x_1 \\ x_2 \end{bmatrix}$, where x_1 and x_2 are real numbers called the **components** of the vector. Mathematically, we write

$$\mathbb{R}^2 = \left\{ \begin{bmatrix} x_1 \\ x_2 \end{bmatrix} \mid x_1, x_2 \in \mathbb{R} \right\}$$

Remark

We shall use the notation $\vec{x} = \begin{bmatrix} x_1 \\ x_2 \end{bmatrix}$ to denote vectors in $\mathbb{R}^2$.

Although we are viewing the elements of $\mathbb{R}^2$ as vectors, we can still interpret these geometrically as points. That is, the vector $\vec{p} = \begin{bmatrix} p_1 \\ p_2 \end{bmatrix}$ can be interpreted as the point $P(p_1, p_2)$. Graphically, this is often represented by drawing an arrow from $(0, 0)$ to (p_1, p_2), as shown in Figure 1.1.2. Note, however, that the points between $(0, 0)$ and (p_1, p_2) should not be thought of as points "on the vector." The representation of a vector as an arrow is particularly common in physics; force and acceleration are vector quantities that can conveniently be represented by an arrow of suitable magnitude and direction.

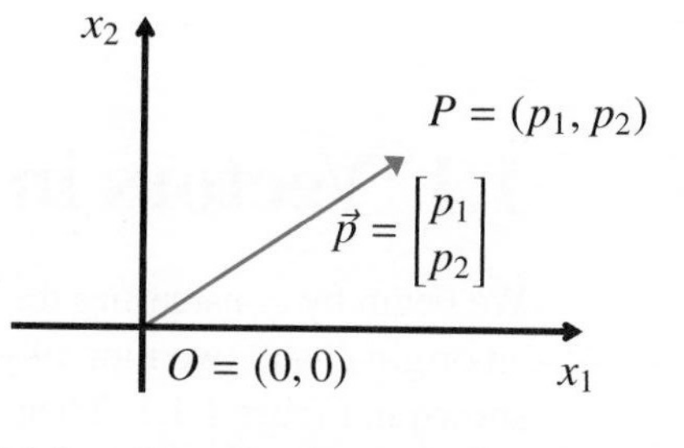

Figure 1.1.2 Graphical representation of a vector.

Definition
Addition and Scalar Multiplication in $\mathbb{R}^2$

If $\vec{x} = \begin{bmatrix} x_1 \\ x_2 \end{bmatrix}$, $\vec{y} = \begin{bmatrix} y_1 \\ y_2 \end{bmatrix}$, and $t \in \mathbb{R}$, then we define **addition** of vectors by

$$\vec{x} + \vec{y} = \begin{bmatrix} x_1 \\ x_2 \end{bmatrix} + \begin{bmatrix} y_1 \\ y_2 \end{bmatrix} = \begin{bmatrix} x_1 + y_1 \\ x_2 + y_2 \end{bmatrix}$$

and the **scalar multiplication** of a vector by a factor of t, called a **scalar**, is defined by

$$t\vec{x} = t\begin{bmatrix} x_1 \\ x_2 \end{bmatrix} = \begin{bmatrix} tx_1 \\ tx_2 \end{bmatrix}$$

The addition of two vectors is illustrated in Figure 1.1.3: construct a parallelogram with vectors $\vec{x}$ and $\vec{y}$ as adjacent sides; then $\vec{x} + \vec{y}$ is the vector corresponding to the vertex of the parallelogram opposite to the origin. Observe that the components really are added according to the definition. This is often called the "parallelogram rule for addition."

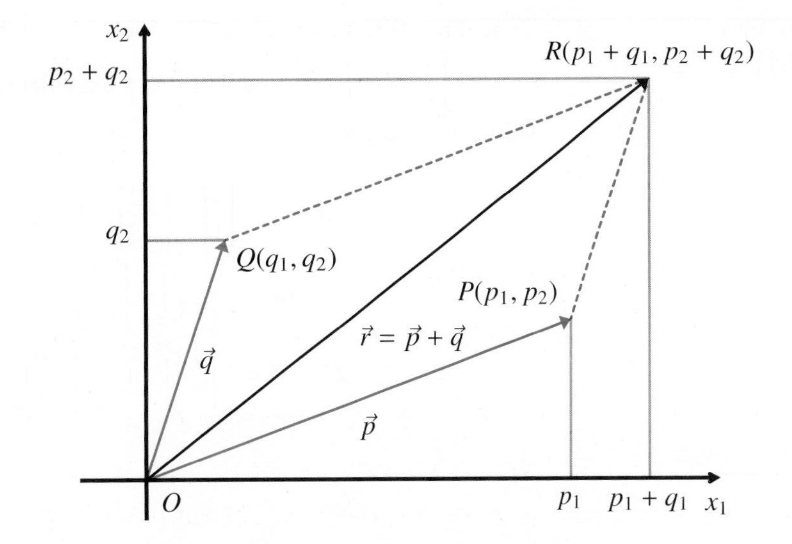

Figure 1.1.3 Addition of vectors $\vec{p}$ and $\vec{q}$.

EXAMPLE 1

Let $\vec{x} = \begin{bmatrix} -2 \\ 3 \end{bmatrix}$ and $\vec{y} = \begin{bmatrix} 5 \\ 1 \end{bmatrix}$. Then

$$\vec{x} + \vec{y} = \begin{bmatrix} -2+5 \\ 3+1 \end{bmatrix} = \begin{bmatrix} 3 \\ 4 \end{bmatrix}$$

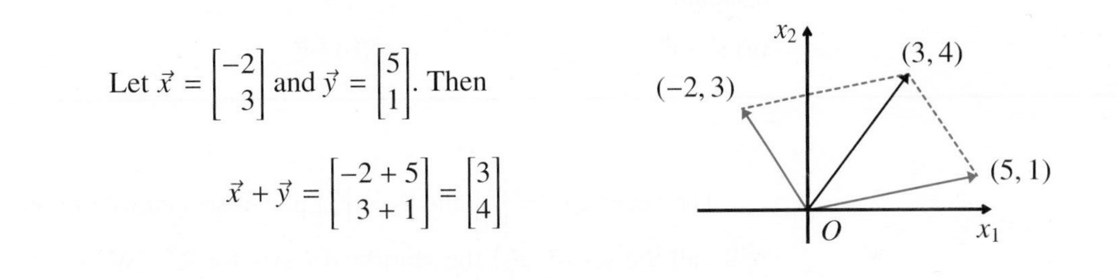

Similarly, scalar multiplication is illustrated in Figure 1.1.4. Observe that multiplication by a negative scalar reverses the direction of the vector. It is important to note that $\vec{x} - \vec{y}$ is to be interpreted as $\vec{x} + (-1)\vec{y}$.

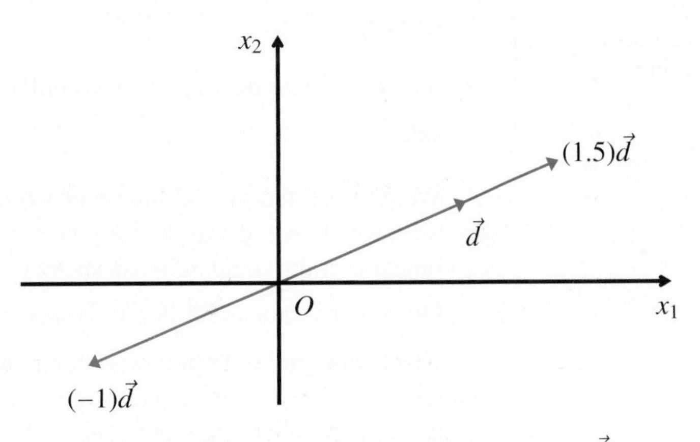

Figure 1.1.4 Scalar multiplication of the vector $\vec{d}$.

EXAMPLE 2

Let $\vec{u} = \begin{bmatrix} 3 \\ 1 \end{bmatrix}$, $\vec{v} = \begin{bmatrix} -2 \\ 3 \end{bmatrix}$, and $\vec{w} = \begin{bmatrix} 0 \\ -1 \end{bmatrix}$. Calculate $\vec{u} + \vec{v}$, $3\vec{w}$, and $2\vec{v} - \vec{w}$.

Solution: We get

$$\vec{u} + \vec{v} = \begin{bmatrix} 3 \\ 1 \end{bmatrix} + \begin{bmatrix} -2 \\ 3 \end{bmatrix} = \begin{bmatrix} 1 \\ 4 \end{bmatrix}$$

$$3\vec{w} = 3\begin{bmatrix} 0 \\ -1 \end{bmatrix} = \begin{bmatrix} 0 \\ -3 \end{bmatrix}$$

$$2\vec{v} - \vec{w} = 2\begin{bmatrix} -2 \\ 3 \end{bmatrix} + (-1)\begin{bmatrix} 0 \\ -1 \end{bmatrix} = \begin{bmatrix} -4 \\ 6 \end{bmatrix} + \begin{bmatrix} 0 \\ 1 \end{bmatrix} = \begin{bmatrix} -4 \\ 7 \end{bmatrix}$$

EXERCISE 1

Let $\vec{u} = \begin{bmatrix} 1 \\ -1 \end{bmatrix}$, $\vec{v} = \begin{bmatrix} 2 \\ 1 \end{bmatrix}$, and $\vec{w} = \begin{bmatrix} 0 \\ 1 \end{bmatrix}$. Calculate each of the following and illustrate with a sketch.

(a) $\vec{u} + \vec{w}$ (b) $-\vec{v}$ (c) $(\vec{u} + \vec{w}) - \vec{v}$

The vectors $\vec{e}_1 = \begin{bmatrix} 1 \\ 0 \end{bmatrix}$ and $\vec{e}_2 = \begin{bmatrix} 0 \\ 1 \end{bmatrix}$ play a special role in our discussion of $\mathbb{R}^2$. We will call the set $\{\vec{e}_1, \vec{e}_2\}$ the **standard basis** for $\mathbb{R}^2$. (We shall discuss the concept of a basis further in Section 1.2.) The basis vectors $\vec{e}_1$ and $\vec{e}_2$ are important because any vector $\vec{v} = \begin{bmatrix} v_1 \\ v_2 \end{bmatrix}$ can be written as a sum of scalar multiples of $\vec{e}_1$ and $\vec{e}_2$ in exactly one way:

$$\vec{v} = \begin{bmatrix} v_1 \\ v_2 \end{bmatrix} = v_1\begin{bmatrix} 1 \\ 0 \end{bmatrix} + v_2\begin{bmatrix} 0 \\ 1 \end{bmatrix} = v_1\vec{e}_1 + v_2\vec{e}_2$$

Remark

In physics and engineering, it is common to use the notation $\mathbf{i} = \begin{bmatrix} 1 \\ 0 \end{bmatrix}$ and $\mathbf{j} = \begin{bmatrix} 0 \\ 1 \end{bmatrix}$ instead.

We will use the phrase **linear combination** to mean "sum of scalar multiples." So, we have shown above that any vector $\vec{x} \in \mathbb{R}^2$ can be written as a unique linear combination of the standard basis vectors.

One other vector in $\mathbb{R}^2$ deserves special mention: the **zero vector**, $\vec{0} = \begin{bmatrix} 0 \\ 0 \end{bmatrix}$. Some important properties of the zero vector, which are easy to verify, are that for any $\vec{x} \in \mathbb{R}^2$,

(1) $\vec{0} + \vec{x} = \vec{x}$
(2) $\vec{x} + (-1)\vec{x} = \vec{0}$
(3) $0\vec{x} = \vec{0}$

The Vector Equation of a Line in $\mathbb{R}^2$

In Figure 1.1.4, it is apparent that the set of all multiples of a vector $\vec{d}$ creates a line through the origin. We make this our definition of a line in $\mathbb{R}^2$: a **line through the origin in** $\mathbb{R}^2$ is a set of the form

$$\{t\vec{d} \mid t \in \mathbb{R}\}$$

Often we do not use formal set notation but simply write the **vector equation** of the line:

$$\vec{x} = t\vec{d}, \quad t \in \mathbb{R}$$

The vector $\vec{d}$ is called the **direction vector** of the line.

Similarly, we define the **line through $\vec{p}$ with direction vector $\vec{d}$** to be the set

$$\{\vec{p} + t\vec{d} \mid t \in \mathbb{R}\}$$

which has the vector equation

$$\vec{x} = \vec{p} + t\vec{d}, \quad t \in \mathbb{R}$$

This line is parallel to the line with equation $\vec{x} = t\vec{d}, t \in \mathbb{R}$ because of the parallelogram rule for addition. As shown in Figure 1.1.5, each point on the line through $\vec{p}$ can be obtained from a corresponding point on the line $\vec{x} = t\vec{d}$ by adding the vector $\vec{p}$. We say that the line has been **translated** by $\vec{p}$. More generally, two lines are parallel if the direction vector of one line is a non-zero multiple of the direction vector of the other line.

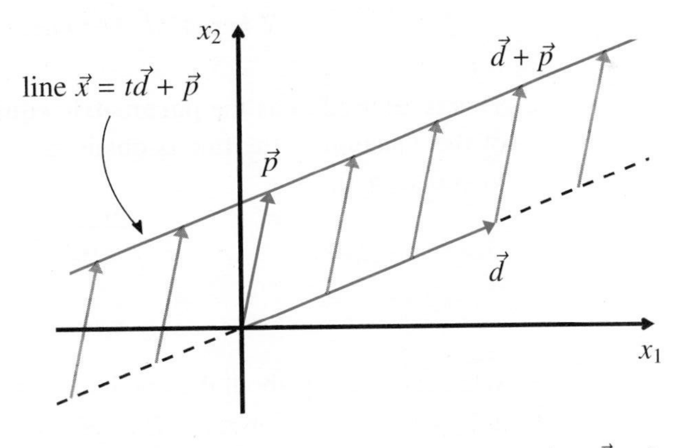

Figure 1.1.5 The line with vector equation $\vec{x} = t\vec{d} + \vec{p}$.

EXAMPLE 3

A vector equation of the line through the point $P(2, -3)$ with direction vector $\begin{bmatrix} -4 \\ 5 \end{bmatrix}$ is

$$\vec{x} = \begin{bmatrix} 2 \\ -3 \end{bmatrix} + t\begin{bmatrix} -4 \\ 5 \end{bmatrix}, \quad t \in \mathbb{R}$$

EXAMPLE 4

Write the vector equation of a line through $P(1, 2)$ parallel to the line with vector equation

$$\vec{x} = t\begin{bmatrix} 2 \\ 3 \end{bmatrix}, \quad t \in \mathbb{R}$$

Solution: Since they are parallel, we can choose the same direction vector. Hence, the vector equation of the line is

$$\vec{x} = \begin{bmatrix} 1 \\ 2 \end{bmatrix} + t\begin{bmatrix} 2 \\ 3 \end{bmatrix}, \quad t \in \mathbb{R}$$

EXERCISE 2

Write the vector equation of a line through $P(0, 0)$ parallel to the line with vector equation

$$\vec{x} = \begin{bmatrix} 4 \\ -3 \end{bmatrix} + t\begin{bmatrix} 3 \\ -1 \end{bmatrix}, \quad t \in \mathbb{R}$$

Sometimes the components of a vector equation are written separately:

$$\vec{x} = \vec{p} + t\vec{d} \text{ becomes } \begin{cases} x_1 = p_1 + td_1 \\ x_2 = p_2 + td_2, \end{cases} \quad t \in \mathbb{R}$$

This is referred to as the **parametric equation** of the line. The familiar **scalar form** of the equation of the line is obtained by eliminating the parameter t. Provided that $d_1 \neq 0, d_2 \neq 0$,

$$\frac{x_1 - p_1}{d_1} = t = \frac{x_2 - p_2}{d_2}$$

or

$$x_2 = p_2 + \frac{d_2}{d_1}(x_1 - p_1)$$

What can you say about the line if $d_1 = 0$ or $d_2 = 0$?

EXAMPLE 5

Write the vector, parametric, and scalar equations of the line passing through the point $P(3, 4)$ with direction vector $\begin{bmatrix} -5 \\ 1 \end{bmatrix}$.

Solution: The vector equation is $\vec{x} = \begin{bmatrix} 3 \\ 4 \end{bmatrix} + t\begin{bmatrix} -5 \\ 1 \end{bmatrix}, \quad t \in \mathbb{R}$.

So, the parametric equation is $\begin{cases} x_1 = 3 - 5t \\ x_2 = 4 + t, \end{cases} \quad t \in \mathbb{R}$.

The scalar equation is $x_2 = 4 - \frac{1}{5}(x_1 - 3)$.

Directed Line Segments For dealing with certain geometrical problems, it is useful to introduce **directed line segments**. We denote the directed line segment from point P to point Q by $\vec{PQ}$. We think of it as an "arrow" starting at P and pointing towards Q. We shall identify directed line segments from the origin O with the corresponding vectors; we write $\vec{OP} = \vec{p}$, $\vec{OQ} = \vec{q}$, and so on. A directed line segment that starts at the origin is called the **position vector** of the point.

For many problems, we are interested only in the direction and length of the directed line segment; we are not interested in the point where it is located. For example, in Figure 1.1.3, we may wish to treat the line segment $\vec{QR}$ as if it were the same as $\vec{OP}$. Taking our cue from this example, for arbitrary points P, Q, R in $\mathbb{R}^2$, we define $\vec{QR}$ to be **equivalent** to $\vec{OP}$ if $\vec{r} - \vec{q} = \vec{p}$. In this case, we have used one directed line segment $\vec{OP}$ starting from the origin in our definition.

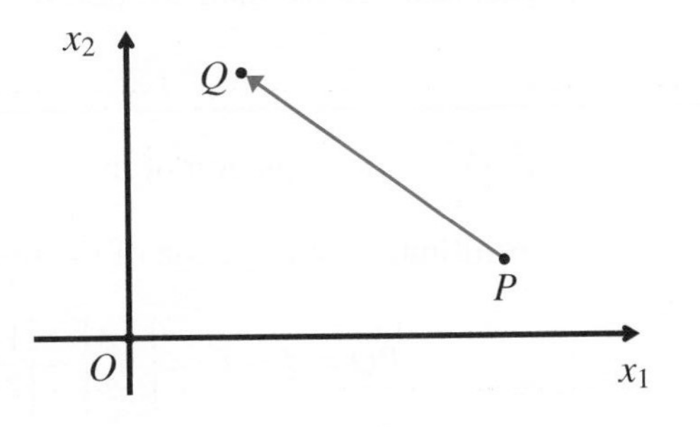

Figure 1.1.6 A directed line segment from P to Q.

More generally, for arbitrary points Q, R, S, and T in $\mathbb{R}^2$, we define $\vec{QR}$ to be equivalent to $\vec{ST}$ if they are both equivalent to the same $\vec{OP}$ for some P. That is, if

$$\vec{r} - \vec{q} = \vec{p} \text{ and } \vec{t} - \vec{s} = \vec{p} \text{ for the same } \vec{p}$$

We can abbreviate this by simply requiring that

$$\vec{r} - \vec{q} = \vec{t} - \vec{s}$$

EXAMPLE 6

For points $Q(1,3)$, $R(6,-1)$, $S(-2,4)$, and $T(3,0)$, we have that $\vec{QR}$ is equivalent to $\vec{ST}$ because

$$\vec{r} - \vec{q} = \begin{bmatrix} 6 \\ -1 \end{bmatrix} - \begin{bmatrix} 1 \\ 3 \end{bmatrix} = \begin{bmatrix} 5 \\ -4 \end{bmatrix} = \begin{bmatrix} 3 \\ 0 \end{bmatrix} - \begin{bmatrix} -2 \\ 4 \end{bmatrix} = \vec{t} - \vec{s}$$

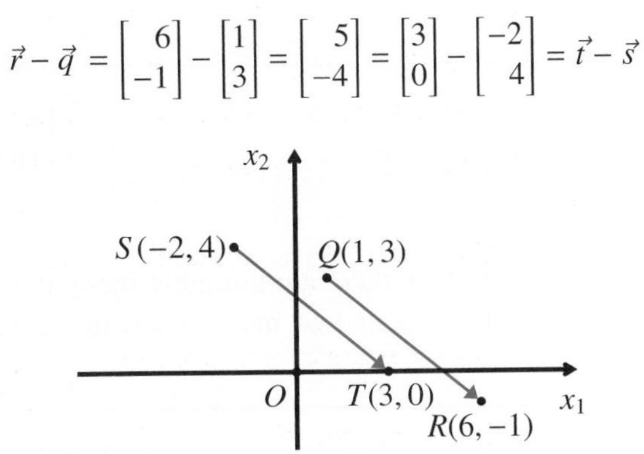

In some problems, where it is not necessary to distinguish between equivalent directed line segments, we "identify" them (that is, we treat them as the same object) and write $\vec{PQ} = \vec{RS}$. Indeed, we identify them with the corresponding line segment starting at the origin, so in Example 6 we write $\vec{QR} = \vec{ST} = \begin{bmatrix} 5 \\ -4 \end{bmatrix}$.

Remark

Writing $\vec{QR} = \vec{ST}$ is a bit sloppy—an abuse of notation—because $\vec{QR}$ is not really the same object as $\vec{ST}$. However, introducing the precise language of "equivalence classes" and more careful notation with directed line segments is not helpful at this stage. By introducing directed line segments, we are encouraged to think about vectors that are located at arbitrary points in space. This is helpful in solving some geometrical problems, as we shall see below.

EXAMPLE 7

Find a vector equation of the line through $P(1, 2)$ and $Q(3, -1)$.

Solution: The direction of the line is

$$\vec{PQ} = \vec{q} - \vec{p} = \begin{bmatrix} 3 \\ -1 \end{bmatrix} - \begin{bmatrix} 1 \\ 2 \end{bmatrix} = \begin{bmatrix} 2 \\ -3 \end{bmatrix}$$

Hence, a vector equation of the line with direction $\vec{PQ}$ that passes through $P(1, 2)$ is

$$\vec{x} = \vec{p} + t\vec{PQ} = \begin{bmatrix} 1 \\ 2 \end{bmatrix} + t \begin{bmatrix} 2 \\ -3 \end{bmatrix}, \quad t \in \mathbb{R}$$

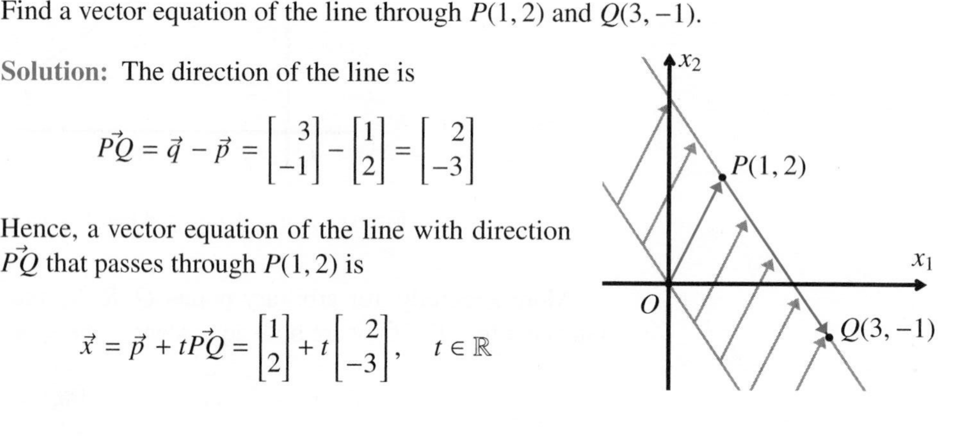

Observe in the example above that we would have the same line if we started at the second point and "moved" toward the first point—or even if we took a direction vector in the opposite direction. Thus, the same line is described by the vector equations

$$\vec{x} = \begin{bmatrix} 3 \\ -1 \end{bmatrix} + r \begin{bmatrix} -2 \\ 3 \end{bmatrix}, \quad r \in \mathbb{R}$$

$$\vec{x} = \begin{bmatrix} 3 \\ -1 \end{bmatrix} + s \begin{bmatrix} 2 \\ -3 \end{bmatrix}, \quad s \in \mathbb{R}$$

$$\vec{x} = \begin{bmatrix} 1 \\ 2 \end{bmatrix} + t \begin{bmatrix} -2 \\ 3 \end{bmatrix}, \quad t \in \mathbb{R}$$

In fact, there are infinitely many descriptions of a line: we may choose any point on the line, and we may use any non-zero multiple of the direction vector.

EXERCISE 3

Find a vector equation of the line through $P(1, 1)$ and $Q(-2, 2)$.

Vectors and Lines in $\mathbb{R}^3$

Everything we have done so far works perfectly well in three dimensions. We choose an origin O and three mutually perpendicular axes, as shown in Figure 1.1.7. The x_1-axis is usually pictured coming out of the page (or blackboard), the x_2-axis to the right, and the x_3-axis towards the top of the picture.

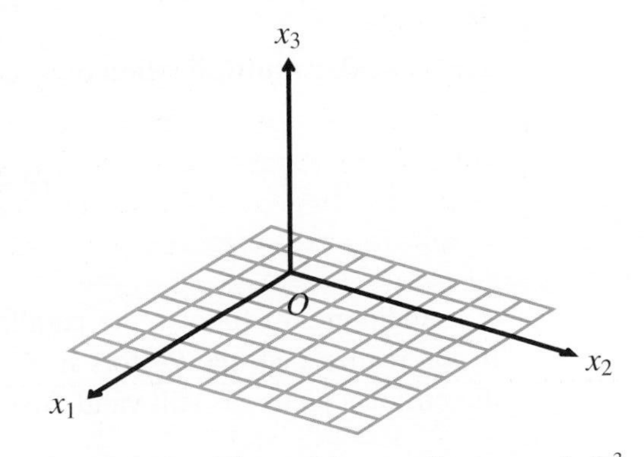

Figure 1.1.7 The positive coordinate axes in $\mathbb{R}^3$.

It should be noted that we are adopting the convention that the coordinate axes form a **right-handed system**. One way to visualize a right-handed system is to spread out the thumb, index finger, and middle finger of your right hand. The thumb is the x_1-axis, the index finger is the x_2-axis, and the middle finger is the x_3-axis. See Figure 1.1.8.

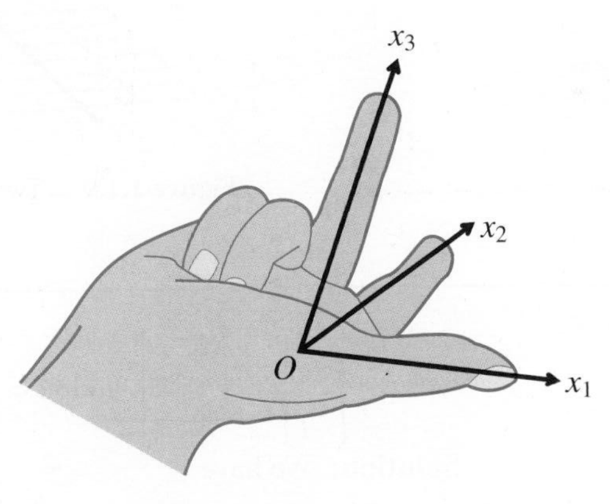

Figure 1.1.8 Identifying a right-handed system.

We now define $\mathbb{R}^3$ to be the three-dimensional analog of $\mathbb{R}^2$.

Definition
$\mathbb{R}^3$

$\mathbb{R}^3$ is the set of all vectors of the form $\begin{bmatrix} x_1 \\ x_2 \\ x_3 \end{bmatrix}$, where $x_1, x_2,$ and x_3 are real numbers.

Mathematically, we write

$$\mathbb{R}^3 = \left\{ \begin{bmatrix} x_1 \\ x_2 \\ x_3 \end{bmatrix} \mid x_1, x_2, x_3 \in \mathbb{R} \right\}$$

Definition
Addition and Scalar Multiplication in $\mathbb{R}^3$

If $\vec{x} = \begin{bmatrix} x_1 \\ x_2 \\ x_3 \end{bmatrix}$, $\vec{y} = \begin{bmatrix} y_1 \\ y_2 \\ y_3 \end{bmatrix}$, and $t \in \mathbb{R}$, then we define **addition** of vectors by

$$\vec{x} + \vec{y} = \begin{bmatrix} x_1 \\ x_2 \\ x_3 \end{bmatrix} + \begin{bmatrix} y_1 \\ y_2 \\ y_3 \end{bmatrix} = \begin{bmatrix} x_1 + y_1 \\ x_2 + y_2 \\ x_3 + y_3 \end{bmatrix}$$

and the **scalar multiplication** of a vector by a factor of t by

$$t\vec{x} = t\begin{bmatrix} x_1 \\ x_2 \\ x_3 \end{bmatrix} = \begin{bmatrix} tx_1 \\ tx_2 \\ tx_3 \end{bmatrix}$$

Addition still follows the parallelogram rule. It may help you to visualize this if you realize that two vectors in $\mathbb{R}^3$ must lie within a plane in $\mathbb{R}^3$ so that the two-dimensional picture is still valid. See Figure 1.1.9.

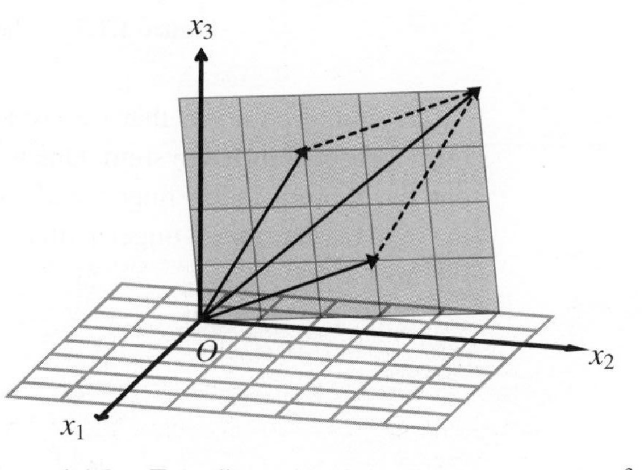

Figure 1.1.9 Two-dimensional parallelogram rule in $\mathbb{R}^3$.

EXAMPLE 8

Let $\vec{u} = \begin{bmatrix} 1 \\ 1 \\ -1 \end{bmatrix}$, $\vec{v} = \begin{bmatrix} -2 \\ 1 \\ 2 \end{bmatrix}$, and $\vec{w} = \begin{bmatrix} 1 \\ 0 \\ 1 \end{bmatrix}$. Calculate $\vec{v} + \vec{u}$, $-\vec{w}$, and $-\vec{v} + 2\vec{w} - \vec{u}$.

Solution: We have

$$\vec{v} + \vec{u} = \begin{bmatrix} -2 \\ 1 \\ 2 \end{bmatrix} + \begin{bmatrix} 1 \\ 1 \\ -1 \end{bmatrix} = \begin{bmatrix} -1 \\ 2 \\ 1 \end{bmatrix}$$

$$-\vec{w} = -\begin{bmatrix} 1 \\ 0 \\ 1 \end{bmatrix} = \begin{bmatrix} -1 \\ 0 \\ -1 \end{bmatrix}$$

$$-\vec{v} + 2\vec{w} - \vec{u} = -\begin{bmatrix} -2 \\ 1 \\ 2 \end{bmatrix} + 2\begin{bmatrix} 1 \\ 0 \\ 1 \end{bmatrix} - \begin{bmatrix} 1 \\ 1 \\ -1 \end{bmatrix} = \begin{bmatrix} 2 \\ -1 \\ -2 \end{bmatrix} + \begin{bmatrix} 2 \\ 0 \\ 2 \end{bmatrix} + \begin{bmatrix} -1 \\ -1 \\ 1 \end{bmatrix} = \begin{bmatrix} 3 \\ -2 \\ 1 \end{bmatrix}$$

It is useful to introduce the standard basis for $\mathbb{R}^3$ just as we did for $\mathbb{R}^2$. Define

$$\vec{e}_1 = \begin{bmatrix} 1 \\ 0 \\ 0 \end{bmatrix}, \quad \vec{e}_2 = \begin{bmatrix} 0 \\ 1 \\ 0 \end{bmatrix}, \quad \vec{e}_3 = \begin{bmatrix} 0 \\ 0 \\ 1 \end{bmatrix}$$

Then any vector $\vec{v} = \begin{bmatrix} v_1 \\ v_2 \\ v_3 \end{bmatrix}$ can be written as the linear combination

$$\vec{v} = v_1 \begin{bmatrix} 1 \\ 0 \\ 0 \end{bmatrix} + v_2 \begin{bmatrix} 0 \\ 1 \\ 0 \end{bmatrix} + v_3 \begin{bmatrix} 0 \\ 0 \\ 1 \end{bmatrix} = v_1\vec{e}_1 + v_2\vec{e}_2 + v_3\vec{e}_3$$

Remark

In physics and engineering, it is common to use the notation $\mathbf{i} = \vec{e}_1, \mathbf{j} = \vec{e}_2$, and $\mathbf{k} = \vec{e}_3$ instead.

The zero vector $\vec{0} = \begin{bmatrix} 0 \\ 0 \\ 0 \end{bmatrix}$ in $\mathbb{R}^3$ has the same properties as the zero vector in $\mathbb{R}^2$.

Directed line segments are the same in three-dimensional space as in the two-dimensional case.

A line through the point P in $\mathbb{R}^3$ (corresponding to a vector $\vec{p}$) with direction vector $\vec{d} \neq \vec{0}$ can be described by a vector equation:

$$\vec{x} = \vec{p} + t\vec{d}, \quad t \in \mathbb{R}$$

It is important to realize that a line in $\mathbb{R}^3$ cannot be described by a single scalar linear equation, as in $\mathbb{R}^2$. We shall see in Section 1.3 that such an equation describes a plane in $\mathbb{R}^3$.

EXAMPLE 9

Find a vector equation of the line that passes through the points $P(1, 5, -2)$ and $Q(4, -1, 3)$.

Solution: A direction vector is $\vec{d} = \vec{q} - \vec{p} = \begin{bmatrix} 3 \\ -6 \\ 5 \end{bmatrix}$. Hence a vector equation of the line is

$$\vec{x} = \begin{bmatrix} 1 \\ 5 \\ -2 \end{bmatrix} + t \begin{bmatrix} 3 \\ -6 \\ 5 \end{bmatrix}, \quad t \in \mathbb{R}$$

Note that the corresponding parametric equations are $x_1 = 1 + 3t$, $x_2 = 5 - 6t$, and $x_3 = -2 + 5t$.

EXERCISE 4

Find a vector equation of the line that passes through the points $P(1, 2, 2)$ and $Q(1, -2, 3)$.

PROBLEMS 1.1

Practice Problems

A1 Compute each of the following linear combinations and illustrate with a sketch.

(a) $\begin{bmatrix}1\\4\end{bmatrix}+\begin{bmatrix}2\\3\end{bmatrix}$ (b) $\begin{bmatrix}3\\2\end{bmatrix}-\begin{bmatrix}4\\1\end{bmatrix}$

(c) $3\begin{bmatrix}-1\\4\end{bmatrix}$ (d) $2\begin{bmatrix}2\\1\end{bmatrix}-2\begin{bmatrix}3\\-1\end{bmatrix}$

A2 Compute each of the following linear combinations.

(a) $\begin{bmatrix}4\\-2\end{bmatrix}+\begin{bmatrix}-1\\3\end{bmatrix}$ (b) $\begin{bmatrix}-3\\-4\end{bmatrix}-\begin{bmatrix}-2\\5\end{bmatrix}$

(c) $-2\begin{bmatrix}3\\-2\end{bmatrix}$ (d) $\frac{1}{2}\begin{bmatrix}2\\6\end{bmatrix}+\frac{1}{3}\begin{bmatrix}4\\3\end{bmatrix}$

(e) $\frac{2}{3}\begin{bmatrix}3\\1\end{bmatrix}-2\begin{bmatrix}1/4\\1/3\end{bmatrix}$ (f) $\sqrt{2}\begin{bmatrix}\sqrt{2}\\\sqrt{3}\end{bmatrix}+3\begin{bmatrix}1\\\sqrt{6}\end{bmatrix}$

A3 Compute each of the following linear combinations.

(a) $\begin{bmatrix}2\\3\\4\end{bmatrix}-\begin{bmatrix}5\\1\\-2\end{bmatrix}$ (b) $\begin{bmatrix}2\\1\\-6\end{bmatrix}+\begin{bmatrix}-3\\1\\-4\end{bmatrix}$

(c) $-6\begin{bmatrix}4\\-5\\-6\end{bmatrix}$ (d) $-2\begin{bmatrix}-5\\1\\1\end{bmatrix}+3\begin{bmatrix}-1\\0\\-1\end{bmatrix}$

(e) $2\begin{bmatrix}2/3\\-1/3\\2\end{bmatrix}+\frac{1}{3}\begin{bmatrix}3\\-2\\1\end{bmatrix}$ (f) $\sqrt{2}\begin{bmatrix}1\\1\\1\end{bmatrix}+\pi\begin{bmatrix}-1\\0\\1\end{bmatrix}$

A4 Let $\vec{v}=\begin{bmatrix}1\\2\\-2\end{bmatrix}$ and $\vec{w}=\begin{bmatrix}2\\-1\\3\end{bmatrix}$. Determine

(a) $2\vec{v}-3\vec{w}$
(b) $-3(\vec{v}+2\vec{w})+5\vec{v}$
(c) $\vec{u}$ such that $\vec{w}-2\vec{u}=3\vec{v}$
(d) $\vec{u}$ such that $\vec{u}-3\vec{v}=2\vec{u}$

A5 Let $\vec{v}=\begin{bmatrix}3\\1\\1\end{bmatrix}$ and $\vec{w}=\begin{bmatrix}5\\-1\\-2\end{bmatrix}$. Determine

(a) $\frac{1}{2}\vec{v}+\frac{1}{2}\vec{w}$
(b) $2(\vec{v}+\vec{w})-(2\vec{v}-3\vec{w})$
(c) $\vec{u}$ such that $\vec{w}-\vec{u}=2\vec{v}$
(d) $\vec{u}$ such that $\frac{1}{2}\vec{u}+\frac{1}{3}\vec{v}=\vec{w}$

A6 Consider the points $P(2,3,1)$, $Q(3,1,-2)$, $R(1,4,0)$, $S(-5,1,5)$. Determine $\vec{PQ}$, $\vec{PR}$, $\vec{PS}$, $\vec{QR}$, and $\vec{SR}$, and verify that $\vec{PQ}+\vec{QR}=\vec{PR}=\vec{PS}+\vec{SR}$.

A7 Write a vector equation of the line passing through the given points with the given direction vector.

(a) $P(3,4)$, $\vec{d}=\begin{bmatrix}-5\\1\end{bmatrix}$

(b) $P(2,3)$, $\vec{d}=\begin{bmatrix}-4\\-6\end{bmatrix}$

(c) $P(2,0,5)$, $\vec{d}=\begin{bmatrix}4\\-2\\-11\end{bmatrix}$

(d) $P(4,1,5)$, $\vec{d}=\begin{bmatrix}-2\\1\\2\end{bmatrix}$

A8 Write a vector equation for the line that passes through the given points.

(a) $P(-1,2)$, $Q(2,-3)$
(b) $P(4,1)$, $Q(-2,-1)$
(c) $P(1,3,-5)$, $Q(-2,1,0)$
(d) $P(-2,1,1)$, $Q(4,2,2)$
(e) $P\left(\frac{1}{2},\frac{1}{4},1\right)$, $Q\left(-1,1,\frac{1}{3}\right)$

A9 For each of the following lines in $\mathbb{R}^2$, determine a vector equation and parametric equations.

(a) $x_2=3x_1+2$
(b) $2x_1+3x_2=5$

A10 (a) A set of points in $\mathbb{R}^n$ is **collinear** if all the points lie on the same line. By considering directed line segments, give a general method for determining whether a given set of three points is collinear.

(b) Determine whether the points $P(1,2)$, $Q(4,1)$, and $R(-5,4)$ are collinear. Show how you decide.

(c) Determine whether the points $S(1,0,1)$, $T(3,-2,3)$, and $U(-3,4,-1)$ are collinear. Show how you decide.

Homework Problems

B1 Compute each of the following linear combinations and illustrate with a sketch.

(a) $\begin{bmatrix} 1 \\ -2 \end{bmatrix} + \begin{bmatrix} -1 \\ 3 \end{bmatrix}$

(b) $\begin{bmatrix} 4 \\ -3 \end{bmatrix} - \begin{bmatrix} 2 \\ 3 \end{bmatrix}$

(c) $-3\begin{bmatrix} 1 \\ -2 \end{bmatrix}$

(d) $-3\begin{bmatrix} 5 \\ 2 \end{bmatrix} - \begin{bmatrix} 3 \\ 3 \end{bmatrix}$

B2 Compute each of the following linear combinations.

(a) $\begin{bmatrix} 6 \\ 3 \end{bmatrix} + \begin{bmatrix} -6 \\ -3 \end{bmatrix}$

(b) $\begin{bmatrix} 6 \\ 3 \end{bmatrix} - \begin{bmatrix} 6 \\ 3 \end{bmatrix}$

(c) $2\begin{bmatrix} -2 \\ -1 \end{bmatrix}$

(d) $\frac{3}{5}\begin{bmatrix} 2 \\ 10 \end{bmatrix} - \frac{3}{10}\begin{bmatrix} 4 \\ 2 \end{bmatrix}$

(e) $\sqrt{2}\begin{bmatrix} \sqrt{3} \\ 1 \end{bmatrix} + \sqrt{3}\begin{bmatrix} 2 \\ -1/2 \end{bmatrix} - \begin{bmatrix} \sqrt{6} \\ \sqrt{3} \end{bmatrix}$

B3 Compute each of the following linear combinations.

(a) $\begin{bmatrix} 5 \\ -4 \\ -3 \end{bmatrix} - \begin{bmatrix} 2 \\ -1 \\ 3 \end{bmatrix}$

(b) $\begin{bmatrix} 4 \\ 1 \\ 2 \end{bmatrix} + \begin{bmatrix} 2 \\ -1 \\ -2 \end{bmatrix}$

(c) $4\begin{bmatrix} 2 \\ -5 \\ -1 \end{bmatrix}$

(d) $0\begin{bmatrix} -3 \\ 3 \\ -2 \end{bmatrix}$

(e) $2\begin{bmatrix} 2/3 \\ -1/3 \\ 2 \end{bmatrix} + \frac{1}{3}\begin{bmatrix} 3 \\ -2 \\ 1 \end{bmatrix}$

(f) $(1+\sqrt{2})\begin{bmatrix} 1-\sqrt{2} \\ 0 \\ \sqrt{2}-1 \end{bmatrix} - \frac{1}{2}\begin{bmatrix} -2 \\ 0 \\ 2 \end{bmatrix}$

B4 Let $\vec{v} = \begin{bmatrix} 3 \\ 5 \\ -1 \end{bmatrix}$ and $\vec{w} = \begin{bmatrix} -5 \\ 2 \\ 1 \end{bmatrix}$. Determine

(a) $2\vec{v} - 3\vec{w}$

(b) $-2(\vec{v} - \vec{w}) - 3\vec{w}$

(c) $\vec{u}$ such that $\vec{w} - 2\vec{u} = 3\vec{v}$

(d) $\vec{u}$ such that $2\vec{u} + 3\vec{w} = \vec{v}$

B5 Let $\vec{v} = \begin{bmatrix} 2 \\ -2 \\ 2 \end{bmatrix}$ and $\vec{w} = \begin{bmatrix} 4 \\ -3 \\ 1 \end{bmatrix}$. Determine

(a) $3\vec{v} - 2\vec{w}$

(b) $-\frac{1}{2}\vec{v} + \frac{1}{4}\vec{w}$

(c) $\vec{u}$ such that $\vec{v} + \vec{u} = \vec{v}$

(d) $\vec{u}$ such that $2\vec{u} - \vec{w} = 2\vec{v}$

B6 (a) Consider the points $P(1,4,1)$, $Q(4,3,-1)$, $R(-1,4,2)$, and $S(8,6,-5)$. Determine $\vec{PQ}$, $\vec{PR}$, $\vec{PS}$, $\vec{QR}$, and $\vec{SR}$, and verify that $\vec{PQ}+\vec{QR} = \vec{PR} = \vec{PS} + \vec{SR}$.

(b) Consider the points $P(3,-2,1)$, $Q(2,7,-3)$, $R(3,1,5)$, and $S(-2,4,-1)$. Determine $\vec{PQ}$, $\vec{PR}$, $\vec{PS}$, $\vec{QR}$, and $\vec{SR}$, and verify that $\vec{PQ}+\vec{QR} = \vec{PR} = \vec{PS} + \vec{SR}$.

B7 Write a vector equation of the line passing through the given points with the given direction vector.

(a) $P(-3,4)$, $\vec{d} = \begin{bmatrix} 4 \\ -3 \end{bmatrix}$

(b) $P(0,0)$, $\vec{d} = \begin{bmatrix} 1 \\ 3 \\ 1 \end{bmatrix}$

(c) $P(2,3,-1)$, $\vec{d} = \begin{bmatrix} 2 \\ -4 \\ 8 \end{bmatrix}$

(d) $P(3,1,2)$, $\vec{d} = \begin{bmatrix} 2 \\ -3 \\ 2 \end{bmatrix}$

B8 Write a vector equation for the line that passes through the given points.

(a) $P(3,1)$, $Q(1,2)$

(b) $P(1,-2,1)$, $Q(0,0,0)$

(c) $P(2,-6,3)$, $Q(-1,5,2)$

(d) $P\left(1,-1,\frac{1}{2}\right)$, $Q\left(\frac{1}{2},\frac{1}{3},1\right)$

B9 For each of the following lines in $\mathbb{R}^2$, determine a vector equation and parametric equations.

(a) $x_2 = -2x_1 + 3$

(b) $x_1 + 2x_2 = 3$

B10 (You will need the solution from Problem A10 (a) to answer this.)

(a) Determine whether the points $P(2,1,1)$, $Q(1,2,3)$, and $R(4,-1,-3)$ are collinear. Show how you decide.

(b) Determine whether the points $S(1,1,0)$, $T(6,2,1)$, and $U(-4,0,-1)$ are collinear. Show how you decide.

Computer Problems

C1 Let $\vec{v}_1 = \begin{bmatrix} -33 \\ -46 \\ 2 \end{bmatrix}$, $\vec{v}_2 = \begin{bmatrix} 21 \\ 31 \\ -7 \end{bmatrix}$, $\vec{v}_3 = \begin{bmatrix} -36 \\ -28 \\ 46 \end{bmatrix}$, and $\vec{v}_4 = \begin{bmatrix} -47 \\ -29 \\ 43 \end{bmatrix}$.

Use computer software to evaluate each of the following.

(a) $17\vec{v}_1 + 5\vec{v}_2 - 3\vec{v}_3 + 42\vec{v}_4$

(b) $-1440\vec{v}_1 - 2341\vec{v}_2 - 919\vec{v}_3 + 669\vec{v}_4$

Conceptual Problems

D1 Let $\vec{v} = \begin{bmatrix} 1 \\ 1 \end{bmatrix}$ and $\vec{w} = \begin{bmatrix} 1 \\ -1 \end{bmatrix}$.

(a) Find real numbers t_1 and t_2 such that $t_1\vec{v}+t_2\vec{w} = \begin{bmatrix} 13 \\ -12 \end{bmatrix}$. Illustrate with a sketch.

(b) Find real numbers t_1 and t_2 such that $t_1\vec{v}+t_2\vec{w} = \begin{bmatrix} x_1 \\ x_2 \end{bmatrix}$ for any $x_1, x_2 \in \mathbb{R}$.

(c) Use your result in part (b) to find real numbers t_1 and t_2 such that $t_1\vec{v}_1 + t_2\vec{v}_2 = \begin{bmatrix} \sqrt{2} \\ \pi \end{bmatrix}$.

D2 Let P, Q, and R be points in $\mathbb{R}^2$ corresponding to vectors $\vec{p}$, $\vec{q}$, and $\vec{r}$ respectively.

(a) Explain in terms of directed line segments why

$$\vec{PQ} + \vec{QR} + \vec{RP} = \vec{0}$$

(b) Verify the equation of part (a) by expressing $\vec{PQ}$, $\vec{QR}$, and $\vec{RP}$ in terms of $\vec{p}$, $\vec{q}$, and $\vec{r}$.

D3 Let $\vec{p}$ and $\vec{d} \neq \vec{0}$ be vectors in $\mathbb{R}^2$. Prove that $\vec{x} = \vec{p} + t\vec{d}$, $t \in \mathbb{R}$, is a line in $\mathbb{R}^2$ passing through the origin if and only if $\vec{p}$ is a scalar multiple of $\vec{d}$.

D4 Let $\vec{x}$ and $\vec{y}$ be vectors in $\mathbb{R}^3$ and $t \in \mathbb{R}$ be a scalar. Prove that

$$t(\vec{x} + \vec{y}) = t\vec{x} + t\vec{y}$$

1.2 Vectors in $\mathbb{R}^n$

We now extend the ideas from the previous section to n-dimensional Euclidean space $\mathbb{R}^n$.

Students sometimes do not see the point in discussing n-dimensional space because it does not seem to correspond to any physical realistic geometry. But, in a number of instances, *more than three dimensions are important. For example, to discuss the motion of a particle, an engineer needs to specify its position (3 variables) and its velocity (3 more variables); the engineer therefore has 6 variables. A scientist working in string theory works with 11 dimensional space-time variables. An economist seeking to model the Canadian economy uses many variables: one standard model has more than 1500 variables. Of course, calculations in such huge models are carried out by computer. Even so, understanding the ideas of geometry and linear algebra is necessary to decide which calculations are required and what the results mean.*

Addition and Scalar Multiplication of Vectors in $\mathbb{R}^n$

Definition
$\mathbb{R}^n$

$\mathbb{R}^n$ is the set of all vectors of the form $\begin{bmatrix} x_1 \\ \vdots \\ x_n \end{bmatrix}$, where $x_i \in \mathbb{R}$. Mathematically,

$$\mathbb{R}^n = \left\{ \begin{bmatrix} x_1 \\ \vdots \\ x_n \end{bmatrix} \mid x_1, \dots, x_n \in \mathbb{R} \right\}$$

Definition
Addition and Scalar Multiplication in $\mathbb{R}^n$

If $\vec{x} = \begin{bmatrix} x_1 \\ \vdots \\ x_n \end{bmatrix}$, $\vec{y} = \begin{bmatrix} y_1 \\ \vdots \\ y_n \end{bmatrix}$, and $t \in \mathbb{R}$, then we define **addition** of vectors by

$$\vec{x} + \vec{y} = \begin{bmatrix} x_1 \\ \vdots \\ x_n \end{bmatrix} + \begin{bmatrix} y_1 \\ \vdots \\ y_n \end{bmatrix} = \begin{bmatrix} x_1 + y_1 \\ \vdots \\ x_n + y_n \end{bmatrix}$$

and the **scalar multiplication** of a vector by a factor of t by

$$t\vec{x} = t\begin{bmatrix} x_1 \\ \vdots \\ x_n \end{bmatrix} = \begin{bmatrix} tx_1 \\ \vdots \\ tx_n \end{bmatrix}$$

Theorem 1

For all $\vec{w}, \vec{x}, \vec{y} \in \mathbb{R}^n$ and $s, t \in \mathbb{R}$ we have

(1) $\vec{x} + \vec{y} \in \mathbb{R}^n$ (closed under addition)
(2) $\vec{x} + \vec{y} = \vec{y} + \vec{x}$ (addition is commutative)
(3) $(\vec{x} + \vec{y}) + \vec{w} = \vec{x} + (\vec{y} + \vec{w})$ (addition is associative)
(4) There exists a vector $\vec{0} \in \mathbb{R}^n$ such that $\vec{z} + \vec{0} = \vec{z}$ for all $\vec{z} \in \mathbb{R}^n$ (zero vector)
(5) For each $\vec{x} \in \mathbb{R}^n$ there exists a vector $-\vec{x} \in \mathbb{R}^n$ such that $\vec{x} + (-\vec{x}) = \vec{0}$ (additive inverses)
(6) $t\vec{x} \in \mathbb{R}^n$ (closed under scalar multiplication)
(7) $s(t\vec{x}) = (st)\vec{x}$ (scalar multiplication is associative)
(8) $(s + t)\vec{x} = s\vec{x} + t\vec{x}$ (a distributive law)
(9) $t(\vec{x} + \vec{y}) = t\vec{x} + t\vec{y}$ (another distributive law)
(10) $1\vec{x} = \vec{x}$ (scalar multiplicative identity)

Proof: We will prove properties (1) and (2) from Theorem 1 and leave the other proofs to the reader.

For (1), by definition,

$$\vec{x}+\vec{y} = \begin{bmatrix} x_1 + y_1 \\ \vdots \\ x_n + y_n \end{bmatrix} \in \mathbb{R}^n$$

since $x_i + y_i \in \mathbb{R}$ for $1 \leq i \leq n$.

For (2),

$$\vec{x}+\vec{y} = \begin{bmatrix} x_1 + y_1 \\ \vdots \\ x_n + y_n \end{bmatrix} = \begin{bmatrix} y_1 + x_1 \\ \vdots \\ y_n + x_n \end{bmatrix} = \vec{y} + \vec{x}$$

■

EXERCISE 1 Prove properties (5), (6), and (7) from Theorem 1.

Observe that properties (2), (3), (7), (8), (9), and (10) from Theorem 1 refer only to the operations of addition and scalar multiplication, while the other properties, (1), (4), (5), and (6), are about the relationship between the operations and the set $\mathbb{R}^n$. These facts should be clear in the proof of Theorem 1. Moreover, we see that the zero vector of $\mathbb{R}^n$ is the vector $\vec{0} = \begin{bmatrix} 0 \\ \vdots \\ 0 \end{bmatrix}$, and the additive inverse of $\vec{x}$ is $-\vec{x} = (-1)\vec{x}$. Note that the zero vector satisfies the same properties as the zero vector in $\mathbb{R}^2$ and $\mathbb{R}^3$.

Students often find properties (1) and (6) a little strange. At first glance, it seems obvious that the sum of two vectors in $\mathbb{R}^n$ or the scalar multiple of a vector in $\mathbb{R}^n$ is another vector in $\mathbb{R}^n$. However, these properties are in fact extremely important. We now look at subsets of $\mathbb{R}^n$ that have both of these properties.

Subspaces

Definition
Subspace

A non-empty subset S of $\mathbb{R}^n$ is called a **subspace** of $\mathbb{R}^n$ if for all vectors $\vec{x}, \vec{y} \in S$ and $t \in \mathbb{R}$:

(1) $\vec{x}+\vec{y} \in S$ (closed under addition)
(2) $t\vec{x} \in S$ (closed under scalar multiplication)

The definition requires that a subspace be non-empty. A subspace always contains at least one vector. In particular, it follows from (2) that if we let $t = 0$, then every subspace of $\mathbb{R}^n$ contains the zero vector. This fact provides an easy method for disqualifying any subsets that do not contain the zero vector as subspaces. For example, a line in $\mathbb{R}^3$ cannot be a subspace if it does not pass through the origin. Thus, when checking to determine if a set S is non-empty, it makes sense to first check if $\vec{0} \in S$.

It is easy to see that the set $\{\vec{0}\}$ consisting of only the zero vector in $\mathbb{R}^n$ is a subspace of $\mathbb{R}^n$; this is called the **trivial subspace**. Additionally, $\mathbb{R}^n$ is a subspace of itself. We will see throughout the text that other subspaces arise naturally in linear algebra.

EXAMPLE 1

Show that $S = \left\{ \begin{bmatrix} x_1 \\ x_2 \\ x_3 \end{bmatrix} \mid x_1 - x_2 + x_3 = 0 \right\}$ is a subspace of $\mathbb{R}^3$.

Solution: We observe that, by definition, S is a subset of $\mathbb{R}^3$ and that $\vec{0} = \begin{bmatrix} 0 \\ 0 \\ 0 \end{bmatrix} \in S$ since taking $x_1 = 0$, $x_2 = 0$, and $x_3 = 0$ satisfies $x_1 - x_2 + x_3 = 0$.

Let $\vec{x} = \begin{bmatrix} x_1 \\ x_2 \\ x_3 \end{bmatrix}, \vec{y} = \begin{bmatrix} y_1 \\ y_2 \\ y_3 \end{bmatrix} \in S$. Then they must satisfy the condition of the set, so $x_1 - x_2 + x_3 = 0$ and $y_1 - y_2 + y_3 = 0$.

To show that S is closed under addition, we must show that $\vec{x} + \vec{y}$ satisfies the condition of S. We have

$$\vec{x} + \vec{y} = \begin{bmatrix} x_1 + y_1 \\ x_2 + y_2 \\ x_3 + y_3 \end{bmatrix}$$

and

$$(x_1 + y_1) - (x_2 + y_2) + (x_3 + y_3) = x_1 - x_2 + x_3 + y_1 - y_2 + y_3 = 0 + 0 = 0$$

Hence, $\vec{x} + \vec{y} \in S$.

Similarly, for any $t \in \mathbb{R}$, we have $t\vec{x} = \begin{bmatrix} tx_1 \\ tx_2 \\ tx_3 \end{bmatrix}$ and

$$(tx_1) - (tx_2) + (tx_3) = t(x_1 - x_2 + x_3) = t(0) = 0$$

So, S is closed under scalar multiplication. Therefore, S is a subspace of $\mathbb{R}^3$.

EXAMPLE 2

Show that $T = \left\{ \begin{bmatrix} x_1 \\ x_2 \end{bmatrix} \mid x_1 x_2 = 0 \right\}$ is not a subspace of $\mathbb{R}^2$.

Solution: To show that T is not a subspace, we just need to give one example showing that T does not satisfy the definition of a subspace. We will show that T is not closed under addition.

Observe that $\vec{x} = \begin{bmatrix} 1 \\ 0 \end{bmatrix}$ and $\vec{y} = \begin{bmatrix} 0 \\ 1 \end{bmatrix}$ are both in T, but $\vec{x} + \vec{y} = \begin{bmatrix} 1 \\ 1 \end{bmatrix} \notin T$, since $1(1) \neq 0$. Thus, T is not a subspace of $\mathbb{R}^2$.

EXERCISE 2

Show that $S = \left\{ \begin{bmatrix} x_1 \\ x_2 \end{bmatrix} \mid 2x_1 = x_2 \right\}$ is a subspace of $\mathbb{R}^2$ and $T = \left\{ \begin{bmatrix} x_1 \\ x_2 \end{bmatrix} \mid x_1 + x_2 = 2 \right\}$ is not a subspace of $\mathbb{R}^2$.

Spanning Sets and Linear Independence

One of the main ways that subspaces arise is as the set of all linear combinations of some **spanning set**. We next present an easy theorem and a bit more vocabulary.

Theorem 2

If $\{\vec{v}_1, \dots, \vec{v}_k\}$ is a set of vectors in $\mathbb{R}^n$ and S is the set of all possible linear combinations of these vectors,

$$S = \{t_1\vec{v}_1 + \cdots + t_k\vec{v}_k \mid t_1, \dots, t_k \in \mathbb{R}\}$$

then S is a subspace of $\mathbb{R}^n$.

Proof: By properties (1) and (6) of Theorem 1, $t_1\vec{v}_1 + \cdots + t_k\vec{v}_k \in \mathbb{R}^n$, so S is a subset of $\mathbb{R}^n$. Taking $t_i = 0$ for $1 \leq i \leq k$, we get $\vec{0} = 0\vec{v}_1 + \cdots + 0\vec{v}_k \in S$, so S is non-empty.

Let $\vec{x}, \vec{y} \in S$. Then, for some real numbers s_i and t_i, $1 \leq i \leq k$, $\vec{x} = s_1\vec{v}_1 + \cdots + s_k\vec{v}_k$ and $\vec{y} = t_1\vec{v}_1 + \cdots + t_k\vec{v}_k$. It follows that

$$\vec{x} + \vec{y} = s_1\vec{v}_1 + \cdots + s_k\vec{v}_k + t_1\vec{v}_1 + \cdots + t_k\vec{v}_k = (s_1 + t_1)\vec{v}_1 + \cdots + (s_k + t_k)\vec{v}_k$$

so, $\vec{x} + \vec{y} \in S$ since $(s_i + t_i) \in \mathbb{R}$. Hence, S is closed under addition.

Similarly, for all $t \in \mathbb{R}$,

$$t\vec{x} = t(s_1\vec{v}_1 + \cdots + s_k\vec{v}_k) = (ts_1)\vec{v}_1 + \cdots (ts_k)\vec{v}_k \in S$$

So, S is closed under scalar multiplication. Therefore, S is a subspace of $\mathbb{R}^n$. ■

Definition
Span
Spanning Set

If S is the subspace of $\mathbb{R}^n$ consisting of all linear combinations of the vectors $\vec{v}_1, \dots, \vec{v}_k \in \mathbb{R}^n$, then S is called the subspace **spanned** by the set of vectors $\mathcal{B} = \{\vec{v}_1, \dots, \vec{v}_k\}$, and we say that the set $\mathcal{B}$ **spans** S. The set $\mathcal{B}$ is called a **spanning set** for the subspace S. We denote S by

$$S = \text{Span}\{\vec{v}_1, \dots, \vec{v}_k\} = \text{Span}\,\mathcal{B}$$

EXAMPLE 3

Let $\vec{v} \in \mathbb{R}^2$ with $\vec{v} \neq \vec{0}$ and consider the line L with vector equation $\vec{x} = t\vec{v}$, $t \in \mathbb{R}$. Then L is the subspace spanned by $\{\vec{v}\}$, and $\{\vec{v}\}$ is a spanning set for L. We write $L = \text{Span}\{\vec{v}\}$.

Similarly, for $\vec{v}_1, \vec{v}_2 \in \mathbb{R}^2$, the set M with vector equation $\vec{x} = t_1\vec{v}_1 + t_2\vec{v}_2$ is a subspace of $\mathbb{R}^2$ with spanning set $\{\vec{v}_1, \vec{v}_2\}$. That is, $M = \text{Span}\{\vec{v}_1, \vec{v}_2\}$.

If $\vec{v} \in \mathbb{R}^2$ with $\vec{v} \neq \vec{0}$, then we can guarantee that $\text{Span}\{\vec{v}\}$ represents a line in $\mathbb{R}^2$ that passes through the origin. However, we see that the geometrical interpretation of $\text{Span}\{\vec{v}_1, \vec{v}_2\}$ depends on the choices of $\vec{v}_1$ and $\vec{v}_2$. We demonstrate this with some examples.

EXAMPLE 4

The set $S_1 = \text{Span}\left\{\begin{bmatrix}0\\0\end{bmatrix}, \begin{bmatrix}1\\0\end{bmatrix}\right\}$ has vector equation

$$\vec{x} = t_1\begin{bmatrix}0\\0\end{bmatrix} + t_2\begin{bmatrix}1\\0\end{bmatrix} = t_2\begin{bmatrix}1\\0\end{bmatrix}, \quad t_2 \in \mathbb{R}$$

Hence, S_1 is a line in $\mathbb{R}^2$ that passes through the origin.

The set $S_2 = \text{Span}\left\{\begin{bmatrix}1\\0\end{bmatrix}, \begin{bmatrix}-2\\0\end{bmatrix}\right\}$ has vector equation

$$\vec{x} = t_1\begin{bmatrix}1\\0\end{bmatrix} + t_2\begin{bmatrix}-2\\0\end{bmatrix} = (t_1 - 2t_2)\begin{bmatrix}1\\0\end{bmatrix} = t\begin{bmatrix}1\\0\end{bmatrix}$$

where $t = t_1 - 2t_2 \in \mathbb{R}$. Hence, S_2 represents the same line as S_1. That is,

$$S_2 = \text{Span}\left\{\begin{bmatrix}1\\0\end{bmatrix}\right\} = S_1$$

The set $S_3 = \text{Span}\left\{\begin{bmatrix}1\\0\end{bmatrix}, \begin{bmatrix}0\\1\end{bmatrix}\right\}$ has vector equation

$$\vec{x} = t_1\begin{bmatrix}1\\0\end{bmatrix} + t_2\begin{bmatrix}0\\1\end{bmatrix} = \begin{bmatrix}t_1\\t_2\end{bmatrix}, \quad t_1, t_2 \in \mathbb{R}$$

Hence, $S_3 = \mathbb{R}^2$. That is, S_3 spans the entire two-dimensional plane.

From these examples, we observe that $\{\vec{v}_1, \vec{v}_2\}$ is a spanning set for $\mathbb{R}^2$ if and only if neither $\vec{v}_1$ nor $\vec{v}_2$ is a scalar multiple of the other. This also means that neither vector can be the zero vector. We now look at this in $\mathbb{R}^3$.

EXAMPLE 5

The set $S_1 = \text{Span}\left\{\begin{bmatrix}0\\-2\\-2\end{bmatrix}, \begin{bmatrix}0\\3\\3\end{bmatrix}, \begin{bmatrix}0\\\sqrt{2}\\\sqrt{2}\end{bmatrix}\right\}$ has vector equation

$$\vec{x} = t_1\begin{bmatrix}0\\-2\\-2\end{bmatrix} + t_2\begin{bmatrix}0\\3\\3\end{bmatrix} + t_3\begin{bmatrix}0\\\sqrt{2}\\\sqrt{2}\end{bmatrix} = (-2t_1 + 3t_2 + \sqrt{2}t_3)\begin{bmatrix}0\\1\\1\end{bmatrix}, \quad t_1, t_2, t_3 \in \mathbb{R}$$

Hence,

$$S_1 = \text{Span}\left\{\begin{bmatrix}0\\1\\1\end{bmatrix}\right\}$$

EXAMPLE 5
(continued)

The set $S_2 = \operatorname{Span}\left\{\begin{bmatrix}1\\0\\-1\end{bmatrix}, \begin{bmatrix}1\\1\\-1\end{bmatrix}, \begin{bmatrix}0\\1\\0\end{bmatrix}\right\}$ has vector equation

$$\vec{x} = t_1\begin{bmatrix}1\\0\\-1\end{bmatrix} + t_2\begin{bmatrix}1\\1\\-1\end{bmatrix} + t_3\begin{bmatrix}0\\1\\0\end{bmatrix}, \quad t_1, t_2, t_3 \in \mathbb{R}$$

which can be written as

$$\vec{x} = t_1\begin{bmatrix}1\\0\\-1\end{bmatrix} + t_2\begin{bmatrix}1\\0\\-1\end{bmatrix} + t_2\begin{bmatrix}0\\1\\0\end{bmatrix} + t_3\begin{bmatrix}0\\1\\0\end{bmatrix} = (t_1 + t_2)\begin{bmatrix}1\\0\\-1\end{bmatrix} + (t_2 + t_3)\begin{bmatrix}0\\1\\0\end{bmatrix}$$

So, $S_2 = \operatorname{Span}\left\{\begin{bmatrix}1\\0\\-1\end{bmatrix}, \begin{bmatrix}0\\1\\0\end{bmatrix}\right\}$.

We extend this to the general case in $\mathbb{R}^n$.

Theorem 3

Let $\vec{v}_1, \dots, \vec{v}_k$ be vectors in $\mathbb{R}^n$. If $\vec{v}_k$ can be written as a linear combination of $\vec{v}_1, \dots, \vec{v}_{k-1}$, then

$$\operatorname{Span}\{\vec{v}_1, \dots, \vec{v}_k\} = \operatorname{Span}\{\vec{v}_1, \dots, \vec{v}_{k-1}\}$$

Proof: We are assuming that there exists $t_1, \dots, t_{k-1} \in \mathbb{R}$ such that

$$t_1\vec{v}_1 + \cdots + t_{k-1}\vec{v}_{k-1} = \vec{v}_k$$

Let $\vec{x} \in \operatorname{Span}\{\vec{v}_1, \dots, \vec{v}_k\}$. Then, there exists $s_1, \dots, s_k \in \mathbb{R}$ such that

$$\begin{aligned}\vec{x} &= s_1\vec{v}_1 + \cdots + s_{k-1}\vec{v}_{k-1} + s_k\vec{v}_k \\ &= s_1\vec{v}_1 + \cdots + s_{k-1}\vec{v}_{k-1} + s_k(t_1\vec{v}_1 + \cdots + t_{k-1}\vec{v}_{k-1}) \\ &= (s_1 + s_kt_1)\vec{v}_1 + \cdots + (s_{k-1} + s_kt_{k-1})\vec{v}_{k-1}\end{aligned}$$

Thus, $\vec{x} \in \operatorname{Span}\{\vec{v}_1, \dots, \vec{v}_{k-1}\}$. Hence, $\operatorname{Span}\{\vec{v}_1, \dots, \vec{v}_k\} \subseteq \operatorname{Span}\{\vec{v}_1, \dots, \vec{v}_{k-1}\}$. Clearly, we have $\operatorname{Span}\{\vec{v}_1, \dots, \vec{v}_{k-1}\} \subseteq \operatorname{Span}\{\vec{v}_1, \dots, \vec{v}_k\}$ and so

$$\operatorname{Span}\{\vec{v}_1, \dots, \vec{v}_k\} = \operatorname{Span}\{\vec{v}_1, \dots, \vec{v}_{k-1}\}$$

as required. ■

In fact, any vector which can be written as a linear combination of the other vectors in the set can be removed without changing the spanned set. It is important in linear algebra to identify when a spanning set can be simplified by removing a vector that can be written as a linear combination of the other vectors. We will call such sets **linearly dependent**. If a spanning set is as simple as possible, then we will call the set **linearly independent**. To identify whether a set is linearly dependent or linearly independent, we require a mathematical definition.

Assume that the set $\{\vec{v}_1, \ldots, \vec{v}_k\}$ is linearly dependent. Then one of the vectors, say $\vec{v}_i$, is equal to a linear combination of some (or all) of the other vectors. Hence, we can find scalars $t_1, \ldots t_k \in \mathbb{R}$ such that

$$-t_i\vec{v}_i = t_1\vec{v}_1 + \cdots + t_{i-1}\vec{v}_{i-1} + t_{i+1}\vec{v}_{i+1} + \cdots + t_k\vec{v}_k$$

where $t_i \neq 0$. Thus, a set is linearly dependent if the equation

$$\vec{0} = t_1\vec{v}_1 + \cdots + t_{i-1}\vec{v}_{i-1} - t_i\vec{v}_i + t_{i+1}\vec{v}_{i+1} + \cdots + t_k\vec{v}_k$$

has a solution where at least one of the coefficients is non-zero. On the other hand, if the set is linearly independent, then the only solution to this equation must be when all the coefficients are 0. For example, if any coefficient is non-zero, say $t_i \neq 0$, then we can write

$$-t_i\vec{v}_i = t_1\vec{v}_1 + \cdots + t_{i-1}\vec{v}_{i-1} + t_{i+1}\vec{v}_{i+1} + \cdots + t_k\vec{v}_k$$

Thus, $\vec{v}_i \in \operatorname{Span}\{\vec{v}_1, \ldots, \vec{v}_{i-1}, \vec{v}_{i+1}, \ldots, \vec{v}_n\}$, and so the set can be simplified by using Theorem 3.

We make this our mathematical definition.

Definition
Linearly Dependent
Linearly Independent

A set of vectors $\{\vec{v}_1, \ldots, \vec{v}_k\}$ is said to be **linearly dependent** if there exist coefficients $t_1, \ldots, t_k$ not all zero such that

$$\vec{0} = t_1\vec{v}_1 + \cdots + t_k\vec{v}_k$$

A set of vectors $\{\vec{v}_1, \ldots, \vec{v}_k\}$ is said to be **linearly independent** if the only solution to

$$\vec{0} = t_1\vec{v}_1 + \cdots + t_k\vec{v}_k$$

is $t_1 = t_2 = \cdots = t_k = 0$. This is called the **trivial solution**.

Theorem 4

If a set of vectors $\{\vec{v}_1, \ldots, \vec{v}_k\}$ contains the zero vector, then it is linearly dependent.

Proof: Assume $\vec{v}_i = \vec{0}$. Then we have

$$0\vec{v}_1 + \cdots + 0\vec{v}_{i-1} + 1\vec{v}_i + 0\vec{v}_{i+1} + \cdots + 0\vec{v}_k = \vec{0}$$

Hence, the equation $\vec{0} = t_1\vec{v}_1 + \cdots + t_k\vec{v}_k$ has a solution with one coefficient, t_i, that is non-zero. So, by definition, the set is linearly dependent. ■

EXAMPLE 6

Show that the set $\left\{\begin{bmatrix} 7 \\ -14 \\ 6 \end{bmatrix}, \begin{bmatrix} -10 \\ 15 \\ 15/14 \end{bmatrix}, \begin{bmatrix} -1 \\ 0 \\ 3 \end{bmatrix}\right\}$ is linearly dependent.

Solution: We consider

$$t_1\begin{bmatrix} 7 \\ -14 \\ 6 \end{bmatrix} + t_2\begin{bmatrix} -10 \\ 15 \\ 15/14 \end{bmatrix} + t_3\begin{bmatrix} -1 \\ 0 \\ 3 \end{bmatrix} = \begin{bmatrix} 0 \\ 0 \\ 0 \end{bmatrix}$$

EXAMPLE 6
(continued)

Using operations on vectors, we get

$$\begin{bmatrix} 7t_1 - 10t_2 - t_3 \\ -14t_1 + 15t_2 \\ 6t_1 + \frac{15}{14}t_2 + 3t_3 \end{bmatrix} = \begin{bmatrix} 0 \\ 0 \\ 0 \end{bmatrix}$$

Since vectors are equal only if their corresponding entries are equal, this gives us three equations in three unknowns

$$\begin{aligned} 7t_1 - 10t_2 - t_3 &= 0 \\ -14t_1 + 15t_2 &= 0 \\ 6t_1 + \frac{15}{14}t_2 + 3t_3 &= 0 \end{aligned}$$

Solving using substitution and elimination, we find that there are in fact infinitely many possible solutions. One is $t_1 = \frac{3}{7}$, $t_2 = \frac{2}{5}$, $t_3 = -1$. Hence, the set is linearly dependent.

EXERCISE 3

Determine whether $\left\{ \begin{bmatrix} 1 \\ 0 \\ 1 \end{bmatrix}, \begin{bmatrix} 0 \\ 1 \\ 1 \end{bmatrix}, \begin{bmatrix} 1 \\ 1 \\ 0 \end{bmatrix} \right\}$ is linearly dependent or linearly independent.

Remark

Observe that determining whether a set $\{\vec{v}_1, \dots, \vec{v}_k\}$ in $\mathbb{R}^n$ is linearly dependent or linearly independent requires determining solutions of the equation $t_1\vec{v}_1 + \cdots + t_k\vec{v}_k = \vec{0}$. However, this equation actually represents n equations (one for each entry of the vectors) in k unknowns $t_1, \dots, t_k$. In the next chapter, we will look at how to efficiently solve such systems of equations.

What we have derived above is that the simplest spanning set for a subspace S is one that is linearly independent. Hence, we make the following definition.

Definition
Basis

If $\{\vec{v}_1, \dots, \vec{v}_k\}$ is a spanning set for a subspace S of $\mathbb{R}^n$ and $\{\vec{v}_1, \dots, \vec{v}_k\}$ is linearly independent, then $\{\vec{v}_1, \dots, \vec{v}_k\}$ is called a **basis** for S.

EXAMPLE 7

Let $\vec{v}_1 = \begin{bmatrix} 1 \\ -2 \\ 0 \end{bmatrix}$, $\vec{v}_2 = \begin{bmatrix} -1 \\ 1 \\ 1 \end{bmatrix}$, $\vec{v}_3 = \begin{bmatrix} 1 \\ -4 \\ 2 \end{bmatrix}$, and let S be the subspace of $\mathbb{R}^3$ given by $S = \operatorname{Span}\{\vec{v}_1, \vec{v}_2, \vec{v}_3\}$. Find a basis for S.

Solution: Observe that $\{\vec{v}_1, \vec{v}_2, \vec{v}_3\}$ is linearly dependent, since

$$3\begin{bmatrix} 1 \\ -2 \\ 0 \end{bmatrix} + 2\begin{bmatrix} -1 \\ 1 \\ 1 \end{bmatrix} - \begin{bmatrix} 1 \\ -4 \\ 2 \end{bmatrix} = \begin{bmatrix} 0 \\ 0 \\ 0 \end{bmatrix}$$

EXAMPLE 7
(continued)

In particular, we can write $\vec{v}_3$ as a linear combination of $\vec{v}_1$ and $\vec{v}_2$. Hence, by Theorem 3,

$$S = \text{Span}\{\vec{v}_1, \vec{v}_2, \vec{v}_3\} = \text{Span}\{\vec{v}_1, \vec{v}_2\}$$

Moreover, observe that the only solution to

$$t_1 \begin{bmatrix} 1 \\ -2 \\ 0 \end{bmatrix} + t_2 \begin{bmatrix} -1 \\ 1 \\ 1 \end{bmatrix} = \begin{bmatrix} 0 \\ 0 \\ 0 \end{bmatrix}$$

is $t_1 = t_2 = 0$ since neither $\vec{v}_1$ nor $\vec{v}_2$ is a scalar multiple of the other. Hence, $\{\vec{v}_1, \vec{v}_2\}$ is linearly independent.
Therefore, $\{\vec{v}_1, \vec{v}_2\}$ is linearly independent and spans S and so it is a basis for S.

Bases (the plural of basis) will be extremely important throughout the remainder of the book. At this point, however, we just define the following very important basis.

Definition
Standard Basis for $\mathbb{R}^n$

In $\mathbb{R}^n$, let $\vec{e}_i$ represent the vector whose i-th component is 1 and all other components are 0. The set $\{\vec{e}_1, \dots, \vec{e}_n\}$ is called the **standard basis for $\mathbb{R}^n$**.

Observe that this definition matches that of the standard basis for $\mathbb{R}^2$ and $\mathbb{R}^3$ given in Section 1.1.

EXAMPLE 8

The standard basis for $\mathbb{R}^3$ is $\{\vec{e}_1, \vec{e}_2, \vec{e}_3\} = \left\{ \begin{bmatrix} 1 \\ 0 \\ 0 \end{bmatrix}, \begin{bmatrix} 0 \\ 1 \\ 0 \end{bmatrix}, \begin{bmatrix} 0 \\ 0 \\ 1 \end{bmatrix} \right\}$.

It is linearly independent since the only solution to

$$\begin{bmatrix} 0 \\ 0 \\ 0 \end{bmatrix} = t_1 \begin{bmatrix} 1 \\ 0 \\ 0 \end{bmatrix} + t_2 \begin{bmatrix} 0 \\ 1 \\ 0 \end{bmatrix} + t_3 \begin{bmatrix} 0 \\ 0 \\ 1 \end{bmatrix} = \begin{bmatrix} t_1 \\ t_2 \\ t_3 \end{bmatrix}$$

is $t_1 = t_2 = t_3 = 0$. Moreover, it is a spanning set for $\mathbb{R}^3$ since every vector $\vec{x} = \begin{bmatrix} x_1 \\ x_2 \\ x_3 \end{bmatrix} \in \mathbb{R}^3$ can be written as a linear combination of the basis vectors. In particular,

$$\begin{bmatrix} x_1 \\ x_2 \\ x_3 \end{bmatrix} = x_1 \begin{bmatrix} 1 \\ 0 \\ 0 \end{bmatrix} + x_2 \begin{bmatrix} 0 \\ 1 \\ 0 \end{bmatrix} + x_3 \begin{bmatrix} 0 \\ 0 \\ 1 \end{bmatrix}$$

Remark

Compare the result of Example 8 with the meaning of point notation $P(a, b, c)$. When we say $P(a, b, c)$ we mean the point P having a amount in the x-direction, b amount in the y-direction, and c amount in the z-direction. So, observe that the standard basis vectors represent our usual coordinate axes.

EXERCISE 4

State the standard basis for $\mathbb{R}^5$. Prove that it is linearly independent and show that it is a spanning set for $\mathbb{R}^5$.

Surfaces in Higher Dimensions

We can now extend our geometrical concepts of lines and planes to $\mathbb{R}^n$ for $n > 3$. To match what we did in $\mathbb{R}^2$ and $\mathbb{R}^3$, we make the following definition.

Definition — Line in $\mathbb{R}^n$

Let $\vec{p}, \vec{v} \in \mathbb{R}^n$ with $\vec{v} \neq \vec{0}$. Then we call the set with vector equation $\vec{x} = \vec{p} + t_1\vec{v}$, $t_1 \in \mathbb{R}$ a **line** in $\mathbb{R}^n$ that passes through $\vec{p}$.

Definition — Plane in $\mathbb{R}^n$

Let $\vec{v}_1, \vec{v}_2, \vec{p} \in \mathbb{R}^n$, with $\{\vec{v}_1, \vec{v}_2\}$ being a linearly independent set. Then the set with vector equation $\vec{x} = \vec{p} + t_1\vec{v}_1 + t_2\vec{v}_2$, $t_1, t_2 \in \mathbb{R}$ is called a **plane** in $\mathbb{R}^n$ that passes through $\vec{p}$.

Definition — Hyperplane in $\mathbb{R}^n$

Let $\vec{v}_1, \dots, \vec{v}_{n-1}, \vec{p} \in \mathbb{R}^n$, with $\{\vec{v}_1, \dots, \vec{v}_{n-1}\}$ being linearly independent. Then the set with vector equation $\vec{x} = \vec{p} + t_1\vec{v}_1 + \cdots + t_{n-1}\vec{v}_{n-1}$, $t_i \in \mathbb{R}$ is called a **hyperplane** in $\mathbb{R}^n$ that passes through $\vec{p}$.

EXAMPLE 9

The set $\operatorname{Span}\left\{\begin{bmatrix}1\\0\\0\\1\end{bmatrix}, \begin{bmatrix}0\\1\\0\\-2\end{bmatrix}, \begin{bmatrix}0\\1\\1\\-1\end{bmatrix}\right\}$ is a hyperplane since $\left\{\begin{bmatrix}1\\0\\0\\1\end{bmatrix}, \begin{bmatrix}0\\1\\0\\-2\end{bmatrix}, \begin{bmatrix}0\\1\\1\\-1\end{bmatrix}\right\}$ is linearly independent in $\mathbb{R}^4$.

EXAMPLE 10

Show that the set $\operatorname{Span}\left\{\begin{bmatrix}1\\2\\1\end{bmatrix}, \begin{bmatrix}-1\\1\\2\end{bmatrix}, \begin{bmatrix}0\\3\\3\end{bmatrix}\right\}$ defines a plane in $\mathbb{R}^3$.

Solution: Observe that the set is linearly dependent since $\begin{bmatrix}1\\2\\1\end{bmatrix} + \begin{bmatrix}-1\\1\\2\end{bmatrix} = \begin{bmatrix}0\\3\\3\end{bmatrix}$. Hence, the simplified vector equation of the set spanned by these vectors is

EXAMPLE 10
(continued)

$$\vec{x} = t_1\begin{bmatrix}1\\2\\1\end{bmatrix} + t_2\begin{bmatrix}-1\\1\\2\end{bmatrix}, \quad t_1, t_2 \in \mathbb{R}$$

Since the set $\left\{\begin{bmatrix}1\\2\\1\end{bmatrix}, \begin{bmatrix}-1\\1\\2\end{bmatrix}\right\}$ is linearly independent, this is the vector equation of a plane passing through the origin in $\mathbb{R}^3$.

PROBLEMS 1.2

Practice Problems

A1 Compute each of the following linear combinations.

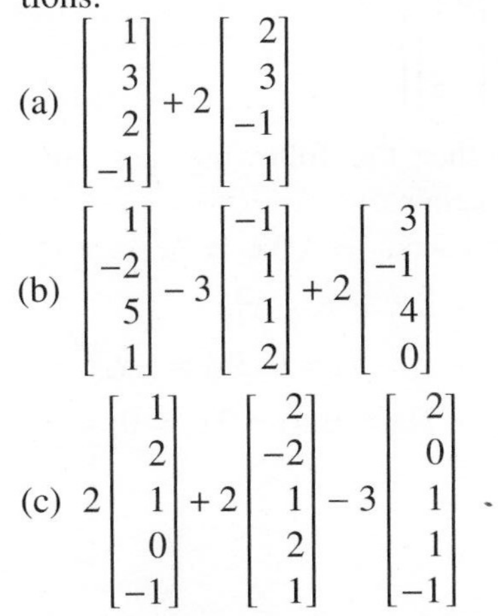

(a) $\begin{bmatrix}1\\3\\2\\-1\end{bmatrix} + 2\begin{bmatrix}2\\3\\-1\\1\end{bmatrix}$

(b) $\begin{bmatrix}1\\-2\\5\\1\end{bmatrix} - 3\begin{bmatrix}-1\\1\\1\\2\end{bmatrix} + 2\begin{bmatrix}3\\-1\\4\\0\end{bmatrix}$

(c) $2\begin{bmatrix}1\\2\\1\\0\\-1\end{bmatrix} + 2\begin{bmatrix}2\\-2\\1\\2\\1\end{bmatrix} - 3\begin{bmatrix}2\\0\\1\\1\\-1\end{bmatrix}$

A2 For each of the following sets, show that the set is or is not a subspace of the appropriate $\mathbb{R}^n$.

(a) $\left\{\begin{bmatrix}x_1\\x_2\\x_3\end{bmatrix} \mid x_1^2 - x_2^2 = x_3\right\}$

(b) $\left\{\begin{bmatrix}x_1\\x_2\\x_3\end{bmatrix} \mid x_1 = x_3\right\}$

(c) $\left\{\begin{bmatrix}x_1\\x_2\end{bmatrix} \mid x_1 + x_2 = 0\right\}$

(d) $\left\{\begin{bmatrix}x_1\\x_2\\x_3\end{bmatrix} \mid x_1x_2 = x_3\right\}$

(e) $\vec{x} = \begin{bmatrix}2\\3\\4\end{bmatrix} + t_1\begin{bmatrix}1\\1\\1\end{bmatrix} + t_2\begin{bmatrix}1\\2\\3\end{bmatrix}$

(f) $\text{Span}\left\{\begin{bmatrix}1\\0\\1\\2\end{bmatrix}\right\}$

A3 Determine whether the following sets are subspaces of $\mathbb{R}^4$. Explain.

(a) $\{\vec{x} \in \mathbb{R}^4 \mid x_1 + x_2 + x_3 + x_4 = 0\}$

(b) $\{\vec{v}_1\}$, where $\vec{v}_1 \neq \vec{0}$.

(c) $\{\vec{x} \in \mathbb{R}^4 \mid x_1 + 2x_3 = 5, x_1 - 3x_4 = 0\}$

(d) $\{\vec{x} \in \mathbb{R}^4 \mid x_1 = x_3x_4, x_2 - x_4 = 0\}$

(e) $\{\vec{x} \in \mathbb{R}^4 \mid 2x_1 = 3x_4, x_2 - 5x_3 = 0\}$

(f) $\{\vec{x} \in \mathbb{R}^4 \mid x_1 + x_2 = -x_4, x_3 = 2\}$

A4 Show that each of the following sets is linearly dependent. Do so by writing a non-trivial linear combination of the vectors that equals the zero vector.

(a) $\left\{\begin{bmatrix}0\\0\\0\end{bmatrix}, \begin{bmatrix}1\\0\\1\end{bmatrix}, \begin{bmatrix}2\\1\\-1\end{bmatrix}\right\}$

(b) $\left\{\begin{bmatrix}2\\-1\\3\end{bmatrix}, \begin{bmatrix}0\\2\\1\end{bmatrix}, \begin{bmatrix}0\\4\\2\end{bmatrix}\right\}$

(c) $\left\{\begin{bmatrix}1\\1\\0\end{bmatrix}, \begin{bmatrix}1\\1\\1\end{bmatrix}, \begin{bmatrix}2\\2\\1\end{bmatrix}\right\}$

(d) $\left\{\begin{bmatrix}1\\1\end{bmatrix}, \begin{bmatrix}1\\2\end{bmatrix}, \begin{bmatrix}1\\3\end{bmatrix}\right\}$

A5 Determine if the following sets represent lines, planes, or hyperplanes in $\mathbb{R}^4$. Give a basis for each.

(a) $\operatorname{Span}\left\{\begin{bmatrix}1\\0\\1\\1\end{bmatrix}, \begin{bmatrix}1\\2\\1\\3\end{bmatrix}\right\}$

(b) $\operatorname{Span}\left\{\begin{bmatrix}1\\0\\0\\0\end{bmatrix}, \begin{bmatrix}0\\1\\0\\0\end{bmatrix}, \begin{bmatrix}0\\0\\0\\1\end{bmatrix}\right\}$

(c) $\operatorname{Span}\left\{\begin{bmatrix}3\\1\\-1\\0\end{bmatrix}, \begin{bmatrix}0\\0\\0\\0\end{bmatrix}, \begin{bmatrix}6\\2\\-2\\0\end{bmatrix}\right\}$

(d) $\operatorname{Span}\left\{\begin{bmatrix}1\\1\\0\\2\end{bmatrix}, \begin{bmatrix}1\\0\\0\\-1\end{bmatrix}, \begin{bmatrix}2\\1\\0\\1\end{bmatrix}\right\}$

A6 Let $\vec{p}, \vec{d} \in \mathbb{R}^n$. Prove that $\vec{x} = \vec{p} + t\vec{d}$, $t \in \mathbb{R}$ is a subspace of $\mathbb{R}^n$ if and only if $\vec{p}$ is a scalar multiple of $\vec{d}$.

A7 Suppose that $\mathcal{B} = \{\vec{v}_1, \dots, \vec{v}_k\}$ is a linearly independent set in $\mathbb{R}^n$. Prove that any non-empty subset of $\mathcal{B}$ is linearly independent.

Homework Problems

B1 Compute each of the following linear combinations.

(a) $\begin{bmatrix}2\\1\\-1\\3\end{bmatrix} - \begin{bmatrix}4\\-2\\1\\-2\end{bmatrix}$

(b) $\begin{bmatrix}2\\-2\\3\\2\\1\end{bmatrix} + 2\begin{bmatrix}2\\1\\2\\3\\0\end{bmatrix} + 2\begin{bmatrix}2\\1\\3\\1\\-1\end{bmatrix}$

(c) $3\begin{bmatrix}1\\1\\0\\0\\2\end{bmatrix} - 2\begin{bmatrix}1\\-2\\3\\-4\\3\end{bmatrix} + \begin{bmatrix}-1\\-7\\6\\-7\\0\end{bmatrix}$

B2 For each of the following sets, show that the set is or is not a subspace of the appropriate $\mathbb{R}^n$.

(a) $\left\{\begin{bmatrix}x_1\\x_2\end{bmatrix} \mid x_1 + x_2 = 1\right\}$

(b) $\left\{\begin{bmatrix}x_1\\x_2\\x_3\end{bmatrix} \mid x_1 + x_2 \geq 0\right\}$

(c) $\left\{\begin{bmatrix}x_1\\x_2\\x_3\end{bmatrix} \mid x_1 - 2x_2 = x_3, x_1 + x_2 = 0\right\}$

(d) $\left\{\begin{bmatrix}x_1\\x_2\\x_3\end{bmatrix} \mid x_1^2 = x_2x_3\right\}$

(e) $\vec{x} = \begin{bmatrix}1\\0\\1\end{bmatrix} + t_1\begin{bmatrix}2\\1\\2\end{bmatrix} + t_2\begin{bmatrix}1\\1\\1\end{bmatrix}$

(f) $\operatorname{Span}\left\{\begin{bmatrix}1\\2\\1\end{bmatrix}, \begin{bmatrix}-1\\1\\3\end{bmatrix}\right\}$

B3 Determine whether the following sets are subspaces of $\mathbb{R}^4$. Explain.

(a) $\left\{\vec{x} \in \mathbb{R}^4 \mid 2x_1 - 5x_4 = 7, 3x_2 = 2x_4\right\}$

(b) $\left\{\vec{x} \in \mathbb{R}^4 \mid x_1^2 + 2x_2^2 = x_3^2 + x_4^2\right\}$

(c) $\left\{\vec{x} \in \mathbb{R}^4 \mid x_1 + x_2 + x_4 = 0, 3x_3 = -x_4\right\}$

(d) $\left\{\vec{x} \in \mathbb{R}^4 \mid x_1 + 2x_3 = 0, x_1 - 3x_4 = 0\right\}$

(e) $\left\{\begin{bmatrix}1\\0\\0\\0\end{bmatrix}, \begin{bmatrix}0\\1\\0\\1\end{bmatrix}\right\}$

(f) $\operatorname{Span}\left\{\begin{bmatrix}1\\2\\1\end{bmatrix}, \begin{bmatrix}-1\\1\\3\end{bmatrix}\right\}$

B4 Show that each of the following sets is linearly dependent by writing a non-trivial linear combination of the vectors that equals the zero vector.

(a) $\left\{\begin{bmatrix}1\\1\\1\end{bmatrix}, \begin{bmatrix}2\\2\\2\end{bmatrix}\right\}$

(b) $\left\{\begin{bmatrix}3\\-2\\4\end{bmatrix}, \begin{bmatrix}1\\2\\-1\end{bmatrix}, \begin{bmatrix}-6\\4\\-8\end{bmatrix}\right\}$

(c) $\left\{\begin{bmatrix}2\\1\\2\end{bmatrix}, \begin{bmatrix}5\\4\\5\end{bmatrix}, \begin{bmatrix}1\\1\\1\end{bmatrix}\right\}$

(d) $\left\{\begin{bmatrix}3\\1\end{bmatrix}, \begin{bmatrix}1\\2\end{bmatrix}, \begin{bmatrix}-1\\2\end{bmatrix}\right\}$

B5 Determine if the following sets represent lines, planes, or hyperplanes in $\mathbb{R}^4$. Give a basis for each.

(a) Span $\left\{\begin{bmatrix}2\\-1\\1\\2\end{bmatrix}, \begin{bmatrix}-2\\1\\-1\\-2\end{bmatrix}\right\}$

(b) Span $\left\{\begin{bmatrix}2\\0\\0\\1\end{bmatrix}, \begin{bmatrix}0\\2\\0\\1\end{bmatrix}, \begin{bmatrix}0\\0\\2\\1\end{bmatrix}\right\}$

(c) Span $\left\{\begin{bmatrix}1\\2\\2\\0\end{bmatrix}, \begin{bmatrix}0\\0\\0\\0\end{bmatrix}, \begin{bmatrix}0\\1\\0\\1\end{bmatrix}\right\}$

(d) Span $\left\{\begin{bmatrix}3\\1\\2\\1\end{bmatrix}, \begin{bmatrix}1\\1\\2\\1\end{bmatrix}, \begin{bmatrix}2\\0\\0\\0\end{bmatrix}\right\}$

Computer Problems

C1 Let $\vec{v}_1 = \begin{bmatrix}1.23\\4.16\\-2.21\\0.34\end{bmatrix}$, $\vec{v}_2 = \begin{bmatrix}4.21\\-3.14\\0\\2.71\end{bmatrix}$, $\vec{v}_3 = \begin{bmatrix}-9.6\\1.01\\2.02\\1.99\end{bmatrix}$,

and $\vec{v}_4 = \begin{bmatrix}0.33\\2.12\\-3.23\\0.89\end{bmatrix}$.

Use computer software to evaluate each of the following.

(a) $3\vec{v}_1 - 2\vec{v}_2 + 5\vec{v}_3 - 3\vec{v}_4$

(b) $2.4\vec{v}_1 - 1.3\vec{v}_2 + \sqrt{2}\vec{v}_3 - \sqrt{3}\vec{v}_4$.

Conceptual Problems

D1 Prove property (8) from Theorem 1.

D2 Prove property (9) from Theorem 1.

D3 Let U and V be subspaces of $\mathbb{R}^n$.

(a) Prove that the intersection of U and V is a subspace of $\mathbb{R}^n$.

(b) Give an example to show that the union of two subspaces of $\mathbb{R}^n$ does not have to be a subspace of $\mathbb{R}^n$.

(c) Define $U + V = \{\vec{u} + \vec{v} \mid \vec{u} \in U, \vec{v} \in V\}$. Prove that $U + V$ is a subspace of $\mathbb{R}^n$.

D4 Pick vectors $\vec{p}, \vec{v}_1, \vec{v}_2$, and $\vec{v}_3$ in $\mathbb{R}^4$ such that the vector equation $\vec{x} = \vec{p} + t_1\vec{v}_1 + t_2\vec{v}_2 + t_3\vec{v}_3$

(a) Is a hyperplane not passing through the origin

(b) Is a plane passing through the origin

(c) Is the point $(1, 3, 1, 1)$

(d) Is a line passing through the origin

D5 Let $\mathcal{B} = \{\vec{v}_1, \dots, \vec{v}_k\}$ be a linearly independent set of vectors in $\mathbb{R}^n$. Prove that every vector in Span $\mathcal{B}$ can be written as a unique linear combination of the vectors in $\mathcal{B}$.

D6 Let $\mathcal{B} = \{\vec{v}_1, \dots, \vec{v}_k\}$ be a linearly independent set of vectors in $\mathbb{R}^n$ and let $\vec{x} \notin \operatorname{Span}\mathcal{B}$. Prove that $\{\vec{v}_1, \dots, \vec{v}_k, \vec{x}\}$ is linearly independent.

D7 Let $\vec{v}_1, \vec{v}_2 \in \mathbb{R}^n$ and let s and t be fixed real numbers. Prove that

$$\operatorname{Span}\{\vec{v}_1, \vec{v}_2\} = \operatorname{Span}\{\vec{v}_1, s\vec{v}_1 + t\vec{v}_2\}$$

D8 Let $\vec{v}_1, \vec{v}_2, \vec{v}_3 \in \mathbb{R}^n$. State whether each of the following statements is true or false. If the statement is true, explain briefly. If the statement is false, give a counterexample.

(a) If $\vec{v}_2 = t\vec{v}_1$ for some real number t, then $\{\vec{v}_1, \vec{v}_2\}$ is linearly dependent.

(b) If $\vec{v}_1$ is not a scalar multiple of $\vec{v}_2$, then $\{\vec{v}_1, \vec{v}_2\}$ is linearly independent.

(c) If $\{\vec{v}_1, \vec{v}_2, \vec{v}_3\}$ is linearly dependent, then $\vec{v}_1$ can be written as a linear combination of $\vec{v}_2$ and $\vec{v}_3$.

(d) If $\vec{v}_1$ can be written as a linear combination of $\vec{v}_2$ and $\vec{v}_3$, then $\{\vec{v}_1, \vec{v}_2, \vec{v}_3\}$ is linearly dependent.

(e) $\{\vec{v}_1\}$ is not a subspace of $\mathbb{R}^n$.

(f) Span$\{\vec{v}_1\}$ is a subspace of $\mathbb{R}^n$.

1.3 Length and Dot Products

In many physical applications, we are given measurements in terms of angles and magnitudes. We must convert this data into vectors so that we can apply the tools of linear algebra to solve problems. For example, we may need to find a vector representing the path (and speed) of a plane flying northwest at 1300 km/h. To do this, we need to identify the length of a vector and the angle between two vectors. In this section, we see how we can calculate both of these quantities with the dot product operator.

Length and Dot Products in $\mathbb{R}^2$, and $\mathbb{R}^3$

The length of a vector in $\mathbb{R}^2$ is defined by the usual distance formula (that is, Pythagoras' Theorem), as in Figure 1.3.10.

Definition
Length in $\mathbb{R}^2$

If $\vec{x} = \begin{bmatrix} x_1 \\ x_2 \end{bmatrix} \in \mathbb{R}^2$, its **length** is defined to be

$$\|\vec{x}\| = \sqrt{x_1^2 + x_2^2}$$

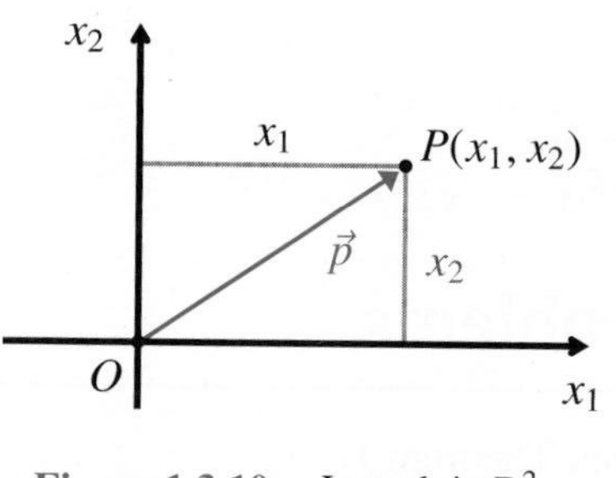

Figure 1.3.10 Length in $\mathbb{R}^2$.

For vectors in $\mathbb{R}^3$, the formula for the length can be obtained from a two-step calculation using the formula for $\mathbb{R}^2$, as shown in Figure 1.3.11. Consider the point $X(x_1, x_2, x_3)$ and let P be the point $P(x_1, x_2, 0)$. Observe that OPX is a right triangle, so that

$$\|\vec{x}\|^2 = \|\vec{OP}\|^2 + \|\vec{PX}\|^2 = (x_1^2 + x_2^2) + x_3^2$$

Definition
Length in $\mathbb{R}^3$

If $\vec{x} = \begin{bmatrix} x_1 \\ x_2 \\ x_3 \end{bmatrix} \in \mathbb{R}^3$, its **length** is defined to be

$$\|\vec{x}\| = \sqrt{x_1^2 + x_2^2 + x_3^2}$$

One immediate application of this formula is to calculate the distance between two points. In particular, if we have points P and Q, then the distance between them is the length of the directed line segment $\vec{PQ}$.

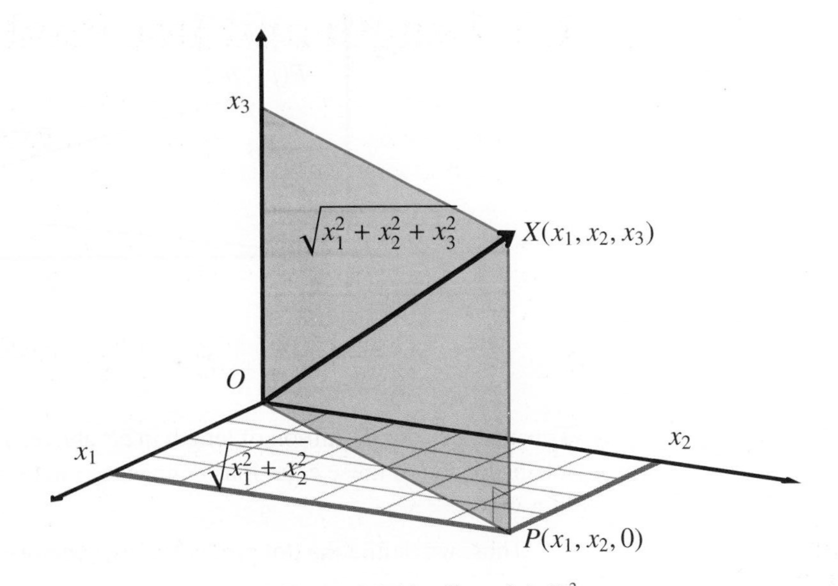

Figure 1.3.11 Length in $\mathbb{R}^3$.

EXAMPLE 1

Find the distance between the points $P(-1, 3, 4)$, $Q(2, -5, 1)$ in $\mathbb{R}^3$.

Solution: We have $\vec{PQ} = \begin{bmatrix} 2-(-1) \\ -5-3 \\ 1-4 \end{bmatrix} = \begin{bmatrix} 3 \\ -8 \\ -3 \end{bmatrix}$. Hence, the distance between the two points is

$$\|\vec{PQ}\| = \sqrt{3^2 + (-8)^2 + (-3)^2} = \sqrt{82}$$

Angles and the Dot Product Determining the angle between two vectors in $\mathbb{R}^2$ leads to the important idea of the **dot product** of two vectors. Consider Figure 1.3.12. The Law of Cosines gives

$$\|\vec{PQ}\|^2 = \|\vec{OP}\|^2 + \|\vec{OQ}\|^2 - 2\|\vec{OP}\| \, \|\vec{OQ}\| \cos\theta \tag{1.1}$$

Substituting $\vec{OP} = \vec{p} = \begin{bmatrix} p_1 \\ p_2 \end{bmatrix}$, $\vec{OQ} = \vec{q} = \begin{bmatrix} q_1 \\ q_2 \end{bmatrix}$, $\vec{PQ} = \vec{p} - \vec{q} = \begin{bmatrix} p_1 - q_1 \\ p_2 - q_2 \end{bmatrix}$ into (1.1) and simplifying gives

$$p_1q_1 + p_2q_2 = \|\vec{p}\| \, \|\vec{q}\| \cos\theta$$

For vectors in $\mathbb{R}^3$, a similar calculation gives

$$p_1q_1 + p_2q_2 + p_3q_3 = \|\vec{p}\| \, \|\vec{q}\| \cos\theta$$

Observe that if $\vec{p} = \vec{q}$, then $\theta = 0$ radians, and we get

$$p_1^2 + p_2^2 + p_3^2 = \|\vec{p}\|^2$$

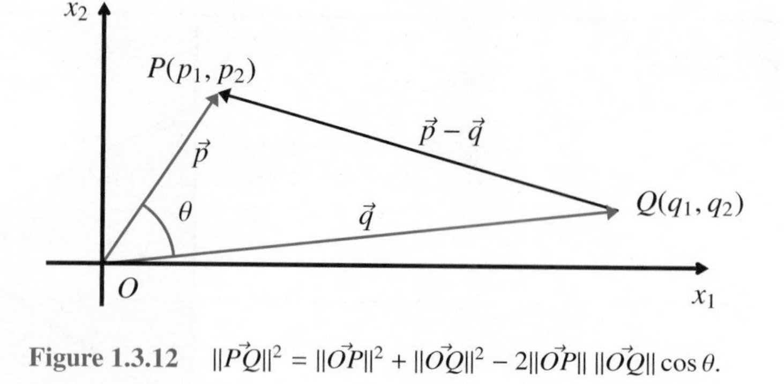

Figure 1.3.12 $\|\vec{PQ}\|^2 = \|\vec{OP}\|^2 + \|\vec{OQ}\|^2 - 2\|\vec{OP}\|\,\|\vec{OQ}\|\cos\theta$.

This matches our definition of length in $\mathbb{R}^3$ above. Thus, we see that the formula on the left-hand side of these equations defines the angle between vectors and also the length of a vector.

Thus, we define the dot product of two vectors $\vec{x} = \begin{bmatrix} x_1 \\ x_2 \end{bmatrix}$, $\vec{y} = \begin{bmatrix} y_1 \\ y_2 \end{bmatrix}$ in $\mathbb{R}^2$ by

$$\vec{x} \cdot \vec{y} = x_1y_1 + x_2y_2$$

Similarly, the dot product of vectors $\vec{x} = \begin{bmatrix} x_1 \\ x_2 \\ x_3 \end{bmatrix}$ and $\vec{y} = \begin{bmatrix} y_1 \\ y_2 \\ y_3 \end{bmatrix}$ in $\mathbb{R}^3$ is defined by

$$\vec{x} \cdot \vec{y} = x_1y_1 + x_2y_2 + x_3y_3$$

Thus, in $\mathbb{R}^2$ and $\mathbb{R}^3$, the cosine of the angle between vectors $\vec{x}$ and $\vec{y}$ can be calculated by means of the formula

$$\cos\theta = \frac{\vec{x} \cdot \vec{y}}{\|\vec{x}\|\,\|\vec{y}\|} \tag{1.2}$$

where θ is always chosen to satisfy $0 \leq \theta \leq \pi$.

EXAMPLE 2

Find the angle in $\mathbb{R}^3$ between $\vec{v} = \begin{bmatrix} 1 \\ 4 \\ -2 \end{bmatrix}$, $\vec{w} = \begin{bmatrix} 3 \\ -1 \\ 4 \end{bmatrix}$.

Solution: We have

$$\vec{v} \cdot \vec{w} = 1(3) + 4(-1) + (-2)(4) = -9$$
$$\|\vec{v}\| = \sqrt{1 + 16 + 4} = \sqrt{21}$$
$$\|\vec{w}\| = \sqrt{9 + 1 + 16} = \sqrt{26}$$

Hence,

$$\cos\theta = \frac{-9}{\sqrt{21}\,\sqrt{26}} \approx -0.38516$$

So $\theta \approx 1.966$ radians. (Note that since $\cos\theta$ is negative, θ is between $\frac{\pi}{2}$ and π.)

EXAMPLE 3

Find the angle in $\mathbb{R}^2$ between $\vec{v} = \begin{bmatrix} 1 \\ -2 \end{bmatrix}$ and $\vec{w} = \begin{bmatrix} 2 \\ 1 \end{bmatrix}$.

Solution: We have $\vec{v} \cdot \vec{w} = 1(2) + (-2)(1) = 0$. Hence, $\cos\theta = \frac{0}{\|\vec{v}\|\,\|\vec{w}\|} = 0$. Thus, $\theta = \frac{\pi}{2}$ radians. That is, $\vec{v}$ and $\vec{w}$ are perpendicular to each other.

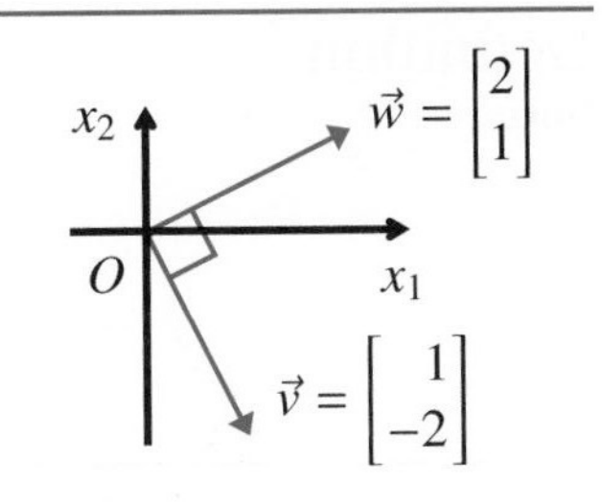

EXERCISE 1

Find the angle in $\mathbb{R}^3$ between $\vec{v} = \begin{bmatrix} 1 \\ 2 \\ -1 \end{bmatrix}$ and $\vec{w} = \begin{bmatrix} 1 \\ -1 \\ -1 \end{bmatrix}$.

Length and Dot Product in $\mathbb{R}^n$

We now extend everything we did in $\mathbb{R}^2$ and $\mathbb{R}^3$ to $\mathbb{R}^n$. We begin by defining the dot product.

Definition
Dot Product

Let $\vec{x} = \begin{bmatrix} x_1 \\ \vdots \\ x_n \end{bmatrix}$ and $\vec{y} = \begin{bmatrix} y_1 \\ \vdots \\ y_n \end{bmatrix}$ be vectors in $\mathbb{R}^n$. Then the **dot product** of $\vec{x}$ and $\vec{y}$ is

$$\vec{x} \cdot \vec{y} = x_1y_1 + x_2y_2 + \cdots + x_ny_n$$

Remark

The dot product is also sometimes called the **scalar product** or the **standard inner product**.

From this definition, some important properties follow.

Theorem 1

Let $\vec{x}, \vec{y}, \vec{z} \in \mathbb{R}^n$ and $t \in \mathbb{R}$. Then,

(1) $\vec{x} \cdot \vec{x} \geq 0$ and $\vec{x} \cdot \vec{x} = 0$ if and only if $\vec{x} = \vec{0}$
(2) $\vec{x} \cdot \vec{y} = \vec{y} \cdot \vec{x}$
(3) $\vec{x} \cdot (\vec{y} + \vec{z}) = \vec{x} \cdot \vec{y} + \vec{x} \cdot \vec{z}$
(4) $(t\vec{x}) \cdot \vec{y} = t(\vec{x} \cdot \vec{y}) = \vec{x} \cdot (t\vec{y})$

Proof: We leave the proof of these properties to the reader.

Because of property (1), we can now define the length of a vector in $\mathbb{R}^n$. The word **norm** is often used as a synonym for **length** when we are speaking of vectors.

Definition
Norm

Let $\vec{x} = \begin{bmatrix} x_1 \\ \vdots \\ x_n \end{bmatrix}$. We define the **norm** or **length** of $\vec{x}$ by

$$\|\vec{x}\| = \sqrt{\vec{x} \cdot \vec{x}} = \sqrt{x_1^2 + \cdots + x_n^2}$$

EXAMPLE 4

Let $\vec{x} = \begin{bmatrix} 2 \\ 1 \\ 3 \\ -1 \end{bmatrix}$ and $\vec{y} = \begin{bmatrix} 1/3 \\ -2/3 \\ 0 \\ -2/3 \end{bmatrix}$. Find $\|\vec{x}\|$ and $\|\vec{y}\|$.

Solution: We have

$$\|\vec{x}\| = \sqrt{2^2 + 1^2 + 3^2 + (-1)^2} = \sqrt{15}$$

$$\|\vec{y}\| = \sqrt{(1/3)^2 + (-2/3)^2 + 0^2 + (-2/3)^2} = \sqrt{1/9 + 4/9 + 0 + 4/9} = 1$$

EXERCISE 2

Let $\vec{x} = \begin{bmatrix} 1 \\ 2 \\ 1 \end{bmatrix}$ and let $\vec{y} = \frac{1}{\|\vec{x}\|}\vec{x}$. Determine $\|\vec{x}\|$ and $\|\vec{y}\|$.

We now give some important properties of the norm in $\mathbb{R}^n$.

Theorem 2

Let $\vec{x}, \vec{y} \in \mathbb{R}^n$ and $t \in \mathbb{R}$. Then

(1) $\|\vec{x}\| \geq 0$ and $\|\vec{x}\| = 0$ if and only if $\vec{x} = \vec{0}$
(2) $\|t\vec{x}\| = |t|\|\vec{x}\|$
(3) $|\vec{x} \cdot \vec{y}| \leq \|\vec{x}\|\,\|\vec{y}\|$, with equality if and only if $\{\vec{x}, \vec{y}\}$ is linearly dependent
(4) $\|\vec{x} + \vec{y}\| \leq \|\vec{x}\| + \|\vec{y}\|$

Proof: Property (1) of Theorem 2 follows immediately from property (1) of Theorem 1.

(2) $\|t\vec{x}\| = \sqrt{(tx_1)^2 + \cdots + (tx_n)^2} = \sqrt{t^2}\sqrt{x_1^2 + \cdots + x_n^2} = |t|\|\vec{x}\|$.

(3) Suppose that $\{\vec{x}, \vec{y}\}$ is linearly dependent. Then, either $\vec{y} = \vec{0}$ or $\vec{x} = t\vec{y}$. If $\vec{y}$ is the zero vector, then both sides are equal to 0. If $\vec{x} = t\vec{y}$, then

$$|\vec{x} \cdot \vec{y}| = |\vec{x} \cdot (t\vec{x})| = |t(\vec{x} \cdot \vec{x})| = |t|\|\vec{x}\|^2 = \|\vec{x}\|\,\|t\vec{x}\| = \|\vec{x}\|\,\|\vec{y}\|$$

Suppose that $\{\vec{x}, \vec{y}\}$ is linearly independent. Then $t\vec{x} + \vec{y} \neq \vec{0}$ for any $t \in \mathbb{R}$. Therefore, by property (1),we have $(t\vec{x} + \vec{y}) \cdot (t\vec{x} + \vec{y}) > 0$ for any $t \in \mathbb{R}$. Use property (3) of Theorem 1 to expand, and we obtain

$$(\vec{x} \cdot \vec{x})t^2 + (2\vec{x} \cdot \vec{y})t + (\vec{y} \cdot \vec{y}) > 0, \quad \text{for all } t \in \mathbb{R} \tag{1.3}$$

Note that $\vec{x} \cdot \vec{x} > 0$ since $\vec{x} \neq \vec{0}$. Now a quadratic expression $At^2 + Bt + C$ with $A > 0$ is always positive if and only if the corresponding equation has no real roots. From the quadratic formula, this is true if and only if $B^2 - 4AC < 0$. Thus, inequality (1.3) implies that

$$4(\vec{x} \cdot \vec{y})^2 - 4(\vec{x} \cdot \vec{x})(\vec{y} \cdot \vec{y}) < 0$$

which can be simplified to the required inequality.

(4) Observe that the required statement is equivalent to

$$\|\vec{x} + \vec{y}\|^2 \leq (\|\vec{x}\| + \|\vec{y}\|)^2$$

The squared form will allow us to use the dot product conveniently. Thus, we consider

$$\begin{aligned} \|\vec{x} + \vec{y}\|^2 - (\|\vec{x}\| + \|\vec{y}\|)^2 &= (\vec{x} + \vec{y}) \cdot (\vec{x} + \vec{y}) - (\|\vec{x}\|^2 + 2\|\vec{x}\|\,\|\vec{y}\| + \|\vec{y}\|^2) \\ &= \vec{x} \cdot \vec{x} + \vec{x} \cdot \vec{y} + \vec{y} \cdot \vec{x} + \vec{y} \cdot \vec{y} - (\vec{x} \cdot \vec{x} + 2\|\vec{x}\|\,\|\vec{y}\| + \vec{y} \cdot \vec{y}) \\ &= 2\vec{x} \cdot \vec{y} - 2\|\vec{x}\|\,\|\vec{y}\| \\ &\leq 0 \qquad \text{by (3)} \end{aligned}$$

■

Remark

Property (3) is called the **Cauchy-Schwarz Inequality** (or Cauchy-Schwarz-Buniakowski). Property (4) is the **Triangle Inequality**.

EXERCISE 3

Prove that the vector $\hat{x} = \frac{1}{\|\vec{x}\|}\vec{x}$ is parallel to $\vec{x}$ and satisfies $\|\hat{x}\| = 1$.

Definition
Unit Vector

A vector $\vec{x} \in \mathbb{R}^n$ such that $\|\vec{x}\| = 1$ is called a **unit vector**.

We will see that unit vectors can be very useful. We often want to find a unit vector that has the same direction as a given vector $\vec{x}$. Using the result in Exercise 3, we see that this is the vector

$$\hat{x} = \frac{1}{\|\vec{x}\|}\vec{x}$$

We could now define the angle between vectors $\vec{x}$ and $\vec{y}$ in $\mathbb{R}^n$ by matching equation (1.2). However, in linear algebra we are generally interested only in whether two vectors are perpendicular. To agree with the result of Example 3, we make the following definition.

Definition
Orthogonal

Two vectors $\vec{x}$ and $\vec{y}$ in $\mathbb{R}^n$ are **orthogonal** to each other if and only if $\vec{x} \cdot \vec{y} = 0$.

Notice that this definition implies that $\vec{0}$ is orthogonal to every vector in $\mathbb{R}^n$.

EXAMPLE 5

Let $\vec{v} = \begin{bmatrix} 1 \\ 0 \\ 3 \\ -2 \end{bmatrix}$, $\vec{w} = \begin{bmatrix} 2 \\ 3 \\ 0 \\ 1 \end{bmatrix}$, and $\vec{z} = \begin{bmatrix} -1 \\ -1 \\ 1 \\ 2 \end{bmatrix}$. Show that $\vec{v}$ is orthogonal to $\vec{w}$ but $\vec{v}$ is not orthogonal to $\vec{z}$.

Solution: We have $\vec{v} \cdot \vec{w} = 1(2) + 0(3) + 3(0) + (-2)(1) = 0$, so they are orthogonal. $\vec{v} \cdot \vec{z} = 1(-1) + 0(-1) + 3(1) + (-2)(2) = -2$, so they are not orthogonal.

The Scalar Equation of Planes and Hyperplanes

We saw in Section 1.2 that a plane can be described by the vector equation $\vec{x} = \vec{p} + t_1\vec{v}_1 + t_2\vec{v}_2$, $t_1, t_2 \in \mathbb{R}$, where $\{\vec{v}_1, \vec{v}_2\}$ is linearly independent. In many problems, it is more useful to have a **scalar equation** that represents the plane. We now look at how to use the dot product to find such an equation.

Suppose that we want to find the equation of a plane that passes through the point $P(p_1, p_2, p_3)$. Suppose that we can find a vector $\vec{n} = \begin{bmatrix} n_1 \\ n_2 \\ n_3 \end{bmatrix}$, called the **normal vector** of the plane, that is orthogonal to any directed line segment $\vec{PQ}$ lying in the plane. (That is, $\vec{n}$ is orthogonal to $\vec{PQ}$ for any point Q in the plane; see Figure 1.3.13.) To find the equation of this plane, let $X(x_1, x_2, x_3)$ be any point on the plane. Then $\vec{n}$ is orthogonal to $\vec{PX}$, so

$$0 = \vec{n} \cdot \vec{PX} = \vec{n} \cdot (\vec{x} - \vec{p}) = n_1(x_1 - p_1) + n_2(x_2 - p_2) + n_3(x_3 - p_3)$$

This equation, which must be satisfied by the coordinates of a point X in the plane, can be written in the form

$$n_1x_1 + n_2x_2 + n_3x_3 = d, \quad \text{where} \quad d = n_1p_1 + n_2p_2 + n_3p_3 = \vec{n} \cdot \vec{p}$$

This is the standard equation of this plane. For computational purposes, the form $\vec{n} \cdot (\vec{x} - \vec{p}) = 0$ is often easiest to use.

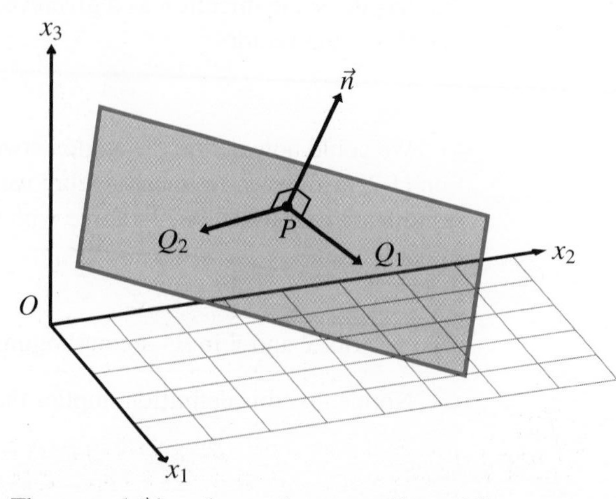

Figure 1.3.13 The normal $\vec{n}$ is orthogonal to every directed line segment lying in the plane.

EXAMPLE 6

Find the scalar equation of the plane that passes through the point $P(2, 3, -1)$ and has normal vector $\vec{n} = \begin{bmatrix} 1 \\ -4 \\ 1 \end{bmatrix}$.

Solution: The equation is

$$\vec{n} \cdot (\vec{x} - \vec{p}) = \begin{bmatrix} 1 \\ -4 \\ 1 \end{bmatrix} \cdot \begin{bmatrix} x_1 - 2 \\ x_2 - 3 \\ x_3 + 1 \end{bmatrix} = 0$$

or

$$x_1 - 4x_2 + x_3 = 1(2) + (-4)(3) + 1(-1) = -11$$

It is important to note that we can reverse the reasoning that leads to the equation of the plane in order to identify the set of points that satisfies an equation of the form $n_1x_1 + n_2x_2 + n_3x_3 = d$. If $n_1 \neq 0$, this can be written as

$$n_1x_1 - d + n_2x_2 + n_3x_3 = 0, \text{ or } \vec{n} \cdot \begin{bmatrix} x_1 - d/n_1 \\ x_2 \\ x_3 \end{bmatrix} = 0$$

where $\vec{n} = \begin{bmatrix} n_1 \\ n_2 \\ n_3 \end{bmatrix}$. This equation describes a plane through the point $P(d/n_1, 0, 0)$, with normal vector $\vec{n}$. If $\vec{n}_2 \neq 0$, we could combine the d with the x_2 term and find that the plane passes through the point $P(0, d/n_2, 0)$. In fact, we could find a point in the plane in many ways, but the normal vector will always be a non-zero scalar multiple of $\vec{n}$.

EXAMPLE 7

Describe the set of points in $\mathbb{R}^3$ that satisfies $5x_1 - 6x_2 + 7x_3 = 11$.

Solution: We wish to rewrite the equation in the form $\vec{n} \cdot (\vec{x} - \vec{p}) = 0$. Using our work above, we get

$$(5x_1 - 11) - 6x_2 + 7x_3 = 0$$

or

$$5\left(x_1 - \frac{11}{5}\right) - 6(x_2 - 0) + 7(x_3 - 0) = 0$$

Thus, we identify the set as a plane with normal vector $\vec{n} = \begin{bmatrix} 5 \\ -6 \\ 7 \end{bmatrix}$, passing through the point $(11/5, 0, 0)$. Alternatively, if $x_1 = x_3 = 0$, we find that $x_2 = -11/6$, so the plane passes through $(0, -11/6, 0)$.

Two planes are defined to be **parallel** if the normal vector to one plane is a non-zero scalar multiple of the normal vector of the other plane. Thus, for example, the plane $x_1 + 2x_2 - x_3 = 1$ is parallel to the plane $2x_1 + 4x_2 - 2x_3 = 7$.

Two planes are **orthogonal** to each other if their normal vectors are orthogonal. For example, the plane $x_1 + x_2 + x_3 = 0$ is orthogonal to the plane $x_1 + x_2 - 2x_3 = 0$ since

$$\begin{bmatrix} 1 \\ 1 \\ 1 \end{bmatrix} \cdot \begin{bmatrix} 1 \\ 1 \\ -2 \end{bmatrix} = 0$$

EXAMPLE 8

Find a scalar equation of the plane that contains the point $P(2, 4, -1)$ and is parallel to the plane $2x_1 + 3x_2 - 5x_3 = 6$.

Solution: An equation of the plane must be of the form $2x_1 + 3x_2 - 5x_3 = d$ since the planes are parallel. The plane must pass through P, so we find that a scalar equation of the plane is

$$2x_1 + 3x_2 - 5x_3 = 2(2) + 3(4) + (-5)(-1) = 21$$

EXAMPLE 9

Find a scalar equation of a plane that contains the point $P(3, -1, 3)$ and is orthogonal to the plane $x_1 - 2x_2 + 4x_3 = 2$.

Solution: The normal vector must be orthogonal to $\begin{bmatrix} 1 \\ -2 \\ 4 \end{bmatrix}$, so we pick $\begin{bmatrix} 0 \\ 2 \\ 1 \end{bmatrix}$. Thus, an equation of this plane must be of the form $2x_2 + x_3 = d$. Since the plane must pass through P, we find that a scalar equation of this plane is

$$2x_2 + x_3 = 0(3) + 2(-1) + (1)(3) = 1$$

EXERCISE 4

Find a scalar equation of the plane that contains the point $P(1, 2, 3)$ and is parallel to the plane $x_1 - 3x_2 - 2x_3 = -5$.

The Scalar Equation of a Hyperplane Repeating what we did above we can find the scalar equation of a hyperplane in $\mathbb{R}^n$. In particular, if we have a vector $\vec{m}$ that is orthogonal to any directed line segment $\vec{PQ}$ lying in the hyperplane, then for any point $X(x_1, \dots, x_n)$ in the hyperplane, we have

$$0 = \vec{m} \cdot \vec{PX}$$

As before, we can rearrange this as

$$\begin{aligned} 0 &= \vec{m} \cdot (\vec{x} - \vec{p}) \\ 0 &= \vec{m} \cdot \vec{x} - \vec{m} \cdot \vec{p} \\ \vec{m} \cdot \vec{x} &= \vec{m} \cdot \vec{p} \\ m_1x_1 + \cdots + m_nx_n &= \vec{m} \cdot \vec{p} \end{aligned}$$

Thus, we see that a single scalar equation in $\mathbb{R}^n$ represents a hyperplane in $\mathbb{R}^n$.

EXAMPLE 10

Find the scalar equation of the hyperplane in $\mathbb{R}^4$ that has normal vector $\vec{m} = \begin{bmatrix} 2 \\ 3 \\ -2 \\ 1 \end{bmatrix}$ and passes through the point $P(1, 0, 2, -1)$.

Solution: The equation is

$$2x_1 + 3x_2 - 2x_3 + x_4 = 2(1) + 3(0) + (-2)(2) + 1(-1) = -3$$

PROBLEMS 1.3

Practice Problems

A1 Calculate the lengths of the given vectors.

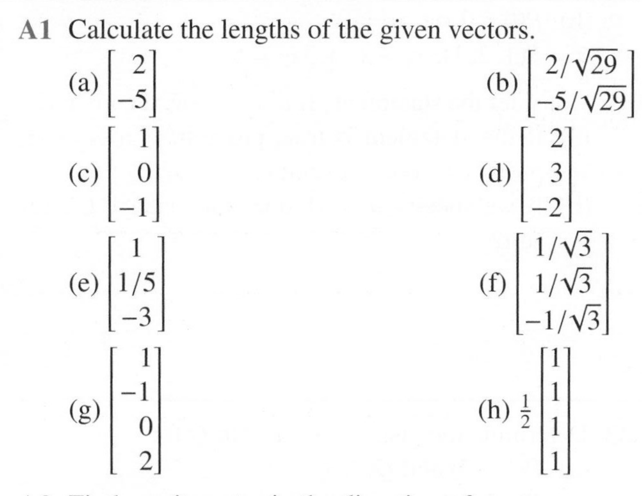

(a) $\begin{bmatrix} 2 \\ -5 \end{bmatrix}$ (b) $\begin{bmatrix} 2/\sqrt{29} \\ -5/\sqrt{29} \end{bmatrix}$

(c) $\begin{bmatrix} 1 \\ 0 \\ -1 \end{bmatrix}$ (d) $\begin{bmatrix} 2 \\ 3 \\ -2 \end{bmatrix}$

(e) $\begin{bmatrix} 1 \\ 1/5 \\ -3 \end{bmatrix}$ (f) $\begin{bmatrix} 1/\sqrt{3} \\ 1/\sqrt{3} \\ -1/\sqrt{3} \end{bmatrix}$

(g) $\begin{bmatrix} 1 \\ -1 \\ 0 \\ 2 \end{bmatrix}$ (h) $\frac{1}{2}\begin{bmatrix} 1 \\ 1 \\ 1 \\ 1 \end{bmatrix}$

A2 Find a unit vector in the direction of

(a) $\begin{bmatrix} 3 \\ -4 \end{bmatrix}$ (b) $\begin{bmatrix} 1 \\ 1 \end{bmatrix}$

(c) $\begin{bmatrix} -1 \\ 0 \\ 2 \end{bmatrix}$ (d) $\begin{bmatrix} 0 \\ -3 \\ 0 \end{bmatrix}$

(e) $\begin{bmatrix} -2 \\ -2 \\ 1 \\ 0 \end{bmatrix}$ (f) $\begin{bmatrix} 1 \\ 0 \\ 0 \\ -1 \end{bmatrix}$

A3 Determine the distance from P to Q for

(a) $P(2, 3)$ and $Q(-4, 1)$

(b) $P(1, 1, -2)$ and $Q(-3, 1, 1)$

(c) $P(4, -6, 1)$ and $Q(-3, 5, 1)$

(d) $P(2, 1, 1, 5)$ and $Q(4, 6, -2, 1)$

A4 Verify the triangle inequality and the Cauchy-Schwarz inequality if

(a) $\vec{x} = \begin{bmatrix} 4 \\ 3 \\ 1 \end{bmatrix}$ and $\vec{y} = \begin{bmatrix} 2 \\ 1 \\ 5 \end{bmatrix}$

(b) $\vec{x} = \begin{bmatrix} 1 \\ -1 \\ 2 \end{bmatrix}$ and $\vec{y} = \begin{bmatrix} -3 \\ 2 \\ 4 \end{bmatrix}$

A5 Determine whether each pair of vectors is orthogonal.

(a) $\begin{bmatrix} 1 \\ 3 \\ 2 \end{bmatrix}, \begin{bmatrix} 2 \\ -2 \\ 2 \end{bmatrix}$ (b) $\begin{bmatrix} -3 \\ 1 \\ 7 \end{bmatrix}, \begin{bmatrix} 2 \\ -1 \\ 1 \end{bmatrix}$

(c) $\begin{bmatrix} 2 \\ 1 \\ 1 \end{bmatrix}, \begin{bmatrix} -1 \\ 4 \\ 2 \end{bmatrix}$ (d) $\begin{bmatrix} 4 \\ 1 \\ 0 \\ -2 \end{bmatrix}, \begin{bmatrix} -1 \\ 4 \\ 3 \\ 0 \end{bmatrix}$

(e) $\begin{bmatrix} 0 \\ 0 \\ 0 \\ 0 \end{bmatrix}, \begin{bmatrix} x_1 \\ x_2 \\ x_3 \\ x_4 \end{bmatrix}$ (f) $\begin{bmatrix} 1/3 \\ 2/3 \\ -1/3 \\ 3 \end{bmatrix}, \begin{bmatrix} 3/2 \\ 0 \\ -3/2 \\ 1 \end{bmatrix}$

A6 Determine all values of k for which each pair of vectors is orthogonal.

(a) $\begin{bmatrix} 3 \\ -1 \end{bmatrix}, \begin{bmatrix} 2 \\ k \end{bmatrix}$ (b) $\begin{bmatrix} 3 \\ -1 \end{bmatrix}, \begin{bmatrix} k \\ k^2 \end{bmatrix}$

(c) $\begin{bmatrix} 1 \\ 2 \\ 3 \end{bmatrix}, \begin{bmatrix} 3 \\ -k \\ k \end{bmatrix}$ (d) $\begin{bmatrix} 1 \\ 2 \\ 3 \\ 4 \end{bmatrix}, \begin{bmatrix} k \\ k \\ -k \\ 0 \end{bmatrix}$

A7 Find a scalar equation of the plane that contains the given point with the given normal vector.

(a) $P(-1,2,-3)$, $\vec{n} = \begin{bmatrix} 2 \\ 4 \\ -1 \end{bmatrix}$

(b) $P(2,5,4)$, $\vec{n} = \begin{bmatrix} 3 \\ 0 \\ 5 \end{bmatrix}$

(c) $P(1,-1,1)$, $\vec{n} = \begin{bmatrix} 3 \\ -4 \\ 1 \end{bmatrix}$

(d) $P(2,1,1)$, $\vec{n} = \begin{bmatrix} -4 \\ -2 \\ -2 \end{bmatrix}$

A8 Determine a scalar equation of the hyperplane that passes through the given point with the given normal vector.

(a) $P(1,1,-1)$, $\vec{n} = \begin{bmatrix} 3 \\ 1 \\ 4 \end{bmatrix}$

(b) $P(2,-2,0,1)$, $\vec{n} = \begin{bmatrix} 0 \\ 1 \\ 3 \\ 3 \end{bmatrix}$

(c) $P(0,0,0,0)$, $\vec{n} = \begin{bmatrix} 1 \\ -4 \\ 5 \\ -2 \end{bmatrix}$

(d) $P(1,0,1,2,1)$, $\vec{n} = \begin{bmatrix} 0 \\ 1 \\ 2 \\ -1 \\ 1 \end{bmatrix}$

A9 Determine a normal vector for the hyperplane.

(a) $2x_1 + x_2 = 3$ in $\mathbb{R}^2$

(b) $3x_1 - 2x_2 + 3x_3 = 7$ in $\mathbb{R}^3$

(c) $-4x_1 + 3x_2 - 5x_3 - 6 = 0$ in $\mathbb{R}^3$

(d) $x_1 - x_2 + 2x_3 - 3x_4 = 5$ in $\mathbb{R}^4$

(e) $x_1 + x_2 - x_3 + 2x_4 - x_5 = 0$ in $\mathbb{R}^5$

A10 Find an equation for the plane through the given point and parallel to the given plane.

(a) $P(1,-3,-1)$, $2x_1 - 3x_2 + 5x_3 = 17$

(b) $P(0,-2,4)$, $x_2 = 0$

(c) $P(1,2,1)$, $x_1 - x_2 + 3x_3 = 5$

A11 Consider the statement "If $\vec{u}\cdot\vec{v} = \vec{u}\cdot\vec{w}$, then $\vec{v} = \vec{w}$."

(a) If the statement is true, prove it. If it is false, provide a counterexample.

(b) If we specify $\vec{u} \neq \vec{0}$, does this change the result?

Homework Problems

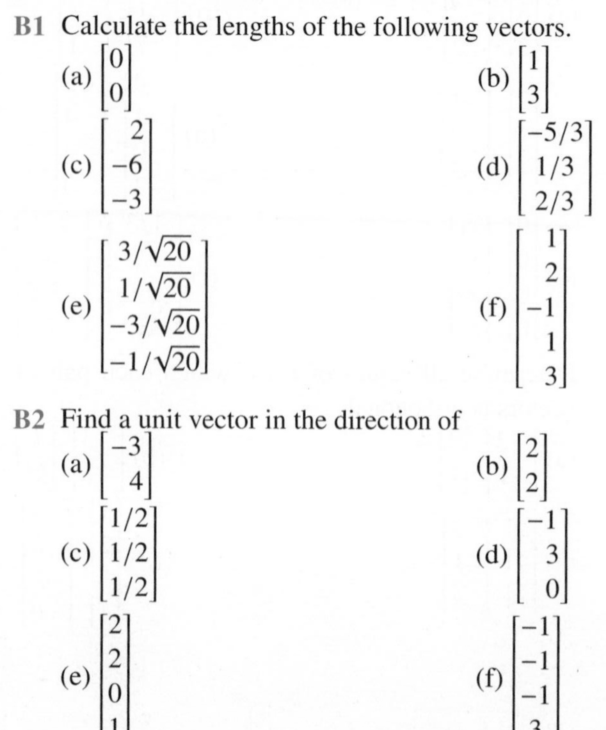

B1 Calculate the lengths of the following vectors.

(a) $\begin{bmatrix} 0 \\ 0 \end{bmatrix}$ (b) $\begin{bmatrix} 1 \\ 3 \end{bmatrix}$

(c) $\begin{bmatrix} 2 \\ -6 \\ -3 \end{bmatrix}$ (d) $\begin{bmatrix} -5/3 \\ 1/3 \\ 2/3 \end{bmatrix}$

(e) $\begin{bmatrix} 3/\sqrt{20} \\ 1/\sqrt{20} \\ -3/\sqrt{20} \\ -1/\sqrt{20} \end{bmatrix}$ (f) $\begin{bmatrix} 1 \\ 2 \\ -1 \\ 1 \\ 3 \end{bmatrix}$

B2 Find a unit vector in the direction of

(a) $\begin{bmatrix} -3 \\ 4 \end{bmatrix}$ (b) $\begin{bmatrix} 2 \\ 2 \end{bmatrix}$

(c) $\begin{bmatrix} 1/2 \\ 1/2 \\ 1/2 \end{bmatrix}$ (d) $\begin{bmatrix} -1 \\ 3 \\ 0 \end{bmatrix}$

(e) $\begin{bmatrix} 2 \\ 2 \\ 0 \\ 1 \end{bmatrix}$ (f) $\begin{bmatrix} -1 \\ -1 \\ -1 \\ 3 \end{bmatrix}$

B3 Determine the distance from P to Q for

(a) $P(1,-3)$ and $Q(2,3)$

(b) $P(-2,-2,5)$ and $Q(-4,1,4)$

(c) $P(3,1,-3)$ and $Q(-1,4,5)$

(d) $P(5,-2,-3,6)$ and $Q(2,5,-4,3)$

B4 Verify the triangle inequality and the Cauchy-Schwarz inequality if

(a) $\vec{x} = \begin{bmatrix} 2 \\ -6 \\ -3 \end{bmatrix}$ and $\vec{y} = \begin{bmatrix} -3 \\ 4 \\ 5 \end{bmatrix}$

(b) $\vec{x} = \begin{bmatrix} 4 \\ 1 \\ -2 \end{bmatrix}$ and $\vec{y} = \begin{bmatrix} 3 \\ 5 \\ 1 \end{bmatrix}$

B5 Determine whether each pair of vectors is orthogonal.

(a) $\begin{bmatrix} 1 \\ 2 \end{bmatrix}, \begin{bmatrix} -2 \\ 2 \end{bmatrix}$ (b) $\begin{bmatrix} 4 \\ 6 \end{bmatrix}, \begin{bmatrix} -3 \\ 2 \end{bmatrix}$

(c) $\begin{bmatrix} 1 \\ 4 \\ 1 \end{bmatrix}, \begin{bmatrix} -4 \\ 1 \\ -4 \end{bmatrix}$ (d) $\begin{bmatrix} 1 \\ 3 \\ 1 \end{bmatrix}, \begin{bmatrix} 3 \\ -1 \\ 0 \end{bmatrix}$

(e) $\begin{bmatrix}1\\2\\1\\2\end{bmatrix}, \begin{bmatrix}-3\\0\\5\\1\end{bmatrix}$ (f) $\begin{bmatrix}2\\1\\-1\\-2\\1\end{bmatrix}, \begin{bmatrix}-2\\1\\-1\\0\\2\end{bmatrix}$

B6 Determine all values of k for which each pair of vectors is orthogonal.

(a) $\begin{bmatrix}1\\2\end{bmatrix}, \begin{bmatrix}k\\k\end{bmatrix}$ (b) $\begin{bmatrix}-2\\1\end{bmatrix}, \begin{bmatrix}k\\k^2\end{bmatrix}$

(c) $\begin{bmatrix}1\\2\\1\end{bmatrix}, \begin{bmatrix}k\\2k\\4\end{bmatrix}$ (d) $\begin{bmatrix}1\\-1\\1\end{bmatrix}, \begin{bmatrix}k\\3k\\k^2\end{bmatrix}$

B7 Find a scalar equation of the plane that contains the given point with the given normal vector.

(a) $P(-3,-3,1), \vec{n} = \begin{bmatrix}-1\\4\\7\end{bmatrix}$

(b) $P(6,-2,5), \vec{n} = \begin{bmatrix}4\\-2\\1\end{bmatrix}$

(c) $P(0,0,0), \vec{n} = \begin{bmatrix}1\\2\\1\end{bmatrix}$

(d) $P(1,1,1), \vec{n} = \begin{bmatrix}1\\1\\1\end{bmatrix}$

B8 Determine a scalar equation of the hyperplane that passes through the given point with the given normal vector.

(a) $P(2,1,1,5), \vec{n} = \begin{bmatrix}3\\-2\\-5\\1\end{bmatrix}$

(b) $P(3,1,0,7), \vec{n} = \begin{bmatrix}2\\-4\\1\\-3\end{bmatrix}$

(c) $P(0,0,0,0), \vec{n} = \begin{bmatrix}1\\0\\0\\0\end{bmatrix}$

(d) $P(1,2,0,1,1), \vec{n} = \begin{bmatrix}0\\1\\-2\\1\\1\end{bmatrix}$

B9 Determine a normal vector for the given hyperplane.

(a) $2x_1 + 3x_2 = 0$ in $\mathbb{R}^2$
(b) $-x_1 - 2x_2 + 5x_3 = 7$ in $\mathbb{R}^3$
(c) $x_1 + 4x_2 - x_4 = 2$ in $\mathbb{R}^4$
(d) $x_1 + x_2 + x_3 - 2x_4 = 5$ in $\mathbb{R}^4$
(e) $x_1 - x_5 = 0$ in $\mathbb{R}^5$
(f) $x_1 + x_2 - 2x_3 + x_4 - x_5 = 1$ in $\mathbb{R}^5$

B10 Find an equation for the plane through the given point and parallel to the given plane.

(a) $P(3,-1,7)$, $5x_1 - x_2 - 2x_3 = 6$
(b) $P(-1,2,-5)$, $2x_2 + 3x_3 = 7$
(c) $P(2,1,1)$, $3x_1 - 2x_2 + 3x_3 = 7$
(d) $P(1,0,1)$, $x_1 - 5x_2 + 3x_3 = 0$

Computer Problems

C1 Let $\vec{v}_1 = \begin{bmatrix}1.12\\2.10\\7.03\\4.15\\6.13\end{bmatrix}$, $\vec{v}_2 = \begin{bmatrix}1.00\\3.12\\-0.45\\-2.21\\2.00\end{bmatrix}$, $\vec{v}_3 = \begin{bmatrix}-3.13\\1.21\\3.31\\1.14\\-0.01\end{bmatrix}$, and

$\vec{v}_4 = \begin{bmatrix}1.12\\0\\2.13\\3.15\\3.40\end{bmatrix}$.

Use computer software to evaluate each of the following.

(a) $\vec{v}_1 \cdot \vec{v}_2$
(b) $\|\vec{v}_3\|$
(c) $\vec{v}_2 \cdot \vec{v}_4$
(d) $\|\vec{v}_2 + \vec{v}_4\|^2$

Conceptual Problems

D1 (a) Using geometrical arguments, what can you say about the vectors $\vec{p}$, $\vec{n}$, and $\vec{d}$ if the line with vector equation $\vec{x} = \vec{p} + t\vec{d}$ and the plane with scalar equation $\vec{n} \cdot \vec{x} = k$ have no point of intersection?

(b) Confirm your answer in part (a) by determining when it is possible to find a value of the parameter t that gives a point of intersection.

D2 Prove that, as a consequence of the triangle inequality, $\big|\|\vec{x}\| - \|\vec{y}\|\big| \leq \|\vec{x} - \vec{y}\|$. (Hint: $\|\vec{x}\| = \|\vec{x} - \vec{y} + \vec{y}\|$.)

D3 Let $\vec{v}_1$ and $\vec{v}_2$ be orthogonal vectors in $\mathbb{R}^n$. Prove that $\|\vec{v}_1 + \vec{v}_2\|^2 = \|\vec{v}_1\|^2 + \|\vec{v}_2\|^2$.

D4 Determine the equation of the set of points in $\mathbb{R}^3$ that are equidistant from points P and Q. Explain why the set is a plane and determine its normal vector.

D5 Find the scalar equation of the plane such that each point of the plane is equidistant from the points $P(2, 2, 5)$ and $Q(-3, 4, 1)$ in two ways.

(a) Write and simplify the equation $\|\vec{PX}\| = \|\vec{QX}\|$.

(b) Determine a point on the plane and the normal vector by geometrical arguments.

D6 Let $\vec{u} \in \mathbb{R}^n$. Prove that the set of all vectors orthogonal to $\vec{u}$ is a subspace of $\mathbb{R}^n$.

D7 Let S be any set of vectors in $\mathbb{R}^n$. Let $S^\perp$ be the set of all vectors that are orthogonal to every vector in S. That is,

$$S^\perp = \{\vec{w} \in \mathbb{R}^n \mid \vec{v} \cdot \vec{w} = 0 \text{ for all } \vec{v} \in S\}$$

Show that $S^\perp$ is a subspace of $\mathbb{R}^n$.

D8 Let $\{\vec{v}_1, \dots, \vec{v}_k\}$ be a set of non-zero vectors in $\mathbb{R}^n$ such that all of the vectors are mutually orthogonal. That is, $\vec{v}_i \cdot \vec{v}_j = 0$ for all $i \neq j$. Prove that $\{\vec{v}_1, \dots, \vec{v}_k\}$ is linearly independent.

D9 (a) Let $\vec{n}$ be a unit vector in $\mathbb{R}^3$. Let α be the angle between $\vec{n}$ and the x_1-axis, let β be the angle between $\vec{n}$ and the x_2-axis, and let γ be the angle between $\vec{n}$ and the x_3-axis. Explain why

$$\vec{n} = \begin{bmatrix} \cos\alpha \\ \cos\beta \\ \cos\gamma \end{bmatrix}$$

(Hint: Take the dot product of $\vec{n}$ with the standard basis vectors.)

Because of this equation, the components n_1, n_2, n_3 are sometimes called the **direction cosines**.

(b) Explain why $\cos^2\alpha + \cos^2\beta + \cos^2\gamma = 1$.

(c) Give a two-dimensional version of the direction cosines and explain the connection to the identity $\cos^2\theta + \sin^2\theta = 1$.

1.4 Projections and Minimum Distance

The idea of a projection is one of the most important applications of the dot product. Suppose that we want to know how much of a given vector $\vec{y}$ is in the direction of some other given vector $\vec{x}$ (see Figure 1.4.14). In elementary physics, this is exactly what is required when a force is "resolved" into its components along certain directions (for example, into its vertical and horizontal components). When we define projections, it is helpful to think of examples in two or three dimensions, but the ideas do not really depend on whether the vectors are in $\mathbb{R}^2$, $\mathbb{R}^3$, or $\mathbb{R}^n$.

Projections

First, let us consider the case where $\vec{x} = \vec{e}_1$ in $\mathbb{R}^2$. How much of an arbitrary vector $\vec{y} = \begin{bmatrix} y_1 \\ y_2 \end{bmatrix}$ points along $\vec{x}$? Clearly, the part of $\vec{y}$ that is in the direction of $\vec{x}$ is $\begin{bmatrix} y_1 \\ 0 \end{bmatrix} = y_1\vec{e}_1 = (\vec{y} \cdot \vec{x})\vec{x}$. This will be called the **projection of $\vec{y}$ onto $\vec{x}$** and is denoted $\text{proj}_{\vec{x}}\, \vec{y}$.

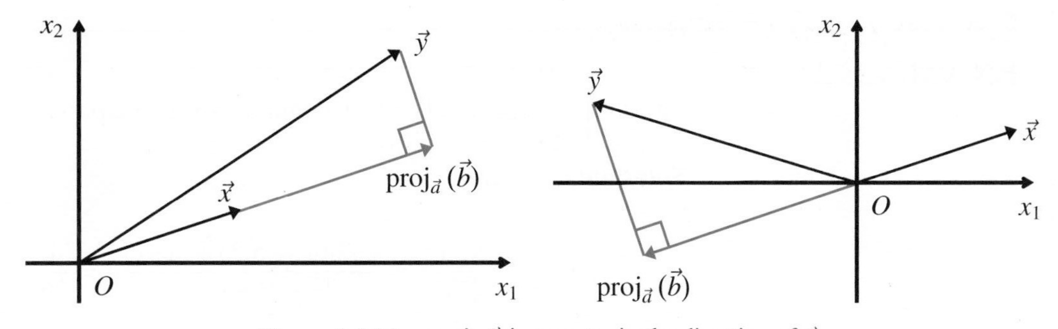

Figure 1.4.14 $\text{proj}_{\vec{x}}\,\vec{y}$ is a vector in the direction of $\vec{x}$.

Next, consider the case where $\vec{x} \in \mathbb{R}^2$ has arbitrary direction and is a unit vector. First, draw the line through the origin with direction vector $\vec{x}$. Now, draw the line perpendicular to this line that passes through the point (y_1, y_2). This forms a right triangle, as in Figure 1.4.14. The projection of $\vec{y}$ onto $\vec{x}$ is the portion of the triangle that lies on the line with direction $\vec{x}$. Thus, the resulting projection is a scalar multiple of $\vec{x}$, that is $\text{proj}_{\vec{x}}\,\vec{y} = k\vec{x}$. We need to determine the value of k. To do this, let $\vec{w}$ denote the vector from $\text{proj}_{\vec{x}}\,\vec{y}$ to $\vec{y}$. Then, by definition, $\vec{z}$ is orthogonal to $\vec{x}$ and we can write

$$\vec{y} = \vec{z} + \text{proj}_{\vec{x}}\,\vec{y} = \vec{z} + k\vec{x}$$

We now employ a very useful and common trick-take the dot product of $\vec{y}$ with $\vec{x}$:

$$\vec{y} \cdot \vec{x} = (\vec{z} + k\vec{x}) \cdot \vec{x} = \vec{z} \cdot \vec{x} + (k\vec{x}) \cdot \vec{x} = 0 + k(\vec{x} \cdot \vec{x}) = k\|\vec{x}\|^2 = k$$

since $\vec{x}$ is a unit vector. Thus,

$$\text{proj}_{\vec{x}}\,\vec{y} = (\vec{y} \cdot \vec{x})\vec{x}$$

EXAMPLE 1

Find the projection of $\vec{u} = \begin{bmatrix} -3 \\ 1 \end{bmatrix}$ onto the unit vector $\vec{v} = \begin{bmatrix} 1/\sqrt{2} \\ 1/\sqrt{2} \end{bmatrix}$.

Solution: We have

$$\begin{aligned} \text{proj}_{\vec{v}}\,\vec{u} &= (\vec{u} \cdot \vec{v})\vec{v} \\ &= \left(\frac{-3}{\sqrt{2}} + \frac{1}{\sqrt{2}} \right) \begin{bmatrix} 1/\sqrt{2} \\ 1/\sqrt{2} \end{bmatrix} \\ &= \frac{-2}{\sqrt{2}} \begin{bmatrix} 1/\sqrt{2} \\ 1/\sqrt{2} \end{bmatrix} \\ &= \begin{bmatrix} -1 \\ -1 \end{bmatrix} \end{aligned}$$

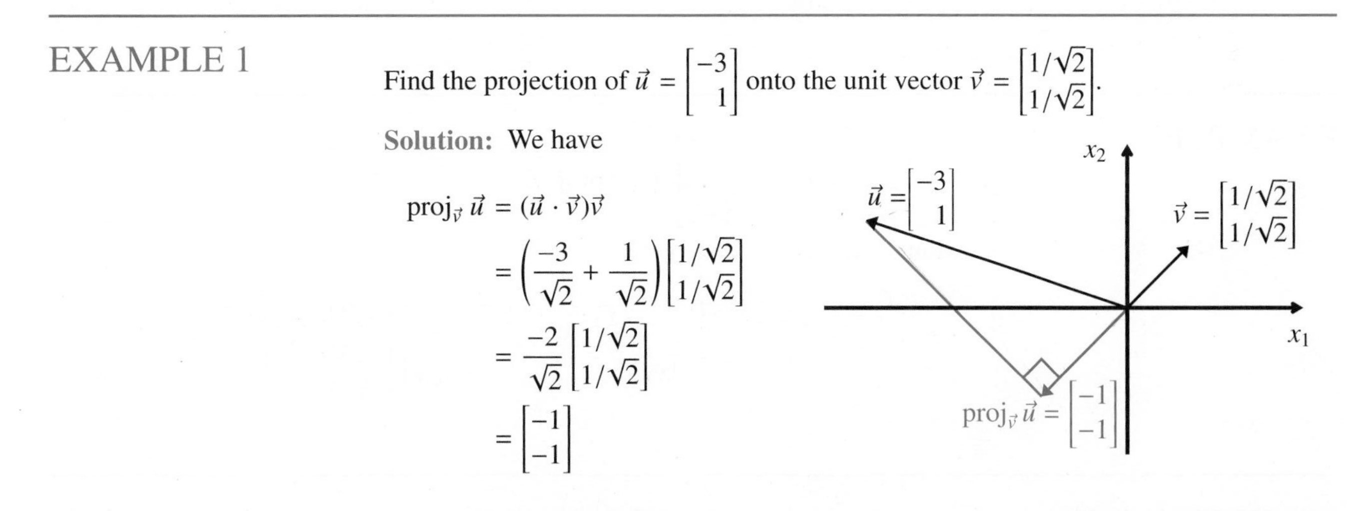

If $\vec{x} \in \mathbb{R}^2$ is an arbitrary non-zero vector, then the unit vector in the direction of $\vec{x}$ is $\hat{x} = \frac{\vec{x}}{\|\vec{x}\|}$. Hence, we find that the projection of $\vec{y}$ onto $\vec{x}$ is

$$\text{proj}_{\vec{x}}\,\vec{y} = \text{proj}_{\hat{x}}\,\vec{y} = (\vec{y} \cdot \hat{x})\hat{x} = \left(\vec{y} \cdot \frac{\vec{x}}{\|\vec{x}\|} \right) \frac{\vec{x}}{\|\vec{x}\|} = \frac{\vec{y} \cdot \vec{x}}{\|\vec{x}\|^2}\vec{x}$$

To match this result, we make the following definition for vectors in $\mathbb{R}^n$.

Definition
Projection

For any vectors $\vec{y}, \vec{x}$ in $\mathbb{R}^n$, with $\vec{x} \neq \vec{0}$, we define the **projection** of $\vec{y}$ onto $\vec{x}$ by

$$\text{proj}_{\vec{x}}\,\vec{y} = \frac{\vec{y} \cdot \vec{x}}{\|\vec{x}\|^2}\vec{x}$$

EXAMPLE 2

Let $\vec{v} = \begin{bmatrix} 4 \\ 3 \\ -1 \end{bmatrix}$ and $\vec{u} = \begin{bmatrix} -2 \\ 5 \\ 3 \end{bmatrix}$. Determine $\text{proj}_{\vec{v}}\, \vec{u}$ and $\text{proj}_{\vec{u}}\, \vec{v}$.

Solution: We have

$$\text{proj}_{\vec{v}}\, \vec{u} = \frac{\vec{u} \cdot \vec{v}}{\|\vec{v}\|^2}\vec{v} = \frac{(-2)(4) + 5(3) + 3(-1)}{4^2 + 3^2 + (-1)^2}\vec{v} = \frac{4}{26}\begin{bmatrix} 4 \\ 3 \\ -1 \end{bmatrix} = \begin{bmatrix} 8/13 \\ 6/13 \\ -2/13 \end{bmatrix}$$

$$\text{proj}_{\vec{u}}\, \vec{v} = \frac{\vec{v} \cdot \vec{u}}{\|\vec{u}\|^2}\vec{u} = \frac{(4)(-2) + 3(5) + (-1)(3)}{(-2)^2 + 5^2 + 3^2}\vec{u} = \frac{4}{38}\begin{bmatrix} -2 \\ 5 \\ 3 \end{bmatrix} = \begin{bmatrix} -4/19 \\ 10/19 \\ 6/19 \end{bmatrix}$$

Remarks

1. This example illustrates that, in general, $\text{proj}_{\vec{x}}\, \vec{y} \neq \text{proj}_{\vec{y}}\, \vec{x}$. Of course, we should not expect equality because $\text{proj}_{\vec{x}}\, \vec{y}$ is in the direction of $\vec{x}$, whereas $\text{proj}_{\vec{y}}\, \vec{x}$ is in the direction of $\vec{y}$.

2. Observe that for any $\vec{x} \in \mathbb{R}^n$, we can consider $\text{proj}_{\vec{x}}$ a function whose domain and codomain are $\mathbb{R}^n$. To indicate this, we can write $\text{proj}_{\vec{x}} : \mathbb{R}^n \to \mathbb{R}^n$. Since the output of this function is a vector, we call it a **vector-valued function**.

EXAMPLE 3

Let $\vec{v} = \begin{bmatrix} 1/\sqrt{3} \\ 1/\sqrt{3} \\ 1/\sqrt{3} \end{bmatrix}$ and $\vec{u} = \begin{bmatrix} 5 \\ 3 \\ -2 \end{bmatrix}$. Find $\text{proj}_{\vec{v}}\, \vec{u}$.

Solution: Since $\|\vec{v}\| = 1$, we get

$$\text{proj}_{\vec{v}}\, \vec{u} = \frac{\vec{u} \cdot \vec{v}}{\|\vec{v}\|^2}\vec{v} = (\vec{u} \cdot \vec{v})\vec{v} = \frac{6}{\sqrt{3}}\begin{bmatrix} 1/\sqrt{3} \\ 1/\sqrt{3} \\ 1/\sqrt{3} \end{bmatrix} = \begin{bmatrix} 2 \\ 2 \\ 2 \end{bmatrix}$$

EXERCISE 1

Let $\vec{v} = \begin{bmatrix} 1 \\ 2 \\ 2 \end{bmatrix}$ and $\vec{u} = \begin{bmatrix} -2 \\ 3 \\ 2 \end{bmatrix}$. Determine $\text{proj}_{\vec{v}}\, \vec{u}$ and $\text{proj}_{\vec{u}}\, \vec{v}$.

The Perpendicular Part

When you resolve a force in physics, you often not only want the component of the force in the direction of a given vector $\vec{x}$, but also the component of the force perpendicular to $\vec{x}$.

We begin by restating the problem. In $\mathbb{R}^n$, given a fixed vector $\vec{x}$ and any other $\vec{y}$, express $\vec{y}$ as the sum of a vector parallel to $\vec{x}$ and a vector orthogonal to $\vec{x}$. That is, write $\vec{y} = \vec{w} + \vec{z}$, where $\vec{w} = c\vec{x}$ for some $c \in \mathbb{R}$ and $\vec{z} \cdot \vec{x} = 0$.

We use the same trick we did $\mathbb{R}^2$. Taking the dot product of $\vec{x}$ and $\vec{y}$ gives

$$\vec{y} \cdot \vec{x} = (\vec{z} + \vec{w}) \cdot \vec{x} = \vec{z} \cdot \vec{x} + (c\vec{x}) \cdot \vec{x} = 0 + c(\vec{x} \cdot \vec{x}) = c\|\vec{x}\|^2$$

Therefore, $c = \dfrac{\vec{y} \cdot \vec{x}}{\|\vec{x}\|^2}$, so in fact, $\vec{w} = c\vec{x} = \operatorname{proj}_{\vec{x}} \vec{y}$, as we might have expected. One bonus of approaching the problem this way is that it is now clear that this is the only way to choose $\vec{w}$ to satisfy the problem.

Next, since $\vec{y} = \operatorname{proj}_{\vec{x}} \vec{y} + \vec{z}$, it follows that $\vec{z} = \vec{y} - \operatorname{proj}_{\vec{x}} \vec{y}$. Is this $\vec{z}$ really orthogonal to $\vec{x}$? To check, calculate

$$\begin{aligned}
\vec{z} \cdot \vec{x} &= (\vec{y} - \operatorname{proj}_{\vec{x}} \vec{y}) \cdot \vec{x} \\
&= \vec{y} \cdot \vec{x} - \left(\frac{\vec{y} \cdot \vec{x}}{\|\vec{x}\|^2}\vec{x}\right) \cdot \vec{x} \\
&= \vec{y} \cdot \vec{x} - \left(\frac{\vec{y} \cdot \vec{x}}{\|\vec{x}\|^2}\right)(\vec{x} \cdot \vec{x}) \\
&= \vec{y} \cdot \vec{x} - \left(\frac{\vec{y} \cdot \vec{x}}{\|\vec{x}\|^2}\right)\|\vec{x}\|^2 \\
&= \vec{y} \cdot \vec{x} - \vec{y} \cdot \vec{x} = 0
\end{aligned}$$

So, $\vec{z}$ is orthogonal to $\vec{x}$, as required. Since it is often useful to construct a vector $\vec{z}$ in this way, we introduce a name for it.

Definition
Perpendicular of a Projection

For any vectors $\vec{x}, \vec{y} \in R^n$, with $\vec{x} \neq \vec{0}$, define the **projection of $\vec{y}$ perpendicular to $\vec{x}$** to be

$$\operatorname{perp}_{\vec{x}} \vec{y} = \vec{y} - \operatorname{proj}_{\vec{x}} \vec{y}$$

Notice that $\operatorname{perp}_{\vec{x}} \vec{y}$ is again a vector-valued function on $\mathbb{R}^n$. Also observe that $\vec{y} = \operatorname{proj}_{\vec{x}} \vec{y} + \operatorname{perp}_{\vec{x}} \vec{y}$. See Figure 1.4.15.

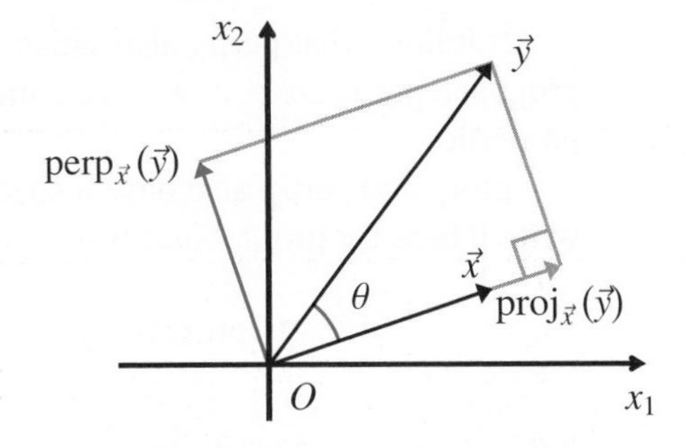

Figure 1.4.15 $\operatorname{perp}_{\vec{x}}(\vec{y})$ is perpendicular to $\vec{x}$, and $\operatorname{proj}_{\vec{x}}(\vec{y}) + \operatorname{perp}_{\vec{x}}(\vec{y}) = \vec{y}$.

EXAMPLE 4

Let $\vec{v} = \begin{bmatrix} 2 \\ 1 \\ 1 \end{bmatrix}$ and $\vec{u} = \begin{bmatrix} 1 \\ 5 \\ 1 \end{bmatrix}$. Determine $\text{proj}_{\vec{v}}\, \vec{u}$ and $\text{perp}_{\vec{v}}\, \vec{u}$.

Solution:

$$\text{proj}_{\vec{v}}\, \vec{u} = \frac{\vec{u} \cdot \vec{v}}{\|\vec{v}\|^2}\vec{v} = \frac{8}{6}\begin{bmatrix} 2 \\ 1 \\ 1 \end{bmatrix} = \begin{bmatrix} 8/3 \\ 4/3 \\ 4/3 \end{bmatrix}$$

$$\text{perp}_{\vec{v}}\, \vec{u} = \vec{u} - \text{proj}_{\vec{v}}\, \vec{u} = \begin{bmatrix} 1 \\ 5 \\ 1 \end{bmatrix} - \begin{bmatrix} 8/3 \\ 4/3 \\ 4/3 \end{bmatrix} = \begin{bmatrix} -5/3 \\ 11/3 \\ -1/3 \end{bmatrix}$$

EXERCISE 2

Let $\vec{v} = \begin{bmatrix} 3 \\ 1 \\ 2 \end{bmatrix}$ and $\vec{u} = \begin{bmatrix} 1 \\ -2 \\ 0 \end{bmatrix}$. Determine $\text{proj}_{\vec{v}}\, \vec{u}$ and $\text{perp}_{\vec{v}}\, \vec{u}$.

Some Properties of Projections

Projections will appear several times in this book, and some of their special properties are important. Two of them are called the **linearity properties**:

(L1) $\text{proj}_{\vec{x}}(\vec{y} + \vec{z}) = \text{proj}_{\vec{x}}\, \vec{y} + \text{proj}_{\vec{x}}\, \vec{z}$ for all $\vec{y}, \vec{z} \in \mathbb{R}^n$

(L2) $\text{proj}_{\vec{x}}(t\vec{y}) = t\, \text{proj}_{\vec{x}}\, \vec{y}$ for all $\vec{y} \in \mathbb{R}^n$ and all $t \in \mathbb{R}$

EXERCISE 3

Verify that properties (L1) and (L2) are true.

It follows that $\text{perp}_{\vec{x}}$ also satisfies the corresponding equations. We shall see that $\text{proj}_{\vec{x}}$ and $\text{perp}_{\vec{x}}$ are just two cases amongst the many functions that satisfy the linearity properties.

$\text{proj}_{\vec{x}}$ and $\text{perp}_{\vec{x}}$ also have a special property called the **projection property**. We write it here for $\text{proj}_{\vec{x}}$, but it is also true for $\text{perp}_{\vec{x}}$:

$$\text{proj}_{\vec{x}}\,(\text{proj}_{\vec{x}}\, \vec{y}) = \text{proj}_{\vec{x}}\, \vec{y}, \qquad \text{for all } \vec{y} \text{ in } \mathbb{R}^n$$

Minimum Distance

What is the distance from a point $Q(q_1, q_2)$ to the line with vector equation $\vec{x} = \vec{p} + t\vec{d}$? In this and similar problems, *distance* always means the minimum distance. Geometrically, we see that the minimum distance is found along a line segment perpendicular to the given line through a point P on the line. A formal proof that minimum distance requires perpendicularity by using Pythagoras' Theorem. (See Problem D3.)

To answer the question, take *any* point on the line $\vec{x} = \vec{p} + t\vec{d}$. The obvious choice is $P(p_1, p_2)$ corresponding to $\vec{p}$. From Figure 1.4.16, we see that the required distance is the length $\text{perp}_{\vec{d}}\, \vec{PQ}$.

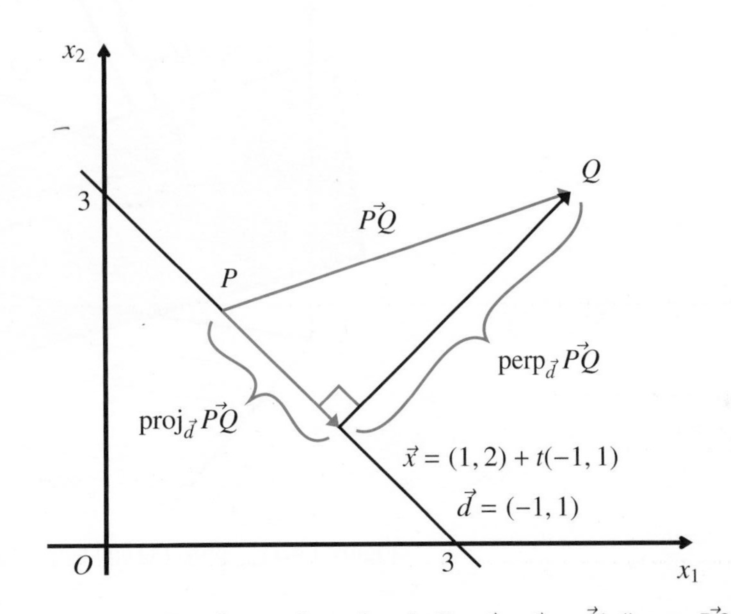

Figure 1.4.16 The distance from Q to the line $\vec{x} = \vec{p} + t\vec{d}$ is $\|\text{perp}_{\vec{d}}\, \vec{PQ}\|$.

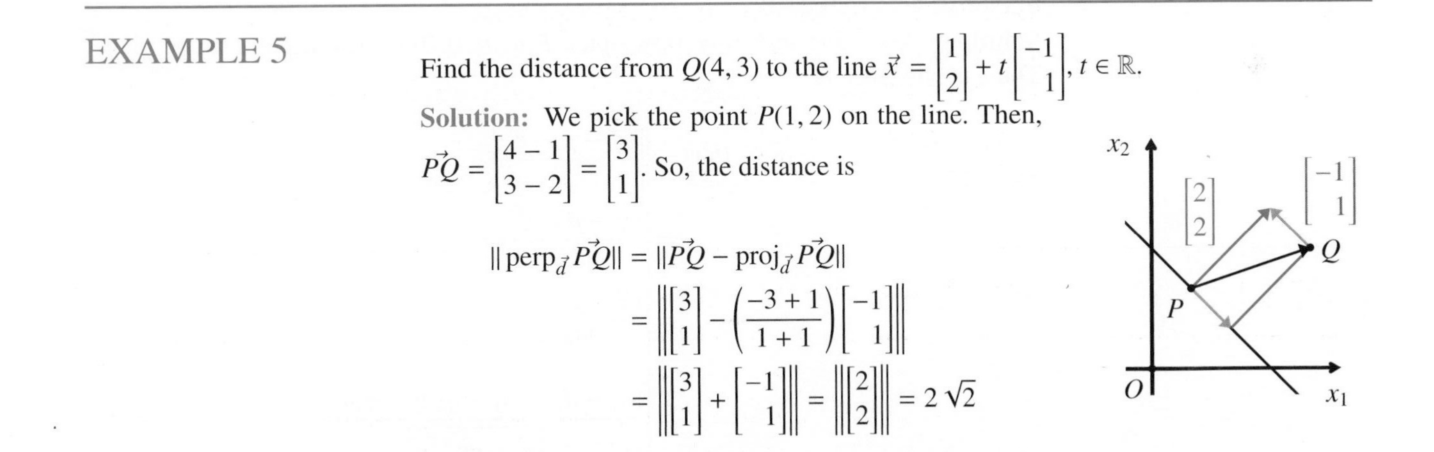

EXAMPLE 5

Find the distance from $Q(4, 3)$ to the line $\vec{x} = \begin{bmatrix} 1 \\ 2 \end{bmatrix} + t\begin{bmatrix} -1 \\ 1 \end{bmatrix}$, $t \in \mathbb{R}$.

Solution: We pick the point $P(1, 2)$ on the line. Then, $\vec{PQ} = \begin{bmatrix} 4-1 \\ 3-2 \end{bmatrix} = \begin{bmatrix} 3 \\ 1 \end{bmatrix}$. So, the distance is

$$\begin{aligned} \|\text{perp}_{\vec{d}}\, \vec{PQ}\| &= \|\vec{PQ} - \text{proj}_{\vec{d}}\, \vec{PQ}\| \\ &= \left\| \begin{bmatrix} 3 \\ 1 \end{bmatrix} - \left(\frac{-3+1}{1+1} \right) \begin{bmatrix} -1 \\ 1 \end{bmatrix} \right\| \\ &= \left\| \begin{bmatrix} 3 \\ 1 \end{bmatrix} + \begin{bmatrix} -1 \\ 1 \end{bmatrix} \right\| = \left\| \begin{bmatrix} 2 \\ 2 \end{bmatrix} \right\| = 2\sqrt{2} \end{aligned}$$

Notice that in this problem and similar problems, we take advantage of the fact that the direction vector $\vec{d}$ can be thought of as "starting" at any point. When $\text{perp}_{\vec{d}}\, \vec{AB}$ is calculated, both vectors are "located" at point P. When projections were originally defined, it was implicitly assumed that all vectors were located at the origin. Now, it is apparent that the definitions make sense as long as all vectors in the calculation are located at the same point.

We now want to look at the similar problem of finding the distance from a point $Q(q_1, q_2, q_3)$ to a plane in $\mathbb{R}^3$ with normal vector $\vec{n}$. If P is any point in the plane, then $\text{proj}_{\vec{n}}\, \vec{PQ}$ is the directed line segment from the plane to the point Q that is perpendicular to the plane. Hence, $\|\text{proj}_{\vec{n}} \vec{PQ}\|$ is the distance from Q to the plane. Moreover, $\text{perp}_{\vec{n}} \vec{PQ}$ is a directed line segment lying in the plane. In particular, it is the projection of $\vec{PQ}$ onto the plane. See Figure 1.4.17.

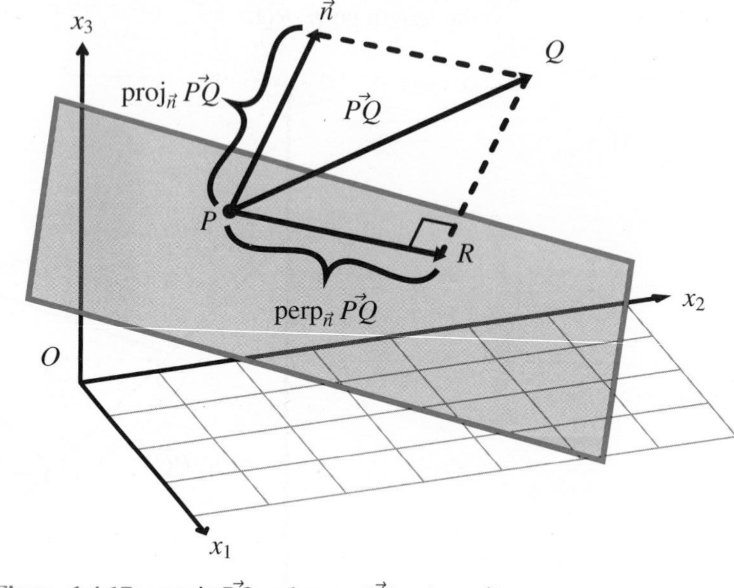

Figure 1.4.17 $\text{proj}_{\vec{n}} \vec{PQ}$ and $\text{perp}_{\vec{n}} \vec{PQ}$, where $\vec{n}$ is normal to the plane.

EXAMPLE 6

What is the distance from $Q(q_1, q_2, q_3)$ to a plane in $\mathbb{R}^3$ with equation $n_1x_1 + n_2x_2 + n_3x_3 = d$?

Solution: Assuming that $n_1 \neq 0$, we pick $P(d/n_1, 0, 0)$ to be our point in the plane. Thus, the distance is

$$\begin{aligned}
\|\text{proj}_{\vec{n}} \vec{PQ}\| &= \left\| \frac{(\vec{q} - \vec{p}) \cdot \vec{n}}{\|\vec{n}\|^2} \vec{n} \right\| \\
&= \left| \frac{(\vec{q} - \vec{p}) \cdot \vec{n}}{\|\vec{n}\|^2} \right| \|\vec{n}\| \\
&= \left| \frac{(\vec{q} - \vec{p}) \cdot \vec{n}}{\|\vec{n}\|} \right| \\
&= \left| \frac{(q_1 - d/n_1)n_1 + q_2n_2 + q_3n_3}{\sqrt{n_1^2 + n_2^2 + n_3^2}} \right| \\
&= \left| \frac{q_1n_1 + q_2n_2 + q_3n_3 - d}{\sqrt{n_1^2 + n_2^2 + n_3^2}} \right|
\end{aligned}$$

This is a standard formula for this distance problem. However, the lengths of projections along or perpendicular to a suitable vector can be used for all of these problems. It is better to learn to use this powerful and versatile idea, as illustrated in the problems above, than to memorize complicated formulas.

Finding the Nearest Point In some applications, we need to determine the point in the plane that is nearest to the point Q. Let us call this point R, as in Figure 1.4.17. Then we can determine R by observing that

$$\vec{OR} = \vec{OP} + \vec{PR} = \vec{OP} + \text{perp}_{\vec{n}}\,\vec{PQ}$$

However, we get an easier calculation if we observe from the figure that

$$\vec{OR} = \vec{OQ} + \vec{QR} = \vec{OQ} + \text{proj}_{\vec{n}}\,\vec{QP}$$

Notice that we need $\vec{QP}$ here instead of $\vec{PQ}$. Problem D4 asks you to check that these two calculations of $\vec{OR}$ are consistent.

If the plane in this problem passes through the origin, then we may take $P = O$, and the point in the plane that is closest to Q is given by $\text{perp}_{\vec{n}}\,\vec{q}$.

EXAMPLE 7

Find the point on the plane $x_1 - 2x_2 + 2x_3 = 5$ that is closest to the point $Q(2, 1, 1)$.

Solution: We pick $P(1, -1, 1)$ to be the point on the plane. Then $\vec{QP} = \begin{bmatrix} -1 \\ -2 \\ 0 \end{bmatrix}$, and we find that the point on the plane closest to Q is

$$\vec{OR} = \vec{q} + \text{proj}_{\vec{n}}\,\vec{QP} = \vec{q} + \frac{\vec{QP}\cdot\vec{n}}{\|\vec{n}\|^2}\vec{n} = \begin{bmatrix} 2 \\ 1 \\ 1 \end{bmatrix} + \frac{3}{9}\begin{bmatrix} 1 \\ -2 \\ 2 \end{bmatrix} = \begin{bmatrix} 7/3 \\ 1/3 \\ 5/3 \end{bmatrix}$$

PROBLEMS 1.4

Practice Problems

A1 For each given pair of vectors $\vec{v}$ and $\vec{u}$, determine $\text{proj}_{\vec{v}}\,\vec{u}$ and $\text{perp}_{\vec{v}}\,\vec{u}$. Check your results by verifying that $\text{proj}_{\vec{v}}\,\vec{u} + \text{perp}_{\vec{v}}\,\vec{u} = \vec{u}$ and $\vec{v} \cdot \text{perp}_{\vec{v}}\,\vec{u} = 0$.

(a) $\vec{v} = \begin{bmatrix} 0 \\ 1 \end{bmatrix}, \vec{u} = \begin{bmatrix} 3 \\ -5 \end{bmatrix}$

(b) $\vec{v} = \begin{bmatrix} 3/5 \\ 4/5 \end{bmatrix}, \vec{u} = \begin{bmatrix} -4 \\ 6 \end{bmatrix}$

(c) $\vec{v} = \begin{bmatrix} 0 \\ 1 \\ 0 \end{bmatrix}, \vec{u} = \begin{bmatrix} -3 \\ 5 \\ 2 \end{bmatrix}$

(d) $\vec{v} = \begin{bmatrix} 1/3 \\ -2/3 \\ 2/3 \end{bmatrix}, \vec{u} = \begin{bmatrix} 4 \\ 1 \\ -3 \end{bmatrix}$

(e) $\vec{v} = \begin{bmatrix} 1 \\ 1 \\ 0 \\ -2 \end{bmatrix}, \vec{u} = \begin{bmatrix} -1 \\ -1 \\ 2 \\ -1 \end{bmatrix}$

(f) $\vec{v} = \begin{bmatrix} 1 \\ 0 \\ 0 \\ 1 \end{bmatrix}, \vec{u} = \begin{bmatrix} 2 \\ 3 \\ 2 \\ -3 \end{bmatrix}$

A2 Determine $\text{proj}_{\vec{v}}\,\vec{u}$ and $\text{perp}_{\vec{v}}\,\vec{u}$ where

(a) $\vec{v} = \begin{bmatrix} 1 \\ 1 \end{bmatrix}, \vec{u} = \begin{bmatrix} 3 \\ -3 \end{bmatrix}$

(b) $\vec{v} = \begin{bmatrix} 2 \\ 3 \\ -2 \end{bmatrix}, \vec{u} = \begin{bmatrix} 4 \\ -1 \\ 3 \end{bmatrix}$

(c) $\vec{v} = \begin{bmatrix} -2 \\ 1 \\ -1 \end{bmatrix}, \vec{u} = \begin{bmatrix} 5 \\ -1 \\ 3 \end{bmatrix}$

(d) $\vec{v} = \begin{bmatrix} 1 \\ 1 \\ -2 \end{bmatrix}, \vec{u} = \begin{bmatrix} 4 \\ 1 \\ -2 \end{bmatrix}$

(e) $\vec{v} = \begin{bmatrix} -1 \\ 2 \\ 1 \\ -3 \end{bmatrix}, \vec{u} = \begin{bmatrix} 2 \\ -1 \\ 2 \\ 1 \end{bmatrix}$

A3 Consider the force represented by the vector $\vec{F} = \begin{bmatrix} 10 \\ 18 \\ -6 \end{bmatrix}$ and let $\vec{u} = \begin{bmatrix} 2 \\ 6 \\ 3 \end{bmatrix}$.

(a) Determine a unit vector in the direction of $\vec{u}$.

(b) Find the projection of $\vec{F}$ onto $\vec{u}$.

(c) Determine the projection of $\vec{F}$ perpendicular to $\vec{u}$.

A4 Follow the same instructions as in Problem A2 but with $\vec{F} = \begin{bmatrix} 3 \\ 11 \\ 2 \end{bmatrix}$ and $\vec{u} = \begin{bmatrix} 3 \\ 1 \\ -2 \end{bmatrix}$.

A5 For the given point and line, find by projection the point on the line that is closest to the given point and use perp to find the distance from the point to the line.

(a) $Q(0,0)$, line $\vec{x} = \begin{bmatrix} 1 \\ 4 \end{bmatrix} + t\begin{bmatrix} -2 \\ 2 \end{bmatrix}, \quad t \in \mathbb{R}$

(b) $Q(2,5)$, line $\vec{x} = \begin{bmatrix} 3 \\ 7 \end{bmatrix} + t\begin{bmatrix} 1 \\ -4 \end{bmatrix}, \quad t \in \mathbb{R}$

(c) $Q(1,0,1)$, line $\vec{x} = \begin{bmatrix} 2 \\ 2 \\ -1 \end{bmatrix} + t\begin{bmatrix} 1 \\ -2 \\ 1 \end{bmatrix}, \quad t \in \mathbb{R}$

(d) $Q(2,3,2)$, line $\vec{x} = \begin{bmatrix} 1 \\ 1 \\ -1 \end{bmatrix} + t\begin{bmatrix} 1 \\ 4 \\ 1 \end{bmatrix}, \quad t \in \mathbb{R}$

A6 Use a projection to find the distance from the point to the plane.

(a) $Q(2,3,1)$, plane $3x_1 - x_2 + 4x_3 = 5$

(b) $Q(-2,3,-1)$, plane $2x_1 - 3x_2 - 5x_3 = 5$

(c) $Q(0,2,-1)$, plane $2x_1 - x_3 = 5$

(d) $Q(-1,-1,1)$, plane $2x_1 - x_2 - x_3 = 4$

A7 For the given point and hyperplane in $\mathbb{R}^4$, determine by a projection the point in the hyperplane that is closest to the given point.

(a) $Q(1,0,0,1)$, hyperplane $2x_1 - x_2 + x_3 + x_4 = 0$

(b) $Q(1,2,1,3)$, hyperplane $x_1 - 2x_2 + 3x_3 = 1$

(c) $Q(2,4,3,4)$, hyperplane $3x_1 - x_2 + 4x_3 + x_4 = 0$

(d) $Q(-1,3,2,-2)$, hyperplane $x_1 + 2x_2 + x_3 - x_4 = 4$

Homework Problems

B1 For each given pair of vectors $\vec{v}$ and $\vec{u}$, determine $\text{proj}_{\vec{v}}\,\vec{u}$ and $\text{perp}_{\vec{v}}\,\vec{u}$. Check your results by verifying that $\text{proj}_{\vec{v}}\,\vec{u} + \text{perp}_{\vec{v}}\,\vec{u} = \vec{u}$ and $\vec{v} \cdot \text{perp}_{\vec{v}}\,\vec{u} = 0$.

(a) $\vec{v} = \begin{bmatrix} 0 \\ 1 \end{bmatrix}, \vec{u} = \begin{bmatrix} 4 \\ 3 \end{bmatrix}$

(b) $\vec{v} = \begin{bmatrix} 4/5 \\ -3/5 \end{bmatrix}, \vec{u} = \begin{bmatrix} -2 \\ 5 \end{bmatrix}$

(c) $\vec{v} = \begin{bmatrix} 0 \\ 0 \\ 1 \end{bmatrix}, \vec{u} = \begin{bmatrix} 2 \\ -4 \\ 7 \end{bmatrix}$

(d) $\vec{v} = \begin{bmatrix} 2 \\ 2 \\ -1 \end{bmatrix}, \vec{u} = \begin{bmatrix} -2 \\ 3 \\ 2 \end{bmatrix}$

(e) $\vec{v} = \begin{bmatrix} 1 \\ 1 \\ 0 \\ 1 \end{bmatrix}, \vec{u} = \begin{bmatrix} 3 \\ 3 \\ -4 \\ 2 \end{bmatrix}$

(f) $\vec{v} = \begin{bmatrix} 1 \\ 2 \\ -1 \\ 3 \end{bmatrix}, \vec{u} = \begin{bmatrix} 2 \\ 3 \\ 2 \\ 1 \end{bmatrix}$

B2 Determine $\text{proj}_{\vec{v}}\,\vec{u}$ and $\text{perp}_{\vec{v}}\,\vec{u}$ where

(a) $\vec{v} = \begin{bmatrix} 2 \\ 3 \end{bmatrix} \vec{u} = \begin{bmatrix} -3 \\ -1 \end{bmatrix}$

(b) $\vec{v} = \begin{bmatrix} 1 \\ 3 \end{bmatrix}, \vec{u} = \begin{bmatrix} 2 \\ 3 \end{bmatrix}$

(c) $\vec{v} = \begin{bmatrix} 1 \\ -2 \\ 1 \end{bmatrix}, \vec{u} = \begin{bmatrix} 5 \\ 1 \\ -2 \end{bmatrix}$

(d) $\vec{v} = \begin{bmatrix} 0 \\ 1 \\ -1 \end{bmatrix}, \vec{u} = \begin{bmatrix} 4 \\ 4 \\ 2 \end{bmatrix}$

(e) $\vec{v} = \begin{bmatrix} 1 \\ 0 \\ -1 \end{bmatrix}, \vec{u} = \begin{bmatrix} 6 \\ 2 \\ 6 \end{bmatrix}$

(f) $\vec{v} = \begin{bmatrix} 2 \\ 0 \\ 1 \\ 1 \end{bmatrix}, \vec{u} = \begin{bmatrix} -1 \\ 2 \\ -1 \\ 2 \end{bmatrix}$

B3 Consider the force represented by the vector $\vec{F} = \begin{bmatrix} 12 \\ -15 \\ 4 \end{bmatrix}$ and let $\vec{u} = \begin{bmatrix} -1 \\ 8 \\ 2 \end{bmatrix}$.

(a) Determine a unit vector in the direction of $\vec{u}$.
(b) Find the projection of $\vec{F}$ onto $\vec{u}$.
(c) Determine the projection of $\vec{F}$ perpendicular to $\vec{u}$.

B4 Follow the same instructions as in Problem A2 but with $\vec{F} = \begin{bmatrix} 5 \\ 13 \\ 3 \end{bmatrix}$ and $\vec{u} = \begin{bmatrix} 2 \\ -4 \\ 3 \end{bmatrix}$.

B5 For the given point and line, find by projection the point on the line that is closest to the given point and use perp to find the distance from the point to the line.

(a) $Q(3,-5)$, line $\vec{x} = \begin{bmatrix} 2 \\ -4 \end{bmatrix} + t\begin{bmatrix} 3 \\ -4 \end{bmatrix}, \quad t \in \mathbb{R}$
(b) $Q(0,0,2)$, line $\vec{x} = \begin{bmatrix} 2 \\ 1 \\ 0 \end{bmatrix} + t\begin{bmatrix} -1 \\ 2 \\ 2 \end{bmatrix}, \quad t \in \mathbb{R}$
(c) $Q(1,-3,0)$, line $\vec{x} = \begin{bmatrix} 1 \\ -1 \\ -1 \end{bmatrix} + t\begin{bmatrix} 1 \\ 1 \\ 1 \end{bmatrix}, \quad t \in \mathbb{R}$
(d) $Q(-1,2,3)$, line $\vec{x} = \begin{bmatrix} 4 \\ 1 \\ -2 \end{bmatrix} + t\begin{bmatrix} 2 \\ 1 \\ 4 \end{bmatrix}, \quad t \in \mathbb{R}$

B6 Use a projection to find the distance from the point to the plane.

(a) $Q(2,-3,-1)$, plane $2x_1 - 3x_2 - 5x_3 = 7$
(b) $Q(0,0,2)$, plane $2x_1 + x_2 - 4x_3 = 5$
(c) $Q(2,-1,2)$, plane $x_1 - x_2 - x_3 = 6$
(d) $Q(0,0,0)$, plane $-x_1 + 2x_2 - x_3 = 5$

B7 For the given point and hyperplane in $\mathbb{R}^4$, determine by a projection the point in the hyperplane that is closest to the given point.

(a) $Q(2,1,0,-1)$, hyperplane $2x_1 - x_3 + 3x_4 = 0$
(b) $Q(1,3,0,-1)$, hyperplane $2x_1 - 2x_2 + x_3 + 3x_4 = 0$
(c) $Q(3,1,2,-6)$, hyperplane $3x_1 - x_2 - x_3 + x_4 = 3$
(d) $Q(5,2,3,7)$, hyperplane $2x_1 + x_2 + 4x_3 + 3x_4 = 40$

Computer Problems

C1 Let $\vec{v}_1 = \begin{bmatrix} 1.12 \\ 2.10 \\ 7.03 \\ 4.15 \\ 6.13 \end{bmatrix}$, $\vec{v}_2 = \begin{bmatrix} 1.00 \\ 3.12 \\ -0.45 \\ -2.21 \\ 2.00 \end{bmatrix}$, $\vec{v}_3 = \begin{bmatrix} 1.00 \\ -1.01 \\ 1.02 \\ -1.03 \\ 1.04 \end{bmatrix}$, and $\vec{v}_4 = \begin{bmatrix} 1.12 \\ 0 \\ 2.13 \\ 3.15 \\ 3.40 \end{bmatrix}$.

Use computer software to evaluate each of the following.

(a) $\text{proj}_{\vec{v}_1} \vec{v}_3$
(b) $\text{perp}_{\vec{v}_1} \vec{v}_3$
(c) $\text{proj}_{\vec{v}_3} \vec{v}_1$
(d) $\text{proj}_{\vec{v}_4} \vec{v}_2$

Conceptual Problems

D1 (a) Given $\vec{u}$ and $\vec{v}$ in $\mathbb{R}^3$ with $\vec{u} \neq \vec{0}$ and $\vec{v} \neq \vec{0}$, verify that the composite map $C : \mathbb{R}^3 \to \mathbb{R}^3$ defined by $C(\vec{x}) = \text{proj}_{\vec{u}}(\text{proj}_{\vec{v}}\, \vec{x})$ also has the linearity properties (L1) and (L2).

(b) (b) Suppose that $C(\vec{x}) = \vec{0}$ for all $\vec{x} \in \mathbb{R}^3$, where C is defined as in part (a). What can you say about $\vec{u}$ and $\vec{v}$? Explain.

D2 By linearity property (L2), we know that $\text{proj}_{\vec{u}}(-\vec{x}) = -\text{proj}_{\vec{u}}\, \vec{x}$. Check, and explain geometrically, that $\text{proj}_{-\vec{u}}\, \vec{x} = \text{proj}_{\vec{u}}\, \vec{x}$.

D3 (a) (Pythagoras' theorem) Use the fact that $\|\vec{x}\|^2 = \vec{x} \cdot \vec{x}$ to prove that $\|\vec{x} + \vec{y}\|^2 = \|\vec{x}\|^2 + \|\vec{y}\|^2$ if and only if $\vec{x} \cdot \vec{y} = 0$.

(b) Let ℓ be the line in $\mathbb{R}^n$ with vector equation $\vec{x} = t\vec{d}$ and let P be any point that is not on ℓ. Prove that for any point Q on the line, the smallest value of $\|\vec{p} - \vec{q}\|^2$ is obtained when $\vec{q} = \text{proj}_{\vec{d}}(\vec{p})$ (that is, when $\vec{p} - \vec{q}$ is perpendicular to $\vec{d}$). (Hint: Consider $\|\vec{p} - \vec{q}\| = \|\vec{p} - \text{proj}_{\vec{d}}(\vec{p}) + \text{proj}_{\vec{d}}(\vec{p}) - \vec{q}\|$.)

D4 By using the definition of $\text{perp}_{\vec{n}}$ and the fact that $\vec{PQ} = -\vec{QP}$, show that

$$\vec{OP} + \text{perp}_{\vec{n}}\, \vec{PQ} = \vec{OQ} + \text{proj}_{\vec{n}}\, \vec{QP}$$

D5 (a) Let $\vec{u} = \begin{bmatrix} 1 \\ 1 \\ -1 \end{bmatrix}$ and $\vec{x} = \begin{bmatrix} 2 \\ 5 \\ 3 \end{bmatrix}$. Show that $\text{proj}_{\vec{u}}(\text{perp}_{\vec{u}}(\vec{x})) = \vec{0}$.

(b) For any $\vec{u} \in \mathbb{R}^3$, prove algebraically that for any $\vec{x} \in \mathbb{R}^3$, $\text{proj}_{\vec{u}}(\text{perp}_{\vec{u}}(\vec{x})) = \vec{0}$.

(c) Explain geometrically why $\text{proj}_{\vec{u}}(\text{perp}_{\vec{u}}(\vec{x})) = \vec{0}$ for every $\vec{x}$.

1.5 Cross-Products and Volumes

Given a pair of vectors $\vec{u}$ and $\vec{v}$ in $\mathbb{R}^3$, how can we find a third vector $\vec{w}$ that is orthogonal to both $\vec{u}$ and $\vec{v}$? This problem arises naturally in many ways. For example, to find the scalar equation of a plane whose vector equation is $\vec{x} = r\vec{u} + s\vec{v}$, we must find the normal vector $\vec{n}$ orthogonal to $\vec{u}$ and $\vec{v}$. In physics, it is observed that the force on an electrically charged particle moving in a magnetic field is in the direction orthogonal to the velocity of the particle and to the vector describing the magnetic field.

Cross-Products

Let $\vec{u}, \vec{v} \in \mathbb{R}^3$. If $\vec{w}$ is orthogonal to both $\vec{u}$ and $\vec{v}$, it must satisfy the equations

$$\vec{u} \cdot \vec{w} = u_1w_1 + u_2w_2 + u_3w_3 = 0$$
$$\vec{v} \cdot \vec{w} = v_1w_1 + v_2w_2 + v_3w_3 = 0$$

In Chapter 2, we shall develop systematic methods for solving such equations for w_1, w_2, w_3. For the present, we simply give a solution:

$$\vec{w} = \begin{bmatrix} u_2v_3 - u_3v_2 \\ u_3v_1 - u_1v_3 \\ u_1v_2 - u_2v_1 \end{bmatrix}$$

EXERCISE 1 Verify that $\vec{w} \cdot \vec{u} = 0$ and $\vec{w} \cdot \vec{v} = 0$.

Also notice from the form of the equations that any multiple of $\vec{w}$ would also be orthogonal to both $\vec{u}$ and $\vec{v}$.

Definition
Cross-Product

The **cross-product** of vectors $\vec{u} = \begin{bmatrix} u_1 \\ u_2 \\ u_3 \end{bmatrix}$ and $\vec{v} = \begin{bmatrix} v_1 \\ v_2 \\ v_3 \end{bmatrix}$ is defined by

$$\vec{u} \times \vec{v} = \begin{bmatrix} u_2v_3 - u_3v_2 \\ u_3v_1 - u_1v_3 \\ u_1v_2 - u_2v_1 \end{bmatrix}$$

EXAMPLE 1

Calculate the cross-product of $\begin{bmatrix} 2 \\ 3 \\ 5 \end{bmatrix}$ and $\begin{bmatrix} -1 \\ 1 \\ 2 \end{bmatrix}$.

Solution: $\begin{bmatrix} 2 \\ 3 \\ 5 \end{bmatrix} \times \begin{bmatrix} -1 \\ 1 \\ 2 \end{bmatrix} = \begin{bmatrix} 6-5 \\ -5-4 \\ 2-(-3) \end{bmatrix} = \begin{bmatrix} 1 \\ -9 \\ 5 \end{bmatrix}$.

Remarks

1. Unlike the dot product of two vectors, which is a scalar, the cross-product of two vectors is itself a new vector.

2. The cross-product is a construction that is defined only in $\mathbb{R}^3$. (There is a generalization to higher dimensions, but it is considerably more complicated, and it will not be considered in this book.)

The formula for the cross-product is a little awkward to remember, but there are many tricks for remembering it. One way is to write the components of $\vec{u}$ in a row above the components of $\vec{v}$:

$$\begin{matrix} u_1 & u_2 & u_3 \\ v_1 & v_2 & v_3 \end{matrix}$$

Then, for the first entry in $\vec{u} \times \vec{v}$, we cover the first column and calculate the difference of the products of the cross-terms:

$$\begin{vmatrix} u_2 & u_3 \\ v_2 & v_3 \end{vmatrix} \Rightarrow u_2v_3 - u_3v_2$$

For the second entry in $\vec{u} \times \vec{v}$, we cover the second column and take the negative of the difference of the products of the cross-terms:

$$-\begin{vmatrix} u_1 & u_3 \\ v_1 & v_3 \end{vmatrix} \Rightarrow -(u_1v_3 - u_3v_1)$$

Similarly, for the third entry, we cover the third column and calculate the difference of the products of the cross-terms:

$$\begin{vmatrix} u_1 & u_2 \\ v_1 & v_2 \end{vmatrix} \Rightarrow u_1v_2 - u_2v_1$$

Note carefully that the second term must be given a minus sign in order for this procedure to provide the correct answer. Since the formula can be difficult to remember, we recommend checking the answer by verifying that it is orthogonal to both $\vec{u}$ and $\vec{v}$.

EXERCISE 2

Calculate the cross-product of $\begin{bmatrix} 3 \\ -2 \\ 1 \end{bmatrix}$ and $\begin{bmatrix} 2 \\ 3 \\ 7 \end{bmatrix}$.

By construction, $\vec{u} \times \vec{v}$ is orthogonal to $\vec{u}$ and $\vec{v}$, so the direction of $\vec{u} \times \vec{v}$ is known except for sign: does it point "up" or "down"? The general rule is as follows: the three vectors $\vec{u}$, $\vec{v}$, and $\vec{u} \times \vec{v}$, taken in this order, form a right-handed system. Let us see how this works for simple cases.

EXERCISE 3

Let $\vec{e}_1$, $\vec{e}_2$, and $\vec{e}_3$ be the standard basis vectors in $\mathbb{R}^3$. Verify that

$$\vec{e}_1 \times \vec{e}_2 = \vec{e}_3, \quad \vec{e}_2 \times \vec{e}_3 = \vec{e}_1, \quad \vec{e}_3 \times \vec{e}_1 = \vec{e}_2$$

but

$$\vec{e}_2 \times \vec{e}_1 = -\vec{e}_3, \quad \vec{e}_3 \times \vec{e}_2 = -\vec{e}_1, \quad \vec{e}_1 \times \vec{e}_3 = -\vec{e}_2$$

Check that in every case, the three vectors taken in order form a right-handed system.

These simple examples also suggest some of the general properties of the cross-product.

Theorem 1

For $\vec{x}, \vec{y}, \vec{z} \in \mathbb{R}^3$ and $t \in \mathbb{R}$, we have

(1) $\vec{x} \times \vec{y} = -\vec{y} \times \vec{x}$
(2) $\vec{x} \times \vec{x} = \vec{0}$
(3) $\vec{x} \times (\vec{y} + \vec{z}) = \vec{x} \times \vec{y} + \vec{x} \times \vec{z}$
(4) $(t\vec{x}) \times \vec{y} = t(\vec{x} \times \vec{y})$

Proof: These properties follow easily from the definition of the cross-product and are left to the reader.

One rule we might expect does not in fact hold. In general,

$$\vec{x} \times (\vec{y} \times \vec{z}) \neq (\vec{x} \times \vec{y}) \times \vec{z}$$

This means that the parentheses cannot be omitted in a cross-product. (There are formulas available for these triple-vector products, but we shall not need them. See Problem F3 in Further Problems at the end of this chapter.)

The Length of the Cross-Product

Given $\vec{u}$ and $\vec{v}$, the direction of their cross-product is known. What is the length of the cross-product of $\vec{u}$ and $\vec{v}$? We give the answer in the following theorem.

Theorem 2

Let $\vec{u}, \vec{v} \in \mathbb{R}$ and θ be the angle between $\vec{u}$ and $\vec{v}$, then $\|\vec{u} \times \vec{v}\| = \|\vec{u}\| \, \|\vec{v}\| \sin \theta$.

Proof: We give an outline of the proof. We have

$$\|\vec{u} \times \vec{v}\|^2 = (u_2v_3 - u_3v_2)^2 + (u_3v_1 - u_1v_3)^2 + (u_1v_2 - u_2v_1)^2$$

Expand by the binomial theorem and then add and subtract the term $(u_1^2v_1^2+u_2^2v_2^2+u_3^2v_3^2)$. The resulting terms can be arranged so as to be seen to be equal to

$$(u_1^2 + u_2^2 + u_3^2)(v_1^2 + v_2^2 + v_3^2) - (u_1v_1 + u_2v_2 + u_3v_3)^2$$

Thus,

$$\begin{aligned}\|\vec{u} \times \vec{v}\|^2 &= \|\vec{u}\|^2\|\vec{v}\|^2 - (\vec{u} \cdot \vec{v})^2 = \|\vec{u}\|^2\|\vec{v}\|^2 - \|\vec{u}\|^2\|\vec{v}\|^2 \cos^2\theta \\ &= \|\vec{u}\|^2\|\vec{v}\|^2(1 - \cos^2\theta) = \|\vec{u}\|^2\|\vec{v}\|^2 \sin^2\theta\end{aligned}$$

and the result follows. ■

To interpret this formula, consider Figure 1.5.18. Assuming that $\vec{u} \times \vec{v} \neq \vec{0}$, the vectors $\vec{u}$ and $\vec{v}$ determine a parallelogram. Take the length of $\vec{u}$ to be the base of the parallelogram; the altitude is the length of $\text{perp}_{\vec{u}}\,\vec{v}$. From trigonometry, we know that this length is $\|\vec{v}\|\sin\theta$, so that the area of the parallelogram is

$$(\text{base}) \times (\text{altitude}) = \|\vec{u}\|\;\|\vec{v}\|\sin\theta = \|\vec{u} \times \vec{v}\|$$

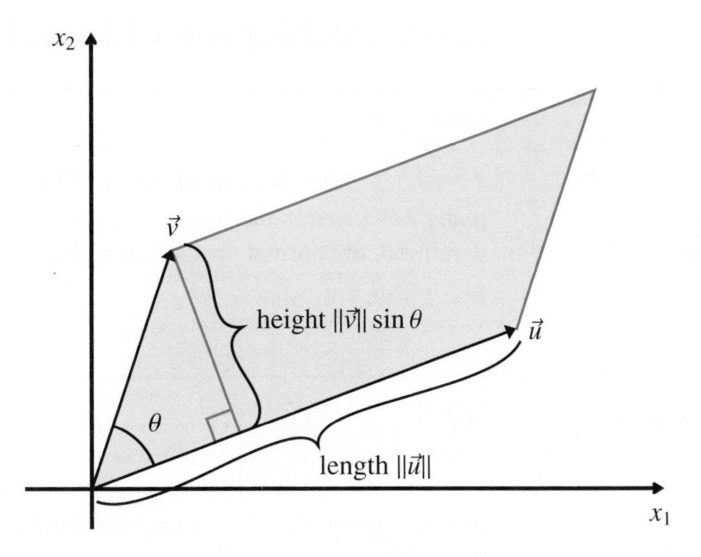

Figure 1.5.18 The area of the parallelogram is $\|\vec{u}\|\;\|\vec{v}\|\sin\theta$.

EXAMPLE 2

Find the area of the parallelogram determined by $\vec{u} = \begin{bmatrix}1\\0\\0\end{bmatrix}$ and $\vec{v} = \begin{bmatrix}1\\1\\0\end{bmatrix}$.

Solution: By Theorem 2, we find that the area is

$$\|\vec{u} \times \vec{v}\| = \left\|\begin{bmatrix}0\\0\\1\end{bmatrix}\right\| = 1$$

EXAMPLE 3

Find the area of the parallelogram determined by $\vec{u} = \begin{bmatrix} 1 \\ -2 \\ 1 \end{bmatrix}$ and $\vec{v} = \begin{bmatrix} -1 \\ 1 \\ 1 \end{bmatrix}$.

Solution: By Theorem 2, the area is

$$\|\vec{u} \times \vec{v}\| = \|(-3, -2, -1)\| = \sqrt{14}$$

EXERCISE 4

Find the area of the parallelogram determined by $\vec{u} = \begin{bmatrix} 0 \\ 2 \\ -1 \end{bmatrix}$ and $\vec{v} = \begin{bmatrix} 1 \\ 2 \\ -2 \end{bmatrix}$.

Some Problems on Lines, Planes, and Distances

The cross-product allows us to answer many questions about lines, planes, and distances in $\mathbb{R}^3$.

Finding the Normal to a Plane In Section 1.2, the vector equation of a plane was given in the form $\vec{x} = \vec{p} + s\vec{u} + t\vec{v}$, where $\{\vec{u}, \vec{v}\}$ is linearly independent. By definition, the normal vector $\vec{n}$ must be perpendicular to $\vec{u}$ and $\vec{v}$. Therefore, it will be given by $\vec{n} = \vec{u} \times \vec{v}$.

EXAMPLE 4

The lines $\vec{x} = \begin{bmatrix} 1 \\ 3 \\ 2 \end{bmatrix} + s\begin{bmatrix} 1 \\ 0 \\ 2 \end{bmatrix}$ and $\vec{x} = \begin{bmatrix} 1 \\ 3 \\ 2 \end{bmatrix} + t\begin{bmatrix} -1 \\ 2 \\ 1 \end{bmatrix}$ must lie in a common plane since they have the point $(1, 3, 2)$ in common. Find a scalar equation of the plane that contains these lines.

Solution: The normal to the plane is

$$\vec{n} = \begin{bmatrix} 1 \\ 0 \\ 2 \end{bmatrix} \times \begin{bmatrix} -1 \\ 2 \\ 1 \end{bmatrix} = \begin{bmatrix} -4 \\ -3 \\ 2 \end{bmatrix}$$

Therefore, since the plane passes through $P(1, 3, 2)$, we find that an equation of the plane is

$$-4x_1 - 3x_2 + 2x_3 = (-4)(1) + (-3)(3) + 2(2) = -9$$

EXAMPLE 5

Find a scalar equation of the plane that contains the three points $P(1,-2,1)$, $Q(2,-2,-1)$, and $R(4,1,1)$.

Solution: Since P, Q, and R lie in the plane, then so do the directed line segments $\vec{PQ}$ and $\vec{PR}$. Hence, the normal to the plane is given by

$$\vec{n} = \vec{PQ} \times \vec{PR} = \begin{bmatrix} 1 \\ 0 \\ -2 \end{bmatrix} \times \begin{bmatrix} 3 \\ 3 \\ 0 \end{bmatrix} = \begin{bmatrix} 6 \\ -6 \\ 3 \end{bmatrix}$$

Since the plane passes through P, we find that an equation of the plane is

$$6x_1 - 6x_2 + 3x_3 = (6)(1) + (-6)(-2) + 1(3) = 21, \quad \text{or} \quad 2x_1 - 2x_2 + x_3 = 7$$

Finding the Line of Intersection of Two Planes Unless two planes in $\mathbb{R}^3$ are parallel, their intersection will be a line. The direction vector of this line lies in both planes, so it is perpendicular to both of the normals. It can therefore be obtained as the cross-product of the two normals. Once we find a point that lies on this line, we can write the vector equation of the line.

EXAMPLE 6

Find a vector equation of the line of intersection of the two planes $x_1 + x_2 - 2x_3 = 3$ and $2x_1 - x_2 + 3x_3 = 6$.

Solution: The normal vectors of the planes are $\begin{bmatrix} 1 \\ 1 \\ -2 \end{bmatrix}$ and $\begin{bmatrix} 2 \\ -1 \\ 3 \end{bmatrix}$. Hence, the direction vector of the line of intersection is

$$\vec{d} = \begin{bmatrix} 1 \\ 1 \\ -2 \end{bmatrix} \times \begin{bmatrix} 2 \\ -1 \\ 3 \end{bmatrix} = \begin{bmatrix} 1 \\ -7 \\ -3 \end{bmatrix}$$

One easy way to find a point on the line is to let $x_3 = 0$ and then solve the remaining equations $x_1 + x_2 = 3$ and $2x_1 - x_2 = 6$. The solution is $x_1 = 3$ and $x_2 = 0$. Hence, a vector equation of the line of intersection is

$$\vec{x} = \begin{bmatrix} 3 \\ 0 \\ 0 \end{bmatrix} + t \begin{bmatrix} 1 \\ -7 \\ -3 \end{bmatrix}, \quad t \in \mathbb{R}$$

EXERCISE 5

Find a vector equation of the line of intersection of the two planes $-x_1 - 2x_2 + x_3 = -2$ and $2x_1 + x_2 - 2x_3 = 1$.

The Scalar Triple Product and Volumes in $\mathbb{R}^3$ The three vectors $\vec{u}$, $\vec{v}$, and $\vec{w}$ in $\mathbb{R}^3$ may be taken to be the three adjacent edges of a parallelepiped (see Figure 1.5.19). Is there an expression for the volume of the parallelepiped in terms of the three vectors? To obtain such a formula, observe that the parallelogram determined by $\vec{u}$ and $\vec{v}$ can be regarded as the base of the solid of the parallelepiped. This base has area $\|\vec{u} \times \vec{v}\|$. With respect to this base, the altitude of the solid is the length of the amount of $\vec{w}$ in the direction of the normal vector $\vec{n} = \vec{u} \times \vec{v}$ to the base. That is,

$$\text{altitude} = \| \text{proj}_{\vec{n}} \vec{w}\| = \frac{|\vec{w} \cdot \vec{n}|}{\|\vec{n}\|} = \frac{|\vec{w} \cdot (\vec{u} \times \vec{v})|}{\|\vec{u} \times \vec{v}\|}$$

Thus, to get the volume of the parallelepiped, multiply this altitude by the area of the base to get

$$\text{volume of the parallelepiped} = \frac{|\vec{w} \cdot (\vec{u} \times \vec{v})|}{\|\vec{u} \times \vec{v}\|} \times \|\vec{u} \times \vec{v}\| = |\vec{w} \cdot (\vec{u} \times \vec{v})|$$

The product $\vec{w} \cdot (\vec{u} \times \vec{v})$ is called the **scalar triple product** of $\vec{w}$, $\vec{u}$, and $\vec{v}$. Notice that the result is a real number (a scalar).

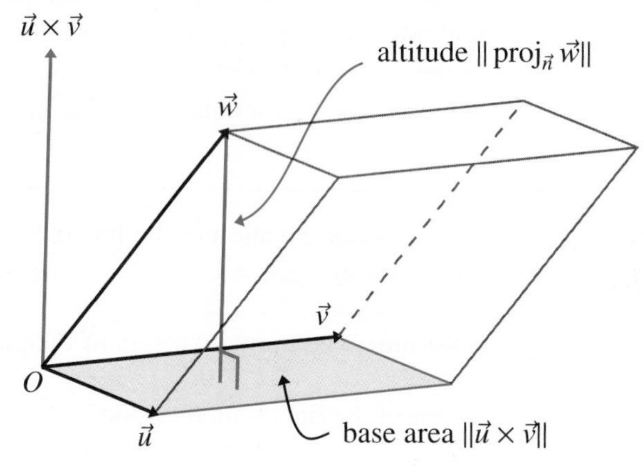

Figure 1.5.19 The parallelepiped with adjacent edges $\vec{u}$, $\vec{v}$, $\vec{w}$ has volume given by $|\vec{w} \cdot (\vec{u} \times \vec{v})|$.

The sign of the scalar triple product also has an interpretation. Recall that the ordered triple of vectors $\{\vec{u}, \vec{v}, \vec{u} \times \vec{v}\}$ is right-handed; we can think of $\vec{u} \times \vec{v}$ as the "upwards" normal vector to the plane with vector equation $\vec{x} = s\vec{u} + t\vec{v}$. Some other vector $\vec{w}$ is then "upwards," and $\{\vec{u}, \vec{v}, \vec{w}\}$ (in that order) is right-handed, if and only if the scalar triple product is positive. If the triple scalar product is negative, then $\{\vec{u}, \vec{v}, \vec{w}\}$ is a left-handed system.

It is often useful to note that

$$\vec{w} \cdot (\vec{u} \times \vec{v}) = \vec{u} \cdot (\vec{v} \times \vec{w}) = \vec{v} \cdot (\vec{w} \times \vec{u})$$

This is straightfoward but tedious to verify.

EXAMPLE 7

Find the volume of the parallelepiped determined by the vectors $\begin{bmatrix}1\\0\\2\end{bmatrix}$, $\begin{bmatrix}2\\1\\1\end{bmatrix}$, and $\begin{bmatrix}-1\\-1\\3\end{bmatrix}$.

Solution: The volume V is

$$V = \left|\begin{bmatrix}1\\0\\2\end{bmatrix} \cdot \left(\begin{bmatrix}2\\1\\1\end{bmatrix} \times \begin{bmatrix}-1\\-1\\3\end{bmatrix}\right)\right| = \left|\begin{bmatrix}1\\0\\2\end{bmatrix} \cdot \begin{bmatrix}4\\-7\\-1\end{bmatrix}\right| = 2$$

PROBLEMS 1.5

Practice Problems

A1 Calculate the following cross-products.

(a) $\begin{bmatrix}1\\-5\\2\end{bmatrix} \times \begin{bmatrix}-2\\1\\5\end{bmatrix}$ (b) $\begin{bmatrix}2\\-3\\-5\end{bmatrix} \times \begin{bmatrix}4\\-2\\7\end{bmatrix}$

(c) $\begin{bmatrix}-1\\0\\-1\end{bmatrix} \times \begin{bmatrix}0\\4\\5\end{bmatrix}$ (d) $\begin{bmatrix}1\\2\\0\end{bmatrix} \times \begin{bmatrix}-1\\-3\\0\end{bmatrix}$

(e) $\begin{bmatrix}4\\-2\\6\end{bmatrix} \times \begin{bmatrix}-2\\1\\-3\end{bmatrix}$ (f) $\begin{bmatrix}3\\1\\3\end{bmatrix} \times \begin{bmatrix}3\\1\\3\end{bmatrix}$

A2 Let $\vec{u} = \begin{bmatrix}-1\\4\\2\end{bmatrix}$, $\vec{v} = \begin{bmatrix}3\\1\\-1\end{bmatrix}$, and $\vec{w} = \begin{bmatrix}2\\-3\\-1\end{bmatrix}$. Check by calculation that the following general properties hold.

(a) $\vec{u} \times \vec{u} = \vec{0}$

(b) $\vec{u} \times \vec{v} = -\vec{v} \times \vec{u}$

(c) $\vec{u} \times 3\vec{w} = 3(\vec{u} \times \vec{w})$

(d) $\vec{u} \times (\vec{v} + \vec{w}) = \vec{u} \times \vec{v} + \vec{u} \times \vec{w}$

(e) $\vec{u} \cdot (\vec{v} \times \vec{w}) = \vec{w} \cdot (\vec{u} \times \vec{v})$

(f) $\vec{u} \cdot (\vec{v} \times \vec{w}) = -\vec{v} \cdot (\vec{u} \times \vec{w})$

A3 Calculate the area of the parallelogram determined by each pair of vectors.

(a) $\begin{bmatrix}1\\2\\1\end{bmatrix}, \begin{bmatrix}2\\3\\-1\end{bmatrix}$ (b) $\begin{bmatrix}1\\0\\1\end{bmatrix}, \begin{bmatrix}1\\1\\4\end{bmatrix}$

(c) $\begin{bmatrix}1\\2\\0\end{bmatrix}, \begin{bmatrix}-2\\5\\0\end{bmatrix}$ (d) $\begin{bmatrix}-3\\1\end{bmatrix}, \begin{bmatrix}4\\3\end{bmatrix}$

(Hint: For (d), think of these vectors as $\begin{bmatrix}-3\\1\\0\end{bmatrix}$ and $\begin{bmatrix}4\\3\\0\end{bmatrix}$ in $\mathbb{R}^3$.)

A4 Determine the scalar equation of the plane with vector equation

(a) $\vec{x} = \begin{bmatrix}1\\4\\7\end{bmatrix} + s\begin{bmatrix}2\\3\\-1\end{bmatrix} + t\begin{bmatrix}4\\1\\0\end{bmatrix}$

(b) $\vec{x} = \begin{bmatrix}2\\3\\-1\end{bmatrix} + s\begin{bmatrix}1\\1\\0\end{bmatrix} + t\begin{bmatrix}-2\\1\\2\end{bmatrix}$

(c) $\vec{x} = \begin{bmatrix}1\\-1\\3\end{bmatrix} + s\begin{bmatrix}2\\-2\\1\end{bmatrix} + t\begin{bmatrix}0\\3\\1\end{bmatrix}$

(d) $\vec{x} = s\begin{bmatrix}1\\3\\2\end{bmatrix} + t\begin{bmatrix}-2\\4\\-3\end{bmatrix}$

A5 Determine the scalar equation of the plane that contains each set of points.

(a) $P(2, 1, 5)$, $Q(4, -3, 2)$, $R(2, 6, -1)$

(b) $P(3, 1, 4)$, $Q(-2, 0, 2)$, $R(1, 4, -1)$

(c) $P(-1, 4, 2)$, $Q(3, 1, -1)$, $R(2, -3, -1)$

(d) $P(1, 0, 1)$, $Q(-1, 0, 1)$, $R(0, 0, 0)$

A6 Determine a vector equation of the line of intersection of the given planes.

(a) $x_1 + 3x_2 - x_3 = 5$ and $2x_1 - 5x_2 + x_3 = 7$

(b) $2x_1 - 3x_3 = 7$ and $x_2 + 2x_3 = 4$

A7 Find the volume of the parallelepiped determined by each set of vectors.

(a) $\begin{bmatrix}1\\0\\1\end{bmatrix}, \begin{bmatrix}0\\1\\1\end{bmatrix}, \begin{bmatrix}0\\0\\1\end{bmatrix}$

(b) $\begin{bmatrix}4\\1\\-1\end{bmatrix}, \begin{bmatrix}-1\\5\\2\end{bmatrix}, \begin{bmatrix}1\\1\\6\end{bmatrix}$

(c) $\begin{bmatrix}-2\\1\\2\end{bmatrix}, \begin{bmatrix}3\\1\\2\end{bmatrix}, \begin{bmatrix}0\\2\\5\end{bmatrix}$

(d) $\begin{bmatrix}1\\5\\-3\end{bmatrix}, \begin{bmatrix}1\\0\\-1\end{bmatrix}, \begin{bmatrix}3\\0\\4\end{bmatrix}$

(e) $\begin{bmatrix}-1\\1\\1\end{bmatrix}, \begin{bmatrix}-2\\4\\0\end{bmatrix}, \begin{bmatrix}2\\2\\2\end{bmatrix}$

A8 What does it mean, geometrically, if $\vec{u} \cdot (\vec{v} \times \vec{w}) = 0$?

A9 Show that $(\vec{u} - \vec{v}) \times (\vec{u} + \vec{v}) = 2\vec{u} \times \vec{v}$.

Homework Problems

B1 Calculate the following cross-products.

(a) $\begin{bmatrix}4\\-2\\4\end{bmatrix}\times\begin{bmatrix}-6\\3\\-6\end{bmatrix}$ (b) $\begin{bmatrix}2\\-1\\6\end{bmatrix}\times\begin{bmatrix}4\\4\\5\end{bmatrix}$

(c) $\begin{bmatrix}1\\2\\1\end{bmatrix}\times\begin{bmatrix}3\\-2\\1\end{bmatrix}$ (d) $\begin{bmatrix}-3\\1\\0\end{bmatrix}\times\begin{bmatrix}1\\5\\2\end{bmatrix}$

(e) $\begin{bmatrix}2\\-4\\1/2\end{bmatrix}\times\begin{bmatrix}-1\\1\\3\end{bmatrix}$ (f) $\begin{bmatrix}1\\3\\-1\end{bmatrix}\times\begin{bmatrix}1\\3\\0\end{bmatrix}$

B2 Let $\vec{u} = \begin{bmatrix}4\\1\\1\end{bmatrix}$, $\vec{v} = \begin{bmatrix}0\\4\\-2\end{bmatrix}$, and $\vec{w} = \begin{bmatrix}1\\-2\\3\end{bmatrix}$. Check by calculation that the following general properties hold.

(a) $\vec{u}\times\vec{u} = \vec{0}$
(b) $\vec{u}\times\vec{v} = -\vec{v}\times\vec{u}$
(c) $\vec{u}\times 2\vec{w} = 2(\vec{u}\times\vec{w})$
(d) $\vec{u}\times(\vec{v}+\vec{w}) = \vec{u}\times\vec{v}+\vec{u}\times\vec{w}$
(e) $\vec{u}\cdot(\vec{v}\times\vec{w}) = \vec{w}\cdot(\vec{u}\times\vec{v})$
(f) $\vec{u}\cdot(\vec{v}\times\vec{w}) = -\vec{v}\cdot(\vec{u}\times\vec{w})$

B3 Calculate the area of the parallelogram determined by each pair of vectors.

(a) $\begin{bmatrix}2\\1\\3\end{bmatrix},\begin{bmatrix}3\\-1\\5\end{bmatrix}$ (b) $\begin{bmatrix}2\\1\\4\end{bmatrix},\begin{bmatrix}5\\-1\\-1\end{bmatrix}$

(c) $\begin{bmatrix}1\\0\\-5\end{bmatrix},\begin{bmatrix}-3\\1\\1\end{bmatrix}$ (d) $\begin{bmatrix}2\\3\end{bmatrix},\begin{bmatrix}-5\\2\end{bmatrix}$

(Hint: For (d), think of these vectors as $\begin{bmatrix}2\\3\\0\end{bmatrix}$ and $\begin{bmatrix}-5\\2\\0\end{bmatrix}$ in $\mathbb{R}^3$.)

B4 Determine the scalar equation of the plane with vector equation

(a) $\vec{x} = \begin{bmatrix}4\\-2\\-5\end{bmatrix}+s\begin{bmatrix}-1\\3\\-2\end{bmatrix}+t\begin{bmatrix}1\\4\\5\end{bmatrix}$

(b) $\vec{x} = \begin{bmatrix}-1\\0\\1\end{bmatrix}+s\begin{bmatrix}3\\1\\1\end{bmatrix}+t\begin{bmatrix}-1\\-1\\3\end{bmatrix}$

(c) $\vec{x} = \begin{bmatrix}3\\4\\5\end{bmatrix}+s\begin{bmatrix}-2\\1\\1\end{bmatrix}+t\begin{bmatrix}-1\\0\\2\end{bmatrix}$

(d) $\vec{x} = s\begin{bmatrix}2\\2\\1\end{bmatrix}+t\begin{bmatrix}1\\-2\\-1\end{bmatrix}$

B5 Determine the scalar equation of the plane that contains each set of points.

(a) $P(5,2,1)$, $Q(-3,2,4)$, $R(8,1,6)$
(b) $P(5,-1,2)$, $Q(-1,3,4)$, $R(3,1,1)$
(c) $P(0,2,1)$, $Q(3,-1,1)$, $R(1,3,0)$
(d) $P(1,5,-3)$, $Q(2,6,-1)$, $R(1,0,1)$

B6 Determine a vector equation of the line of intersection of the given planes.

(a) $x_1+4x_2+x_3 = 5$ and $3x_1-7x_2-x_3 = 6$
(b) $x_1-3x_2-2x_3 = 4$ and $3x_1+2x_2+x_3 = 2$

B7 Find the volume of the parallelepiped determined by each set of vectors.

(a) $\begin{bmatrix}5\\-1\\1\end{bmatrix},\begin{bmatrix}1\\4\\-1\end{bmatrix},\begin{bmatrix}1\\2\\-6\end{bmatrix}$

(b) $\begin{bmatrix}1\\1\\4\end{bmatrix},\begin{bmatrix}1\\3\\4\end{bmatrix},\begin{bmatrix}-2\\1\\-5\end{bmatrix}$

(c) $\begin{bmatrix}3\\1\\0\end{bmatrix},\begin{bmatrix}2\\3\\3\end{bmatrix},\begin{bmatrix}1\\4\\-1\end{bmatrix}$

(d) $\begin{bmatrix}6\\1\\5\end{bmatrix},\begin{bmatrix}7\\0\\5\end{bmatrix},\begin{bmatrix}2\\2\\-4\end{bmatrix}$

Conceptual Problems

D1 Show that if X is a point on the line through P and Q, then $\vec{x}\times(\vec{q}-\vec{p}) = \vec{p}\times\vec{q}$, where $\vec{x} = \overrightarrow{OX}$, $\vec{p} = \overrightarrow{OP}$, and $\vec{q} = \overrightarrow{OQ}$.

D2 Consider the following statement: "If $\vec{u}\neq\vec{0}$, and $\vec{u}\times\vec{v} = \vec{u}\times\vec{w}$, then $\vec{v} = \vec{w}$." If the statement is true, prove it. If it is false, give a counterexample.

D3 Explain why $\vec{u} \times (\vec{v} \times \vec{w})$ must be a vector that satisfies the vector equation $\vec{x} = s\vec{v} + t\vec{w}$.

D4 Give an example of distinct vectors $\vec{u}$, $\vec{v}$, and $\vec{w}$ in $\mathbb{R}^3$ such that

(a) $\vec{u} \times (\vec{v} \times \vec{w}) = (\vec{u} \times \vec{v}) \times \vec{w}$

(b) $\vec{u} \times (\vec{v} \times \vec{w}) \neq (\vec{u} \times \vec{v}) \times \vec{w}$

CHAPTER REVIEW

Suggestions for Student Review

Organizing your own review is an important step towards mastering new material. It is much more valuable than memorizing someone else's list of key ideas. To retain new concepts as useful tools, you must be able to state definitions and make connections between various ideas and techniques. You should also be able to give (or, even better, create) instructive examples. The suggestions below are not intended to be an exhaustive checklist; instead, they suggest the kinds of activities and questioning that will help you gain a confident grasp of the material.

1 Find some person or persons to talk with about mathematics. There's lots of evidence that this is the best way to learn. Be sure you do your share of asking and answering. Note that a little bit of embarrassment is a small price for learning. Also, be sure to get lots of practice in writing answers independently.

2 Draw pictures to illustrate addition of vectors, subtraction of vectors, and multiplication of a vector by a scalar (general case). (Section 1.1)

3 Explain how you find a vector equation for a line and make up examples to show why the vector equation of a line is not unique. (Albert Einstein once said, "If you can't explain it simply, you don't understand it well enough.") (Section 1.1)

4 State the definition of a subspace of $\mathbb{R}^n$. Give examples of subspaces in $\mathbb{R}^3$ that are lines, planes, and all of $\mathbb{R}^3$. Show that there is only one subspace in $\mathbb{R}^3$ that does not have infinitely many vectors in it. (Section 1.2)

5 Show that the subspace spanned by three vectors in $\mathbb{R}^3$ can either be a point, a line, a plane, or all of $\mathbb{R}^3$, by giving examples. Explain how this relates with the concept of linear independence. (Section 1.2)

6 Let $\{\vec{v}_1, \vec{v}_2\}$ be a linearly independent spanning set for a subspace S of $\mathbb{R}^3$. Explain how you could construct other spanning sets and other linearly independent spanning sets for S. (Section 1.2)

7 State the formal definition of linear independence. Explain the connection between the formal definition of linear dependence and an intuitive geometric understanding of linear dependence. Why is linear independence important when looking at spanning sets? (Section 1.2)

8 State the relation (or relations) between the length in $\mathbb{R}^3$ and the dot product in $\mathbb{R}^3$. Use examples to illustrate. (Section 1.3)

9 Explain how projection onto a vector $\vec{v}$ is defined in terms of the dot product. Illustrate with a picture. Define the part of a vector $\vec{x}$ perpendicular to $\vec{v}$ and verify (in the general case) that it is perpendicular to $\vec{v}$. (Section 1.4)

10 Explain with a picture how projections help us to solve the minimum distance problem. (Section 1.4)

11 Discuss the role of the normal vector to a plane in determining the scalar equation of the plane. Explain how you can get from a scalar equation of a plane to a vector equation for the plane and from a vector equation of the plane to the scalar equation. (Sections 1.3 and 1.5)

12 State the important algebraic and geometric properties of the cross-product. (Section 1.5)

Chapter Quiz

Note: Your instructor may have different ideas of an appropriate level of difficulty for a test on this material.

E1 Determine a vector equation of the line passing through points $P(-2, 1, -4)$ and $Q(5, -2, 1)$.

E2 Determine the scalar equation of the plane that contains the points $P(1, -1, 0)$, $Q(3, 1, -2)$, and $R(-4, 1, 6)$.

E3 Show that $\left\{\begin{bmatrix}1\\2\end{bmatrix}, \begin{bmatrix}-1\\2\end{bmatrix}\right\}$ is a basis for $\mathbb{R}^2$.

E4 Prove that $S = \left\{\begin{bmatrix}x_1\\x_2\\x_3\end{bmatrix} \in \mathbb{R}^3 \mid a_1x_2 + a_2x_2 + a_3x_3 = d\right\}$ is a subspace of $\mathbb{R}^3$ for any real numbers a_1, a_2, a_3 if and only if $d = 0$.

E5 Determine the cosine of the angle between $\vec{v} = \begin{bmatrix}2\\-3\\1\end{bmatrix}$ and each of the coordinate axes.

E6 Find the point on the line $\vec{x} = t\begin{bmatrix}3\\-2\\3\end{bmatrix}, t \in \mathbb{R}$ that is closest to the point $P(2, 3, 4)$. Illustrate your method of calculation with a sketch.

E7 Find the point on the hyperplane $x_1 + x_2 + x_3 + x_4 = 1$ that is closest to the point $P(3, -2, 0, 2)$ and determine the distance from the point to the plane.

E8 Determine a non-zero vector that is orthogonal to both $\begin{bmatrix}1\\2\\0\end{bmatrix}$ and $\begin{bmatrix}-3\\1\\1\end{bmatrix}$.

E9 Prove that the volume of the parallelepiped determined by $\vec{u}$, $\vec{v}$, and $\vec{w}$ has the same volume as the parallelepiped determined by $(\vec{u} + k\vec{v})$, $\vec{v}$, and $\vec{w}$.

E10 Each of the following statements is to be interpreted in $\mathbb{R}^3$. Determine whether each statement is true, and if so, explain briefly. If false, give a counterexample.

(i) Any three distinct points lie in exactly one plane.

(ii) The subspace spanned by a single non-zero vector is a line passing through the origin.

(iii) The set $\text{Span}\{\vec{v}_1, \dots, \vec{v}_k\}$ is linearly dependent.

(iv) The dot product of a vector with itself cannot be zero.

(v) For any vectors $\vec{x}$ and $\vec{y}$, $\text{proj}_{\vec{x}}\, \vec{y} = \text{proj}_{\vec{y}}\, \vec{x}$.

(vi) For any vectors $\vec{x}$ and $\vec{y}$, the set $\{\text{proj}_{\vec{x}}\, \vec{y}, \text{perp}_{\vec{x}}\, \vec{y}\}$ is linearly independent.

(vii) The area of the parallelogram determined by $\vec{u}$ and $\vec{v}$ is the same as the area of the parallelogram determined by $\vec{u}$ and $(\vec{v} + 3\vec{u})$.

Further Problems

These problems are intended to be a little more challenging than the problems at the end of each section. Some explore topics beyond the material discussed in the text.

F1 Consider the statement "If $\vec{u} \neq \vec{0}$, and both $\vec{u} \cdot \vec{v} = \vec{u} \cdot \vec{w}$ and $\vec{u} \times \vec{v} = \vec{u} \times \vec{w}$, then $\vec{v} = \vec{w}$." Either prove the statement or give a counterexample.

F2 Suppose that $\vec{u}$ and $\vec{v}$ are orthogonal unit vectors in $\mathbb{R}^3$. Prove that for every $\vec{x} \in \mathbb{R}^3$,

$$\text{perp}_{\vec{u}\times\vec{v}}(\vec{x}) = \text{proj}_{\vec{u}}(\vec{x}) + \text{proj}_{\vec{v}}(\vec{x})$$

F3 In Problem 1.5.D3, you were asked to show that $\vec{u} \times (\vec{v} \times \vec{w}) = s\vec{v} + t\vec{w}$ for some $s, t \in \mathbb{R}$.

(a) By direct calculation, prove that $\vec{u} \times (\vec{v} \times \vec{w}) = (\vec{u} \cdot \vec{w})\vec{v} - (\vec{u} \cdot \vec{v})\vec{w}$.

(b) Prove that $\vec{u} \times (\vec{v} \times \vec{w}) + \vec{v} \times (\vec{w} \times \vec{u}) + \vec{w} \times (\vec{u} \times \vec{v}) = \vec{0}$.

F4 Prove that

(a) $\vec{u} \cdot \vec{v} = \frac{1}{4}\|\vec{u} + \vec{v}\|^2 - \frac{1}{4}\|\vec{u} - \vec{v}\|^2$

(b) $\|\vec{u} + \vec{v}\|^2 + \|\vec{u} - \vec{v}\|^2 = 2\|\vec{u}\|^2 + 2\|\vec{v}\|^2$

(c) Interpret (a) and (b) in terms of a parallelogram determined by vectors $\vec{u}$ and $\vec{v}$.

F5 Show that if P, Q, and R are collinear points and $\vec{OP} = \vec{p}$, $\vec{OQ} = \vec{q}$, and $\vec{OR} = \vec{r}$, then

$$(\vec{p} \times \vec{q}) + (\vec{q} \times \vec{r}) + (\vec{r} \times \vec{p}) = \vec{0}$$

F6 In $\mathbb{R}^2$, two lines fail to have a point of intersection only if they are parallel. However, in $\mathbb{R}^3$, a pair of lines can fail to have a point of intersection even if they are not parallel. Two such lines in $\mathbb{R}^3$ are called **skew**.

(a) Observe that if two lines are skew, then they do not lie in a common plane. Show that two skew lines do lie in parallel planes.

(b) Find the distance between the skew lines

$$\vec{x} = \begin{bmatrix} 1 \\ 4 \\ 2 \end{bmatrix} + s\begin{bmatrix} 2 \\ 0 \\ 1 \end{bmatrix}, s \in \mathbb{R} \text{ and } \vec{x} = \begin{bmatrix} 2 \\ -3 \\ 1 \end{bmatrix} + t\begin{bmatrix} 1 \\ 1 \\ 3 \end{bmatrix}, \quad t \in \mathbb{R}$$

Visit the text's website at www.pearsoncanada.ca/norman for practice quizzes, additional applications, and an essay on linearity and superposition in physics.

CHAPTER 2

Systems of Linear Equations

CHAPTER OUTLINE

In a few places in Chapter 1, we needed to find a vector $\vec{x}$ in $\mathbb{R}^n$ that simultaneously satisfied several linear equations. In such cases, we used a **system of linear equations***. Such systems arise frequently in almost every conceivable area where mathematics is applied: in analyzing stresses in complicated structures; in allocating resources or managing inventory; in determining appropriate controls to guide aircraft or robots; and as a fundamental tool in the numerical analysis of the flow of fluids or heat.*

The standard method of solving systems of linear equations is elimination. Elimination can be represented by row reduction of a matrix to its reduced row echelon form. This is a fundamental procedure in linear algebra. Obtaining and interpreting the reduced row echelon form of a matrix will play an important role in almost everything we do in the rest of this book.

2.1 Systems of Linear Equations and Elimination

Definition
Linear Equation

A **linear equation** in n variables $x_1, \dots, x_n$ is an equation that can be written in the form

$$a_1x_1 + a_2x_2 + a_3x_3 + \cdots + a_nx_n = b \tag{2.1}$$

The numbers $a_1, \dots, a_n$ are called the **coefficients** of the equation, and b is usually referred to as "the right-hand side," or "the constant term." The x_i are the unknowns or variables to be solved for.

EXAMPLE 1

The equations

$$x_1 + 2x_2 = 4 \tag{2.2}$$

$$x_1 - 3x_2 + \sqrt{3}x_3 = \pi x_4 \tag{2.3}$$

are both linear equations since they both can be written in the form of equation (2.1).

The equation $x_1^2 - x_2 = 1$ is not a linear equation.

Definition
Solution

A vector $\begin{bmatrix} s_1 \\ \vdots \\ s_n \end{bmatrix}$ in $\mathbb{R}^n$ is called a **solution** of equation (2.1) if the equation is satisfied when we make the substitution $x_1 = s_1, x_2 = s_2, \ldots, x_n = s_n$.

EXAMPLE 2

A few solutions of equation (2.2) are $\begin{bmatrix} 2 \\ 1 \end{bmatrix}$, $\begin{bmatrix} 3 \\ 0.5 \end{bmatrix}$, and $\begin{bmatrix} 6 \\ -1 \end{bmatrix}$ since

$$2 + 2(1) = 4$$
$$3 + 2(0.5) = 4$$
$$6 + 2(-1) = 4$$

The vector $\begin{bmatrix} 0 \\ 0 \\ 0 \\ 0 \end{bmatrix}$ is clearly a solution of $x_1 - 3x_2 + \sqrt{3}x_3 = \pi x_4$.

A general system of m linear equations in n variables is written in the form

$$\begin{aligned} a_{11}x_1 + a_{12}x_2 + \cdots + a_{1n}x_n &= b_1 \\ a_{21}x_1 + a_{22}x_2 + \cdots + a_{2n}x_n &= b_2 \\ &\vdots \\ a_{m1}x_1 + a_{m2}x_2 + \cdots + a_{mn}x_n &= b_m \end{aligned}$$

Note that for each coefficient, the first index indicates in which equation the coefficient appears. The second index indicates which variable the coefficient multiplies. That is, a_{ij} is the coefficient of x_j in the i-th equation. The indices on the right-hand side indicate which equation the constant appears in.

We want to establish a standard procedure for determining all the solutions of such a system—*if there are any solutions*! It will be convenient to speak of the **solution set** of a system to mean the set of all solutions of the system.

The standard procedure for solving a system of linear equations is **elimination**. By multiplying and adding some of the original equations, we can eliminate *some* of the variables from *some* of the equations. The result will be a simpler system of equations that has the same solution set as the original system, but is easier to solve.

We say that two systems of equations are **equivalent** if they have the same solution set. *In elimination, each elimination step must be reversible and must leave the solution set unchanged. Every system produced during an elimination procedure is equivalent to the original system.* We begin with an example and explain the general rules as we proceed.

EXAMPLE 3

Find all solutions of the system of linear equations

$$\begin{aligned} x_1 + x_2 - 2x_3 &= 4 \\ x_1 + 3x_2 - x_3 &= 7 \\ 2x_1 + x_2 - 5x_3 &= 7 \end{aligned}$$

Solution: To solve this system by elimination, we begin by eliminating x_1 from all equations except the first one.

Add (−1) times the first equation to the second equation. The first and third equations are unchanged, so the system is now

$$\begin{aligned} x_1 + x_2 - 2x_3 &= 4 \\ 2x_2 + x_3 &= 3 \\ 2x_1 + x_2 - 5x_3 &= 7 \end{aligned}$$

Note two important things about this step. First, if x_1, x_2, x_3 satisfy the original system, then they certainly satisfy the revised system after the step. This follows from the rule of arithmetic that if $P = Q$ and $R = S$, then $P + R = Q + S$. So, when we add two equations and both are satisfied, the resulting sum equation is satisfied. Thus, the revised system is equivalent to the original system.

Second, the step is reversible: to get back to the original system, we just add (1) times the first equation to the revised second equation.

Add (−2) times the first equation to the third equation.

$$\begin{aligned} x_1 + x_2 - 2x_3 &= 4 \\ 2x_2 + x_3 &= 3 \\ -x_2 - x_3 &= -1 \end{aligned}$$

Again, note that this step is reversible and does not change the solution set. Also note that x_1 has been eliminated from all equations except the first one, so now we leave the first equation and turn our attention to x_2.

Although we will not modify or use the first equation in the next several steps, we keep writing the entire system after each step. This is important because it leads to a good general procedure for dealing with large systems.

It is convenient, but not necessary, to work with an equation in which x_2 has the coefficient 1. We could multiply the second equation by $1/2$, but to avoid fractions, follow the steps on the next page.

EXAMPLE 3
(continued)

Interchange the second and third equations. This is another reversible step that does not change the solution set:

$$\begin{aligned} x_1 + x_2 - 2x_3 &= 4 \\ -x_2 - x_3 &= -1 \\ 2x_2 + x_3 &= 3 \end{aligned}$$

Multiply the second equation by (−1). This step is reversible and does not change the solution set:

$$\begin{aligned} x_1 + x_2 - 2x_3 &= 4 \\ x_2 + x_3 &= 1 \\ 2x_2 + x_3 &= 3 \end{aligned}$$

Add (−2) times the second equation to the third equation.

$$\begin{aligned} x_1 + x_2 - 2x_3 &= 4 \\ x_2 + x_3 &= 1 \\ -x_3 &= 1 \end{aligned}$$

In the third equation, all variables except x_3 have been eliminated; by elimination, we have solved for x_3. Using similar steps, we could continue and eliminate x_3 from the second and first equations and x_2 from the first equation. However, it is often a much simpler task to complete the solution process by **back-substitution**.

First, observe that $x_3 = -1$. Substitute this value into the second equation and find that

$$x_2 = 1 - x_3 = 1 - (-1) = 2$$

Next, substitute these values back into the first equation to obtain

$$x_1 = 4 - x_2 + 2x_3 = 4 - 2 + 2(-1) = 0$$

Thus, the only solution of this system is $\begin{bmatrix} x_1 \\ x_2 \\ x_3 \end{bmatrix} = \begin{bmatrix} 0 \\ 2 \\ -1 \end{bmatrix}$. Since the final system is equivalent to the original system, this solution is also the unique solution of the problem.

Observe that we can easily check that $\begin{bmatrix} 0 \\ 2 \\ -1 \end{bmatrix}$ satisfies the original system of equations:

$$\begin{aligned} 0 + 2 - 2(-1) &= 4 \\ 0 + 3(2) - (-1) &= 7 \\ 2(0) + 2 - 5(-1) &= 7 \end{aligned}$$

It is important to observe the form of the equations in our final system. The first variable with a non-zero coefficient in each equation, called a **leading variable**, does not appear in any equation below it. Also, the leading variable in the second equation is to the right of the leading variable in the first, and the leading variable in the third is to the right of the leading variable in the second.

The system solved in Example 3 is a particularly simple one. However, the solution procedure introduces all the steps that are needed in the process of elimination. They are worth reviewing.

Types of Steps in Elimination

(1) Multiply one equation by a non-zero constant.

(2) Interchange two equations.

(3) Add a multiple of one equation to another equation.

Warning! Do *not* combine steps of type (1) and type (3) into one step of the form "Add a multiple of one equation to a multiple of another equation." Although such a combination would not lead to errors in this chapter, it would lead to errors when we apply these ideas in Chapter 5.

EXAMPLE 4

Determine the solution set of the system of linear equations

$$\begin{aligned} x_1 + 2x_3 + x_4 &= 14 \\ x_1 + 3x_3 + 3x_4 &= 19 \end{aligned}$$

Remark

Notice that neither equation contains x_2. This may seem peculiar, but it happens in some applications that one of the variables of interest does not appear in the linear equations. If it truly is one of the variables of the problem, ignoring it is incorrect. Rewrite the equations to make it explicit:

$$\begin{aligned} x_1 + 0x_2 + 2x_3 + x_4 &= 14 \\ x_1 + 0x_2 + 3x_3 + 3x_4 &= 19 \end{aligned}$$

Solution: As in Example 3, we want our leading variable in the first equation to be to the left of the leading variable in the second equation, and we want the leading variable to be eliminated from the second equation. Thus, we use a type (3) step to eliminate x_1 from the second equation.

Add (−1) times the first equation to the second equation:

$$\begin{aligned} x_1 + 0x_2 + 2x_3 + x_4 &= 14 \\ x_3 + 2x_4 &= 5 \end{aligned}$$

Observe that x_2 is not shown in the second equation because the leading variable must have a non-zero coefficient. Moreover, we have already finished our elimination procedure as we have our desired form. The solution can now be completed by back-substitution.

Note that the equations do not completely determine both x_3 and x_4: one of them can be chosen arbitrarily, and the equations can still be satisfied. For consistency, we always choose the variables that do not appear as a leading variable in any equation to be the ones that will be chosen arbitrarily. We will call these **free variables**.

EXAMPLE 4
(continued)

Thus, in the revised system, we see that neither x_2 nor x_4 appears as a leading variable in any equation. Therefore, x_2 and x_4 are the free variables and may be chosen arbitrarily (for example, $x_4 = t \in \mathbb{R}$ and $x_2 = s \in \mathbb{R}$). Then the second equation can be solved for the leading variable x_3:

$$x_3 = 5 - 2x_4 = 5 - 2t$$

Now, solve the first equation for its leading variable x_1:

$$x_1 = 14 - 2x_3 - x_4 = 14 - 2(5 - 2t) - t = 4 + 3t$$

Thus, the solution set of the system is

$$\begin{bmatrix} x_1 \\ x_2 \\ x_3 \\ x_4 \end{bmatrix} = \begin{bmatrix} 4 + 3t \\ s \\ 5 - 2t \\ t \end{bmatrix}, \quad s, t \in \mathbb{R}$$

In this case, there are infinitely many solutions because for each value of s and each value of t that we choose, we get a different solution. We say that this equation is the **general solution** of the system, and we call s and t the **parameters** of the general solution. For many purposes, it is useful to recognize that this solution can be split into a constant part, a part in t, and a part in s:

$$\begin{bmatrix} x_1 \\ x_2 \\ x_3 \\ x_4 \end{bmatrix} = \begin{bmatrix} 4 \\ 0 \\ 5 \\ 0 \end{bmatrix} + s\begin{bmatrix} 0 \\ 1 \\ 0 \\ 0 \end{bmatrix} + t\begin{bmatrix} 3 \\ 0 \\ -2 \\ 1 \end{bmatrix}$$

This will be the standard format for displaying general solutions. It is acceptable to leave x_2 in the place of s and x_4 in the place of t, but then you *must* say $x_2, x_4 \in \mathbb{R}$. Observe that one immediate advantage of this form is that we can instantly see the geometric interpretation of the solution. The intersection of the two hyperplanes $x_1 + 2x_3 + x_4 = 14$ and $x_1 + 3x_3 + 3x_4 = 19$ in $\mathbb{R}^4$ is a plane in $\mathbb{R}^4$ that passes through $P(4, 0, 5, 0)$.

The solution procedure we have introduced is known as **Gaussian elimination with back-substitution**. A slight variation of this procedure is introduced in the next section.

EXERCISE 1

Find the general solution to the system of linear equations

$$\begin{aligned} 2x_1 + 4x_2 + 0x_3 &= 12 \\ x_1 + 2x_2 - x_3 &= 4 \end{aligned}$$

Use the general solution to find three different solutions of the system.

The Matrix Representation of a System of Linear Equations

After you have solved a few systems of equations using elimination, you may realize that you could write the solution faster if you could omit the letters x_1, x_2, and so on—as long as you could keep the coefficients lined up properly. To do this, we write out the coefficients in a rectangular array called a **matrix**. A general linear system of m equations in n unknowns will be represented by the matrix

$$\left[\begin{array}{cccccc|c} a_{11} & a_{12} & \cdots & a_{1j} & \cdots & a_{1n} & b_1 \\ a_{21} & a_{22} & \cdots & a_{2j} & \cdots & a_{2n} & b_2 \\ \vdots & \vdots & & \vdots & & \vdots & \vdots \\ a_{i1} & a_{i2} & \cdots & a_{ij} & \cdots & a_{in} & b_i \\ \vdots & \vdots & & \vdots & & \vdots & \vdots \\ a_{m1} & a_{m2} & \cdots & a_{mj} & \cdots & a_{mn} & b_m \end{array}\right]$$

where the coefficient a_{ij} appears in the i-th row and j-th column of the coefficient matrix. This is called the **augmented matrix** of the system; it is augmented because it includes as its last column the right-hand side of the equations. The matrix without this last column is called the **coefficient matrix** of the system:

$$\begin{bmatrix} a_{11} & a_{12} & \cdots & a_{1j} & \cdots & a_{1n} \\ a_{21} & a_{22} & \cdots & a_{2j} & \cdots & a_{2n} \\ \vdots & \vdots & & \vdots & & \vdots \\ a_{i1} & a_{i2} & \cdots & a_{ij} & \cdots & a_{in} \\ \vdots & \vdots & & \vdots & & \vdots \\ a_{m1} & a_{m2} & \cdots & a_{mj} & \cdots & a_{mn} \end{bmatrix}$$

For convenience, we sometimes denote the augmented matrix of a system with coefficient matrix A and right-hand side $\vec{b} = \begin{bmatrix} b_1 \\ \vdots \\ b_m \end{bmatrix}$ by $\left[A \mid \vec{b}\,\right]$. In Chapter 3, we will develop another way of representing a system of linear equations.

EXAMPLE 5

Write the coefficient matrix and augmented matrix for the following system:

$$\begin{aligned} 3x_1 + 8x_2 - 18x_3 + x_4 &= 35 \\ x_1 + 2x_2 \;\; - 4x_3 \qquad &= 11 \\ x_1 + 3x_2 \;\; - 7x_3 + x_4 &= 10 \end{aligned}$$

Solution: The coefficient matrix is formed by writing the coefficients of each equation as the rows of the matrix. Thus, we get the matrix

$$A = \begin{bmatrix} 3 & 8 & -18 & 1 \\ 1 & 2 & -4 & 0 \\ 1 & 3 & -7 & 1 \end{bmatrix}$$

EXAMPLE 5
(continued)

For the augmented matrix, we just add the right-hand side as the last column. We get

$$\left[\begin{array}{cccc|c} 3 & 8 & -18 & 1 & 35 \\ 1 & 2 & -4 & 0 & 11 \\ 1 & 3 & -7 & 1 & 10 \end{array}\right]$$

EXAMPLE 6

Write the system of linear equations that has the augmented matrix

$$\left[\begin{array}{ccc|c} 1 & 0 & 2 & 3 \\ 0 & -1 & 1 & 1 \\ 0 & 0 & 1 & -2 \end{array}\right]$$

Solution: The rows of the matrix tell us the coefficients and *right-hand side* of each equation. We get the system

$$\begin{aligned} x_1 \qquad + 2x_3 &= 3 \\ -x_2 + x_3 &= 1 \\ x_3 &= -2 \end{aligned}$$

Remark

Another way to view the coefficient matrix is to see that the j-th column of the matrix is the vector containing all the coefficients of x_j. This view will become very important in Chapter 3 and beyond.

Since each row in the augmented matrix corresponds to an equation in the system of linear equations, performing operations on the equations of the system corresponds to performing the same operations on the rows of the matrix. Thus, the steps in elimination correspond to the following elementary row operations.

Types of Elementary Row Operations

(1) Multiply one row by a non-zero constant.

(2) Interchange two rows.

(3) Add a multiple of one row to another.

As with the steps in elimination, we do not combine operations of type (1) and type (3) into one operation.

The process of performing elementary row operations on a matrix to bring it into some simpler form is called **row reduction**.

Recall that if a system of equations is obtained from another system by one or more of the elimination steps, the systems are said to be **equivalent**. For matrices, if the matrix M is row reduced into a matrix N by a sequence of elementary row operations, then we say that M is **row equivalent** to N. Just as elimination steps are reversible, so are elementary row operations. It follows that if M is row equivalent to N, then N is row equivalent to M, so we may say that M and N are row equivalent. It also follows that if A is row equivalent to B and B is row equivalent to C, then A is row equivalent to C.

Let us see how the elimination in Example 3 appears in matrix notation. To do this, we introduce notation to indicate this elementary row operation. We write $R_i + cR_j$

to indicate adding c times row j to row i, $R_i \updownarrow R_j$ to indicate interchanging row i and row j. We write cR_i to indicate multiplying row i by a non-zero scalar c. Additionally, at each step we will use $\sim$ to indicate that the matrices are row equivalent. Note that it would be incorrect to use $=$ or $\Rightarrow$ instead of $\sim$. As one becomes confident with elementary row operations, one may omit these indicators of which elementary row operations were used. However, including them can make checking the steps easier, and instructors may require them in work submitted for grading.

EXAMPLE 7

The augmented matrix for the system in Example 3 is

$$\left[\begin{array}{ccc|c} 1 & 1 & -2 & 4 \\ 1 & 3 & -1 & 7 \\ 2 & 1 & -5 & 7 \end{array}\right]$$

The first step in the elimination was to add (–1) times the first equation to the second. Here we add (–1) multiplied by the first row to the second. We write

$$\left[\begin{array}{ccc|c} 1 & 1 & -2 & 4 \\ 1 & 3 & -1 & 7 \\ 2 & 1 & -5 & 7 \end{array}\right] R_2 + (-1)R_1 \sim \left[\begin{array}{ccc|c} 1 & 1 & -2 & 4 \\ 0 & 2 & 1 & 3 \\ 2 & 1 & -5 & 7 \end{array}\right]$$

The remaining steps are

$$\left[\begin{array}{ccc|c} 1 & 1 & -2 & 4 \\ 0 & 2 & 1 & 3 \\ 2 & 1 & -5 & 7 \end{array}\right] \begin{array}{c} \\ \\ R_3 - 2R_1 \end{array} \sim \left[\begin{array}{ccc|c} 1 & 1 & -2 & 4 \\ 0 & 2 & 1 & 3 \\ 0 & -1 & -1 & -1 \end{array}\right] R_2 \updownarrow R_3$$

$$\sim \left[\begin{array}{ccc|c} 1 & 1 & -2 & 4 \\ 0 & -1 & -1 & -1 \\ 0 & 2 & 1 & 3 \end{array}\right] (-1)R_2 \sim \left[\begin{array}{ccc|c} 1 & 1 & -2 & 4 \\ 0 & 1 & 1 & 1 \\ 0 & 2 & 1 & 3 \end{array}\right] \begin{array}{c} \\ \\ R_3 - 2R_2 \end{array} \sim \left[\begin{array}{ccc|c} 1 & 1 & -2 & 4 \\ 0 & 1 & 1 & 1 \\ 0 & 0 & -1 & 1 \end{array}\right]$$

All the elementary row operations corresponding to the elimination in Example 3 have been performed. Observe that the final matrix is the augmented matrix for the final system of linear equations that we obtained in Example 3.

EXAMPLE 8

The matrix representation of the elimination in Example 4 is

$$\left[\begin{array}{cccc|c} 1 & 0 & 2 & 1 & 14 \\ 1 & 0 & 3 & 3 & 19 \end{array}\right] R_2 + (-1)R_1 \sim \left[\begin{array}{cccc|c} 1 & 0 & 2 & 1 & 14 \\ 0 & 0 & 1 & 2 & 5 \end{array}\right]$$

EXERCISE 2

Write out the matrix representation of the elimination used in Exercise 1.

In the next example, we will solve a system of linear equations using Gaussian elimination with back-substitution entirely in matrix form.

EXAMPLE 9 Find the general solution of the system

$$\begin{aligned} 3x_1 + 8x_2 - 18x_3 + x_4 &= 35 \\ x_1 + 2x_2 - 4x_3 &= 11 \\ x_1 + 3x_2 - 7x_3 + x_4 &= 10 \end{aligned}$$

Solution: Write the augmented matrix of the system and row reduce:

$$\left[\begin{array}{cccc|c} 3 & 8 & -18 & 1 & 35 \\ 1 & 2 & -4 & 0 & 11 \\ 1 & 3 & -7 & 1 & 10 \end{array}\right] R_1 \updownarrow R_2 \sim \left[\begin{array}{cccc|c} 1 & 2 & -4 & 0 & 11 \\ 3 & 8 & -18 & 1 & 35 \\ 1 & 3 & -7 & 1 & 10 \end{array}\right] \begin{array}{c} \\ R_2 - 3R_1 \\ \\ \end{array} \sim$$

$$\left[\begin{array}{cccc|c} 1 & 2 & -4 & 0 & 11 \\ 0 & 2 & -6 & 1 & 2 \\ 1 & 3 & -7 & 1 & 10 \end{array}\right] \begin{array}{c} \\ \\ R_3 - R_1 \end{array} \sim \left[\begin{array}{cccc|c} 1 & 2 & -4 & 0 & 11 \\ 0 & 2 & -6 & 1 & 2 \\ 0 & 1 & -3 & 1 & -1 \end{array}\right] R_2 \updownarrow R_3 \sim$$

$$\left[\begin{array}{cccc|c} 1 & 2 & -4 & 0 & 11 \\ 0 & 1 & -3 & 1 & -1 \\ 0 & 2 & -6 & 1 & 2 \end{array}\right] \begin{array}{c} \\ \\ R_3 - 2R_2 \end{array} \sim \left[\begin{array}{cccc|c} 1 & 2 & -4 & 0 & 11 \\ 0 & 1 & -3 & 1 & -1 \\ 0 & 0 & 0 & -1 & 4 \end{array}\right]$$

To find the general solution, we now interpret the final matrix as the augmented matrix of the equivalent system. We get the system

$$\begin{aligned} x_1 + 2x_2 - 4x_3 &= 11 \\ x_2 - 3x_3 + x_4 &= -1 \\ -x_4 &= 4 \end{aligned}$$

We see that x_3 is a free variable, so we let $x_3 = t \in \mathbb{R}$. Then we use back-substitution to get

$$\begin{aligned} x_4 &= -4 \\ x_2 &= -1 + 3x_3 - x_4 = 3 + 3t \\ x_1 &= 11 - 2x_2 + 4x_3 = 5 - 2t \end{aligned}$$

Thus, the general solution is

$$\begin{bmatrix} x_1 \\ x_2 \\ x_3 \\ x_4 \end{bmatrix} = \begin{bmatrix} 5 - 2t \\ 3 + 3t \\ t \\ -4 \end{bmatrix} = \begin{bmatrix} 5 \\ 3 \\ 0 \\ -4 \end{bmatrix} + t \begin{bmatrix} -2 \\ 3 \\ 1 \\ 0 \end{bmatrix}, \quad t \in \mathbb{R}$$

Check this solution by substituting these values for x_1, x_2, x_3, x_4 into the original equations.

Observe that there are many different ways that we could choose to row reduce the augmented matrix in any of these examples. For instance, in Example 9 we could interchange row 1 and row 3 instead of interchanging row 1 and row 2. Alternatively, we could use the elementary row operations $R_2 - \frac{1}{3}R_1$ and $R_3 - \frac{1}{3}R_1$ to eliminate the non-zero entries beneath the first leading variable. It is natural to ask if there is a way of determining which elementary row operations will work the best. Unfortunately, there is no such algorithm for doing these by hand. However, we will give a basic algorithm for row reducing a matrix into the "proper" form. We start by defining this form.

Row Echelon Form

Based on how we used elimination to solve the system of equations, we define the following form of a matrix.

Definition
Row Echelon Form (REF)

A matrix is in **row echelon form (REF)** if

(1) When all entries in a row are zeros, this row appears below all rows that contain a non-zero entry.
(2) When two non-zero rows are compared, the first non-zero entry, called the leading entry, in the upper row is to the left of the leading entry in the lower row.

Remark

It follows from these properties that all entries in a column beneath a leading entry must be 0. For otherwise, (1) or (2) would be violated.

EXAMPLE 10

Determine which of the following matrices are in row echelon form. For each matrix that is not in row echelon form, explain why it is not in row echelon form.

$$\text{(a)} \begin{bmatrix} 1 & 1 & -2 & 4 \\ 0 & 1 & 1 & 1 \\ 0 & 0 & 1 & -1 \end{bmatrix} \qquad \text{(b)} \begin{bmatrix} 0 & 1 & 0 & 1 & 2 \\ 0 & 0 & 0 & -3 & 1 \\ 0 & 0 & 0 & 0 & 0 \end{bmatrix}$$

$$\text{(c)} \begin{bmatrix} 1 & 0 & 0 & 0 \\ 0 & 0 & 2 & -3 \\ 0 & 2 & 3 & 3 \end{bmatrix} \qquad \text{(d)} \begin{bmatrix} 1 & 1 & 2 & -1 & 1 \\ 1 & 3 & 1 & 4 & -2 \end{bmatrix}$$

Solution: The matrices in (a) and (b) are both in REF. The matrix in (c) is not in REF since the leading entry in the second row is to the right of the leading entry in the third row. The matrix in (d) is not in REF since the leading entry in the second row is beneath the leading entry in the first row.

Any matrix can be row reduced to row echelon form by using the following steps First, consider the first column of the matrix; if it consists entirely of zero entries, move to the next column. If it contains some non-zero entry, interchange rows (if necessary) so that the top entry in the column is non-zero. Of course, the column may contain multiple non-zero entries. You can use any of these non-zero entries, but some choices will make your calculations considerably easier than others; see the Remarks below. We will call this entry a **pivot**. Use elementary row operations of type (3) to make all entries beneath the pivot into zeros. Next, consider the submatrix consisting of all columns to the right of the column we have just worked on and all rows below the row with the most recently obtained leading entry. Repeat the procedure described for this submatrix to obtain the next pivot with zeros below it. Keep repeating the procedure until we have "used up" all rows and columns of the original matrix. The resulting matrix is in row echelon form.

EXAMPLE 11

Row reduce the augmented matrix of the following system to row echelon form and use it to determine all solutions of the system:

$$\begin{aligned} x_1 + x_2 \quad\quad &= 1 \\ x_2 + x_3 &= 2 \\ x_1 + 2x_2 + x_3 &= -2 \end{aligned}$$

Solution: Write the augmented matrix of the system and row reduce:

$$\left[\begin{array}{ccc|c} 1 & 1 & 0 & 1 \\ 0 & 1 & 1 & 2 \\ 1 & 2 & 1 & -2 \end{array}\right] R_3 - R_1 \sim \left[\begin{array}{ccc|c} 1 & 1 & 0 & 1 \\ 0 & 1 & 1 & 2 \\ 0 & 1 & 1 & -3 \end{array}\right] R_3 - R_2 \sim \left[\begin{array}{ccc|c} 1 & 1 & 0 & 1 \\ 0 & 1 & 1 & 2 \\ 0 & 0 & 0 & -5 \end{array}\right]$$

Observe that when we write the system of linear equations represented by the augmented matrix, we get

$$\begin{aligned} x_1 + x_2 \quad\quad &= 1 \\ x_2 + x_3 &= 2 \\ 0 &= -5 \end{aligned}$$

Clearly, the last equation is impossible. This means we cannot find values of x_1, x_2, x_3 that satisfy all three equations. Hence, this system has no solution.

Remarks

1. Although the previous algorithm will always work, it is not necessarily the fastest or easiest method to use for any particular matrix. In principle, it does not matter which non-zero entry is chosen as the pivot in the procedure just described. In practice, it can have considerable impact on the amount of work required and on the accuracy of the result. The ability to row reduce a general matrix to REF by hand quickly and efficiently comes only with a lot of practice. Note that for hand calculations on simple integer examples, it is sensible to go to some trouble to postpone fractions because avoiding fractions may reduce both the effort required and the chance of making errors.

2. Observe that the row echelon form for a given matrix A is not unique. In particular, every non-zero matrix has infinitely many row echelon forms that are all row equivalent. However, it can be shown that any two row echelon forms for the same matrix A must agree on the position of the leading entries. (This fact may seem obvious, but it is not easy to prove. It follows from Problem F6 in the Chapter 4 Further Problems.)

We have now seen that a system of linear equations may have exactly one solution, infinitely many solutions, or no solution. We will now discuss this in greater detail.

Consistent Systems and Unique Solutions

We shall see that it is often important to be able to recognize whether a given system has a solution and, if so, how many solutions. A system that has at least one solution is called **consistent**, and a system that does not have any solutions is called **inconsistent**.

To illustrate the possibilities, consider a system of three linear equations in three unknowns. Each equation can be considered as the equation of a plane in $\mathbb{R}^3$. A solution of the system determines a point of intersection of the three planes that we call P_1, P_2, P_3. Figure 2.1.1 illustrates an inconsistent system: there is no point common to all three planes. Figure 2.1.2 illustrates a unique solution: all three planes intersect in exactly one point. Figure 2.1.3 demonstrates a case where there are infinitely many solutions.

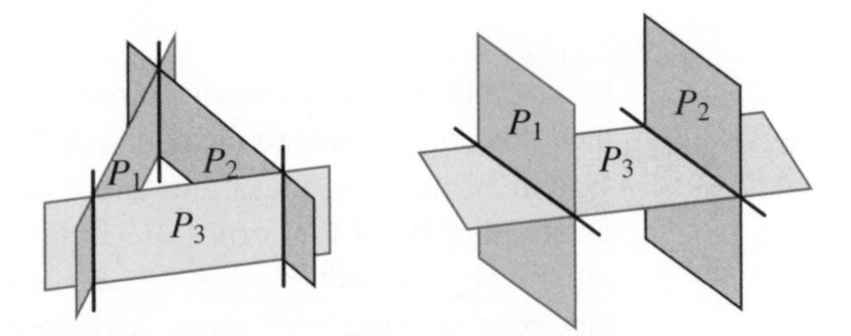

Figure 2.1.1 Two cases where three planes have no common point of intersection: the corresponding system is inconsistent.

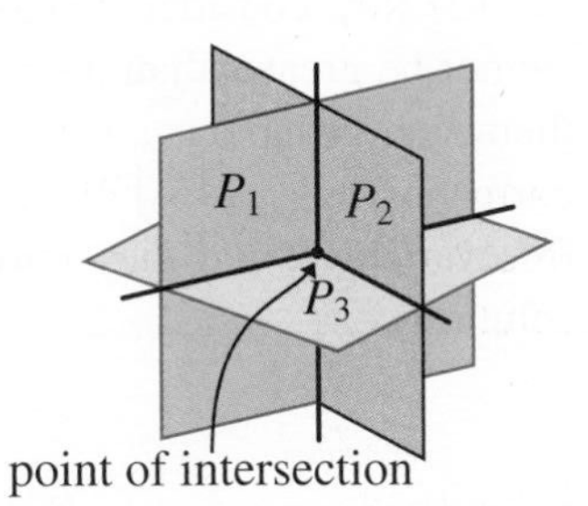

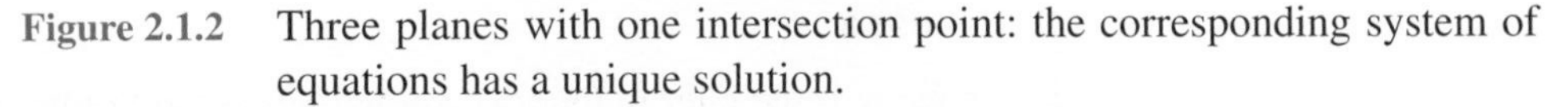

Figure 2.1.2 Three planes with one intersection point: the corresponding system of equations has a unique solution.

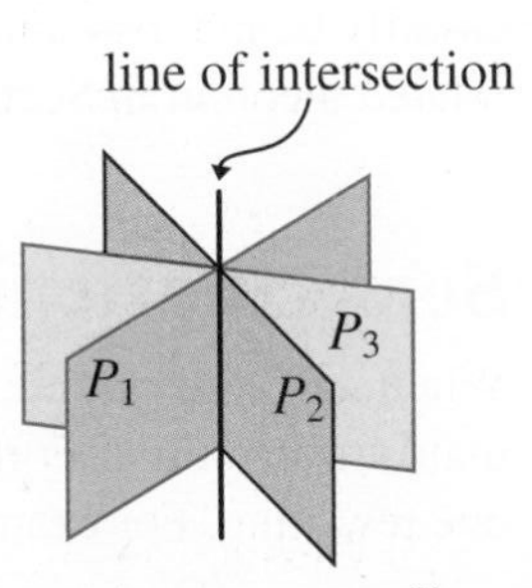

Figure 2.1.3 Three planes that meet in a common line: the corresponding system has infinitely many solutions.

Row echelon form allows us to answer questions about consistency and uniqueness. In particular, we have the following theorem.

Theorem 1

Suppose that the augmented matrix $\left[A \mid \vec{b}\right]$ of a system of linear equations is row equivalent to $\left[S \mid \vec{c}\right]$, which is in row echelon form.

(1) The given system is inconsistent if and only if some row of $\left[S \mid \vec{c}\right]$ is of the form $\left[\begin{array}{cccc|c} 0 & 0 & \cdots & 0 & c \end{array}\right]$, with $c \neq 0$.

(2) If the system is consistent, there are two possibilities. Either the number of pivots in S is equal to the number of variables in the system and the system has a unique solution, or the number of pivots is less than the number of variables and the system has infinitely many solutions.

Proof: (1) If $\left[S \mid \vec{c}\right]$ contains a row of the form $\left[\begin{array}{cccc|c} 0 & 0 & \cdots & 0 & c \end{array}\right]$, where $c \neq 0$, then this corresponds to the equation $0 = c$, which clearly has no solution. Hence, the system is inconsistent. On the other hand, if it contains no such row, then each row must either be of the form $\left[\begin{array}{cccc|c} 0 & 0 & \cdots & 0 & 0 \end{array}\right]$, which corresponds to an equation satisfied by any values of $x_1, \dots, x_n$, or else contains a pivot. We may ignore the rows that consist entirely of zeros, leaving only rows with pivots. In the latter case, the corresponding system can be solved by assigning arbitrary values to the free variables and then determining the remaining variables by back-substitution. Thus, if there is no row of the form $\left[\begin{array}{cccc|c} 0 & 0 & \cdots & 0 & c \end{array}\right]$, the system cannot be inconsistent.

(2) Now consider the case of a consistent system. The number of leading variables cannot be greater than the number of columns in the coefficient matrix; if it is equal, then each variable is a leading variable and thus is determined uniquely by the system corresponding to $\left[S \mid \vec{c}\right]$. If some variables are not leading variables, then they are free variables, and they may be chosen arbitrarily. Hence, there are infinitely many solutions. ■

Remark

As we will see later in the text, sometimes we are only interested in whether a system is consistent or inconsistent or in how many solutions a system has. We may not necessarily be interested in finding a particular solution. In these cases, Theorem 1 or the related theorem in Section 2.2 is very useful.

Some Shortcuts and Some Bad Moves

When carrying out elementary row operations, you may get weary of rewriting the matrix every time. Fortunately, we can combine some elementary row operations in one rewriting. For example,

$$\left[\begin{array}{ccc|c} 1 & 1 & -2 & 4 \\ 1 & 3 & -1 & 7 \\ 2 & 1 & -5 & 7 \end{array}\right] \begin{array}{l} \\ R_2 - R_1 \\ R_3 - 2R_1 \end{array} \sim \left[\begin{array}{ccc|c} 1 & 1 & -2 & 4 \\ 0 & 2 & 1 & 3 \\ 0 & -1 & -1 & -1 \end{array}\right]$$

Choosing one particular row (in this case, the first row) and adding multiples of it to several other rows is perfectly acceptable. There are other elementary row operations that can be combined, but these should not be used until one is extremely comfortable with row reducing. This is because some combinations of steps do cause errors. For example,

$$\begin{bmatrix} 1 & 1 & 3 \\ 1 & 2 & 4 \end{bmatrix} \begin{matrix} R_1 - R_2 \\ R_2 - R_1 \end{matrix} \sim \begin{bmatrix} 0 & -1 & -1 \\ 0 & 1 & 1 \end{bmatrix} \qquad (\textit{WRONG!})$$

This is nonsense because the final matrix should have a leading 1 in the first column. By performing one elementary row operation, we change one row; thereafter we must use that row in its new changed form. Thus, when performing multiple elementary row operations in one step, make sure that you are not modifying a row that you are using in another elementary row operation.

A Word Problem

To illustrate the application of systems of equations to word problems, we give a simple example. More interesting applications usually require a fair bit of background. In Section 2.4 we discuss two applications from physics/engineering.

EXAMPLE 12

A boy has a jar full of coins. Altogether there are 180 nickels, dimes, and quarters. The number of dimes is one-half of the total number of nickels and quarters. The value of the coins is $16.00. How many of each kind of coin does he have?

Solution: Let n be the number of nickels, d the number of dimes, and q the number of quarters. Then

$$n + d + q = 180$$

The second piece of information we are given is that

$$d = \frac{1}{2}(n + q)$$

We rewrite this into standard form for a linear equation:

$$n - 2d + q = 0$$

Finally, we have the value of the coins, in cents:

$$5n + 10d + 25q = 1600$$

Thus, n, d, and q satisfy the system of linear equations:

$$\begin{aligned} n \quad + d \quad + q &= 180 \\ n \quad - 2d \quad + q &= 0 \\ 5n + 10d + 25q &= 1600 \end{aligned}$$

Write the augmented matrix and row reduce:

$$\left[\begin{array}{ccc|c} 1 & 1 & 1 & 180 \\ 1 & -2 & 1 & 0 \\ 5 & 10 & 25 & 1600 \end{array}\right] \begin{matrix} \\ R_2 - R_1 \\ R_3 - 5R_1 \end{matrix} \sim \left[\begin{array}{ccc|c} 1 & 1 & 1 & 180 \\ 0 & -3 & 0 & -180 \\ 0 & 5 & 20 & 700 \end{array}\right] \begin{matrix} \\ (-1/3)R_2 \\ (1/5)R_3 \end{matrix} \sim$$

$$\left[\begin{array}{ccc|c} 1 & 1 & 1 & 180 \\ 0 & 1 & 0 & 60 \\ 0 & 1 & 4 & 140 \end{array}\right] \begin{matrix} \\ \\ R_3 - R_2 \end{matrix} \sim \left[\begin{array}{ccc|c} 1 & 1 & 1 & 180 \\ 0 & 1 & 0 & 60 \\ 0 & 0 & 4 & 80 \end{array}\right]$$

EXAMPLE 12
(continued)

According to Theorem 1, the system is consistent with a unique solution. In particular, writing the final augmented matrix as a system of equations, we get

$$\begin{aligned} n + d + q &= 180 \\ d &= 60 \\ 4q &= 80 \end{aligned}$$

So, by back-substitution, we get $q = 20$, $d = 60$, $n = 180 - d - q = 100$. Hence, the boy has 100 nickels, 60 dimes, and 20 quarters.

A Remark on Computer Calculations

In computer calculations, the choice of the pivot may affect accuracy. The problem is that real numbers are represented to only a finite number of digits in a computer, so inevitably some round-off or truncation errors occur. When you are doing a large number of arithmetic operations, these errors can accumulate, and they can be particularly serious if at some stage you subtract two nearly equal numbers. The following example gives some idea of the difficulties that might be encountered.

The system

$$\begin{aligned} 0.1000x_1 + 0.9990x_2 &= 1.000 \\ 0.1000x_1 + 1.000x_2 &= 1.006 \end{aligned}$$

is easily found to have solution $x_2 = 6.000$, $x_1 = -49.94$. Notice that the coefficients were given to four digits. Suppose all entries are rounded to three digits. The system becomes

$$\begin{aligned} 0.100x_1 + 0.999x_2 &= 1.00 \\ 0.100x_1 + 1.00x_2 &= 1.01 \end{aligned}$$

The solution is now $x_2 = 10$, $x_1 = -89.9$. Notice that despite the fact that there was only a small change in one term on the right-hand side, the resulting solution is not close to the solution of the original problem. Geometrically, this can be understood by observing that the solution is the intersection point of two nearly parallel lines; therefore, a small displacement of one line causes a major shift of the intersection point. Difficulties of this kind may arise in higher-dimensional systems of equations in real applications.

Carefully choosing pivots in computer programs can reduce the error caused by these sorts of problems. However, some matrices are **ill conditioned**; even with high-precision calculations, the solutions produced by computers with such matrices may be unreliable. In applications, the entries in the matrices may be experimentally determined, and small errors in the entries may result in large errors in calculated solutions, no matter how much precision is used in computation. To understand this problem better, you need to know something about sources of error in numerical computation—and more linear algebra. We shall not discuss it further in this book, but you should be aware of the difficulty if you use computers to solve systems of linear equations.

PROBLEMS 2.1

Practice Problems

A1 Solve each of the following systems by back-substitution and write the general solution in standard form.

(a)
$$\begin{aligned} x_1 - 3x_2 &= 5 \\ x_2 &= 4 \end{aligned}$$

(b)
$$\begin{aligned} x_1 + 2x_2 - x_3 &= 7 \\ x_3 &= 6 \end{aligned}$$

(c)
$$\begin{aligned} x_1 + 3x_2 - 2x_3 &= 4 \\ x_2 + 5x_3 &= 2 \\ x_3 &= 2 \end{aligned}$$

(d)
$$\begin{aligned} x_1 - 2x_2 + x_3 + 4x_4 &= 7 \\ x_2 \qquad - x_4 &= -3 \\ x_3 + x_4 &= 2 \end{aligned}$$

A2 Which of the following matrices are in row echelon form? For each matrix not in row echelon form, explain why it is not.

(a) $A = \begin{bmatrix} 1 & 2 & 3 & 4 \\ 0 & 1 & -2 & -3 \\ 0 & 0 & 0 & 3 \end{bmatrix}$

(b) $B = \begin{bmatrix} 0 & 1 & 2 & 3 \\ 0 & 0 & 1 & 1 \\ 0 & 0 & 0 & 0 \end{bmatrix}$

(c) $C = \begin{bmatrix} 1 & -1 & -2 & -3 \\ 0 & 1 & 2 & 0 \\ 0 & 1 & 0 & 3 \end{bmatrix}$

(d) $D = \begin{bmatrix} 1 & 0 & 2 & 1 \\ 0 & 0 & 0 & 1 \\ 0 & 0 & 1 & 1 \end{bmatrix}$

A3 Row reduce the following matrices to obtain a row equivalent matrix in row echelon form. Show your steps.

(a) $\begin{bmatrix} 4 & 1 & 1 \\ 1 & -3 & 2 \end{bmatrix}$

(b) $\begin{bmatrix} 2 & -2 & 5 & 8 \\ 1 & -1 & 2 & 3 \\ -1 & 1 & 0 & 2 \end{bmatrix}$

(c) $\begin{bmatrix} 1 & -1 & -1 \\ 2 & -1 & -2 \\ 5 & 0 & 0 \\ 3 & 4 & 5 \end{bmatrix}$

(d) $\begin{bmatrix} 2 & 0 & 2 & 0 \\ 1 & 2 & 3 & 4 \\ 1 & 4 & 9 & 16 \\ 3 & 6 & 13 & 20 \end{bmatrix}$

(e) $\begin{bmatrix} 0 & 1 & 2 & 1 \\ 1 & 2 & 1 & 1 \\ 3 & -1 & -4 & 1 \\ 2 & 1 & 3 & 6 \end{bmatrix}$

(f) $\begin{bmatrix} 3 & 1 & 8 & 2 & 4 \\ 1 & 0 & 3 & 0 & 1 \\ 0 & 2 & -2 & 4 & 3 \\ -4 & 1 & 11 & 3 & 8 \end{bmatrix}$

A4 Each of the following is an augmented matrix of a system of linear equations. Determine whether the system is consistent. If it is, determine the general solution.

(a) $\left[\begin{array}{ccc|c} 1 & 2 & -1 & 2 \\ 0 & 1 & 3 & 4 \\ 0 & 0 & 0 & -5 \end{array}\right]$

(b) $\left[\begin{array}{ccc|c} 1 & 0 & 0 & 2 \\ 0 & 0 & 1 & 3 \\ 0 & 0 & 0 & 0 \end{array}\right]$

(c) $\left[\begin{array}{cccc|c} 1 & 0 & 1 & 0 & 1 \\ 0 & 1 & 1 & 1 & 2 \\ 0 & 0 & 0 & 1 & 3 \end{array}\right]$

(d) $\left[\begin{array}{cccc|c} 1 & 1 & -1 & 3 & 1 \\ 0 & 0 & 2 & 1 & 3 \\ 0 & 0 & 0 & 1 & -2 \end{array}\right]$

(e) $\left[\begin{array}{cccc|c} 1 & 0 & 1 & -1 & 0 \\ 0 & 1 & 0 & 0 & 0 \\ 0 & 0 & 0 & 0 & 0 \\ 0 & 0 & 0 & 0 & 0 \end{array}\right]$

A5 For each of the following systems of linear equations:

(i) Write the augmented matrix.

(ii) Obtain a row equivalent matrix in row echelon form.

(iii) Determine whether the system is consistent or inconsistent. If it is consistent, determine the number of parameters in the general solution.

(iv) If the system is consistent, write its general solution in standard form.

(a)
$$\begin{aligned} 3x_1 - 5x_2 &= 2 \\ x_1 + 2x_2 &= 4 \end{aligned}$$

(b)
$$\begin{aligned} x_1 + 2x_2 + x_3 &= 5 \\ 2x_1 - 3x_2 + 2x_3 &= 6 \end{aligned}$$

(c)
$$\begin{aligned} x_1 + 2x_2 - 3x_3 &= 8 \\ x_1 + 3x_2 - 5x_3 &= 11 \\ 2x_1 + 5x_2 - 8x_3 &= 19 \end{aligned}$$

(d)
$$\begin{aligned} -3x_1 + 6x_2 + 16x_3 &= 36 \\ x_1 - 2x_2 - 5x_3 &= -11 \\ 2x_1 - 3x_2 - 8x_3 &= -17 \end{aligned}$$

(e)
$$\begin{aligned} x_1 + 2x_2 - x_3 &= 4 \\ 2x_1 + 5x_2 + x_3 &= 10 \\ 4x_1 + 9x_2 - x_3 &= 19 \end{aligned}$$

(f)
$$\begin{aligned} x_1 + 2x_2 - 3x_3 &= -5 \\ 2x_1 + 4x_2 - 6x_3 + x_4 &= -8 \\ 6x_1 + 13x_2 - 17x_3 + 4x_4 &= -21 \end{aligned}$$

(g)
$$\begin{aligned} 2x_2 - 2x_3 + x_5 &= 2 \\ x_1 + 2x_2 - 3x_3 + x_4 + 4x_5 &= 1 \\ 2x_1 + 4x_2 - 5x_3 + 3x_4 + 8x_5 &= 3 \\ 2x_1 + 5x_2 - 7x_3 + 3x_4 + 10x_5 &= 5 \end{aligned}$$

A6 Each of the following matrices is an augmented matrix of a system of linear equations. Determine the values of a, b, c, d for which the systems are consistent. If a system is consistent, determine whether the system has a unique solution.

(a) $\left[\begin{array}{ccc|c} 2 & 4 & -3 & 6 \\ 0 & b & 7 & 2 \\ 0 & 0 & a & a \end{array}\right]$

(b) $\left[\begin{array}{cccc|c} 1 & -1 & 4 & -2 & 5 \\ 0 & 1 & 2 & 3 & 4 \\ 0 & 0 & d & 5 & 7 \\ 0 & 0 & 0 & cd & c \end{array}\right]$

A7 A fruit seller has apples, bananas, and oranges. Altogether he has 1500 pieces of fruit. On average, each apple weighs 120 grams, each banana weighs 140 grams, and each orange weighs 160 grams. He can sell apples for 25 cents each, bananas for 20 cents each, and oranges for 30 cents each. If the fruit weighs 208 kilograms, and the total selling price is \$380, how many of each kind of fruit does the fruit seller have?

A8 A student is taking courses in algebra, calculus, and physics at a college where grades are given in percentages. To determine her standing for a physics prize, a weighted average is calculated based on 50% of the student's physics grades, 30% of her calculus grade, and 20% of her algebra grade; the weighted average is 84. For an applied mathematics prize, a weighted average based on one-third of each of the three grades is calculated to be 83. For a pure mathematics prize, her average based on 50% of her calculus grade and 50% of her algebra grade is 82.5. What are her grades in the individual courses?

Homework Problems

B1 Solve each of the following systems by back-substitution and write the general solution in standard form.

(a)
$$\begin{aligned} x_1 - 2x_2 - x_3 &= 5 \\ x_2 + 3x_3 &= 4 \\ x_3 &= -2 \end{aligned}$$

(b)
$$\begin{aligned} x_1 - 3x_2 + x_3 &= 1 \\ x_2 + 2x_3 &= -1 \end{aligned}$$

(c)
$$\begin{aligned} x_1 + 3x_2 - x_3 + 2x_4 &= -1 \\ x_2 - x_3 + 2x_4 &= -1 \\ x_3 + 3x_4 &= 3 \end{aligned}$$

(d)
$$\begin{aligned} x_1 + 3x_2 - 2x_3 - 2x_4 + 2x_5 &= -2 \\ x_2 - 2x_3 - 2x_4 + 3x_5 &= 4 \\ x_3 + x_4 - 2x_5 &= -3 \end{aligned}$$

B2 Which of the following matrices are in row echelon form? For each matrix not in row echelon form, explain why it is not.

(a) $A = \begin{bmatrix} 0 & 1 & -2 & 5 \\ 0 & 0 & 1 & 2 \\ 0 & 0 & 0 & 1 \end{bmatrix}$

(b) $B = \begin{bmatrix} 1 & 0 & 1 & 1 \\ 0 & 3 & 0 & 6 \\ 0 & 0 & 1 & 2 \end{bmatrix}$

(c) $C = \begin{bmatrix} 1 & 2 & 2 & 4 \\ 1 & 0 & 0 & 3 \\ 0 & 0 & 0 & 1 \end{bmatrix}$

(d) $D = \begin{bmatrix} 1 & 0 & 2 & 1 \\ 0 & 1 & 2 & 1 \\ 0 & 0 & 1 & -3 \end{bmatrix}$

B3 Row reduce the following matrices to obtain a row equivalent matrix in row echelon form. Show your steps.

(a) $\begin{bmatrix} 2 & 4 & 2 & -1 \\ 1 & 3 & 3 & 0 \\ 2 & 5 & 5 & -4 \end{bmatrix}$

(b) $\begin{bmatrix} 5 & 7 & -2 & 7 & 4 \\ 1 & 1 & 0 & 1 & 0 \\ 3 & 0 & 3 & -6 & 5 \end{bmatrix}$

(c) $\begin{bmatrix} 2 & 1 & 0 & 7 \\ 1 & 1 & -1 & 2 \\ 3 & 2 & 0 & 12 \\ 6 & 4 & -1 & 25 \end{bmatrix}$

(d) $\begin{bmatrix} 1 & 2 & 1 & 3 & 0 \\ 2 & 5 & 2 & 6 & 1 \\ 3 & 7 & 4 & 9 & 3 \\ 2 & 6 & 2 & 6 & 5 \end{bmatrix}$

B4 Each of the following is an augmented matrix of a system of linear equations. Determine whether the system is consistent. If it is, determine the general solution.

(a) $\left[\begin{array}{cccc|c} 1 & -1 & 1 & 0 & 1 \\ 0 & 1 & 0 & -1 & 2 \\ 0 & 0 & 2 & 4 & -3 \end{array}\right]$

(b) $\left[\begin{array}{cccc|c} 1 & -2 & 1 & 0 & 1 \\ 0 & 0 & 1 & 2 & 3 \\ 0 & 0 & 0 & 0 & 1 \end{array}\right]$

(c) $\left[\begin{array}{cccc|c} 2 & 1 & 0 & 1 & 2 \\ 0 & 1 & 0 & 1 & 1 \\ 0 & 0 & 1 & 1 & 2 \end{array}\right]$

(d) $\left[\begin{array}{cccc|c} 1 & 0 & 1 & 0 & -1 \\ 0 & 0 & 1 & -1 & 0 \\ 0 & 0 & 0 & 0 & 0 \\ 0 & 0 & 0 & 0 & 0 \end{array}\right]$

B5 For each of the following systems of linear equations:

(i) Write the augmented matrix.

(ii) Obtain a row equivalent matrix in row echelon form.

(iii) Determine whether the system is consistent or inconsistent. If it is consistent, then determine the number of parameters in the general solution.

(iv) If the system is consistent, write its general solution in standard form.

(a)
$$\begin{aligned} 2x_1 + x_2 + 5x_3 &= -4 \\ x_1 + x_2 + x_3 &= -2 \end{aligned}$$

(b)
$$\begin{aligned} 2x_1 + x_2 - x_3 &= 6 \\ x_1 - 2x_2 - 2x_3 &= 1 \\ -x_1 + 12x_2 + 8x_3 &= 7 \end{aligned}$$

(c)
$$\begin{aligned} x_2 + x_3 &= 2 \\ x_1 + x_2 + x_3 &= 3 \\ 2x_1 + 3x_2 + 3x_3 &= 9 \end{aligned}$$

(d)
$$\begin{aligned} x_1 + x_2 &= -7 \\ 2x_1 + 4x_2 + x_3 &= -16 \\ x_1 + 2x_2 + x_3 &= 9 \end{aligned}$$

(e)
$$\begin{aligned} x_1 + x_2 + 2x_3 + x_4 &= 3 \\ x_1 + 2x_2 + 4x_3 + x_4 &= 7 \\ x_1 + x_4 &= -21 \end{aligned}$$

(f)
$$\begin{aligned} x_1 + x_2 + 2x_3 + x_4 &= 1 \\ x_1 + 2x_2 + 4x_3 + x_4 &= -1 \\ x_1 + x_4 &= 3 \end{aligned}$$

B6 Each of the following matrices is an augmented matrix of a system of linear equations. Determine the values of a, b, c, d for which the systems are consistent. In cases where the system is consistent, determine whether the system has a unique solution.

(a) $\left[\begin{array}{ccc|c} 1 & -2 & 4 & 7 \\ 0 & a^2-1 & a & 3 \\ 0 & 0 & b & -3 \end{array}\right]$

(b) $\left[\begin{array}{cccc|c} 1 & 0 & 2 & 5 & 2 \\ 0 & c & c & 0 & 1 \\ 0 & 0 & c & 0 & c \\ 0 & 0 & 0 & cd & c+d \end{array}\right]$

B7 A bookkeeper is trying to determine the prices that a manufacturer was charging. He examines old sales slips that show the number of various items shipped and the total price. He finds that 20 armchairs, 10 sofa beds, and 8 double beds cost $15,200; 15 armchairs, 12 sofa beds, and 10 double beds cost $15,700; and 12 armchairs, 20 sofa beds, and 10 double beds cost $19,600. Determine the cost for each item or explain why the sales slips must be in error.

B8 (Requires knowledge of forces and moments.) A rod 10 m long is pivoted at its centre; it swings in the horizontal plane. Forces of magnitude F_1, F_2, F_3 are applied perpendicular to the rod in the directions indicated by the arrows in the diagram below; F_1 is applied to the left end of the rod, F_2 is applied at a point 2 m to the right of centre and F_3 at a point 4 m to the right. The total force on the pivot is zero, the moment about the centre is zero, and the sum of the magnitudes of forces is 80 newtons. Write a system of three equations for F_1, F_2, and F_3; write the corresponding augmented matrix; and use the standard procedure to find F_1, F_2, and F_3.

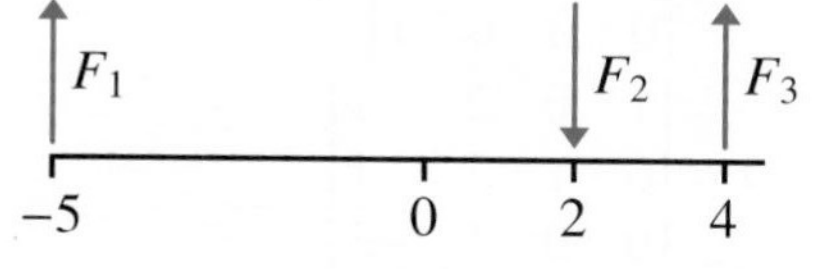

B9 Students at Linear University write a linear algebra examination. An average mark is computed for 100 students in business, an average is computed for 300 students in liberal arts, and an average is computed for 200 students in science. The average of these three averages is 85%. However, the overall average for the 600 students is 86%. Also, the average for the 300 students in business and science is 4 marks higher than the average for the students in liberal arts. Determine the average for each group of students by solving a system of linear equations.

Computer Problems

C1 Use computer software to determine a matrix in row echelon form that is row equivalent to each of the following matrices.

(a) $A = \begin{bmatrix} 35 & 45 & 18 & 13 \\ 17 & 65 & -61 & 7 \\ 23 & 19 & 6 & 41 \end{bmatrix}$

(b) $B = \begin{bmatrix} -25 & -36 & 37 & 41 & 22 \\ 50 & -38 & 49 & 13 & 45 \\ 27 & -23 & 6 & -21 & 27 \end{bmatrix}$

C2 Redo Problems A3, A5, B3, and B5 using a computer.

C3 Suppose that a system of linear equations has the augmented matrix.

$$\left[\begin{array}{ccc|c} 1.121 & -2.015 & 2.131 & 4.612 \\ 2.501 & 3.214 & 4.130 & 3.115 \\ -1.639 & -12.473 & -1.827 & 8.430 \end{array}\right]$$

(a) Determine the solution of this system.

(b) Change the entry in the second row and third column from 4.130 to 4.080 and find the solution.

Conceptual Problems

D1 Consider the linear system in x, y, z, and w:

$$\begin{aligned} x + y \qquad + w &= b \\ 2x + 3y + z + 5w &= 6 \\ z + w &= 4 \\ 2y + 2z + aw &= 1 \end{aligned}$$

For what values of the constants a and b is the system

(a) Inconsistent?

(b) Consistent with a unique solution?

(c) Consistent with infinitely many solutions?

D2 Recall that in $\mathbb{R}^3$, two planes $\vec{n} \cdot \vec{x} = c$ and $\vec{m} \cdot \vec{x} = d$ are parallel if and only if the normal vector $\vec{m}$ is a non-zero multiple of the normal vector $\vec{n}$. Row reduce a suitable augmented matrix to explain why two parallel planes must either coincide or else have no points in common.

2.2 Reduced Row Echelon Form, Rank, and Homogeneous Systems

To determine the solution of a system of linear equations, elimination with back-substitution, as described in Section 2.1, is the standard basic procedure. In some situations and applications, however, it is advantageous to carry the elimination steps (elementary row operations) as far as possible to avoid the need for back-substitution.

To see what further elementary row operations might be worthwhile, recall that the Gaussian elimination procedure proceeds by selecting a pivot and using elementary row operations to create zeros beneath the pivot. The only further elimination steps that simplify the system are steps that create zeros above the pivot.

EXAMPLE 1

In Example 2.1.7, we row reduced the augmented matrix for the original system to a row equivalent matrix in row echelon form. That is, we found that

$$\left[\begin{array}{ccc|c} 1 & 1 & -2 & 4 \\ 1 & 3 & -1 & 7 \\ 2 & 1 & -5 & 7 \end{array}\right] \sim \left[\begin{array}{ccc|c} 1 & 1 & -2 & 4 \\ 0 & 1 & 1 & 1 \\ 0 & 0 & -1 & 1 \end{array}\right]$$

Instead of using back-substitution to solve the system as we did in Example 2.1.7, we instead perform the following elementary row operations:

$$\left[\begin{array}{ccc|c} 1 & 1 & -2 & 4 \\ 0 & 1 & 1 & 1 \\ 0 & 0 & -1 & 1 \end{array}\right] \begin{array}{c} \\ \\ (-1)R_3 \end{array} \sim \left[\begin{array}{ccc|c} 1 & 1 & -2 & 4 \\ 0 & 1 & 1 & 1 \\ 0 & 0 & 1 & -1 \end{array}\right] \begin{array}{c} \\ R_2 - R_3 \\ \\ \end{array} \sim$$

$$\left[\begin{array}{ccc|c} 1 & 1 & -2 & 4 \\ 0 & 1 & 0 & 2 \\ 0 & 0 & 1 & -1 \end{array}\right] \begin{array}{c} R_1 + 2R_3 \\ \\ \\ \end{array} \sim \left[\begin{array}{ccc|c} 1 & 1 & 0 & 2 \\ 0 & 1 & 0 & 2 \\ 0 & 0 & 1 & -1 \end{array}\right] \begin{array}{c} R_1 - R_2 \\ \\ \\ \end{array} \sim \left[\begin{array}{ccc|c} 1 & 0 & 0 & 0 \\ 0 & 1 & 0 & 2 \\ 0 & 0 & 1 & -1 \end{array}\right]$$

This is the augmented matrix for the system $x_1 = 0$, $x_2 = 2$, and $x_3 = -1$, which gives us the solution we found in Example 2.1.7.

This system has been solved by **complete elimination**. The leading variable in the j-th equation has been eliminated from every other equation. This procedure is often called **Gauss-Jordan elimination** to distinguish it from Gaussian elimination with back-substitution. Observe that the elementary row operations used in Example 1 are exactly the same as the operations performed in the back-substitution in Example 2.1.7.

A matrix corresponding to a system on which Gauss-Jordan elimination has been carried out is in a special kind of row echelon form.

Definition

Reduced Row Echelon Form (RREF)

A matrix R is said to be in **reduced row echelon form (RREF)** if

(1) It is in row echelon form.
(2) All leading entries are 1, called a **leading 1**.
(3) In a column with a leading 1, all the other entries are zeros.

As in the case of row echelon form, it is easy to see that every matrix is row equivalent to a matrix in reduced row echelon form via Gauss-Jordan elimination. However, in this case we get a stronger result.

Theorem 1

For any given matrix A there is a unique matrix in reduced row echelon form that is row equivalent to A.

Proof: You are asked to prove that there is only one matrix in reduced row echelon form that is row equivalent to A in Problem F6 in the Chapter 4 Further Problem. ■

EXAMPLE 2

Obtain the matrix in reduced row echelon form that is row equivalent to the matrix

$$A = \begin{bmatrix} 1 & 1 & 2 & -2 & 2 \\ 3 & 3 & 5 & 0 & 2 \end{bmatrix}$$

Solution: Row reducing the matrix, we get

$$\begin{bmatrix} 1 & 1 & 2 & -2 & 2 \\ 3 & 3 & 5 & 0 & 2 \end{bmatrix} \begin{matrix} \\ R_2 - 3R_1 \end{matrix} \sim \begin{bmatrix} 1 & 1 & 2 & -2 & 2 \\ 0 & 0 & -1 & 6 & -4 \end{bmatrix} \begin{matrix} R_1 + 2R_2 \\ \\ \end{matrix} \sim$$
$$\begin{bmatrix} 1 & 1 & 0 & 10 & -6 \\ 0 & 0 & -1 & 6 & -4 \end{bmatrix} \begin{matrix} \\ (-1)R_2 \end{matrix} \sim \begin{bmatrix} 1 & 1 & 0 & 10 & -6 \\ 0 & 0 & 1 & -6 & 4 \end{bmatrix}$$

This final matrix is in reduced row echelon form.

EXERCISE 1

Row reduce the matrices in Examples 2.1.9 and 2.1.12 into reduced row echelon form.

Because of the uniqueness of the reduced row echelon form, we often speak of the reduced row echelon form of a matrix or of reducing a matrix to its reduced row echelon form.

Remarks

1. In general, reducing an augmented matrix to reduced row echelon form to solve a system is not more efficient than the method used in Section 2.1. As previously mentioned, both methods are essentially equivalent for solving small systems by hand.

2. When row reducing to reduced row echelon form by hand, it seems more natural not to obtain a row echelon form first. Instead, you might obtain zeros below and above any leading 1 before moving on to the next leading 1. However, for programming a computer to row reduce a matrix, this is a poor strategy because it requires more multiplications and additions than the previous strategy. See Problem F2 at the end of the chapter.

Rank of a Matrix

We saw in Theorem 2.1.1 that the number of leading entries in a row echelon form of the augmented matrix of a system determines whether the system is consistent or inconsistent. It also determines how many solutions (one or infinitely many) the system has if it is consistent. Thus, we make the following definition.

Definition
Rank

The **rank** of a matrix A is the number of leading 1s in its reduced row echelon form and is denoted by $\text{rank}(A)$.

The rank of A is also equal to the number of leading entries in any row echelon form of A. However, since the row echelon form is not unique, it is more tiresome to give clear arguments in terms of row echelon form. In Section 3.4 we shall see a more conceptual way of describing rank.

EXAMPLE 3

The rank of the matrix in Example 1 is 3 since the RREF of the matrix has three leading 1s. The rank of the matrix in Example 2 is 2 as the RREF of the matrix has two leading 1s.

EXERCISE 2

Determine the rank of each of the following matrices:

(a) $A = \begin{bmatrix} 1 & 0 & 1 & 0 \\ 0 & 0 & 1 & 1 \\ 0 & 0 & 0 & 0 \end{bmatrix}$ (b) $B = \begin{bmatrix} 1 & 1 & 0 & 1 \\ 0 & 0 & 0 & 0 \\ 0 & 0 & 0 & 3 \\ 0 & 0 & 0 & 2 \end{bmatrix}$

Theorem 2

Let $\left[A \mid \vec{b} \right]$ be a system of m linear equations in n variables.

(1) The system is consistent if and only if the rank of the coefficient matrix A is equal to the rank of the augmented matrix $\left[A \mid \vec{b} \right]$.

Theorem 2 (continued)

(2) If the system is consistent, then the number of parameters in the general solution is the number of variables minus the rank of the matrix:

$$\#\text{ of parameters } = n - \operatorname{rank}(A)$$

Proof: Notice that the first n columns of the reduced row echelon form of $\left[A \mid \vec{b}\right]$ consists of the reduced row echelon form of A. By Theorem 2.1.1, the system is inconsistent if and only if the reduced row echelon form of $\left[A \mid \vec{b}\right]$ contains a row of the form $\left[\begin{array}{ccc|c} 0 & \cdots & 0 & 1 \end{array}\right]$. But this is true if and only if the rank of $\left[A \mid \vec{b}\right]$ is greater than the rank of A.

If the system is consistent, then the free variables are the variables that are not leading variables of any equation in a row echelon form of the matrix. Thus, by definition, there are $n - \operatorname{rank}(A)$ free variables and hence $n - \operatorname{rank}(A)$ parameters in the general solution. ■

Corollary 3

Let $\left[A \mid \vec{b}\right]$ be a system of m linear equations in n variables. Then $\left[A \mid \vec{b}\right]$ is consistent for all $\vec{b} \in \mathbb{R}^n$ if and only if $\operatorname{rank}(A) = m$.

Proof: This follows immediately from Theorem 2.

Homogeneous Linear Equations

Frequently systems of linear equations appear where all of the terms on the right-hand side are zero.

Definition
Homogeneous

A linear equation is **homogeneous** if the right-hand side is zero. A system of linear equations is **homogeneous** if all of the equations of the system are homogeneous.

Since a homogeneous system is a special case of the systems already discussed, no new tools or techniques are needed to solve them. However, we normally work only with the coefficient matrix of a homogeneous system since the last column of the augmented matrix consists entirely of zeros.

EXAMPLE 4

Find the general solution of the homogeneous system

$$\begin{aligned} 2x_1 + x_2 \phantom{{}- x_3} &= 0 \\ x_1 + x_2 - x_3 &= 0 \\ -x_2 + 2x_3 &= 0 \end{aligned}$$

EXAMPLE 4
(continued)

Solution: We row reduce the coefficient matrix of the system to RREF:

$$\begin{bmatrix} 2 & 1 & 0 \\ 1 & 1 & -1 \\ 0 & -1 & 2 \end{bmatrix} R_1 \updownarrow R_2 \sim \begin{bmatrix} 1 & 1 & -1 \\ 2 & 1 & 0 \\ 0 & -1 & 2 \end{bmatrix} R_2 - 2R_1 \sim$$

$$\begin{bmatrix} 1 & 1 & -1 \\ 0 & -1 & 2 \\ 0 & -1 & 2 \end{bmatrix} \begin{matrix} R_1 + R_2 \\ \\ R_3 - R_2 \end{matrix} \sim \begin{bmatrix} 1 & 0 & 1 \\ 0 & -1 & 2 \\ 0 & 0 & 0 \end{bmatrix} (-1)R_2 \sim \begin{bmatrix} 1 & 0 & 1 \\ 0 & 1 & -2 \\ 0 & 0 & 0 \end{bmatrix}$$

This corresponds to the homogeneous system

$$\begin{aligned} x_1 \quad\quad + x_3 &= 0 \\ x_2 - 2x_3 &= 0 \end{aligned}$$

Hence, x_3 is a free variable, so we let $x_3 = t \in \mathbb{R}$. Then $x_1 = -x_3 = -t$, $x_2 = 2x_3 = 2t$, and the general solution is

$$\begin{bmatrix} x_1 \\ x_2 \\ x_3 \end{bmatrix} = \begin{bmatrix} -t \\ 2t \\ t \end{bmatrix} = t \begin{bmatrix} -1 \\ 2 \\ 1 \end{bmatrix}$$

Observe that every homogeneous system is consistent as the zero vector $\vec{0}$ will certainly be a solution. We call $\vec{0}$ the **trivial solution**. Thus, as we will see frequently throughout the text, when dealing with homogeneous systems, we are often mostly interested in how many parameters are in the general solution. Of course, for this we can apply Theorem 2.

EXAMPLE 5

Determine the number of parameters in the general solution of the homogeneous system

$$\begin{aligned} x_1 + 2x_2 + 2x_3 + x_4 + 4x_5 &= 0 \\ 3x_1 + 7x_2 + 7x_3 + 3x_4 + 13x_5 &= 0 \\ 2x_1 + 5x_2 + 5x_3 + 2x_4 + 9x_5 &= 0 \end{aligned}$$

Solution: We row reduce the coefficient matrix:

$$\begin{bmatrix} 1 & 2 & 2 & 1 & 4 \\ 3 & 7 & 7 & 3 & 13 \\ 2 & 5 & 5 & 2 & 9 \end{bmatrix} \begin{matrix} \\ R_2 - 3R_1 \\ R_3 - 2R_1 \end{matrix} \sim \begin{bmatrix} 1 & 2 & 2 & 1 & 4 \\ 0 & 1 & 1 & 0 & 1 \\ 0 & 1 & 1 & 0 & 1 \end{bmatrix} \begin{matrix} R_1 - 2R_2 \\ \\ R_3 - R_2 \end{matrix} \sim \begin{bmatrix} 1 & 0 & 0 & 1 & 2 \\ 0 & 1 & 1 & 0 & 1 \\ 0 & 0 & 0 & 0 & 0 \end{bmatrix}$$

The rank of the coefficient matrix is 2 and the number of variables is 5. Thus, by Theorem 2, there are $5 - 2 = 3$ parameters in the general solution.

EXERCISE 3

Find the general solution of the system in Example 5.

PROBLEMS 2.2

Practice Problems

A1 Determine the RREF and rank of the following matrices.

(a) $\begin{bmatrix} 2 & 1 \\ 1 & -1 \\ 3 & 2 \end{bmatrix}$

(b) $\begin{bmatrix} 2 & 0 & 1 \\ 0 & 1 & 2 \\ 1 & 1 & 1 \end{bmatrix}$

(c) $\begin{bmatrix} 1 & 2 & 3 \\ 2 & 1 & 2 \\ 2 & 3 & 4 \end{bmatrix}$

(d) $\begin{bmatrix} 1 & 0 & -2 \\ 2 & 1 & 2 \\ 2 & 3 & 4 \end{bmatrix}$

(e) $\begin{bmatrix} 1 & 2 & 1 \\ 1 & 2 & 3 \\ -1 & -2 & 3 \\ 2 & 4 & 3 \end{bmatrix}$

(f) $\begin{bmatrix} 1 & 1 & 1 & 1 \\ 1 & 1 & 1 & 0 \\ 1 & 1 & 0 & 0 \end{bmatrix}$

(g) $\begin{bmatrix} 2 & -1 & 2 & 8 \\ 1 & -1 & 0 & 2 \\ 3 & -2 & 3 & 13 \end{bmatrix}$

(h) $\begin{bmatrix} 1 & 1 & 0 & 1 \\ 0 & 1 & 1 & 2 \\ 2 & 3 & 1 & 4 \\ 1 & 2 & 3 & 4 \end{bmatrix}$

(i) $\begin{bmatrix} 0 & 1 & 0 & 2 & 5 \\ 3 & 1 & 8 & 5 & 3 \\ 1 & 0 & 3 & 2 & 1 \\ 2 & 1 & 6 & 7 & 1 \end{bmatrix}$

A2 Suppose that each of the following is the coefficient matrix of a homogeneous system already in RREF. For each matrix, determine the number of parameters in the general solution and write out the general solution in standard form.

(a) $\begin{bmatrix} 1 & 0 & 2 & 0 \\ 0 & 1 & -1 & 0 \\ 0 & 0 & 0 & 1 \end{bmatrix}$

(b) $\begin{bmatrix} 0 & 1 & 2 & 0 \\ 0 & 0 & 0 & 1 \\ 0 & 0 & 0 & 0 \end{bmatrix}$

(c) $\begin{bmatrix} 1 & -3 & 2 & 0 \\ 0 & 0 & 0 & 1 \\ 0 & 0 & 0 & 0 \\ 0 & 0 & 0 & 0 \end{bmatrix}$

(d) $\begin{bmatrix} 1 & 0 & 2 & 0 & 0 \\ 0 & 1 & -1 & 0 & -2 \\ 0 & 0 & 0 & 1 & 1 \end{bmatrix}$

(e) $\begin{bmatrix} 1 & 0 & 0 & 4 & 0 \\ 0 & 0 & 1 & -5 & 0 \\ 0 & 0 & 0 & 0 & 1 \end{bmatrix}$

(f) $\begin{bmatrix} 1 & 0 & 0 & 0 & 0 \\ 0 & 1 & 1 & 0 & 0 \\ 0 & 0 & 0 & 1 & 0 \\ 0 & 0 & 0 & 0 & 1 \end{bmatrix}$

A3 For each homogeneous system, write the coefficient matrix and determine the rank and number of parameters in the general solution. Then determine the general solution.

(a)
$$\begin{aligned} 2x_2 - 5x_3 &= 0 \\ x_1 + 2x_2 + 3x_3 &= 0 \\ x_1 + 4x_2 - 3x_3 &= 0 \end{aligned}$$

(b)
$$\begin{aligned} 3x_1 + x_2 - 9x_3 &= 0 \\ x_1 + x_2 - 5x_3 &= 0 \\ 2x_1 + x_2 - 7x_3 &= 0 \end{aligned}$$

(c)
$$\begin{aligned} x_1 - x_2 + 2x_3 - 3x_4 &= 0 \\ 3x_1 - 3x_2 + 8x_3 - 5x_4 &= 0 \\ 2x_1 - 2x_2 + 5x_3 - 4x_4 &= 0 \\ 3x_1 - 3x_2 + 7x_3 - 7x_4 &= 0 \end{aligned}$$

(d)
$$\begin{aligned} x_2 + 2x_3 + 2x_4 &= 0 \\ x_1 + 2x_2 + 5x_3 + 3x_4 - x_5 &= 0 \\ 2x_1 + x_2 + 5x_3 + x_4 - 3x_5 &= 0 \\ x_1 + x_2 + 4x_3 + 2x_4 - 2x_5 &= 0 \end{aligned}$$

A4 Solve the following systems of linear equations by row reducing the coefficient matrix to RREF. Compare your steps with your solutions from the Problem 2.1.A5.

(a)
$$\begin{aligned} 3x_1 - 5x_2 &= 2 \\ x_1 + 2x_2 &= 4 \end{aligned}$$

(b)
$$\begin{aligned} x_1 + 2x_2 + x_3 &= 5 \\ 2x_1 - 3x_2 + 2x_3 &= 6 \end{aligned}$$

(c)
$$\begin{aligned} x_1 + 2x_2 - 3x_3 &= 8 \\ x_1 + 3x_2 - 5x_3 &= 11 \\ 2x_1 + 5x_2 - 8x_3 &= 19 \end{aligned}$$

(d)
$$\begin{aligned} -3x_1 + 6x_2 + 16x_3 &= 36 \\ x_1 - 2x_2 - 5x_3 &= -11 \\ 2x_1 - 3x_2 - 8x_3 &= -17 \end{aligned}$$

(e)
$$\begin{aligned} x_1 + 2x_2 - x_3 &= 4 \\ 2x_1 + 5x_2 + x_3 &= 10 \\ 4x_1 + 9x_2 - x_3 &= 19 \end{aligned}$$

(f)
$$\begin{aligned} x_1 + 2x_2 - 3x_3 &= -5 \\ 2x_1 + 4x_2 - 6x_3 + x_4 &= -8 \\ 6x_1 + 13x_2 - 17x_3 + 4x_4 &= -21 \end{aligned}$$

(g)
$$\begin{aligned} 2x_2 - 2x_3 + x_5 &= 2 \\ x_1 + 2x_2 - 3x_3 + x_4 + 4x_5 &= 1 \\ 2x_1 + 4x_2 - 5x_3 + 3x_4 + 8x_5 &= 3 \\ 2x_1 + 5x_2 - 7x_3 + 3x_4 + 10x_5 &= 5 \end{aligned}$$

A5 Solve the system $\left[A \mid \vec{b}\right]$ by row reducing the coefficient matrix to RREF. Then, without any further operations, find the general solution to the homogeneous $\left[A \mid \vec{0}\right]$.

(a) $A = \begin{bmatrix} 2 & -1 & 4 \\ 1 & 3 & 0 \\ 1 & 1 & 2 \end{bmatrix}, \vec{b} = \begin{bmatrix} 1 \\ 0 \\ 2 \end{bmatrix}$

(b) $A = \begin{bmatrix} 1 & 7 & 5 \\ 1 & 0 & 5 \\ -1 & 2 & -5 \end{bmatrix}, \vec{b} = \begin{bmatrix} 5 \\ -2 \\ 4 \end{bmatrix}$

(c) $A = \begin{bmatrix} 0 & -1 & 5 & -2 \\ -1 & -1 & -4 & -1 \end{bmatrix}, \vec{b} = \begin{bmatrix} -1 \\ 4 \end{bmatrix}$

(d) $A = \begin{bmatrix} 1 & 0 & -1 & -1 \\ 4 & 3 & 2 & -4 \\ -1 & -4 & -3 & 5 \end{bmatrix}, \vec{b} = \begin{bmatrix} 3 \\ 3 \\ 5 \end{bmatrix}$

Homework Problems

B1 Determine the RREF and rank of the following matrices.

(a) $\begin{bmatrix} 2 & 3 \\ 3 & -1 \\ 3 & 2 \end{bmatrix}$

(b) $\begin{bmatrix} 2 & 5 & 3 \\ 1 & 2 & 2 \\ 1 & 3 & 2 \end{bmatrix}$

(c) $\begin{bmatrix} 2 & 4 & 8 \\ 1 & 1 & 3 \\ 1 & -1 & 1 \end{bmatrix}$

(d) $\begin{bmatrix} 1 & -1 & 2 \\ -1 & 1 & 4 \\ 4 & -3 & 3 \\ 3 & -2 & 3 \end{bmatrix}$

(e) $\begin{bmatrix} 1 & 1 & 2 & 1 \\ 2 & 1 & 4 & 3 \\ 0 & 3 & 2 & 1 \end{bmatrix}$

(f) $\begin{bmatrix} 2 & 1 & 1 & 1 \\ 3 & 2 & 1 & 1 \\ 4 & 3 & 2 & 1 \\ 1 & 0 & -1 & 0 \end{bmatrix}$

(g) $\begin{bmatrix} 1 & 1 & 2 & 1 & -2 \\ 2 & 2 & 4 & 3 & -6 \\ 0 & 1 & 2 & 2 & -4 \\ 3 & 2 & 4 & 2 & -4 \end{bmatrix}$

B2 Suppose that each of the following is the coefficient matrix of a homogeneous system already in RREF. For each matrix, determine the number of parameters in the general solution and write out the general solution in standard form.

(a) $\begin{bmatrix} 1 & 3 & 0 & -1 \\ 0 & 0 & 1 & 2 \\ 0 & 0 & 0 & 0 \end{bmatrix}$

(b) $\begin{bmatrix} 1 & 1 & 0 & -2 & 0 \\ 0 & 0 & 1 & 1 & 0 \\ 0 & 0 & 0 & 0 & 1 \end{bmatrix}$

(c) $\begin{bmatrix} 1 & 2 & 0 & 0 & 0 & -3 \\ 0 & 0 & 1 & -5 & 0 & 4 \\ 0 & 0 & 0 & 0 & 1 & 1 \\ 0 & 0 & 0 & 0 & 0 & 0 \end{bmatrix}$

B3 For each homogeneous system, write the coefficient matrix and determine the rank and number of parameters in the general solution. Then determine the general solution.

(a)
$$\begin{aligned} x_1 + 5x_2 - 3x_3 &= 0 \\ 3x_1 + 5x_2 - 9x_3 &= 0 \\ x_1 + x_2 - 3x_3 &= 0 \end{aligned}$$

(b)
$$\begin{aligned} x_1 + 4x_2 - 2x_3 &= 0 \\ 2x_1 \qquad - 3x_3 &= 0 \\ 4x_1 + 8x_2 - 7x_3 &= 0 \end{aligned}$$

(c)
$$\begin{aligned} x_1 + x_2 + x_3 - 2x_4 &= 0 \\ 2x_1 + 7x_2 \qquad - 14x_4 &= 0 \\ x_1 + 3x_2 \qquad - 6x_4 &= 0 \\ x_1 + 4x_2 \qquad - 8x_4 &= 0 \end{aligned}$$

(d)
$$\begin{aligned} x_1 + 3x_2 + x_3 + x_4 + 2x_5 &= 0 \\ 2x_2 + x_3 \qquad - x_5 &= 0 \\ x_1 + 2x_2 + 2x_3 + x_4 \qquad &= 0 \\ x_1 + 2x_2 + x_3 + x_4 + x_5 &= 0 \end{aligned}$$

B4 Solve the system $\left[A \mid \vec{b}\,\right]$ by row reducing the coefficient matrix to RREF. Then, without any further operations, find the general solution to the homogeneous system $\left[A \mid \vec{0}\,\right]$.

(a) $A = \begin{bmatrix} 4 & 0 & 6 \\ 6 & 6 & 3 \\ -2 & 1 & -4 \end{bmatrix}, \vec{b} = \begin{bmatrix} 0 \\ -6 \\ -1 \end{bmatrix}$

(b) $A = \begin{bmatrix} 1 & 2 & -4 \\ -1 & -5 & -5 \\ -4 & 1 & 9 \\ -5 & -4 & 0 \end{bmatrix}, \vec{b} = \begin{bmatrix} 10 \\ -1 \\ 1 \\ 8 \end{bmatrix}$

(c) $A = \begin{bmatrix} 1 & -1 & 4 & -1 \\ -1 & -2 & 5 & -2 \\ -4 & -1 & 2 & 2 \\ 5 & 4 & 1 & 8 \end{bmatrix}, \vec{b} = \begin{bmatrix} 4 \\ 5 \\ -4 \\ 5 \end{bmatrix}$

(d) $A = \begin{bmatrix} 1 & 1 & 3 & 1 & 4 \\ 4 & 4 & 6 & -8 & 4 \\ 1 & 1 & 4 & -2 & 1 \\ 3 & 3 & 2 & -4 & 5 \end{bmatrix}, \vec{b} = \begin{bmatrix} 2 \\ -4 \\ -6 \\ 6 \end{bmatrix}$

Computer Problems

C1 Determine the RREF of the augmented matrix of the system of linear equations

$$\begin{aligned} 2.01x + 3.45y + 2.23z &= 4.13 \\ 1.57x + 2.03y - 3.11z &= 6.11 \\ 2.23x + 7.10y - 4.28z &= 0.47 \end{aligned}$$

If the system is consistent, determine the general solution.

C2 Determine the RREF of the matrices in Problem 2.1.C1.

Conceptual Problems

D1 We want to find a vector $\vec{x} \neq \vec{0}$ in $\mathbb{R}^3$ that is simultaneously orthogonal to given vectors $\vec{a}, \vec{b}, \vec{c} \in \mathbb{R}^3$.

(a) Write equations that must be satisfied by $\vec{x}$.

(b) What condition must be satisfied by the rank of the matrix $A = \begin{bmatrix} a_1 & a_2 & a_3 \\ b_1 & b_2 & b_3 \\ c_1 & c_2 & c_3 \end{bmatrix}$ if there are to be non-trivial solutions? Explain.

D2 (a) Suppose that $\begin{bmatrix} 1 & 0 & 2 \\ 0 & 1 & -1 \end{bmatrix}$ is the coefficient matrix of a homogeneous system of linear equations. Find the general solution of the system and indicate why it describes a line through the origin.

(b) Suppose that a matrix A with two rows and three columns is the coefficient matrix of a homogeneous system. If A has rank 2, then explain why the solution set of the homogeneous system is a line through the origin. What could you say if $\text{rank}(A) = 1$?

(c) Let $\vec{u}$, $\vec{v}$, and $\vec{w}$ be three vectors in $\mathbb{R}^4$. Write conditions on a vector $\vec{x} \in \mathbb{R}^4$ such that $\vec{x}$ is orthogonal to $\vec{u}$, $\vec{v}$, and $\vec{w}$. (This should lead to a homogeneous system with coefficient matrix C, whose rows are $\vec{u}$, $\vec{v}$, and $\vec{w}$.) What does the rank of C tell us about the set of vectors $\vec{x}$ that are orthogonal to $\vec{u}$, $\vec{v}$, and $\vec{w}$?

D3 What can you say about the consistency of a system of m linear equations in n variables and the number of parameters in the general solution if:

(a) $m = 5$, $n = 7$, the rank of the coefficient matrix is 4?

(b) $m = 3$, $n = 6$, the rank of the coefficient matrix is 3?

(c) $m = 5$, $n = 4$, the rank of the augmented matrix is 4?

D4 A system of linear equations has augmented matrix $\left[\begin{array}{ccc|c} 1 & a & b & 1 \\ 1 & 1 & 0 & a \\ 1 & 0 & 1 & b \end{array}\right]$. For which values of a and b is the system consistent? Are there values for which there is a unique solution? Determine the general solution.

2.3 Application to Spanning and Linear Independence

As discussed at the beginning of this chapter, solving systems of linear equations will play an important role in much of what we do in the rest of the text. Here we will show how to use the methods described in this chapter to solve some of the problems we encountered in Chapter 1.

Spanning Problems

Recall that a vector $\vec{v} \in \mathbb{R}^n$ is in the span of a set $\{\vec{v}_1, \dots, \vec{v}_k\}$ of vectors in $\mathbb{R}^n$ if and only if there exists scalars $t_1, \dots, t_k \in \mathbb{R}$ such that

$$t_1\vec{v}_1 + \cdots + t_k\vec{v}_k = \vec{v}$$

Observe that this vector equation actually represents n equations (one for each component of the vectors) in the k unknowns $t_1, \dots, t_k$. Thus, it is easy to establish whether a vector is in the span of a set; we just need to determine whether the corresponding system of linear equations is consistent or not.

EXAMPLE 1

Determine whether the vector $\vec{v} = \begin{bmatrix} -2 \\ -3 \\ 1 \end{bmatrix}$ is in the set Span $\left\{ \begin{bmatrix} 1 \\ 1 \\ 1 \end{bmatrix}, \begin{bmatrix} 1 \\ -1 \\ 5 \end{bmatrix}, \begin{bmatrix} 2 \\ 1 \\ 4 \end{bmatrix}, \begin{bmatrix} -1 \\ -3 \\ 3 \end{bmatrix} \right\}$.

Solution: Consider the vector equation

$$t_1 \begin{bmatrix} 1 \\ 1 \\ 1 \end{bmatrix} + t_2 \begin{bmatrix} 1 \\ -1 \\ 5 \end{bmatrix} + t_3 \begin{bmatrix} 2 \\ 1 \\ 4 \end{bmatrix} + t_4 \begin{bmatrix} -1 \\ -3 \\ 3 \end{bmatrix} = \begin{bmatrix} -2 \\ -3 \\ 1 \end{bmatrix}$$

Simplifying the left-hand side using vector operations, we get

$$\begin{bmatrix} t_1 + t_2 + 2t_3 - t_4 \\ t_1 - t_2 + t_3 - 3t_4 \\ t_1 + 5t_2 + 4t_3 + 3t_4 \end{bmatrix} = \begin{bmatrix} -2 \\ -3 \\ 1 \end{bmatrix}$$

Comparing corresponding entries gives the system of linear equations

$$\begin{aligned} t_1 + t_2 + 2t_3 - t_4 &= -2 \\ t_1 - t_2 + t_3 - 3t_4 &= -3 \\ t_1 + 5t_2 + 4t_3 + 3t_4 &= 1 \end{aligned}$$

We row reduce the augmented matrix:

$$\left[\begin{array}{cccc|c} 1 & 1 & 2 & -1 & -2 \\ 1 & -1 & 1 & -3 & -3 \\ 1 & 5 & 4 & 3 & 1 \end{array}\right] \begin{array}{l} \\ R_2 - R_1 \\ R_3 - R_1 \end{array} \sim \left[\begin{array}{cccc|c} 1 & 1 & 2 & -1 & -2 \\ 0 & -2 & -1 & -2 & -1 \\ 0 & 4 & 2 & 4 & 3 \end{array}\right] \begin{array}{l} \\ \\ R_3 + 2R_2 \end{array}$$

$$\sim \left[\begin{array}{cccc|c} 1 & 1 & 2 & -1 & -2 \\ 0 & -2 & -1 & -2 & -1 \\ 0 & 0 & 0 & 0 & 1 \end{array}\right]$$

By Theorem 2.1.1, the system is inconsistent. Hence, there do not exist values of t_1, t_2, t_3, and t_4 that satisfy the system of equations, so $\vec{v}$ is not in the span of the vectors.

EXAMPLE 2

Let $\vec{v}_1 = \begin{bmatrix} 1 \\ 2 \\ 1 \end{bmatrix}$, $\vec{v}_2 = \begin{bmatrix} -2 \\ 1 \\ 0 \end{bmatrix}$, and $\vec{v}_3 = \begin{bmatrix} 1 \\ 1 \\ 1 \end{bmatrix}$. Write $\vec{v} = \begin{bmatrix} -1 \\ 1 \\ -1 \end{bmatrix}$ as a linear combination of $\vec{v}_1$, $\vec{v}_2$, and $\vec{v}_3$.

Solution: We need to find scalars t_1, t_2, and t_3 such that

$$t_1\vec{v}_1 + t_2\vec{v}_2 + t_3\vec{v}_3 = \vec{v}$$

Simplifying the left-hand side using vector operations, we get

$$\begin{bmatrix} t_1 - 2t_2 + t_3 \\ 2t_1 + t_2 + t_3 \\ t_1 + t_3 \end{bmatrix} = \begin{bmatrix} -1 \\ 1 \\ -1 \end{bmatrix}$$

EXAMPLE 2
(continued)

This gives the system of linear equations:

$$\begin{aligned} t_1 - 2t_2 + t_3 &= -1 \\ 2t_1 + t_2 + t_3 &= 1 \\ t_1 \qquad + t_3 &= -1 \end{aligned}$$

Row reducing the augmented matrix to RREF gives

$$\left[\begin{array}{ccc|c} 1 & -2 & 1 & -1 \\ 2 & 1 & 1 & 1 \\ 1 & 0 & 1 & -1 \end{array}\right] \sim \left[\begin{array}{ccc|c} 1 & 0 & 0 & 2 \\ 0 & 1 & 0 & 0 \\ 0 & 0 & 1 & -3 \end{array}\right]$$

The solution is $t_1 = 2$, $t_2 = 0$, and $t_3 = -3$. Hence, we get $2\vec{v}_1 + 0\vec{v}_2 - 3\vec{v}_3 = \vec{v}$.

EXERCISE 1

Determine whether $\vec{v} = \begin{bmatrix} 1 \\ 3 \\ 1 \end{bmatrix}$ is in the set Span $\left\{ \begin{bmatrix} 1 \\ -3 \\ -3 \end{bmatrix}, \begin{bmatrix} 2 \\ -2 \\ 1 \end{bmatrix}, \begin{bmatrix} -2 \\ 2 \\ -3 \end{bmatrix} \right\}$.

EXAMPLE 3

Consider Span $\left\{ \begin{bmatrix} 1 \\ 2 \\ 1 \\ 1 \end{bmatrix}, \begin{bmatrix} 1 \\ 1 \\ 3 \\ 1 \end{bmatrix}, \begin{bmatrix} 3 \\ 5 \\ 5 \\ 3 \end{bmatrix} \right\}$. Find a homogeneous system of linear equations that defines this set.

Solution: A vector $\vec{x} = \begin{bmatrix} x_1 \\ x_2 \\ x_3 \\ x_4 \end{bmatrix}$ is in this set if and only if for some t_1, t_2, and t_3,

$$t_1 \begin{bmatrix} 1 \\ 2 \\ 1 \\ 1 \end{bmatrix} + t_2 \begin{bmatrix} 1 \\ 1 \\ 3 \\ 1 \end{bmatrix} + t_3 \begin{bmatrix} 3 \\ 5 \\ 5 \\ 3 \end{bmatrix} = \begin{bmatrix} x_1 \\ x_2 \\ x_3 \\ x_4 \end{bmatrix}$$

Simplifying the left-hand side gives us a system of equations with augmented matrix

$$\left[\begin{array}{ccc|c} 1 & 1 & 3 & x_1 \\ 2 & 1 & 5 & x_2 \\ 1 & 3 & 5 & x_3 \\ 1 & 1 & 3 & x_4 \end{array}\right]$$

Row reducing this matrix to RREF gives

$$\left[\begin{array}{ccc|c} 1 & 1 & 3 & x_1 \\ 2 & 1 & 5 & x_2 \\ 1 & 3 & 5 & x_3 \\ 1 & 1 & 3 & x_4 \end{array}\right] \sim \left[\begin{array}{ccc|c} 1 & 1 & 3 & x_1 \\ 0 & 1 & 1 & 2x_1 - x_2 \\ 0 & 0 & 0 & -5x_1 + 2x_2 + x_3 \\ 0 & 0 & 0 & -x_1 + x_4 \end{array}\right]$$

The system is consistent if and only if $-5x_1 + 2x_2 + x_3 = 0$ and $-x_1 + x_4 = 0$. Thus, this system of linear equations defines the set.

EXAMPLE 4

Show that every vector $\vec{v} \in \mathbb{R}^3$ can be written as a linear combination of the vectors $\vec{v}_1 = \begin{bmatrix} -3 \\ 1 \\ -2 \end{bmatrix}$, $\vec{v}_2 = \begin{bmatrix} 1 \\ 3 \\ -3 \end{bmatrix}$, and $\vec{v}_3 = \begin{bmatrix} 2 \\ -1 \\ 1 \end{bmatrix}$.

Solution: To show that every vector $\vec{v} \in \mathbb{R}^3$ can be written as a linear combination of the vectors $\vec{v}_1$, $\vec{v}_2$, and $\vec{v}_3$, we need to show that the system

$$t_1\vec{v}_1 + t_2\vec{v}_2 + t_3\vec{v}_3 = \vec{v}$$

is consistent for all $\vec{v} \in \mathbb{R}^3$.

Simplifying the left-hand side gives us a system of equations with coefficient matrix

$$\begin{bmatrix} -3 & 1 & 2 \\ 1 & 3 & -1 \\ -2 & -3 & 1 \end{bmatrix}$$

Row reducing the coefficient matrix to RREF gives

$$\begin{bmatrix} -3 & 1 & 2 \\ 1 & 3 & -1 \\ -2 & -3 & 1 \end{bmatrix} \sim \begin{bmatrix} 1 & 0 & 0 \\ 0 & 1 & 0 \\ 0 & 0 & 1 \end{bmatrix}$$

Hence, the rank of the matrix is 3, which equals the number of rows (equations). Hence, by Theorem 2.2.2, the system is consistent for all $\vec{v} \in \mathbb{R}^3$, as required.

We generalize the method used in Example 4 to get the following important results.

Lemma 1

A set of k vectors $\{\vec{v}_1, \dots, \vec{v}_k\}$ in $\mathbb{R}^n$ spans $\mathbb{R}^n$ if and only if the rank of the coefficient matrix of the system $t_1\vec{v}_1 + \cdots + t_k\vec{v}_k = \vec{v}$ is n.

Proof: If $\text{Span}\{\vec{v}_1, \dots, \vec{v}_k\} = \mathbb{R}^n$, then every vector $\vec{b} \in \mathbb{R}^n$ can be written as a linear combination of the vectors $\{\vec{v}_1, \dots, \vec{v}_k\}$. That is, the system of linear equations

$$t_1\vec{v}_1 + \cdots + t_k\vec{v}_k = \vec{b}$$

has a solution for every $\vec{b} \in \mathbb{R}^n$. By Theorem 2.2.2, this means that the rank of the coefficient matrix of the system equals n (the number of equations). On the other hand, if the rank of the coefficient matrix of the system $t_1\vec{v}_1 + \cdots + t_k\vec{v}_k = \vec{v}$ is n, then the system is consistent for all $\vec{v} \in \mathbb{R}^n$ by Theorem 2.2.2. Hence, $\text{Span}\{\vec{v}_1, \dots, \vec{v}_k\} = \mathbb{R}^n$. ■

Theorem 2 Let $\{\vec{v}_1, \dots, \vec{v}_k\}$ be a set of k vectors in $\mathbb{R}^n$. If $\text{Span}\{\vec{v}_1, \dots, \vec{v}_k\} = \mathbb{R}^n$, then $k \geq n$.

Proof: By Lemma 1, if $\text{Span}\{\vec{v}_1, \dots, \vec{v}_k\} = \mathbb{R}^n$, then the rank of the coefficient matrix is n. But, if we have n leading 1s, then there must be at least n columns in the matrix to contain the leading 1s. Hence, the number of columns, k, must be greater than or equal to n. ■

Linear Independence Problems

Recall that a set of vectors $\{\vec{v}_1, \dots, \vec{v}_k\}$ in $\mathbb{R}^n$ is said to be linearly independent if and only if the only solution to the vector equation

$$t_1\vec{v}_1 + \cdots + t_k\vec{v}_k = \vec{0}$$

is the solution $t_i = 0$ for $1 \leq i \leq k$. From our work above, we see that this is true when the corresponding homogeneous system of n equations in k unknowns has a unique solution (the trivial solution).

EXAMPLE 5

Determine whether the set $\left\{ \begin{bmatrix} 1 \\ 1 \\ 1 \end{bmatrix}, \begin{bmatrix} 1 \\ -1 \\ 5 \end{bmatrix}, \begin{bmatrix} 2 \\ 1 \\ 4 \end{bmatrix}, \begin{bmatrix} -1 \\ -3 \\ 3 \end{bmatrix} \right\}$ is linearly independent or dependent.

Solution: Consider

$$t_1 \begin{bmatrix} 1 \\ 1 \\ 1 \end{bmatrix} + t_2 \begin{bmatrix} 1 \\ -1 \\ 5 \end{bmatrix} + t_3 \begin{bmatrix} 2 \\ 1 \\ 4 \end{bmatrix} + t_4 \begin{bmatrix} -1 \\ -3 \\ 3 \end{bmatrix} = \begin{bmatrix} 0 \\ 0 \\ 0 \end{bmatrix}$$

Simplifying as above, this gives the homogeneous system with coefficient matrix

$$\begin{bmatrix} 1 & 1 & 2 & -1 \\ 1 & -1 & 1 & -3 \\ 1 & 5 & 4 & 3 \end{bmatrix}$$

Notice that we do not even need to row reduce this matrix. By Theorem 2.2.2, the number of parameters in the general solution equals the number of variables minus the rank of the matrix. There are four variables, but the maximum the rank can be is 3 since there are only three rows. Hence, the number of parameters is at least one, so the system has infinitely many solutions. Therefore, the set is linearly dependent.

EXAMPLE 6

Let $\vec{v}_1 = \begin{bmatrix} 1 \\ 2 \\ 1 \end{bmatrix}$, $\vec{v}_2 = \begin{bmatrix} -2 \\ 1 \\ 0 \end{bmatrix}$, and $\vec{v}_3 = \begin{bmatrix} 1 \\ 1 \\ 1 \end{bmatrix}$. Determine whether the set $\{\vec{v}_1, \vec{v}_2, \vec{v}_3\}$ is linearly independent or dependent.

EXAMPLE 6 (continued)

Solution: Consider $t_1\vec{v}_1 + t_2\vec{v}_2 + t_3\vec{v}_3 = \vec{0}$. As above, we find that the coefficient matrix of the corresponding system is $\begin{bmatrix} 1 & -2 & 1 \\ 2 & 1 & 1 \\ 1 & 0 & 1 \end{bmatrix}$. Using the same elementary row operations as in Example 2, we get

$$\begin{bmatrix} 1 & -2 & 1 \\ 2 & 1 & 1 \\ 1 & 0 & 1 \end{bmatrix} \sim \begin{bmatrix} 1 & 0 & 0 \\ 0 & 1 & 0 \\ 0 & 0 & 1 \end{bmatrix}$$

Therefore, the set is linearly independent since the system has a unique solution (the trivial solution).

EXERCISE 2

Determine whether the set $\left\{ \begin{bmatrix} -1 \\ 1 \\ -3 \end{bmatrix}, \begin{bmatrix} -2 \\ -3 \\ -3 \end{bmatrix}, \begin{bmatrix} 1 \\ -1 \\ 3 \end{bmatrix} \right\}$ is linearly independent or dependent.

Again, we can generalize the method used in Examples 5 and 6 to prove some important results.

Lemma 3

A set of vectors $\{\vec{v}_1, \dots, \vec{v}_k\}$ in $\mathbb{R}^n$ is linearly independent if and only if the rank of the coefficient matrix of the homogeneous system $t_1\vec{v}_1 + \cdots + t_k\vec{v}_k = \vec{0}$ is k.

Proof: If $\{\vec{v}_1, \dots, \vec{v}_k\}$ is linearly independent, then the system of linear equations

$$c_1\vec{v}_1 + \cdots + c_k\vec{v}_k = \vec{0}$$

has a unique solution. Thus, the rank of the coefficient matrix equals the number of unknowns k by Theorem 2.2.2.

On the other hand, if the rank of the coefficient matrix equals k, then the homogeneous system has $k - k = 0$ parameters. Therefore, it has the unique solution $t_1 = \cdots = t_k = 0$, and so the set is linearly independent. ■

Theorem 4

If $\{\vec{v}_1, \dots, \vec{v}_k\}$ is a linearly independent set of vectors in $\mathbb{R}^n$, then $k \leq n$.

Proof: By Lemma 3, if $\{\vec{v}_1, \dots, \vec{v}_k\}$ is linearly independent, then the rank of the coefficient matrix is k. Hence, there must be at least k rows in the matrix to contain the leading 1s. Therefore, the number of rows n must be greater than or equal to k. ■

Bases of Subspaces

Recall from Section 1.2 that we defined a basis of a subspace S of $\mathbb{R}^n$ to be a linearly independent set that spans S. Thus, with our previous tools, we can now easily identify a basis for a subspace. In particular, to show that a set $\mathcal{B}$ of vectors in $\mathbb{R}^n$ is a basis for a subspace S, we just need to show that Span $\mathcal{B} = S$ and $\mathcal{B}$ is linearly independent. We demonstrate this with a couple examples.

EXAMPLE 7

Let $\mathcal{B} = \left\{ \begin{bmatrix} 1 \\ 1 \\ 2 \end{bmatrix}, \begin{bmatrix} 5 \\ -2 \\ 2 \end{bmatrix}, \begin{bmatrix} -2 \\ 3 \\ 1 \end{bmatrix} \right\}$. Prove that $\mathcal{B}$ is a basis for $\mathbb{R}^3$.

Solution: To show that every vector $\vec{v} \in \mathbb{R}^3$ can be written as a linear combination of the vectors in $\mathcal{B}$, we just need to show that the system

$$t_1 \begin{bmatrix} 1 \\ 1 \\ 2 \end{bmatrix} + t_2 \begin{bmatrix} 5 \\ -2 \\ 2 \end{bmatrix} + t_3 \begin{bmatrix} -2 \\ 3 \\ 1 \end{bmatrix} = \vec{v}$$

is consistent for all $\vec{v} \in \mathbb{R}^3$.

Row reducing the coefficient matrix that corresponds to this system to RREF gives

$$\begin{bmatrix} 1 & 5 & -2 \\ 1 & -2 & 3 \\ 2 & 2 & 1 \end{bmatrix} \sim \begin{bmatrix} 1 & 0 & 0 \\ 0 & 1 & 0 \\ 0 & 0 & 1 \end{bmatrix}$$

The rank of the matrix is 3, which equals the number of rows (equations). Hence, by Theorem 2.2.2, the system is consistent for all $\vec{v} \in \mathbb{R}^3$, as required.

Moreover, to determine whether $\mathcal{B}$ is linearly independent, we would perform the same elementary row operations on the same coefficient matrix. So, we see that the rank of the matrix also equals the number of columns (variables). By Theorem 2.2.2, the system has a unique solution, and hence $\mathcal{B}$ is also linearly independent. Thus, $\mathcal{B}$ is a basis for $\mathbb{R}^3$.

EXAMPLE 8

Show that $\mathcal{B} = \left\{ \begin{bmatrix} 1 \\ 2 \\ -1 \end{bmatrix}, \begin{bmatrix} 1 \\ 1 \\ 1 \end{bmatrix} \right\}$ is a basis for the plane $-3x_1 + 2x_2 + x_3 = 0$.

Solution: We first observe that $\mathcal{B}$ is clearly linearly independent since neither vector is a scalar multiple of the other. Thus, we need to show that every vector in the plane can be written as a linear combination of the vectors in $\mathcal{B}$. To do this, observe that any vector $\vec{x}$ in the plane must satisfy the condition of the plane. Hence, every vector in the plane has the form

$$\vec{x} = \begin{bmatrix} x_1 \\ x_2 \\ 3x_1 - 2x_2 \end{bmatrix}$$

since $x_3 = 3x_1 - 2x_2$. Therefore, we now just need to show that the equation

$$t_1 \begin{bmatrix} 1 \\ 2 \\ -1 \end{bmatrix} + t_2 \begin{bmatrix} 1 \\ 1 \\ 1 \end{bmatrix} = \begin{bmatrix} x_1 \\ x_2 \\ 3x_1 - 2x_2 \end{bmatrix}$$

EXAMPLE 8 (continued)

is always consistent. Row reducing the corresponding augmented matrix gives

$$\left[\begin{array}{cc|c} 1 & 1 & x_1 \\ 2 & 1 & x_2 \\ -1 & 1 & 3x_1 - 2x_2 \end{array}\right] \sim \left[\begin{array}{cc|c} 1 & 1 & x_1 \\ 0 & 1 & 2x_1 - x_2 \\ 0 & 0 & 0 \end{array}\right]$$

So, the system is consistent and hence $\mathcal{B}$ is a basis for the plane.

Theorem 5

A set of vectors $\{\vec{v}_1, \dots, \vec{v}_n\}$ is a basis for $\mathbb{R}^n$ if and only if the rank of the coefficient matrix of $t_1\vec{v}_1 + \cdots + t_n\vec{v}_n = \vec{v}$ is n.

Proof: If $\{\vec{v}_1, \dots, \vec{v}_n\}$ is a basis for $\mathbb{R}^n$, then it is linearly independent. Hence, by Lemma 3, the rank of the coefficient matrix is n.

If the rank of the coefficient matrix is n, then the set is linearly independent and spans $\mathbb{R}^n$ by Lemma 1 and Lemma 3. ■

Theorem 5 gives us a condition to test whether a set of n vectors in $\mathbb{R}^n$ is a basis for $\mathbb{R}^n$. Moreover, Lemma 1 and Lemma 3 tell us that a basis of $\mathbb{R}^n$ must contain n vectors. We now want to prove that every basis of a subspace S of $\mathbb{R}^n$ must contain the same number of vectors.

Lemma 6

Suppose that S is a non-trivial subspace of $\mathbb{R}^n$ and $\operatorname{Span}\{\vec{v}_1, \dots, \vec{v}_\ell\} = S$. If $\{\vec{u}_1, \dots, \vec{u}_k\}$ is a linearly independent set of vectors in S, then $k \leq \ell$.

Proof: Since each $\vec{u}_i$, $1 \leq i \leq k$, is a vector in S, by definition of spanning, it can be written as a linear combination of the vectors $\vec{v}_1, \dots, \vec{v}_\ell$. We get

$$\begin{aligned} \vec{u}_1 &= a_{11}\vec{v}_1 + a_{21}\vec{v}_2 + \cdots + a_{\ell 1}\vec{v}_\ell \\ \vec{u}_2 &= a_{12}\vec{v}_1 + a_{22}\vec{v}_2 + \cdots + a_{\ell 2}\vec{v}_\ell \\ &\vdots \\ \vec{u}_k &= a_{1k}\vec{v}_1 + a_{2k}\vec{v}_2 + \cdots + a_{\ell k}\vec{v}_\ell \end{aligned}$$

Consider the equation

$$\begin{aligned} \vec{0} &= t_1\vec{u}_1 + \cdots + t_k\vec{u}_k \\ &= t_1(a_{11}\vec{v}_1 + a_{21}\vec{v}_2 + \cdots + a_{\ell 1}\vec{v}_\ell) + \cdots + t_k(a_{1k}\vec{v}_1 + a_{2k}\vec{v}_2 + \cdots + a_{\ell k}\vec{v}_\ell) \\ &= (a_{11}t_1 + \cdots + a_{1k}t_k)\vec{v}_1 + \cdots + (a_{\ell 1}t_1 + \cdots + a_{\ell k}t_k)\vec{v}_\ell \end{aligned}$$

Since $\{\vec{v}_1, \dots, \vec{v}_\ell\}$ is linearly independent, the only solution to this equation is

$$\begin{aligned} a_{11}t_1 + \cdots + a_{1k}t_k &= 0 \\ \vdots \qquad & \quad \vdots \\ a_{\ell 1}t_1 + \cdots + a_{\ell k}t_k &= 0 \end{aligned}$$

This gives a homogeneous system of ℓ equations in the k unknowns $t_1, \dots, t_k$. If $k > \ell$, then this system would have a non-trivial solution, which would imply that $\{\vec{u}_1, \dots, \vec{u}_k\}$ is linearly dependent. But we assumed that $\{\vec{u}_1, \dots, \vec{u}_k\}$ is linearly independent, so we must have $k \leq \ell$. ■

Theorem 7

If $\{\vec{v}_1, \dots, \vec{v}_\ell\}$ and $\{\vec{u}_1, \dots, \vec{u}_k\}$ are both bases of a non-trivial subspace S of $\mathbb{R}^n$, then $k = \ell$.

Proof: We know that $\{\vec{v}_1, \dots, \vec{v}_\ell\}$ is a basis for S, so it is linearly independent. Also, $\{\vec{u}_1, \dots, \vec{u}_k\}$ is a basis for S, so $\mathrm{Span}\{\vec{u}_1, \dots, \vec{u}_k\} = S$. Thus, by Lemma 6, we get $\ell \leq k$. Similarly, $\{\vec{u}_1, \dots, \vec{u}_k\}$ is linearly independent as it is a basis for S, and $\mathrm{Span}\{\vec{v}_1, \dots, \mathbf{v}_\ell\} = S$, so $\ell \geq k$. Therefore, $\ell = k$, as required. ■

This theorem justifies the following definition.

Definition
Dimension

If S is a non-trivial subspace of $\mathbb{R}^n$ with a basis containing k vectors, then we say that the **dimension** of S is k and write

$$\dim S = k$$

So that we can talk about the basis of any subspace of $\mathbb{R}^n$, we define the empty set to be a basis for the trivial subspace $\{\vec{0}\}$ of $\mathbb{R}^n$ and thus say that the dimension of the trivial vector space is 0.

EXAMPLE 9

By definition, a plane in $\mathbb{R}^n$ that passes through the origin is spanned by a set of two linearly independent vectors. Thus, such a plane is 2-dimensional since every basis of the plane will have two vectors. This result fits with our geometric understanding of a plane.

PROBLEMS 2.3

Practice Problems

A1 Let $B = \left\{ \begin{bmatrix} 1 \\ 0 \\ 1 \\ 1 \end{bmatrix}, \begin{bmatrix} 2 \\ 1 \\ 0 \\ 1 \end{bmatrix}, \begin{bmatrix} -1 \\ 1 \\ 2 \\ 1 \end{bmatrix} \right\}$. For each of the following vectors, either express it as a linear combination of the vectors of B or show that it is not a vector in Span B. (a) $\begin{bmatrix} -3 \\ 2 \\ 8 \\ 4 \end{bmatrix}$ (b) $\begin{bmatrix} 5 \\ 4 \\ 6 \\ 7 \end{bmatrix}$ (c) $\begin{bmatrix} 2 \\ -2 \\ 1 \\ 1 \end{bmatrix}$

A2 Let $B = \left\{ \begin{bmatrix} 1 \\ -1 \\ 1 \\ 0 \end{bmatrix}, \begin{bmatrix} -1 \\ 1 \\ 0 \\ 2 \end{bmatrix}, \begin{bmatrix} 1 \\ 1 \\ -1 \\ -1 \end{bmatrix} \right\}$. For each of the following vectors, either express it as a linear combination of the vectors of B or show that it is not a vector in Span B.

(a) $\begin{bmatrix} 3 \\ 2 \\ -1 \\ -1 \end{bmatrix}$ (b) $\begin{bmatrix} -7 \\ 3 \\ 0 \\ 8 \end{bmatrix}$ (c) $\begin{bmatrix} 1 \\ 1 \\ 1 \\ 1 \end{bmatrix}$

A3 Using the procedure in Example 3, find a homogeneous system that defines the given set.

(a) Span $\left\{ \begin{bmatrix} 1 \\ 0 \\ 0 \end{bmatrix}, \begin{bmatrix} 0 \\ 1 \\ 0 \end{bmatrix} \right\}$

(b) Span $\left\{ \begin{bmatrix} 2 \\ 1 \\ 0 \end{bmatrix} \right\}$

(c) Span $\left\{ \begin{bmatrix} 1 \\ 1 \\ 2 \end{bmatrix}, \begin{bmatrix} 3 \\ -1 \\ 0 \end{bmatrix} \right\}$

(d) Span $\left\{ \begin{bmatrix} 2 \\ 1 \\ -1 \end{bmatrix}, \begin{bmatrix} 1 \\ -2 \\ 1 \end{bmatrix} \right\}$

(e) Span $\left\{ \begin{bmatrix} 1 \\ 0 \\ 1 \\ 0 \end{bmatrix}, \begin{bmatrix} 2 \\ -1 \\ 1 \\ 1 \end{bmatrix} \right\}$

(f) Span $\left\{ \begin{bmatrix} 1 \\ -1 \\ 1 \\ 2 \end{bmatrix}, \begin{bmatrix} 0 \\ 1 \\ 3 \\ -2 \end{bmatrix}, \begin{bmatrix} -2 \\ 0 \\ 4 \\ -3 \end{bmatrix} \right\}$

A4 Using the procedure in Example 8, determine whether the given set is a basis for the given plane or hyperplane.

(a) $\left\{ \begin{bmatrix} 1 \\ 1 \\ 2 \end{bmatrix}, \begin{bmatrix} 1 \\ 0 \\ 1 \end{bmatrix} \right\}$ for $x_1 + x_2 - x_3 = 0$

(b) $\left\{ \begin{bmatrix} 1 \\ 1 \\ 1 \end{bmatrix}, \begin{bmatrix} 1 \\ 2 \\ -3 \end{bmatrix} \right\}$ for $2x_1 - 3x_2 + x_3 = 0$

(c) $\left\{ \begin{bmatrix} 1 \\ 1 \\ 0 \\ 0 \end{bmatrix}, \begin{bmatrix} 0 \\ 0 \\ 1 \\ 1 \end{bmatrix}, \begin{bmatrix} 3 \\ 1 \\ 0 \\ 1 \end{bmatrix} \right\}$ for $x_1 - x_2 + 2x_3 - 2x_4 = 0$

A5 Determine whether the following sets are linearly independent. If the set is linearly dependent, find all linear combinations of the vectors that equal the zero vector.

(a) $\left\{ \begin{bmatrix} 1 \\ 2 \\ 1 \\ -1 \end{bmatrix}, \begin{bmatrix} 1 \\ 2 \\ 3 \\ 1 \end{bmatrix}, \begin{bmatrix} 1 \\ -3 \\ 2 \\ 1 \end{bmatrix} \right\}$

(b) $\left\{ \begin{bmatrix} 1 \\ 0 \\ 1 \\ 0 \end{bmatrix}, \begin{bmatrix} 0 \\ 1 \\ 1 \\ 1 \end{bmatrix}, \begin{bmatrix} 0 \\ 0 \\ 1 \\ 1 \end{bmatrix}, \begin{bmatrix} 3 \\ 2 \\ 6 \\ 3 \end{bmatrix} \right\}$

(c) $\left\{ \begin{bmatrix} 1 \\ 1 \\ 0 \\ 1 \\ 1 \end{bmatrix}, \begin{bmatrix} 2 \\ 3 \\ 1 \\ 3 \\ 3 \end{bmatrix}, \begin{bmatrix} 0 \\ 1 \\ 1 \\ 1 \\ 1 \end{bmatrix} \right\}$

A6 Determine all values of k such that the given set is linearly independent.

(a) $\left\{ \begin{bmatrix} 1 \\ 0 \\ 1 \\ 0 \end{bmatrix}, \begin{bmatrix} 0 \\ 1 \\ 1 \\ 1 \end{bmatrix}, \begin{bmatrix} 2 \\ -3 \\ -1 \\ k \end{bmatrix} \right\}$

(b) $\left\{ \begin{bmatrix} 1 \\ 1 \\ 1 \\ 2 \end{bmatrix}, \begin{bmatrix} 1 \\ -1 \\ 2 \\ 0 \end{bmatrix}, \begin{bmatrix} -1 \\ 2 \\ k \\ 1 \end{bmatrix} \right\}$

A7 Determine whether the given set is a basis for $\mathbb{R}^3$.

(a) $\left\{\begin{bmatrix}1\\1\\2\end{bmatrix},\begin{bmatrix}1\\-1\\-1\end{bmatrix},\begin{bmatrix}2\\1\\1\end{bmatrix}\right\}$

(b) $\left\{\begin{bmatrix}-2\\2\\1\end{bmatrix},\begin{bmatrix}3\\-1\\2\end{bmatrix}\right\}$

(c) $\left\{\begin{bmatrix}1\\0\\1\end{bmatrix},\begin{bmatrix}-1\\2\\1\end{bmatrix},\begin{bmatrix}1\\3\\5\end{bmatrix},\begin{bmatrix}2\\-1\\-4\end{bmatrix}\right\}$

(d) $\left\{\begin{bmatrix}1\\-1\\1\end{bmatrix},\begin{bmatrix}1\\2\\-1\end{bmatrix},\begin{bmatrix}3\\0\\1\end{bmatrix}\right\}$

Homework Problems

B1 Let $B=\left\{\begin{bmatrix}1\\1\\0\\1\end{bmatrix},\begin{bmatrix}2\\0\\0\\2\end{bmatrix},\begin{bmatrix}0\\2\\-1\\1\end{bmatrix}\right\}$. For each of the following vectors, either express it as a linear combination of the vectors of B or show that it is not a vector in Span B.

(a) $\begin{bmatrix}-4\\-2\\2\\-6\end{bmatrix}$ (b) $\begin{bmatrix}6\\0\\0\\3\end{bmatrix}$ (c) $\begin{bmatrix}3\\-1\\2\\1\end{bmatrix}$

B2 Let $B=\left\{\begin{bmatrix}1\\2\\1\\0\end{bmatrix},\begin{bmatrix}1\\1\\3\\1\end{bmatrix},\begin{bmatrix}1\\-1\\1\\-1\end{bmatrix}\right\}$. For each of the following vectors, either express it as a linear combination of the vectors of B or show that it is not a vector in Span B.

(a) $\begin{bmatrix}0\\1\\4\\2\end{bmatrix}$ (b) $\begin{bmatrix}6\\4\\10\\3\end{bmatrix}$ (c) $\begin{bmatrix}2\\3\\1\\-1\end{bmatrix}$

B3 Using the procedure in Example 3, find a homogeneous system that defines the given set.

(a) Span $\left\{\begin{bmatrix}1\\-1\\2\end{bmatrix},\begin{bmatrix}1\\2\\5\end{bmatrix}\right\}$

(b) Span $\left\{\begin{bmatrix}1\\-3\\1\end{bmatrix}\right\}$

(c) Span $\left\{\begin{bmatrix}2\\1\\-1\end{bmatrix},\begin{bmatrix}-4\\2\\2\end{bmatrix},\begin{bmatrix}2\\3\\-1\end{bmatrix}\right\}$

(d) Span $\left\{\begin{bmatrix}1\\3\\0\\-2\end{bmatrix},\begin{bmatrix}0\\-1\\3\\4\end{bmatrix}\right\}$

(e) Span $\left\{\begin{bmatrix}1\\1\\1\\-1\end{bmatrix},\begin{bmatrix}-1\\3\\3\\1\end{bmatrix},\begin{bmatrix}0\\-2\\-5\\2\end{bmatrix}\right\}$

(f) Span $\left\{\begin{bmatrix}1\\0\\1\\-1\\0\end{bmatrix},\begin{bmatrix}0\\1\\2\\-5\\0\end{bmatrix},\begin{bmatrix}0\\0\\1\\-1\\1\end{bmatrix},\begin{bmatrix}0\\0\\0\\0\\3\end{bmatrix}\right\}$

B4 Using the procedure in Example 8, determine whether the given set is a basis for the given plane or hyperplane.

(a) $\left\{\begin{bmatrix}-1\\1\\1\end{bmatrix},\begin{bmatrix}1\\1\\-3\end{bmatrix}\right\}$ for $2x_1+x_2+x_3=0$

(b) $\left\{\begin{bmatrix}1\\-2\\1\end{bmatrix},\begin{bmatrix}1\\0\\4\end{bmatrix}\right\}$ for $4x_1+2x_2-x_3=0$

(c) $\left\{\begin{bmatrix}1\\1\\0\\1\end{bmatrix},\begin{bmatrix}1\\-1\\1\\0\end{bmatrix},\begin{bmatrix}0\\0\\1\\0\end{bmatrix}\right\}$ for $2x_1+3x_2-5x_4=0$

(d) $\left\{\begin{bmatrix}6\\1\\-3\\0\end{bmatrix},\begin{bmatrix}0\\3\\0\\1\end{bmatrix},\begin{bmatrix}-2\\0\\1\\1\end{bmatrix}\right\}$ for $x_1+2x_3=0$

B5 Determine whether the following sets are linearly independent. If the set is linearly dependent, find all linear combinations of the vectors that are $\vec{0}$.

(a) $\left\{\begin{bmatrix}1\\0\\1\\0\end{bmatrix},\begin{bmatrix}1\\2\\0\\1\end{bmatrix},\begin{bmatrix}0\\1\\2\\3\end{bmatrix}\right\}$

(b) $\left\{\begin{bmatrix}1\\0\\1\\1\\0\end{bmatrix},\begin{bmatrix}1\\2\\0\\0\\1\end{bmatrix},\begin{bmatrix}0\\1\\1\\2\\-2\end{bmatrix},\begin{bmatrix}1\\-3\\1\\0\\1\end{bmatrix}\right\}$

B6 Determine all values of k such that the given set is linearly independent.

(a) $\left\{\begin{bmatrix}1\\1\\2\\1\end{bmatrix}, \begin{bmatrix}-1\\1\\-1\\2\end{bmatrix}, \begin{bmatrix}3\\1\\5\\k\end{bmatrix}\right\}$

(b) $\left\{\begin{bmatrix}1\\-1\\3\\1\end{bmatrix}, \begin{bmatrix}-1\\1\\2\\1\end{bmatrix}, \begin{bmatrix}-1\\1\\k\\5\end{bmatrix}\right\}$

B7 Determine whether the given set is a basis for $\mathbb{R}^3$.

(a) $\left\{\begin{bmatrix}1\\-1\\3\end{bmatrix}, \begin{bmatrix}2\\1\\-1\end{bmatrix}, \begin{bmatrix}1\\2\\-4\end{bmatrix}\right\}$

(b) $\left\{\begin{bmatrix}1\\3\\1\end{bmatrix}, \begin{bmatrix}-1\\0\\1\end{bmatrix}, \begin{bmatrix}5\\1\\0\end{bmatrix}, \begin{bmatrix}0\\0\\3\end{bmatrix}\right\}$

(c) $\left\{\begin{bmatrix}1\\2\\1\end{bmatrix}, \begin{bmatrix}-1\\-1\\0\end{bmatrix}, \begin{bmatrix}1\\1\\7\end{bmatrix}\right\}$

(d) $\left\{\begin{bmatrix}1\\0\\1\end{bmatrix}, \begin{bmatrix}0\\1\\0\end{bmatrix}\right\}$

Computer Problems

C1 Let $B = \left\{\begin{bmatrix}2\\-1\\1\\2\\4\\-2\\1\end{bmatrix}, \begin{bmatrix}1\\-2\\1\\0\\6\\2\\3\end{bmatrix}, \begin{bmatrix}1\\0\\-1\\5\\-1\\8\\2\end{bmatrix}, \begin{bmatrix}3\\-1\\0\\-4\\3\\6\\3\end{bmatrix}, \begin{bmatrix}1\\0\\1\\-2\\1\\3\\5\end{bmatrix}\right\}$ and $\vec{v} = \begin{bmatrix}0\\1\\0\\0\\0\\0\\0\end{bmatrix}$.

(a) Determine whether B is linearly independent or dependent.

(b) Determine whether $\vec{v}$ is in Span B.

Conceptual Problems

D1 Let $B = \{\vec{e}_1, \dots, \vec{e}_n\}$ be the standard basis for $\mathbb{R}^n$. Prove that Span $B = \mathbb{R}^n$ and that B is linearly independent.

D2 Let $B = \{\vec{v}_1, \dots, \vec{v}_k\}$ be vectors in $\mathbb{R}^n$.

(a) Prove that if $k < n$, then there exists a vector $\vec{v} \in \mathbb{R}^n$ such that $\vec{v} \notin$ Span B.

(b) Prove that if $k > n$, then B must be linearly dependent.

(c) Prove that if $k = n$, then Span $B = \mathbb{R}^n$ if and only if B is linearly independent.

2.4 Applications of Systems of Linear Equations

Resistor Circuits in Electricity

The flow of electrical current in simple electrical circuits is described by simple linear laws. In an electrical circuit, the **current** has a direction and therefore has a sign attached to it; **voltage** is also a signed quantity; **resistance** is a positive scalar. The laws for electrical circuits are discussed next.

Ohm's Law If an electrical current of magnitude I amperes is flowing through a resistor with resistance R ohms, then the drop in the voltage across the resistor is $V = IR$, measured in volts. The filament in a light bulb and the heating element of an electrical heater are familiar examples of electrical resistors. (See Figure 2.4.4.)

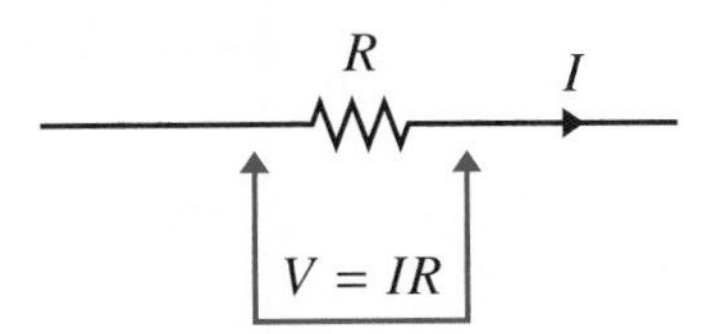

Figure 2.4.4 Ohm's law: the voltage across the resistor is $V = IR$.

Kirchhoff's Laws (1) At a node or junction where several currents enter, the signed sum of the currents entering the node is zero. (See Figure 2.4.5.)

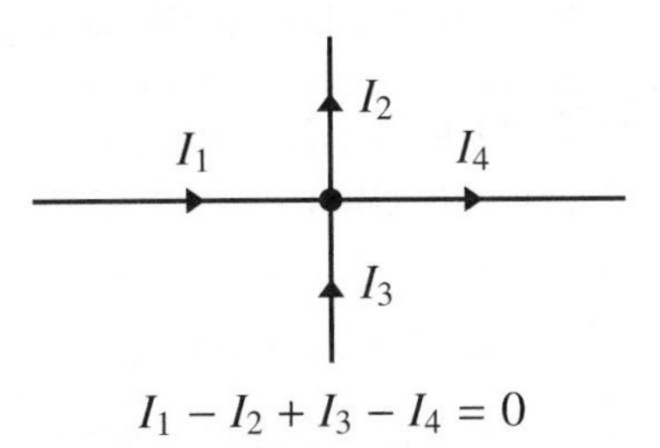

Figure 2.4.5 One of Kirchhoff's laws: $I_1 - I_2 + I_3 - I_4 = 0$.

(2) In a closed loop consisting of only resistors and an **electromotive force** E (for example, E might be due to a battery), the sum of the voltage drops across resistors is equal to E. (See Figure 2.4.6.)

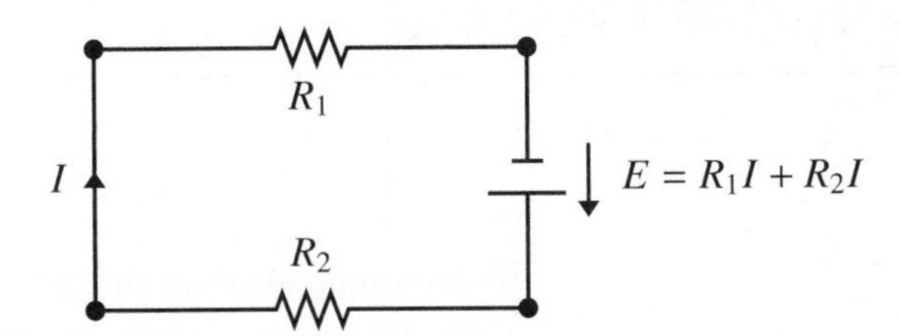

Figure 2.4.6 Kirchhoff's other law: $E = R_1 I + R_2 I$.

Note that we adopt the convention of drawing an arrow to show the direction of I or of E. These arrows can be assigned arbitrarily, and then the circuit laws will determine whether the quantity has a positive or negative sign. It is important to be consistent in using these assigned directions when you write down Kirchhoff's law for loops.

Sometimes it is necessary to determine the current flowing in each of the loops of a network of loops, as shown in Figure 2.4.7. (If the sources of electromotive force are distributed in various places, it will not be sufficient to deal with the problems as a collection of resistors "in parallel and/or in series.") In such problems, it is convenient to introduce the idea of the "current in the loop," which will be denoted i. The true current across any circuit element is given as the algebraic (signed) sum of the "loop currents" flowing through that circuit element. For example, in Figure 2.4.7, the circuit consists of four loops, and a loop current has been indicated in each loop. Across the resistor R_1 in the figure, the true current is simply the loop current i_1; however, across the resistor R_2, the true current (directed from top to bottom) is $i_1 - i_2$. Similarly, across R_4, the true current (from right to left) is $i_1 - i_3$.

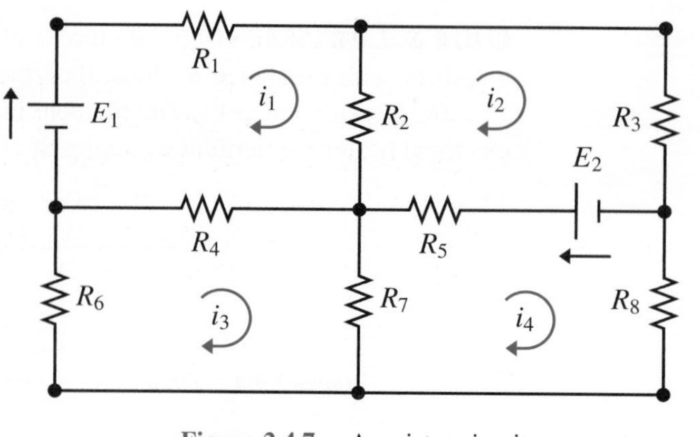

Figure 2.4.7 A resistor circuit.

The reason for introducing these loop currents for our present problem is that there are fewer loop currents than there are currents through individual elements. Moreover, Kirchhoff's law at the nodes is automatically satisfied, so we do not have to write nearly so many equations.

To determine the currents in the loops, it is necessary to use Kirchhoff's second law with Ohm's law describing the voltage drops across the resistors. For Figure 2.4.7, the resulting equations for each loop are:

- The top-left loop: $R_1i_1 + R_2(i_1 - i_2) + R_4(i_1 - i_3) = E_1$
- The top-right loop: $R_3i_2 + R_5(i_2 - i_4) + R_2(i_2 - i_1) = E_2$
- The bottom-left loop: $R_6i_3 + R_4(i_3 - i_1) + R_7(i_3 - i_4) = 0$
- The bottom-right loop: $R_8i_4 + R_7(i_4 - i_3) + R_5(i_4 - i_2) = -E_2$

Multiply out and collect terms to display the equations as a system in the variables i_1, i_2, i_3, and i_4. The augmented matrix of the system is

$$\left[\begin{array}{cccc|c} R_1 + R_2 + R_4 & -R_2 & -R_4 & 0 & E_1 \\ -R_2 & R_2 + R_3 + R_5 & 0 & -R_5 & E_2 \\ -R_4 & 0 & R_4 + R_6 + R_7 & -R_7 & 0 \\ 0 & -R_5 & -R_7 & R_5 + R_7 + R_8 & -E_2 \end{array}\right]$$

To determine the loop currents, this augmented matrix must be reduced to row echelon form. There is no particular purpose in finding an explicit solution for this general problem, and in a linear algebra course, there is no particular value in plugging in particular values for E_1, E_2, and the seven resistors. Instead, the point of this example is to show that even for a fairly simple electrical circuit with the most basic elements (resistors), the analysis requires you to be competent in dealing with large systems of linear equations. Systematic, efficient methods of solution are essential.

Obviously, as the number of loops in the network grows, so does the number of variables and so does the number of equations. For larger systems, it is important to know whether you have the correct number of equations to determine the unknowns. Thus, the theorems in Sections 2.1 and 2.2, the idea of rank, and the idea of linear independence are all important.

The Moral of This Example Linear algebra is an essential tool for dealing with large systems of linear equations that may arise in dealing with circuits; really interesting examples cannot be given without assuming greater knowledge of electrical circuits and their components.

Planar Trusses

It is common to use trusses, such as the one shown in Figure 2.4.8, in construction. For example, many bridges employ some variation of this design. When designing such structures, it is necessary to determine the **axial forces** in each **member** of the structure (that is, the force along the long axis of the member). To keep this simple, only two-dimensional trusses with hinged joints will be considered; it will be assumed that any displacements of the joints under loading are small enough to be negligible.

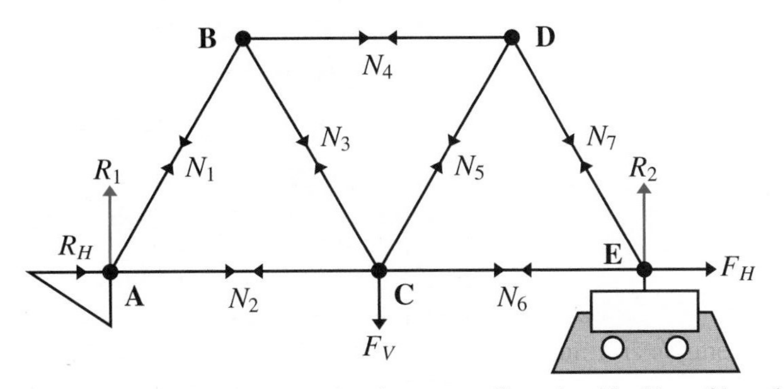

Figure 2.4.8 A planar truss. All triangles are equilateral, with sides of length s.

The external loads (such as vehicles on a bridge, or wind or waves) are assumed to be given. The reaction forces at the supports (shown as R_1, R_2, and R_H in the figure) are also external forces; these forces must have values such that the total external force on the structure is zero. To get enough information to design a truss for a particular application, we must determine the forces in the members under various loadings. To illustrate the kinds of equations that arise, we shall consider only the very simple case of a vertical force F_V acting at C and a horizontal force F_H acting at E. Notice that in this figure, the right-hand end of the truss is allowed to undergo small horizontal displacements; it turns out that if a reaction force were applied here as well, the equations would not uniquely determine all the unknown forces (the structure would be "statically indeterminate"), and other considerations would have to be introduced.

The geometry of the truss is assumed given: here it will be assumed that the triangles are equilateral, with all sides equal to s metres.

First consider the equations that indicate that the total force on the structure is zero and that the total moment about some convenient point due to those forces is zero. Note that the axial force along the members does not appear in this first set of equations.

- Total horizontal force: $R_H + F_H = 0$
- Total vertical force: $R_1 + R_2 - F_V = 0$
- Moment about A: $-F_V(s) + R_2(2s) = 0$, so $R_2 = \frac{1}{2}F_V = R_1$

Next, we consider the system of equations obtained from the fact that the sum of the forces at each joint must be zero. The moments are automatically zero because the forces along the members act through the joints.

At a joint, each member at that joint exerts a force in the direction of the axis of the member. It will be assumed that each member is in tension, so it is "pulling" away from the joint; if it were compressed, it would be "pushing" at the joint. As indicated in the figure, the force exerted on this joint A by the upper-left-hand member has magnitude N_1; with the conventions that forces to the right are positive and forces up are positive, the force vector exerted by this member on the joint A is $\begin{bmatrix} N_1/2 \\ \sqrt{3}N_1/2 \end{bmatrix}$. On the joint B, the same member will exert a force $\begin{bmatrix} -N_1/2 \\ -\sqrt{3}N_1/2 \end{bmatrix}$. If N_1 is positive, the force is a tension force; if N_1 is negative, there is compression.

For each of the joints A, B, C, D, and E, there are two equations—the first for the sum of horizontal forces and the second for the sum of the vertical forces:

$$
\begin{array}{llllllll}
A1 & N_1/2 & +N_2 & & & & +R_H = & 0 \\
A2 & \sqrt{3}N_1/2 & & & & & +R_1 = & 0 \\
B1 & -N_1/2 & +N_3/2 & +N_4 & & & = & 0 \\
B2 & -\sqrt{3}N_1/2 & -\sqrt{3}N_3/2 & & & & = & 0 \\
C1 & & -N_2 - N_3/2 & +N_5/2 & +N_6 & & = & 0 \\
C2 & & \sqrt{3}N_3/2 & +\sqrt{3}N_5/2 & & & = & F_V \\
D1 & & & -N_4 - N_5/2 & & +N_7/2 & = & 0 \\
D2 & & & -\sqrt{3}N_5/2 & & -\sqrt{3}N_7/2 & = & 0 \\
E1 & & & & -N_6 - N_7/2 & & = & -F_H \\
E2 & & & & & \sqrt{3}N_7/2 & +R_2 = & 0
\end{array}
$$

Notice that if the reaction forces are treated as unknowns, this is a system of 10 equations in 10 unknowns. The geometry of the truss and its supports determines the coefficient matrix of this system, and it could be shown that the system is necessarily consistent with a unique solution. Notice also that if the horizontal force equations (A1, B1, C1, D1, and E1) are added together, the sum is the total horizontal force equation, and similarly the sum of the vertical force equations is the total vertical force equation. A suitable combination of the equations would also produce the moment equation, so if those three equations are solved as above, then the 10 joint equations will still be a consistent system for the remaining 7 axial force variables.

For this particular truss, the system of equations is quite easy to solve, since some of the variables are already leading variables. For example, if $F_H = 0$, from A2 and E2 it follows that $N_1 = N_7 = -\frac{1}{\sqrt{3}}F_V$ and then B2, C2, and D2 give $N_3 = N_5 = \frac{1}{\sqrt{3}}F_V$; then A1 and E1 imply that $N_2 = N_6 = \frac{1}{2\sqrt{3}}F_V$, and B1 implies that $N_4 = -\frac{1}{\sqrt{3}}F_V$. Note that the members AC, BC, CD, and CE are under tension, and AB, BD, and DE experience compression, which makes intuitive sense.

This is a particularly simple truss. In the real world, trusses often involve many more members and use more complicated geometry; trusses may also be three-dimensional. Therefore, the systems of equations that arise may be considerably larger and more complicated. It is also sometimes essential to introduce considerations other than the equations of equilibrium of forces in statics. To study these questions, you need to know the basic facts of linear algebra.

It is worth noting that in the system of equations above, each of the quantities N_1, $N_2, \ldots, N_7$ appears with a non-zero coefficient in only some of the equations. Since each member touches only two joints, this sort of special structure will often occur in the equations that arise in the analysis of trusses. A deeper knowledge of linear algebra is important in understanding how such special features of linear equations may be exploited to produce efficient solution methods.

Linear Programming

Linear programming is a procedure for deciding the best way to allocate resources. "Best" may mean fastest, most profitable, least expensive, or best by whatever criterion is appropriate. For linear programming to be applicable, the problem must have some special features. These will be illustrated by an example.

In a primitive economy, a man decides to earn a living by making hinges and gate latches. He is able to obtain a supply of 25 kilograms a week of suitable metal at a price of 2 cowrie shells per kilogram. His design requires 500 grams to make a hinge and 250 grams to make a gate latch. With his primitive tools, he finds that he can make a hinge in 1 hour, and it takes 3/4 hour to make a gate latch. He is willing to work 60 hours a week. The going price is 3 cowrie shells for a hinge and 2 cowrie shells for a gate latch. How many hinges and how many gate latches should he produce each week in order to maximize his net income?

To analyze the problem, let x be the number of hinges produced per week and let y be the number of gate latches. Then the amount of metal used is $(0.5x + 0.25y)$ kilograms. Clearly, this must be less than or equal to 25 kilograms:

$$0.5x + 0.25y \leq 25$$

or

$$2x + y \leq 100$$

Such an inequality is called a **constraint** on x and y; it is a linear constraint because the corresponding equation is linear.

Our producer also has a time constraint: the time taken making hinges plus the time taken making gate latches cannot exceed 60 hours. Therefore,

$$1x + 0.75y \leq 60$$

or

$$4x + 3y \leq 240$$

Obviously, also $x \geq 0$ and $y \geq 0$.

The producer's net revenue for selling x hinges and y gate latches is $R(x, y) = 3x + 2y - 2(25)$ cowrie shells. This is called the **objective function** for the problem. The mathematical problem can now be stated as follows: Find the point (x, y) that maximizes the objective function $R(x, y) = 3x + 2y - 50$, subject to the linear constraints $x \geq 0$, $y \geq 0$, $2x + y \leq 100$, and $4x + 3y \leq 240$.

This is a **linear programming problem** because it asks for the maximum (or minimum) of a *linear* objective function, subject to *linear* constraints. It is useful to introduce one piece of special vocabulary: the **feasible** set for the problem is the set of (x, y) satisfying all of the constraints. The solution procedure relies on the fact that the feasible set for a linear programming problem has a special kind of shape. (See Figure 2.4.9 for the feasible set for this particular problem.) *Any line that meets the feasible set either meets the set in a single line segment or only touches the set on its boundary.* In particular, because of the way the feasible set is defined in terms of linear inequalities, it turns out that it is impossible for one line to meet the feasible set in two separate pieces.

For example, the shaded region if Figure 2.4.10 cannot possibly be the feasible set for a linear programming problem, because some lines meet the region in two line segments. (This property of feasible sets is not difficult to prove, but since this is only a brief illustration, the proof is omitted.)

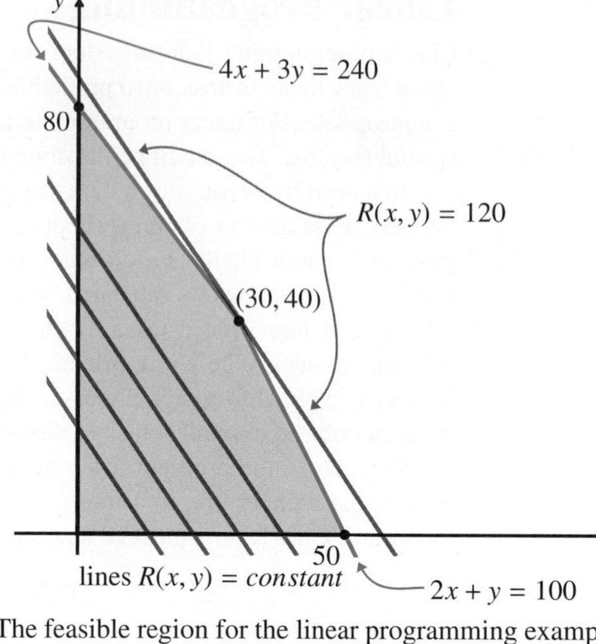

Figure 2.4.9 The feasible region for the linear programming example. The grey lines are level sets of the objective function R.

Now consider sets of the form $R(x, y) = k$, where k is a constant; these are called the **level sets** of R. These sets obviously form a family of parallel lines, and some of them are shown in Figure 2.4.9. Choose some point in the feasible set: check that $(20, 20)$ is such a point. Then the line

$$R(x, y) = R(20, 20) = 50$$

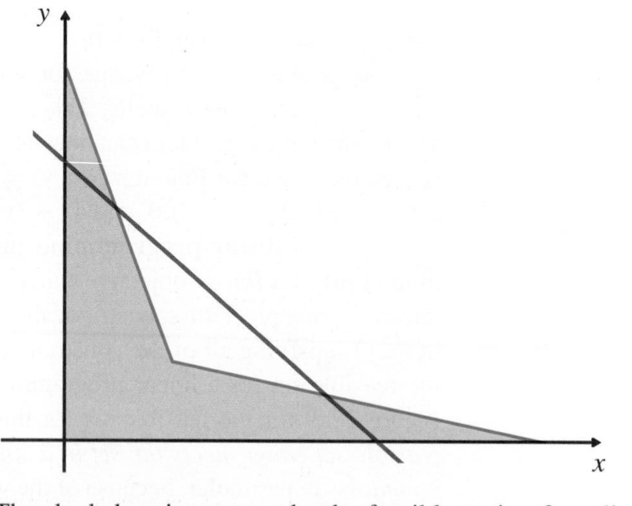

Figure 2.4.10 The shaded region cannot be the feasible region for a linear programming problem because it meets a line in two segments.

meets the feasible set in a line segment. $(30, 30)$ is also a feasible point (check), and

$$R(x, y) = R(30, 30) = 100$$

also meets the feasible set in a line segment. You can tell that $(30, 30)$ is not a boundary point of the feasible set because it satisfies all the constraints with strict inequality; boundary points must satisfy one of the constraints with equality.

As we move further from the origin into the first quadrant, $R(x, y)$ increases. The biggest possible value for $R(x, y)$ will occur at a point where the set $R(x, y) = k$ (for some constant k to be determined) just touches the feasible set. For larger values of $R(x, y)$, the set $R(x, y) = k$ does not meet the feasible set at all, so there are no feasible points that give such bigger values of R. The touching must occur at a **vertex**—that is, at an intersection point of two of the boundary lines. (In general, the line $R(x, y) = k$ for the largest possible constant could touch the feasible set along a line segment that makes up part of the boundary. But such a line segment has two vertices as endpoints, so it is correct to say that the touching occurs at a vertex.)

For this particular problem, the vertices of the feasible set are easily found to be $(0, 0)$, $(50, 0)$, and $(0, 80)$, and the solution of the system of equations is

$$\begin{aligned} 2x + y &= 100 \\ 4x + 3y &= 240 \end{aligned}$$

For this particular problem, the vertices of the feasible set are $(0, 0)$, $(50, 0)$, $(0, 80)$, and $(30, 80)$. Now compare the values of $R(x, y)$ at all of these vertices: $R(0, 0)$, $R(50, 0)$, $R(0, 80) = 110$, and $R(30, 40) = 120$. The vertex $(30, 40)$ gives the best net revenue, so the producer should make 30 hinges and 40 gate latches each week.

General Remarks Problems involving allocation of resources are common in business and government. Problems such as scheduling ship transits through a canal can be analyzed this way. Oil companies must make choices about the grades of crude oil to use in their refineries, and about the amounts of various refined products to produce. Such problems often involve tens or even hundreds of variables—and similar numbers of constraints. The boundaries of the feasible set are hyperplanes in some $\mathbb{R}^n$, where n is large. Although the basic principles of the solution method remain the same as in this example (look for the best vertex), the problem is much more complicated because there are so many vertices. In fact, it is a challenge to find vertices; simply solving all possible combinations of systems of boundary equations is not good enough. Note in the simple two-dimensional example that the point $(60, 0)$ is the intersection point of two of the lines ($y = 0$ and $4x + 3y = 240$) that make up the boundary, but it is not a vertex of the feasible region because it fails to satisfy the constraint $2x + y \leq 100$. For higher-dimension problems, drawing pictures is not good enough, and an organized approach is called for.

The standard method for solving linear programming problems has been the **simplex method**, which finds an initial vertex and then prescribes a method (very similar to row reduction) for moving to another vertex, improving the value of the objective function with each step.

Again, it has been possible to hint at major application areas for linear algebra, but to pursue one of these would require the development of specialized mathematical tools and information from specialized disciplines.

PROBLEMS 2.4

Practice Problems

A1 Determine the system of equations for the reaction forces and axial forces in members for the truss shown in the diagram.

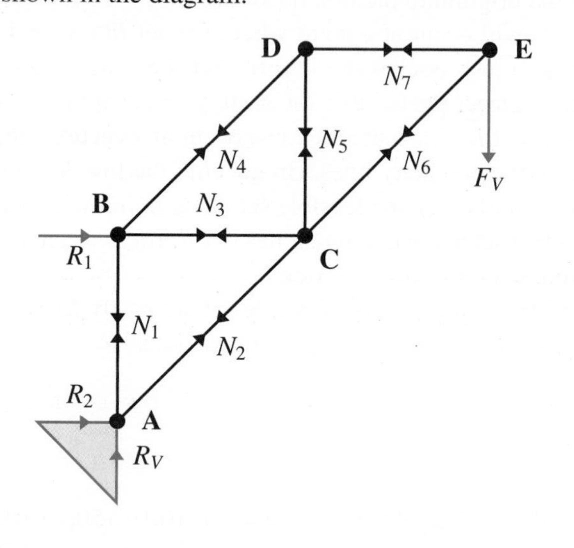

A2 Determine the augmented matrix of the system of linear equations, and determine the loop currents indicated in the diagram.

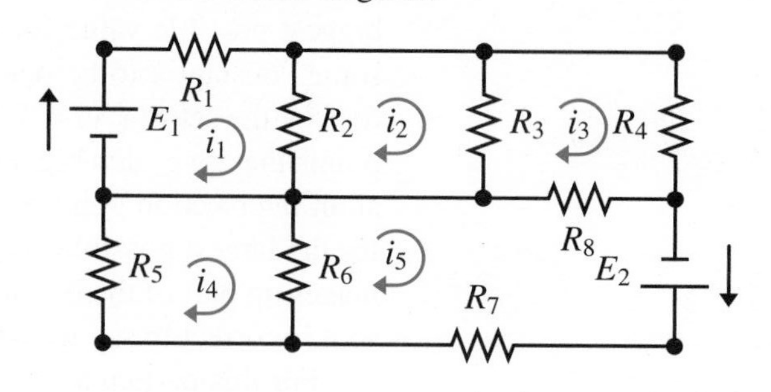

A3 Find the maximum value of the objective function $x+y$ subject to the constraints $0 \leq x \leq 100, 0 \leq y \leq 80$, and $4x + 5y \leq 600$. Sketch the feasible region.

CHAPTER REVIEW

Suggestions for Student Review

1 Explain why elimination works as a method for solving systems of linear equations. (Section 2.1)

2 When you row reduce an augmented matrix $\left[A \mid \vec{b}\right]$ to solve a system of linear equations, why can you stop when the matrix is in row echelon form? How do you use this form to decide if the system is consistent and if it has a unique solution? (Section 2.1)

3 How is reduced row echelon form different from row echelon form? (Section 2.2)

4 (a) Write the augmented matrix of a consistent non-homogeneous system of three linear equations in four variables, such that the coefficient matrix is in row echelon form (but not reduced row echelon form) and of rank 3.

(b) Determine the general solution of your system.

(c) Perform the following sequence of elementary row operations on your augmented matrix:

(i) Interchange the first and second rows.

(ii) Add the (new) second row to the first row.

(iii) Add twice the second row to the third row.

(iv) Add the third row to the second.

(d) Regard the result of (c) as the augmented matrix of a system and solve that system directly. (Don't just use the reverse operation in (c).) Check that your general solution agrees with (b).

5 For homogeneous systems, how can you use the row echelon form to determine whether there are non-trivial solutions and, if there are, how many parameters there are in the general solution? Is there any case where we know (by inspection) that a homogeneous system has non-trivial solutions? (Section 2.2)

6 Write a short explanation of how you use information about consistency of systems and uniqueness of solutions in testing for linear independence and in determining whether a vector $\vec{x}$ belongs to a given subspace of $\mathbb{R}^n$. (Section 2.3)

7 Explain how to determine whether a set of vectors $\{\vec{v}_1, \dots, \vec{v}_k\}$ in $\mathbb{R}^n$ is both linearly independent and a spanning set for a subspace S of $\mathbb{R}^n$. What form must the reduced row echelon form of the coefficient matrix of the vector equation $t_1\vec{v}_1 + \cdots + t_k\vec{v}_k = \vec{b}$ have if the set is a linearly independent spanning set? (Section 2.3)

Chapter Quiz

E1 Determine whether the following system is consistent by row reducing its augmented matrix:

$$\begin{aligned} x_2 - 2x_3 + x_4 &= 2 \\ 2x_1 - 2x_2 + 4x_3 - x_4 &= 10 \\ x_1 - x_2 + x_3 &= 2 \\ x_1 + x_3 &= 9 \end{aligned}$$

If it is consistent, determine the general solution. Show your steps clearly.

E2 Find a matrix in reduced row echelon form that is row equivalent to

$$A = \begin{bmatrix} 0 & 3 & 3 & 0 & -1 \\ 1 & 1 & 3 & 3 & 1 \\ 2 & 4 & 9 & 6 & 1 \\ -2 & -4 & -6 & -3 & -1 \end{bmatrix}$$

Show your steps clearly.

E3 The matrix $A = \left[\begin{array}{cccc|c} 1 & 2 & 3 & a & 2 \\ 0 & 2 & 1 & 0 & -3 \\ 0 & 0 & b+2 & 0 & b \\ 0 & 0 & 0 & c^2-1 & c+1 \end{array}\right]$ is the augmented matrix of a system of linear equations.

(a) Determine all values of (a, b, c) such that the system is consistent and all values of (a, b, c) such that the system is inconsistent.

(b) Determine all values of (a, b, c) such that the system has a unique solution.

E4 (a) Determine all vectors in $\mathbb{R}^5$ that are orthogonal to $\begin{bmatrix} 1 \\ 1 \\ 3 \\ 1 \\ 4 \end{bmatrix}, \begin{bmatrix} 2 \\ 1 \\ 5 \\ 0 \\ 0 \end{bmatrix}$, and $\begin{bmatrix} 3 \\ 2 \\ 8 \\ 5 \\ 9 \end{bmatrix}$.

(b) Let $\vec{u}$, $\vec{v}$, and $\vec{w}$ be three vectors in $\mathbb{R}^5$. Explain why there must be non-zero vectors orthogonal to all of $\vec{u}$, $\vec{v}$, and $\vec{w}$.

E5 Determine whether $\mathcal{B} = \left\{ \begin{bmatrix} 3 \\ 1 \\ 2 \end{bmatrix}, \begin{bmatrix} 1 \\ 1 \\ 6 \end{bmatrix}, \begin{bmatrix} 4 \\ 1 \\ 5 \end{bmatrix} \right\}$ is a basis for $\mathbb{R}^3$.

E6 Indicate whether the following statements are true or false. In each case, justify your answer with a brief explanation or counterexample.

(a) A consistent system must have a unique solution.

(b) If there are more equations than variables in a non-homogeneous system of linear equations, then the system must be inconsistent.

(c) Some homogeneous systems of linear equations have unique solutions.

(d) If there are more variables than equations in a system of linear equations, then the system cannot have a unique solution.

Further Problems

These problems are intended to be challenging. They may not be of interest to all students.

F1 The purpose of this exercise is to explore the relationship between the general solution of the system $\left[A \mid \vec{b}\right]$ and the general solution of the **corresponding homogeneous** system $\left[A \mid \vec{0}\right]$. This relation will be studied with different tools in Section 3.4. We begin by considering some examples where the coefficient matrix is in reduced row echelon form.

(a) Let $R = \begin{bmatrix} 1 & 0 & r_{13} \\ 0 & 1 & r_{23} \end{bmatrix}$. Show that the general solution of the homogeneous system $\left[R \mid \vec{0}\right]$ is

$$\vec{x}_H = t\vec{v}, \quad t \in \mathbb{R}$$

where $\vec{v}$ is expressed in terms of r_{13} and r_{23}. Show that the general solution of the non-homogeneous system $[R \mid \vec{c}\,]$ is

$$\vec{x}_N = \vec{p} + \vec{x}_H$$

where $\vec{p}$ is expressed in terms of $\vec{c}$, and $\vec{x}_H$ is as above.

(b) Let $R = \begin{bmatrix} 1 & r_{12} & 0 & 0 & r_{15} \\ 0 & 0 & 1 & 0 & r_{25} \\ 0 & 0 & 0 & 1 & r_{35} \end{bmatrix}$. Show that the general solution of the homogeneous system $\left[R \mid \vec{0}\right]$ is

$$\vec{x}_H = t_1\vec{v}_1 + t_2\vec{v}_2, \quad t_i \in \mathbb{R}$$

where each of $\vec{v}_1$ and $\vec{v}_2$ can be expressed in terms of the entries r_{ij}. Express each $\vec{v}_i$ explicitly. Then show that the general solution of $[R \mid \vec{c}\,]$ can be written as

$$\vec{x}_N = \vec{p} + \vec{x}_H$$

where $\vec{p}$ is expressed in terms of the components $\vec{c}$, and $\vec{x}_H$ is the solution of the corresponding homogeneous system.

The pattern should now be apparent; if it is not, try again with another special case of R. In the next part of this exercise, create an effective labelling system so that you can clearly indicate what you want to say.

(c) Let R be a matrix in reduced row echelon form, with m rows, n columns, and rank k. Show that the general solution of the homogeneous system $\left[R \mid \vec{0}\right]$ is

$$\vec{x}_H = t_1\vec{v}_1 + \cdots + t_{n-k}\vec{v}_{n-k}, \quad t_i \in \mathbb{R}$$

where each $\vec{v}_i$ is expressed in terms of the entries in R. Suppose that the system $[R \mid \vec{c}\,]$ is consistent and show that the general solution is

$$\vec{x}_N = \vec{p} + \vec{x}_H$$

where $\vec{p}$ is expressed in terms of the components of $\vec{c}$, and $\vec{x}_H$ is the solution of the corresponding homogeneous system.

(d) Use the result of (c) to discuss the relationship between the general solution of the consistent system $\left[A \mid \vec{b}\right]$ and the corresponding homogeneous system $\left[A \mid \vec{0}\right]$.

F2 *This problem involves comparing the efficiency of row reduction procedures.*

When we use a computer to solve large systems of linear equations, we want to keep the number of arithmetic operations as small as possible. This reduces the time taken for calculations, which is important in many industrial and commercial applications. It also tends to improve accuracy: every arithmetic operation is an opportunity to *lose* accuracy through truncation or round-off, subtraction of two nearly equal numbers, and so on.

We want to count the number of multiplications and/or divisions in solving a system by elimination. We focus on these operations because they are more time-consuming than addition or subtraction, and the number of additions is approximately the same as the number of multiplications. We make certain assumptions: the system $\left[A \mid \vec{b}\right]$ has n equations and n variables, and it is consistent with a unique solution. (Equivalently, A has n rows, n columns, and rank n.) We assume for simplicity that no row interchanges are required. (If row interchanges are required, they can be handled by renaming "addresses" in the computer.)

(a) How many multiplications and divisions are required to reduce $\left[A \mid \vec{b}\right]$ to a form $\left[C \mid \vec{d}\right]$ such that C is in row echelon form?

Hints

(1) To carry out the obvious first elementary row operation, compute $\frac{a_{21}}{a_{11}}$—one division. Since we know what will happen in the first column, we do not multiply a_{11} by $\frac{a_{21}}{a_{11}}$, but we must multiply every other element of the first row of $\left[A \mid \vec{b}\right]$ by this factor and subtract the product from the corresponding element of the second row—n multiplications.

(2) Obtain zeros in the remaining entries in the first column, then move to the $(n-1)$ by n blocks consisting of the reduced version of $\left[A \mid \vec{b}\right]$ with the first row and first column deleted.

(3) Note that $\sum_{i=1}^{n} i = \frac{n(n+1)}{2}$ and $\sum_{i=1}^{n} i^2 = \frac{n(n+1)(2n+1)}{6}$.

(4) The biggest term in your answer should be $n^3/3$. Note that n^3 is much greater than n^2 when n is large.

(b) Determine how many multiplications and divisions are required to solve the system with the augmented matrix $\left[C \mid \vec{d}\right]$ of part (a) by back-substitution.

(c) Show that the number of multiplications and divisions required to row reduce $[R \mid \vec{c}\,]$ to reduced row echelon form is the same as the number used in solving the system by back-substitution. Conclude that the Gauss-Jordan procedure is as efficient as Gaussian elimination with back-substitution. For large n, the number of multiplications and divisions is roughly $\frac{n^3}{3}$.

(d) Suppose that we do a "clumsy" Gauss-Jordan procedure. We do not first obtain row echelon form; instead we obtain zeros in all entries above and below a pivot before moving on to the next column. Show that the number of multiplications and divisions required in this procedure is roughly $\frac{n^3}{2}$, so that this procedure requires approximately 50% more operations than the more efficient procedures.

Visit the text's website at www.pearsoncanada.ca/norman for practice quizzes, additional applications, and an essay on linearity and superposition in physics.

CHAPTER 3

Matrices, Linear Mappings, and Inverses

CHAPTER OUTLINE

In many applications of linear algebra, we use vectors in $\mathbb{R}^n$ to represent quantities, such as forces, and then use the tools of Chapters 1 and 2 to solve various problems. However, there are many times when it is useful to translate a problem into other linear algebra objects. In this chapter, we look at two of these fundamental objects: matrices and linear mappings. We now explore the properties of these objects and show how they are tied together with the material from Chapters 1 and 2.

3.1 Operations on Matrices

We used matrices essentially as bookkeeping devices in Chapter 2. Matrices also possess interesting algebraic properties, so they have wider and more powerful applications than is suggested by their use in solving systems of equations. We now look at some of these algebraic properties.

Equality, Addition, and Scalar Multiplication of Matrices

We have seen that matrices are useful in solving systems of linear equations. However, we shall see that matrices show up in different kinds of problems, and *it is important to be able to think of matrices as "things" that are worth studying and playing with—and these things may have no connection with a system of equations.* A **matrix** is a rectangular array of numbers. A typical matrix has the form

$$A = \begin{bmatrix} a_{11} & a_{12} & \cdots & a_{1j} & \cdots & a_{1n} \\ a_{21} & a_{22} & \cdots & a_{2j} & \cdots & a_{2n} \\ \vdots & \vdots & & \vdots & & \vdots \\ a_{i1} & a_{i2} & \cdots & a_{ij} & \cdots & a_{in} \\ \vdots & \vdots & & \vdots & & \vdots \\ a_{m1} & a_{m2} & \cdots & a_{mj} & \cdots & a_{mn} \end{bmatrix}$$

We say that A is an $m \times n$ matrix when A has m rows and n columns. Two matrices A and B are **equal** if and only if they have the same size and their corresponding entries are equal. That is, if $a_{ij} = b_{ij}$ for $1 \leq i \leq m$, $1 \leq j \leq n$.

For now, we will consider only matrices whose entries a_{ij} are real numbers. We will look at matrices whose entries are complex numbers in Chapter 9.

Remark

We sometimes denote the ij-th entry of a matrix A by

$$(A)_{ij} = a_{ij}$$

This may seem pointless for a single matrix, but it is useful when dealing with multiple matrices.

Several special types of matrices arise frequently in linear algebra.

Definition Square Matrix

An $n \times n$ matrix (where the number of rows of the matrix is equal to the number of columns) is called a **square matrix**.

Definition Upper Triangular, Lower Triangular

A square matrix U is said to be **upper triangular** if the entries beneath the main diagonal are all zero—that is, $u_{ij} = 0$ whenever $i > j$. A square matrix L is said to be **lower triangular** if the entries above the main diagonal are all zero—in particular, $l_{ij} = 0$ whenever $i < j$.

EXAMPLE 1

The matrices $\begin{bmatrix} 2 & 3 \\ 0 & 1 \end{bmatrix}$ and $\begin{bmatrix} 3 & 1 & 2 \\ 0 & 0 & 2 \\ 0 & 0 & 1 \end{bmatrix}$ are upper triangular, while $\begin{bmatrix} -3 & 0 \\ 1 & 2 \end{bmatrix}$ and $\begin{bmatrix} 0 & 0 & 0 \\ -2 & 3 & 0 \\ 0 & -2 & 1 \end{bmatrix}$ are lower triangular. The matrix $\begin{bmatrix} 2 & 0 & 0 \\ 0 & 3 & 0 \\ 0 & 0 & 1 \end{bmatrix}$ is both upper and lower triangular.

Definition Diagonal Matrix

A matrix D that is both upper and lower triangular is called a **diagonal matrix**—that is, $d_{ij} = 0$ for all $i \neq j$. We denote an $n \times n$ diagonal matrix by

$$D = \text{diag}(d_{11}, d_{22}, \cdots, d_{nn})$$

EXAMPLE 2

We denote the diagonal matrix $D = \begin{bmatrix} \sqrt{3} & 0 \\ 0 & -2 \end{bmatrix}$ by $D = \text{diag}(\sqrt{3}, -2)$, while diag(0, 3, 1) is the diagonal matrix $\begin{bmatrix} 0 & 0 & 0 \\ 0 & 3 & 0 \\ 0 & 0 & 1 \end{bmatrix}$.

Also, we can think of a vector in $\mathbb{R}^n$ as an $n \times 1$ matrix, called a **column matrix**. For this reason, it makes sense to define operations on matrices to match those with vectors in $\mathbb{R}^n$.

Definition
Addition and Scalar Multiplication of Matrices

Let A and B be $m \times n$ matrices and $t \in \mathbb{R}$ a scalar. We define **addition** of matrices by

$$(A + B)_{ij} = (A)_{ij} + (B)_{ij}$$

and the **scalar multiplication** of matrices by

$$(tA)_{ij} = t(A)_{ij}$$

EXAMPLE 3

Perform the following operations.

(a) $\begin{bmatrix} 2 & 3 \\ 4 & 1 \end{bmatrix} + \begin{bmatrix} 5 & 1 \\ -2 & 7 \end{bmatrix}$

Solution: $\begin{bmatrix} 2 & 3 \\ 4 & 1 \end{bmatrix} + \begin{bmatrix} 5 & 1 \\ -2 & 7 \end{bmatrix} = \begin{bmatrix} 2+5 & 3+1 \\ 4+(-2) & 1+7 \end{bmatrix} = \begin{bmatrix} 7 & 4 \\ 2 & 8 \end{bmatrix}$

(b) $\begin{bmatrix} 3 & 0 \\ 1 & -5 \end{bmatrix} - \begin{bmatrix} 1 & -1 \\ -2 & 0 \end{bmatrix}$

Solution: $\begin{bmatrix} 3 & 0 \\ 1 & -5 \end{bmatrix} - \begin{bmatrix} 1 & -1 \\ -2 & 0 \end{bmatrix} = \begin{bmatrix} 3-1 & 0-(-1) \\ 1-(-2) & -5-0 \end{bmatrix} = \begin{bmatrix} 2 & 1 \\ 3 & -5 \end{bmatrix}$

(c) $5\begin{bmatrix} 2 & 3 \\ 4 & 1 \end{bmatrix}$

Solution: $5\begin{bmatrix} 2 & 3 \\ 4 & 1 \end{bmatrix} = \begin{bmatrix} 5(2) & 5(3) \\ 5(4) & 5(1) \end{bmatrix} = \begin{bmatrix} 10 & 15 \\ 20 & 5 \end{bmatrix}$

(d) $2\begin{bmatrix} 1 & 3 \\ 0 & -1 \end{bmatrix} + 3\begin{bmatrix} 4 & 0 \\ 1 & 2 \end{bmatrix}$

Solution: $2\begin{bmatrix} 1 & 3 \\ 0 & -1 \end{bmatrix} + 3\begin{bmatrix} 4 & 0 \\ 1 & 2 \end{bmatrix} = \begin{bmatrix} 2 & 6 \\ 0 & -2 \end{bmatrix} + \begin{bmatrix} 12 & 0 \\ 3 & 6 \end{bmatrix} = \begin{bmatrix} 2+12 & 6+0 \\ 0+3 & -2+6 \end{bmatrix} = \begin{bmatrix} 14 & 6 \\ 3 & 4 \end{bmatrix}$

Note that matrix addition is defined only if the matrices are the same size.

Properties of Matrix Addition and Multiplication by Scalars We now look at the properties of addition and scalar multiplication of matrices. It is very important to notice that these are the exact same ten properties discussed in Section 1.2 for addition and scalar multiplication of vectors in $\mathbb{R}^n$.

Theorem 1

Let A, B, C be $m \times n$ matrices and let s, t be real scalars. Then

(1) $A + B$ is an $m \times n$ matrix (closed under addition)
(2) $A + B = B + A$ (addition is commutative)
(3) $(A + B) + C = A + (B + C)$ (addition is associative)
(4) There exists a matrix, denoted by $O_{m,n}$, such that $A + O_{m,n} = A$; in particular, $O_{m,n}$ is the $m \times n$ matrix with all entries as zero and is called the **zero matrix** (zero vector)
(5) For each matrix A, there exists an $m \times n$ matrix $(-A)$, with the property that $A+(-A) = O_{m,n}$; in particular, $(-A)$ is defined by $(-A)_{ij} = -(A)_{ij}$ (additive inverses)
(6) sA is an $m \times n$ matrix (closed under scalar multiplication)
(7) $s(tA) = (st)A$ (scalar multiplication is associative)
(8) $(s + t)A = sA + tA$ (a distributive law)
(9) $s(A + B) = sA + sB$ (another distributive law)
(10) $1A = A$ (scalar multiplicative identity)

These properties follow easily from the definitions of addition and multiplication by scalars. The proofs are left to the reader.

Since we can now compute linear combinations of matrices, it makes sense to look at the set of all possible linear combinations of a set of matrices. And, as with vectors in $\mathbb{R}^n$, this goes hand-in-hand with the concept of linear independence. We mimic the definitions we had for vectors in $\mathbb{R}^n$.

Definition
Span

Let $\mathcal{B} = \{A_1, \ldots, A_k\}$ be a set of $m \times n$ matrices. Then the **span** of $\mathcal{B}$ is defined as

$$\operatorname{Span} \mathcal{B} = \{t_1A_1 + \cdots + t_kA_k \mid t_1, \ldots, t_k \in \mathbb{R}\}$$

Definition
Linearly Independent
Linearly Dependent

Let $\mathcal{B} = \{A_1, \ldots, A_\ell\}$ be a set of $m \times n$ matrices. Then $\mathcal{B}$ is said to be **linearly independent** if the only solution to the equation

$$t_1A_1 + \cdots + t_\ell A_\ell = O_{m,n}$$

is $t_1 = \cdots = t_\ell = 0$; otherwise, $\mathcal{B}$ is said to be **linearly dependent**.

EXAMPLE 4

Determine if $\begin{bmatrix} 1 & 2 \\ 3 & 4 \end{bmatrix}$ is in the span of

$$\mathcal{B} = \left\{ \begin{bmatrix} 1 & 1 \\ 0 & 0 \end{bmatrix}, \begin{bmatrix} 1 & 0 \\ 0 & 1 \end{bmatrix}, \begin{bmatrix} 0 & 1 \\ 1 & 0 \end{bmatrix}, \begin{bmatrix} 0 & 1 \\ 0 & 1 \end{bmatrix} \right\}$$

Solution: We want to find if there are $t_1, t_2, t_3,$ and t_4 such that

$$\begin{bmatrix} 1 & 2 \\ 3 & 4 \end{bmatrix} = t_1 \begin{bmatrix} 1 & 1 \\ 0 & 0 \end{bmatrix} + t_2 \begin{bmatrix} 1 & 0 \\ 0 & 1 \end{bmatrix} + t_3 \begin{bmatrix} 0 & 1 \\ 1 & 0 \end{bmatrix} + t_4 \begin{bmatrix} 0 & 1 \\ 0 & 1 \end{bmatrix} = \begin{bmatrix} t_1 + t_2 & t_1 + t_3 + t_4 \\ t_3 & t_2 + t_4 \end{bmatrix}$$

EXAMPLE 4
(continued)

Since two matrices are equal if and only if their corresponding entries are equal, this gives the system of linear equations

$$\begin{aligned} t_1 + t_2 &= 1 \\ t_1 + t_3 + t_4 &= 2 \\ t_3 &= 3 \\ t_2 + t_4 &= 4 \end{aligned}$$

Row reducing the augmented matrix gives

$$\left[\begin{array}{cccc|c} 1 & 1 & 0 & 0 & 1 \\ 1 & 0 & 1 & 1 & 2 \\ 0 & 0 & 1 & 0 & 3 \\ 0 & 1 & 0 & 1 & 4 \end{array}\right] \sim \left[\begin{array}{cccc|c} 1 & 0 & 0 & 0 & -2 \\ 0 & 1 & 0 & 0 & 3 \\ 0 & 0 & 1 & 0 & 3 \\ 0 & 0 & 0 & 1 & 1 \end{array}\right]$$

We see that the system is consistent. Therefore, $\begin{bmatrix} 1 & 2 \\ 3 & 4 \end{bmatrix}$ is in the span of $\mathcal{B}$. In particular, we have $t_1 = -2$, $t_2 = 3$, $t_3 = 3$, and $t_4 = 1$.

EXAMPLE 5

Determine if the set $\mathcal{B} = \left\{ \begin{bmatrix} 1 & 2 \\ 2 & -1 \end{bmatrix}, \begin{bmatrix} 3 & 2 \\ 1 & 1 \end{bmatrix}, \begin{bmatrix} 0 & 0 \\ 2 & 2 \end{bmatrix} \right\}$ is linearly dependent or linearly independent.

Solution: We consider the equation

$$\begin{bmatrix} 0 & 0 \\ 0 & 0 \end{bmatrix} = t_1 \begin{bmatrix} 1 & 2 \\ 2 & -1 \end{bmatrix} + t_2 \begin{bmatrix} 3 & 2 \\ 1 & 1 \end{bmatrix} + t_3 \begin{bmatrix} 0 & 0 \\ 2 & 2 \end{bmatrix} = \begin{bmatrix} t_1 + 3t_2 & 2t_1 + 2t_2 \\ 2t_1 + t_2 + 2t_3 & -t_1 + t_2 + 2t_3 \end{bmatrix}$$

Row reducing the coefficient matrix of the corresponding homogeneous system gives

$$\begin{bmatrix} 1 & 3 & 0 \\ 2 & 2 & 0 \\ 2 & 1 & 2 \\ -1 & 1 & 2 \end{bmatrix} \sim \begin{bmatrix} 1 & 0 & 0 \\ 0 & 1 & 0 \\ 0 & 0 & 1 \\ 0 & 0 & 0 \end{bmatrix}$$

The only solution is $t_1 = t_2 = t_3 = 0$, so S is linearly independent.

EXERCISE 1

Determine if $\mathcal{B} = \left\{ \begin{bmatrix} 1 & 2 \\ 1 & 1 \end{bmatrix}, \begin{bmatrix} 1 & 1 \\ 3 & 1 \end{bmatrix}, \begin{bmatrix} 3 & 5 \\ 5 & 3 \end{bmatrix}, \begin{bmatrix} 0 & -1 \\ -2 & 0 \end{bmatrix} \right\}$ is linearly dependent or linearly independent. Is $X = \begin{bmatrix} 1 & 5 \\ -5 & 1 \end{bmatrix}$ in the span of $\mathcal{B}$?

EXERCISE 2

Consider $\mathcal{B} = \left\{ \begin{bmatrix} 1 & 0 \\ 0 & 0 \end{bmatrix}, \begin{bmatrix} 0 & 1 \\ 0 & 0 \end{bmatrix}, \begin{bmatrix} 0 & 0 \\ 1 & 0 \end{bmatrix}, \begin{bmatrix} 0 & 0 \\ 0 & 1 \end{bmatrix} \right\}$. Prove that $\mathcal{B}$ is linearly independent and show that Span $\mathcal{B}$ is the set of all 2×2 matrices. Compare $\mathcal{B}$ with the standard basis for $\mathbb{R}^4$.

The Transpose of a Matrix

Definition
Transpose

Let A be an $m \times n$ matrix. Then the **transpose** of A is the $n \times m$ matrix denoted A^T, whose ij-th entry is the ji-th entry of A. That is,

$$(A^T)_{ij} = (A)_{ji}$$

EXAMPLE 6

Determine the transpose of $A = \begin{bmatrix} -1 & 6 & -4 \\ 3 & 5 & 2 \end{bmatrix}$ and $B = \begin{bmatrix} 1 \\ 0 \\ -1 \end{bmatrix}$.

Solution: $A^T = \begin{bmatrix} -1 & 6 & -4 \\ 3 & 5 & 2 \end{bmatrix}^T = \begin{bmatrix} -1 & 3 \\ 6 & 5 \\ -4 & 2 \end{bmatrix}$ and $B^T = \begin{bmatrix} 1 \\ 0 \\ -1 \end{bmatrix}^T = \begin{bmatrix} 1 & 0 & -1 \end{bmatrix}$.

Observe that taking the transpose of a matrix turns its rows into columns and its columns into rows.

Some Properties of the Transpose How does the operation of transposition combine with addition and scalar multiplication?

Theorem 2

For any matrices A and B and scalar $s \in \mathbb{R}$, we have
(1) $(A^T)^T = A$
(2) $(A + B)^T = A^T + B^T$
(3) $(sA)^T = sA^T$

Proof: 1. $((A^T)^T)_{ij} = (A^T)_{ji} = A_{ij}$.

2. $((A + B)^T)_{ij} = (A + B)_{ji} = (A)_{ji} + (B)_{ji} = (A^T)_{ij} + (B^T)_{ij} = (A^T + B^T)_{ij}$.

3. $((sA)^T)_{ij} = (sA)_{ji} = s(A)_{ji} = s(A^T)_{ij} = (sA^T)_{ij}$.

■

EXERCISE 3

Let $A = \begin{bmatrix} 2 & 3 & 1 \\ -1 & 0 & 5 \end{bmatrix}$. Verify that $(A^T)^T = A$ and $(3A)^T = 3A^T$.

Remark

Since we always represent a vector in $\mathbb{R}^n$ as a column matrix, to represent the row of a matrix as a vector, we will write $\vec{a}^T$. For now, this will be our main use of the transpose; however, it will become much more important later in the book.

An Introduction to Matrix Multiplication

The operations are so far very natural and easy. Multiplication of two matrices is more complicated, but a simple example illustrates that there is one useful natural definition.

Suppose that we wish to change variables; that is, we wish to work with new variables y_1 and y_2 that are defined in terms of the original variables x_1 and x_2 by the equations

$$\begin{aligned} y_1 &= a_{11}x_1 + a_{12}x_2 \\ y_2 &= a_{21}x_1 + a_{22}x_2 \end{aligned}$$

This is a system of linear equations like those in Chapter 2.

It is convenient to write these equations in matrix form. Let $\vec{x} = \begin{bmatrix} x_1 \\ x_2 \end{bmatrix}$, $\vec{y} = \begin{bmatrix} y_1 \\ y_2 \end{bmatrix}$, and $A = \begin{bmatrix} a_{11} & a_{12} \\ a_{21} & a_{22} \end{bmatrix}$ be the coefficient matrix. Then the change of variables equations can be written in the form $\vec{y} = A\vec{x}$, provided that we define the **product of A and $\vec{x}$** according to the following rule:

$$A\vec{x} = \begin{bmatrix} a_{11} & a_{12} \\ a_{21} & a_{22} \end{bmatrix} \begin{bmatrix} x_1 \\ x_2 \end{bmatrix} = \begin{bmatrix} a_{11}x_1 + a_{12}x_2 \\ a_{21}x_1 + a_{22}x_2 \end{bmatrix} \tag{3.1}$$

It is instructive to rewrite these entries in the right-hand matrix as dot products. Let $\vec{a}_1 = \begin{bmatrix} a_{11} \\ a_{12} \end{bmatrix}$ and $\vec{a}_2 = \begin{bmatrix} a_{21} \\ a_{22} \end{bmatrix}$ so that $\vec{a}_1^T = \begin{bmatrix} a_{11} & a_{12} \end{bmatrix}$ and $\vec{a}_2^T = \begin{bmatrix} a_{21} & a_{22} \end{bmatrix}$ are the *rows* of A. Then the equation becomes

$$A\vec{x} = \begin{bmatrix} \vec{a}_1^T \\ \vec{a}_2^T \end{bmatrix} \begin{bmatrix} x_1 \\ x_2 \end{bmatrix} = \begin{bmatrix} \vec{a}_1 \cdot \vec{x} \\ \vec{a}_2 \cdot \vec{x} \end{bmatrix}$$

Thus, in order for the right-hand side of the original equations to be represented correctly by the matrix product $A\vec{x}$, the entry in the first row of $A\vec{x}$ must be the dot product of the first row of A (as a column vector) with the vector $\vec{x}$; the entry in the second row must be the dot product of the second row of A (as a column vector) with $\vec{x}$.

Suppose there is a second change of variables from $\vec{y}$ to $\vec{z}$:

$$\begin{aligned} z_1 &= b_{11}y_1 + b_{12}y_2 \\ z_2 &= b_{21}y_1 + b_{22}y_2 \end{aligned}$$

In matrix form, this is written $\vec{z} = B\vec{y}$. Now suppose that these changes are performed one after the other. The values for y_1 and y_2 from the first change of variables are substituted into the second pair of equations:

$$z_1 = b_{11}(a_{11}x_1 + a_{12}x_2) + b_{12}(a_{21}x_1 + a_{22}x_2)$$
$$z_2 = b_{21}(a_{11}x_1 + a_{12}x_2) + b_{22}(a_{21}x_1 + a_{22}x_2)$$

After simplification, this can be written as

$$\vec{z} = \begin{bmatrix} b_{11}a_{11} + b_{12}a_{21} & b_{11}a_{12} + b_{12}a_{22} \\ b_{21}a_{11} + b_{22}a_{21} & b_{21}a_{12} + b_{22}a_{22} \end{bmatrix} \vec{x}$$

We want this to be equivalent to $\vec{z} = B\vec{y} = BA\vec{x}$. Therefore, the product BA must be

$$BA = \begin{bmatrix} b_{11}a_{11} + b_{12}a_{21} & b_{11}a_{12} + b_{12}a_{22} \\ b_{21}a_{11} + b_{22}a_{21} & b_{21}a_{12} + b_{22}a_{22} \end{bmatrix} \tag{3.2}$$

Thus, the product BA must be defined by the following rules:

- $(BA)_{11}$ is the dot product of the first row of B and the first column of A.
- $(BA)_{12}$ is the dot product of the first row of B and the second column of A.
- $(BA)_{21}$ is the dot product of the second row of B and the first column of A.
- $(BA)_{22}$ is the dot product of the second row of B and the second column of A.

EXAMPLE 7

$$\begin{bmatrix} 2 & 3 \\ 4 & 1 \end{bmatrix} \begin{bmatrix} 5 & 1 \\ -2 & 7 \end{bmatrix} = \begin{bmatrix} 2(5) + 3(-2) & 2(1) + 3(7) \\ 4(5) + 1(-2) & 4(1) + 1(7) \end{bmatrix} = \begin{bmatrix} 4 & 23 \\ 18 & 11 \end{bmatrix}$$

We now want to generalize matrix multiplication. It will be convenient to use $\vec{b}_i^T$ to represent the i-th row of B and $\vec{a}_j$ to represent the j-th column of A. Observe from our work above that we want the ij-th entry of BA to be the dot product of the i-th row of B and the j-th column of A. However, for this to be defined, $\vec{b}_i^T$ must have the same number of entries as $\vec{a}_j$. Hence, the number of entries in the rows of the matrix B (that is, the number of columns of B) must be equal to the number of entries in the columns of A (that is, the number of rows of A). We can now make a precise definition.

Definition
Matrix Multiplication

Let B be an $m \times n$ matrix with rows $\vec{b}_1^T, \dots, \vec{b}_m^T$ and A be an $n \times p$ matrix with columns $\vec{a}_1, \dots, \vec{a}_p$. Then, we define BA to be the matrix whose ij-th entry is

$$(BA)_{ij} = \vec{b}_i \cdot \vec{a}_j$$

Remark

If B is an $m \times n$ matrix and A is a $p \times q$ matrix, then BA is defined only if $n = p$. Moreover, if $n = p$, then the resulting matrix is $m \times q$.

More simply stated, multiplication of two matrices can be performed only if the number of columns in the first matrix is equal the number of rows in the second.

EXAMPLE 8

Perform the following operations.

(a) $\begin{bmatrix} 2 & 3 & 0 & 1 \\ 4 & -1 & 2 & -1 \end{bmatrix} \begin{bmatrix} 3 & 1 \\ 1 & 2 \\ 2 & 3 \\ 0 & 5 \end{bmatrix}$

Solution: $\begin{bmatrix} 2 & 3 & 0 & 1 \\ 4 & -1 & 2 & -1 \end{bmatrix} \begin{bmatrix} 3 & 1 \\ 1 & 2 \\ 2 & 3 \\ 0 & 5 \end{bmatrix} = \begin{bmatrix} 9 & 13 \\ 15 & 3 \end{bmatrix}$

(b) $\begin{bmatrix} 1 & 1 & 2 \\ -2 & -1 & 3 \\ 0 & 0 & 1 \end{bmatrix} \begin{bmatrix} 5 & 6 \\ 4 & 7 \\ 2 & 5 \end{bmatrix}$

Solution: $\begin{bmatrix} 1 & 1 & 2 \\ -2 & -1 & 3 \\ 0 & 0 & 1 \end{bmatrix} \begin{bmatrix} 5 & 6 \\ 4 & 7 \\ 2 & 5 \end{bmatrix} = \begin{bmatrix} 13 & 23 \\ -8 & -4 \\ 2 & 5 \end{bmatrix}$

EXAMPLE 9

$\begin{bmatrix} 2 & 3 \\ 1 & -3 \end{bmatrix} \begin{bmatrix} 2 & -3 \\ 4 & 1 \\ 5 & 7 \end{bmatrix}$ is not defined because the first matrix has two columns and the second matrix has three rows.

EXERCISE 4

Let $A = \begin{bmatrix} 1 & 2 & -1 \\ 2 & 3 & 1 \end{bmatrix}$ and $B = \begin{bmatrix} 2 & 1 \\ 1 & 0 \end{bmatrix}$. Calculate the following or explain why they are not defined.

(a) AB (b) BA (c) A^TA (d) BB^T

EXAMPLE 10

Let $A = \begin{bmatrix} 1 \\ 2 \\ 3 \end{bmatrix}$ and $B = \begin{bmatrix} 6 \\ 5 \\ 4 \end{bmatrix}$. Compute A^TB.

Solution: $A^TB = \begin{bmatrix} 1 & 2 & 3 \end{bmatrix} \begin{bmatrix} 6 \\ 5 \\ 4 \end{bmatrix} = [1(6) + 2(5) + 3(4)] = [28]$.

Observe that if we let $\vec{x} = \begin{bmatrix} 1 \\ 2 \\ 3 \end{bmatrix}$ and $\vec{y} = \begin{bmatrix} 6 \\ 5 \\ 4 \end{bmatrix}$ be vectors in $\mathbb{R}^3$, then $\vec{x} \cdot \vec{y} = 28$, This matches the result in Example 10. This result should not be surprising since we have defined matrix multiplication in terms of the dot product. More generally, for any $\vec{x}, \vec{y} \in \mathbb{R}^n$, we have

$$\vec{x}^T \vec{y} = \vec{x} \cdot \vec{y}$$

where we interpret the 1×1 matrix on the right-hand side as a scalar. This formula will be used frequently later in the book.

Defining matrix multiplication with the dot product fits our view that the rows of the coefficient matrix of a system of linear equations are the coefficients from each equation. We now look at how we could define matrix multiplication by using our alternate view of the coefficient matrix; in that case, the columns of the coefficient matrix are the coefficients of each variable.

Observe that we can write equations (3.1) as

$$A\vec{x} = \begin{bmatrix} a_{11}x_1 + a_{12}x_2 \\ a_{21}x_1 + a_{22}x_2 \end{bmatrix} = \begin{bmatrix} a_{11} \\ a_{21} \end{bmatrix} x_1 + \begin{bmatrix} a_{12} \\ a_{22} \end{bmatrix} x_2$$

That is, we can view $A\vec{x}$ as giving a linear combination of the columns of A. So, for an $m \times n$ matrix A with columns $\vec{a}_1, \dots, \vec{a}_n$ and vector $\vec{x} \in \mathbb{R}^n$, we have

$$A\vec{x} = \begin{bmatrix} \vec{a}_1 & \cdots & \vec{a}_n \end{bmatrix} \begin{bmatrix} x_1 \\ \vdots \\ x_n \end{bmatrix} = x_1\vec{a}_1 + \cdots + x_n\vec{a}_n$$

Using this, observe that (3.2) can be written as

$$BA = \begin{bmatrix} B\vec{a}_1 & B\vec{a}_2 \end{bmatrix}$$

Hence, in general, if A is an $m \times n$ matrix and B is a $p \times m$ matrix, then BA is the $p \times n$ matrix given by

$$BA = \begin{bmatrix} B\vec{a}_1 & \cdots & B\vec{a}_n \end{bmatrix} \tag{3.3}$$

Both interpretations of matrix multiplication will be very useful, so it is important to know and understand both of them.

Remark

We now see that linear combinations of vectors (and hence concepts such as spanning and linear independence), solving systems of linear equations, and matrix multiplication are all closely tied together. We will continue to see these connections later in this chapter and throughout the book.

Summation Notation and Matrix Multiplication Some calculations with matrix products are better described in terms of summation notation:

$$\sum_{k=1}^{n} a_k = a_1 + a_2 + \cdots + a_n$$

Let A be an $m \times n$ matrix with i-th row $\vec{a}_i^T = \begin{bmatrix} a_{i1} & \cdots & a_{in} \end{bmatrix}$ and let B be an $n \times p$ matrix with j-th column $\vec{b}_j = \begin{bmatrix} b_{1j} \\ \vdots \\ b_{nj} \end{bmatrix}$. Then the ij-th entry of AB is

$$(AB)_{ij} = \vec{a}_i \cdot \vec{b}_j = \vec{a}_i^T \vec{b}_j = a_{i1}b_{1j} + a_{i2}b_{2j} + \cdots + a_{in}b_{nj} = \sum_{k=1}^{n} (A)_{ik}(B)_{kj}$$

We can use this notation to help prove some properties of matrix multiplication.

Theorem 3 If A, B, and C are matrices of the correct size so that the required products are defined, and $t \in \mathbb{R}$, then

(1) $A(B + C) = AB + AC$
(2) $t(AB) = (tA)B = A(tB)$
(3) $A(BC) = (AB)C$
(4) $(AB)^T = B^T A^T$

Each of these properties follows easily from the definition of matrix multiplication and properties of summation notation. However, the proofs are not particularly illuminating and so are omitted.

Important Facts **The matrix product is not commutative:** That is, in general, $AB \neq BA$. In fact, if BA is defined, it is not necessarily true that AB is even defined. For example, if B is 2×2 and A is 2×3, then BA is defined, but AB is not. However, even if both AB and BA are defined, they are usually not equal. $AB = BA$ is true only in very special circumstances.

EXAMPLE 11

Show that if $A = \begin{bmatrix} 2 & 3 \\ 4 & -1 \end{bmatrix}$ and $B = \begin{bmatrix} 5 & 1 \\ -2 & 7 \end{bmatrix}$, then $AB \neq BA$.

Solution: $AB = \begin{bmatrix} 2 & 3 \\ 4 & -1 \end{bmatrix} \begin{bmatrix} 5 & 1 \\ -2 & 7 \end{bmatrix} = \begin{bmatrix} 4 & 23 \\ 22 & -3 \end{bmatrix}$,

but

$$BA = \begin{bmatrix} 5 & 1 \\ -2 & 7 \end{bmatrix} \begin{bmatrix} 2 & 3 \\ 4 & -1 \end{bmatrix} = \begin{bmatrix} 14 & 14 \\ 24 & -13 \end{bmatrix}$$

Therefore, $AB \neq BA$.

The cancellation law is almost never valid for matrix multiplication: Thus, if $AB = AC$, then we cannot guarantee that $B = C$.

EXAMPLE 12

Let $A = \begin{bmatrix} 0 & 0 \\ 0 & 1 \end{bmatrix}$, $B = \begin{bmatrix} 5 & 6 \\ 7 & 8 \end{bmatrix}$, and $C = \begin{bmatrix} 2 & 3 \\ 7 & 8 \end{bmatrix}$. Then,

$$AB = \begin{bmatrix} 0 & 0 \\ 0 & 1 \end{bmatrix} \begin{bmatrix} 5 & 6 \\ 7 & 8 \end{bmatrix} = \begin{bmatrix} 0 & 0 \\ 7 & 8 \end{bmatrix} = \begin{bmatrix} 0 & 0 \\ 0 & 1 \end{bmatrix} \begin{bmatrix} 2 & 3 \\ 7 & 8 \end{bmatrix} = AC$$

so $AB = AC$ but $B \neq C$.

Remark

The fact that we do not have the cancellation law for matrix multiplication comes from the fact that **we do not have division for matrices**.

We must distinguish carefully between a general cancellation law and the following theorem, which we will use many times.

Theorem 4 If A and B are $m \times n$ matrices such that $A\vec{x} = B\vec{x}$ for every $\vec{x} \in \mathbb{R}^n$, then $A = B$.

Note that it is the assumption that equality holds for *every* $\vec{x}$ that distinguishes this from a cancellation law.

Proof: You are asked to prove Theorem 4, with hints, in Problem D1.

Identity Matrix

We have seen that the zero matrix $O_{m,n}$ is the additive identity for addition of $m \times n$ matrices. However, since we also have multiplication of matrices, it is important to determine if we have a multiplicative identity. If we do, we need to determine what the multiplicative identity is. First, we observe that for there to exist a matrix A and a matrix I such that $AI = A = IA$, both A and I must be $n \times n$ matrices. The multiplicative identity I is the $n \times n$ matrix that has this property for all $n \times n$ matrices A.

To find how to define I, we begin with a simple case. Let $A = \begin{bmatrix} a & b \\ c & d \end{bmatrix}$. We want to find a matrix $I = \begin{bmatrix} e & f \\ g & h \end{bmatrix}$ such that $AI = A$. By matrix multiplication, we get

$$\begin{bmatrix} a & b \\ c & d \end{bmatrix} = \begin{bmatrix} a & b \\ c & d \end{bmatrix}\begin{bmatrix} e & f \\ g & h \end{bmatrix} = \begin{bmatrix} ae + bg & af + bh \\ ce + dg & cf + dh \end{bmatrix}$$

Thus, we must have $a = ae + bg$, $b = af + bh$, $c = ce + dg$, and $d = cf + dh$. Although this system of equations is not linear, it is still easy to solve. We find that we must have $e = 1 = h$ and $f = g = 0$. Thus,

$$I = \begin{bmatrix} 1 & 0 \\ 0 & 1 \end{bmatrix} = \text{diag}(1, 1)$$

It is easy to verify that I also satisfies $IA = A$. Hence, I is the multiplicative identity for 2×2 matrices. We now extend this definition to the $n \times n$ case.

Definition
Identity Matrix

The $n \times n$ matrix $I = \text{diag}(1, 1, \dots, 1)$ is called the **identity matrix**.

EXAMPLE 13

The 3×3 identity matrix is $I = \text{diag}(1, 1, 1) = \begin{bmatrix} 1 & 0 & 0 \\ 0 & 1 & 0 \\ 0 & 0 & 1 \end{bmatrix}$.

The 4×4 identity matrix is $I = \text{diag}(1, 1, 1, 1) = \begin{bmatrix} 1 & 0 & 0 & 0 \\ 0 & 1 & 0 & 0 \\ 0 & 0 & 1 & 0 \\ 0 & 0 & 0 & 1 \end{bmatrix}$.

Remarks

1. In general, the size of I (the value of n) is clear from context. However, in some cases, we stress the size of the identity matrix by denoting it by I_n. For example, I_2 is the 2×2 identity matrix, and I_m is the $m \times m$ identity matrix.

2. The columns of the identity matrix should seem familiar. If $\{\vec{e}_1, \dots, \vec{e}_n\}$ is the standard basis for $\mathbb{R}^n$, then

$$I_n = \begin{bmatrix} \vec{e}_1 & \cdots & \vec{e}_n \end{bmatrix}$$

Theorem 5 If A is any $m \times n$ matrix, then $I_m A = A = AI_n$.

You are asked to prove this theorem in Problem D2. Note that it immediately implies that I_n is the multiplicative identity for the set of $n \times n$ matrices.

Block Multiplication

Observe that in our second interpretation of matrix multiplication, equation (3.3), we calculated the product BA in **blocks**. That is, we computed the smaller matrix products $B\vec{a}_1, B\vec{a}_2, \dots, B\vec{a}_n$ and put these in the appropriate positions to create BA. This is a very simple example of **block multiplication**. Observe that we could also regard the rows of B as blocks and write

$$BA = \begin{bmatrix} \vec{b}_1^T A \\ \vdots \\ \vec{b}_p^T A \end{bmatrix}$$

There are more general statements about the products of two matrices, each of which have been partitioned into blocks. In addition to clarifying the meaning of some calculations, block multiplication is used in organizing calculations with very large matrices.

Roughly speaking, as long as the sizes of the blocks are chosen so that the products of the blocks are defined and fit together as required, block multiplication is defined by an extension of the usual rules of matrix multiplication. We will demonstrate this with an example.

EXAMPLE 14 Suppose that A is an $m \times n$ matrix, B is an $n \times p$ matrix, and A and B are **partitioned** into blocks as indicated:

$$A = \begin{bmatrix} A_1 \\ A_2 \end{bmatrix}, \qquad B = \begin{bmatrix} B_1 & B_2 \end{bmatrix}$$

Say that A_1 is $r \times n$ so that A_2 is $(m - r) \times n$, while B_1 is $n \times q$ and B_2 is $n \times (p - q)$. Now, the product of a 2×1 matrix and a 1×2 matrix is given by

$$\begin{bmatrix} a_1 \\ a_2 \end{bmatrix} \begin{bmatrix} b_1 & b_2 \end{bmatrix} = \begin{bmatrix} a_1b_1 & a_1b_2 \\ a_2b_1 & a_2b_2 \end{bmatrix}$$

EXAMPLE 14 (continued)

So, for the partitioned block matrices, we have

$$\begin{bmatrix} A_1 \\ A_2 \end{bmatrix}\begin{bmatrix} B_1 & B_2 \end{bmatrix} = \begin{bmatrix} A_1B_1 & A_1B_2 \\ A_2B_1 & A_2B_2 \end{bmatrix}$$

Observe that all the products are defined and the size of the resulting matrix is $m \times p$, as desired.

PROBLEMS 3.1

Practice Problems

A1 Perform the indicated operation.

(a) $\begin{bmatrix} 2 & -2 & 3 \\ 4 & 1 & -1 \end{bmatrix} + \begin{bmatrix} -3 & -4 & 1 \\ 2 & -5 & 3 \end{bmatrix}$

(b) $(-3)\begin{bmatrix} 1 & -2 \\ 2 & 1 \\ 4 & -2 \end{bmatrix}$

(c) $\begin{bmatrix} 2 & 3 \\ 1 & -2 \end{bmatrix} - 3\begin{bmatrix} 1 & -2 \\ 4 & 5 \end{bmatrix}$

A2 Calculate each of the following products or explain why a product is not defined.

(a) $\begin{bmatrix} -2 & 3 \\ 3 & 4 \end{bmatrix}\begin{bmatrix} 4 & 5 & -3 \\ -1 & 3 & 2 \end{bmatrix}$

(b) $\begin{bmatrix} 2 & 0 & 3 \\ 1 & 1 & 1 \\ -1 & 3 & 2 \end{bmatrix}\begin{bmatrix} 6 & -2 \\ 3 & 1 \\ 0 & 5 \end{bmatrix}$

(c) $\begin{bmatrix} 2 & 3 \\ -1 & -1 \\ 5 & 3 \end{bmatrix}\begin{bmatrix} 4 & 3 & 2 & 1 \\ -4 & 0 & 3 & -2 \end{bmatrix}$

(d) $\begin{bmatrix} 2 & 3 \\ -4 & 2 \end{bmatrix}\begin{bmatrix} 1 & 2 \\ 1 & 0 \\ 0 & -3 \end{bmatrix}$

A3 Check that $A(B + C) = AB + AC$ and that $A(3B) = 3(AB)$ for the given matrices.

(a) $A = \begin{bmatrix} 2 & 3 \\ -3 & 1 \end{bmatrix}$, $B = \begin{bmatrix} -3 & 2 \\ 1 & 0 \end{bmatrix}$, $C = \begin{bmatrix} -2 & -3 \\ -2 & 4 \end{bmatrix}$

(b) $A = \begin{bmatrix} 2 & -1 & -2 \\ 3 & 2 & 1 \end{bmatrix}$, $B = \begin{bmatrix} -1 & 2 \\ -3 & 4 \\ 0 & 3 \end{bmatrix}$,

$C = \begin{bmatrix} 5 & -2 \\ -3 & 0 \\ 4 & 3 \end{bmatrix}$

A4 For the given matrices A and B, check whether $A + B$ and AB are defined. If so, check that $(A + B)^T = A^T + B^T$ and $(AB)^T = B^TA^T$.

(a) $A = \begin{bmatrix} 1 & 2 \\ 1 & 3 \\ -2 & 1 \end{bmatrix}$, $B = \begin{bmatrix} -4 & -3 \\ 1 & -1 \\ 3 & 2 \end{bmatrix}$

(b) $A = \begin{bmatrix} 2 & -4 & 5 \\ 4 & 1 & -3 \end{bmatrix}$, $B = \begin{bmatrix} -3 & -4 \\ 5 & -2 \\ 1 & 3 \end{bmatrix}$

A5 Let $A = \begin{bmatrix} 2 & 5 \\ -1 & 3 \end{bmatrix}$, $B = \begin{bmatrix} -1 & 3 & -4 \\ 3 & 5 & 2 \end{bmatrix}$, $C = \begin{bmatrix} 1 & 4 \\ 1 & 3 \\ 4 & -3 \end{bmatrix}$,

and $D = \begin{bmatrix} 4 & 3 & 2 & 1 \\ -1 & 0 & 1 & 2 \\ 2 & 1 & 0 & 3 \end{bmatrix}$.

Determine the following products or state why they do not exist.

(a) AB (b) BA (c) AC
(d) DC (e) CD (f) C^TD
(g) Verify that $A(BC) = (AB)C$
(h) Verify that $(AB)^T = B^TA^T$
(i) Without doing any arithmetic, determine D^TC

A6 Let $A = \begin{bmatrix} 2 & 3 & 1 \\ 3 & -1 & 4 \\ -1 & 0 & 1 \end{bmatrix}$, $\vec{x} = \begin{bmatrix} 1 \\ 2 \\ 4 \end{bmatrix}$, $\vec{y} = \begin{bmatrix} 3 \\ 1 \\ -1 \end{bmatrix}$,

and $\vec{z} = \begin{bmatrix} 0 \\ -1 \\ 1 \end{bmatrix}$.

(a) Determine $A\vec{x}$, $A\vec{y}$, and $A\vec{z}$.

(b) Use the result of (a) to determine $A\begin{bmatrix}1 & 3 & 0\\ 2 & 1 & -1\\ 4 & -1 & 1\end{bmatrix}$.

A7 Let $A = \begin{bmatrix}2 & 3 & 1\\ 3 & -1 & 4\\ -1 & 0 & 1\end{bmatrix}$ and consider $A\vec{x}$ as a linear combination of columns of A to determine $\vec{x}$ if $A\vec{x} = \vec{b}$ where

(a) $\vec{b} = \begin{bmatrix}2\\ 3\\ -1\end{bmatrix}$ (b) $\vec{b} = \begin{bmatrix}4\\ 6\\ -2\end{bmatrix}$

(c) $\vec{b} = \begin{bmatrix}3\\ 12\\ 3\end{bmatrix}$ (d) $\vec{b} = \begin{bmatrix}5\\ 2\\ -1\end{bmatrix}$

A8 Calculate the following products.

(a) $\begin{bmatrix}-3\\ 1\end{bmatrix}\begin{bmatrix}2 & 6\end{bmatrix}$ (b) $\begin{bmatrix}2 & 6\end{bmatrix}\begin{bmatrix}-3\\ 1\end{bmatrix}$

(c) $\begin{bmatrix}2\\ -1\\ 3\end{bmatrix}\begin{bmatrix}5 & 4 & -3\end{bmatrix}$ (d) $\begin{bmatrix}5 & 4 & -3\end{bmatrix}\begin{bmatrix}2\\ -1\\ 3\end{bmatrix}$

A9 Verify the following case of block multiplication by calculating both sides of the equation and comparing.

$$\left[\begin{array}{cc|cc}2 & 3 & -4 & 5\\ -4 & 1 & 2 & 1\end{array}\right]\left[\begin{array}{cc}6 & 3\\ -2 & 4\\ \hline 1 & 3\\ -3 & 2\end{array}\right]$$

$$= \begin{bmatrix}2 & 3\\ -4 & 1\end{bmatrix}\begin{bmatrix}6 & 3\\ -2 & 4\end{bmatrix} + \begin{bmatrix}-4 & 5\\ 2 & 1\end{bmatrix}\begin{bmatrix}1 & 3\\ -3 & [illegible]\end{bmatrix}$$

A10 Determine if $\begin{bmatrix}2 & 3\\ 2 & -3\end{bmatrix}$ is in the span of

$$B = \left\{\begin{bmatrix}1 & 2\\ 1 & 0\end{bmatrix}, \begin{bmatrix}0 & 1\\ -1 & 2\end{bmatrix}, \begin{bmatrix}1 & 1\\ 3 & -1\end{bmatrix}\right\}$$

A11 Determine if the set

$$S = \left\{\begin{bmatrix}-1 & 1\\ 1 & -1\end{bmatrix}, \begin{bmatrix}0 & 1\\ 2 & -2\end{bmatrix}, \begin{bmatrix}1 & -1\\ 3 & -3\end{bmatrix}\right\}$$

is linearly dependent or linearly independent.

A12 Let $A = \begin{bmatrix}\vec{a}_1 & \cdots & \vec{a}_n\end{bmatrix}$ and let $\{\vec{e}_1, \dots, \vec{e}_n\}$ denote the standard basis vectors for $\mathbb{R}^n$. Prove that $A\vec{e}_i = \vec{a}_i$.

Homework Problems

B1 Perform the indicated operation.

(a) $\begin{bmatrix}3 & -2\\ -4 & 1\\ 3 & 7\end{bmatrix} + \begin{bmatrix}5 & 4\\ 1 & 4\\ -6 & -9\end{bmatrix}$

(b) $(-5)\begin{bmatrix}2 & 3 & -6 & -2\\ -7 & 1 & 0 & 5\end{bmatrix}$

(c) $\begin{bmatrix}4 & 2 & 3\\ -2 & 1 & 5\end{bmatrix} - 4\begin{bmatrix}-2 & -1 & 5\\ 6 & 7 & 1\end{bmatrix}$

B2 Calculate each of the following products or explain why a product is not defined.

(a) $\begin{bmatrix}-3 & 2\\ 5 & -1\end{bmatrix}\begin{bmatrix}3 & 1 & -2\\ 2 & -3 & -1\end{bmatrix}$

(b) $\begin{bmatrix}0 & 3 & -1\\ -1 & 2 & -1\\ 1 & 1 & 3\end{bmatrix}\begin{bmatrix}7 & -3\\ 2 & -1\\ 5 & 0\end{bmatrix}$

(c) $\begin{bmatrix}3 & -1\\ 2 & 4\\ 2 & 7\end{bmatrix}\begin{bmatrix}3 & 1 & -2 & 2\\ -2 & 1 & -2 & 3\end{bmatrix}$

(d) $\begin{bmatrix}1 & 2 & -1\\ 2 & 3 & 1\end{bmatrix}\begin{bmatrix}-4 & 1 & [illegible]\\ 6 & 3 & [illegible]\end{bmatrix}$

B3 Check that $A(B + C) = AB$ [illegible] $3(AB)$ for the given matrice[illegible]

(a) $A = \begin{bmatrix}3 & 5\\ -2 & -1\end{bmatrix}$, $B = \begin{bmatrix}4 & [illegible] & [illegible]\\ 1 & -6 & 8\end{bmatrix}$,

$C = \begin{bmatrix}-2 & 1 & 4\\ 2 & 2 & -3\end{bmatrix}$

(b) $A = \begin{bmatrix}3 & 4\\ 1 & 0\\ -3 & 5\end{bmatrix}$, $B = \begin{bmatrix}2 & -3\\ 1 & 5\end{bmatrix}$, $C = \begin{bmatrix}-4 & -6\\ 3 & 1\end{bmatrix}$

B4 For the given matrices A and B, check whether $A + B$ and AB are defined. If so, check that $(A + B)^T = A^T + B^T$ and $(AB)^T = B^T A^T$.

(a) $A = \begin{bmatrix}1 & 1 & 4\\ -2 & 1 & 6\end{bmatrix}$, $B = \begin{bmatrix}5 & 4\\ -2 & 3\\ 0 & -1\end{bmatrix}$

(b) $A = \begin{bmatrix} 1 & 3 & 1 \\ -2 & 1 & 4 \\ 1 & 0 & -3 \end{bmatrix}$, $B = \begin{bmatrix} 6 & -2 & 2 \\ 1 & 1 & 3 \\ 2 & -3 & -4 \end{bmatrix}$

B5 Let $A = \begin{bmatrix} 2 & 1 \\ -1 & -2 \\ 4 & -3 \end{bmatrix}$, $B = \begin{bmatrix} 3 & -4 \\ 4 & 5 \end{bmatrix}$,

$C = \begin{bmatrix} -2 & 5 & -2 \\ -2 & 1 & 0 \end{bmatrix}$, and $D = \begin{bmatrix} 1 & 3 & -5 \\ 0 & 2 & 1 \\ -3 & 2 & 1 \\ 1 & 1 & -1 \end{bmatrix}$.

Determine the following products or state why they do not exist.

(a) AB (b) BA (c) AC
(d) DC (e) DA (f) CD^T
(g) Verify that $B(CA) = (BC)A$
(h) Verify that $(AB)^T = B^TA^T$
(i) Without doing any arithmetic, determine DC^T

B6 Let $A = \begin{bmatrix} 1 & 0 & 2 \\ -1 & 2 & 3 \\ -1 & 1 & 0 \end{bmatrix}$, $\vec{x} = \begin{bmatrix} 4 \\ 0 \\ -2 \end{bmatrix}$, $\vec{y} = \begin{bmatrix} -1 \\ 1 \\ 2 \end{bmatrix}$, and $\vec{z} = \begin{bmatrix} 5 \\ 1 \\ 3 \end{bmatrix}$.

(a) Determine $A\vec{x}$, $A\vec{y}$, and $A\vec{z}$.

(b) Use the result of (a) to determine $A\begin{bmatrix} 4 & 1 & -5 \\ 0 & -1 & -1 \\ -2 & -2 & -3 \end{bmatrix}$.

B7 Let $A = \begin{bmatrix} 3 & 3 & -5 \\ 0 & -1 & -1 \\ -2 & -4 & -3 \end{bmatrix}$ and consider $A\vec{x}$ as a linear combination of columns of A to determine $\vec{x}$ if $A\vec{x} = \vec{b}$ where

(a) $\vec{b} = \begin{bmatrix} 3 \\ -1 \\ -4 \end{bmatrix}$ (b) $\vec{b} = \begin{bmatrix} -6 \\ 0 \\ 4 \end{bmatrix}$

(c) $\vec{b} = \begin{bmatrix} 15 \\ 3 \\ 9 \end{bmatrix}$ (d) $\vec{b} = \begin{bmatrix} 1 \\ -2 \\ -9 \end{bmatrix}$

B8 Calculate the following products.

(a) $\begin{bmatrix} 3 \\ 2 \end{bmatrix}\begin{bmatrix} -2 & 4 \end{bmatrix}$ (b) $\begin{bmatrix} -2 & 4 \end{bmatrix}\begin{bmatrix} 3 \\ 2 \end{bmatrix}$

(c) $\begin{bmatrix} 2 \\ 1 \\ 5 \end{bmatrix}\begin{bmatrix} -3 & 1 & 2 \end{bmatrix}$ (d) $\begin{bmatrix} -3 & 1 & 2 \end{bmatrix}\begin{bmatrix} 2 \\ 1 \\ 5 \end{bmatrix}$

B9 Verify the following case of block multiplication by calculating both sides of the equation and comparing.

$$\left[\begin{array}{cc|cc} 1 & -1 & -2 & 4 \\ 3 & 2 & 1 & 3 \end{array}\right]\left[\begin{array}{cc} 4 & 3 \\ 2 & 5 \\ \hline 6 & 1 \\ -1 & 3 \end{array}\right]$$
$$= \begin{bmatrix} 1 & -1 \\ 3 & 2 \end{bmatrix}\begin{bmatrix} 4 & 3 \\ 2 & 5 \end{bmatrix} + \begin{bmatrix} -2 & 4 \\ 1 & 3 \end{bmatrix}\begin{bmatrix} 6 & 1 \\ -1 & 3 \end{bmatrix}$$

B10 Determine if $\begin{bmatrix} 2 & -1 \\ 1 & 1 \end{bmatrix}$ is in the span of

$$B = \left\{\begin{bmatrix} 1 & -1 \\ 0 & 2 \end{bmatrix}, \begin{bmatrix} -2 & 3 \\ 0 & -1 \end{bmatrix}, \begin{bmatrix} -1 & 2 \\ 1 & 1 \end{bmatrix}\right\}$$

B11 Determine if the set

$$S = \left\{\begin{bmatrix} 1 & 0 \\ -1 & 0 \end{bmatrix}, \begin{bmatrix} -3 & 0 \\ 4 & -2 \end{bmatrix}, \begin{bmatrix} 2 & -3 \\ 0 & -1 \end{bmatrix}, \begin{bmatrix} -3 & -1 \\ 3 & -3 \end{bmatrix}\right\}$$

is linearly dependent or linearly independent.

Computer Problems

C1 Use a computer to check your results for Problems A2 and A5.

C2 Use a computer to determine the following matrix products and the transpose of the products. Note that one of the matrices appears twice. You should not have to enter it twice.

(a) $\begin{bmatrix} 2.12 & 5.35 \\ -1.97 & 3.56 \end{bmatrix}\begin{bmatrix} -1.02 & 3.47 & -4.94 \\ 3.33 & 5.83 & 2.29 \end{bmatrix}$

(b) $\begin{bmatrix} -1.02 & 3.47 & -4.94 \\ 3.33 & 5.83 & 2.29 \end{bmatrix}\begin{bmatrix} 1.88 & 4.25 \\ 1.55 & 3.38 \\ 4.67 & -3.73 \end{bmatrix}$

Conceptual Problems

D1 Prove Theorem 4, using the following hints. To prove $A = B$, prove that $A - B = O_{m,n}$; note that $A\vec{x} = B\vec{x}$ for every $\vec{x} \in \mathbb{R}^n$ if and only if $(A - B)\vec{x} = \vec{0}$ for every $\vec{x} \in \mathbb{R}^n$. Now, suppose that $C\vec{x} = \vec{0}$ for every $\vec{x} \in \mathbb{R}^n$. Consider the case where $\vec{x} = \vec{e}_i$ and conclude that each column of C must be the zero vector.

D2 Let A be an $m \times n$ matrix. Show that $I_m A = A$ and $AI_n = A$.

D3 Find a formula to calculate the ij-th entry of AA^T and of A^TA. Explain why it follows that if AA^T or A^TA is the zero matrix, then A is the zero matrix.

D4 (a) Construct a 2×2 matrix A that is not the zero matrix yet satisfies $A^2 = O_{2,2}$.

(b) Find 2×2 matrices A and B with $A \neq B$ and neither $A = O_{2,2}$ nor $B = O_{2,2}$, such that $A^2 - AB - BA + B^2 = O_{2,2}$.

D5 Find as many 2×2 matrices as you can that satisfy $A^2 = I$.

3.2 Matrix Mappings and Linear Mappings

Functions are a fundamental concept in mathematics. Recall that a **function** f *is a rule that assigns to every element* x *of an initial set called the* **domain** *of the function a unique value* y *in another set called the* **codomain** *of* f*. We say that* f **maps** x *to* y *or that* y *is the* **image** *of* x *under* f*, and we write* f(x) = y*. If* f *is a function with domain* U *and codomain* V*, then we say that* f *maps* U *to* V *and denote this by* $f : U \to V$*. In your earlier mathematics, you met functions* $f : \mathbb{R} \to \mathbb{R}$ *such as* $f(x) = x^2$ *and looked at their various properties. In this section, we will start by looking at more general functions* $f : \mathbb{R}^n \to \mathbb{R}^m$*, commonly called mappings or transformations. We will also look at a class of functions called linear mappings that are very important in linear algebra.*

Matrix Mappings

Using the rule for matrix multiplication introduced in the preceding section, observe that for any $m \times n$ matrix A and vector $\vec{x} \in \mathbb{R}^n$, the product $A\vec{x}$ is a vector in $\mathbb{R}^m$. We see that this is behaving like a function whose domain is $\mathbb{R}^n$ and whose codomain is $\mathbb{R}^m$.

Definition
Matrix Mapping

For any $m \times n$ matrix A, we define a function $f_A : \mathbb{R}^n \to \mathbb{R}^m$ called the **matrix mapping**, corresponding to A by

$$f_A(\vec{x}) = A\vec{x}$$

for any $\vec{x} \in \mathbb{R}^n$.

Remark

Although a matrix mapping sends vectors to vectors, it is much more common to view functions as mapping points to points. Thus, when dealing with mappings in this text, we will often write

$$f(x_1, \dots, x_n) = (y_1, \dots, y_m), \quad \text{or} \quad f(x_1, \dots, x_n) = \begin{bmatrix} y_1 \\ \vdots \\ y_m \end{bmatrix}$$

when, technically, it would be more correct to write $f\left(\begin{bmatrix} x_1 \\ \vdots \\ x_n \end{bmatrix}\right) = \begin{bmatrix} y_1 \\ \vdots \\ y_m \end{bmatrix}$.

EXAMPLE 1

Let $A = \begin{bmatrix} 2 & 3 \\ -1 & 4 \\ 0 & 1 \end{bmatrix}$. Find $f_A(1, 2)$ and $f_A(-1, 4)$.

Solution: We have

$$f_A(1,2) = \begin{bmatrix} 2 & 3 \\ -1 & 4 \\ 0 & 1 \end{bmatrix}\begin{bmatrix} 1 \\ 2 \end{bmatrix} = \begin{bmatrix} 8 \\ 7 \\ 2 \end{bmatrix} \qquad f_A(-1,4) = \begin{bmatrix} 2 & 3 \\ -1 & 4 \\ 0 & 1 \end{bmatrix}\begin{bmatrix} -1 \\ 4 \end{bmatrix} = \begin{bmatrix} 10 \\ 17 \\ 4 \end{bmatrix}$$

EXERCISE 1

Let $A = \begin{bmatrix} 1 & 2 \\ 3 & -1 \end{bmatrix}$. Find $f_A(1, 0)$, $f_A(0, 1)$, and $f_A(2, 3)$.

What is the relationship between the value of $f_A(2, 3)$ and the values of $f_A(1, 0)$ and $f_A(0, 1)$?

EXERCISE 2

Let $A = \begin{bmatrix} 1 & 2 & -1 & 1 \\ 2 & 0 & 2 & 6 \\ 3 & 2 & 1 & 7 \end{bmatrix}$. Find $f_A(-1, 1, 1, 0)$ and $f_A(-3, 1, 0, 1)$.

Based on our results in Exercise 1, is such a relationship always true? A good way to explore this is to look at a more general example.

EXAMPLE 2

Let $A = \begin{bmatrix} a_{11} & a_{12} \\ a_{21} & a_{22} \end{bmatrix}$ and find the values of $f_A(1, 0)$, $f_A(0, 1)$, and $f_A(x_1, x_2)$.

EXAMPLE 2
(continued)

Solution: We have

$$f_A(1,0) = \begin{bmatrix} a_{11} & a_{12} \\ a_{21} & a_{22} \end{bmatrix}\begin{bmatrix} 1 \\ 0 \end{bmatrix} = \begin{bmatrix} a_{11} \\ a_{21} \end{bmatrix}$$
$$f_A(0,1) = \begin{bmatrix} a_{11} & a_{12} \\ a_{21} & a_{22} \end{bmatrix}\begin{bmatrix} 0 \\ 1 \end{bmatrix} = \begin{bmatrix} a_{12} \\ a_{22} \end{bmatrix}$$

Then we get

$$f_A(x_1,x_2) = \begin{bmatrix} a_{11} & a_{12} \\ a_{21} & a_{22} \end{bmatrix}\begin{bmatrix} x_1 \\ x_2 \end{bmatrix} = \begin{bmatrix} a_{11}x_1 + a_{12}x_2 \\ a_{21}x_1 + a_{22}x_2 \end{bmatrix} = x_1\begin{bmatrix} a_{11} \\ a_{21} \end{bmatrix} + x_2\begin{bmatrix} a_{12} \\ a_{22} \end{bmatrix}$$

We can now clearly see the relationship between the image of the standard basis vectors in $\mathbb{R}^2$ and the image of any vector $\vec{x}$. We suspect that this works for any $m \times n$ matrix A.

Theorem 1

Let $\vec{e}_1, \vec{e}_2, \ldots, \vec{e}_n$ be the standard basis vectors of $\mathbb{R}^n$, let A be an $m \times n$ matrix, and let $f_A : \mathbb{R}^n \to \mathbb{R}^m$ be the corresponding matrix mapping. Then, for any vector $\vec{x} \in \mathbb{R}^n$, we have

$$f_A(\vec{x}) = x_1 f_A(\vec{e}_1) + x_2 f_A(\vec{e}_2) + \cdots + x_n f_A(\vec{e}_n)$$

Proof: Let $A = \begin{bmatrix} \vec{a}_1 & \vec{a}_2 & \cdots & \vec{a}_n \end{bmatrix}$. Since $\vec{e}_i$ has 1 as its i-th entry and 0s elsewhere, we get $f_A(\vec{e}_i) = A\vec{e}_i = \vec{a}_i$. Thus, we have

$$f_A(\vec{x}) = A\vec{x} = x_1\vec{a}_1 + \cdots + x_n\vec{a}_n = x_1 f_A(\vec{e}_1) + \cdots x_n f_A(\vec{e}_n)$$

as required. ■

Since the images of the standard basis vectors are just the columns of A, we see that the image of any vector $\vec{x} \in \mathbb{R}^n$ is a linear combination of the columns of A. This should not be surprising as this is one of our interpretations of matrix multiplication. However, it does make us think about how a matrix mapping will affect a linear combination of vectors in $\mathbb{R}^n$. For simplicity, we look at how it affects these separately.

Theorem 2

Let A be an $m \times n$ matrix with corresponding matrix mapping $f_A : \mathbb{R}^n \to \mathbb{R}^m$. Then, for any $\vec{x}, \vec{y} \in \mathbb{R}^n$ and any $t \in \mathbb{R}$, we have
(L1) $f_A(\vec{x} + \vec{y}) = f_A(\vec{x}) + f_A(\vec{y})$
(L2) $f_A(t\vec{x}) = t f_A(\vec{x})$

Proof: Let $\vec{x}, \vec{y} \in \mathbb{R}^n$ and $t \in \mathbb{R}$. Using properties of matrix multiplication, we get

$$f_A(\vec{x} + \vec{y}) = A(\vec{x} + \vec{y}) = A\vec{x} + A\vec{y} = f_A(\vec{x}) + f_A(\vec{y})$$

and

$$f_A(t\vec{x}) = A(t\vec{x}) = tA\vec{x} = t f_A(\vec{x})$$

■

A function that satisfies property (L1) of Theorem 2 is said to **preserve addition**. Similarly, a function satisfying property (L2) of Theorem 2 is said to **preserve scalar multiplication**. Notice that a function that satisfies both properties will in fact **preserve linear combinations**—that is,

$$f_A(t_1\vec{x}_1 + \cdots + t_n\vec{x}_n) = t_1 f_A(\vec{x}_1) + \cdots + t_n f_A(\vec{x}_n)$$

We call such functions **linear mappings**.

Linear Mappings

Definition
Linear Mapping

A function $L : \mathbb{R}^n \to \mathbb{R}^m$ is called a ***linear mapping*** (or linear transformation) if for every $\vec{x}, \vec{y} \in \mathbb{R}^n$ and $t \in \mathbb{R}$ it satisfies the following properties:
(L1) $L(\vec{x} + \vec{y}) = L(\vec{x}) + L(\vec{y})$
(L2) $L(t\vec{x}) = tL(\vec{x})$

Definition
Linear Operator

A **linear operator** is a linear mapping whose domain and codomain are the same.

Theorem 2 shows that every matrix mapping is a linear mapping. In Section 1.4, we showed that $\text{proj}_{\vec{v}}$ and $\text{perp}_{\vec{v}}$ are linear operators from $\mathbb{R}^n$ to $\mathbb{R}^n$.

Remarks

1. *Linear transformation* and *linear mapping* mean exactly the same thing. Some people prefer one or the other, but we shall use both.
2. Since a linear operator L has the same domain and codomain, we often speak of a **linear operator** L **on** $\mathbb{R}^n$ to indicate that L is a linear mapping from $\mathbb{R}^n$ to $\mathbb{R}^n$.
3. For the time being, we have defined only linear mappings whose domain is $\mathbb{R}^n$ and whose codomain is $\mathbb{R}^m$. In Chapter 4, we will look at other sets that can be the domain and/or codomain of linear mappings.

EXAMPLE 3

Show that the mapping $f : \mathbb{R}^2 \to \mathbb{R}^2$ defined by $f(x_1, x_2) = (2x_1 + x_2, -3x_1 + 5x_2)$ is linear.

Solution: To prove that it is linear, we must show that it preserves both addition and scalar multiplication. For any $\vec{y}, \vec{z} \in \mathbb{R}^2$, we have

$$\begin{aligned} f(\vec{y} + \vec{z}) &= f(y_1 + z_1, y_2 + z_2) \\ &= \begin{bmatrix} 2(y_1 + z_1) + (y_2 + z_2) \\ -3(y_1 + z_1) + 5(y_2 + z_2) \end{bmatrix} \\ &= \begin{bmatrix} 2y_1 + y_2 \\ -3y_1 + 5y_2 \end{bmatrix} + \begin{bmatrix} 2z_1 + z_2 \\ -3z_1 + 5z_2 \end{bmatrix} \\ &= f(\vec{y}) + f(\vec{z}) \end{aligned}$$

EXAMPLE 3
(continued)

Thus, f preserves addition. Let $\vec{y} \in \mathbb{R}^2$ and $t \in \mathbb{R}$ be a scalar. Then we have

$$\begin{aligned} f(t\vec{y}) &= f(ty_1, ty_2) \\ &= \begin{bmatrix} 2(ty_1) + (ty_2) \\ -3(ty_1) + 5(ty_2) \end{bmatrix} \\ &= t\begin{bmatrix} 2y_1 + y_2 \\ -3y_1 + 5y_2 \end{bmatrix} \\ &= tf(\vec{y}) \end{aligned}$$

Hence, f also preserves scalar multiplication, and therefore f is a linear operator.

In the previous solution, notice that the proofs for addition and scalar multiplication are very similar. A natural question to ask is whether we can combine these into one step. The answer is *yes*! We can combine the two linearity properties into one statement:

$L : \mathbb{R}^n \to \mathbb{R}^m$ is a linear mapping if for every $\vec{x}$ and $\vec{y}$ in the domain and for any scalar $t \in \mathbb{R}$,

$$L(t\vec{x} + \vec{y}) = tL(\vec{x}) + L(\vec{y})$$

The proof that the definitions are equivalent is left for Problem D1.

EXAMPLE 4

Determine if the mapping $f : \mathbb{R}^3 \to \mathbb{R}$ defined by $f(\vec{x}) = ||\vec{x}||$ is linear.

Solution: We must test whether f preserves addition and scalar multiplication. Let $\vec{x}, \vec{y} \in \mathbb{R}^3$ and consider

$$f(\vec{x} + \vec{y}) = ||\vec{x} + \vec{y}|| \quad \text{and} \quad f(\vec{x}) + f(\vec{y}) = ||\vec{x}|| + ||\vec{y}||$$

Are these equal? By the triangle inequality, we get

$$||\vec{x} + \vec{y}|| \leq ||\vec{x}|| + ||\vec{y}||$$

and we expect equality only when one of $\vec{x}, \vec{y}$ is a multiple of the other. Therefore, we believe that these are not always equal, and consequently f does not preserve addition.

To demonstrate this, we give a counterexample: if $\vec{x} = \begin{bmatrix} 1 \\ 0 \\ 0 \end{bmatrix}$ and $\vec{y} = \begin{bmatrix} 0 \\ 1 \\ 0 \end{bmatrix}$, then

$$f(\vec{x} + \vec{y}) = f(1, 1, 0) = \left\| \begin{bmatrix} 1 \\ 1 \\ 0 \end{bmatrix} \right\| = \sqrt{2}$$

EXAMPLE 4 (continued)

but

$$f(\vec{x}) + f(\vec{y}) = \left\| \begin{bmatrix} 1 \\ 0 \\ 0 \end{bmatrix} \right\| + \left\| \begin{bmatrix} 0 \\ 1 \\ 0 \end{bmatrix} \right\| = 1 + 1 = 2$$

Thus, $f(\vec{x} + \vec{y}) \neq f(\vec{x}) + f(\vec{y})$ for any pair of vectors $\vec{x}, \vec{y}$ in $\mathbb{R}^3$, hence f is not linear.

Note that one counterexample is enough to disqualify a mapping from being linear.

EXERCISE 3

Determine whether the following mappings are linear.

(a) $f : \mathbb{R}^2 \to \mathbb{R}^2$ defined by $f(x_1, x_2) = (x_1^2, x_1 + x_2)$

(b) $g : \mathbb{R}^2 \to \mathbb{R}^2$ defined by $g(x_1, x_2) = (x_2, x_1 - x_2)$

Is Every Linear Mapping a Matrix Mapping?

We saw that every matrix determines a corresponding linear mapping. It is natural to ask whether every linear mapping can be represented as a matrix mapping.

To see if this might be true, let's look at the linear mapping f from Example 3, defined by

$$f(x_1, x_2) = (2x_1 + x_2, -3x_1 + 5x_2)$$

If this is to be a matrix mapping, then from our work with matrix mappings, we know that the columns of the matrix must be the images of the standard basis vectors. We have $f(1,0) = \begin{bmatrix} 2 \\ -3 \end{bmatrix}$, $f(0,1) = \begin{bmatrix} 1 \\ 5 \end{bmatrix}$. So, taking these as columns, we get the matrix $A = \begin{bmatrix} 2 & 1 \\ -3 & 5 \end{bmatrix}$. Now observe that

$$f_A(x_1, x_2) = \begin{bmatrix} 2 & 1 \\ -3 & 5 \end{bmatrix} \begin{bmatrix} x_1 \\ x_2 \end{bmatrix} = \begin{bmatrix} 2x_1 + x_2 \\ -3x_1 + 5x_2 \end{bmatrix} = f(x_1, x_2)$$

Hence, f can be represented as a matrix mapping.

This example not only gives us a good reason to believe it is always true but indicates how we can find the matrix for a given linear mapping L.

Theorem 3

If $L : \mathbb{R}^n \to \mathbb{R}^m$ is a linear mapping, then L can be represented as a matrix mapping, with the corresponding $m \times n$ matrix $[L]$ given by

$$[L] = \begin{bmatrix} L(\vec{e}_1) & L(\vec{e}_2) & \cdots & L(\vec{e}_n) \end{bmatrix}$$

Proof: Since L is linear, for any $\vec{x} \in \mathbb{R}^n$ we have

$$\begin{aligned} L(\vec{x}) &= L(x_1\vec{e}_1 + x_2\vec{e}_2 + \cdots + x_n\vec{e}_n) \\ &= x_1 L(\vec{e}_1) + x_2 L(\vec{e}_2) + \cdots + x_n L(\vec{e}_n) \\ &= \begin{bmatrix} L(\vec{e}_1) & L(\vec{e}_2) & \cdots & L(\vec{e}_n) \end{bmatrix} \begin{bmatrix} x_1 \\ \vdots \\ x_n \end{bmatrix} \\ &= [L]\vec{x} \end{aligned}$$

as required. ■

Remarks

1. Combining this with Theorem 2, we have proven that a mapping is linear if and only if it is a matrix mapping.
2. The matrix $[L]$ determined in this theorem depends on the standard basis. Consequently, the resulting matrix is often called the **standard matrix** of the linear mapping. Later, we shall see that we can use other bases and get different matrices corresponding to the same linear mapping.

EXAMPLE 5

Let $\vec{v} = \begin{bmatrix} 3 \\ 4 \end{bmatrix}$ and $\vec{u} = \begin{bmatrix} 1 \\ 2 \end{bmatrix}$. Find the standard matrix of the mapping $\text{proj}_{\vec{v}} : \mathbb{R}^2 \to \mathbb{R}^2$ and use it to find $\text{proj}_{\vec{v}}\, \vec{u}$.

Solution: Since $\text{proj}_{\vec{v}}$ is linear, we can apply Theorem 3 to find $[\text{proj}_{\vec{v}}]$. The first column of the matrix is the image of the first standard basis vector under $\text{proj}_{\vec{v}}$. Using the methods of Section 1.4, we get

$$\text{proj}_{\vec{v}}\, \vec{e}_1 = \frac{\vec{e}_1 \cdot \vec{v}}{\|\vec{v}\|^2}\vec{v} = \frac{1(3) + 0(4)}{3^2 + 4^2}\begin{bmatrix} 3 \\ 4 \end{bmatrix} = \begin{bmatrix} 9/25 \\ 12/25 \end{bmatrix}$$

Similarly, the second column is the image of the second basis vector:

$$\text{proj}_{\vec{v}}\, \vec{e}_2 = \frac{\vec{e}_2 \cdot \vec{v}}{\|\vec{v}\|^2}\vec{v} = \frac{0(3) + 1(4)}{25}\begin{bmatrix} 3 \\ 4 \end{bmatrix} = \begin{bmatrix} 12/25 \\ 16/25 \end{bmatrix}$$

Hence, the standard matrix of the linear mapping is

$$[\text{proj}_{\vec{v}}] = \begin{bmatrix} \text{proj}_{\vec{v}}\, \vec{e}_1 & \text{proj}_{\vec{v}}\, \vec{e}_2 \end{bmatrix} = \begin{bmatrix} 9/25 & 12/25 \\ 12/25 & 16/25 \end{bmatrix}$$

Therefore, we have

$$\text{proj}_{\vec{v}}\, \vec{u} = [\text{proj}_{\vec{v}}]\vec{u} = \begin{bmatrix} 9/25 & 12/25 \\ 12/25 & 16/25 \end{bmatrix}\begin{bmatrix} 1 \\ 2 \end{bmatrix} = \begin{bmatrix} 33/25 \\ 44/25 \end{bmatrix}$$

EXAMPLE 6

Let $G : \mathbb{R}^3 \to \mathbb{R}^2$ be defined by $G(x_1, x_2, x_3) = (x_1, x_2)$. Prove that G is linear and find the standard matrix of G.

Solution: We first prove that G is linear. For any $\vec{y}, \vec{z} \in \mathbb{R}^3$ and $t \in \mathbb{R}$, we have

$$\begin{aligned} G(t\vec{y} + \vec{z}) &= G(ty_1 + z_1, ty_2 + z_2, ty_3 + z_3) \\ &= \begin{bmatrix} ty_1 + z_1 \\ ty_2 + z_2 \end{bmatrix} = \begin{bmatrix} ty_1 \\ ty_2 \end{bmatrix} + \begin{bmatrix} z_1 \\ z_2 \end{bmatrix} \\ &= t\begin{bmatrix} y_1 \\ y_2 \end{bmatrix} + \begin{bmatrix} z_1 \\ z_2 \end{bmatrix} = tG(\vec{y}) + G(\vec{z}) \end{aligned}$$

Hence, G is linear. Thus, we can apply Theorem 3 to find its standard matrix. The images of the standard basis vectors are

$$G(1,0,0) = \begin{bmatrix} 1 \\ 0 \end{bmatrix}, \qquad G(0,1,0) = \begin{bmatrix} 0 \\ 1 \end{bmatrix}, \qquad G(0,0,1) = \begin{bmatrix} 0 \\ 0 \end{bmatrix}$$

Thus, we have

$$[G] = \begin{bmatrix} G(1,0,0) & G(0,1,0) & G(0,0,1) \end{bmatrix} = \begin{bmatrix} 1 & 0 & 0 \\ 0 & 1 & 0 \end{bmatrix}$$

Did we really need to prove that G was linear first? Couldn't we have just constructed $[G]$ using the image of the standard basis vectors and then said that because G is a matrix mapping, it is linear? *No*! You must always check the conditions of the theorem before using it. Theorem 3 says that *if* f is linear, *then* $[f]$ can be constructed from the images of the standard basis vectors. The converse is not true!

For example, consider the mapping $f(x_1, x_2) = (x_1 x_2, 0)$. The images of the standard basis vectors are $f(1,0) = \begin{bmatrix} 0 \\ 0 \end{bmatrix}$ and $f(0,1) = \begin{bmatrix} 0 \\ 0 \end{bmatrix}$, so we can construct the matrix $\begin{bmatrix} 0 & 0 \\ 0 & 0 \end{bmatrix}$. But this matrix does not represent the mapping! In particular, observe that $f(1,1) = \begin{bmatrix} 1 \\ 0 \end{bmatrix}$ but $\begin{bmatrix} 0 & 0 \\ 0 & 0 \end{bmatrix}\begin{bmatrix} 1 \\ 1 \end{bmatrix} = \begin{bmatrix} 0 \\ 0 \end{bmatrix}$. Hence, even though we can create a matrix using the images of the standard basis vectors, it does not imply that the matrix will represent that mapping, unless we already know the mapping is linear.

EXERCISE 4

Let $H : \mathbb{R}^4 \to \mathbb{R}^2$ be defined by $H(x_1, x_2, x_3, x_4) = (x_3 + x_4, x_1)$. Prove that H is linear and find the standard matrix of H.

Compositions and Linear Combinations of Linear Mappings

We now look at the usual operations on functions and how these affect linear mappings.

Definition
Operations on Linear Mappings

Let L and M be linear mappings from $\mathbb{R}^n$ to $\mathbb{R}^m$ and let $t \in \mathbb{R}$ be a scalar. We define $(L + M)$ to be the mapping from $\mathbb{R}^n$ to $\mathbb{R}^m$ such that

$$(L + M)(\vec{x}) = L(\vec{x}) + M(\vec{x})$$

We define (tL) to be the mapping from $\mathbb{R}^n$ to $\mathbb{R}^m$ such that

$$(tL)(\vec{x}) = tL(\vec{x})$$

Theorem 4

If L and M are linear mappings from $\mathbb{R}^n$ to $\mathbb{R}^m$ and $t \in \mathbb{R}$, then $(L + M)$ and (tL) are linear mappings.

Proof: Let $\vec{x}, \vec{y} \in \mathbb{R}^n$ and $s \in \mathbb{R}$. Then,

$$(tL)(s\vec{x} + \vec{y}) = tL(s\vec{x} + \vec{y}) = t[sL(\vec{x}) + L(\vec{y})] = stL(\vec{x}) + tL(\vec{y}) = s(tL)(\vec{x}) + (tL)(\vec{y})$$

Hence, (tL) is linear. Determining that $L + M$ is linear is left for Problem D2. ■

The properties we proved in Theorem 4 should seem familiar. They are properties (1) and (6) of Theorem 1.2.1 and Theorem 3.1.1. That is, the set of all possible linear mappings from $\mathbb{R}^n$ to $\mathbb{R}^m$ is closed under scalar multiplication and addition. It can be shown that the other eight properties in these theorems also hold.

Definition
Composition of Linear Mappings

Let $L : \mathbb{R}^n \to \mathbb{R}^m$ and $M : \mathbb{R}^m \to \mathbb{R}^p$ be linear mappings. The **composition** $M \circ L : \mathbb{R}^n \to \mathbb{R}^p$ is defined by

$$(M \circ L)(\vec{x}) = M(L(\vec{x}))$$

for all $\vec{x} \in \mathbb{R}^n$.

Note that the definition makes sense only if the domain of the second map M contains the codomain of the first map L, as we are evaluating M at $L(\vec{x})$. Moreover, observe that the order of the mappings in the definition is important.

EXERCISE 5

Prove that if $L : \mathbb{R}^n \to \mathbb{R}^m$ and $M : \mathbb{R}^m \to \mathbb{R}^p$ are linear mappings, then $M \circ L$ is a linear mapping.

Since compositions and linear combinations of linear mappings are linear mappings, it is natural to ask about the standard matrix of these new linear mappings.

Theorem 5

Let $L : \mathbb{R}^n \to \mathbb{R}^m$, $M : \mathbb{R}^n \to \mathbb{R}^m$, and $N : \mathbb{R}^m \to \mathbb{R}^p$ be linear mappings and $t \in \mathbb{R}$. Then,

$$[L + M] = [L] + [M], \qquad [tL] = t[L], \qquad [N \circ L] = [N][L]$$

Proof: We will prove $[tL] = t[L]$ and $[N \circ L] = [N][L]$. You are asked to prove that $[L + M] = [L] + [M]$ in Problem D2.

Observe that for any $\vec{x} \in \mathbb{R}^n$, we have

$$[tL]\vec{x} = (tL)(\vec{x}) = tL(\vec{x}) = t[L]\vec{x}$$

Hence, $[tL] = t[L]$ by Theorem 3.1.4.

We first observe that since $[N]$ is $p \times m$ and $[L]$ is $m \times n$, the matrix product $[N][L]$ is defined. For any $\vec{x} \in \mathbb{R}^n$,

$$[N \circ L]\vec{x} = (N \circ L)(\vec{x}) = N(L(\vec{x})) = N([L]\vec{x}) = [N][L]\vec{x}$$

Thus, $[N \circ L] = [N][L]$ by Theorem 3.1.4. ■

EXAMPLE 7

Let L and M be the linear operators on $\mathbb{R}^2$ defined by

$$L(x_1, x_2) = (2x_1 + x_2, x_1 + x_2) \quad \text{and} \quad M(x_1, x_2) = (x_1 - x_2, -x_1 + 2x_2)$$

Determine $[M \circ L]$.

Solution: We find that

$$L(1, 0) = \begin{bmatrix} 2 \\ 1 \end{bmatrix}, \quad L(0, 1) = \begin{bmatrix} 1 \\ 1 \end{bmatrix}, \quad M(1, 0) = \begin{bmatrix} 1 \\ -1 \end{bmatrix}, \quad M(0, 1) = \begin{bmatrix} -1 \\ 2 \end{bmatrix}$$

Thus,

$$[M \circ L] = \begin{bmatrix} 1 & -1 \\ -1 & 2 \end{bmatrix} \begin{bmatrix} 2 & 1 \\ 1 & 1 \end{bmatrix} = \begin{bmatrix} 1 & 0 \\ 0 & 1 \end{bmatrix}$$

In Example 7, $M \circ L$ is the mapping such that $(M \circ L)(\vec{x}) = \vec{x}$ for all $\vec{x} \in \mathbb{R}^n$. Moreover, the standard matrix of the mapping is the identity matrix. We thus make the following definition.

Definition
Identity Mapping

The linear mapping $\text{Id} : \mathbb{R}^n \to \mathbb{R}^n$ defined by

$$\text{Id}(\vec{x}) = \vec{x}$$

is called the **identity mapping**.

Remark

Since the mappings L and M in Example 7 satisfy $M \circ L = \text{Id}$, they are called *inverses*, as are the matrices $[M]$ and $[L]$. We will look at inverse mappings and matrices in Section 3.5.

PROBLEMS 3.2

Practice Problems

A1 Let $A = \begin{bmatrix} -2 & 3 \\ 3 & 0 \\ 1 & 5 \\ 4 & -6 \end{bmatrix}$ and let f_A be the corresponding matrix mapping.

(a) Determine the domain and codomain of f_A.

(b) Determine $f_A(2, -5)$ and $f_A(-3, 4)$.

(c) Find the images of the standard basis vectors for the domain under f_A.

(d) Determine $f_A(\vec{x})$.

(e) Check your answers in (c) and (d) by calculating $[f_A(\vec{x})]$ using Theorem 3.

A2 Let $A = \begin{bmatrix} 1 & 2 & -3 & 0 \\ 2 & -1 & 0 & 3 \\ 1 & 0 & 2 & -1 \end{bmatrix}$ and let f_A be the corresponding matrix mapping.

(a) Determine the domain and codomain of f_A.

(b) Determine $f_A(2, -2, 3, 1)$ and $f_A(-3, 1, 4, 2)$.

(c) Find the images of the standard basis vectors for the domain under f_A.

(d) Determine $f_A(\vec{x})$.

(e) Check your answers in (c) and (d) by calculating $[f_A(\vec{x})]$ using Theorem 3.

A3 For each of the following mappings, state the domain and codomain. Determine whether the mapping is linear and either prove that it is linear or give a counterexample to show why it cannot be linear.

(a) $f(x_1, x_2) = (\sin x_1, e^{x_2^2})$

(b) $g(x_1, x_2) = (2x_1 + 3x_2, x_1 - x_2)$

(c) $h(x_1, x_2) = (2x_1 + 3x_2, x_1 - x_2, x_1x_2)$

(d) $k(x_1, x_2, x_3) = (x_1 + x_2, 0, x_2 - x_3)$

(e) $\ell(x_1, x_2, x_3) = (x_2, |x_1|)$

(f) $m(x_1) = (x_1, 1, 0)$

A4 For each of the following linear mappings, determine the domain, the codomain, and the standard matrix of the mapping.

(a) $L(x_1, x_2, x_3) = (2x_1 - 3x_2 + x_3, x_2 - 5x_3)$

(b) $K(x_1, x_2, x_3, x_4) = (5x_1 + 3x_3 - x_4, x_2 - 7x_3 + 3x_4)$

(c) $M(x_1, x_2, x_3, x_4) = (x_1 - x_3 + x_4, x_1 + 2x_2 - x_3 - 3x_4, x_2 + x_3, x_1 - x_2 + x_3 - x_4)$

A5 Suppose that S and T are linear mappings with matrices

$$[S] = \begin{bmatrix} 2 & 1 & 3 \\ -1 & 0 & 2 \end{bmatrix}, \quad [T] = \begin{bmatrix} 1 & 2 & -1 \\ 2 & 2 & 3 \end{bmatrix}$$

(a) Determine the domain and codomain of each mapping.

(b) Determine the matrices that represent $(S + T)$ and $(2S - 3T)$.

A6 Suppose that S and T are linear mappings with matrices

$$[S] = \begin{bmatrix} -3 & -3 & 0 & 1 \\ 0 & 2 & 4 & 2 \end{bmatrix}, \quad [T] = \begin{bmatrix} 1 & 4 \\ -2 & 1 \\ 2 & -1 \\ 3 & -4 \end{bmatrix}$$

(a) Determine the domain and codomain of each mapping.

(b) Determine the matrices that represent $S \circ T$ and $T \circ S$.

A7 Let L, M, and N be linear mappings with matrices $[L] = \begin{bmatrix} 2 & 3 \\ -1 & 4 \\ 0 & 1 \end{bmatrix}$, $[M] = \begin{bmatrix} 1 & 1 & 2 \\ 3 & -2 & -1 \end{bmatrix}$, and $[N] = \begin{bmatrix} 2 & 1 \\ 3 & 1 \\ -3 & 0 \\ 1 & 4 \end{bmatrix}$. Determine which of the following compositions are defined and determine the domain and codomain of those that are defined.

(a) $L \circ M$ (b) $M \circ L$
(c) $L \circ N$ (d) $N \circ L$
(e) $M \circ N$ (f) $N \circ M$

A8 Let $\vec{v} = \begin{bmatrix} -2 \\ 1 \end{bmatrix}$. Determine the matrix $[\text{proj}_{\vec{v}}]$.

A9 Let $\vec{v} = \begin{bmatrix} 1 \\ 4 \end{bmatrix}$. Determine the matrix $[\text{perp}_{\vec{v}}]$.

A10 Let $\vec{v} = \begin{bmatrix} 2 \\ 2 \\ -1 \end{bmatrix}$. Determine the matrix $[\text{proj}_{\vec{v}}]$.

Homework Problems

B1 Let $A = \begin{bmatrix} -1 & 4 & -2 \\ -5 & 3 & 1 \end{bmatrix}$ and let f_A be the corresponding matrix mapping.

(a) Determine the domain and codomain of f_A.
(b) Determine $f_A(3, 4, -5)$ and $f_A(-2, 1, -4)$.
(c) Find the images of the standard basis vectors for the domain under f_A.
(d) Determine $f_A(\vec{x})$.
(e) Check your answers in (c) and (d) by calculating $[f_A(\vec{x})]$ using Theorem 3.

B2 Let $A = \begin{bmatrix} 2 & 1 & 0 \\ 0 & 2 & 3 \\ 5 & 7 & 9 \\ 2 & 4 & 8 \end{bmatrix}$ and let f_A be the corresponding matrix mapping.

(a) Determine the domain and codomain of f_A.
(b) Determine $f_A(-4, 2, 1)$ and $f_A(3, -3, 2)$.
(c) Find the images of the standard basis vectors for the domain under f_A.
(d) Determine $f_A(\vec{x})$.
(e) Check your answers in (c) and (d) by calculating $[f_A(\vec{x})]$ using Theorem 3.

B3 For each of the following mappings, state the domain and codomain. Determine whether the mapping is linear and either prove that it is linear or give a counterexample to show why it cannot be linear.

(a) $f(x_1, x_2) = (2x_2, x_1 - x_2)$
(b) $g(x_1, x_2) = (\cos x_2, x_1 x_2^3)$
(c) $h(x_1, x_2, x_3) = (0, 0, x_1 + x_2 + x_3)$
(d) $k(x_1, x_2, x_3) = (0, 0, 0)$
(e) $\ell(x_1, x_2, x_3, x_4) = (x_1, 1, x_4)$
(f) $m(x_1, x_2, x_3, x_4) = (x_1 + x_2 - x_3)$

B4 For each of the following linear mappings, determine the domain, the codomain, and the standard matrix of the mapping.

(a) $L(x_1, x_2, x_3) = (2x_1 - 3x_2, x_2, 4x_1 - 5x_2)$
(b) $K(x_1, x_2, x_3, x_4) = (2x_1 - x_3 + 3x_4, -x_1 - 2x_2 + 2x_3 + x_4, 3x_2 + x_3)$
(c) $M(x_1, x_2, x_3) = (x_3 - x_1, 0, 5x_1 + x_2)$

B5 Suppose that S and T are linear mappings with matrices

$$[S] = \begin{bmatrix} 2 & 1 \\ 1 & -2 \\ 3 & 1 \end{bmatrix} \quad \text{and} \quad [T] = \begin{bmatrix} 3 & 1 \\ 1 & -2 \\ 1 & 0 \end{bmatrix}$$

(a) Determine the domain and codomain of each mapping.
(b) Determine the matrices that represent $(T + S)$ and $(-S + 2T)$.

B6 Suppose that S and T are linear mappings with matrices

$$[S] = \begin{bmatrix} 4 & -3 \\ 1 & 1 \\ 5 & -3 \\ -2 & 0 \end{bmatrix}, [T] = \begin{bmatrix} 4 & 0 & 2 & 3 \\ -2 & 1 & 3 & 0 \end{bmatrix}$$

(a) Determine the domain and codomain of each mapping.
(b) Determine the matrices that represent $S \circ T$ and $T \circ S$.

B7 Let L, M, N be linear mappings with matrices $[L] = \begin{bmatrix} 1 & 4 \\ 3 & 2 \\ -1 & 0 \end{bmatrix}$, $[M] = \begin{bmatrix} 3 & 2 & 2 & 0 \\ 1 & -1 & -1 & 2 \end{bmatrix}$, and

$[N] = \begin{bmatrix} 2 & 1 & 1 \\ 3 & 1 & 4 \\ -3 & 0 & 7 \\ 1 & -4 & 1 \end{bmatrix}$. Determine which of the following compositions are defined and determine the domain and codomain of those that are defined.

(a) $L \circ M$ (b) $M \circ L$
(c) $L \circ N$ (d) $N \circ L$
(e) $M \circ N$ (f) $N \circ M$

B8 Let $\vec{v} = \begin{bmatrix} 1 \\ -3 \end{bmatrix}$. Determine the matrix $[\text{proj}_{\vec{v}}]$.

B9 Let $\vec{v} = \begin{bmatrix} 2 \\ -1 \end{bmatrix}$. Determine the matrix $[\text{perp}_{\vec{v}}]$.

B10 Let $\vec{v} = \begin{bmatrix} 1 \\ -1 \\ 4 \end{bmatrix}$. Determine the matrix $[\text{proj}_{\vec{v}}]$.

B11 Let $\vec{v} = \begin{bmatrix} 1 \\ -2 \\ 2 \end{bmatrix}$. Determine the matrix $[\text{perp}_{\vec{v}}]$.

Conceptual Problems

D1 Let $L : \mathbb{R}^n \to \mathbb{R}^m$. Show that for any $\vec{x}, \vec{y} \in \mathbb{R}^n$ and $t \in \mathbb{R}$, L satisfies $L(\vec{x}+\vec{y}) = L(\vec{x})+L(\vec{y})$ and $L(t\vec{x}) = tL(\vec{x})$ if and only if $L(t\vec{x} + \vec{y}) = tL(\vec{x}) + L(\vec{y})$.

D2 Let $L : \mathbb{R}^n \to \mathbb{R}^m$ and $M : \mathbb{R}^n \to \mathbb{R}^m$ be linear mappings. Prove that $(L+M)$ is linear and that $[L+M] = [L] + [M]$.

D3 Let $\vec{v} \in \mathbb{R}^n$ be a fixed vector and define a mapping $\text{DOT}_{\vec{v}}$ by $\text{DOT}_{\vec{v}}\, \vec{x} = \vec{v} \cdot \vec{x}$. Verify that $\text{DOT}_{\vec{v}}$ is a linear mapping. What is its codomain? Verify that the matrix of this linear mapping can be written as $\vec{v}^T$.

D4 If $\vec{u}$ is a unit vector, show that $[\text{proj}_{\vec{u}}] = \vec{u}\vec{u}^T$.

D5 Let $\vec{v} \in \mathbb{R}^3$ be a fixed vector and define a mapping $\text{CROSS}_{\vec{v}}$ by $\text{CROSS}_{\vec{v}}\, \vec{x} = \vec{v} \times \vec{x}$. Verify that $\text{CROSS}_{\vec{v}}$ is a linear mapping and determine its codomain and standard matrix.

3.3 Geometrical Transformations

Geometrical transformations have long been of great interest to mathematicians. They also have many important applications. Physicists and engineers often rely on simple geometrical transformations to gain understanding of the properties of materials or structures they wish to examine. For example, structural engineers use stretches, shears, and rotations to understand the deformation of materials. Material scientists use rotations and reflections to analyse crystals and other fine structures. Many of these simple geometrical transformations in $\mathbb{R}^2$ and $\mathbb{R}^3$ are linear. The following is a brief partial catalogue of some of these transformations and their matrix representations. ($\text{proj}_{\vec{v}}$ and $\text{perp}_{\vec{v}}$ belong to the list of geometrical transformations, too, but they were discussed in Chapter 1 and so are not included here.)

Rotations in the Plane

$R_\theta : \mathbb{R}^2 \to \mathbb{R}^2$ is defined to be the transformation that rotates $\vec{x}$ counterclockwise through angle θ to the image $R_\theta(\vec{x})$. See Figure 3.3.1. Is R_θ linear? Since a rotation does not change lengths, we get the same result if we first multiply a vector $\vec{x}$ by a scalar t and then rotate through angle θ, or if we first rotate through θ and then multiply by t. Thus, $R_\theta(t\vec{x}) = tR_\theta(\vec{x})$ for any $t \in \mathbb{R}$ and any $\vec{x} \in \mathbb{R}^2$. Since the shape of a parallelogram is not altered by a rotation, the picture for the parallelogram rule of addition should be unchanged under a rotation, so $R_\theta(\vec{x} + \vec{y}) = R_\theta(\vec{x}) + R_\theta(\vec{y})$. Thus,

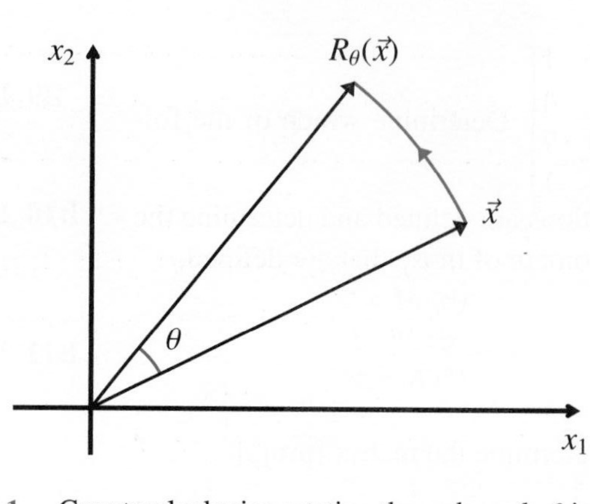

Figure 3.3.1 Counterclockwise rotation through angle θ in the plane.

R_θ is linear. A more formal proof is obtained by showing that R_θ can be represented as a matrix mapping.

Assuming that R_θ is linear, we can use the definition of the standard matrix to calculate $[R_\theta]$. From Figure 3.3.2, we see that

$$R_\theta(1, 0) = \begin{bmatrix} \cos\theta \\ \sin\theta \end{bmatrix}$$

$$R_\theta(0, 1) = \begin{bmatrix} -\sin\theta \\ \cos\theta \end{bmatrix}$$

Hence,

$$[R_\theta] = \begin{bmatrix} \cos\theta & -\sin\theta \\ \sin\theta & \cos\theta \end{bmatrix}$$

It follows from the calculation of the product $[R_\theta]\vec{x}$ that

$$R_\theta(\vec{x}) = \begin{bmatrix} x_1\cos\theta - x_2\sin\theta \\ x_1\sin\theta + x_2\cos\theta \end{bmatrix}$$

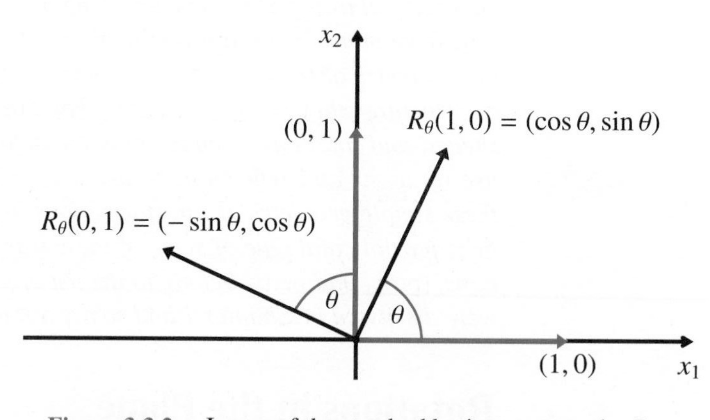

Figure 3.3.2 Image of the standard basis vectors under R_θ.

EXAMPLE 1

What is the matrix of rotation of $\mathbb{R}^2$ through angle $2\pi/3$?

Solution: Since $\cos\frac{2\pi}{3} = -\frac{1}{2}$ and $\sin\frac{2\pi}{3} = \frac{\sqrt{3}}{2}$,

$$[R_{2\pi/3}] = \begin{bmatrix} -1/2 & -\sqrt{3}/2 \\ \sqrt{3}/2 & -1/2 \end{bmatrix}$$

EXERCISE 1

Determine $[R_{\pi/4}]$ and use it to calculate $R_{\pi/4}(1, 1)$. Illustrate with a sketch.

Rotation Through Angle θ About the x_3-axis in $\mathbb{R}^3$

Figure 3.3.3 demonstrates a counterclockwise rotation with respect to the right-handed standard basis. This rotation leaves x_3 unchanged, so that if the transformation is denoted R, then $R(0, 0, 1) = (0, 0, 1)$. Together with the previous case, this tells us that the matrix of this rotation is

$$[R] = \begin{bmatrix} \cos\theta & -\sin\theta & 0 \\ \sin\theta & \cos\theta & 0 \\ 0 & 0 & 1 \end{bmatrix}$$

These ideas can be adapted to give rotations about the other coordinate axes. Is it possible to determine the matrix of a rotation about an arbitrary axis in $\mathbb{R}^3$? We shall see how to do this in Chapter 7.

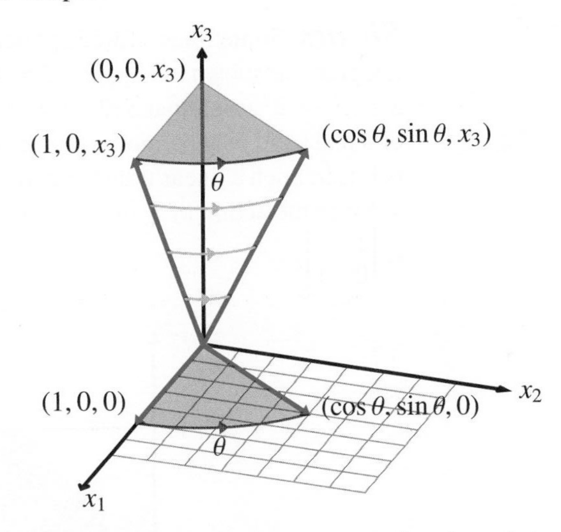

Figure 3.3.3 A right-handed counterclockwise rotation about the x_3-axis in $\mathbb{R}^3$.

Stretches Imagine that all lengths in the x_1-direction in the plane are stretched by a scalar factor $t > 0$, while lengths in the x_2-direction are left unchanged (Figure 3.3.4). This linear transformation, called a "stretch by factor t in the x_1-direction," has matrix $\begin{bmatrix} t & 0 \\ 0 & 1 \end{bmatrix}$. (If $t < 1$, you might prefer to call this a *shrink*.) It should be obvious that stretches can also be defined in the x_2-direction and in higher dimensions. Stretches are important in understanding the deformation of solids.

Contractions and Dilations If a linear operator $T : \mathbb{R}^2 \to \mathbb{R}^2$ has matrix $\begin{bmatrix} t & 0 \\ 0 & t \end{bmatrix}$, with $t > 0$, then for any $\vec{x}$, $T(\vec{x}) = t\vec{x}$, so that this transformation stretches vectors in all directions by the same factor. Thus, for example, a circle of radius 1 centred at the origin is mapped to a circle of radius t at the origin. If $0 < t < 1$, such a transformation is called a **contraction**; if $t > 1$, it is a **dilation**.

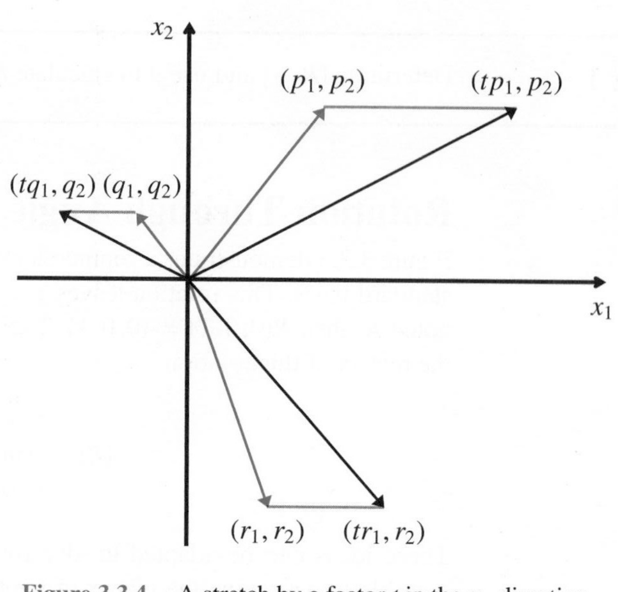

Figure 3.3.4 A stretch by a factor t in the x_1-direction.

Shears Sometimes a force applied to a rectangle will cause it to deform into a parallelogram, as shown in Figure 3.3.5. The change can be described by the transformation $S : \mathbb{R}^2 \to \mathbb{R}^2$, such that $S(2, 0) = (2, 0)$ and $S(0, 1) = (s, 1)$. Although the deformation of a real solid may be more complicated, it is usual to assume that the transformation S is linear. Such a linear transformation is called a **shear** in the direction of x_1 by amount s. Since the action of S on the standard basis vectors is known, we find that its matrix is $\begin{bmatrix} 1 & s \\ 0 & 1 \end{bmatrix}$.

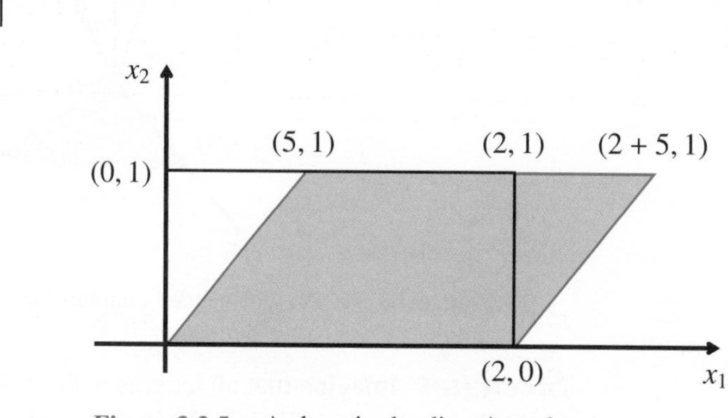

Figure 3.3.5 A shear in the direction of x_1 by amount s.

Reflections in Coordinate Axes in $\mathbb{R}^2$ or Coordinates Planes in $\mathbb{R}^3$ Let $R : \mathbb{R}^2 \to \mathbb{R}^2$ be a reflection in the x_1-axis (see Figure 3.3.6). Then each vector corresponding to a point above the axis is mapped by R to the mirror image vector below. Hence, $R(x_1, x_2) = (x_1, -x_2)$, and it follows that this transformation is linear with matrix $\begin{bmatrix} 1 & 0 \\ 0 & -1 \end{bmatrix}$. Similarly, a reflection in the x_2-axis has the matrix $\begin{bmatrix} -1 & 0 \\ 0 & 1 \end{bmatrix}$.

Next, consider the reflection $T : \mathbb{R}^3 \to \mathbb{R}^3$ that reflects in the x_1x_2-plane (that is, the plane $x_3 = 0$). Points above the plane are reflected to points below the plane. The matrix of this reflection is $\begin{bmatrix} 1 & 0 & 0 \\ 0 & 1 & 0 \\ 0 & 0 & -1 \end{bmatrix}$.

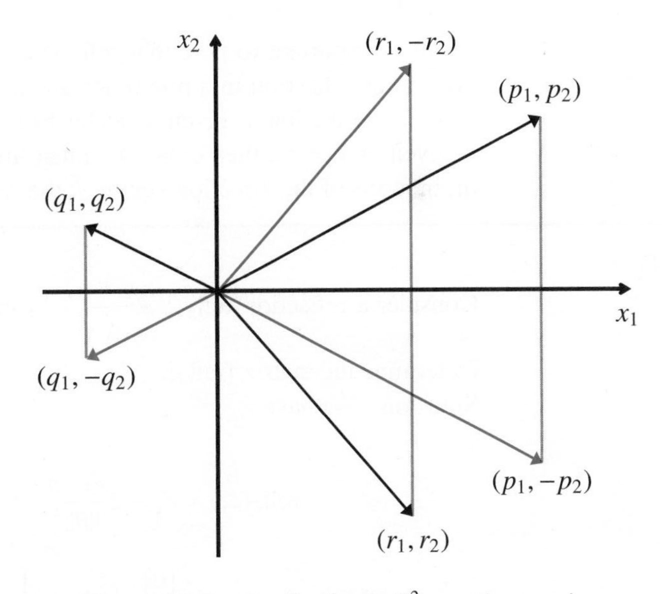

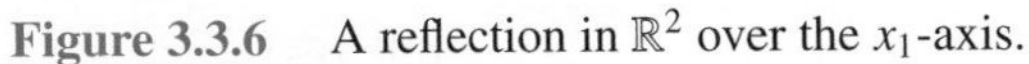
Figure 3.3.6 A reflection in $\mathbb{R}^2$ over the x_1-axis.

EXERCISE 2 Write the matrices for the reflections in the other two coordinate planes in $\mathbb{R}^3$.

General Reflections We consider only reflections in (or "across") lines in $\mathbb{R}^2$ or planes in $\mathbb{R}^3$ that pass through the origin. Reflections in lines or planes not containing the origin involve translations (which are not linear) as well as linear mappings.

Consider the plane in $\mathbb{R}^3$ with equation $\vec{n} \cdot \vec{x} = 0$. Since a reflection is related to $\text{proj}_{\vec{n}}$, a **reflection in the plane with normal vector** $\vec{n}$ will be denoted $\text{refl}_{\vec{n}}$. If a vector $\vec{p}$ corresponds to a point P that does not lie in the plane, its image under $\text{refl}_{\vec{n}}$ is the vector that corresponds to the point on the opposite side of the plane, lying on a line through P perpendicular to the plane of reflection, at the same distance from the plane as P. Figure 3.3.7 shows reflection in a line. From the figure, we see that

$$\text{refl}_{\vec{n}}(\vec{p}) = \vec{p} - 2\,\text{proj}_{\vec{n}}(\vec{p})$$

Since $\text{proj}_{\vec{n}}$ is linear, it is easy to see that $\text{refl}_{\vec{n}}$ is also linear.

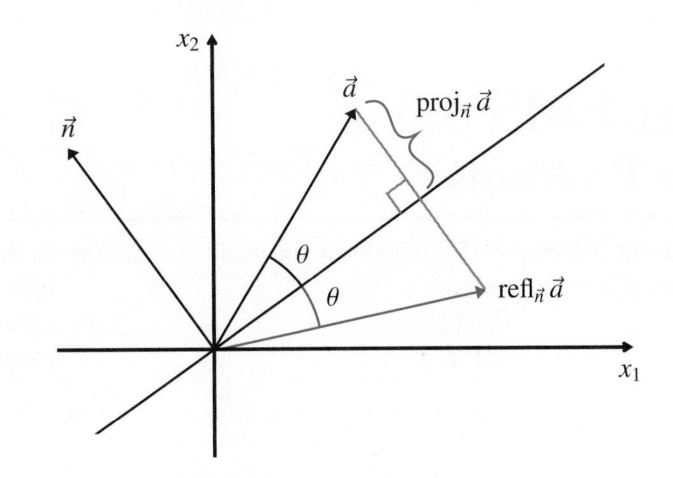

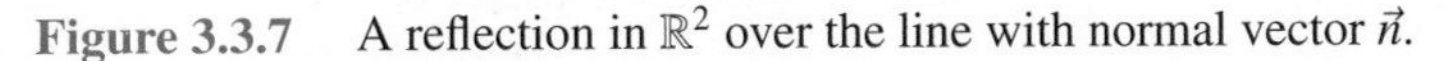
Figure 3.3.7 A reflection in $\mathbb{R}^2$ over the line with normal vector $\vec{n}$.

It is important to note that $\text{refl}_{\vec{n}}$ is a reflection with normal vector $\vec{n}$. The calculations for reflection in a line in $\mathbb{R}^2$ are similar to those for a plane, provided that the equation of the line is given in scalar form $\vec{n} \cdot \vec{x} = 0$. If the vector equation of the line is given as $\vec{x} = t\vec{d}$, then either we must find a normal vector $\vec{n}$ and proceed as above, or, in terms of the direction vector $\vec{d}$, the reflection will map $\vec{p}$ to $(\vec{p} - 2\,\text{perp}_{\vec{d}}\,\vec{p})$.

EXAMPLE 2

Consider a reflection $\text{refl}_{\vec{n}} : \mathbb{R}^3 \to \mathbb{R}^3$ over the plane with normal vector $\vec{n} = \begin{bmatrix} 1 \\ -1 \\ 2 \end{bmatrix}$.

Determine the matrix $[\text{refl}_{\vec{n}}]$.

Solution: We have

$$\text{refl}_{\vec{n}}(\vec{e}_1) = \vec{e}_1 - 2\frac{\vec{e}_1 \cdot \vec{n}}{\|\vec{n}\|^2}\vec{n} = \begin{bmatrix} 1 \\ 0 \\ 0 \end{bmatrix} - 2\left(\frac{1}{6}\right)\begin{bmatrix} 1 \\ -1 \\ 2 \end{bmatrix} = \begin{bmatrix} 2/3 \\ 1/3 \\ -2/3 \end{bmatrix}$$

$$\text{refl}_{\vec{n}}(\vec{e}_2) = \begin{bmatrix} 0 \\ 1 \\ 0 \end{bmatrix} - 2\left(\frac{-1}{6}\right)\begin{bmatrix} 1 \\ -1 \\ 2 \end{bmatrix} = \begin{bmatrix} 1/3 \\ 2/3 \\ 2/3 \end{bmatrix}$$

$$\text{refl}_{\vec{n}}(\vec{e}_3) = \begin{bmatrix} 0 \\ 0 \\ 1 \end{bmatrix} - 2\left(\frac{2}{6}\right)\begin{bmatrix} 1 \\ -1 \\ 2 \end{bmatrix} = \begin{bmatrix} -2/3 \\ 2/3 \\ -1/3 \end{bmatrix}$$

Hence,

$$[\text{refl}_{\vec{n}}] = \begin{bmatrix} 2/3 & 1/3 & -2/3 \\ 1/3 & 2/3 & 2/3 \\ -2/3 & 2/3 & -1/3 \end{bmatrix}$$

Note that we could have also computed $[\text{refl}_{\vec{n}}]$ in the following way. The equation for $\text{refl}_{\vec{n}}(\vec{p})$ can be written as

$$\text{refl}_{\vec{n}}(\vec{p}) = \text{Id}(\vec{p}) - 2\,\text{proj}_{\vec{n}}(\vec{p}) = (\,\text{Id} + (-2)\,\text{proj}_{\vec{n}}\,)(\vec{p})$$

Thus,

$$[\text{refl}_{\vec{n}}] = [\text{Id} - 2\,\text{proj}_{\vec{n}}] = I - 2[\text{proj}_{\vec{n}}]$$

PROBLEMS 3.3

Practice Problems

A1 Determine the matrices of the rotations in the plane through the following angles.

(a) $\frac{\pi}{2}$ (b) π
(c) $-\frac{\pi}{4}$ (d) $\frac{2\pi}{5}$

A2 (a) In the plane, what is the matrix of a stretch S by a factor 5 in the x_2-direction?

(b) Calculate the composition of S *followed by* a rotation through angle θ.

(c) Calculate the composition of S *following* a rotation through angle θ.

A3 Determine the matrices of the following reflections in $\mathbb{R}^2$.

(a) R is a reflection in the line $x_1 + 3x_2 = 0$.
(b) S is a reflection in the line $2x_1 = x_2$.

A4 Determine the matrix of the reflections in the following plane in $\mathbb{R}^3$.

(a) $x_1 + x_2 + x_3 = 0$
(b) $2x_1 - 2x_2 - x_3 = 0$

A5 (a) Let $D : \mathbb{R}^3 \to \mathbb{R}^3$ be the dilation with factor $t = 5$ and let $\text{inj} : \mathbb{R}^3 \to \mathbb{R}^4$ be defined by $\text{inj}(x_1, x_2, x_3) = (x_1, x_2, 0, x_3)$. Determine the matrix of $\text{inj} \circ D$.

(b) Let $P : \mathbb{R}^3 \to \mathbb{R}^2$ be defined by $P(x_1, x_2, x_3) = (x_2, x_3)$ and let S be the shear in $\mathbb{R}^3$ such that $S(x_1, x_2, x_3) = (x_1, x_2, x_3 + 2x_1)$. Determine the matrix of $P \circ S$.

(c) Can you define a shear $T : \mathbb{R}^2 \to \mathbb{R}^2$ such that $T \circ P = P \circ S$, where P and S are as in part (b)?

(d) Let $Q : \mathbb{R}^3 \to \mathbb{R}^2$ be defined by $Q(x_1, x_2, x_3) = (x_1, x_2)$. Determine the matrix of $Q \circ S$, where S is the mapping in part (b).

Homework Problems

B1 Determine the matrices of the rotations in the plane through the following angles.

(a) $-\frac{\pi}{2}$ (b) $-\pi$
(c) $\frac{3\pi}{4}$ (d) $-\frac{\pi}{5}$

B2 (a) In the plane, what is the matrix of a stretch S by a factor 0.6 in the x_2-direction?

(b) Calculate the composition of S *followed by* a rotation through angle θ.

(c) Calculate the composition of S *following* a rotation through angle $\frac{\pi}{4}$.

B3 Determine the matrices of the following reflections in $\mathbb{R}^2$.

(a) R is a reflection in the line $x_1 - 5x_2 = 0$.
(b) S is a reflection in the line $3x_1 + 4x_2 = 0$.

B4 Determine the matrix of the reflections in the following plane in $\mathbb{R}^3$.

(a) $x_1 - 3x_2 - x_3 = 0$
(b) $2x_1 + x_2 - x_3 = 0$

B5 (a) Let $C : \mathbb{R}^3 \to \mathbb{R}^3$ be contraction with factor $1/3$ and let $\text{inj} : \mathbb{R}^3 \to \mathbb{R}^5$ be defined by $\text{inj}(x_1, x_2, x_3) = (0, x_1, 0, x_2, x_3)$. Determine the matrix of $\text{inj} \circ C$.

(b) Let $S : \mathbb{R}^3 \to \mathbb{R}^2$ be the shear defined by $S(x_1, x_2, x_3) = (x_1, x_2 - 2x_3, x_3)$. Determine the matrices $C \circ S$ and $S \circ C$, where C is the contraction in part (a).

(c) Let $T : \mathbb{R}^3 \to \mathbb{R}^3$ be the shear defined by $T(x_1, x_2, x_3) = (x_1 + 3x_2, x_2, x_3)$. Determine the matrix of $S \circ T$ and $T \circ S$, where S is the mapping in part (b).

Conceptual Problems

D1 Verify that for rotations in the plane $[R_\alpha \circ R_\theta] = [R_\alpha][R_\theta] = [R_{\alpha+\theta}]$.

D2 In Problem A3, $[R] = \begin{bmatrix} 4/5 & -3/5 \\ -3/5 & -4/5 \end{bmatrix}$ and $[S] = \begin{bmatrix} -3/5 & 4/5 \\ 4/5 & 3/5 \end{bmatrix}$ are reflection matrices. Calculate $[R \circ S]$ and verify that it can be identified as the matrix of a rotation. Determine the angle of the rotation. Draw a picture illustrating how the composition of these reflections is a rotation.

D3 In $\mathbb{R}^3$, calculate the matrix of the composition of a reflection in the x_2x_3-plane followed by a reflection in the x_1x_2-plane and identify it as a rotation about some coordinate axis. What is the angle of the rotation?

D4 (a) Construct a 2×2 matrix $A \neq I$ such that $A^3 = I$. (Hint: Think geometrically.)
(b) Construct a 2×2 matrix $A \neq I$ such that $A^5 = I$.

D5 From geometrical considerations, we know that $\text{refl}_{\vec{n}} \circ \text{refl}_{\vec{n}} = \text{Id}$. Verify the corresponding matrix equation. (Hint: $[\text{refl}_{\vec{n}}] = I - 2[\text{proj}_{\vec{n}}]$ and $\text{proj}_{\vec{n}}$ satisfied the projection property from Section 1.4.)

3.4 Special Subspaces for Systems and Mappings: Rank Theorem

In the preceeding two sections, we have seen how to represent any linear mapping as a matrix mapping. We now use subspaces of $\mathbb{R}^n$ to explore this further by examining the connection between the properties of L *and its standard matrix* [L] = A. *This will also allow us to show how the solutions of the systems of equations* $A\vec{x} = \vec{b}$ *and* $A\vec{x} = \vec{0}$ *are related.*

Recall from Section 1.2 that a subspace of $\mathbb{R}^n$ is a non-empty subset of $\mathbb{R}^n$ that is closed under addition and closed under scalar multiplication. Moreover, we proved that $\text{Span}\{\vec{v}_1, \dots, \vec{v}_k\}$ is a subspace of $\mathbb{R}^n$. Throughout this section, L will always denote a linear mapping from $\mathbb{R}^n$ to $\mathbb{R}^m$, and A will denote the standard matrix of L.

Solution Space and Nullspace

Theorem 1 Let A be an $m \times n$ matrix. The set $S = \{\vec{x} \in \mathbb{R}^n \mid A\vec{x} = \vec{0}\}$ of all solutions to a homogeneous system $A\vec{x} = \vec{0}$ is a subspace of $\mathbb{R}^n$.

Proof: We have $\vec{0} \in S$ since $A\vec{0} = \vec{0}$.
Let $\vec{x}, \vec{y} \in S$. Then $A(\vec{x} + \vec{y}) = A\vec{x} + A\vec{y} = \vec{0} + \vec{0} = \vec{0}$, so we have $\vec{x} + \vec{y} \in S$.
Let $\vec{x} \in S$ and $t \in \mathbb{R}$. Then $A(t\vec{x}) = tA\vec{x} = t\vec{0} = \vec{0}$. Therefore, $t\vec{x} \in S$.
So, by definition, S is a subspace of $\mathbb{R}^n$. ■

From this result, we make the following definition.

Definition
Solution Space

The set $S = \{\vec{x} \in \mathbb{R}^n \mid A\vec{x} = \vec{0}\}$ of all solutions to a homogeneous system $A\vec{x} = \vec{0}$ is called the **solution space** of the system $A\vec{x} = \vec{0}$.

EXAMPLE 1

Find the solution space of the homogeneous system $x_1 + 2x_2 - 3x_3 = 0$.

Solution: We can solve this very easily by using the methods of Chapter 2. In particular, we find that the general solution is

$$\vec{x} = s\begin{bmatrix}-2\\1\\0\end{bmatrix} + t\begin{bmatrix}3\\0\\1\end{bmatrix}, \quad s, t \in \mathbb{R}$$

or

$$\vec{x} \in \operatorname{Span}\left\{\begin{bmatrix}-2\\1\\0\end{bmatrix}, \begin{bmatrix}3\\0\\1\end{bmatrix}\right\}$$

EXERCISE 1

Let $A = \begin{bmatrix}1 & 0 & 2 & 0 & 4\\0 & 1 & 3 & 0 & 5\\0 & 0 & 0 & 1 & 6\end{bmatrix}$. Find the solution space of $A\vec{x} = \vec{0}$.

Notice that in both of these problems, the solution space is displayed automatically as the span of a set of linearly independent vectors.

For the linear mapping L with standard matrix $A = [L]$, we see that $L(\vec{x}) = A\vec{x}$, by definition of the standard matrix. Hence, the vectors $\vec{x}$ such that $L(\vec{x}) = \vec{0}$ are exactly the same as the vectors satisfying $A\vec{x} = \vec{0}$. Thus, the set of all vectors $\vec{x}$ such that $L(\vec{x}) = \vec{0}$ also forms a subspace of $\mathbb{R}^n$.

Definition
Nullspace

The **nullspace** of a linear mapping L is the set of all vectors whose image under L is the zero vector $\vec{0}$. We write

$$\operatorname{Null}(L) = \{\vec{x} \in \mathbb{R}^n \mid L(\vec{x}) = \vec{0}\}$$

Remark

The word **kernel**—and the notation $\ker(L) = \{\vec{x} \in \mathbb{R}^n \mid L(\vec{x}) = \vec{0}\}$—is often used in place of *nullspace*.

EXAMPLE 2

Let $\vec{v} = \begin{bmatrix}2\\-1\\3\end{bmatrix}$. Find the nullspace of $\operatorname{proj}_{\vec{v}} : \mathbb{R}^3 \to \mathbb{R}^3$.

Solution: Since vectors orthogonal to $\vec{v}$ will be mapped to $\vec{0}$ by $\operatorname{proj}_{\vec{v}}$, the nullspace of $\operatorname{proj}_{\vec{v}}$ is the set of all vectors orthogonal to $\vec{v}$. That is, it is the plane passing through the origin with normal vector $\vec{v}$. Hence, we get

$$\operatorname{Null}(\operatorname{proj}_{\vec{v}}) = \left\{\begin{bmatrix}x_1\\x_2\\x_3\end{bmatrix} \in \mathbb{R}^3 \mid 2x_1 - x_2 + 3x_3 = 0\right\}$$

EXAMPLE 3

Let $L : \mathbb{R}^2 \to \mathbb{R}^3$ be defined by $L(x_1, x_2) = (2x_1 - x_2, 0, x_1 + x_2)$. Find Null($L$).

Solution: We have $\begin{bmatrix} x_1 \\ x_2 \end{bmatrix} \in \text{Null}(L)$ if $L(x_1, x_2) = (0, 0, 0)$. For this to be true, we must have

$$\begin{aligned} 2x_1 - x_2 &= 0 \\ x_1 + x_2 &= 0 \end{aligned}$$

We see that the only solution to this homogeneous system is $\vec{x} = \vec{0}$. Thus $\text{Null}(L) = \{\vec{0}\}$.

EXERCISE 2

Let $L : \mathbb{R}^3 \to \mathbb{R}^2$ be defined by $L(x_1, x_2, x_3) = (x_1 - x_2, -2x_1 + 2x_2 + x_3)$. Find Null($L$).

To match our work with linear mappings, we make the following definition for matrices.

Definition
Nullspace

Let A be an $m \times n$ matrix. Then the **nullspace** of A is

$$\text{Null}(A) = \{\vec{x} \in \mathbb{R}^n \mid A\vec{x} = \vec{0}\}$$

It should be clear that for any linear mapping $L : \mathbb{R}^n \to \mathbb{R}^m$, $\text{Null}(L) = \text{Null}([L])$.

Solution Set of $A\vec{x} = \vec{b}$

Next, we want to consider solutions for a non-homogeneous system $A\vec{x} = \vec{b}$, $\vec{b} \neq \vec{0}$ and compare this solution set with the solution space for the **corresponding homogeneous system** $A\vec{x} = \vec{0}$ (that is, the system with the same coefficient matrix A).

EXAMPLE 4

Find the general solution of the system $x_1 + 2x_2 - 3x_3 = 5$.

Solution: The general solution of $x_1 + 2x_2 - 3x_3 = 5$ is

$$\vec{x} = \begin{bmatrix} 5 \\ 0 \\ 0 \end{bmatrix} + s\begin{bmatrix} -2 \\ 1 \\ 0 \end{bmatrix} + t\begin{bmatrix} 3 \\ 0 \\ 1 \end{bmatrix}, \quad s, t \in \mathbb{R}$$

EXERCISE 3

Let $A = \begin{bmatrix} 1 & 0 & 2 & 0 & 4 \\ 0 & 1 & 3 & 0 & 5 \\ 0 & 0 & 0 & 1 & 6 \end{bmatrix}$ and $\vec{b} = \begin{bmatrix} 7 \\ 8 \\ 9 \end{bmatrix}$. Find the general solution of $A\vec{x} = \vec{b}$.

Observe that in Example 4 and Exercise 3, the general solution is obtained from the solution space of the corresponding homogeneous problem (Example 1 and Exercise 1, respectively) by a translation. We prove this result in the following theorem.

Theorem 2

Let $\vec{p}$ be a solution of the system of linear equations $A\vec{x} = \vec{b}$, $\vec{b} \neq \vec{0}$.

(1) If $\vec{v}$ is any other solution of the same system, then $A(\vec{p} - \vec{v}) = \vec{0}$, so that $\vec{p} - \vec{v}$ is a solution of the corresponding homogeneous system $A\vec{x} = \vec{0}$.

(2) If $\vec{h}$ is any solution of the corresponding system $A\vec{x} = \vec{0}$, then $\vec{p} + \vec{h}$ is a solution of the system $A\vec{x} = \vec{b}$.

Proof: (i) Suppose that $A\vec{v} = \vec{b}$. Then $A(\vec{p} - \vec{v}) = A\vec{p} - A\vec{v} = \vec{b} - \vec{b} = \vec{0}$.

(ii) Suppose that $A\vec{h} = \vec{0}$. Then $A(\vec{p} + \vec{h}) = A\vec{p} + A\vec{h} = \vec{b} + \vec{0} = \vec{b}$.

■

The solution $\vec{p}$ of the non-homogeneous system is sometimes called a **particular solution** of the system. Theorem 2 can thus be restated as follows: any solution of the non-homogeneous system can be obtained by adding a solution of the corresponding homogeneous system to a particular solution.

Range of L and Columnspace of A

Definition
Range

The **range** of a linear mapping $L: \mathbb{R}^n \to \mathbb{R}^m$ is defined to be the set

$$\text{Range}(L) = \{L(\vec{x}) \in \mathbb{R}^m \mid \vec{x} \in \mathbb{R}^n\}$$

EXAMPLE 5

Let $\vec{v} = \begin{bmatrix} 1 \\ 1 \\ 3 \end{bmatrix}$ and consider the linear mapping $\text{proj}_{\vec{v}} : \mathbb{R}^3 \to \mathbb{R}^3$. By definition, every image of this mapping is a multiple of $\vec{v}$, so the range of the mapping is the set of all multiples of $\vec{v}$. On the other hand, the range of $\text{perp}_{\vec{v}}$ is the set of all vectors orthogonal to $\vec{v}$. Note that in each of these cases, the range is a subset of the codomain.

EXAMPLE 6

If L is a rotation, reflection, contraction, or dilation in $\mathbb{R}^3$, then, because of the geometry of the mapping, it is easy to see that the range of L is all of $\mathbb{R}^3$.

EXAMPLE 7

Let $L : \mathbb{R}^2 \to \mathbb{R}^3$ be defined by $L(x_1, x_2) = (2x_1 - x_2, 0, x_1 + x_2)$. Find Range($L$).

Solution: By definition of the range, if $L(\vec{x})$ is any vector in the range, then $L(\vec{x}) = \begin{bmatrix} 2x_1 - x_2 \\ 0 \\ x_1 + x_2 \end{bmatrix}$. Using vector operations, we can write this as

$$L(\vec{x}) = x_1 \begin{bmatrix} 2 \\ 0 \\ 1 \end{bmatrix} + x_2 \begin{bmatrix} -1 \\ 0 \\ 1 \end{bmatrix}$$

This is for any $x_1, x_2 \in \mathbb{R}$, and so $\text{Range}(L) = \text{Span}\left\{ \begin{bmatrix} 2 \\ 0 \\ 1 \end{bmatrix}, \begin{bmatrix} -1 \\ 0 \\ 1 \end{bmatrix} \right\}$.

EXERCISE 4

Let $L : \mathbb{R}^3 \to \mathbb{R}^2$ be defined by $L(x_1, x_2, x_3) = (x_1 - x_2, -2x_1 + 2x_2 + x_3)$. Find Range($L$).

It is natural to ask whether the range of L can easily be described in terms of the matrix A of L. Observe that

$$L(\vec{x}) = A\vec{x} = \begin{bmatrix} \vec{a}_1 & \cdots & \vec{a}_n \end{bmatrix} \begin{bmatrix} x_1 \\ \vdots \\ x_n \end{bmatrix} = x_1\vec{a}_1 + \cdots + x_n\vec{a}_n$$

Thus, the image of $\vec{x}$ under L is a linear combination of the columns of the matrix A.

Definition
Columnspace

The **columnspace** of an $m \times n$ matrix A is the set Col(A) defined by

$$\text{Col}(A) = \{A\vec{x} \in \mathbb{R}^m \mid \vec{x} \in \mathbb{R}^n\}$$

Notice that our second interpretation of matrix-vector multiplication tells us that $A\vec{x}$ is a linear combination of the columns of A. Thus, the columnspace of A is the set of all possible linear combinations of A. In particular, it is the subspace of $\mathbb{R}^m$ spanned by the columns of A. Moreover, if $L : \mathbb{R}^n \to \mathbb{R}^m$ is a linear mapping, then Range(L) = Col(A).

EXAMPLE 8

Let $A = \begin{bmatrix} 1 & 2 & 3 \\ 2 & 1 & -1 \end{bmatrix}$ and $B = \begin{bmatrix} 1 & 3 \\ 2 & -1 \\ 0 & 1 \end{bmatrix}$. Then

$$\text{Col}(A) = \text{Span}\left\{ \begin{bmatrix} 1 \\ 2 \end{bmatrix}, \begin{bmatrix} 2 \\ 1 \end{bmatrix}, \begin{bmatrix} 3 \\ -1 \end{bmatrix} \right\} \quad \text{and} \quad \text{Col}(B) = \text{Span}\left\{ \begin{bmatrix} 1 \\ 2 \\ 0 \end{bmatrix}, \begin{bmatrix} 3 \\ -1 \\ 1 \end{bmatrix} \right\}$$

EXAMPLE 9

If L is the mapping with standard matrix $A = \begin{bmatrix} 1 & 1 \\ 2 & 1 \\ 1 & 3 \end{bmatrix}$, then

$$\text{Range}(L) = \text{Col}(A) = \text{Span}\left\{ \begin{bmatrix} 1 \\ 2 \\ 1 \end{bmatrix}, \begin{bmatrix} 1 \\ 1 \\ 3 \end{bmatrix} \right\}$$

EXERCISE 5

Find the standard matrix A of $L(x_1, x_2, x_3) = (x_1 - x_2, -2x_1 + 2x_2 + x_3)$ and show that $\text{Range}(L) = \text{Col}(A)$.

The range of a linear mapping L with standard matrix A is also related to the system of equations $A\vec{x} = \vec{b}$.

Theorem 3

The system of equations $A\vec{x} = \vec{b}$ is consistent if and only if $\vec{b}$ is in the range of the linear mapping L with standard matrix A (or, equivalently, if and only if $\vec{b}$ is in the columnspace of A).

Proof: If there exists a vector $\vec{x}$ such that $A\vec{x} = \vec{b}$, then $\vec{b} = A\vec{x} = L(\vec{x})$ and hence $\vec{b}$ is in the range of L. Similarly, if $\vec{b}$ is in the range of L, then there exists a vector $\vec{x}$ such that $\vec{b} = L(\vec{x}) = A\vec{x}$. ■

EXAMPLE 10

Suppose that L is a linear mapping with matrix $A = \begin{bmatrix} 1 & 1 \\ 2 & 1 \\ 1 & 3 \end{bmatrix}$. Determine whether $\vec{c} = \begin{bmatrix} 1 \\ 3 \\ -1 \end{bmatrix}$ and $\vec{d} = \begin{bmatrix} 2 \\ 1 \\ 9 \end{bmatrix}$ are in the range of L.

Solution: $\vec{c}$ is in the range of L if and only if the system $A\vec{x} = \vec{c}$ is consistent. Similarly, $\vec{d}$ is in the range of L if and only if $A\vec{x} = \vec{d}$ is consistent. Since the coefficient matrix is the same for the two systems, we can answer both questions simultaneously by row reducing the doubly augmented matrix $\left[\begin{array}{c|c|c} A & \vec{c} & \vec{d} \end{array} \right]$:

$$\left[\begin{array}{cc|c|c} 1 & 1 & 1 & 2 \\ 2 & 1 & 3 & 1 \\ 1 & 3 & -1 & 9 \end{array} \right] \sim \left[\begin{array}{cc|c|c} 1 & 1 & 1 & 2 \\ 0 & 1 & -1 & 3 \\ 0 & 0 & 0 & 1 \end{array} \right]$$

EXAMPLE 10
(continued)

By considering the reduced matrix corresponding to $\left[\begin{array}{c|c} A & \vec{c} \end{array}\right]$ (ignore the last column), we see that the system $A\vec{x} = \vec{c}$ is consistent, so $\vec{c}$ is in the range of L. The reduced matrix corresponding to $\left[\begin{array}{c|c} A & \vec{d} \end{array}\right]$ shows that $A\vec{x} = \vec{d}$ is inconsistent and hence $\vec{d}$ is not in the range of L.

Rowspace of A

The idea of the rowspace of a matrix A is similar to the idea of the columnspace.

Definition
Rowspace

Given an $m \times n$ matrix A, the **rowspace** of A is the subspace spanned by the rows of A (regarded as vectors) and is denoted Row(A).

EXAMPLE 11

Let $A = \begin{bmatrix} 1 & 2 & 3 \\ 2 & 1 & -1 \end{bmatrix}$ and $B = \begin{bmatrix} 1 & 3 \\ 2 & -1 \\ 0 & 1 \end{bmatrix}$. Then

$$\text{Row}(A) = \text{Span}\left\{\begin{bmatrix} 1 \\ 2 \\ 3 \end{bmatrix}, \begin{bmatrix} 2 \\ 1 \\ -1 \end{bmatrix}\right\} \quad \text{and} \quad \text{Row}(B) = \text{Span}\left\{\begin{bmatrix} 1 \\ 3 \end{bmatrix}, \begin{bmatrix} 2 \\ -1 \end{bmatrix}, \begin{bmatrix} 0 \\ 1 \end{bmatrix}\right\}$$

To write a mathematical definition of the rowspace of A, we require linear combinations of the rows of A. But, matrix-vector multiplication only gives us a linear combination of the columns of a matrix. However, we recall that the transpose of a matrix turns rows into columns. Thus, we can precisely define the rowspace of A by

$$\text{Row}(A) = \{A^T\vec{x} \in \mathbb{R}^n \mid \vec{x} \in \mathbb{R}^m\}$$

We now prove an important result about the rowspaces of row equivalent matrices.

Theorem 4

If the $m \times n$ matrix A is row equivalent to the matrix B, then Row(A) = Row(B).

Proof: We will show that applying each of the three elementary row operations does not change the rowspace. Let the rows of A be denoted $\vec{a}_1^T, \dots, \vec{a}_m^T$ and the rows of B be denoted by $\vec{b}_1^T, \dots, \vec{b}_m^T$.

Suppose that B is obtained from A by interchanging two rows of A. Then, except for the order, the rows of A are the same as the rows of B; hence Row(A) = Row(B).

Suppose that B is obtained from A by multiplying the i-th row of A by a non-zero constant t. Then,

$$\begin{aligned} \text{Row}(B) &= \text{Span}\{\vec{b}_1, \dots, \vec{b}_m\} = \text{Span}\{\vec{a}_1, \dots, t\vec{a}_i, \dots, \vec{a}_m\} \\ &= \{c_1\vec{a}_1 + \cdots + c_i(t\vec{a}_i) + \cdots + c_m\vec{a}_m \mid c_i \in \mathbb{R}\} \\ &= \{c_1\vec{a}_1 + \cdots + c_it\vec{a}_i + \cdots + c_m\vec{a}_m \mid c_i \in \mathbb{R}\} \\ &= \text{Span}\{\vec{a}_1, \dots, \vec{a}_m\} = \text{Row}(A) \end{aligned}$$

Now, suppose that B is obtained from A by adding t times the i-th row to the j-th row. Then,

$$\begin{aligned}\text{Row}(B) &= \text{Span}\{\vec{b}_1, \dots, \vec{b}_m\} = \{\vec{a}_1, \dots, \vec{a}_j + t\vec{a}_i, \dots, \vec{a}_m\}\\ &= \{c_1\vec{a}_1 + \cdots + c_i\vec{a}_i + \cdots + c_j(\vec{a}_j + t\vec{a}_i) + \cdots + c_m\vec{a}_m \mid c_i \in \mathbb{R}\}\\ &= \{c_1\vec{a}_1 + \cdots + (c_i + c_jt)\vec{a}_i + \cdots + c_j\vec{a}_j + \cdots + c_m\vec{a}_m \mid c_i \in \mathbb{R}\}\\ &= \text{Span}\{\vec{a}_1, \dots, \vec{a}_m\} = \text{Row}(A)\end{aligned}$$

By considering a sequence of elementary row operations, we see that row equivalent matrices must have the same rowspace. ■

Bases for Row(A), Col(A), and Null(A)

Recall from Section 1.2 that we always want to write a spanning set with as few vectors in the set as possible. We saw this in Section 1.2 when the set was linearly independent. Thus, we defined a **basis** for a subspace S to be a linearly independent spanning set. Moreover in Section 2.3, we defined the **dimension** of the subspace to be the number of vectors in any basis.

Basis of the Rowspace of a Matrix We now determine how to find bases for the rowspace, columnspace, and nullspace of a matrix.

Theorem 5

Let B be the reduced row echelon form of an $m \times n$ matrix A. Then the non-zero rows of B form a basis for Row(A), and hence the dimension of Row(A) equals the rank of A.

Proof: By Theorem 4, we have Row(B) = Row(A). Hence, the non-zero rows of B form a spanning set for the rowspace of A. Thus, we just need to prove that these non-zero rows are linearly independent. Suppose that B has r non-zero rows $\vec{b}_1^T, \dots, \vec{b}_r^T$ and consider

$$t_1\vec{b}_1 + \cdots + t_r\vec{b}_r = \vec{0} \tag{3.4}$$

Observe that if B has a leading 1 in the j-th row, then by definition of reduced row echelon form, all other rows must have a zero as their j-th coordinate. Hence, we must have $t_j = 0$ in (3.4). It follows that the rows are linearly independent and thus form a basis for Row(B) = Row(A). Moreover, since there are r non-zero rows in B, the rank of A is r, and so the dimension of the rowspace equals the rank of A. ■

EXAMPLE 12

Let $A = \begin{bmatrix} 1 & 1 & 1\\ 1 & -1 & 3\\ 3 & 1 & 5\end{bmatrix}$. Find a basis for Row($A$).

Solution: Row reducing A gives

$$\begin{bmatrix} 1 & 1 & 1\\ 1 & -1 & 3\\ 3 & 1 & 5\end{bmatrix} \sim \begin{bmatrix} 1 & 0 & 2\\ 0 & 1 & -1\\ 0 & 0 & 0\end{bmatrix}$$

EXAMPLE 12
(continued)

Hence, by Theorem 5, a basis for Row(A) is $\left\{ \begin{bmatrix} 1 \\ 0 \\ 2 \end{bmatrix}, \begin{bmatrix} 0 \\ 1 \\ -1 \end{bmatrix} \right\}$.

EXERCISE 6

Let $A = \begin{bmatrix} 1 & 1 & 3 \\ 2 & 1 & 5 \\ 1 & 3 & 5 \\ 1 & 1 & 3 \end{bmatrix}$. Find a basis for Row($A$).

Basis of the Columnspace of a Matrix What about a basis for the columnspace of a matrix A? It is remarkable that the same row reduction that gives the basis for the rowspace of A also indicates how to find a basis for the columnspace of A. However, the method is more subtle and requires a little more attention.

Again, let B be the reduced row echelon form of A. Recall that the whole point of the method of row reduction is that a vector $\vec{x}$ satisfies $A\vec{x} = \vec{0}$ if and only if it satisfies $B\vec{x} = \vec{0}$. That is, if we let $\vec{a}_1, \dots, \vec{a}_n$ denote the columns of A and $\vec{b}_1, \dots, \vec{b}_n$ denote the columns of B, then

$$x_1\vec{a}_1 + \cdots + x_n\vec{a}_n = \vec{0}$$

if and only if

$$x_1\vec{b}_1 + \cdots + x_n\vec{b}_n = \vec{0}$$

So, any statement about the linear dependence of the columns of A is true if and only if the same statement is true for the corresponding columns of B.

Theorem 6

Suppose that B is the reduced row echelon form of A. Then the columns of A that correspond to the columns of B with leading 1s form a basis of the columnspace of A. Hence, the dimension of the columnspace equals the rank of A.

Proof: For any $m \times n$ matrix $B = \begin{bmatrix} \vec{b}_1 & \cdots & \vec{b}_n \end{bmatrix}$ in reduced row echelon form, the set of columns containing leading 1s is linearly independent as they are standard basis vectors from $\mathbb{R}^m$. Moreover, if $\vec{b}_i$ is a column of B that does not contain a leading 1, then, by definition of the reduced row echelon form, it can be written as a linear combination of the columns containing leading 1s. Therefore, from our argument above, the corresponding column $\vec{a}_i$ in A can be written as a linear combination of the columns in A that correspond to the columns containing leading 1s in B. Thus, we can remove this column from the spanning set without changing the set it spans by Theorem 1.2.3. We continue to do this until we have removed all the columns of A that correspond to columns of B that do not have leading 1s, and we get a basis for the columnspace of A. ■

EXAMPLE 13

Let $A = \begin{bmatrix} 1 & 2 & 1 & 1 \\ 1 & 2 & 2 & 1 \\ 2 & 4 & 2 & 3 \\ 3 & 6 & 4 & 3 \end{bmatrix}$. Find a basis for Col($A$).

Solution: By row reducing A, we get

$$\begin{bmatrix} 1 & 2 & 1 & 1 \\ 1 & 2 & 2 & 1 \\ 2 & 4 & 2 & 3 \\ 3 & 6 & 4 & 3 \end{bmatrix} \sim \begin{bmatrix} 1 & 2 & 0 & 0 \\ 0 & 0 & 1 & 0 \\ 0 & 0 & 0 & 1 \\ 0 & 0 & 0 & 0 \end{bmatrix}$$

The first, third, and fourth columns of the reduced row echelon form of A are linearly independent. Therefore, by Theorem 2, the first, third, and fourth columns of matrix A form a basis for Col(A). Thus, a basis for Col(A) is $\left\{ \begin{bmatrix} 1 \\ 1 \\ 2 \\ 3 \end{bmatrix}, \begin{bmatrix} 1 \\ 2 \\ 2 \\ 4 \end{bmatrix}, \begin{bmatrix} 1 \\ 1 \\ 3 \\ 3 \end{bmatrix} \right\}$.

Notice in Example 13 that every vector in the columnspace of the reduced row echelon form of A has a last coordinate 0, which is not true in the columnspace of A, so the two columnspaces are not equal. Thus, the first, third, and fourth columns of the reduced row echelon form of A *do not* form a basis for Col(A).

EXERCISE 7

Let $A = \begin{bmatrix} 1 & 1 & 2 & 0 & 3 \\ 1 & -1 & 0 & 2 & -3 \\ -1 & 2 & 1 & -3 & -2 \end{bmatrix}$. Find a basis for Col($A$).

There is an alternative procedure for finding a basis for the columnspace of matrix A, which uses the fact that the columnspace of a matrix A is equal to the rowspace of A^T. However, the basis obtained in this manner is sometimes not as useful, as it may not consist of the columns of A.

EXERCISE 8

Find a basis for the columnspace of the matrix A from Example 13 by finding a basis for the rowspace of A^T.

Basis of the Nullspace of a Matrix In Section 2.2 we saw that if the rank of A was r, then the general solution of the homogeneous system $A\vec{x} = \vec{0}$ was automatically expressed as a spanning set of $n - r$ vectors. We can now show that these spanning vectors are linearly independent, so that the dimension of the nullspace of A is $n - r$. Since this quantity is important, we make the following definition.

Definition
Nullity

Let A be an $m \times n$ matrix. We call the dimension of the nullspace of A the **nullity** of A and denote it by nullity(A).

EXAMPLE 14

Consider the homogeneous system $A\vec{x} = \vec{0}$, where the coefficient matrix

$$A = \begin{bmatrix} 1 & 2 & 0 & 3 & 4 \\ 0 & 0 & 1 & 5 & 6 \end{bmatrix}$$

is already in reduced row echelon form. By finding a basis for Null(A), determine the nullity of A and relate it to rank(A).

Solution: The general solution is

$$\vec{x} = t_1 \begin{bmatrix} -4 \\ 0 \\ -6 \\ 0 \\ 1 \end{bmatrix} + t_2 \begin{bmatrix} -3 \\ 0 \\ -5 \\ 1 \\ 0 \end{bmatrix} + t_3 \begin{bmatrix} -2 \\ 1 \\ 0 \\ 0 \\ 0 \end{bmatrix}$$

Thus, $\left\{ \begin{bmatrix} -4 \\ 0 \\ -6 \\ 0 \\ 1 \end{bmatrix}, \begin{bmatrix} -3 \\ 0 \\ -5 \\ 1 \\ 0 \end{bmatrix}, \begin{bmatrix} -2 \\ 1 \\ 0 \\ 0 \\ 0 \end{bmatrix} \right\}$ is a spanning set for the nullspace of A. We now check for linear independence.

Let us look closely at the coordinates of the general solution $\vec{x}$ corresponding to the free variables (x_2, x_4, and x_5 in this example):

$$\vec{x} = t_1 \begin{bmatrix} * \\ 0 \\ * \\ 0 \\ 1 \end{bmatrix} + t_2 \begin{bmatrix} * \\ 0 \\ * \\ 1 \\ 0 \end{bmatrix} + t_3 \begin{bmatrix} * \\ 1 \\ * \\ 0 \\ 0 \end{bmatrix} = \begin{bmatrix} * \\ t_3 \\ * \\ t_2 \\ t_1 \end{bmatrix}$$

Clearly, this linear combination is the zero vector only if $t_1 = t_2 = t_3 = 0$, so the spanning vectors are linearly independent and hence form a basis for the nullspace of A. It follows that nullity(A) $= 3 =$ (# of columns) $-$ rank(A). In particular, it is the number of free variables in the system.

Following the method in Example 14, we prove the following theorem.

Theorem 7

Let A be an $m \times n$ matrix with rank(A) $= r$. Then the spanning set for the general solution of the homogeneous system $A\vec{x} = \vec{0}$ obtained by the method in Chapter 2 is a basis for Null(A) and the nullity of A is $n - r$.

Proof: Let $\{\vec{v}_1, \dots, \vec{v}_{n-r}\}$ be a spanning set for the general solution of $A\vec{x} = \vec{0}$ obtained in the usual manner and consider

$$t_1\vec{v}_1 + \cdots + t_{n-r}\vec{v}_{n-r} = \vec{0} \tag{3.5}$$

Then, the coefficients t_i are just the parameters to the free variables. Thus, the coordinate associated with the i-th parameter is non-zero only in the i-th vector. Hence, the only possible solution of (3.5) is $t_1 = \cdots = t_{n-r} = 0$. Therefore, the set is a linearly independent spanning set for Null(A) and thus forms a basis for Null(A). Thus, the nullity of A is $n - r$, as required. ■

Putting Theorem 6 and Theorem 7 together gives the following important result.

Theorem 8

[Rank Theorem]
If A is any $m \times n$ matrix, then

$$\text{rank}(A) + \text{nullity}(A) = n$$

EXAMPLE 15

Find a basis for the rowspace, columnspace, and nullspace of $A = \begin{bmatrix} 1 & 2 & 3 \\ -1 & 3 & 2 \\ 0 & 1 & 1 \end{bmatrix}$ and verify the Rank Theorem in this case.

Solution: Row reducing A gives

$$\begin{bmatrix} 1 & 2 & 3 \\ -1 & 3 & 2 \\ 0 & 1 & 1 \end{bmatrix} \sim \begin{bmatrix} 1 & 0 & 1 \\ 0 & 1 & 1 \\ 0 & 0 & 0 \end{bmatrix}$$

Thus, a basis for the rowspace of A is $\left\{ \begin{bmatrix} 1 \\ 0 \\ 1 \end{bmatrix}, \begin{bmatrix} 0 \\ 1 \\ 1 \end{bmatrix} \right\}$. Also, the first and second columns of the reduced row echelon form of A have leading 1s, so the corresponding columns from A form a basis for the columnspace of A. That is, a basis for the columnspace of A is $\left\{ \begin{bmatrix} 1 \\ -1 \\ 0 \end{bmatrix}, \begin{bmatrix} 2 \\ 3 \\ 1 \end{bmatrix} \right\}$. Thus, since the rank of A is equal to the dimension of the columnspace (or rowspace), the rank of A is 2.

By back-substitution, we find that the general solution is $\vec{x} = t \begin{bmatrix} -1 \\ -1 \\ 1 \end{bmatrix}, \quad t \in \mathbb{R}$.

Hence, a basis for the nullspace of A is $\left\{ \begin{bmatrix} -1 \\ -1 \\ 1 \end{bmatrix} \right\}$. Thus, we have nullity($A$) = 1 and

$$\text{rank}(A) + \text{nullity}(A) = 2 + 1 = 3$$

as predicted by the Rank Theorem.

EXERCISE 9

Find a basis for the rowspace, columnspace, and nullspace of $A = \begin{bmatrix} 1 & 1 & -3 & 1 \\ 2 & 3 & -8 & 4 \\ 0 & 1 & -2 & 3 \end{bmatrix}$ and verify the Rank Theorem in this case.

A Summary of Facts About Rank

For an $m \times n$ matrix A:

$$\begin{aligned} \text{the rank of } A &= \text{the number of leading 1s in the reduced row echelon form of } A \\ &= \text{the number of non-zero rows in any row echelon form of } A \\ &= \dim \operatorname{Row}(A) \\ &= \dim \operatorname{Col}(A) \\ &= n - \dim \operatorname{Null}(A) \end{aligned}$$

PROBLEMS 3.4

Practice Problems

A1 Let L be the linear mapping with matrix $\begin{bmatrix} 1 & 0 & -1 \\ 0 & 1 & 3 \\ 1 & 2 & 1 \end{bmatrix}$, $\vec{y}_1 = \begin{bmatrix} 3 \\ 1 \\ 6 \\ 1 \end{bmatrix}$, and $\vec{y}_2 = \begin{bmatrix} 3 \\ -5 \\ 1 \\ 5 \end{bmatrix}$.

(a) Is $\vec{y}_1$ in the range of L? If so, find $\vec{x}$ such that $L(\vec{x}) = \vec{y}_1$.

(b) Is $\vec{y}_2$ in the range of L? If so, find $\vec{x}$ such that $L(\vec{x}) = \vec{y}_2$.

A2 Find a basis for the range and a basis for the nullspace of each of the following linear mappings.

(a) $L(x_1, x_2, x_3) = (2x_1, -x_2 + 2x_3)$

(b) $M(x_1, x_2, x_3, x_4) = (x_4, x_3, 0, x_2, x_1 + x_2 - x_3)$

A3 Determine a matrix of a linear mapping $L : \mathbb{R}^2 \to \mathbb{R}^3$ whose nullspace is Span $\left\{ \begin{bmatrix} 1 \\ 1 \end{bmatrix} \right\}$ and whose range is Span $\left\{ \begin{bmatrix} 1 \\ 2 \\ 3 \end{bmatrix} \right\}$.

A4 Determine a matrix of a linear mapping $L : \mathbb{R}^2 \to \mathbb{R}^3$ whose nullspace is Span $\left\{ \begin{bmatrix} 1 \\ -2 \end{bmatrix} \right\}$ and whose range is Span $\left\{ \begin{bmatrix} 1 \\ 1 \\ 1 \end{bmatrix} \right\}$.

A5 Suppose that each of the following matrices is the coefficient matrix of a homogeneous system of equations. State the following.

(i) The number of variables in the system

(ii) The rank of the matrix

(iii) The dimension of the solution space

(a) $A = \begin{bmatrix} 1 & 2 & 1 & 3 \\ 0 & 0 & 1 & -2 \end{bmatrix}$

(b) $B = \begin{bmatrix} 1 & -2 & 0 & 0 & 5 \\ 0 & 1 & 3 & 4 & -1 \\ 0 & 0 & 0 & 1 & 2 \end{bmatrix}$

(c) $C = \begin{bmatrix} 1 & 0 & 2 & 1 & 1 \\ 0 & 0 & 1 & 5 & -2 \\ 0 & 0 & 0 & 0 & 0 \end{bmatrix}$

(d) $D = \begin{bmatrix} 1 & 0 & 3 & -5 & 1 & -1 \\ 0 & 1 & 2 & -4 & 2 & 1 \\ 0 & 0 & 0 & 1 & 1 & 4 \\ 0 & 0 & 0 & 0 & 0 & 0 \end{bmatrix}$

A6 For each of the following matrices, determine a basis for the rowspace, a subset of the columns that form a basis for the columnspace, and a basis for the nullspace. Verify the Rank Theorem.

(a) $\begin{bmatrix} 1 & 2 & 8 \\ 1 & 1 & 5 \\ 1 & 0 & -2 \end{bmatrix}$

(b) $\begin{bmatrix} 1 & 1 & -3 & 1 \\ 2 & 3 & -8 & 4 \\ 0 & 1 & -2 & 3 \end{bmatrix}$

(c) $\begin{bmatrix} 1 & 2 & 0 & 3 & 0 \\ 1 & 2 & 1 & 7 & 1 \\ 2 & 4 & 0 & 6 & 1 \\ 3 & 6 & 1 & 13 & 2 \end{bmatrix}$

A7 By geometrical arguments, give a basis for the nullspace and a basis for the range of the following linear mappings.

(a) $\text{proj}_{\vec{v}} : \mathbb{R}^3 \to \mathbb{R}^3$, where $\vec{v} = \begin{bmatrix} 1 \\ -2 \\ 2 \end{bmatrix}$

(b) $\text{perp}_{\vec{v}} : \mathbb{R}^3 \to \mathbb{R}^3$, where $\vec{v} = \begin{bmatrix} 3 \\ 1 \\ 2 \end{bmatrix}$

(c) $\text{refl}_{\vec{v}} : \mathbb{R}^3 \to \mathbb{R}^3$, where $\vec{v} = \begin{bmatrix} 0 \\ 1 \\ 0 \end{bmatrix}$

A8 The matrix $A = \begin{bmatrix} 1 & 1 & 1 & 1 & 5 \\ 2 & 3 & 1 & 2 & 11 \\ 1 & 1 & 1 & 3 & 7 \\ 1 & 2 & 0 & -1 & 4 \end{bmatrix}$ has reduced row echelon form $R = \begin{bmatrix} 1 & 0 & 2 & 0 & 3 \\ 0 & 1 & -1 & 0 & 1 \\ 0 & 0 & 0 & 1 & 1 \\ 0 & 0 & 0 & 0 & 0 \end{bmatrix}$.

(a) If the rowspace of A is a subspace of some $\mathbb{R}^n$, what is n?

(b) Without calculation, give a basis for Row(A); outline the theory that explains why this is a basis.

(c) If the columnspace of A is a subspace of some $\mathbb{R}^m$, what is m?

(d) Without calculation, give a basis for the columnspace of A. Why is this the required basis?

(e) Determine the general solution of the system $A\vec{x} = \vec{0}$ and, hence, obtain a spanning set for the solution space.

(f) Explain why the spanning set in (e) is in fact a basis for the solution space.

(g) Verify that the rank of A plus the dimension of the solution space of the system $A\vec{x} = \vec{0}$ is equal to the number of variables in the system.

Homework Problems

B1 Let L be the linear mapping with matrix $\begin{bmatrix} 1 & 2 & 1 \\ 1 & 0 & 3 \\ 2 & 1 & -1 \\ 0 & 2 & 2 \end{bmatrix}$, $\vec{y}_1 = \begin{bmatrix} 5 \\ 5 \\ 4 \\ 4 \end{bmatrix}$, and $\vec{y}_2 = \begin{bmatrix} 1 \\ -3 \\ 2 \\ 1 \end{bmatrix}$.

(a) Is $\vec{y}_1$ in the range of L? If so, find $\vec{x}$ such that $L(\vec{x}) = \vec{y}_1$.

(b) Is $\vec{y}_2$ in the range of L? If so, find $\vec{x}$ such that $L(\vec{x}) = \vec{y}_2$.

B2 Find a basis for the range and a basis for the nullspace of each of the following linear mappings.

(a) $L(x_1, x_2) = (x_1, x_1 + 2x_2, 3x_2)$

(b) $M(x_1, x_2, x_3, x_4) = (x_1 + x_4, x_2 - 2x_3, x_1 - 2x_2 + 3x_3)$

B3 Determine a matrix of a linear mapping $L : \mathbb{R}^2 \to \mathbb{R}^2$ whose nullspace is Span $\left\{\begin{bmatrix}2\\1\end{bmatrix}\right\}$ and whose range is Span $\left\{\begin{bmatrix}2\\1\end{bmatrix}\right\}$.

B4 Determine a matrix of a linear mapping $L : \mathbb{R}^2 \to \mathbb{R}^3$ whose nullspace is Span $\left\{\begin{bmatrix}1\\-1\end{bmatrix}\right\}$ and whose range is Span $\left\{\begin{bmatrix}0\\3\\5\end{bmatrix}\right\}$.

B5 Suppose that each of the following matrices is the coefficient matrix of a homogeneous system of equations. State the following.

(i) The number of variables in the system

(ii) The rank of the matrix

(iii) The dimension of the solution space

(a) $A = \begin{bmatrix}1 & 2 & 0 & 1 & -2\\0 & 1 & 1 & 2 & 0\\0 & 0 & 0 & 1 & 0\end{bmatrix}$

(b) $B = \begin{bmatrix}1 & 0 & 2 & 0\\0 & 1 & -1 & 3\\0 & 0 & 0 & 1\\0 & 0 & 0 & 0\end{bmatrix}$

(c) $C = \begin{bmatrix}1 & 5 & 0 & 2 & -3\\0 & 1 & 3 & -2 & 1\\0 & 0 & 1 & 4 & 2\\0 & 0 & 0 & 0 & 1\end{bmatrix}$

(d) $D = \begin{bmatrix}1 & 6 & 0 & 2 & -1 & 0\\0 & 0 & 1 & -2 & 1 & 2\\0 & 0 & 0 & 0 & 1 & 3\\0 & 0 & 0 & 0 & 0 & 0\end{bmatrix}$

B6 For each of the following matrices, determine a basis for the rowspace, a subset of the columns that form a basis for the columnspace, and a basis for the nullspace. Verify the Rank Theorem.

(a) $\begin{bmatrix}3 & 6 & 1\\2 & 4 & 1\\1 & 2 & 0\end{bmatrix}$

(b) $\begin{bmatrix}0 & 1 & 0 & -2\\1 & 2 & 1 & -1\\2 & 4 & 3 & -1\end{bmatrix}$

(c) $\begin{bmatrix}1 & 1 & 1 & 1 & 1\\0 & 1 & 2 & 3 & 4\\1 & 0 & 1 & 3 & 3\\1 & 1 & 3 & 6 & 8\end{bmatrix}$

B7 By geometrical arguments, give a basis for the nullspace and a basis for the range of the following linear mappings.

(a) $\text{proj}_{\vec{v}} : \mathbb{R}^3 \to \mathbb{R}^3$, where $\vec{v} = \begin{bmatrix}1\\0\\-1\end{bmatrix}$

(b) $\text{perp}_{\vec{v}} : \mathbb{R}^3 \to \mathbb{R}^3$, where $\vec{v} = \begin{bmatrix}-1\\1\\3\end{bmatrix}$

(c) $\text{refl}_{\vec{v}} : \mathbb{R}^3 \to \mathbb{R}^3$, where $\vec{v} = \begin{bmatrix}1\\1\\1\end{bmatrix}$

B8 The matrix $A = \begin{bmatrix}1 & 2 & 0 & 0 & 3 & 0 & -1\\1 & 2 & 1 & 0 & 2 & 0 & -3\\1 & 2 & 0 & 1 & 1 & 1 & 4\\1 & 2 & 1 & 0 & 2 & 1 & 1\\3 & 6 & 2 & 1 & 5 & 2 & 2\end{bmatrix}$ has reduced row echelon form

$$R = \begin{bmatrix}1 & 2 & 0 & 0 & 3 & 0 & -1\\0 & 0 & 1 & 0 & -1 & 0 & -2\\0 & 0 & 0 & 1 & -2 & 0 & 1\\0 & 0 & 0 & 0 & 0 & 1 & 4\\0 & 0 & 0 & 0 & 0 & 0 & 0\end{bmatrix}.$$

(a) If the rowspace of A is a subspace of some $\mathbb{R}^n$, what is n?

(b) Without calculation, give a basis for Row(A); outline the theory that explains why this is a basis.

(c) If the columnspace of A is a subspace of some $\mathbb{R}^m$, what is m?

(d) Without calculation, give a basis for the columnspace of A. Why is this the required basis?

(e) Determine the general solution of the system $A\vec{x} = \vec{0}$ and, hence, obtain a spanning set for the solution space.

(f) Explain why the spanning set in (e) is in fact a basis for the solution space.

(g) Verify that the rank of A plus the dimension of the solution space of the system $A\vec{x} = \vec{0}$ is equal to the number of variables in the system.

Conceptual Problems

D1 Let $L : \mathbb{R}^n \to \mathbb{R}^m$ be a linear mapping. Prove that

$$\dim(\text{Range}(L)) + \dim(\text{Null}(L)) = n$$

D2 Suppose that $L : \mathbb{R}^n \to \mathbb{R}^m$ and $M : \mathbb{R}^m \to \mathbb{R}^p$ are linear mappings.

(a) Show that the range of $M \circ L$ is a subspace of the range of M.

(b) Give an example such that the range of $M \circ L$ is not equal to the range of M.

(c) Show that the nullspace of L is a subspace of the nullspace of $M \circ L$.

D3 (a) If A is a 5×7 matrix and $\text{rank}(A) = 4$, then what is the nullity of A, and what is the dimension of the columnspace of A?

(b) If A is a 5×4 matrix, then what is the largest possible dimension of the nullspace of A? What is the largest possible rank of A?

(c) If A is a 4×5 matrix and $\text{nullity}(A) = 3$, then what is the dimension of the rowspace of A?

D4 Let A be an $n \times n$ matrix such that $A^2 = O_{n,n}$. Prove that the columnspace of A is a subset of the nullspace of A.

3.5 Inverse Matrices and Inverse Mappings

In this section we will look at inverses of matrices and linear mappings. We will make many connections with the material we have covered so far and provide useful tools for the material contained in the rest of the book.

Definition

Inverse

Let A be an $n \times n$ matrix. If there exists an $n \times n$ matrix B such that $AB = I = BA$, then A is said to be **invertible**, and B is called the **inverse** of A (and A is the inverse of B). The inverse of A is denoted A^{-1}.

EXAMPLE 1

The matrix $\begin{bmatrix} 2 & -1 \\ -1 & 1 \end{bmatrix}$ is the inverse of the matrix $\begin{bmatrix} 1 & 1 \\ 1 & 2 \end{bmatrix}$ because

$$\begin{bmatrix} 1 & 1 \\ 1 & 2 \end{bmatrix}\begin{bmatrix} 2 & -1 \\ -1 & 1 \end{bmatrix} = \begin{bmatrix} 1 & 0 \\ 0 & 1 \end{bmatrix} = I$$

and

$$\begin{bmatrix} 2 & -1 \\ -1 & 1 \end{bmatrix}\begin{bmatrix} 1 & 1 \\ 1 & 2 \end{bmatrix} = \begin{bmatrix} 1 & 0 \\ 0 & 1 \end{bmatrix} = I$$

Notice that in the definition, B is *the* inverse of A. This depends on the easily proven fact that the inverse is unique.

Theorem 1 Let A be a square matrix and suppose that $BA = AB = I$ and $CA = AC = I$. Then $B = C$.

Proof: We have $B = BI = B(AC) = (BA)C = IC = C$. ■

Remark

Note that the proof uses less than the full assumptions of the theorem: we have proven that if $BA = I = AC$, then $B = C$. Sometimes we say that if $BA = I$, then B is a "left inverse" of A. Similarly, if $AC = I$, then C is a "right inverse" of A. The proof shows that for a square matrix, any left inverse must be equal to any right inverse. However, non-square matrices may have only a right inverse or a left inverse, but not both (see Problem D4). We will now show that for square matrices, a right inverse is automatically a left inverse.

Theorem 2 Suppose that A and B are $n \times n$ matrices such that $AB = I$. Then $BA = I$, so that $B = A^{-1}$. Moreover, B and A have rank n.

Proof: We first show, by contradiction, that $\text{rank}(B) = n$. Suppose that B has rank less than n. Then, by Theorem 2.2.2, the homogeneous system $B\vec{x} = \vec{0}$ has non-trivial solutions. But this means that for some non-zero $\vec{x}$, $AB\vec{x} = A(B\vec{x}) = A\vec{0} = \vec{0}$. So, AB is certainly not equal to I, which contradicts our assumption. Hence, B must have rank n.

Since B has rank n, the non-homogeneous system $\vec{y} = B\vec{x}$ is consistent for every $\vec{y} \in \mathbb{R}^n$ by Theorem 2.2.2. Now consider

$$BA\vec{y} = BA(B\vec{x}) = B(AB)\vec{x} = BI\vec{x} = B\vec{x} = \vec{y}$$

Thus, $BA\vec{y} = \vec{y}$ for every $\vec{y} \in \mathbb{R}^n$, so $BA = I$ by Theorem 3.1.4. Therefore, $AB = I$ and $BA = I$, so that $B = A^{-1}$.

Since we have $BA = I$, we see that $\text{rank}(A) = n$, by the same argument we used to prove $\text{rank}(B) = n$. ■

Theorem 2 makes it very easy to prove some useful properties of the matrix inverse. In particular, to show that $A^{-1} = B$, we only need to show that $AB = I$.

Theorem 3 Suppose that A and B are invertible matrices and that t is a non-zero real number.

(1) $(tA)^{-1} = \frac{1}{t}A^{-1}$
(2) $(AB)^{-1} = B^{-1}A^{-1}$
(3) $(A^T)^{-1} = (A^{-1})^T$

Proof: We have

$$(tA)\left(\frac{1}{t}A^{-1}\right) = \left(\frac{t}{t}\right)AA^{-1} = 1I = I$$

$$(AB)(B^{-1}A^{-1}) = A(BB^{-1})A^{-1} = AIA^{-1} = AA^{-1} = I$$

$$(A^T)(A^{-1})^T = (A^{-1}A)^T = I^T = I$$

■

A Procedure for Finding the Inverse of a Matrix

For any given square matrix A, we would like to determine whether it has an inverse and, if so, what the inverse is. Fortunately, one procedure answers both questions. We begin by trying to solve the matrix equation $AX = I$ for the unknown square matrix X. If a solution X can be found, then $X = A^{-1}$ by Theorem 2. On the other hand, if no such matrix X can be found, then A is not invertible.

To keep it simple, the procedure is examined in the case where A is 3×3, but it should be clear that it can be applied to any square matrix. Write the matrix equation $AX = I$ in the form

$$\begin{bmatrix} A\vec{x}_1 & A\vec{x}_2 & A\vec{x}_3 \end{bmatrix} = \begin{bmatrix} \vec{e}_1 & \vec{e}_2 & \vec{e}_3 \end{bmatrix}$$

Hence, we have

$$A\vec{x}_1 = \vec{e}_1, \quad A\vec{x}_2 = \vec{e}_2, \quad A\vec{x}_3 = \vec{e}_3$$

So, it is necessary to solve three systems of equations, one for each column of X. Note that each system has a different standard basis vector as its right-hand side, but all have the same coefficient matrix. Since the solution procedure for systems of equations requires that we row reduce the coefficient matrix, we might as well write out a "triple-augmented matrix" and solve all three systems at once. Therefore, write

$$\left[\begin{array}{c|ccc} A & \vec{e}_1 & \vec{e}_2 & \vec{e}_3 \end{array}\right] = \left[\begin{array}{c|c} A & I \end{array}\right]$$

and row reduce to reduced row echelon form to solve.

Suppose that A **is row equivalent to** I, and call the resulting block on the right B so that the reduction gives

$$\left[\begin{array}{c|c} A & I \end{array}\right] \sim \left[\begin{array}{c|c} I & B \end{array}\right] = \left[\begin{array}{c|ccc} I & \vec{b}_1 & \vec{b}_2 & \vec{b}_3 \end{array}\right]$$

Now, we must interpret the final matrix by breaking it up into three systems:

$$I\vec{x}_1 = \vec{b}_1, \quad I\vec{x}_2 = \vec{b}_2, \quad I\vec{x}_3 = \vec{b}_3$$

In particular, $A\vec{b}_1 = \vec{e}_1$, $A\vec{b}_2 = \vec{e}_2$, and $A\vec{b}_3 = \vec{e}_3$. It follows that the first column of the desired matrix X is $\vec{b}_1$, the second column of X is $\vec{b}_2$, and so on. Thus, A is invertible and $B = A^{-1}$.

If the reduced row echelon form of A **is not** I, then $\text{rank}(A) < n$. Hence, A is not invertible, since Theorem 2 tells us that if A is invertible, then $\text{rank}(A) = n$.

First, we summarize the procedure and give an example.

Algorithm 1

Finding A^{-1}

To find the inverse of a square matrix A,

(1) Row reduce the multi-augmented matrix $\left[\begin{array}{c|c} A & I \end{array}\right]$ so that the left block is in reduced row echelon form.

(2) If the reduced row echelon form is $\left[\begin{array}{c|c} I & B \end{array}\right]$, then $A^{-1} = B$.

(3) If the reduced row echelon form of A is not I, then A is not invertible.

EXAMPLE 2

Determine whether $A = \begin{bmatrix} 1 & 1 & 2 \\ 1 & 2 & 2 \\ 2 & 4 & 3 \end{bmatrix}$ is invertible, and if it is, determine its inverse.

Solution: Write the matrix $\left[\begin{array}{c|c} A & I \end{array}\right]$ and row reduce:

$$\left[\begin{array}{ccc|ccc} 1 & 1 & 2 & 1 & 0 & 0 \\ 1 & 2 & 2 & 0 & 1 & 0 \\ 2 & 4 & 3 & 0 & 0 & 1 \end{array}\right] \sim \left[\begin{array}{ccc|ccc} 1 & 1 & 2 & 1 & 0 & 0 \\ 0 & 1 & 0 & -1 & 1 & 0 \\ 0 & 2 & -1 & -2 & 0 & 1 \end{array}\right] \sim$$

$$\left[\begin{array}{ccc|ccc} 1 & 0 & 2 & 2 & -1 & 0 \\ 0 & 1 & 0 & -1 & 1 & 0 \\ 0 & 0 & -1 & 0 & -2 & 1 \end{array}\right] \sim \left[\begin{array}{ccc|ccc} 1 & 0 & 0 & 2 & -5 & 2 \\ 0 & 1 & 0 & -1 & 1 & 0 \\ 0 & 0 & 1 & 0 & 2 & -1 \end{array}\right]$$

Hence, A is invertible and $A^{-1} = \begin{bmatrix} 2 & -5 & 2 \\ -1 & 1 & 0 \\ 0 & 2 & -1 \end{bmatrix}$.

You should check that the inverse has been correctly calculated by verifying that $AA^{-1} = I$.

EXAMPLE 3

Determine whether $A = \begin{bmatrix} 1 & 2 \\ 2 & 4 \end{bmatrix}$ is invertible, and if it is, determine its inverse.

Solution: Write the matrix $\left[\begin{array}{c|c} A & I \end{array}\right]$ and row reduce:

$$\left[\begin{array}{cc|cc} 1 & 2 & 1 & 0 \\ 2 & 4 & 0 & 1 \end{array}\right] \sim \left[\begin{array}{cc|cc} 1 & 2 & 1 & 0 \\ 0 & 0 & -2 & 1 \end{array}\right]$$

Hence, A is not invertible.

EXERCISE 1

Determine whether $A = \begin{bmatrix} 2 & 3 \\ 4 & 5 \end{bmatrix}$ is invertible, and if it is, determine its inverse.

Some Facts About Square Matrices and Solutions of Linear Systems

In Theorem 2 and in the description of the procedure for finding the inverse matrix, we used some facts about systems of equations with square matrices. It is worth stating them clearly as a theorem. Most of the conclusions are simply special cases of previous results.

Theorem 4

[Invertible Matrix Theorem]
Suppose that A is an $n \times n$ matrix. Then the following statements are equivalent (that is, one is true if and only if each of the others is true).

Theorem 4
(continued)

(1) A is invertible.
(2) $\text{rank}(A) = n$.
(3) The reduced row echelon form of A is I.
(4) For all $\vec{b} \in \mathbb{R}^n$, the system $A\vec{x} = \vec{b}$ is consistent and has a unique solution.
(5) The columns of A are linearly independent.
(6) The columnspace of A is $\mathbb{R}^n$.

Proof: (We use the "implication arrow": $P \Rightarrow Q$ means "if P, then Q." It is common in proving a theorem such as this to prove (1) $\Rightarrow$ (2) $\Rightarrow$ (3) $\Rightarrow$ (4) $\Rightarrow$ (5) $\Rightarrow$ (6) $\Rightarrow$ (1), so that any statement implies any other.)
(1) $\Rightarrow$ (2): This is the second part of Theorem 2.
(2) $\Rightarrow$ (3): This follows immediately from the definition of rank and the fact that A is $n \times n$.
(3) $\Rightarrow$ (4): This follows immediately from Theorem 2.2.2.
(4) $\Rightarrow$ (5): Assume that the only solution to $A\vec{x} = \vec{0}$ is the trivial solution. Hence, if $A = \begin{bmatrix} \vec{a}_1 & \cdots \vec{a}_n \end{bmatrix}$, then $\vec{0} = A\vec{x} = x_1\vec{a}_1 + \cdots + x_n\vec{a}_n$ has only the solution $x_1 = \cdots = x_n = 0$. Thus, the columns $\vec{a}_1, \dots, \vec{a}_n$ of A are linearly independent.
(5) $\Rightarrow$ (6): If the columns of A are linearly independent, then $A\vec{x} = \vec{0}$ has a unique solution, so the rank of A is n. Thus, $A\vec{x} = \vec{b}$ is consistent for all $\vec{b} \in \mathbb{R}^n$ and so $\text{Col}(A) = \mathbb{R}^n$.
(6) $\Rightarrow$ (1): If $\text{Col}(A) = \mathbb{R}^n$, then $A\vec{x}_i = \vec{e}_i$ is consistent for $1 \leq i \leq n$. Thus, A is invertible. ■

Amongst other things, this theorem tells us that if a matrix A is invertible, then the system $A\vec{x} = \vec{b}$ is consistent with a unique solution. However, the way we proved the theorem does not immediately tell us how to find the unique solution. We now demonstrate this.

Let A be an invertible square matrix and consider the system $A\vec{x} = \vec{b}$. Multiplying both sides of the equation by A^{-1} gives $A^{-1}A\vec{x} = A^{-1}\vec{b}$ and hence $\vec{x} = A^{-1}\vec{b}$.

EXAMPLE 4

Let $A = \begin{bmatrix} 1 & 1 \\ 1 & 2 \end{bmatrix}$. Find the solution of $A\vec{x} = \begin{bmatrix} 2 \\ 4 \end{bmatrix}$.

Solution: By Example 1, $A^{-1} = \begin{bmatrix} 2 & -1 \\ -1 & 1 \end{bmatrix}$. Thus, the solution of $A\vec{x} = \begin{bmatrix} 2 \\ 4 \end{bmatrix}$ is

$$\vec{x} = A^{-1}\vec{b} = \begin{bmatrix} 2 & -1 \\ -1 & 1 \end{bmatrix}\begin{bmatrix} 2 \\ 4 \end{bmatrix} = \begin{bmatrix} 0 \\ 2 \end{bmatrix}$$

EXERCISE 2

Let $A = \begin{bmatrix} 1 & 3 \\ -2 & 1 \end{bmatrix}$ and $\vec{b} = \begin{bmatrix} 7 \\ 14 \end{bmatrix}$. Determine A^{-1} and use it to find the solution of $A\vec{x} = \vec{b}$.

It likely seems very inefficient to solve Exercise 2 by the method described. One would think that simply row reducing the augmented matrix of the system would make more sense. However, observe that if we wanted to solve many systems of equations with the same coefficient matrix A, we would need to compute A^{-1} once, and then each system can be solved by the problem of solving the system to simple matrix multiplication.

Remark

It might seem surprising at first that we can solve a system of linear equations without performing any elementary row operations and instead just using matrix multiplication. Of course, with some thought, one realizes that the elementary row operations are "contained" inside the inverse of the matrix (which we obtained by row reducing). In the next section, we will see more of the connection between matrix multiplication and row reducing.

Inverse Linear Mappings

It is useful to introduce the inverse of a linear mapping here because many geometrical transformations provide nice examples of inverses. Note that just as the inverse matrix is defined only for square matrices, the inverse of a linear mapping is defined only for linear operators. Recall that the identity transformation Id is the linear mapping defined by $\mathrm{Id}(\vec{x}) = \vec{x}$ for all $\vec{x}$.

Definition
Inverse Mapping

If $L : \mathbb{R}^n \to \mathbb{R}^n$ is a linear mapping and there exists another linear mapping $M : \mathbb{R}^n \to \mathbb{R}^n$ such that $M \circ L = \mathrm{Id} = L \circ M$, then L is said to be **invertible**, and M is called the **inverse** of L, usually denoted L^{-1}.

Observe that if M is the inverse of L and $L(\vec{v}) = \vec{w}$, then

$$M(\vec{w}) = M(L(\vec{v})) = (M \circ L)(\vec{v}) = \mathrm{Id}(\vec{v}) = \vec{v}$$

Similarly, if $M(\vec{w}) = \vec{v}$, then

$$L(\vec{v}) = L(M(\vec{w})) = (L \circ M)(\vec{w}) = \mathrm{Id}(\vec{w}) = \vec{w}$$

So we have $L(\vec{v}) = \vec{w}$ if and only if $M(\vec{w}) = \vec{v}$.

Theorem 5

Suppose that $L : \mathbb{R}^n \to \mathbb{R}^n$ is a linear mapping with standard matrix $[L] = A$ and that $M : \mathbb{R}^n \to \mathbb{R}^n$ is a linear mapping with standard matrix $[M] = B$. Then M is the inverse of L if and only if B is the inverse of A.

Proof: By Theorem 3.2.5, $[M \circ L] = [M][L]$. Hence, $L \circ M = \mathrm{Id} = M \circ L$ if and only if $AB = I = BA$. ■

For many of the geometrical transformations of Section 3.3, an inverse transformation is easily found by geometrical arguments, and these provide many examples of inverse matrices.

EXAMPLE 5

For each of the following geometrical transformations, determine the inverse transformation. Verify that the product of the standard matrix of the transformation and its inverse is the identity matrix.

(a) The rotation R_θ of the plane

(b) In the plane, a stretch T by a factor of t in the x_1-direction

Solution: (a) The inverse transformation is to just rotate by $-\theta$. That is, $(R_\theta)^{-1} = R_{-\theta}$. We have

$$[R_{-\theta}] = \begin{bmatrix} \cos(-\theta) & -\sin(-\theta) \\ \sin(-\theta) & \cos(-\theta) \end{bmatrix} = \begin{bmatrix} \cos\theta & \sin\theta \\ -\sin\theta & \cos\theta \end{bmatrix}$$

since $\sin(-\theta) = -\sin\theta$ and $\cos(-\theta) = \cos\theta$. Hence,

$$\begin{aligned}[R_\theta][R_{-\theta}] &= \begin{bmatrix} \cos\theta & -\sin\theta \\ \sin\theta & \cos\theta \end{bmatrix}\begin{bmatrix} \cos\theta & \sin\theta \\ -\sin\theta & \cos\theta \end{bmatrix} \\ &= \begin{bmatrix} \cos^2\theta + \sin^2\theta & \cos\theta\sin\theta - \cos\theta\sin\theta \\ -\sin\theta\cos\theta + \sin\theta\cos\theta & \cos^2\theta + \sin^2\theta \end{bmatrix} = \begin{bmatrix} 1 & 0 \\ 0 & 1 \end{bmatrix}\end{aligned}$$

(b) The inverse transformation T^{-1} is a stretch by a factor of $\frac{1}{t}$ in the x_1-direction:

$$[T][T^{-1}] = \begin{bmatrix} t & 0 \\ 0 & 1 \end{bmatrix}\begin{bmatrix} 1/t & 0 \\ 0 & 1 \end{bmatrix} = \begin{bmatrix} 1 & 0 \\ 0 & 1 \end{bmatrix}$$

EXERCISE 3

For each of the following geometrical transformations, determine the inverse transformation. Verify that the product of the standard matrix of the transformation and its inverse is the identity matrix.

(a) A reflection over the line $x_2 = x_1$ in the plane

(b) A shear in the plane by a factor of t in the x_1-direction

Observe that if $\vec{y} \in \mathbb{R}^n$ is in the domain of the inverse M, then it must be in the range of the original L. Therefore, it follows that if L has an inverse, the range of L must be all of the codomain $\mathbb{R}^n$. Moreover, if $L(\vec{x}_1) = \vec{y} = L(\vec{x}_2)$, then by applying M to both sides, we have

$$M(L(\vec{x}_1)) = M(L(\vec{x}_2)) \Rightarrow x_1 = x_2$$

Hence, we have shown that for any $\vec{y} \in \mathbb{R}^n$, there exists a unique $\vec{x} \in \mathbb{R}^n$ such that $L(\vec{x}) = \vec{y}$. This property is the linear mapping version of statement (4) of Theorem 4 about square matrices.

Theorem 6 [Invertible Matrix Theorem, cont.]
Suppose that $L : \mathbb{R}^n \to \mathbb{R}^n$ is a linear mapping with standard matrix $A = [L]$. Then, the following statements are equivalent to each other and to the statements of Theorem 4.

(7) L is invertible.
(8) $\text{Range}(L) = \mathbb{R}^n$.
(9) $\text{Null}(L) = \{\vec{0}\}$.

Proof: (We use $P \Leftrightarrow Q$ to mean "P if and only if Q.")
(1) $\Leftrightarrow$ (7): This is Theorem 5.
(4) $\Leftrightarrow$ (8): Since $L(\vec{x}) = A\vec{x}$, we know that for every $\vec{b} \in \mathbb{R}^n$ there exists a $\vec{x} \in \mathbb{R}^n$ such that $L(\vec{x}) = \vec{b}$ if and only if $A\vec{x} = \vec{b}$ is consistent for every $\vec{b} \in \mathbb{R}^n$.
(7) $\Rightarrow$ (9): Assume that L is invertible. Hence, there is a linear mapping L^{-1} such that $L^{-1}(L(\vec{x})) = \vec{x}$. If $\vec{x} \in \text{Null}(L)$, then $L(\vec{x}) = \vec{0}$ and then $L^{-1}L(\vec{x}) = L^{-1}(\vec{0})$. But, this gives

$$\vec{x} = L^{-1}(\vec{0}) = \vec{0}$$

since L^{-1} is linear. Thus, $\text{Null}(L) = \{\vec{0}\}$.
(9) $\Rightarrow$ (3): Assume that $\text{Null}(L) = \{\vec{0}\}$. Then $L(\vec{x}) = A\vec{x} = \vec{0}$ has a unique solution. Thus, $\text{rank}(A) = n$, by Theorem 2.2.2. ■

Remark

It is possible to give many alternate proofs of the equivalences in Theorem 4 and Theorem 6. You should be able to start with any one of the nine statements and show that it implies any of the other eight.

EXAMPLE 6

Prove that the linear mapping $\text{proj}_{\vec{v}}$ is not invertible for any $\vec{v} \in \mathbb{R}^n$, $n \geq 2$.
Solution: By definition, $\text{proj}_{\vec{v}}(\vec{x}) = t\vec{v}$, for some $t \in \mathbb{R}$. Hence, any vector $\vec{y} \in \mathbb{R}^n$ that is not a scalar multiple of $\vec{v}$ is not in the range of $\text{proj}_{\vec{v}}$. Thus, $\text{Range}(\text{proj}_{\vec{v}}) \neq \mathbb{R}^n$, hence $\text{proj}_{\vec{v}}$ is not invertible, by Theorem 6.

EXAMPLE 7

Prove that the linear mapping $L : \mathbb{R}^3 \to \mathbb{R}^3$ defined by

$$L(x_1, x_2, x_3) = (2x_1 + x_2, x_3, x_2 - 2x_3)$$

is invertible.
Solution: Assume that $\vec{x}$ is in the nullspace of L. Then $L(\vec{x}) = \vec{0}$, so by definition of L, we have

$$\begin{aligned} 2x_1 + x_2 &= 0 \\ x_3 &= 0 \\ x_2 - 2x_3 &= 0 \end{aligned}$$

The only solution to this system is $x_1 = x_2 = x_3 = 0$. Thus, $\text{Null}(L) = \{\vec{0}\}$ and hence L is invertible, by Theorem 6.

Finally, recall that the matrix condition $AB = BA = I$ implies that the matrix inverse can be defined only for square matrices. Here is an example that illustrates for linear mappings that the domain and codomain of L must be the same if it is to have an inverse.

EXAMPLE 8

Consider the linear mappings $P : \mathbb{R}^4 \to \mathbb{R}^3$ defined by $P(x_1, x_2, x_3, x_4) = (x_1, x_2, x_3)$ and $\text{inj} : \mathbb{R}^3 \to \mathbb{R}^4$ defined by $\text{inj}(x_1, x_2, x_3) = (x_1, x_2, x_3, 0)$.

It is easy to see that $P \circ \text{inj} = \text{Id}$ but that $\text{inj} \circ P \neq \text{Id}$. Thus, P is not an inverse for inj. Notice that P satisfies the condition that its range is all of its codomain, but it fails the condition that its nullspace is trivial. On the other hand, inj satisfies the condition that its nullspace is trivial but fails the condition that its range is all of its codomain.

PROBLEMS 3.5

Practice Problems

A1 For each of the following matrices, either show that the matrix is not invertible or find its inverse. Check by multiplication.

(a) $\begin{bmatrix} 3 & -4 \\ 2 & 5 \end{bmatrix}$

(b) $\begin{bmatrix} 1 & 0 & 1 \\ 2 & 1 & 3 \\ 1 & 0 & 2 \end{bmatrix}$

(c) $\begin{bmatrix} 1 & 0 & 2 \\ 1 & 1 & 3 \\ 3 & 1 & 7 \end{bmatrix}$

(d) $\begin{bmatrix} 0 & 0 & 1 \\ 0 & 1 & 1 \\ 1 & 1 & 1 \end{bmatrix}$

(e) $\begin{bmatrix} 1 & 1 & 3 & 1 \\ 0 & 2 & 1 & 0 \\ 2 & 2 & 7 & 1 \\ 0 & 6 & 3 & 1 \end{bmatrix}$

(f) $\begin{bmatrix} 1 & 0 & 1 & 0 & 1 \\ 0 & 1 & 0 & 1 & 0 \\ 0 & 0 & 1 & 1 & 1 \\ 0 & 0 & 0 & 1 & 2 \\ 0 & 0 & 0 & 0 & 1 \end{bmatrix}$

A2 Let $B = \begin{bmatrix} 1 & 0 & 1 \\ 2 & 1 & 3 \\ 1 & 0 & 2 \end{bmatrix}$. Use B^{-1} to find the solutions of the following.

(a) $B\vec{x} = \begin{bmatrix} 1 \\ 1 \\ 1 \end{bmatrix}$

(b) $B\vec{x} = \begin{bmatrix} -1 \\ 0 \\ 1 \end{bmatrix}$

(c) $B\vec{x} = \left(\begin{bmatrix} 1 \\ 1 \\ 1 \end{bmatrix} + \begin{bmatrix} -1 \\ 0 \\ 1 \end{bmatrix} \right)$

A3 Let $A = \begin{bmatrix} 2 & 1 \\ 3 & 2 \end{bmatrix}$ and $B = \begin{bmatrix} 1 & 2 \\ 3 & 5 \end{bmatrix}$.

(a) Find A^{-1} and B^{-1}.

(b) Calculate AB and $(AB)^{-1}$ and check that $(AB)^{-1} = B^{-1}A^{-1}$.

(c) Calculate $(3A)^{-1}$ and check that it equals $\frac{1}{3}A^{-1}$.

(d) Calculate $(A^T)^{-1}$ and check that $A^T(A^T)^{-1} = I$.

A4 By geometrical arguments, determine the inverse of each of the following matrices.

(a) The matrix of the rotation $R_{\pi/6}$ in the plane.

(b) $\begin{bmatrix} 1 & -3 \\ 0 & 1 \end{bmatrix}$

(c) $\begin{bmatrix} 5 & 0 \\ 0 & 1 \end{bmatrix}$

(d) $\begin{bmatrix} 1 & 0 & 0 \\ 0 & -1 & 0 \\ 0 & 0 & 1 \end{bmatrix}$

A5 The mappings in this problem are from $\mathbb{R}^2 \to \mathbb{R}^2$.

(a) Determine the matrix of the shear S by a factor of 2 in the x_2-direction and the matrix of S^{-1}.

(b) Determine the matrix of the reflection R in the line $x_1 - x_2 = 0$ and the matrix of R^{-1}.

(c) Determine the matrix of $(R \circ S)^{-1}$ and the matrix of $(S \circ R)^{-1}$ (without determining the matrices of $R \circ S$ and $S \circ R$).

A6 Suppose that $L : \mathbb{R}^n \to \mathbb{R}^n$ is a linear mapping and that $M : \mathbb{R}^n \to \mathbb{R}^n$ is a function (not assumed to be linear) such that $\vec{x} = M(\vec{y})$ if and only if $\vec{y} = L(\vec{x})$. Show that M is also linear.

Homework Problems

B1 For each of the following matrices, either show that the matrix is not invertible or find its inverse. Check by multiplication.

(a) $\begin{bmatrix} 1 & 4 \\ -2 & 7 \end{bmatrix}$

(b) $\begin{bmatrix} 1 & -1 & 2 \\ 3 & 1 & 5 \\ 2 & 2 & 3 \end{bmatrix}$

(c) $\begin{bmatrix} 1 & 2 & 0 \\ 2 & 2 & 5 \\ 1 & -1 & 3 \end{bmatrix}$

(d) $\begin{bmatrix} 1 & 1 & -2 \\ 2 & 1 & 5 \\ 4 & 3 & 1 \end{bmatrix}$

(e) $\begin{bmatrix} 2 & -1 & 3 \\ 1 & 2 & 2 \\ 1 & 0 & 1 \end{bmatrix}$

(f) $\begin{bmatrix} 1 & -1 & 0 & 2 \\ 0 & 1 & 1 & 0 \\ 2 & -2 & 3 & 5 \\ 1 & 0 & 1 & 3 \end{bmatrix}$

(g) $\begin{bmatrix} 0 & 2 & 2 & 5 \\ 0 & 1 & 0 & 3 \\ 1 & 3 & 1 & 3 \\ 3 & 6 & 0 & 3 \end{bmatrix}$

(h) $\begin{bmatrix} 1 & 0 & 0 & 0 & 1 \\ 0 & 0 & 0 & 1 & 0 \\ 0 & 0 & 1 & 0 & 0 \\ 0 & 1 & 0 & 0 & 0 \\ 1 & 0 & 0 & 0 & 0 \end{bmatrix}$

(i) $\begin{bmatrix} 0 & 0 & 0 & 0 & 1 \\ 0 & 0 & 0 & 1 & 0 \\ 0 & 0 & 1 & 1 & 1 \\ 0 & 1 & 0 & 0 & 0 \\ 1 & 0 & 0 & 0 & 0 \end{bmatrix}$

(j) $\begin{bmatrix} 1 & 0 & 0 & 0 & 0 \\ 0 & 1 & 0 & 0 & 0 \\ 0 & 0 & 1 & 0 & 0 \\ 0 & 0 & 0 & 1 & 0 \\ 1 & 1 & 1 & 1 & 1 \end{bmatrix}$

B2 Let $A = \begin{bmatrix} 2 & -1 & 1 \\ 0 & 1 & 1 \\ 1 & -1 & -1 \end{bmatrix}$.

(a) Find A^{-1}.

(b) Use (a) to solve $A\vec{x} = \vec{b}$ if

(i) $\vec{b} = \begin{bmatrix} 2 \\ 3 \\ -1 \end{bmatrix}$

(ii) $\vec{b} = 3\begin{bmatrix} 2 \\ 3 \\ -1 \end{bmatrix}$

(iii) $\vec{b} = \begin{bmatrix} 4 \\ -2 \\ 3 \end{bmatrix}$

B3 Let $A = \begin{bmatrix} 2 & 5 \\ 1 & 2 \end{bmatrix}$ and $B = \begin{bmatrix} 1 & 2 \\ 3 & 7 \end{bmatrix}$.

(a) Find A^{-1} and B^{-1}.

(b) Calculate AB and $(AB)^{-1}$ and check that $(AB)^{-1} = B^{-1}A^{-1}$.

(c) Calculate $(5A)^{-1}$ and check that it equals $\frac{1}{5}A^{-1}$.

(d) Calculate $(A^T)^{-1}$ and check that $A^T(A^T)^{-1} = I$.

B4 By geometrical arguments, determine the inverse of each of the following matrices.

(a) The matrix of the rotation $R_{\pi/4}$ in the plane.

(b) $\begin{bmatrix} 1 & 0 \\ 2 & 1 \end{bmatrix}$

(c) $\begin{bmatrix} 1 & 0 \\ 0 & -1/3 \end{bmatrix}$

(d) $\begin{bmatrix} -1 & 0 & 0 \\ 0 & -1 & 0 \\ 0 & 0 & -1 \end{bmatrix}$

B5 The mappings in this problem are from $\mathbb{R}^2 \to \mathbb{R}^2$.

(a) Determine the matrix of the stretch S by a factor of 3 in the x_2-direction and the matrix of S^{-1}.

(b) Determine the matrix of the reflection R in the line $x_1 + x_2 = 0$ and the matrix of R^{-1}.

(c) Determine the matrix of $(R \circ S)^{-1}$ and the matrix of $(S \circ R)^{-1}$ (without determining the matrices of $R \circ S$ and $S \circ R$).

B6 For each of the following pairs of linear mappings from $\mathbb{R}^3 \to \mathbb{R}^3$, determine the matrices $[R^{-1}]$, $[S^{-1}]$, and $[(R \circ S)^{-1}]$.

(a) R is the rotation about the x_1-axis through angle $\pi/2$, and S is the stretch by a factor of 0.5 in the x_3-direction.

(b) R is the reflection in the plane $x_1 - x_3 = 0$, and S is a shear by a factor of 0.4 in the x_3-direction in the x_1x_3-plane.

Computer Problems

C1 Let $A = \begin{bmatrix} 1.23 & 3.11 & 1.01 & 0.00 \\ 2.01 & -2.56 & 3.03 & 0.04 \\ 1.11 & 0.03 & -5.11 & 2.56 \\ 2.14 & -1.90 & 4.05 & 1.88 \end{bmatrix}$.

(a) Use computer software to calculate A^{-1}.

(b) Use computer software to calculate the inverse of A^{-1}. Explain your answer.

Conceptual Problems

D1 Determine an expression in terms of A^{-1} and B^{-1} for $((AB)^T)^{-1}$.

D2 (a) Suppose that A is an $n \times n$ matrix such that $A^3 = I$. Find an expression for A^{-1} in terms of A. (Hint: Find X such that $AX = I$.)

(b) Suppose that B satisfies $B^5 + B^3 + B = I$. Find an expression for B^{-1} in terms of B.

D3 Prove that if A and B are square matrices such that AB is invertible, then A and B are invertible.

D4 Let $A = \begin{bmatrix} 1 & -2 & 0 \\ 1 & -1 & 0 \end{bmatrix}$.

(a) Show that A has a right inverse by finding a matrix B such that $AB = I$.

(b) Show that there cannot exist a matrix C such that $CA = I$. Hence, A cannot have a left inverse.

D5 Prove that the following are equivalent for an $n \times n$ matrix A.

(1) A is invertible.

(2) $\text{Null}(A) = \{\vec{0}\}$.

(3) The rows of A are linearly independent.

(4) A^T is invertible.

3.6 Elementary Matrices

In Sections 3.1 and 3.5, we saw some connections between matrix-vector multiplication and systems of linear equations. In Section 3.2, we observed the connection between linear mappings and matrix-vector multiplication. Since matrix multiplication is a direct extension of matrix-vector multiplication, it should not be surprising that there is a connection between matrix multiplication, systems of linear equations, and linear mappings. We examine this connection through the use of elementary matrices.

Definition
Elementary Matrix

A matrix that can be obtained from the identity matrix by a single elementary row operation is called an **elementary matrix**.

Note that it follows from the definition that an elementary matrix must be square.

EXAMPLE 1

$E_1 = \begin{bmatrix} 1 & t \\ 0 & 1 \end{bmatrix}$ is the elementary matrix obtained from I_2 by adding the product of t times the second row to the first—a single elementary row operation. Observe that E_1 is the matrix of a shear in the x_1-direction by a factor of t.

$E_2 = \begin{bmatrix} 1 & 0 \\ 0 & t \end{bmatrix}$ is the elementary matrix obtained from I_2 by multiplying row 2 by the non-zero scalar t. E_2 is the matrix of stretch in the x_2-direction by a factor of t.

$E_3 = \begin{bmatrix} 0 & 1 \\ 1 & 0 \end{bmatrix}$ is the elementary matrix obtained from I_2 by swapping row 1 and row 2, and it is the matrix of a reflection over $x_2 = x_1$ in the plane.

As in Example 1, it can be shown that every $n \times n$ elementary matrix is the standard matrix of a shear, a stretch, or a reflection. The following theorem tells us that elementary matrices also represent elementary row operations. Hence, performing shears, stretches, and reflections; multiplying by elementary matrices; and using elementary row operations all accomplish the same thing.

Theorem 1

If A is an $n \times n$ matrix and E is the elementary matrix obtained from I_n by a certain elementary row operation, then the product EA is the matrix obtained from A by performing the same elementary row operation.

It would be tedious to write the proof in the general $n \times n$ case. Instead, we illustrate why this works by verifying the conclusion for some simple cases for 3×3 matrices.
Case 1. Consider the elementary row operation of adding k times row 3 to row 1. The corresponding elementary matrix is $E = \begin{bmatrix} 1 & 0 & k \\ 0 & 1 & 0 \\ 0 & 0 & 1 \end{bmatrix}$. Then,

$$\begin{bmatrix} a_{11} & a_{12} & a_{13} \\ a_{21} & a_{22} & a_{23} \\ a_{31} & a_{32} & a_{33} \end{bmatrix} \begin{matrix} R_1 + kR_3 \\ \sim \\ \\ \end{matrix} \begin{bmatrix} a_{11} + ka_{31} & a_{12} + ka_{32} & a_{13} + ka_{33} \\ a_{21} & a_{22} & a_{23} \\ a_{31} & a_{32} & a_{33} \end{bmatrix}$$

while

$$EA = \begin{bmatrix} 1 & 0 & k \\ 0 & 1 & 0 \\ 0 & 0 & 1 \end{bmatrix} \begin{bmatrix} a_{11} & a_{12} & a_{13} \\ a_{21} & a_{22} & a_{23} \\ a_{31} & a_{32} & a_{33} \end{bmatrix} = \begin{bmatrix} a_{11} + ka_{31} & a_{12} + ka_{32} & a_{13} + ka_{33} \\ a_{21} & a_{22} & a_{23} \\ a_{31} & a_{32} & a_{33} \end{bmatrix}$$

Case 2. Consider the elementary row operation of swapping row 2 with row 3, which has the corresponding elementary matrix $E = \begin{bmatrix} 1 & 0 & 0 \\ 0 & 0 & 1 \\ 0 & 1 & 0 \end{bmatrix}$:

$$\begin{bmatrix} a_{11} & a_{12} & a_{13} \\ a_{21} & a_{22} & a_{23} \\ a_{31} & a_{32} & a_{33} \end{bmatrix} \begin{matrix} \\ R_2 \updownarrow R_3 \\ \end{matrix} \sim \begin{bmatrix} a_{11} & a_{12} & a_{13} \\ a_{31} & a_{32} & a_{33} \\ a_{21} & a_{22} & a_{23} \end{bmatrix}$$

while

$$EA = \begin{bmatrix} 1 & 0 & 0 \\ 0 & 0 & 1 \\ 0 & 1 & 0 \end{bmatrix} \begin{bmatrix} a_{11} & a_{12} & a_{13} \\ a_{21} & a_{22} & a_{23} \\ a_{31} & a_{32} & a_{33} \end{bmatrix} = \begin{bmatrix} a_{11} & a_{12} & a_{13} \\ a_{31} & a_{32} & a_{33} \\ a_{21} & a_{22} & a_{23} \end{bmatrix}$$

Again, the conclusion of Theorem 1 is verified.

EXERCISE 1

Verify that Theorem 1 also holds for the elementary row operation of multiplying the second row by a non-zero constant for 3×3 matrices.

Theorem 2

For any $m \times n$ matrix A, there exists a sequence of elementary matrices, $E_1, E_2, \dots, E_k$, such that $E_k \cdots E_2 E_1 A$ is equal to the reduced row echelon form of A.

Proof: From our work in Chapter 2, we know that there is a sequence of elementary row operations to bring A into its reduced row echelon form. Call the elementary matrix corresponding to the first operation E_1, the elementary matrix corresponding to the second operation E_2, and so on, until the final elementary row operation corresponds to E_k. Then, by Theorem 1, $E_1 A$ is the matrix obtained by performing the first elementary row operation on A, $E_2 E_1 A$ is the matrix obtained by performing the second elementary row operation on $E_1 A$ (that is, performing the first two elementary row operations on A), and $E_k \cdots E_2 E_1 A$ is the matrix obtained after performing all of the elementary row operations on A, in the specified order. ■

EXAMPLE 2

Let $A = \begin{bmatrix} 1 & 2 & 1 \\ 2 & 4 & 4 \end{bmatrix}$. Find a sequence of elementary matrices $E_1, \dots, E_k$ such that $E_k \cdots E_1 A$ is the reduced row echelon form of A.

Solution: We row reduce A keeping track of our elementary row operations:

$$\begin{bmatrix} 1 & 2 & 1 \\ 2 & 4 & 4 \end{bmatrix} \begin{matrix} \\ R_2 - 2R_1 \end{matrix} \sim \begin{bmatrix} 1 & 2 & 1 \\ 0 & 0 & 2 \end{bmatrix} \begin{matrix} \\ \frac{1}{2}R_2 \end{matrix} \sim \begin{bmatrix} 1 & 2 & 1 \\ 0 & 0 & 1 \end{bmatrix} \begin{matrix} R_1 - R_2 \\ \\ \end{matrix} \sim \begin{bmatrix} 1 & 2 & 0 \\ 0 & 0 & 1 \end{bmatrix}$$

The first elementary row operation is $R_2 - 2R_1$, so $E_1 = \begin{bmatrix} 1 & 0 \\ -2 & 1 \end{bmatrix}$.

The second elementary row operation is $\frac{1}{2}R_2$, so $E_2 = \begin{bmatrix} 1 & 0 \\ 0 & 1/2 \end{bmatrix}$.

EXAMPLE 2 (continued)

The third elementary row operation is $R_1 - R_2$, so $E_3 = \begin{bmatrix} 1 & -1 \\ 0 & 1 \end{bmatrix}$.

Thus, $E_3E_2E_1A = \begin{bmatrix} 1 & -1 \\ 0 & 1 \end{bmatrix}\begin{bmatrix} 1 & 0 \\ 0 & 1/2 \end{bmatrix}\begin{bmatrix} 1 & 0 \\ -2 & 1 \end{bmatrix}\begin{bmatrix} 1 & 2 & 1 \\ 2 & 4 & 4 \end{bmatrix} = \begin{bmatrix} 1 & 2 & 0 \\ 0 & 0 & 1 \end{bmatrix}$.

Remark

We know that the elementary matrices in Example 2 must be 2×2 for two reasons. First, we had only two rows in A to perform elementary row operations on, so this must be the same with the corresponding elementary matrices. Second, for the matrix multiplication E_1A to be defined, we know that the number of columns in E_1 must be equal to the number of rows in A. Also, E_1 is square since it is elementary.

EXERCISE 2

Let $A = \begin{bmatrix} 1 & 1 \\ 2 & 2 \\ 3 & 2 \end{bmatrix}$. Find a sequence of elementary matrices $E_1, \dots, E_k$ such that $E_k \cdots E_1A$ is the reduced row echelon form of A.

In the special case where A is an invertible square matrix, the reduced row echelon form of A is I. Hence, by Theorem 2, there exists a sequence of elementary row operations such that $E_k \cdots E_1A = I$. Thus, the matrix $B = E_k \cdots E_1$ satisfies $BA = I$, so B is the inverse of A. Observe that this result corresponds exactly to two facts we observed in Section 3.5. First, it demonstrates our procedure for finding the inverse of a matrix by row reducing $\left[\begin{array}{c|c} A & I \end{array}\right]$. Second, it shows us that solving a system $A\vec{x} = \vec{b}$ by row reducing or by computing $\vec{x} = A^{-1}\vec{b}$ yields the same result.

Finally, observe that elementary row operations are invertible since they are reversible, and thus reflections, shears, and stretches are invertible. Moreover, since the reverse operation of an elementary row operation is another elementary row operation, the inverse of an elementary matrix is another elementary matrix. We use this to prove the following important result.

Theorem 3 If an $n \times n$ matrix A has rank n, then it may be represented as a product of elementary matrices.

Proof: By Theorem 2, there exists a sequence of elementary row operations such that $E_k \cdots E_1A = I$. Since E_k is invertible, we can multiply both sides on the left by $(E_k)^{-1}$ to get

$$(E_k)^{-1}E_kE_{k-1} \cdots E_1A = (E_k)^{-1}I \quad \text{or} \quad E_{k-1} \cdots E_1A = E_k^{-1}$$

Next, we multiply both sides by E_{k-1}^{-1} to get

$$E_{k-2} \cdots E_1A = E_{k-1}^{-1}E_k^{-1}$$

We continue to multiply by the inverse of the elementary matrix on the left until the equation becomes

$$A = E_1^{-1}E_2^{-1} \cdots E_k^{-1}$$

Thus, since the inverse of an elementary matrix is elementary, we have written A as a product of elementary matrices. ■

Remark

Observe that writing A as a product of simpler matrices is kind of like factoring a polynomial (although it is definitely not the same). This is an example of a **matrix decomposition**. There are many very important matrix decompositions in linear algebra. We will look at a useful matrix decomposition in the next section and a couple more of them later in the book.

EXAMPLE 3

Let $A = \begin{bmatrix} 0 & 2 \\ 1 & 1 \end{bmatrix}$. Write A and A^{-1} as a product of elementary matrices.

Solution: We row reduce A to I, keeping track of the elementary row operations used:

$$\begin{bmatrix} 0 & 2 \\ 1 & 1 \end{bmatrix} \begin{matrix} R_2 \updownarrow R_1 \end{matrix} \sim \begin{bmatrix} 1 & 1 \\ 0 & 2 \end{bmatrix} \begin{matrix} \\ \frac{1}{2}R_2 \end{matrix} \sim \begin{bmatrix} 1 & 1 \\ 0 & 1 \end{bmatrix} \begin{matrix} R_1 - R_2 \\ \end{matrix} \sim \begin{bmatrix} 1 & 0 \\ 0 & 1 \end{bmatrix}$$

Hence, we have

$$E_1 = \begin{bmatrix} 0 & 1 \\ 1 & 0 \end{bmatrix}, \quad E_2 = \begin{bmatrix} 1 & 0 \\ 0 & 1/2 \end{bmatrix}, \quad E_3 = \begin{bmatrix} 1 & -1 \\ 0 & 1 \end{bmatrix}$$

Thus,

$$A^{-1} = E_3E_2E_1$$

and

$$A = E_1^{-1}E_2^{-1}E_3^{-1} = \begin{bmatrix} 0 & 1 \\ 1 & 0 \end{bmatrix}\begin{bmatrix} 1 & 0 \\ 0 & 2 \end{bmatrix}\begin{bmatrix} 1 & 1 \\ 0 & 1 \end{bmatrix}$$

PROBLEMS 3.6

Practice Problems

A1 Write a 3×3 elementary matrix that corresponds to each of the following elementary row operations. Multiply each of the elementary matrices by $A = \begin{bmatrix} 1 & 2 & 3 \\ -1 & 3 & 4 \\ 4 & 2 & 0 \end{bmatrix}$ and verify that the product EA is the matrix obtained from A by the elementary row operation.

(a) Add (–5) times the second row to the first row.
(b) Swap the second and third rows.
(c) Multiply the third row by (–1).
(d) Multiply the second row by 6.
(e) Add 4 times the first row to the third.

A2 Write a 4×4 elementary matrix that corresponds to each of the following elementary row operations.

(a) Add (–3) times the third row to the fourth row.
(b) Swap the second and fourth rows.
(c) Multiply the third row by (–3).
(d) Add 2 times the first row to the third row.
(e) Multiply the first row by 3.
(f) Swap the first and third rows.

A3 For each of the following matrices, either state that it is an elementary matrix and state the corresponding elementary row operation or explain why it is not elementary.

(a) $\begin{bmatrix} 1 & 0 & 0 \\ 0 & 1 & 0 \\ 0 & -4 & 1 \end{bmatrix}$ (b) $\begin{bmatrix} -1 & 0 & 0 \\ 0 & 1 & 0 \\ 0 & 0 & -1 \end{bmatrix}$

(c) $\begin{bmatrix} 3 & 0 & 1 \\ 0 & 1 & 0 \\ 0 & 0 & 1 \end{bmatrix}$ (d) $\begin{bmatrix} 0 & 0 & 1 \\ 0 & 1 & 0 \\ 1 & 0 & 0 \end{bmatrix}$

(e) $\begin{bmatrix} 0 & 1 & 0 \\ 0 & 0 & 1 \\ 1 & 0 & 0 \end{bmatrix}$ (f) $\begin{bmatrix} 1 & 0 & 0 \\ 0 & 1 & 0 \\ 0 & 0 & 1 \end{bmatrix}$

A4 For each of the following matrices:
(i) Find a sequence of elementary matrices $E_k, \dots, E_1$ such that $E_k \cdots E_1 A = I$.
(ii) Determine A^{-1} by computing $E_k \cdots E_1$.
(iii) Write A as a product of elementary matrices.

(a) $A = \begin{bmatrix} 1 & 3 & 4 \\ 0 & 0 & 2 \\ 0 & 1 & 0 \end{bmatrix}$

(b) $A = \begin{bmatrix} 1 & 2 & 2 \\ 0 & 1 & 3 \\ 2 & 4 & 5 \end{bmatrix}$

(c) $A = \begin{bmatrix} 1 & 0 & -1 \\ -2 & 0 & -2 \\ -4 & 1 & 4 \end{bmatrix}$

(d) $A = \begin{bmatrix} 1 & -2 & 4 & 1 \\ -1 & 3 & -4 & -1 \\ 0 & 1 & 2 & 0 \\ -2 & 4 & -8 & -1 \end{bmatrix}$

Homework Problems

B1 Write a 3×3 elementary matrix that corresponds to each of the following elementary row operations. Multiply each of the elementary matrices by $A = \begin{bmatrix} 2 & 1 & -1 \\ 2 & 0 & 5 \\ 1 & -3 & -2 \end{bmatrix}$ and verify that the product EA is the matrix obtained from A by the elementary row operation.
(a) Add 4 times the second row to the first row.
(b) Swap the first and third rows.
(c) Multiply the second row by (–3).
(d) Multiply the first row by 2.
(e) Add (–2) times the first row to the third.
(f) Swap the first and second rows.

B2 Write a 4×4 elementary matrix that corresponds to each of the following elementary row operations.
(a) Add 6 times the fourth row to the second row.
(b) Multiply the second row by 5.
(c) Swap the first and fourth rows.
(d) Swap the third and fourth rows.
(e) Add (–2) times the third row to the first row.
(f) Multiply the fourth row by (–2).

B3 For each of the following matrices, either state that it is an elementary matrix and state the corresponding elementary row operation or explain why it is not elementary.

(a) $\begin{bmatrix} 0 & 0 & 1 \\ 1 & 0 & 0 \\ 0 & 1 & 0 \end{bmatrix}$ (b) $\begin{bmatrix} 1 & 0 & 0 \\ 0 & 1 & 0 \\ 0 & 3 & 1 \end{bmatrix}$

(c) $\begin{bmatrix} -1 & 0 & 0 \\ 0 & 0 & 1 \\ 0 & 1 & 0 \end{bmatrix}$ (d) $\begin{bmatrix} 1 & 0 & 1 \\ 0 & 1 & 0 \\ 1 & 0 & 1 \end{bmatrix}$

(e) $\begin{bmatrix} 2 & 0 & 0 \\ 0 & 1 & 0 \\ 0 & 0 & 1 \end{bmatrix}$ (f) $\begin{bmatrix} 1 & 0 & 0 \\ -1 & 1 & 0 \\ 0 & 0 & 1 \end{bmatrix}$

B4 For each of the following matrices:
(i) Find a sequence of elementary matrices $E_k, \dots, E_1$ such that $E_k \cdots E_1 A = I$.
(ii) Determine A^{-1} by computing $E_k \cdots E_1$.
(iii) Write A as a product of elementary matrices.

(a) $A = \begin{bmatrix} 1 & 2 & -1 \\ 0 & 1 & 2 \\ 2 & 4 & 0 \end{bmatrix}$

(b) $A = \begin{bmatrix} 1 & 0 & 0 \\ 2 & 3 & 0 \\ 1 & 4 & 1 \end{bmatrix}$

(c) $A = \begin{bmatrix} 0 & 1 & 2 \\ -4 & -3 & -3 \\ 1 & 1 & 1 \end{bmatrix}$

(d) $A = \begin{bmatrix} 1 & -2 & 1 & 1 \\ 3 & -4 & 2 & 2 \\ -2 & 4 & -1 & -3 \\ 3 & -5 & 1 & 5 \end{bmatrix}$

Conceptual Problems

D1 (a) Let $L : \mathbb{R}^2 \to \mathbb{R}^2$ be the invertible linear operator with standard matrix $A = \begin{bmatrix} 0 & -2 \\ 1 & -4 \end{bmatrix}$. By writing A as a product of elementary matrices, show that L can be written as a composition of shears, stretches, and reflections.

(b) Explain how we know that every invertible linear operator $L : \mathbb{R}^n \to \mathbb{R}^n$ can be written as a composition of shears, stretches, and reflections.

D2 For 2×2 matrices, verify that Theorem 1 holds for the elementary row operations add t times row 1 to row 2 and multiply row 2 by a factor of $t \neq 0$.

D3 Let $A = \begin{bmatrix} 1 & 2 \\ 2 & 5 \end{bmatrix}$ and $\vec{b} = \begin{bmatrix} 3 \\ 5 \end{bmatrix}$.

(a) Determine elementary matrices E_1 and E_2 such that $E_2E_1A = I$.

(b) Since A is invertible, we know that the system $A\vec{x} = \vec{b}$ has the unique solution $\vec{x} = A^{-1}\vec{b} = E_2E_1\vec{b}$. Instead of using matrix multiplication, calculate the solution $\vec{x}$ in the following way. First, compute $E_1\vec{b}$ by performing the elementary row operation associated with E_1 on the matrix $\vec{b}$. Then compute $\vec{x} = E_2E_1\vec{b}$ by performing the elementary row operation associated with E_2 on the result for $E_1\vec{b}$.

(c) Solve the system $A\vec{x} = \vec{b}$ by row reducing $\left[\begin{array}{c|c} A & \vec{b} \end{array}\right]$. Observe the operations that you use on the augmented part of the system and compare with part (b).

3.7 *LU*-Decomposition

One of the most basic and useful ideas in mathematics is the concept of a factorization of an object. You have already seen that it can be very useful to factor a number into primes or to factor a polynomial. Similarly, in many applications of linear algebra, we may want to decompose a matrix into factors that have certain properties.

In applied linear algebra, we often need to quickly solve multiple systems $A\vec{x} = \vec{b}$, where the coefficient matrix A remains the same but the vector $\vec{b}$ changes. The goal of this section is to derive a matrix factorization called the LU-decomposition, which is commonly used in computer algorithms to solve such problems.

We now start our look at the LU-decomposition by recalling the definition of upper-triangular and lower-triangular matrices.

Definition
Upper Triangular
Lower Triangular

An $n \times n$ matrix U is said to be **upper triangular** if the entries beneath the main diagonal are all zero—that is, $(U)_{ij} = 0$ whenever $i > j$. An $n \times n$ matrix L is said to be **lower triangular** if the entries above the main diagonal are all zero—in particular, $(L)_{ij} = 0$ whenever $i < j$.

Remark

By definition, a matrix in row echelon form is upper triangular. This fact is very important for the LU-decomposition.

Observe that for each such system $A\vec{x} = \vec{b}$, we can use the same row operations to row reduce $\left[\begin{array}{c|c} A & \vec{b} \end{array}\right]$ to row echelon form and then solve the system using back-substitution. The only difference between the systems will then be the effect of the row operations on $\vec{b}$. In particular, we see that the two important pieces of information we require are the row echelon form of A and the elementary row operation used.

For our purposes, we will assume that our $n \times n$ coefficient matrix A can be brought into row echelon form using only elementary row operations of the form add a multiple of one row to another. Since we can row reduce a matrix to a row echelon form without multiplying a row by a non-zero constant, omitting this row operation is not a problem. However, omitting row interchanges may seem rather serious: without row interchanges, we cannot bring a matrix such as $\begin{bmatrix} 0 & 1 \\ 1 & 2 \end{bmatrix}$ into row echelon form. However, we only omit row interchanges because it is difficult to keep track of them by hand. A computer can keep track of row interchanges without physically moving entries from one location to another. At the end of the section, we will comment on the case where swapping rows is required.

Thus, for such a matrix A, to row reduce A to a row echelon form, we will only use row operations of the form $R_i + sR_j$, where $i > j$. Each such row operation will have a corresponding elementary matrix that is lower triangular and has 1s along the main diagonal. So, under our assumption, there are elementary matrices $E_1, \dots, E_k$ that are all lower triangular such that

$$E_k \cdots E_1 A = U$$

where U is a row echelon form of A. Since $E_k \cdots E_1$ is invertible, we can write $A = (E_k \cdots E_1)^{-1} U$ and define

$$L = (E_k \cdots E_1)^{-1} = E_1^{-1} \cdots E_k^{-1}$$

Since the inverse of a lower-triangular elementary matrix is lower triangular, and a product of lower-triangular matrices is lower triangular, L is lower triangular. (You are asked to prove this in Problem D1.) Therefore, this gives us the matrix decomposition $A = LU$, where U is upper triangular and L is lower triangular. Moreover, L contains the information about the row operations used to bring A to U.

Theorem 1

If A is an $n \times n$ matrix that can be row reduced to row echelon form without swapping rows, then there exists an upper triangular matrix U and lower triangular matrix L such that $A = LU$.

Definition
LU-Decomposition

Writing an $n \times n$ matrix A as a product LU, where L is lower triangular and U is upper triangular, is called an ***LU*-decomposition** of A.

Our derivation has given an algorithm for finding the LU-decomposition of such a matrix.

EXAMPLE 1

Find an LU-decomposition of $A = \begin{bmatrix} 2 & -1 & 4 \\ 4 & -1 & 6 \\ -1 & -1 & 3 \end{bmatrix}$.

Solution: Row-reducing and keeping track of our row-operations gives

$$\begin{bmatrix} 2 & -1 & 4 \\ 4 & -1 & 6 \\ -1 & -1 & 3 \end{bmatrix} \begin{matrix} \\ R_2 - 2R_1 \\ R_3 + \frac{1}{2}R_1 \end{matrix} \sim \begin{bmatrix} 2 & -1 & 4 \\ 0 & 1 & -2 \\ 0 & -3/2 & 5 \end{bmatrix} \begin{matrix} \\ \\ R_3 + \frac{3}{2}R_2 \end{matrix}$$

$$\sim \begin{bmatrix} 2 & -1 & 4 \\ 0 & 1 & -2 \\ 0 & 0 & 2 \end{bmatrix} = U$$

We have $E_3E_2E_1A = U$, so $A = E_1^{-1}E_2^{-1}E_3^{-1}U$, where

$$E_1 = \begin{bmatrix} 1 & 0 & 0 \\ -2 & 1 & 0 \\ 0 & 0 & 1 \end{bmatrix}, E_2 = \begin{bmatrix} 1 & 0 & 0 \\ 0 & 1 & 0 \\ 1/2 & 0 & 1 \end{bmatrix}, E_3 = \begin{bmatrix} 1 & 0 & 0 \\ 0 & 1 & 0 \\ 0 & 3/2 & 1 \end{bmatrix}$$

Hence, we let

$$L = E_1^{-1}E_2^{-1}E_3^{-1} = \begin{bmatrix} 1 & 0 & 0 \\ 2 & 1 & 0 \\ 0 & 0 & 1 \end{bmatrix} \begin{bmatrix} 1 & 0 & 0 \\ 0 & 1 & 0 \\ -1/2 & 0 & 1 \end{bmatrix} \begin{bmatrix} 1 & 0 & 0 \\ 0 & 1 & 0 \\ 0 & -3/2 & 1 \end{bmatrix}$$

$$= \begin{bmatrix} 1 & 0 & 0 \\ 2 & 1 & 0 \\ 0 & 0 & 1 \end{bmatrix} \begin{bmatrix} 1 & 0 & 0 \\ 0 & 1 & 0 \\ -1/2 & -3/2 & 1 \end{bmatrix} = \begin{bmatrix} 1 & 0 & 0 \\ 2 & 1 & 0 \\ -1/2 & -3/2 & 1 \end{bmatrix}$$

And we get $A = LU$.

Observe from this example that the entries in L are just the negative of the multipliers used to put a zero in the corresponding entry. To see why this is the case, observe that if $E_k \cdots E_1A = U$, then

$$(E_k \cdots E_1)L = (E_k \cdots E_1)(E_k \cdots E_1)^{-1} = I$$

Hence, the same row operations that reduce A to U will reduce L to I.

This makes the LU-decomposition extremely easy to find.

EXAMPLE 2

Find an LU-decomposition of $B = \begin{bmatrix} 2 & 1 & -1 \\ -4 & 3 & 3 \\ 6 & 8 & -3 \end{bmatrix}$.

EXAMPLE 2 (continued)

Solution: By row-reducing, we get

$$\begin{bmatrix} 2 & 1 & -1 \\ -4 & 3 & 3 \\ 6 & 8 & -3 \end{bmatrix} \begin{matrix} \\ R_2 + 2R_1 \\ R_3 - 3R_1 \end{matrix} \sim \begin{bmatrix} 2 & 1 & -1 \\ 0 & 5 & 1 \\ 0 & 5 & 0 \end{bmatrix} \quad \Rightarrow \quad L = \begin{bmatrix} 1 & 0 & 0 \\ -2 & 1 & 0 \\ 3 & * & 1 \end{bmatrix}$$

$$\begin{bmatrix} 2 & 1 & -1 \\ 0 & 5 & 1 \\ 0 & 5 & 0 \end{bmatrix} \begin{matrix} \\ \\ R_3 - R_2 \end{matrix} \sim \begin{bmatrix} 2 & 1 & -1 \\ 0 & 5 & 1 \\ 0 & 0 & -1 \end{bmatrix} \quad \Rightarrow \quad L = \begin{bmatrix} 1 & 0 & 0 \\ -2 & 1 & 0 \\ 3 & 1 & 1 \end{bmatrix}$$

Therefore, we have

$$B = LU = \begin{bmatrix} 1 & 0 & 0 \\ -2 & 1 & 0 \\ 3 & 1 & 1 \end{bmatrix} \begin{bmatrix} 2 & 1 & -1 \\ 0 & 5 & 1 \\ 0 & 0 & -1 \end{bmatrix}$$

EXAMPLE 3

Find an LU-decomposition of $C = \begin{bmatrix} 1 & 2 & -3 \\ 2 & 2 & 3 \\ -4 & -2 & 1 \end{bmatrix}$.

Solution: By row-reducing, we get

$$\begin{bmatrix} 1 & 2 & -3 \\ 2 & 2 & 3 \\ -4 & -2 & 1 \end{bmatrix} \begin{matrix} \\ R_2 - 2R_1 \\ R_3 + 4R_1 \end{matrix} \sim \begin{bmatrix} 1 & 2 & -3 \\ 0 & -2 & 9 \\ 0 & 6 & -11 \end{bmatrix} \quad \Rightarrow \quad L = \begin{bmatrix} 1 & 0 & 0 \\ 2 & 1 & 0 \\ -4 & * & 1 \end{bmatrix}$$

$$\begin{bmatrix} 1 & 2 & -3 \\ 0 & -2 & 9 \\ 0 & 6 & -11 \end{bmatrix} \begin{matrix} \\ \\ R_3 + 3R_2 \end{matrix} \sim \begin{bmatrix} 1 & 2 & -3 \\ 0 & -2 & 9 \\ 0 & 0 & 16 \end{bmatrix} \quad \Rightarrow \quad L = \begin{bmatrix} 1 & 0 & 0 \\ 2 & 1 & 0 \\ -4 & -3 & 1 \end{bmatrix}$$

Therefore, we have

$$C = LU = \begin{bmatrix} 1 & 0 & 0 \\ 2 & 1 & 0 \\ -4 & -3 & 1 \end{bmatrix} \begin{bmatrix} 1 & 2 & -3 \\ 0 & -2 & 9 \\ 0 & 0 & 16 \end{bmatrix}$$

EXERCISE 1

Find an LU-decomposition of $A = \begin{bmatrix} -1 & 1 & 2 \\ 4 & -1 & -3 \\ -3 & -3 & 1 \end{bmatrix}$.

Solving Systems with the *LU*-Decomposition

We now look at how to use the *LU*-decomposition to solve the system $A\vec{x} = \vec{b}$. If $A = LU$, the system can be written as

$$LU\vec{x} = \vec{b}$$

Letting $\vec{y} = U\vec{x}$, we can write $LU\vec{x} = \vec{b}$ as two systems:

$$L\vec{y} = \vec{b} \qquad \text{and} \qquad U\vec{x} = \vec{y}$$

which both have triangular coefficient matrices. This allows us to solve both systems immediately, using substitution. In particular, since L is lower triangular, we use forward-substitution to solve $\vec{y}$ and then solve $U\vec{x} = \vec{y}$ for $\vec{x}$ using back-substitution.

Remark

Observe that the first system is really calculating how performing the row operations on A would have affected $\vec{b}$.

EXAMPLE 4

Let $B = \begin{bmatrix} 2 & 1 & -1 \\ -4 & 3 & 3 \\ 6 & 8 & -3 \end{bmatrix}$ and $\vec{b} = \begin{bmatrix} 3 \\ -13 \\ 4 \end{bmatrix}$. Use an *LU*-decomposition of B to solve $B\vec{x} = \vec{b}$.

Solution: In Example 2 we found an *LU*-decomposition of B. We write $B\vec{x} = \vec{b}$ as $LU\vec{x} = \vec{b}$ and take $\vec{y} = U\vec{x}$. Writing out the system $L\vec{y} = \vec{b}$, we get

$$\begin{aligned} y_1 &= 3 \\ -2y_1 + y_2 &= -13 \\ 3y_1 + y_2 + y_3 &= 4 \end{aligned}$$

Using forward-substitution, we find that $y_1 = 3$, so $y_2 = -13 + 2(3) = -7$ and $y_3 = 4 - 3(3) - (-7) = 2$. Hence, $\vec{y} = \begin{bmatrix} 3 \\ -7 \\ 2 \end{bmatrix}$.

Thus, our system $U\vec{x} = \vec{y}$ is

$$\begin{aligned} 2x_1 + x_2 - x_3 &= 3 \\ 5x_2 + x_3 &= -7 \\ -x_3 &= 2 \end{aligned}$$

Using back-substitution, we get $x_3 = -2$, $5x_2 = -7 - (-2) \Rightarrow x_2 = -1$ and $2x_1 = 3 - (-1) + (-2) \Rightarrow x_1 = 1$. Thus, the solution is $\vec{x} = \begin{bmatrix} 1 \\ -1 \\ -2 \end{bmatrix}$.

EXAMPLE 5

Use LU-decomposition to solve $\begin{bmatrix} 1 & 1 & 1 \\ -1 & -2 & 3 \\ -2 & -4 & 6 \end{bmatrix} \vec{x} = \begin{bmatrix} 1 \\ 6 \\ 12 \end{bmatrix}$.

Solution: We first find an LU-decomposition for $\begin{bmatrix} 1 & 1 & 1 \\ -1 & -2 & 3 \\ -2 & -4 & 6 \end{bmatrix}$. Row reducing gives

$$\begin{bmatrix} 1 & 1 & 1 \\ -1 & -2 & 3 \\ -2 & -4 & 6 \end{bmatrix} \begin{matrix} \\ R_2 + R_1 \\ R_3 + 2R_1 \end{matrix} \sim \begin{bmatrix} 1 & 1 & 1 \\ 0 & -1 & 4 \\ 0 & -2 & 8 \end{bmatrix} \quad \Rightarrow \quad L = \begin{bmatrix} 1 & 0 & 0 \\ -1 & 1 & 0 \\ -2 & * & 1 \end{bmatrix}$$

$$\begin{bmatrix} 1 & 1 & 1 \\ 0 & -1 & 4 \\ 0 & -2 & 8 \end{bmatrix} \begin{matrix} \\ \\ R_3 - 2R_2 \end{matrix} \sim \begin{bmatrix} 1 & 1 & 1 \\ 0 & -1 & 4 \\ 0 & 0 & 0 \end{bmatrix} \quad \Rightarrow \quad L = \begin{bmatrix} 1 & 0 & 0 \\ -1 & 1 & 0 \\ -2 & 2 & 1 \end{bmatrix}$$

Thus, $U = \begin{bmatrix} 1 & 1 & 1 \\ 0 & -1 & 4 \\ 0 & 0 & 0 \end{bmatrix}$. We let $\vec{y} = U\vec{x}$ and solve $L\vec{y} = \vec{b}$. This gives

$$\begin{aligned} y_1 &= 1 \\ -y_1 + y_2 &= 6 \\ -2y_1 + 2y_2 + y_3 &= 12 \end{aligned}$$

So, $y_1 = 1$, $y_2 = 6 + 1 = 7$, and $y_3 = 12 + 2 - 14 = 0$. Then we solve $U\vec{x} = \begin{bmatrix} 1 \\ 7 \\ 0 \end{bmatrix}$, which is

$$\begin{aligned} x_1 + x_2 + x_3 &= 1 \\ -x_2 + 4x_3 &= 7 \\ 0x_3 &= 0 \end{aligned}$$

This gives $x_3 = t \in \mathbb{R}$, $-x_2 = 7 - 4t$, so $x_2 = -7 + 4t$ and $x_1 = 1 + (7 - 4t) - t = 8 - 5t$. Hence,

$$\vec{x} = \begin{bmatrix} 8 - 5t \\ -7 + 4t \\ t \end{bmatrix} = \begin{bmatrix} 8 \\ -7 \\ 0 \end{bmatrix} + t \begin{bmatrix} -5 \\ 4 \\ 1 \end{bmatrix}$$

EXERCISE 2

Let $A = \begin{bmatrix} -1 & 1 & 2 \\ 4 & -1 & -3 \\ -3 & -3 & 1 \end{bmatrix}$. Use the LU-decomposition of A that you found in Exercise 1 to solve the system $A\vec{x} = \vec{b}$, where:

(a) $\vec{b} = \begin{bmatrix} 3 \\ 2 \\ 6 \end{bmatrix}$ (b) $\vec{b} = \begin{bmatrix} 8 \\ 2 \\ -9 \end{bmatrix}$

A Comment About Swapping Rows

It can be shown that for any $n \times n$ matrix A, we can first rearrange the rows of A to get a matrix that has an LU-factorization. In particular, for every matrix A, there exists a matrix P, called a permutation matrix, that can be obtained by only performing row swaps on the identity matrix such that

$$PA = LU$$

Then, to solve $A\vec{x} = \vec{b}$, we use the LU-decomposition as before to solve the equivalent system

$$PA\vec{x} = P\vec{b}$$

PROBLEMS 3.7

Practice Problems

A1 Find an LU-decomposition of each of the following matrices.

(a) $\begin{bmatrix} -2 & -1 & 5 \\ -4 & 0 & -2 \\ 2 & 1 & 3 \end{bmatrix}$ (b) $\begin{bmatrix} 1 & -2 & 4 \\ 3 & -2 & 4 \\ 2 & 2 & -5 \end{bmatrix}$

(c) $\begin{bmatrix} 2 & -4 & 5 \\ 2 & 5 & 2 \\ 2 & -1 & 5 \end{bmatrix}$ (d) $\begin{bmatrix} 1 & 5 & 3 & 4 \\ -2 & -6 & -1 & 3 \\ 0 & 2 & -1 & -1 \\ 0 & 0 & 0 & 0 \end{bmatrix}$

(e) $\begin{bmatrix} 1 & -2 & 1 & 1 \\ 0 & -3 & -2 & 1 \\ 3 & -3 & 2 & -1 \\ 0 & 4 & -3 & 0 \end{bmatrix}$ (f) $\begin{bmatrix} -2 & -1 & 2 & 0 \\ 4 & 3 & -2 & 2 \\ 3 & 3 & 4 & 3 \\ 2 & -1 & 2 & -4 \end{bmatrix}$

A2 For each matrix, find an LU-decomposition and use it to solve $A\vec{x} = \vec{b}_i$, for $i = 1, 2$.

(a) $A = \begin{bmatrix} 1 & 0 & 3 \\ -2 & 1 & -3 \\ -1 & 4 & 5 \end{bmatrix}, \vec{b}_1 = \begin{bmatrix} 3 \\ -4 \\ -3 \end{bmatrix}, \vec{b}_2 = \begin{bmatrix} 2 \\ -5 \\ -2 \end{bmatrix}$

(b) $A = \begin{bmatrix} 1 & 0 & -2 \\ -1 & -4 & 4 \\ 3 & -4 & -1 \end{bmatrix}, \vec{b}_1 = \begin{bmatrix} -1 \\ -7 \\ -5 \end{bmatrix}, \vec{b}_2 = \begin{bmatrix} 2 \\ 0 \\ -1 \end{bmatrix}$

(c) $A = \begin{bmatrix} 1 & 0 & 1 \\ -3 & 2 & -1 \\ -3 & 4 & 2 \end{bmatrix}, \vec{b}_1 = \begin{bmatrix} 3 \\ -5 \\ -1 \end{bmatrix}, \vec{b}_2 = \begin{bmatrix} -4 \\ 4 \\ -5 \end{bmatrix}$

(d) $A = \begin{bmatrix} -1 & 2 & -3 & 0 \\ 0 & -1 & 3 & 1 \\ 3 & -8 & 3 & 2 \\ 1 & -2 & 3 & 1 \end{bmatrix}, \vec{b}_1 = \begin{bmatrix} -6 \\ 7 \\ -4 \\ 5 \end{bmatrix}, \vec{b}_2 = \begin{bmatrix} 5 \\ -3 \\ 3 \\ -5 \end{bmatrix}$

Homework Problems

B1 Find an LU-decomposition of each of the following matrices.

(a) $\begin{bmatrix} 1 & -1 \\ 5 & 1 \end{bmatrix}$ (b) $\begin{bmatrix} -1 & -2 & 1 \\ -4 & -5 & 3 \\ 2 & 1 & -1 \end{bmatrix}$

(c) $\begin{bmatrix} 1 & -1 & 3 \\ 0 & -4 & 2 \\ -1 & -4 & -2 \end{bmatrix}$ (d) $\begin{bmatrix} 2 & 2 & -1 \\ 4 & 3 & 0 \\ -4 & -3 & -1 \end{bmatrix}$

(e) $\begin{bmatrix} 1 & 2 & 0 & -2 \\ 2 & 3 & -1 & -4 \\ 1 & -3 & -4 & 0 \\ 4 & 5 & 0 & 0 \end{bmatrix}$ (f) $\begin{bmatrix} 2 & -1 & 3 & -1 \\ -2 & 2 & -3 & 2 \\ 4 & 5 & 5 & 5 \\ -3 & 4 & -5 & 4 \end{bmatrix}$

B2 For each matrix, find an LU-decomposition and use it to solve $A\vec{x} = \vec{b}_i$, for $i = 1, 2$.

(a) $A = \begin{bmatrix} -1 & 0 & -2 \\ 4 & 1 & 2 \\ 3 & 0 & 3 \end{bmatrix}, \vec{b}_1 = \begin{bmatrix} -6 \\ 9 \\ 9 \end{bmatrix}, \vec{b}_2 = \begin{bmatrix} 2 \\ 5 \\ 3 \end{bmatrix}$

(b) $A = \begin{bmatrix} -2 & -4 & 2 \\ 2 & 2 & -2 \\ 5 & 6 & -4 \end{bmatrix}, \vec{b}_1 = \begin{bmatrix} -2 \\ 4 \\ 1 \end{bmatrix}, \vec{b}_2 = \begin{bmatrix} 4 \\ 2 \\ 0 \end{bmatrix}$

(c) $A = \begin{bmatrix} 1 & -2 & 4 \\ 1 & 3 & 9 \\ -4 & 1 & -8 \end{bmatrix}, \vec{b}_1 = \begin{bmatrix} 11 \\ -9 \\ -16 \end{bmatrix}, \vec{b}_2 = \begin{bmatrix} -6 \\ -1 \\ 2 \end{bmatrix}$

(d) $A = \begin{bmatrix} 1 & -1 & -2 & 0 \\ 2 & -5 & -3 & -2 \\ -2 & 5 & 5 & -2 \\ 4 & -4 & -4 & -3 \end{bmatrix}, \vec{b}_1 = \begin{bmatrix} -3 \\ -4 \\ 8 \\ -4 \end{bmatrix}, \vec{b}_2 = \begin{bmatrix} 0 \\ 4 \\ -2 \\ 5 \end{bmatrix}$

Conceptual Problems

D1 (a) Prove that the inverse of a lower-triangular elementary matrix is lower triangular.

(b) Prove that a product of lower-triangular matrices is lower triangular.

CHAPTER REVIEW

Suggestions for Student Review

Try to answer all of these questions before checking answers at the suggested locations. In particular, try to invent your own examples. These review suggestions are intended to help you carry out your review. They may not cover every idea you need to master. Working in small groups may improve your efficiency.

1 State the rules for determining the product of two matrices A and B. What condition(s) must be satisfied by the sizes of A and B for the product to be defined? How do each of these rules correspond to writing a system of linear equations? (Section 3.1)

2 Explain clearly the relationship between a matrix and the corresponding matrix mapping. Explain how you determine the matrix of a given linear mapping. Pick some vector $\vec{v} \in \mathbb{R}^2$ and determine the standard matrices of the linear mappings $\text{proj}_{\vec{v}}$, $\text{perp}_{\vec{v}}$, and $\text{refl}_{\vec{v}}$. Check that your answers are correct by using each matrix to determine the image of $\vec{v}$ and a vector orthogonal to $\vec{v}$ under the mapping. (Section 3.2)

3 Determine the image of the vector $\begin{bmatrix} 1 \\ 1 \end{bmatrix}$ under the rotation of the plane through the angle $\frac{\pi}{6}$. Check that the image has the same length as the original vector. (Section 3.3)

4 Outline the relationships between the solution set for the homogeneous system $A\vec{x} = \vec{0}$ and solutions for the corresponding non-homogeneous system $A\vec{x} = \vec{b}$. Illustrate by giving some specific examples where A is 2×3 and is in reduced row echelon form. Also discuss connections between these sets and special subspaces for linear mappings. (Section 3.4)

5 (a) How many ways can you recognize the rank of a matrix? State them all. (Section 3.4)

(b) State the connection between the rank of a matrix A and the dimension of the solution space of $A\vec{x} = \vec{0}$. (Section 3.4)

(c) Illustrate your answers to (a) and (b) by constructing examples of 4×5 matrices in row echelon form of (i) rank 4; (ii) rank 3; and (iii) rank 2. In each case, actually determine the general solution of the system $A\vec{x} = \vec{0}$ and check that the solution space has the correct dimension. (Section 3.4)

6 (a) Outline the procedure for determining the inverse of a matrix. Indicate why it might not produce an inverse for some matrix A. Use the matrices of some geometric linear mappings to give two or three examples of matrices that have inverses and two examples of square matrices that do not have inverses. (Section 3.5)

(b) Pick a fairly simple 3×3 matrix (that has not too many zeros) and try to find its inverse. If it is not invertible, try another. When you have an inverse, check its correctness by multiplication. (Section 3.5)

7 For 3×3 matrices, choose one elementary row operation of each of the three types, call these E_1, E_2, E_3. Choose an arbitrary 3×3 matrix A and check that E_iA is the matrix obtained from A by the appropriate elementary row operations. (Section 3.6)

Chapter Quiz

E1 Let $A = \begin{bmatrix} 2 & -5 & -3 \\ -3 & 4 & -7 \end{bmatrix}$ and $B = \begin{bmatrix} 2 & -1 & 4 \\ 3 & 0 & 2 \\ 1 & -1 & 5 \end{bmatrix}$.

Either determine the following products or explain why they are not defined.

(a) AB (b) BA (c) BA^T

E2 (a) Let $A = \begin{bmatrix} -3 & 0 & 4 \\ 2 & -4 & -1 \end{bmatrix}$, let f_A be the matrix mapping with matrix A, and let $\vec{u} = \begin{bmatrix} 1 \\ 1 \\ -2 \end{bmatrix}$ and $\vec{v} = \begin{bmatrix} 4 \\ -2 \\ -1 \end{bmatrix}$. Determine $f_A(\vec{u})$ and $f_A(\vec{v})$.

(b) Use the result of part (a) to calculate $A\begin{bmatrix} 4 & 1 \\ -2 & 1 \\ -1 & -2 \end{bmatrix}$.

E3 Let R be the rotation through angle $\frac{\pi}{3}$ about the x_3-axis in $\mathbb{R}^3$ and let M be a reflection in $\mathbb{R}^3$ in the plane with equation $-x_1 - x_2 + 2x_3 = 0$. Determine:

(a) The matrix of R

(b) The matrix of M

(c) The matrix of $[R \circ M]$

E4 Let $A = \begin{bmatrix} 1 & 0 & 2 & 1 & 0 \\ 2 & 1 & 3 & 2 & 0 \\ 1 & 1 & 1 & 1 & 1 \end{bmatrix}$ and $\vec{b} = \begin{bmatrix} 5 \\ 16 \\ 18 \end{bmatrix}$. Determine the solution set of $A\vec{x} = \vec{b}$ and the solution space of $A\vec{x} = \vec{0}$ and discuss the relationship between the two sets.

E5 Let $B = \begin{bmatrix} 1 & 2 & 0 \\ -1 & -1 & -1 \\ 1 & 3 & 0 \\ 0 & 2 & -1 \end{bmatrix}$, $\vec{u} = \begin{bmatrix} 4 \\ -3 \\ 5 \\ 3 \end{bmatrix}$, and $\vec{v} = \begin{bmatrix} -5 \\ 6 \\ -7 \\ -1 \end{bmatrix}$.

(a) Using only one sequence of elementary row operations, determine whether $\vec{u}$ is in the columnspace of B and whether $\vec{v}$ is in the range of the linear mapping f_B with matrix B.

(b) Determine from your calculation in part (a) a vector $\vec{x}$ such that $f_B(\vec{x}) = \vec{v}$.

(c) Determine a vector $\vec{y}$ such that $f_B(\vec{y}) = \begin{bmatrix} 2 \\ -1 \\ 3 \\ 2 \end{bmatrix}$ (the second column of B).

E6 You are given the matrix A below and a row echelon form of A. Determine a basis for the rowspace, columnspace, and nullspace of A.

$$A = \begin{bmatrix} 1 & 0 & 1 & 1 & 1 \\ 2 & 1 & 1 & 2 & 5 \\ 0 & 2 & -2 & 1 & 8 \\ 3 & 3 & 0 & 4 & 14 \end{bmatrix} \sim \begin{bmatrix} 1 & 0 & 1 & 1 & 1 \\ 0 & 1 & -1 & 0 & 3 \\ 0 & 0 & 0 & 1 & 2 \\ 0 & 0 & 0 & 0 & 0 \end{bmatrix}$$

E7 Determine the inverse of the matrix $A = \begin{bmatrix} 1 & 0 & 0 & -1 \\ 0 & 0 & 1 & 0 \\ 0 & 2 & 0 & 1 \\ 1 & 0 & 0 & 2 \end{bmatrix}$.

E8 Determine all values of p such that the matrix $\begin{bmatrix} 1 & 0 & p \\ 1 & 1 & 0 \\ 2 & 1 & 1 \end{bmatrix}$ is invertible and determine its inverse.

E9 Prove that the range of a linear mapping $L : \mathbb{R}^n \to \mathbb{R}^m$ is a subspace of the codomain.

E10 Let $\{\vec{v}_1, \dots, \vec{v}_k\}$ be a linearly independent set in $\mathbb{R}^n$ and let $L : \mathbb{R}^n \to \mathbb{R}^m$ be a linear mapping. Prove that if $\text{Null}(L) = \{\vec{0}\}$, then $\{L(\vec{v}_1), \dots, L(\vec{v}_k)\}$ is a linearly independent set in $\mathbb{R}^m$.

E11 Let $A = \begin{bmatrix} 1 & 0 & -2 \\ 0 & 2 & -3 \\ 0 & 0 & 4 \end{bmatrix}$.

(a) Determine a sequence of elementary matrices $E_1, \dots, E_k$, such that $E_k \cdots E_1 A = I$.

(b) By inverting the elementary matrices in part (a), write A as a product of elementary matrices.

E12 For each of the following, either give an example or explain (in terms of theorems or definitions) why no such example can exist.

(a) A matrix K such that $KM = MK$ for all 3×3 matrices M.

(b) A matrix K such that $KM = MK$ for all 3×4 matrices M.

(c) The matrix of a linear mapping $L:\mathbb{R}^2 \to \mathbb{R}^3$ whose range is $\text{Span}\left\{\begin{bmatrix} 1 \\ 1 \end{bmatrix}\right\}$ and whose nullspace is $\text{Span}\left\{\begin{bmatrix} 2 \\ 3 \end{bmatrix}\right\}$.

(d) The matrix of a linear mapping $L:\mathbb{R}^2 \to \mathbb{R}^3$ whose range is $\text{Span}\left\{\begin{bmatrix} 1 \\ 1 \\ 2 \end{bmatrix}\right\}$ and whose nullspace is $\text{Span}\left\{\begin{bmatrix} 2 \\ 3 \end{bmatrix}\right\}$.

(e) A linear mapping $L : \mathbb{R}^3 \to \mathbb{R}^3$ such that the range of L is all of $\mathbb{R}^3$ and the nullspace of L is $\text{Span}\left\{\begin{bmatrix} 1 \\ -1 \\ 1 \end{bmatrix}\right\}$.

(f) An invertible 4×4 matrix of rank 3.

Further Problems

These problems are intended to be challenging. They may not be of interest to all students.

F1 We say that matrix C commutes with matrix D if $CD = DC$. Show that the set of matrices that commute with $A = \begin{bmatrix} 3 & 2 \\ 0 & 1 \end{bmatrix}$ is the set of matrices of the form $pI + qA$, where p and q are arbitrary scalars.

F2 Let A be some fixed $n \times n$ matrix. Show that the set $C(A)$ of matrices that commutes with A is closed under addition, scalar multiplication, and matrix multiplication.

F3 A square matrix A is said to be **nilpotent** if some power of A is equal to the zero matrix. Show that the matrix $\begin{bmatrix} 0 & a_{12} & a_{13} \\ 0 & 0 & a_{23} \\ 0 & 0 & 0 \end{bmatrix}$ is nilpotent. Generalize.

F4 (a) Suppose that ℓ is a line in $\mathbb{R}^2$ passing through the origin and making an angle θ with the positive x_1-axis. Let refl_θ denote a reflection in this line. Determine the matrix $[\text{refl}_\theta]$ in terms of functions of θ.

(b) Let refl_α denote a reflection in a second line, and by considering the matrix $[\text{refl}_\alpha \circ \text{refl}_\theta]$, show that the composition of two reflections in the plane is a rotation. Express the angle of the rotation in terms of α and θ.

F5 (Isometries of $\mathbb{R}^2$) A linear transformation $L : \mathbb{R}^2 \to \mathbb{R}^2$ is an **isometry** of $\mathbb{R}^2$ if L preserves lengths (that is, if $\|L(\vec{x})\| = \|\vec{x}\|$ for every $\vec{x} \in \mathbb{R}^2$).

(a) Show that an isometry preserves the dot product (that is, $L(\vec{x}) \cdot L(\vec{y}) = \vec{x} \cdot \vec{y}$ for every $\vec{x}, \vec{y} \in \mathbb{R}^2$). (Hint: Consider $L(\vec{x} + \vec{y})$.)

(b) Show that the columns of that matrix $[L]$ must be orthogonal to each other and of length 1. Deduce that any isometry of $\mathbb{R}^2$ must be the composition of a reflection and a rotation. (Hint: You may find it helpful to use the result of Problem F4 (a).)

F6 (a) Suppose that A and B are $n \times n$ matrices such that $A + B$ and $A - B$ are invertible and that C and D are arbitrary $n \times n$ matrices. Show that there are $n \times n$ matrices X and Y satisfying the system

$$AX + BY = C$$
$$BX + AY = D$$

(b) With the same assumptions as in part (a), give a careful explanation of why the matrix $\begin{bmatrix} A & B \\ B & A \end{bmatrix}$ must be invertible and obtain an expression for its inverse in terms of $(A + B)^{-1}$ and $(A - B)^{-1}$.

Visit the text's website at www.pearsoncanada.ca/norman for practice quizzes, additional applications, and an essay on linearity and superposition in physics.

CHAPTER 4

Vector Spaces

CHAPTER OUTLINE

This chapter explores some of the most important ideas in linear algebra. Some of these ideas have appeared as special cases before, but here we give definitions and examine them in more general settings.

4.1 Spaces of Polynomials

We now compare sets of polynomials under standard addition and scalar multiplication to sets of vectors in $\mathbb{R}^n$ and sets of matrices.

Addition and Scalar Multiplication of Polynomials

Recall that if $p(x) = a_0 + a_1x + \cdots + a_nx^n$, $q(x) = b_0 + b_1x + \cdots + b_nx^n$, and $t \in \mathbb{R}$, then

$$(p + q)(x) = (a_0 + b_0) + (a_1 + b_1)x + \cdots + (a_n + b_n)x^n$$

and

$$(tp)(x) = ta_0 + (ta_1)x + \cdots + (ta_n)x^n$$

Moreover, two polynomials p and q are equal if and only if $a_i = b_i$ for $0 \leq i \leq n$.

EXAMPLE 1

Perform the following operations.

(a) $(2 + 3x + 4x^2 + x^3) + (5 + x - 2x^2 + 7x^3)$

Solution:
$$\begin{aligned}(2+3x+4x^2+x^3)+(5+x-2x^2+7x^3) &= 2+5+(3+1)x+(4-2)x^2+(1+7)x^3 \\ &= 7 + 4x + 2x^2 + 8x^3\end{aligned}$$

(b) $(3 + x^2 - 5x^3) - (1 - x - 2x^2)$

Solution:
$$\begin{aligned}(3+x^2-5x^3)-(1-x-2x^2) &= 3-1+[0-(-1)]x+[1-(-2)]x^2+[-5-0]x^3 \\ &= 2 + x + 3x^2 - 5x^3\end{aligned}$$

EXAMPLE 1 (continued)

(c) $5(2 + 3x + 4x^2 + x^3)$

Solution: $5(2+3x+4x^2+x^3) = 5(2)+5(3)x+5(4)x^2+5(1)x^3 = 10+15x+20x^2+5x^3$

(d) $2(1 + 3x - x^3) + 3(4 + x^2 + 2x^3)$

Solution:
$$\begin{aligned} 2(1 + 3x - x^3) + 3(4 + x^2 + 2x^3) &= 2(1) + 2(3)x + 2(0)x^2 + 2(-1)x^3 \\ &\quad + 3(4) + 3(0)x + 3(1)x^2 + 3(2)x^3 \\ &= 2 + 12 + (6 + 0)x + (0 + 3)x^2 + (-2 + 6)x^3 \\ &= 14 + 6x + 3x^2 + 4x^3 \end{aligned}$$

Properties of Polynomial Addition and Scalar Multiplication

Theorem 1

Let $p(x)$, $q(x)$ and $r(x)$ be polynomials of degree at most n and let $s, t \in \mathbb{R}$. Then

(1) $p(x) + q(x)$ is a polynomial of degree at most n
(2) $(p(x) + q(x)) + r(x) = p(x) + (q(x) + r(x))$
(3) The polynomial $0 = 0 + 0x + \cdots + 0x^n$, called the **zero polynomial**, satisfies $p(x) + 0 = p(x) = 0 + p(x)$ for any polynomial $p(x)$
(4) For each polynomial $p(x)$, there exists an additive inverse, denoted $(-p)(x)$, with the property that $p(x) + (-p)(x) = 0$; in particular, $(-p)(x) = -p(x)$
(5) $p(x) + q(x) = q(x) + p(x)$
(6) $sp(x)$ is a polynomial of degree at most n
(7) $s(tp(x)) = (st)p(x)$
(8) $(s + t)p(x) = sp(x) + tp(x)$
(9) $s(p(x) + q(x)) = sp(x) + sq(x)$
(10) $1p(x) = p(x)$

Remarks

1. These properties follow easily from the definitions of addition and scalar multiplication and are very similar to those for vectors in $\mathbb{R}^n$. Thus, the proofs are left to the reader.

2. Observe that these are the same 10 properties we had for addition and scalar multiplication of vectors in $\mathbb{R}^n$ (Theorem 1.2.1) and of matrices (Theorem 3.1.1).

3. When we look at polynomials in this way, it is the coefficients of the polynomials that are important.

As with vectors in $\mathbb{R}^n$ and matrices, we can also consider linear combinations of polynomials. We make the following definition.

Definition
Span

Let $\mathcal{B} = \{p_1(x), \ldots, p_k(x)\}$ be a set of polynomials of degree at most n. Then the **span** of $\mathcal{B}$ is defined as

$$\operatorname{Span} \mathcal{B} = \{t_1 p_1(x) + \cdots + t_k p_k(x) \mid t_1, \ldots, t_k \in \mathbb{R}\}$$

Definition
Linearly Independent

The set $\mathcal{B} = \{p_1(x), \dots, p_k(x)\}$ is said to be **linearly independent** if the only solution to the equation

$$t_1 p_1(x) + \cdots + t_k p_k(x) = 0$$

is $t_1 = \cdots = t_k = 0$; otherwise, $\mathcal{B}$ is said to be **linearly dependent**.

EXAMPLE 2

Determine if $1 + 2x + 3x^2 + 4x^3$ is in the span of $\mathcal{B} = \{1 + x, 1 + x^3, x + x^2, x + x^3\}$.

Solution: We want to determine if there are t_1, t_2, t_3, t_4 such that

$$\begin{aligned} 1 + 2x + 3x^2 + 4x^3 &= t_1(1 + x) + t_2(1 + x^3) + t_3(x + x^2) + t_4(x + x^3) \\ &= (t_1 + t_2) + (t_1 + t_3 + t_4)x + t_3x^2 + (t_2 + t_4)x^3 \end{aligned}$$

By comparing the coefficients of the different powers of x on both sides of the equation, we get the system of linear equations

$$\begin{aligned} t_1 + t_2 \qquad\qquad &= 1 \\ t_1 \qquad + t_3 + t_4 &= 2 \\ t_3 \qquad &= 3 \\ t_2 \qquad + t_4 &= 4 \end{aligned}$$

Row reducing the augmented matrix gives

$$\left[\begin{array}{cccc|c} 1 & 1 & 0 & 0 & 1 \\ 1 & 0 & 1 & 1 & 2 \\ 0 & 0 & 1 & 0 & 3 \\ 0 & 1 & 0 & 1 & 4 \end{array}\right] \sim \left[\begin{array}{cccc|c} 1 & 0 & 0 & 0 & -2 \\ 0 & 1 & 0 & 0 & 3 \\ 0 & 0 & 1 & 0 & 3 \\ 0 & 0 & 0 & 1 & 1 \end{array}\right]$$

We see that the system is consistent; therefore, $1 + 2x + 3x^2 + 4x^3$ is in the span of $\mathcal{B}$. In particular, we have $t_1 = -2$, $t_2 = 3$, $t_3 = 3$, and $t_4 = 1$.

EXAMPLE 3

Determine if the set $\mathcal{B} = \{1 + 2x + 2x^2 - x^3, 3 + 2x + x^2 + x^3, 2x^2 + 2x^3\}$ is linearly dependent or linearly independent.

Solution: Consider

$$\begin{aligned} 0 &= t_1(1 + 2x + 2x^2 - x^3) + t_2(3 + 2x + x^2 + x^3) + t_3(2x^2 + 2x^3) \\ &= (t_1 + 3t_2) + (2t_1 + 2t_2)x + (2t_1 + t_2 + 2t_3)x^2 + (-t_1 + t_2 + 2t_3)x^3 \end{aligned}$$

Comparing coefficients of the powers of x, we get a homogeneous system of linear equations. Row reducing the associated coefficient matrix gives

$$\begin{bmatrix} 1 & 3 & 0 \\ 2 & 2 & 0 \\ 2 & 1 & 2 \\ -1 & 1 & 2 \end{bmatrix} \sim \begin{bmatrix} 1 & 0 & 0 \\ 0 & 1 & 0 \\ 0 & 0 & 1 \\ 0 & 0 & 0 \end{bmatrix}$$

The only solution is $t_1 = t_2 = t_3 = 0$. Hence $\mathcal{B}$ is linearly independent.

EXERCISE 1 Determine if $\mathcal{B} = \{1 + 2x + x^2 + x^3, 1 + x + 3x^2 + x^3, 3 + 5x + 5x^2 - 3x^3, -x - 2x^2\}$ is linearly dependent or linearly independent. Is $p(x) = 1 + 5x - 5x^2 + x^3$ in the span of $\mathcal{B}$?

EXERCISE 2 Consider $\mathcal{B} = \{1, x, x^2, x^3\}$. Prove that $\mathcal{B}$ is linearly independent and show that Span $\mathcal{B}$ is the set of all polynomials of degree less than or equal to 3.

PROBLEMS 4.1

Practice Problems

A1 Calculate the following.
(a) $(2 - 2x + 3x^2 + 4x^3) + (-3 - 4x + x^2 + 2x^3)$
(b) $(-3)(1 - 2x + 2x^2 + x^3 + 4x^4)$
(c) $(2 + 3x + x^2 - 2x^3) - 3(1 - 2x + 4x^2 + 5x^3)$
(d) $(2 + 3x + 4x^2) - (5 + x - 2x^2)$
(e) $-2(-5 + x + x^2) + 3(-1 - x^2)$
(f) $2\left(\frac{2}{3} - \frac{1}{3}x + 2x^2\right) + \frac{1}{3}\left(3 - 2x + x^2\right)$
(g) $\sqrt{2}(1 + x + x^2) + \pi(-1 + x^2)$

A2 Let $\mathcal{B} = \{1 + x^2 + x^3, 2 + x + x^3, -1 + x + 2x^2 + x^3\}$. For each of the following polynomials, either express it as a linear combination of the polynomials in $\mathcal{B}$ or show that it is not in Span $\mathcal{B}$.
(a) 0
(b) $2 + 4x + 3x^2 + 4x^3$
(c) $-x + 2x^2 + x^3$
(d) $-4 - x + 3x^2$
(e) $-1 + 7x + 5x^2 + 4x^3$
(f) $2 + x + 5x^3$

A3 Determine which of the following sets are linearly independent. If a set is linearly dependent, find all linear combinations of the polynomials that equal the zero polynomial.
(a) $\left\{1 + 2x + x^2 - x^3, 5x + x^2, 1 - 3x + 2x^2 + x^3\right\}$
(b) $\left\{1 + x + x^2, x, x^2 + x^3, 3 + 2x + 2x^2 - x^3\right\}$
(c) $\left\{3 + x + x^2, 4 + x - x^2, 1 + 2x + x^2 + 2x^3, -1 + 5x^2 + x^3\right\}$
(d) $\left\{1 + x + x^3 + x^4, 2 + x - x^2 + x^3 + x^4, x + x^2 + x^3 + x^4\right\}$

A4 Prove that the set $\mathcal{B} = \{1, x - 1, (x - 1)^2\}$ is linearly independent and show that Span $\mathcal{B}$ is the set of all polynomials of degree less than or equal to 2.

Homework Problems

B1 Calculate the following.
(a) $(3 + 4x - 2x^2 + 5x^3) - (1 - 2x + 5x^3)$
(b) $(-2)(2 + x + x^2 + 3x^3 - x^4)$
(c) $(-1)(2 + x + 4x^2 + 2x^3) - 2(-1 - 2x - 2x^2 - x^3)$
(d) $3(1 + x + x^3) + 2(x - x^2 + x^3)$
(e) $0(1 + 3x^3 - 4x^4)$
(f) $\frac{1}{4}\left(3 - \frac{2}{3}x + x^2\right) + \frac{1}{2}\left(2 + 4x + x^2\right)$
(g) $\left(1 + \sqrt{2}\right)\left(1 - \sqrt{2} + (\sqrt{2} - 1)x^2\right) - \frac{1}{2}\left(-2 + 2x^2\right)$

B2 Let $\mathcal{B} = \{1 + x, x + x^2, 1 - x^3\}$. For each of the following polynomials, either express it as a linear combination of the polynomials in $\mathcal{B}$ or show that it is not in Span $\mathcal{B}$.
(a) $p(x) = 1$
(b) $p(x) = 5x + 2x^2 + 3x^3$
(c) $q(x) = 3 + x^2 - 4x^3$
(d) $q(x) = 1 + x^3$

B3 Determine which of the following sets are linearly independent. If a set is linearly dependent, find all linear combinations of the polynomials that equal the zero polynomial.
(a) $\left\{x^2, x^3, x^2 + x^3 + x^4\right\}$
(b) $\left\{1 + \frac{x^2}{2}, 1 - \frac{x^2}{2}, x + \frac{x^3}{6}, x - \frac{x^3}{6}\right\}$
(c) $\left\{1 + x + x^3, x + x^3 + x^5, 1 - x^5\right\}$
(d) $\{1 - 2x + x^4, x - 2x^2 + x^5, 1 - 3x + x^3\}$
(e) $\{1 + 2x + x^2 - x^3, 2 + 3x - x^2 + x^3 + x^4, 1 + x - 2x^2 + 2x^3 + x^4, 1 + 2x + x^2 + x^3 - 3x^4, 4 + 6x - 2x^2 + 5x^4\}$

B4 Prove that the set $\mathcal{B} = \{1, x - 2, (x - 2)^2, (x - 2)^3\}$ is linearly independent and show that Span $\mathcal{B}$ is the set of all polynomials of degree less than or equal to 3.

Conceptual Problems

D1 Let $\mathcal{B} = \{p_1(x), \dots, p_k(x)\}$ be a set of polynomials of degree at most n.
  (a) Prove that if $k < n + 1$, then there exists a polynomial $q(x)$ of degree at most n such that $q(x) \notin \operatorname{Span} \mathcal{B}$.
  (b) Prove that if $k > n + 1$, then $\mathcal{B}$ must be linearly dependent.

4.2 Vector Spaces

We have now seen that addition and scalar multiplication of matrices and polynomials satisfy the same 10 properties as vectors in $\mathbb{R}^n$. Moreover, we commented in Section 3.2 that addition and scalar multiplication of linear mappings also satisfy these same properties. In fact, many other mathematical objects also have these important properties. Instead of analysing each of these objects separately, it is useful to define one abstract concept that encompasses them all.

Vector Spaces

Definition
Vector Space over $\mathbb{R}$

A **vector space over** $\mathbb{R}$ is a set $\mathbb{V}$ together with an operation of **addition**, usually denoted $\mathbf{x} + \mathbf{y}$ for any $\mathbf{x}, \mathbf{y} \in \mathbb{V}$, and an operation of **scalar multiplication**, usually denoted $s\mathbf{x}$ for any $\mathbf{x} \in \mathbb{V}$ and $s \in \mathbb{R}$, such that for any $\mathbf{x}, \mathbf{y}, \mathbf{z} \in \mathbb{V}$ and $s, t \in \mathbb{R}$ we have all of the following properties:

V1 $\mathbf{x} + \mathbf{y} \in \mathbb{V}$ (closed under addition)
V2 $(\mathbf{x} + \mathbf{y}) + \mathbf{z} = \mathbf{x} + (\mathbf{y} + \mathbf{z})$ (addition is associative)
V3 There is an element $\mathbf{0} \in \mathbb{V}$, called the **zero vector**, such that

$$\mathbf{x} + \mathbf{0} = \mathbf{x} = \mathbf{0} + \mathbf{x} \quad \text{(additive identity)}$$

V4 For each $\mathbf{x} \in \mathbb{V}$ there exists an element $-\mathbf{x} \in \mathbb{V}$ such that

$$\mathbf{x} + (-\mathbf{x}) = \mathbf{0} \quad \text{(additive inverse)}$$

V5 $\mathbf{x} + \mathbf{y} = \mathbf{y} + \mathbf{x}$ (addition is commutative)
V6 $s\mathbf{x} \in \mathbb{V}$ (closed under scalar multiplication)
V7 $s(t\mathbf{x}) = (st)\mathbf{x}$ (scalar multiplication is associative)
V8 $(s + t)\mathbf{x} = s\mathbf{x} + t\mathbf{x}$ (scalar addition is distributive)
V9 $s(\mathbf{x} + \mathbf{y}) = s\mathbf{x} + s\mathbf{y}$ (scalar multiplication is distributive)
V10 $1\mathbf{x} = \mathbf{x}$ (scalar multiplicative identity)

Remarks

1. We will call the elements of a vector space **vectors**. Note that these can be very different objects than vectors in $\mathbb{R}^n$. Thus, we will always denote, as in the definition above, a vector in a general vector space in boldface (for example, $\mathbf{x}$). However, in vector spaces such as $\mathbb{R}^n$, matrix spaces, or polynomial spaces, we will often use the notation we introduced earlier.

2. Some people prefer to denote the operations of addition and scalar multiplication in general vector spaces by $\oplus$ and $\odot$, respectively, to stress the fact that these do not need to be "standard" addition and scalar multiplication.

3. Since every vector space contains a zero vector by V3, the empty set cannot be a vector space.

4. When working with multiple vector spaces, we sometimes use a subscript to denote the vector space to which the zero vector belongs. For example, $\mathbf{0}_{\mathbb{V}}$ would represent the zero vector in the vector space $\mathbb{V}$.

5. Vector spaces can be defined using other number systems as the scalars. For example, note that the definition makes perfect sense if rational numbers are used instead of the real numbers. Vector spaces over the complex numbers are discussed in Chapter 9. Until Chapter 9, "vector space" means "vector space over $\mathbb{R}$."

6. We define vector spaces to have the same structure as $\mathbb{R}^n$. The study of vector spaces is the study of this common structure. However, it is possible that vectors in individual vector spaces have other aspects not common to all vector spaces, such as matrix multiplication or factorization of polynomials.

EXAMPLE 1

$\mathbb{R}^n$ is a vector space with addition and scalar multiplication defined in the usual way. We call these standard addition and scalar multiplication of vectors in $\mathbb{R}^n$.

EXAMPLE 2

P_n, the set of all polynomials of degree at most n, is a vector space with standard addition and scalar multiplication of polynomials.

EXAMPLE 3

$M(m, n)$, the set of all $m \times n$ matrices, is a vector space with standard addition and scalar multiplication of matrices.

EXAMPLE 4

Consider the set of polynomials of degree n. Is this a vector space with standard addition and scalar multiplication? No, since it does not contain the zero polynomial. Note also that the sum of two polynomials of degree n may be of degree lower than n. For example, $(1 + x^n) - (1 - x^n) = 2$, which is of degree 0. Thus, the set is also not closed under addition.

EXAMPLE 5

Let $\mathcal{F}(a, b)$ denote the set of all functions $f : (a, b) \to \mathbb{R}$. If $f, g \in \mathcal{F}(a, b)$, then the sum is defined by $(f + g)(x) = f(x) + g(x)$, and multiplication by a scalar $t \in \mathbb{R}$ is defined by $(tf)(x) = tf(x)$. With these definitions, $\mathcal{F}(a, b)$ is a vector space.

EXAMPLE 6

Let $C(a, b)$ denote the set of all functions that are continuous on the interval (a, b). Since the sum of continuous functions is continuous and a scalar multiple of a continuous function is continuous, $C(a, b)$ is a vector space. See Figure 4.2.1.

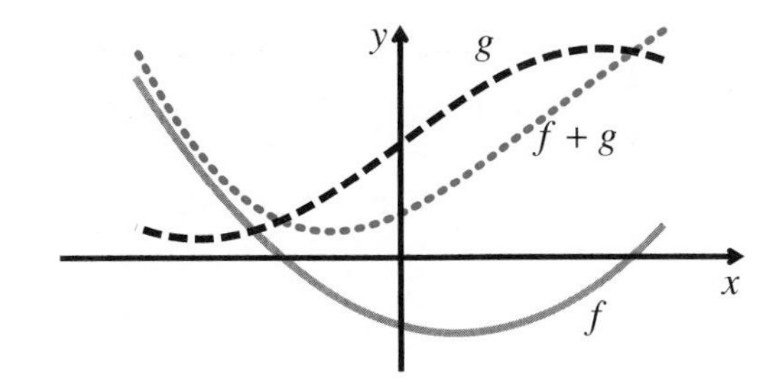

Figure 4.2.1 The sum of continuous functions f and g is a continuous function.

EXAMPLE 7

Let $\mathbb{T}$ be the set of all solutions to $x_1 + 2x_2 = 1$, $2x_1 + 3x_2 = 0$. Is $\mathbb{T}$ a vector space with standard addition and scalar multiplication? No. This set with these operations does not satisfy many of the vector space axioms. For example, V1 does not hold since $\begin{bmatrix} -3 \\ 2 \end{bmatrix}$ is in $\mathbb{T}$; it is a solution of this system of linear equations. But $\begin{bmatrix} -3 \\ 2 \end{bmatrix} + \begin{bmatrix} -3 \\ 2 \end{bmatrix} = \begin{bmatrix} -6 \\ 4 \end{bmatrix}$ is not a solution of the system and hence not in $\mathbb{T}$.

EXAMPLE 8

Consider $\mathbb{V} = \{(x, y) \mid x, y \in \mathbb{R}\}$ with addition defined by $(x_1, y_1) + (x_2, y_2) = (x_1 + x_2, y_1 + y_2)$ and scalar multiplication defined by $k(x, y) = (ky, kx)$. Is $\mathbb{V}$ a vector space? No, since $1(2, 3) = (3, 2) \neq (2, 3)$, it does not satisfy V10. Note that it also does not satisfy V7.

EXAMPLE 9

Let $\mathbb{S} = \left\{ \begin{bmatrix} x_1 \\ x_2 \\ x_1 + x_2 \end{bmatrix} \mid x_1, x_2 \in \mathbb{R} \right\}$. Is $\mathbb{S}$ a vector space with standard addition and scalar multiplication in $\mathbb{R}^3$? Yes. Let's verify the axioms.

First, observe that axioms V2, V5, V7, V8, V9, and V10 refer only to the operations of addition and scalar multiplication. Thus, we know that these operations must satisfy all the axioms as they are the operations of the vector space $\mathbb{R}^3$.

Let $\vec{x} = \begin{bmatrix} x_1 \\ x_2 \\ x_1 + x_2 \end{bmatrix}$ and $\vec{y} = \begin{bmatrix} y_1 \\ y_2 \\ y_1 + y_2 \end{bmatrix}$ be vectors in $\mathbb{S}$.

V1 We have

$$\vec{x} + \vec{y} = \begin{bmatrix} x_1 \\ x_2 \\ x_1 + x_2 \end{bmatrix} + \begin{bmatrix} y_1 \\ y_2 \\ y_1 + y_2 \end{bmatrix} = \begin{bmatrix} x_1 + y_1 \\ x_2 + y_2 \\ x_1 + y_1 + x_2 + y_2 \end{bmatrix}$$

Observe that if we let $z_1 = x_1 + y_1$ and $z_2 = x_2 + y_2$, then $z_1 + z_2 = x_1 + y_1 + x_2 + y_2$ and hence

$$\vec{x} + \vec{y} = \begin{bmatrix} z_1 \\ z_2 \\ z_1 + z_2 \end{bmatrix} \in S$$

since it satisfies the conditions of the set. Therefore, $\mathbb{S}$ is closed under addition.

V3 The vector $\vec{0} = \begin{bmatrix} 0 \\ 0 \\ 0 \end{bmatrix}$ satisfies $\vec{x} + \vec{0} = \vec{x} = \vec{0} + \vec{x}$ and is in $\mathbb{S}$ since it satisfies the conditions of $\mathbb{S}$.

EXAMPLE 9
(continued)

V4 The additive inverse of $\vec{x} = \begin{bmatrix} x_1 \\ x_2 \\ x_1 + x_2 \end{bmatrix}$ is $(-\vec{x}) = \begin{bmatrix} -x_1 \\ -x_2 \\ -x_1 - x_2 \end{bmatrix}$, which is in $\mathbb{S}$ since it satisfies the conditions of $\mathbb{S}$.

V6 $t\vec{x} = t\begin{bmatrix} x_1 \\ x_2 \\ x_1 + x_2 \end{bmatrix} = \begin{bmatrix} tx_1 \\ tx_2 \\ tx_1 + tx_2 \end{bmatrix} \in \mathbb{S}$. Therefore, $\mathbb{S}$ is closed under scalar multiplication.

Thus, $\mathbb{S}$ with these operators is a vector space as it satisfies all 10 axioms.

EXERCISE 1

Prove that the set $\mathbb{S} = \left\{\begin{bmatrix} x_1 \\ x_2 \end{bmatrix} \mid x_1, x_2 \in \mathbb{Z}\right\}$ is not a vector space using standard addition and scalar multiplication of vectors in $\mathbb{R}^2$.

EXERCISE 2

Let $\mathbb{S} = \left\{\begin{bmatrix} a_1 & 0 \\ 0 & a_2 \end{bmatrix} \mid a_1, a_2 \in \mathbb{R}\right\}$. Prove that $\mathbb{S}$ is a vector space using standard addition and scalar multiplication of matrices. This is the vector space of 2×2 diagonal matrices.

Again, one advantage of having the abstract concept of a vector space is that when we prove a result about a general vector space, it instantly applies to all of the examples of vector spaces. To demonstrate this, we give three additional properties that follow easily from the vector space axioms.

Theorem 1

Let $\mathbb{V}$ be a vector space. Then
(1) $0\mathbf{x} = \mathbf{0}$ for all $\mathbf{x} \in \mathbb{V}$
(2) $(-1)\mathbf{x} = -\mathbf{x}$ for all $\mathbf{x} \in \mathbb{V}$
(3) $t\mathbf{0} = \mathbf{0}$ for all $t \in \mathbb{R}$

Proof: We will prove (1). You are asked to prove (2) and (3) in Problem D1. For any $\mathbf{x} \in \mathbb{V}$ we have

$$\begin{aligned}
0\mathbf{x} &= 0\mathbf{x} + \mathbf{0} && \text{by V3} \\
&= 0\mathbf{x} + [\mathbf{x} + (-\mathbf{x})] && \text{by V4} \\
&= 0\mathbf{x} + [1\mathbf{x} + (-\mathbf{x})] && \text{by V10} \\
&= [0\mathbf{x} + 1\mathbf{x}] + (-\mathbf{x}) && \text{by V2} \\
&= (0 + 1)\mathbf{x} + (-\mathbf{x}) && \text{by V8} \\
&= 1\mathbf{x} + (-\mathbf{x}) && \text{operation of numbers in } \mathbb{R} \\
&= \mathbf{x} + (-\mathbf{x}) && \text{by V10} \\
&= \mathbf{0} && \text{by V4}
\end{aligned}$$

■

Thus, if we know that $\mathbb{V}$ is a vector space, we can determine the zero vector of $\mathbb{V}$ by finding $0\mathbf{x}$ for any $\mathbf{x} \in \mathbb{V}$. Similarly, we can determine the additive inverse of any vector $\mathbf{x} \in \mathbb{V}$ by computing $(-1)\mathbf{x}$.

EXAMPLE 10

Let $\mathbb{V} = \{(a, b) \mid a, b \in \mathbb{R}, b > 0\}$ and define addition by $(a, b) \oplus (c, d) = (ad + bc, bd)$ and define scalar multiplication by $t \odot (a, b) = (tab^{t-1}, b^t)$. Use Theorem 1 to show that axioms V3 and V4 hold for $\mathbb{V}$ with these operations. (Note that we are using $\oplus$ and $\odot$ to represent the operations of addition and scalar multiplication in the vector space to help distinguish the difference between these and the operations of addition and multiplication of real numbers.)

Solution: We do not know if $\mathbb{V}$ is a vector space. If it is, then by Theorem 1 we must have

$$\mathbf{0} = 0 \odot (a, b) = (0ab^{-1}, b^0) = (0, 1)$$

Observe that $(0, 1) \in \mathbb{V}$ and for any $(a, b) \in \mathbb{V}$ we have

$$(a, b) \oplus (0, 1) = (a(1) + b(0), b(1)) = (a, b) = (0(b) + 1(a), 1(b)) = (0, 1) \oplus (a, b)$$

So, $\mathbb{V}$ satisfies V3 using $\mathbf{0} = (0, 1)$.

Similarly, if $\mathbb{V}$ is a vector space, then by Theorem 1 for any $\mathbf{x} = (a, b) \in \mathbb{V}$ we must have

$$(-\mathbf{x}) = (-1) \odot (a, b) = (-ab^{-2}, b^{-1})$$

Observe that for any $(a, b) \in \mathbb{V}$ we have $(-ab^{-2}, b^{-1}) \in \mathbb{V}$ since $b^{-1} > 0$ whenever $b > 0$. Also,

$$(a, b) \oplus (-ab^{-2}, b^{-1}) = (ab^{-1} + b(-ab^{-2}), bb^{-1}) = (ab^{-1} - ab^{-1}, 1) = (0, 1)$$

So, $\mathbb{V}$ satisfies V4 using $-(a, b) = (-ab^{-2}, b^{-1})$.

You are asked to complete the proof that $\mathbb{V}$ is indeed a vector space in Problem D2.

Subspaces

In Example 9 we showed that $\mathbb{S}$ is a vector space that is contained inside the vector space $\mathbb{R}^3$. Observe that, by the definition of a subspace of $\mathbb{R}^n$ in Section 1.2, $\mathbb{S}$ is actually a subspace of $\mathbb{R}^3$. We now generalize these ideas to general vector spaces.

Definition
Subspace

Suppose that $\mathbb{V}$ is a vector space. A non-empty subset $\mathbb{U}$ of $\mathbb{V}$ is a **subspace** of $\mathbb{V}$ if it satisfies the following two properties:

S1 $\mathbf{x} + \mathbf{y} \in \mathbb{U}$ for all $\mathbf{x}, \mathbf{y} \in \mathbb{U}$ ($\mathbb{U}$ is closed under addition)

S2 $t\mathbf{x} \in \mathbb{U}$ for all $\mathbf{x} \in \mathbb{U}$ and $t \in \mathbb{R}$ ($\mathbb{U}$ is closed under scalar multiplication)

Equivalent Definition. If $\mathbb{U}$ is a subset of a vector space $\mathbb{V}$ and $\mathbb{U}$ is also a vector space using the same operations as $\mathbb{V}$, then $\mathbb{U}$ is a **subspace** of $\mathbb{V}$.

To prove that both of these definitions are equivalent, we first observe that if $\mathbb{U}$ is a vector space, then it satisfies properties S1 and S2 as these are vector space axioms V1 and V6. On the other hand, as in Example 4.2.9, we know the operations must satisfy axioms V2, V5, V7, V8, V9, and V10 since $\mathbb{V}$ is a vector space. For the remaining axioms we have

V1 Follows from property S1

V3 Follows from Theorem 1 and property S2 because for any $\mathbf{u} \in \mathbb{U}$ we have $\mathbf{0} = 0\vec{u} \in \mathbb{U}$

V4 Follows from Theorem 1 and property S2 because for any $\mathbf{u} \in \mathbb{U}$, the additive inverse of $\mathbf{u}$ is $(-\mathbf{u}) = (-1)\mathbf{u} \in \mathbb{U}$

V6 Follows from property S2

Hence, all 10 axioms are satisfied. Therefore, $\mathbb{U}$ is also a vector space under the operations of $\mathbb{V}$.

Remarks

1. When proving that a set $\mathbb{U}$ is a subspace of a vector space $\mathbb{V}$, it is important not to forget to show that $\mathbb{U}$ is actually a subset of $\mathbb{V}$.
2. As with subspaces of $\mathbb{R}^n$ in Section 1.2, we typically show that the subset is non-empty by showing that it contains the zero vector of $\mathbb{V}$.

EXAMPLE 11

In Exercise 2 you proved that $\mathbb{S} = \left\{ \begin{bmatrix} a_1 & 0 \\ 0 & a_2 \end{bmatrix} \mid a_1, a_2 \in \mathbb{R} \right\}$ is a vector space. Thus, since $\mathbb{S}$ is a subset of $M(2,2)$, it is a subspace of $M(2,2)$.

EXAMPLE 12

Let $\mathbb{U} = \{p(x) \in P_3 \mid p(3) = 0\}$. Show that $\mathbb{U}$ is a subspace of P_3.

Solution: By definition, $\mathbb{U}$ is a subset of P_3. The zero vector in P_3 maps x to 0 for all x; hence it maps 3 to 0. Therefore, the zero vector of P_3 is in $\mathbb{U}$, and hence $\mathbb{U}$ is non-empty.

Let $p(x), q(x) \in \mathbb{U}$ and $s \in \mathbb{R}$. Then $p(3) = 0$ and $q(3) = 0$.

S1 $(p+q)(3) = p(3) + q(3) = 0 + 0 = 0$ so $p(x) + q(x) \in \mathbb{U}$

S2 $(sp)(3) = sp(3) = s0 = 0$, so $sp(x) \in \mathbb{U}$

Hence, $\mathbb{U}$ is a subspace of P_3. Note that this also implies that $\mathbb{U}$ is itself a vector space.

EXAMPLE 13

Define the **trace** of a 2×2 matrix by $\operatorname{tr}\left(\begin{bmatrix} a_{11} & a_{12} \\ a_{21} & a_{22} \end{bmatrix}\right) = a_{11} + a_{22}$. Prove that $\mathbb{S} = \{A \in M(2,2) \mid \operatorname{tr}(A) = 0\}$ is a subspace of $M(2,2)$.

Solution: By definition, $\mathbb{S}$ is a subset of $M(2,2)$. The zero vector of $M(2,2)$ is $O_{2,2} = \begin{bmatrix} 0 & 0 \\ 0 & 0 \end{bmatrix}$. Clearly, $\operatorname{tr}(O_{2,2}) = 0$, so $O_{2,2} \in \mathbb{S}$.

Let $A, B \in \mathbb{S}$ and $s \in \mathbb{R}$. Then $a_{11} + a_{22} = \operatorname{tr}(A) = 0$ and $b_{11} + b_{22} = \operatorname{tr}(B) = 0$.

S1 $\operatorname{tr}(A+B) = \operatorname{tr}\left(\begin{bmatrix} a_{11}+b_{11} & a_{12}+b_{12} \\ a_{21}+b_{21} & a_{22}+b_{22} \end{bmatrix}\right) = a_{11} + a_{22} + b_{11} + b_{22} = 0 + 0 = 0$, so $A + B \in \mathbb{S}$

EXAMPLE 13 (continued)

S2 $\operatorname{tr}(sA) = \operatorname{tr}\left(\begin{bmatrix} sa_{11} & sa_{12} \\ sa_{21} & sa_{22} \end{bmatrix}\right) = sa_{11} + sa_{22} = s(a_{11} + a_{22}) = s(0) = 0$, so $sA \in \mathbb{S}$

Hence, $\mathbb{S}$ is a subspace of $M(2,2)$.

EXAMPLE 14

The vector space $\mathbb{R}^2$ is *not* a subspace of $\mathbb{R}^3$, since $\mathbb{R}^2$ is not a subset of $\mathbb{R}^3$. That is, if we take any vector $\vec{x} = \begin{bmatrix} x_1 \\ x_2 \end{bmatrix} \in \mathbb{R}^2$, this is not a vector in $\mathbb{R}^3$, since a vector in $\mathbb{R}^3$ has three components.

EXERCISE 3

Prove that $\mathbb{U} = \{a + bx + cx^2 \in P_2 \mid b + c = a\}$ is a subspace of P_2.

EXERCISE 4

Let $\mathbb{V}$ be a vector space. Prove that $\{\mathbf{0}\}$ is also a vector space, called the **trivial vector space**, under the same operations as $\mathbb{V}$ by proving it is a subspace of $\mathbb{V}$.

In the previous exercise you proved that $\{\mathbf{0}\}$ is a subspace of any vector space $\mathbb{V}$. Furthermore, by definition, $\mathbb{V}$ is a subspace of itself. We now prove that the set of all possible linear combinations of a set of vectors in a vector space $\mathbb{V}$ is also a subspace.

Theorem 2

If $\{\mathbf{v}_1, \dots, \mathbf{v}_k\}$ is a set of vectors in a vector space $\mathbb{V}$ and $\mathbb{S}$ is the set of all possible linear combinations of these vectors,

$$\mathbb{S} = \{t_1\mathbf{v}_1 + \cdots + t_k\mathbf{v}_k \mid t_1, \dots, t_k \in \mathbb{R}\}$$

then $\mathbb{S}$ is a subspace of $\mathbb{V}$.

Proof: By V1 and V6, $t_1\mathbf{v}_1 + \cdots + t_k\mathbf{v}_k \in \mathbb{V}$. Hence, $\mathbb{S}$ is a subset of $\mathbb{V}$. Also, by taking $t_i = 0$ for $1 \leq i \leq k$, we get

$$0\mathbf{v}_1 + \cdots + 0\mathbf{v}_k = 0(\mathbf{v}_1 + \cdots + \mathbf{v}_k) = \mathbf{0}_{\mathbb{V}}$$

by V9 and Theorem 1, so $\mathbf{0}_{\mathbb{V}} \in \mathbb{S}$.

Let $\mathbf{x}, \mathbf{y} \in \mathbb{S}$. Then for some real numbers s_i and t_i, $1 \leq i \leq k$, $\mathbf{x} = s_1\mathbf{v}_1 + \cdots + s_k\mathbf{v}_k$ and $\mathbf{y} = t_1\mathbf{v}_1 + \cdots + t_k\mathbf{v}_k$. It follows that

$$\mathbf{x} + \mathbf{y} = s_1\mathbf{v}_1 + \cdots + s_k\mathbf{v}_k + t_1\mathbf{v}_1 + \cdots + t_k\mathbf{v}_k = (s_1 + t_1)\mathbf{v}_1 + \cdots + (s_k + t_k)\mathbf{v}_k$$

using V8. So, $\mathbf{x} + \mathbf{y} \in \mathbb{S}$ since $(s_i + t_i) \in \mathbb{R}$. Hence, $\mathbb{S}$ is closed under addition.

Similarly, for all $r \in \mathbb{R}$,

$$r\mathbf{x} = r(s_1\mathbf{v}_1 + \cdots + s_k\mathbf{v}_k) = (rs_1)\mathbf{v}_1 + \cdots + (rs_k)\mathbf{v}_k \in \mathbb{S}$$

by V7. Thus, $\mathbb{S}$ is also closed under scalar multiplication. Therefore, $\mathbb{S}$ is a subspace of $\mathbb{V}$. ■

To match what we did in Sections 1.2, 3.1, and 4.1, we make the following definition.

Definition
Span
Spanning Set

If $\mathbb{S}$ is the subspace of the vector space $\mathbb{V}$ consisting of all linear combinations of the vectors $\mathbf{v}_1, \dots, \mathbf{v}_k \in \mathbb{V}$, then $\mathbb{S}$ is called the subspace **spanned** by $\mathcal{B} = \{\mathbf{v}_1, \dots, \mathbf{v}_k\}$, and we say that the set $\mathcal{B}$ **spans** $\mathbb{S}$. The set $\mathcal{B}$ is called a **spanning set** for the subspace $\mathbb{S}$. We denote $\mathbb{S}$ by

$$\mathbb{S} = \text{Span}\{\mathbf{v}_1, \dots, \mathbf{v}_k\} = \text{Span}\,\mathcal{B}$$

In Sections 1.2, 3.1, and 4.1, we saw that the concept of spanning is closely related to the concept of linear independence.

Definition
Linearly Independent
Linearly Dependent

If $\mathcal{B} = \{\mathbf{v}_1, \dots, \mathbf{v}_k\}$ is a set of vectors in a vector space $\mathbb{V}$, then $\mathcal{B}$ is said to be **linearly independent** if the only solution to the equation

$$t_1\mathbf{v}_1 + \cdots + t_k\mathbf{v}_k = \mathbf{0}$$

is $t_1 = \cdots = t_k = 0$; otherwise, $\mathcal{B}$ is said to be **linearly dependent**.

Remark

The procedure for determining if a vector is in a span or if a set is linearly independent in a general vector space is exactly the same as we saw for $\mathbb{R}^n$, $M(m,n)$, and P_n in Sections 2.3, 3.1, and 4.1. We will see further examples of this in the next section.

PROBLEMS 4.2

Practice Problems

A1 Determine, with proof, which of the following sets are subspaces of the given vector space.

(a) $\left\{\begin{bmatrix} x_1 \\ x_2 \\ x_3 \\ x_4 \end{bmatrix} \mid x_1 + 2x_2 = 0, x_1, x_2, x_3, x_4 \in \mathbb{R}\right\}$ of $\mathbb{R}^4$

(b) $\left\{\begin{bmatrix} a_1 & a_2 \\ a_3 & a_4 \end{bmatrix} \mid a_1 + 2a_2 = 0, a_1, a_2, a_3, a_4 \in \mathbb{R}\right\}$ of $M(2,2)$

(c) $\{a_0 + a_1x + a_2x^2 + a_3x^3 \mid a_0 + 2a_1 = 0, a_0, a_1, a_2, a_3 \in \mathbb{R}\}$ of P_3

(d) $\left\{\begin{bmatrix} a_1 & a_2 \\ a_3 & a_4 \end{bmatrix} \mid a_1, a_2, a_3, a_4 \in \mathbb{Z}\right\}$ of $M(2,2)$

(e) $\left\{\begin{bmatrix} a_1 & a_2 \\ a_3 & a_4 \end{bmatrix} \mid a_1a_4 - a_2a_3 = 0, a_1, a_2, a_3, a_4 \in \mathbb{R}\right\}$ of $M(2,2)$

(f) $\left\{\begin{bmatrix} a_1 & a_2 \\ 0 & 0 \end{bmatrix} \mid a_1 = a_2, a_1, a_2 \in \mathbb{R}\right\}$ of $M(2,2)$.

A2 Determine, with proof, whether the following subsets of $M(n,n)$ are subspaces.

(a) The subset of diagonal matrices

(b) The subset of matrices that are in row echelon form

(c) The subset of **symmetric** matrices (A matrix A is **symmetric** if $A^T = A$ or, equivalently, if $a_{ij} = a_{ji}$ for all i and j.)

(d) The subset of upper triangular matrices

A3 Determine, with proof, whether the following subsets of P_5 are subspaces.

(a) $\{p(x) \in P_5 \mid p(-x) = p(x)$ for all $x \in \mathbb{R}\}$ (the subset of even polynomials)

(b) $\{(1 + x^2)p(x) \mid p(x) \in P_3\}$

(c) $\{a_0, a_1x + \cdots + a_4x^4 \mid a_0 = a_4, a_1 = a_3, a_i \in \mathbb{R}\}$

(d) $\{p(x) \in P_5 \mid p(0) = 1\}$

(e) $\{a_0 + a_1x + a_2x^2 \mid a_0, a_1, a_2 \in \mathbb{R}\}$

A4 Let $\mathcal{F}$ be the vector space of all real-valued functions of a real variable. Determine, with proof, which of the following subsets of $\mathcal{F}$ are subspaces.
(a) $\{f \in \mathcal{F} \mid f(3) = 0\}$
(b) $\{f \in \mathcal{F} \mid f(3) = 1\}$
(c) $\{f \in \mathcal{F} \mid f(-x) = f(x) \text{ for all } x \in \mathbb{R}\}$
(d) $\{f \in \mathcal{F} \mid f(x) \geq 0 \text{ for all } x \in \mathbb{R}\}$

A5 Show that any set of vectors in a vector space $\mathbb{V}$ that contains the zero vector is linearly dependent.

Homework Problems

B1 Determine, with proof, which of the following sets are subspaces of the given vector space.
(a) $\left\{\begin{bmatrix} x_1 \\ x_2 \\ x_3 \\ x_4 \end{bmatrix} \mid x_1 + x_3 = x_4, x_1, x_2, x_3, x_4 \in \mathbb{R}\right\}$ of $\mathbb{R}^4$
(b) $\left\{\begin{bmatrix} a_1 & a_2 \\ a_3 & a_4 \end{bmatrix} \mid a_1 + a_3 = a_4, a_1, a_2, a_3, a_4 \in \mathbb{R}\right\}$ of $M(2,2)$
(c) $\{a_0 + a_1x + a_2x^2 + a_3x^3 \mid a_0 + a_2 = a_3, a_0, a_1, a_2, a_3 \in \mathbb{R}\}$ of P_3
(d) $\{a_0 + a_1x^2 \mid a_0, a_1 \in \mathbb{R}\}$ of P_4
(e) $\left\{\begin{bmatrix} 1 & 2 \\ 3 & 4 \end{bmatrix}\right\}$ of $M(2,2)$
(f) $\left\{\begin{bmatrix} a_1 & a_2 \\ 0 & a_3 \end{bmatrix} \mid a_1 - a_3 = 1, a_1, a_2, a_3 \in \mathbb{R}\right\}$ of $M(2,2)$.

B2 Determine, with proof, whether the following subsets of $M(3,3)$ are subspaces.
(a) $\{A \in M(3,3) \mid \text{tr}(A) = 0\}$
(b) The subset of invertible 3×3 matrices
(c) The subset of 3×3 matrices A such that $A\begin{bmatrix} 1 \\ 2 \\ 3 \end{bmatrix} = \begin{bmatrix} 0 \\ 0 \\ 0 \end{bmatrix}$
(d) The subset of 3×3 matrices A such that $A\begin{bmatrix} 4 \\ 5 \\ 6 \end{bmatrix} = \begin{bmatrix} 1 \\ 2 \\ 3 \end{bmatrix}$
(e) The subset of **skew-symmetric** matrices. (A matrix A is skew-symmetric if $A^T = -A$; that is, $a_{ij} = -a_{ji}$ for all i and j.)

B3 Determine, with proof, whether the following subsets of P_5 are subspaces.
(a) $\{p(x) \in P_5 \mid p(-x) = -p(x) \text{ for all } x \in \mathbb{R}\}$ (the subset of odd polynomials)
(b) $\{(p(x))^2 \mid p(x) \in P_2\}$
(c) $\{a_0 + a_1x + \cdots + a_4x^4 \mid a_1a_4 = 1 \in \mathbb{R}\}$
(d) $\{x^3p(x) \mid p(x) \in P_2\}$
(e) $\{p(x) \mid p(1) = 0\}$

B4 Let $\mathcal{F}$ be the vector space of all real-valued functions of a real variable. Determine, with proof, which of the following subsets of $\mathcal{F}$ are subspaces.
(a) $\{f \in \mathcal{F} \mid f(3) + f(5) = 0\}$
(b) $\{f \in \mathcal{F} \mid f(1) + f(2) = 1\}$
(c) $\{f \in \mathcal{F} \mid |f(x)| \leq 1\}$
(d) $\{f \in \mathcal{F} \mid f \text{ is increasing on } \mathbb{R}\}$

Conceptual Problems

D1 Let $\mathbb{V}$ be a vector space.
(a) Prove that $-\mathbf{x} = (-1)\mathbf{x}$ for every $\mathbf{x} \in \mathbb{V}$.
(b) Prove that the zero vector in $\mathbb{V}$ is unique.
(c) Prove that $t\mathbf{0} = \mathbf{0}$ for every $t \in \mathbb{R}$.

D2 Let $\mathbb{V} = \{(a,b) \mid a, b \in \mathbb{R}, b > 0\}$ and define addition by $(a,b) \oplus (c,d) = (ad + bc, bd)$ and scalar multiplication by $t \odot (a,b) = (tab^{t-1}, b^t)$ for any $t \in \mathbb{R}$. Prove that $\mathbb{V}$ is a vector space with these operations.

D3 Let $\mathbb{V} = \{x \in \mathbb{R} \mid x > 0\}$ and define addition by $x \oplus y = xy$ and scalar multiplication by $t \odot x = x^t$ for any $t \in \mathbb{R}$. Prove that $\mathbb{V}$ is a vector space with these operations.

D4 Let $\mathbb{L}$ denote the set of all linear operators $L : \mathbb{R}^n \to \mathbb{R}^n$ with standard addition and scalar multiplication of linear mappings. Prove that $\mathbb{L}$ is a vector space under these operations.

D5 Suppose that $\mathbb{U}$ and $\mathbb{V}$ are vector spaces over $\mathbb{R}$. The **Cartesian product** of $\mathbb{U}$ and $\mathbb{V}$ is defined to be

$$\mathbb{U} \times \mathbb{V} = \{(\mathbf{u}, \mathbf{v}) \mid \mathbf{u} \in \mathbb{U}, \mathbf{v} \in \mathbb{V}\}$$

(a) In $\mathbb{U} \times \mathbb{V}$ define addition and scalar multiplication by

$$(\mathbf{u}_1, \mathbf{v}_1) \oplus (\mathbf{u}_2, \mathbf{v}_2) = (\mathbf{u}_1 + \mathbf{u}_2, \mathbf{v}_1 + \mathbf{v}_2)$$
$$t \odot (\mathbf{u}_1, \mathbf{v}_1) = (t\mathbf{u}_1, t\mathbf{v}_1)$$

Verify that with these operations that $\mathbb{U} \times \mathbb{V}$ is a vector space.

(b) Verify that $\mathbb{U} \times \{\mathbf{0}_{\mathbb{V}}\}$ is a subspace of $\mathbb{U} \times \mathbb{V}$.

(c) Suppose instead that scalar multiplication is defined by $t \odot (\mathbf{u}, \mathbf{v}) = (t\mathbf{u}, \mathbf{v})$, while addition is defined as in part (a). Is $\mathbb{U} \times \mathbb{V}$ a vector space with these operations?

4.3 Bases and Dimensions

In Chapters 1 and 3, much of the discussion was dependant on the use of the standard basis in $\mathbb{R}^n$. For example, the dot product of two vectors $\vec{a}$ and $\vec{b}$ was defined in terms of the standard components of the vectors. As another example, the standard matrix [L] *of a linear mapping was determined by calculating the images of the standard basis vectors. Therefore, it would be useful to define the same concept for any vector space.*

Bases

Recall from Section 1.2 that the two important properties of the standard basis in $\mathbb{R}^n$ were that it spanned $\mathbb{R}^n$ and it was linearly independent. It is clear that we should want a basis $\mathcal{B}$ for a vector space $\mathbb{V}$ to be a spanning set so that every vector in $\mathbb{V}$ can be written as a linear combination of the vectors in $\mathcal{B}$. Why would it be important that the set $\mathcal{B}$ be linearly independent? The following theorem answers this question.

Theorem 1 **Unique Representation Theorem**
Let $\mathcal{B} = \{\mathbf{v}_1, \dots, \mathbf{v}_n\}$ be a spanning set for a vector space $\mathbb{V}$. Then every vector in $\mathbb{V}$ can be expressed in a *unique* way as a linear combination of the vectors of $\mathcal{B}$ if and only if the set $\mathcal{B}$ is linearly independent.

Proof: Let $\mathbf{x}$ be any vector in $\mathbb{V}$. Since Span $\mathcal{B} = \mathbb{V}$, we have that $\mathbf{x}$ can be written as a linear combination of the vectors in $\mathcal{B}$. Assume that there are linear combinations

$$\mathbf{x} = a_1\mathbf{v}_1 + \cdots + a_n\mathbf{v}_n \quad \text{and} \quad \mathbf{x} = b_1\mathbf{v}_1 + \cdots + b_n\mathbf{v}_n$$

This gives

$$a_1\mathbf{v}_1 + \cdots + a_n\mathbf{v}_n = \mathbf{x} = b_1\mathbf{v}_1 + \cdots + b_n\mathbf{v}_n$$

which implies

$$\mathbf{0} = \mathbf{x} - \mathbf{x} = (a_1\mathbf{v}_1 + \cdots + a_n\mathbf{v}_n) - (b_1\mathbf{v}_1 + \cdots + b_n\mathbf{v}_n) = (a_1 - b_1)\mathbf{v}_1 + \cdots + (a_n - b_n)\mathbf{v}_n$$

If $\mathcal{B}$ is linearly independent, then we must have $a_i - b_i = 0$, so $a_i = b_i$ for $1 \leq i \leq n$. Hence, $\mathbf{x}$ has a unique representation.

On the other hand, if $\mathcal{B}$ is linearly dependent, then

$$\mathbf{0} = t_1\mathbf{v}_1 + \cdots + t_n\mathbf{v}_n$$

has a solution where at least one of the coefficients is non-zero. But

$$\mathbf{0} = 0\mathbf{v}_1 + \cdots + 0\mathbf{v}_n$$

Hence, $\mathbf{0}$ can be expressed as a linear combination of the vectors in $\mathcal{B}$ in multiple ways. ■

Thus, if $\mathcal{B}$ is a linearly independent spanning set for a vector space $\mathbb{V}$, then every vector in $\mathbb{V}$ can be written as a unique linear combination of the vectors in $\mathcal{B}$.

Definition
Basis

A set $\mathcal{B}$ of vectors in a vector space $\mathbb{V}$ is a **basis** if it is a linearly independent spanning set for $\mathbb{V}$.

Remark

According to this definition, the trivial vector space $\{\mathbf{0}\}$ does not have a basis since any set of vectors containing the zero vector in a vector space $\mathbb{V}$ is linearly dependent. However, we would like every vector space to have a basis, so we define the empty set to be a basis for the trivial vector space.

EXAMPLE 1

The set of vectors $\left\{\begin{bmatrix}1 & 0\\0 & 0\end{bmatrix}, \begin{bmatrix}0 & 1\\0 & 0\end{bmatrix}, \begin{bmatrix}0 & 0\\1 & 0\end{bmatrix}, \begin{bmatrix}0 & 0\\0 & 1\end{bmatrix}\right\}$ in Exercise 3.1.2 is a basis for $M(2,2)$. It is called the standard basis for $M(2,2)$.

The set of vectors $\{1, x, x^2, x^3\}$ in Exercise 4.1.2 is a basis for P_3. In particular, the set $\{1, x, \dots, x^n\}$ is called the standard basis for P_n.

EXAMPLE 2

Prove that the set $C = \left\{\begin{bmatrix}1\\1\\0\end{bmatrix}, \begin{bmatrix}1\\1\\1\end{bmatrix}, \begin{bmatrix}1\\0\\1\end{bmatrix}\right\}$ is a basis for $\mathbb{R}^3$.

Solution: We need to show that $\operatorname{Span} C = \mathbb{R}^3$ and that C is linearly independent. To prove that $\operatorname{Span} C = \mathbb{R}^3$, we need to show that every vector $\vec{x} \in \mathbb{R}^3$ can be written as a linear combination of the vectors in C. Consider

$$\begin{bmatrix}x_1\\x_2\\x_3\end{bmatrix} = t_1\begin{bmatrix}1\\1\\0\end{bmatrix} + t_2\begin{bmatrix}1\\1\\1\end{bmatrix} + t_3\begin{bmatrix}1\\0\\1\end{bmatrix} = \begin{bmatrix}t_1 + t_2 + t_3\\t_1 + t_2\\t_2 + t_3\end{bmatrix}$$

Row reducing the corresponding coefficient matrix gives

$$\begin{bmatrix}1 & 1 & 1\\1 & 1 & 0\\0 & 1 & 1\end{bmatrix} \sim \begin{bmatrix}1 & 0 & 0\\0 & 1 & 0\\0 & 0 & 1\end{bmatrix}$$

Observe that the rank of the coefficient matrix equals the number of rows, so by Theorem 2.2.2, the system is consistent for every $\vec{x} \in \mathbb{R}^3$. Hence, $\operatorname{Span} C = \mathbb{R}^3$. Moreover, since the rank of the coefficient matrix equals the number of columns, there are no parameters in the general solution. Therefore, we have a unique solution when we take $\vec{x} = \vec{0}$, so C is also linearly independent. Hence, it is a basis for $\mathbb{R}^3$.

EXAMPLE 3

Is the set $\mathcal{B} = \left\{ \begin{bmatrix} 1 & 2 \\ -1 & 1 \end{bmatrix}, \begin{bmatrix} 0 & 1 \\ 3 & 1 \end{bmatrix}, \begin{bmatrix} 2 & 5 \\ 1 & 3 \end{bmatrix} \right\}$ a basis for the subspace Span $\mathcal{B}$ of $M(2,2)$?

Solution: Since $\mathcal{B}$ is a spanning set for Span $\mathcal{B}$, we just need to check if the vectors in $\mathcal{B}$ are linearly independent. Consider the equation

$$\begin{bmatrix} 0 & 0 \\ 0 & 0 \end{bmatrix} = t_1 \begin{bmatrix} 1 & 2 \\ -1 & 1 \end{bmatrix} + t_2 \begin{bmatrix} 0 & 1 \\ 3 & 1 \end{bmatrix} + t_3 \begin{bmatrix} 2 & 5 \\ 1 & 3 \end{bmatrix} = \begin{bmatrix} t_1 + 2t_3 & 2t_1 + t_2 + 5t_3 \\ -t_1 + 3t_2 + t_3 & t_1 + t_2 + 3t_3 \end{bmatrix}$$

Row reducing the coefficient matrix of the corresponding system gives

$$\begin{bmatrix} 1 & 0 & 2 \\ 2 & 1 & 5 \\ -1 & 3 & 1 \\ 1 & 1 & 3 \end{bmatrix} \sim \begin{bmatrix} 1 & 0 & 2 \\ 0 & 1 & 1 \\ 0 & 0 & 0 \\ 0 & 0 & 0 \end{bmatrix}$$

Observe that this implies that there are non-trivial solutions to the system. For example, one non-trivial solution is given by $t_1 = -2$, $t_2 = -1$, and $t_3 = 1$, and you can verify that

$$(-2)\begin{bmatrix} 1 & 2 \\ -1 & 1 \end{bmatrix} + (-1)\begin{bmatrix} 0 & 1 \\ 3 & 1 \end{bmatrix} + (1)\begin{bmatrix} 2 & 5 \\ 1 & 3 \end{bmatrix} = \begin{bmatrix} 0 & 0 \\ 0 & 0 \end{bmatrix}$$

Therefore, the given vectors are linearly dependent and do not form a basis for Span $\mathcal{B}$.

EXAMPLE 4

Is the set $C = \{3 + 2x + 2x^2, 1 + x^2, 1 + x + x^2\}$ a basis for P_2?

Solution: Consider the equation

$$\begin{aligned} a_0 + a_1x + a_2x^2 &= t_1(3 + 2x + 2x^2) + t_2(1 + x^2) + t_3(1 + x + x^2) \\ &= (3t_1 + t_2 + t_3) + (2t_1 + t_3)x + (2t_1 + t_2 + t_3)x^2 \end{aligned}$$

Row reducing the coefficient matrix of the corresponding system gives

$$\begin{bmatrix} 3 & 1 & 1 \\ 2 & 0 & 1 \\ 2 & 1 & 1 \end{bmatrix} \sim \begin{bmatrix} 1 & 0 & 0 \\ 0 & 1 & 0 \\ 0 & 0 & 1 \end{bmatrix}$$

Observe that this implies that the system is consistent and has a unique solution for all $a_0 + a_1x + a_2x^2 \in P_2$. Thus, C is a basis for P_2.

EXERCISE 1

Prove that the set $\mathcal{B} = \{1 + 2x + x^2, 1 + x^2, 1 + x\}$ is a basis for P_2.

EXAMPLE 5

Determine a basis for the subspace $\mathbb{S} = \{p(x) \in P_2 \mid p(1) = 0\}$ of P_2.

Solution: We first find a spanning set for $\mathbb{S}$ and then show that it is linearly independent. By the Factor Theorem, if $p(1) = 0$, then $(x - 1)$ is a factor of $p(x)$. That is, every polynomial $p(x) \in \mathbb{S}$ can be written in the form

$$p(x) = (x - 1)(ax + b) = a(x^2 - x) + b(x - 1)$$

Thus, we see that $\mathbb{S} = \text{Span}\{x^2 - x, x - 1\}$. Consider

$$t_1(x^2 - x) + t_2(x - 1) = 0$$

The only solution is $t_1 = t_2 = 0$. Hence, $\{x^2 - x, x - 1\}$ is linearly independent. Thus, $\{x^2 - x, x - 1\}$ is a linearly independent spanning set of $\mathbb{S}$ and hence a basis.

Obtaining a Basis from an Arbitrary Finite Spanning Set

Many times throughout the rest of this book, we will need to determine a basis for a vector space. One standard way of doing this is to first determine a spanning set for the vector space and then to remove vectors from the spanning set until we have a basis. We now outline this procedure.

Suppose that $\mathcal{T} = \{\mathbf{v}_1, \dots, \mathbf{v}_k\}$ is a spanning set for a non-trivial vector space $\mathbb{V}$. We want to choose a subset of $\mathcal{T}$ that is a basis for $\mathbb{V}$. If $\mathcal{T}$ is linearly independent, then $\mathcal{T}$ is a basis for $\mathbb{V}$, and we are done. If $\mathcal{T}$ is linearly dependent, then $t_1\mathbf{v}_1 + \cdots + t_k\mathbf{v}_k = \mathbf{0}$ has a solution where at least one of the coefficients is non-zero, say $t_i \neq 0$. Then, we can solve the equation for $\mathbf{v}_i$ to get

$$\mathbf{v}_i = -\frac{1}{t_i}(t_1\mathbf{v}_1 + \cdots + t_{i-1}\mathbf{v}_{i-1} + t_{i+1}\mathbf{v}_{i+1} + \cdots + t_k\mathbf{v}_k)$$

So, for any $\mathbf{x} \in \mathbb{V}$ we have

$$\begin{aligned}\mathbf{x} &= a_1\mathbf{v}_1 + \cdots + a_{i-1}\mathbf{v}_{i-1} + a_i\mathbf{v}_i + a_{i+1}\mathbf{v}_{i+1} + \cdots + a_k\mathbf{v}_k \\ &= a_1\mathbf{v}_1 + \cdots + a_{i-1}\mathbf{v}_{i-1} + a_i\left[-\frac{1}{t_i}(t_1\mathbf{v}_1 + \cdots + t_{i-1}\mathbf{v}_{i-1} + t_{i+1}\mathbf{v}_{i+1} + \cdots + t_k\mathbf{v}_k)\right] \\ &\quad + a_{i+1}\mathbf{v}_{i+1} + \cdots + a_k\mathbf{v}_k\end{aligned}$$

Thus, any $\mathbf{x} \in \mathbb{V}$ can be expressed as a linear combination of the set $\mathcal{T}\backslash\{\mathbf{v}_i\}$. This shows that $\mathcal{T}\backslash\{\mathbf{v}_i\}$ is a spanning set for $\mathbb{V}$. If $\mathcal{T}\backslash\{\mathbf{v}_i\}$ is linearly independent, it is a basis for $\mathbb{V}$, and the procedure is finished. Otherwise, we repeat the procedure to omit a second vector, say $\mathbf{v}_j$, and get $\mathcal{T}\backslash\{\mathbf{v}_i, \mathbf{v}_j\}$, which still spans $\mathbb{V}$. In this fashion, we must eventually get a linearly independent set. (Certainly, if there is only one non-zero vector left, it forms a linearly independent set.) Thus, we obtain a subset of $\mathcal{T}$ that is a basis for $\mathbb{V}$.

EXAMPLE 6

If $\mathcal{T} = \left\{ \begin{bmatrix} 1 \\ 1 \\ -2 \end{bmatrix}, \begin{bmatrix} 2 \\ -1 \\ 1 \end{bmatrix}, \begin{bmatrix} 1 \\ -2 \\ 3 \end{bmatrix}, \begin{bmatrix} 1 \\ 5 \\ 3 \end{bmatrix} \right\}$, determine a subset of $\mathcal{T}$ that is a basis for Span $\mathcal{T}$.

Solution: Consider

$$\begin{bmatrix} 0 \\ 0 \\ 0 \end{bmatrix} = t_1 \begin{bmatrix} 1 \\ 1 \\ -2 \end{bmatrix} + t_2 \begin{bmatrix} 2 \\ -1 \\ 1 \end{bmatrix} + t_3 \begin{bmatrix} 1 \\ -2 \\ 3 \end{bmatrix} + t_4 \begin{bmatrix} 1 \\ 5 \\ 3 \end{bmatrix} = \begin{bmatrix} t_1 + 2t_2 + t_3 + t_4 \\ t_1 - t_2 - 2t_3 + 5t_4 \\ -2t_1 + t_2 + 3t_3 + 3t_4 \end{bmatrix}$$

We row reduce the corresponding coefficent matrix

$$\begin{bmatrix} 1 & 2 & 1 & 1 \\ 1 & -1 & -2 & 5 \\ -2 & 1 & 3 & 3 \end{bmatrix} \sim \begin{bmatrix} 1 & 0 & -1 & 0 \\ 0 & 1 & 1 & 0 \\ 0 & 0 & 0 & 1 \end{bmatrix}$$

The general solution is $\begin{bmatrix} t_1 \\ t_2 \\ t_3 \\ t_4 \end{bmatrix} = s \begin{bmatrix} 1 \\ -1 \\ 1 \\ 0 \end{bmatrix}$, $s \in \mathbb{R}$. Taking $s = 1$, we get $\begin{bmatrix} t_1 \\ t_2 \\ t_3 \\ t_4 \end{bmatrix} = \begin{bmatrix} 1 \\ -1 \\ 1 \\ 0 \end{bmatrix}$, which gives

$$\begin{bmatrix} 1 \\ 1 \\ -2 \end{bmatrix} - \begin{bmatrix} 2 \\ -1 \\ 1 \end{bmatrix} + \begin{bmatrix} 1 \\ -2 \\ 3 \end{bmatrix} = \begin{bmatrix} 0 \\ 0 \\ 0 \end{bmatrix} \quad \text{or} \quad \begin{bmatrix} 1 \\ -2 \\ 3 \end{bmatrix} = -\begin{bmatrix} 1 \\ 1 \\ -2 \end{bmatrix} + \begin{bmatrix} 2 \\ -1 \\ 1 \end{bmatrix}$$

Thus, we can omit $\begin{bmatrix} 1 \\ -2 \\ 3 \end{bmatrix}$ from $\mathcal{T}$ and consider $\mathcal{T} \setminus \left\{ \begin{bmatrix} 1 \\ -2 \\ 3 \end{bmatrix} \right\} = \left\{ \begin{bmatrix} 1 \\ 1 \\ -2 \end{bmatrix}, \begin{bmatrix} 2 \\ -1 \\ 1 \end{bmatrix}, \begin{bmatrix} 1 \\ 5 \\ 3 \end{bmatrix} \right\}$

Now consider

$$\begin{bmatrix} 0 \\ 0 \\ 0 \end{bmatrix} = t_1 \begin{bmatrix} 1 \\ 1 \\ -2 \end{bmatrix} + t_2 \begin{bmatrix} 2 \\ -1 \\ 1 \end{bmatrix} + t_3 \begin{bmatrix} 1 \\ 5 \\ 3 \end{bmatrix}$$

The matrix is the same as above except that the third column is omitted, so the same row operations give

$$\begin{bmatrix} 1 & 2 & 1 \\ 1 & -1 & 5 \\ -2 & 1 & 3 \end{bmatrix} \sim \begin{bmatrix} 1 & 0 & 0 \\ 0 & 1 & 0 \\ 0 & 0 & 1 \end{bmatrix}$$

Hence, the only solution is $t_1 = t_2 = t_3 = 0$ and we conclude that $\left\{ \begin{bmatrix} 1 \\ 1 \\ -2 \end{bmatrix}, \begin{bmatrix} 2 \\ -1 \\ 1 \end{bmatrix}, \begin{bmatrix} 1 \\ 5 \\ 3 \end{bmatrix} \right\}$ is linearly independent and thus a basis for Span $\mathcal{T}$.

EXERCISE 2 Let $\mathcal{B} = \{1 - x, 2 + 2x + x^2, x + x^2, 1 + x^2\}$. Determine a subset of $\mathcal{B}$ that is a basis for Span $\mathcal{B}$.

Dimension

We saw in Section 2.3 that every basis of a subspace $\mathbb{S}$ of $\mathbb{R}^n$ contains the same number of vectors. We now prove that this result holds for general vector spaces. Observe that the proof of this result is essentially identical to that in Section 2.3.

Lemma 2 Suppose that $\mathbb{V}$ is a vector space and Span$\{\mathbf{v}_1, \dots, \mathbf{v}_n\} = \mathbb{V}$. If $\{\mathbf{u}_1, \dots, \mathbf{u}_k\}$ is a linearly independent set in $\mathbb{V}$, then $k \leq n$.

Proof: Since each $\mathbf{u}_i$, $1 \leq i \leq k$, is a vector in $\mathbb{V}$, it can be written as a linear combination of the $\mathbf{v}_j$'s. We get

$$\begin{aligned}
\mathbf{u}_1 &= a_{11}\mathbf{v}_1 + a_{21}\mathbf{v}_2 + \cdots + a_{n1}\mathbf{v}_n \\
\mathbf{u}_2 &= a_{12}\mathbf{v}_1 + a_{22}\mathbf{v}_2 + \cdots + a_{n2}\mathbf{v}_n \\
&\ \ \vdots \\
\mathbf{u}_k &= a_{1k}\mathbf{v}_1 + a_{2k}\mathbf{v}_2 + \cdots + a_{nk}\mathbf{v}_n
\end{aligned}$$

Consider the equation

$$\begin{aligned}
\mathbf{0} &= t_1\mathbf{u}_1 + \cdots + t_k\mathbf{u}_k \\
&= t_1(a_{11}\mathbf{v}_1 + a_{21}\mathbf{v}_2 + \cdots + a_{n1}\mathbf{v}_n) + \cdots + t_k(a_{1k}\mathbf{v}_1 + a_{2k}\mathbf{v}_2 + \cdots + a_{nk}\mathbf{v}_n) \\
&= (a_{11}t_1 + \cdots + a_{1k}t_k)\mathbf{v}_1 + \cdots + (a_{n1}t_1 + \cdots + a_{nk}t_k)\mathbf{v}_n
\end{aligned}$$

Since $\mathbf{0} = 0\mathbf{v}_1 + \cdots + 0\mathbf{v}_n$ we can write this as

$$0\mathbf{v}_1 + \cdots + 0\mathbf{v}_n = (a_{11}t_1 + \cdots + a_{1k}t_k)\mathbf{v}_1 + \cdots + (a_{n1}t_1 + \cdots + a_{nk}t_k)\mathbf{v}_n$$

Comparing coefficients of $\mathbf{v}_i$ we get the homogeneous system

$$\begin{aligned}
a_{11}t_1 + \cdots + a_{1k}t_k &= 0 \\
\vdots \qquad\quad & \ \ \vdots \\
a_{n1}t_1 + \cdots + a_{nk}t_k &= 0
\end{aligned}$$

of n equations in k unknowns $t_1, \dots, t_k$. If $k > n$, then this system would have a non-trivial solution, which would imply that $\{\mathbf{u}_1, \dots, \mathbf{u}_k\}$ is linearly dependent. But, we assumed that $\{\mathbf{u}_1, \dots, \mathbf{u}_k\}$ is linearly independent, so we must have $k \leq n$. ■

Theorem 3 If $\mathcal{B} = \{\mathbf{v}_1, \dots, \mathbf{v}_n\}$ and $C = \{\mathbf{u}_1, \dots, \mathbf{u}_k\}$ are both bases of a vector space $\mathbb{V}$, then $k = n$.

Proof: On one hand, $\mathcal{B}$ is a basis for $\mathbb{V}$, so it is linearly independent. Also, C is a basis for $\mathbb{V}$, so Span $C = \mathbb{V}$. Thus, by Lemma 2, we get that $n \leq k$. Similarly, C is linearly independent as it is a basis for $\mathbb{V}$, and Span $\mathcal{B} = \mathbb{V}$, since $\mathcal{B}$ is a basis for $\mathbb{V}$. So Lemma 2 gives $n \geq k$. Therefore, $n = k$, as required. ■

As in Section 2.3, this theorem justifies the following definition of the dimension of a vector space.

Definition
Dimension

If a vector space $\mathbb{V}$ has a basis with n vectors, then we say that the **dimension** of $\mathbb{V}$ is n and write

$$\dim \mathbb{V} = n$$

If a vector space $\mathbb{V}$ does not have a basis with finitely many elements, then $\mathbb{V}$ is called **infinite-dimensional**. The dimension of the trivial vector space is defined to be 0.

Remark

Properties of infinite-dimensional spaces are beyond the scope of this book.

EXAMPLE 7

(a) $\mathbb{R}^n$ is n-dimensional because the standard basis contains n vectors.

(b) The vector space $M(m,n)$ is $(m \times n)$-dimensional since the standard basis has $m \times n$ vectors.

(c) The vector space P_n is $(n + 1)$-dimensional as it has the standard basis $\{1, x, x^2, \dots, x^n\}$.

(d) The vector space $C(a,b)$ is infinite-dimensional as it contains all polynomials (along with many other types of functions). Most function spaces are infinite-dimensional.

EXAMPLE 8

Let $\mathbb{S} = \operatorname{Span}\left\{\begin{bmatrix}1\\3\\2\end{bmatrix}, \begin{bmatrix}-1\\-2\\-1\end{bmatrix}, \begin{bmatrix}-1\\2\\3\end{bmatrix}, \begin{bmatrix}-1\\3\\4\end{bmatrix}\right\}$. Show that $\dim \mathbb{S} = 2$.

Solution: Consider

$$\begin{bmatrix}0\\0\\0\end{bmatrix} = t_1\begin{bmatrix}1\\3\\2\end{bmatrix} + t_2\begin{bmatrix}-1\\-2\\-1\end{bmatrix} + t_3\begin{bmatrix}-1\\2\\3\end{bmatrix} + t_4\begin{bmatrix}-1\\3\\4\end{bmatrix} = \begin{bmatrix}t_1 - t_2 - t_3 - t_4\\3t_1 - 2t_2 + 2t_3 + 3t_4\\2t_1 - t_2 + 3t_3 + 4t_4\end{bmatrix}$$

We row reduce the corresponding coefficent matrix

$$\begin{bmatrix}1 & -1 & -1 & -1\\3 & -2 & 2 & 3\\2 & -1 & 3 & 4\end{bmatrix} \sim \begin{bmatrix}1 & 0 & 4 & 5\\0 & 1 & 5 & 6\\0 & 0 & 0 & 0\end{bmatrix}$$

Observe that this implies that $\begin{bmatrix}-1\\2\\3\end{bmatrix}$ and $\begin{bmatrix}-1\\3\\4\end{bmatrix}$ can be written as linear combinations of the first two vectors. Thus, $\mathbb{S} = \operatorname{Span}\left\{\begin{bmatrix}1\\3\\2\end{bmatrix}, \begin{bmatrix}-1\\-2\\-1\end{bmatrix}\right\}$. Moreover, $\mathcal{B} = \left\{\begin{bmatrix}1\\3\\2\end{bmatrix}, \begin{bmatrix}-1\\-2\\-1\end{bmatrix}\right\}$ is

EXAMPLE 8 (continued)

clearly linearly independent since neither vector is a scalar multiple of the other, hence $\mathcal{B}$ is a basis for $\mathbb{S}$. Thus, $\dim \mathbb{S} = 2$.

EXAMPLE 9

Let $\mathbb{S} = \left\{ \begin{bmatrix} a & b \\ c & d \end{bmatrix} \in M(2,2) \mid a + b = d \right\}$. Determine the dimension of $\mathbb{S}$.

Solution: Since $d = a + b$, observe that every matrix in $\mathbb{S}$ has the form

$$\begin{bmatrix} a & b \\ c & d \end{bmatrix} = \begin{bmatrix} a & b \\ c & a+b \end{bmatrix} = a\begin{bmatrix} 1 & 0 \\ 0 & 1 \end{bmatrix} + b\begin{bmatrix} 0 & 1 \\ 0 & 1 \end{bmatrix} + c\begin{bmatrix} 0 & 0 \\ 1 & 0 \end{bmatrix}$$

Thus,

$$\mathbb{S} = \operatorname{Span}\left\{ \begin{bmatrix} 1 & 0 \\ 0 & 1 \end{bmatrix}, \begin{bmatrix} 0 & 1 \\ 0 & 1 \end{bmatrix}, \begin{bmatrix} 0 & 0 \\ 1 & 0 \end{bmatrix} \right\}$$

It is easy to show that this spanning set $\left\{ \begin{bmatrix} 1 & 0 \\ 0 & 1 \end{bmatrix}, \begin{bmatrix} 0 & 1 \\ 0 & 1 \end{bmatrix}, \begin{bmatrix} 0 & 0 \\ 1 & 0 \end{bmatrix} \right\}$ for $\mathbb{S}$ is also linearly independent and hence is a basis for $\mathbb{S}$. Thus, $\dim \mathbb{S} = 3$.

EXERCISE 3

Find the dimension of $\mathbb{S} = \{a + bx + cx^2 + dx^3 \in P_3 \mid a + b + c + d = 0\}$.

Extending a Linearly Independent Subset to a Basis

Sometimes a linearly independent subset $\mathcal{T} = \{\mathbf{v}_1, \dots, \mathbf{v}_k\}$ is given in an n-dimensional vector space $\mathbb{V}$, and it is necessary to include these in a basis for $\mathbb{V}$. If $\operatorname{Span}\mathcal{T} \neq \mathbb{V}$, then there exists some vector $\mathbf{w}_{k+1}$ that is in $\mathbb{V}$ but not in $\operatorname{Span}\mathcal{T}$. Now consider

$$t_1\mathbf{v}_1 + \cdots + t_k\mathbf{v}_k + t_{k+1}\mathbf{w}_{k+1} = \mathbf{0} \tag{4.1}$$

If $t_{k+1} \neq 0$, then we have

$$\mathbf{w}_{k+1} = -\frac{t_1}{t_{k+1}}\mathbf{v}_1 - \cdots - \frac{t_k}{t_{k+1}}\mathbf{v}_k$$

and so $\mathbf{w}_{k+1}$ can be written as a linear combination of the vectors in $\mathcal{T}$, which cannot be since $\mathbf{w}_{k+1} \notin \operatorname{Span}\mathcal{T}$. Therefore, we must have $t_{k+1} = 0$. In this case, (4.1) becomes

$$\mathbf{0} = t_1\mathbf{v}_1 + \cdots + t_k\mathbf{v}_k$$

But, $\mathcal{T}$ is linearly independent, which implies that $t_1 = \cdots = t_k = 0$. Thus, $t_1\mathbf{v}_1 + \cdots + t_k\mathbf{v}_k + t_{k+1}\mathbf{w}_{k+1} = \mathbf{0}$ implies that $t_1 = \cdots = t_{k+1} = 0$, and hence $\{\mathbf{v}_1, \dots, \mathbf{v}_k, \mathbf{w}_{k+1}\}$ is linearly independent. Now, if $\operatorname{Span}\{\mathbf{v}_1, \dots, \mathbf{v}_k, \mathbf{w}_{k+1}\} = \mathbb{V}$, then it is a basis for $\mathbb{V}$. If not, we repeat the procedure to add another vector $\mathbf{w}_{k+2}$ to get $\{\mathbf{v}_1, \dots, \mathbf{v}_k, \mathbf{w}_{k+1}, \mathbf{w}_{k+2}\}$, which is linearly independent. In this fashion, we must eventually get a basis, since according to Lemma 2, there cannot be more than n linearly independent vectors in an n-dimensional vector space.

EXAMPLE 10

Let $C = \left\{ \begin{bmatrix} 1 & 1 \\ 0 & 1 \end{bmatrix}, \begin{bmatrix} -2 & -1 \\ 1 & 1 \end{bmatrix} \right\}$. Extend C to a basis for $M(2,2)$.

Solution: We first want to determine whether C is a spanning set for $M(2,2)$. Consider

$$\begin{bmatrix} b_1 & b_2 \\ b_3 & b_4 \end{bmatrix} = t_1 \begin{bmatrix} 1 & 1 \\ 0 & 1 \end{bmatrix} + t_2 \begin{bmatrix} -2 & -1 \\ 1 & 1 \end{bmatrix} = \begin{bmatrix} t_1 - 2t_2 & t_1 - t_2 \\ t_2 & t_1 + t_2 \end{bmatrix}$$

Row reducing the augmented matrix of the associated system gives

$$\left[\begin{array}{cc|c} 1 & -2 & b_1 \\ 1 & -1 & b_2 \\ 0 & 1 & b_3 \\ 1 & 1 & b_4 \end{array}\right] \sim \left[\begin{array}{cc|c} 1 & -2 & b_1 \\ 0 & 1 & b_2 - b_1 \\ 0 & 0 & b_1 - b_2 + b_3 \\ 0 & 0 & 2b_1 - 3b_2 + b_4 \end{array}\right]$$

Hence, $\mathcal{B}$ is not a spanning set of $M(2,2)$ since any matrix $\begin{bmatrix} b_1 & b_2 \\ b_3 & b_4 \end{bmatrix}$ with $b_1 - b_2 + b_3 \neq 0$ (or $2b_1 - 3b_2 + b_4 \neq 0$) is not in Span $\mathcal{B}$. In particular, $\begin{bmatrix} 0 & 0 \\ 1 & 0 \end{bmatrix}$ is not in the span of $\mathcal{B}$. Hence, by the procedure above, we should add this matrix to the set. We let

$$\mathcal{B}_1 = \left\{ \begin{bmatrix} 1 & 1 \\ 0 & 1 \end{bmatrix}, \begin{bmatrix} -2 & -1 \\ 1 & 1 \end{bmatrix}, \begin{bmatrix} 0 & 0 \\ 1 & 0 \end{bmatrix} \right\}$$

and repeat the procedure. Consider

$$\begin{bmatrix} b_1 & b_2 \\ b_3 & b_4 \end{bmatrix} = t_1 \begin{bmatrix} 1 & 1 \\ 0 & 1 \end{bmatrix} + t_2 \begin{bmatrix} -2 & -1 \\ 1 & 1 \end{bmatrix} + t_3 \begin{bmatrix} 0 & 0 \\ 1 & 0 \end{bmatrix}$$
$$= \begin{bmatrix} t_1 - 2t_2 & t_1 - t_2 \\ t_2 + t_3 & t_1 + t_2 \end{bmatrix}$$

Row reducing the augmented matrix of the associated system gives

$$\left[\begin{array}{ccc|c} 1 & -2 & 0 & b_1 \\ 1 & -1 & 0 & b_2 \\ 0 & 1 & 1 & b_3 \\ 1 & 1 & 0 & b_4 \end{array}\right] \sim \left[\begin{array}{ccc|c} 1 & -2 & 0 & b_1 \\ 0 & 1 & 0 & b_2 - b_1 \\ 0 & 0 & 1 & b_1 - b_2 + b_3 \\ 0 & 0 & 0 & 2b_1 - 3b_2 + b_4 \end{array}\right]$$

So, any matrix $\begin{bmatrix} b_1 & b_2 \\ b_3 & b_4 \end{bmatrix}$ with $2b_1 - 3b_2 + b_4 \neq 0$ is not in Span $\mathcal{B}_1$. For example, $\begin{bmatrix} 0 & 0 \\ 0 & 1 \end{bmatrix}$ is not in the span of $\mathcal{B}_1$ and thus $\mathcal{B}_1$ is not a basis for $M(2,2)$. Adding $\begin{bmatrix} 0 & 0 \\ 0 & 1 \end{bmatrix}$ to $\mathcal{B}_1$ we get

$$\mathcal{B}_2 = \left\{ \begin{bmatrix} 1 & 1 \\ 0 & 1 \end{bmatrix}, \begin{bmatrix} -2 & -1 \\ 1 & 1 \end{bmatrix}, \begin{bmatrix} 0 & 0 \\ 1 & 0 \end{bmatrix}, \begin{bmatrix} 0 & 0 \\ 0 & 1 \end{bmatrix} \right\}$$

By construction $\mathcal{B}_2$ is a linearly independent. Moreover, we can show that it spans $M(2,2)$. Thus, it is a basis for $M(2,2)$.

EXERCISE 4

Extend the set $\mathcal{T} = \left\{ \begin{bmatrix} 1 \\ 1 \\ 1 \end{bmatrix} \right\}$ to a basis for $\mathbb{R}^3$.

Knowing the dimension of a finite dimensional vector space $\mathbb{V}$ is very useful when trying to invent a basis for $\mathbb{V}$, as the next theorem demonstrates.

Theorem 4

Let $\mathbb{V}$ be an n-dimensional vector space. Then
(1) A set of more than n vectors in $\mathbb{V}$ must be linearly dependent.
(2) A set of fewer than n vectors cannot span $\mathbb{V}$.
(3) A set with n elements of $\mathbb{V}$ is a spanning set for $\mathbb{V}$ if and only if it is linearly independent.

Proof: (1) This is Lemma 2 above.

(2) Suppose that $\mathbb{V}$ can be spanned by a set of $k < n$ vectors. If $\mathbb{V}$ is n-dimensional, then it has a basis containing n vectors. This means we have a set of n linearly independent vectors in a set spanned by $k < n$ vectors which contradicts (1).

(3) If $\mathcal{B}$ is a linearly independent set of n vectors that does not span $\mathbb{V}$, then it can be extended to a basis for $\mathbb{V}$ by the procedure above. But, this would give a linearly independent set of more than n vectors in $\mathbb{V}$, which contradicts (1).

Similarly, if $\mathcal{B}$ is a spanning set for $\mathbb{V}$ that is not linearly independent, then by the procedure above, there is a linearly independent proper subset of $\mathcal{B}$ that spans $\mathbb{V}$. But, then $\mathbb{V}$ would be spanned by a set of fewer that n vectors, which contradicts (2). ■

EXAMPLE 11

1. Produce a basis $\mathcal{B}$ for the plane $\mathcal{P}$ in $\mathbb{R}^3$ with equation $x_1 + 2x_2 - x_3 = 0$.

2. Extend the basis $\mathcal{B}$ to obtain a basis C for $\mathbb{R}^3$.

Solution: (a) We know that a plane in $\mathbb{R}^3$ has dimension 2. By Theorem 4, we just need to pick two linearly independent vectors that lie in the plane. Observe that $\vec{v}_1 = \begin{bmatrix} 1 \\ 0 \\ 1 \end{bmatrix}$ and $\vec{v}_2 = \begin{bmatrix} 0 \\ 1 \\ 2 \end{bmatrix}$ both satisfy the equation of the plane and neither is a scalar multiple of the other. Thus they are linearly independent, and $\mathcal{B} = \{\vec{v}_1, \vec{v}_2\}$ is a basis for the plane $\mathcal{P}$.

(b) From the procedure above, we need to add a vector that is not in the span of $\{\vec{v}_1, \vec{v}_2\}$. But, Span$\{\vec{v}_1, \vec{v}_2\}$ is in the plane, so we need to pick any vector not in the plane. Observe that $\vec{v}_3 = \begin{bmatrix} 1 \\ 0 \\ 0 \end{bmatrix}$ does not satisfy the equation of the plane and hence is not in the plane. Thus, $\{\vec{v}_1, \vec{v}_2, \vec{v}_3\}$ is a linearly independent set of three vectors in $\mathbb{R}^3$ and therefore is a basis for $\mathbb{R}^3$, according to Theorem 4.

EXERCISE 5 Produce a basis for the hyperplane in $\mathbb{R}^4$ with equation $x_1 - x_2 + x_3 - 2x_4 = 0$ and extend the basis to obtain a basis for $\mathbb{R}^4$.

PROBLEMS 4.3

Practice Problems

A1 Determine whether each set is a basis for $\mathbb{R}^3$.

(a) $\left\{\begin{bmatrix}1\\1\\2\end{bmatrix}, \begin{bmatrix}1\\-1\\-1\end{bmatrix}, \begin{bmatrix}2\\1\\1\end{bmatrix}\right\}$

(b) $\left\{\begin{bmatrix}-2\\2\\1\end{bmatrix}, \begin{bmatrix}3\\-1\\2\end{bmatrix}\right\}$

(c) $\left\{\begin{bmatrix}1\\0\\1\end{bmatrix}, \begin{bmatrix}-1\\2\\1\end{bmatrix}, \begin{bmatrix}1\\3\\5\end{bmatrix}, \begin{bmatrix}2\\-1\\-4\end{bmatrix}\right\}$

(d) $\left\{\begin{bmatrix}-1\\3\\5\end{bmatrix}, \begin{bmatrix}2\\4\\0\end{bmatrix}, \begin{bmatrix}1\\4\\2\end{bmatrix}\right\}$

(e) $\left\{\begin{bmatrix}1\\-1\\1\end{bmatrix}, \begin{bmatrix}1\\2\\-1\end{bmatrix}, \begin{bmatrix}3\\0\\1\end{bmatrix}\right\}$

A2 Let $\mathcal{B} = \left\{\begin{bmatrix}1\\1\\1\\2\end{bmatrix}, \begin{bmatrix}2\\3\\1\\3\end{bmatrix}, \begin{bmatrix}3\\3\\4\\11\end{bmatrix}, \begin{bmatrix}2\\3\\1\\4\end{bmatrix}\right\}$. Prove that $\mathcal{B}$ is a basis for $\mathbb{R}^4$.

A3 Select a basis for Span $\mathcal{B}$ from each of the following sets $\mathcal{B}$ and determine the dimension of Span $\mathcal{B}$.

(a) $\mathcal{B} = \left\{\begin{bmatrix}1\\-2\\1\end{bmatrix}, \begin{bmatrix}0\\1\\2\end{bmatrix}, \begin{bmatrix}2\\0\\10\end{bmatrix}, \begin{bmatrix}1\\1\\7\end{bmatrix}\right\}$

(b) $\mathcal{B} = \left\{\begin{bmatrix}1\\3\\2\end{bmatrix}, \begin{bmatrix}-2\\-6\\-4\end{bmatrix}, \begin{bmatrix}-1\\-1\\2\end{bmatrix}, \begin{bmatrix}0\\4\\8\end{bmatrix}, \begin{bmatrix}0\\1\\1\end{bmatrix}\right\}$

A4 Select a basis for Span $\mathcal{B}$ from each of the following sets $\mathcal{B}$ and determine the dimension of Span $\mathcal{B}$.

(a) $\mathcal{B} = \left\{\begin{bmatrix}1 & 1\\-1 & 1\end{bmatrix}, \begin{bmatrix}0 & 1\\3 & -1\end{bmatrix}, \begin{bmatrix}1 & -1\\2 & -3\end{bmatrix}, \begin{bmatrix}2 & 1\\4 & -3\end{bmatrix}\right\}$

(b) $\mathcal{B} = \left\{\begin{bmatrix}1 & -1\\-1 & -1\end{bmatrix}, \begin{bmatrix}1 & 1\\1 & -1\end{bmatrix}, \begin{bmatrix}0 & 1\\2 & 1\end{bmatrix}, \begin{bmatrix}3 & 0\\1 & -2\end{bmatrix}, \begin{bmatrix}0 & 1\\0 & 1\end{bmatrix}\right\}$

(c) $\mathcal{B} = \left\{\begin{bmatrix}1 & 0\\0 & 1\end{bmatrix}, \begin{bmatrix}0 & 1\\1 & 0\end{bmatrix}, \begin{bmatrix}0 & 1\\0 & -1\end{bmatrix}, \begin{bmatrix}1 & 1\\1 & 0\end{bmatrix}\right\}$

A5 Determine the dimension of the vector space of polynomials spanned by $\mathcal{B}$.

(a) $\mathcal{B} = \{1 + x, 1 + x + x^2, 1 + x^3\}$

(b) $\mathcal{B} = \{1 + x, 1 - x, 1 + x^3, 1 - x^3\}$

(c) $\mathcal{B} = \{1 + x + x^2, 1 - x^3, 1 - 2x + 2x^2 - x^3, 1 - x^2 + 2x^3, x^2 + x^3\}$

A6 (a) Using the method in Example 11, determine a basis for the plane $2x_1 - x_2 - x_3 = 0$ in $\mathbb{R}^3$.

(b) Extend the basis of part (a) to obtain a basis for $\mathbb{R}^3$.

A7 (a) Using the method in Example 11, determine a basis for the hyperplane $x_1 - x_2 + x_3 - x_4 = 0$ in $\mathbb{R}^4$.

(b) Extend the basis of part (a) to obtain a basis for $\mathbb{R}^4$.

A8 Obtain a basis for each of the following vector spaces and determine the dimension.

(a) $\mathbb{S} = \{a + bx + cx^2 \in P_2 \mid a = -c\}$

(b) $\mathbb{S} = \left\{\begin{bmatrix}a & b\\0 & c\end{bmatrix} \in M(2,2) \mid a, b, c \in \mathbb{R}\right\}$

(c) $\mathbb{S} = \left\{\begin{bmatrix}x_1\\x_2\\x_3\end{bmatrix} \in \mathbb{R}^3 \mid \begin{bmatrix}x_1\\x_2\\x_3\end{bmatrix} \cdot \begin{bmatrix}1\\1\\1\end{bmatrix} = 0\right\}$

(d) $\mathbb{S} = \{p(x) \in P_2 \mid p(2) = 0\}$

(e) $\mathbb{S} = \left\{\begin{bmatrix}a & b\\c & d\end{bmatrix} \in M(2,2) \mid a = -c\right\}$

Homework Problems

B1 Determine whether each set is a basis for $\mathbb{R}^3$.

(a) $\left\{\begin{bmatrix}1\\-1\\3\end{bmatrix},\begin{bmatrix}2\\1\\-1\end{bmatrix},\begin{bmatrix}1\\2\\-4\end{bmatrix}\right\}$

(b) $\left\{\begin{bmatrix}1\\3\\1\end{bmatrix},\begin{bmatrix}-1\\0\\1\end{bmatrix},\begin{bmatrix}5\\1\\0\end{bmatrix},\begin{bmatrix}0\\0\\3\end{bmatrix}\right\}$

(c) $\left\{\begin{bmatrix}1\\2\\1\end{bmatrix},\begin{bmatrix}-1\\-1\\0\end{bmatrix},\begin{bmatrix}1\\1\\7\end{bmatrix}\right\}$

(d) $\left\{\begin{bmatrix}1\\0\\1\end{bmatrix},\begin{bmatrix}0\\1\\0\end{bmatrix}\right\}$

(e) $\left\{\begin{bmatrix}1\\2\\-4\end{bmatrix},\begin{bmatrix}2\\1\\-1\end{bmatrix},\begin{bmatrix}1\\2\\-4\end{bmatrix}\right\}$

B2 Let $\mathcal{B} = \left\{\begin{bmatrix}1 & -2\\1 & 1\end{bmatrix},\begin{bmatrix}2 & -3\\3 & 4\end{bmatrix},\begin{bmatrix}3 & -6\\4 & 5\end{bmatrix},\begin{bmatrix}3 & -6\\4 & 6\end{bmatrix}\right\}$. Prove that $\mathcal{B}$ is a basis for $M(2,2)$.

B3 Determine whether each set is a basis for P_2.

(a) $\{1 + x + x^2, 1 - x^2\}$

(b) $\{1 + x^2, 2 - x + 2x^2, -2 + 2x\}$

(c) $\{1 + x^2, 1 + x + x^2, -x - x^2, 1 + 2x^2\}$

(d) $\{3 + 2x + 2x^2, 1 + x + x^2, 1 - x - x^2\}$

B4 Select a basis for Span $\mathcal{B}$ from each of the following sets $\mathcal{B}$ and determine the dimension of Span $\mathcal{B}$.

(a) $\mathcal{B} = \left\{\begin{bmatrix}1\\2\\1\end{bmatrix},\begin{bmatrix}2\\3\\-2\end{bmatrix},\begin{bmatrix}-1\\0\\7\end{bmatrix}\right\}$

(b) $\mathcal{B} = \left\{\begin{bmatrix}1\\1\\-1\end{bmatrix},\begin{bmatrix}1\\-1\\2\end{bmatrix},\begin{bmatrix}2\\0\\1\end{bmatrix},\begin{bmatrix}0\\2\\-3\end{bmatrix},\begin{bmatrix}3\\-3\\-2\end{bmatrix}\right\}$

B5 Select a basis for Span $\mathcal{B}$ from each of the following sets $\mathcal{B}$ and determine the dimension of Span $\mathcal{B}$.

(a) $\mathcal{B} = \left\{\begin{bmatrix}1 & 1\\2 & 1\end{bmatrix},\begin{bmatrix}-2 & -2\\-4 & -2\end{bmatrix},\begin{bmatrix}1 & -1\\1 & 0\end{bmatrix},\begin{bmatrix}3 & 1\\5 & 2\end{bmatrix}\right\}$

(b) $\mathcal{B} = \left\{\begin{bmatrix}1 & 2\\-1 & 1\end{bmatrix},\begin{bmatrix}1 & 2\\2 & 1\end{bmatrix},\begin{bmatrix}3 & 6\\0 & 3\end{bmatrix},\begin{bmatrix}0 & 1\\0 & -2\end{bmatrix},\begin{bmatrix}1 & 3\\2 & -1\end{bmatrix}\right\}$

B6 Determine the dimension of the vector space of polynomials spanned by $\mathcal{B}$.

(a) $\mathcal{B} = \{1 + x + x^2, x + x^2 + x^3, 1 - x^3\}$

(b) $\mathcal{B} = \{1 + x + x^2, x + x^2 + x^3, 1 + x^2 + x^3\}$

B7 (a) Using the method in Example 11, determine a basis for the plane $x_1 + 3x_2 + 4x_3 = 0$ in $\mathbb{R}^3$.

(b) Extend the basis of part (a) to obtain a basis for $\mathbb{R}^3$.

B8 (a) Using the method in Example 11, determine a basis for the hyperplane $x_1 + x_2 + 2x_3 + x_4 = 0$ in $\mathbb{R}^4$.

(b) Extend the basis of part (a) to obtain a basis for $\mathbb{R}^4$.

B9 Obtain a basis for each of the following vector spaces and determine the dimension.

(a) $\mathbb{S} = \left\{\begin{bmatrix}a & 0\\0 & b\end{bmatrix} \in M(2,2) \mid a, b \in \mathbb{R}\right\}$

(b) $\mathbb{S} = \left\{(1 + x^2)p(x) \mid p(x) \in P_2\right\}$

(c) $\mathbb{S} = \left\{\begin{bmatrix}x_1\\x_2\\x_3\end{bmatrix} \in \mathbb{R}^3 \mid \begin{bmatrix}x_1\\x_2\\x_3\end{bmatrix} \cdot \begin{bmatrix}1\\0\\-1\end{bmatrix} = 0\right\}$

(d) $\mathbb{S} = \left\{\begin{bmatrix}a & b\\c & d\end{bmatrix} \in M(2,2) \mid a - b = 0,\ b - c = 0, c - a = 0 \in \mathbb{R}\right\}$

(e) $\mathbb{S} = \{p(x) \in P_5 \mid p(-x) = p(x) \text{ for all } x\}$

Conceptual Problems

D1 (a) It may be said that "a basis for a finite dimensional vector space $\mathbb{V}$ is a maximal (largest possible) linearly independent set in $\mathbb{V}$." Explain why this makes sense in terms of statements in this section.

(b) It may be said that "a basis for a finite dimensional vector space $\mathbb{V}$ is a minimal spanning set for $\mathbb{V}$." Explain why this makes sense in terms of statements in this section.

D2 Let $\mathbb{V}$ be an n-dimensional vector space. Prove that if $\mathbb{S}$ is a subspace of $\mathbb{V}$ and $\dim \mathbb{S} = n$, then $\mathbb{S} = \mathbb{V}$.

D3 (a) Show that if $\{\mathbf{v}_1, \mathbf{v}_2\}$ is a basis for a vector space $\mathbb{V}$, then for any real number t, $\{\mathbf{v}_1, \mathbf{v}_2 + t\mathbf{v}_1\}$ is also a basis for $\mathbb{V}$.

(b) Show that if $\{\mathbf{v}_1, \mathbf{v}_2, \mathbf{v}_3\}$ is a basis for a vector space $\mathbb{V}$, then for any $s, t \in \mathbb{R}$, $\{\mathbf{v}_1, \mathbf{v}_2, \mathbf{v}_3 + t\mathbf{v}_1 + s\mathbf{v}_2\}$ is also a basis for $\mathbb{V}$.

D4 In Problem 4.2.D2 you proved that $\mathbb{V} = \{(a, b) \mid a, b \in \mathbb{R}, b > 0\}$, with addition defined by $(a, b) \oplus (c, d) = (ad + bc, bd)$ and scalar multiplication defined by $t \odot (a, b) = (tab^{t-1}, b^t)$, was a vector space over $\mathbb{R}$. Find, with justification, a basis for $\mathbb{V}$ and hence determine the dimension of $\mathbb{V}$.

4.4 Coordinates with Respect to a Basis

In Section 4.3, we saw that a vector space may have many different bases. Why might we want a basis other than the standard basis? In some problems, it is much more convenient to use a different basis than the standard basis. For example, a stretch by a factor of 3 in $\mathbb{R}^2$ in the direction of the vector $\begin{bmatrix}1\\2\end{bmatrix}$ is geometrically easy to understand. However, it would be awkward to determine the standard matrix of this stretch and then determine its effect on any other vector. It would be much better to have a basis that takes advantage of the direction $\begin{bmatrix}1\\2\end{bmatrix}$ and of the direction $\begin{bmatrix}-2\\1\end{bmatrix}$, which remain unchanged under the stretch.

Alternatively, consider the reflection in the plane $x_1 + 2x_2 - 3x_3 = 0$ in $\mathbb{R}^3$. It is easy to describe this by saying that it reverses the normal vector $\begin{bmatrix}1\\2\\-3\end{bmatrix}$ to $\begin{bmatrix}-1\\-2\\3\end{bmatrix}$ and leaves unchanged any vectors lying in the plane (such as the vectors $\begin{bmatrix}2\\-1\\0\end{bmatrix}$ and $\begin{bmatrix}3\\0\\1\end{bmatrix}$). See Figure 4.4.2. Describing this reflection in terms of these vectors gives more geometrical information than describing it in terms of the standard basis vectors.

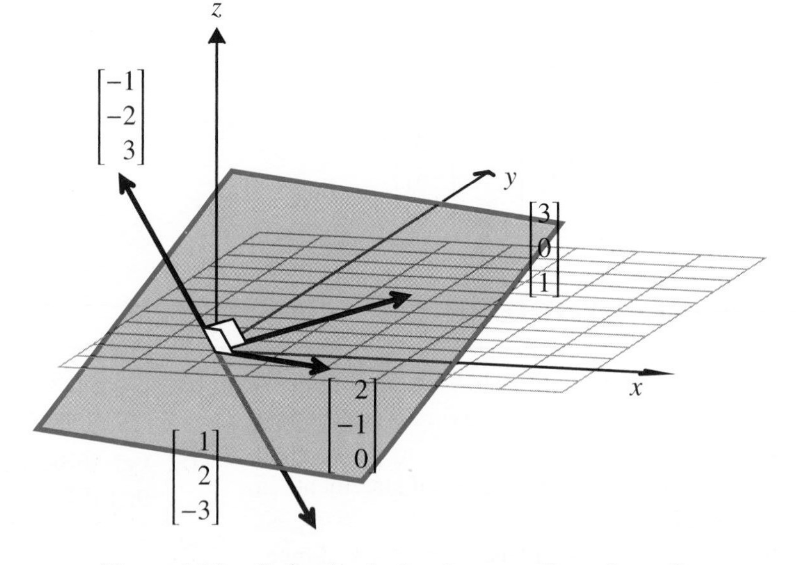

Figure 4.4.2 Reflection in the plane $x_1 + 2x_2 - 3x_3 = 0$.

Notice that in these examples, the geometry itself provides us with a preferred basis for the appropriate space. However, to make use of these preferred bases, we need to know how to represent an arbitrary vector in a vector space $\mathbb{V}$ in terms of a basis $\mathcal{B}$ for $\mathbb{V}$.

Definition
Coordinate Vector

Suppose that $\mathcal{B} = \{\mathbf{v}_1, \dots, \mathbf{v}_n\}$ is a basis for the vector space $\mathbb{V}$. If $\mathbf{x} \in \mathbb{V}$ with $\mathbf{x} = x_1\mathbf{v}_1 + x_2\mathbf{v}_2 + \cdots + x_n\mathbf{v}_n$, then the **coordinate vector** of $\mathbf{x}$ with respect to the basis $\mathcal{B}$ is

$$[\mathbf{x}]_{\mathcal{B}} = \begin{bmatrix} x_1 \\ x_2 \\ \vdots \\ x_n \end{bmatrix}$$

Remarks

1. This definition makes sense because of the Unique Representation Theorem (Theorem 4.3.1).
2. Observe that the coordinate vector $[\mathbf{x}]_{\mathcal{B}}$ depends on the order in which the basis vectors appear. In this book, "basis" always means **ordered basis**; that is, it is always assumed that a basis is specified in the order in which the basis vectors are listed.
3. We often say "the coordinates of $\mathbf{x}$ with respect to $\mathcal{B}$" or "the $\mathcal{B}$-coordinates of $\mathbf{x}$" instead of the coordinate vector.

EXAMPLE 1

The set $\mathcal{B} = \left\{\begin{bmatrix}1\\0\end{bmatrix}, \begin{bmatrix}1\\1\end{bmatrix}\right\}$ is a basis for $\mathbb{R}^2$. Find the coordinates of $\vec{a} = \begin{bmatrix}3\\2\end{bmatrix}$ and $\vec{b} = \begin{bmatrix}1\\-2\end{bmatrix}$ with respect to the basis $\mathcal{B}$.

Solution: For $\vec{a}$, we must find a_1 and a_2 such that $a_1\begin{bmatrix}1\\0\end{bmatrix} + a_2\begin{bmatrix}1\\1\end{bmatrix} = \begin{bmatrix}3\\2\end{bmatrix}$. By inspection, we see that a solution is $a_1 = 1$ and $a_2 = 2$. Thus,

$$[\vec{a}]_{\mathcal{B}} = \begin{bmatrix}3\\2\end{bmatrix}_{\mathcal{B}} = \begin{bmatrix}1\\2\end{bmatrix}$$

Similarly, we see that a solution of $b_1\begin{bmatrix}1\\0\end{bmatrix} + b_2\begin{bmatrix}1\\1\end{bmatrix} = \begin{bmatrix}1\\-2\end{bmatrix}$ is $b_1 = 3$, $b_2 = -2$. Hence

$$[\vec{b}]_{\mathcal{B}} = \begin{bmatrix}1\\-2\end{bmatrix}_{\mathcal{B}} = \begin{bmatrix}3\\-2\end{bmatrix}$$

Figure 4.4.3 shows $\mathbb{R}^2$ with this basis and $\vec{a}$. Notice that the use of the basis $\mathcal{B}$ means that the space is covered with two families of parallel coordinate lines, one with direction vector $\begin{bmatrix}1\\0\end{bmatrix}$ and the other with direction vector $\begin{bmatrix}1\\1\end{bmatrix}$. Coordinates are established relative to these two families. The axes of this new coordinate system are obviously not orthogonal to each other. Such non-orthogonal coordinate systems arise naturally in the study of some crystalline structures in material science.

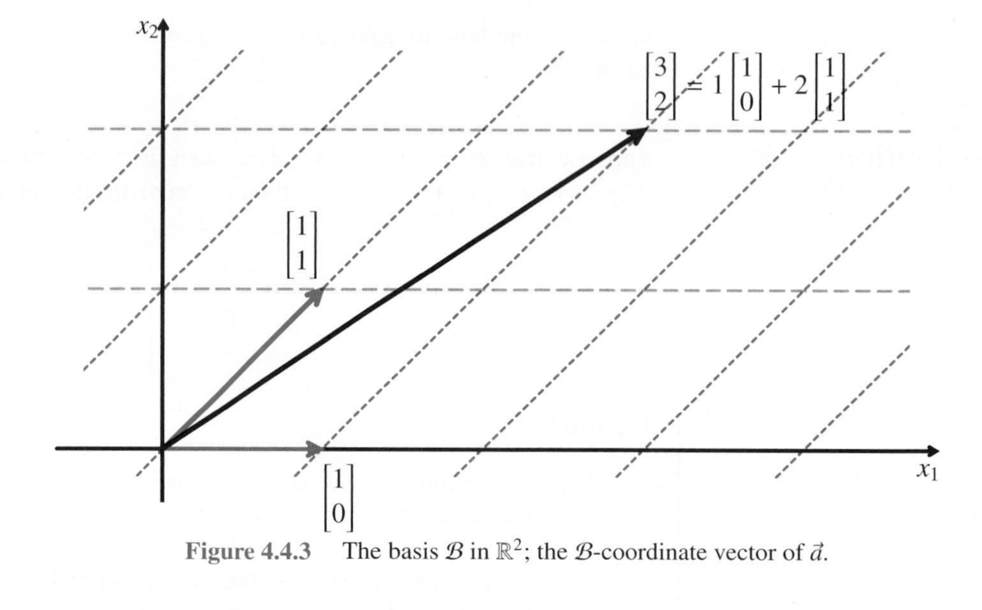

Figure 4.4.3 The basis $\mathcal{B}$ in $\mathbb{R}^2$; the $\mathcal{B}$-coordinate vector of $\vec{a}$.

EXAMPLE 2

The set $\mathcal{B} = \left\{\begin{bmatrix}3 & 2\\2 & 2\end{bmatrix}, \begin{bmatrix}1 & 0\\1 & 1\end{bmatrix}, \begin{bmatrix}1 & 1\\1 & 0\end{bmatrix}\right\}$ is a basis for the subspace Span $\mathcal{B}$. Determine whether the matrices $A = \begin{bmatrix}1 & -1\\0 & 3\end{bmatrix}$ and $B = \begin{bmatrix}1 & 4\\0 & 3\end{bmatrix}$ are in Span $\mathcal{B}$. If they are, determine their $\mathcal{B}$-coordinate vector.

Solution: We are required to determine whether there are numbers u_1, u_2, u_3 and/or v_1, v_2, v_3, such that

$$u_1\begin{bmatrix}3 & 2\\2 & 2\end{bmatrix} + u_2\begin{bmatrix}1 & 0\\1 & 1\end{bmatrix} + u_3\begin{bmatrix}1 & 1\\1 & 0\end{bmatrix} = \begin{bmatrix}1 & -1\\0 & 3\end{bmatrix}$$
$$v_1\begin{bmatrix}3 & 2\\2 & 2\end{bmatrix} + v_2\begin{bmatrix}1 & 0\\1 & 1\end{bmatrix} + v_3\begin{bmatrix}1 & 1\\1 & 0\end{bmatrix} = \begin{bmatrix}1 & 4\\0 & 3\end{bmatrix}$$

Since the two systems have the same coefficient matrix, augment the coefficient matrix twice by adjoining both right-hand sides and row reduce

$$\left[\begin{array}{ccc|cc}3 & 1 & 1 & 1 & 1\\2 & 0 & 1 & -1 & 4\\2 & 1 & 1 & 0 & 0\\2 & 1 & 0 & 3 & 3\end{array}\right] \sim \left[\begin{array}{ccc|cc}1 & 0 & 0 & 1 & 6\\0 & 1 & 0 & 1 & -9\\0 & 0 & 1 & -3 & -8\\0 & 0 & 0 & 0 & 5\end{array}\right]$$

It is now clear that the system with the second matrix B is not consistent, so that B is not in the subspace spanned by $\mathcal{B}$. On the other hand, we see that the system with the first matrix A has the unique solution $u_1 = 1$, $u_2 = 1$, and $u_3 = -3$. Thus,

$$[A]_{\mathcal{B}} = \begin{bmatrix}1\\1\\-3\end{bmatrix}$$

Note that there are only three $\mathcal{B}$-coordinates because the basis $\mathcal{B}$ has only three vectors. Also note that there is an immediate check available as we can verify that

EXAMPLE 2
(continued)

$$\begin{bmatrix} 1 & -1 \\ 0 & 3 \end{bmatrix} = 1\begin{bmatrix} 3 & 2 \\ 2 & 2 \end{bmatrix} + 1\begin{bmatrix} 1 & 0 \\ 1 & 1 \end{bmatrix} - 3\begin{bmatrix} 1 & 1 \\ 1 & 0 \end{bmatrix}$$

EXERCISE 1

Determine the coordinate vector of $\begin{bmatrix} 0 \\ 2 \\ 2 \end{bmatrix}$ with respect to the basis $\mathcal{B} = \left\{ \begin{bmatrix} 2 \\ 1 \\ -1 \end{bmatrix}, \begin{bmatrix} -3 \\ -1 \\ 2 \end{bmatrix} \right\}$ of Span $\mathcal{B}$.

EXAMPLE 3

Suppose that you have written a computer program to perform certain operations with polynomials in P_2. You need to include some method of inputting the polynomial you are considering. If you use the standard basis $\{1, x, x^2\}$ for P_2, to input the polynomial $3 - 5x + 2x^2$, you would surely write your program in such a way that you would type "3, −5, 2", the standard coordinate vector, as the input.

On the other hand, for some problems in differential equations, you might prefer the basis $\mathcal{B} = \{1 - x^2, x, 1 + x^2\}$. To find the $\mathcal{B}$-coordinates of $3 - 5x + 2x^2$, we must find t_1, t_2 and t_3 such that

$$3 - 5x + 2x^2 = t_1(1 - x^2) + t_2x + t_3(1 + x^2) = (t_1 + t_3) + t_2x + (t_3 - t_1)x^2$$

Row reducing the corresponding augmented matrix gives

$$\left[\begin{array}{ccc|c} 1 & 0 & 1 & 3 \\ 0 & 1 & 0 & -5 \\ -1 & 0 & 1 & 2 \end{array}\right] \sim \left[\begin{array}{ccc|c} 1 & 0 & 0 & 1/2 \\ 0 & 1 & 0 & -5 \\ 0 & 0 & 1 & 5/2 \end{array}\right]$$

It follows that the coordinates of $3 - 5x + 2x^2$ with respect to $\mathcal{B}$ are

$$[3 - 5x + 2x^2]_{\mathcal{B}} = \begin{bmatrix} 1/2 \\ -5 \\ 5/2 \end{bmatrix}$$

Thus, if your computer program is written to work in the basis $\mathcal{B}$, then you would input "0.5, −5, 2.5".

We might need to input several polynomials into the computer program written to work in the basis $\mathcal{B}$. In this case, we would want a much faster way of converting standard coordinates to coordinates with respect to the basis $\mathcal{B}$. We now develop a method for doing this.

Theorem 1

Let $\mathcal{B}$ be a basis for a finite dimensional vector space $\mathbb{V}$. Then, for any $\mathbf{x}, \mathbf{y} \in \mathbb{V}$ and $t \in \mathbb{R}$ we have

$$[t\mathbf{x} + \mathbf{y}]_{\mathcal{B}} = t[\mathbf{x}]_{\mathcal{B}} + [\mathbf{y}]_{\mathcal{B}}$$

Proof: Let $\mathcal{B} = \{\mathbf{v}_1, \dots, \mathbf{v}_n\}$. Then, for any vectors $\mathbf{x} = x_1\mathbf{v}_1 + \cdots + x_n\mathbf{v}_n$ and $\mathbf{y} = y_1\mathbf{v}_1 + \cdots + y_n\mathbf{v}_n$ in $\mathbb{V}$ and any $t \in \mathbb{R}$, we have $t\mathbf{x}+\mathbf{y} = (tx_1+y_1)\mathbf{v}_1 + \cdots + (tx_n+y_n)\mathbf{v}_n$, so

$$[t\mathbf{x} + \mathbf{y}]_{\mathcal{B}} = \begin{bmatrix} tx_1 + y_1 \\ \vdots \\ tx_n + y_n \end{bmatrix} = t\begin{bmatrix} x_1 \\ \vdots \\ x_n \end{bmatrix} + \begin{bmatrix} y_1 \\ \vdots \\ y_n \end{bmatrix} = t[\mathbf{x}]_{\mathcal{B}} + [\mathbf{y}]_{\mathcal{B}}$$

■

Let $\mathcal{B}$ be a basis for an n-dimensional vector space $\mathbb{V}$ and let $C = \{\mathbf{w}_1, \dots, \mathbf{w}_n\}$ be another basis for $\mathbb{V}$. Consider $\mathbf{x} \in \mathbb{V}$. Writing $\mathbf{x}$ as a linear combination of the vectors in C gives

$$\mathbf{x} = x_1\mathbf{w}_1 + \cdots + x_n\mathbf{w}_n$$

Taking $\mathcal{B}$-coordinates gives

$$[\mathbf{x}]_B = [x_1\mathbf{w}_1 + \cdots + x_n\mathbf{w}_n]_{\mathcal{B}} = x_1[\mathbf{w}_1]_{\mathcal{B}} + \cdots + x_n[\mathbf{w}_n]_{\mathcal{B}} = \begin{bmatrix} [\mathbf{w}_1]_{\mathcal{B}} \cdots [\mathbf{w}_n]_{\mathcal{B}} \end{bmatrix} \begin{bmatrix} x_1 \\ \vdots \\ x_n \end{bmatrix}$$

Since $\begin{bmatrix} x_1 \\ \vdots \\ x_n \end{bmatrix} = [\mathbf{x}]_C$, we see that this equation gives a formula for calculating the $\mathcal{B}$-coordinates of $\mathbf{x}$ from the C-coordinates of $\mathbf{x}$ using simple matrix-vector multiplication. We call this equation the **change of coordinates equation** and make the following definition.

Definition

Change of Coordinates Matrix

Let $\mathcal{B}$ and $C = \{\mathbf{w}_1, \dots, \mathbf{w}_n\}$ both be bases for a vector space $\mathbb{V}$. The matrix $P = \begin{bmatrix} [\mathbf{w}_1]_{\mathcal{B}} & \cdots & [\mathbf{w}_n]_{\mathcal{B}} \end{bmatrix}$ is called the **change of coordinates matrix** from C-coordinates to $\mathcal{B}$-coordinates and satisfies

$$[\mathbf{x}]_{\mathcal{B}} = P[\mathbf{x}]_C$$

Of course, we could exchange the roles of $\mathcal{B}$ and C to find the change of coordinates matrix Q from $\mathcal{B}$-coordinates to C-coordinates.

Theorem 2

Let $\mathcal{B}$ and C both be bases for a finite-dimensional vector space $\mathbb{V}$. Let P be the change of coordinates matrix from C-coordinates to $\mathcal{B}$-coordinates. Then, P is invertible and P^{-1} is the change of coordinates matrix from $\mathcal{B}$-coordinates to C-coordinates.

The proof is left to Problem D4.

EXAMPLE 4

Let $\mathcal{S} = \{\vec{e}_1, \vec{e}_2, \vec{e}_3\}$ be the standard basis for $\mathbb{R}^3$ and let $\mathcal{B} = \left\{ \begin{bmatrix} 1 \\ 3 \\ -1 \end{bmatrix}, \begin{bmatrix} 2 \\ 1 \\ 1 \end{bmatrix}, \begin{bmatrix} 3 \\ 4 \\ 1 \end{bmatrix} \right\}$. Find the change of coordinates matrix Q from $\mathcal{B}$-coordinates to $\mathcal{S}$-coordinates. Find the change of coordinates matrix P from $\mathcal{S}$-coordinates to $\mathcal{B}$-coordinates. Verify that $PQ = I$.

EXAMPLE 4
(continued)

Solution: To find the change of coordinates matrix Q, we need to find the coordinates of the vectors in $\mathcal{B}$ with respect to the standard basis $\mathcal{S}$. We get

$$Q = \begin{bmatrix} \begin{bmatrix} 1 \\ 3 \\ -1 \end{bmatrix}_{\mathcal{S}} & \begin{bmatrix} 2 \\ 1 \\ 1 \end{bmatrix}_{\mathcal{S}} & \begin{bmatrix} 3 \\ 4 \\ 1 \end{bmatrix}_{\mathcal{S}} \end{bmatrix} = \begin{bmatrix} 1 & 2 & 3 \\ 3 & 1 & 4 \\ -1 & 1 & 1 \end{bmatrix}$$

To find the change of coordinates matrix P, we need to find the coordinates of the standard basis vectors with respect to the basis $\mathcal{B}$. To do this, we solve the augmented systems

$$\left[\begin{array}{ccc|c} 1 & 2 & 3 & 1 \\ 3 & 1 & 4 & 0 \\ -1 & 1 & 1 & 0 \end{array}\right], \quad \left[\begin{array}{ccc|c} 1 & 2 & 3 & 0 \\ 3 & 1 & 4 & 1 \\ -1 & 1 & 1 & 0 \end{array}\right], \quad \left[\begin{array}{ccc|c} 1 & 2 & 3 & 0 \\ 3 & 1 & 4 & 0 \\ -1 & 1 & 1 & 1 \end{array}\right]$$

To make this easier, we row reduce the triple-augmented matrix for the system:

$$\left[\begin{array}{ccc|ccc} 1 & 2 & 3 & 1 & 0 & 0 \\ 3 & 1 & 4 & 0 & 1 & 0 \\ -1 & 1 & 1 & 0 & 0 & 1 \end{array}\right] \sim \left[\begin{array}{ccc|ccc} 1 & 0 & 0 & 3/5 & -1/5 & -1 \\ 0 & 1 & 0 & 7/5 & -4/5 & -1 \\ 0 & 0 & 1 & -4/5 & 3/5 & 1 \end{array}\right]$$

Thus,

$$P = \begin{bmatrix} 3/5 & -1/5 & -1 \\ 7/5 & -4/5 & -1 \\ -4/5 & 3/5 & 1 \end{bmatrix}$$

We now see that

$$\begin{bmatrix} 3/5 & -1/5 & -1 \\ 7/5 & -4/5 & -1 \\ -4/5 & 3/5 & 1 \end{bmatrix} \begin{bmatrix} 1 & 2 & 3 \\ 3 & 1 & 4 \\ -1 & 1 & 1 \end{bmatrix} = \begin{bmatrix} 1 & 0 & 0 \\ 0 & 1 & 0 \\ 0 & 0 & 1 \end{bmatrix}$$

EXAMPLE 5

Let $\mathcal{B} = \{1 - x^2, x, 1 + x^2\}$. Find $[a + bx + cx^2]_{\mathcal{B}}$.

Solution: If we find the change of coordinates matrix P from the standard basis $\mathcal{S} = \{1, x, x^2\}$ of P_2 to $\mathcal{B}$, then we will have

$$[a + bx + cx^2]_{\mathcal{B}} = P[a + bx + cx^2]_{\mathcal{S}} = P \begin{bmatrix} a \\ b \\ c \end{bmatrix}$$

Hence, we just need to find the coordinates of the standard basis vectors with respect to the basis $\mathcal{B}$. To do this, we solve the three systems of linear equations given by

$$t_{11}(1 - x^2) + t_{12}(x) + t_{13}(1 + x^2) = 1$$
$$t_{21}(1 - x^2) + t_{22}(x) + t_{23}(1 + x^2) = x$$
$$t_{31}(1 - x^2) + t_{32}(x) + t_{33}(1 + x^2) = x^2$$

EXAMPLE 5
(continued)

We row reduce the triple-augmented matrix to get

$$\left[\begin{array}{ccc|ccc} 1 & 0 & 1 & 1 & 0 & 0 \\ 0 & 1 & 0 & 0 & 1 & 0 \\ -1 & 0 & 1 & 0 & 0 & 1 \end{array}\right] \sim \left[\begin{array}{ccc|ccc} 1 & 0 & 0 & 1/2 & 0 & -1/2 \\ 0 & 1 & 0 & 0 & 1 & 0 \\ 0 & 0 & 1 & 1/2 & 0 & 1/2 \end{array}\right]$$

Hence,

$$[a + bx + cx^2]_{\mathcal{B}} = \begin{bmatrix} 1/2 & 0 & -1/2 \\ 0 & 1 & 0 \\ 1/2 & 0 & 1/2 \end{bmatrix} \begin{bmatrix} a \\ b \\ c \end{bmatrix} = \begin{bmatrix} \frac{1}{2}a - \frac{1}{2}c \\ b \\ \frac{1}{2}a + \frac{1}{2}c \end{bmatrix}$$

EXERCISE 2

Let $\mathcal{S} = \{\vec{e}_1, \vec{e}_2, \vec{e}_3\}$ be the standard basis for $\mathbb{R}^3$ and let $\mathcal{B} = \left\{ \begin{bmatrix} 1 \\ 1 \\ 2 \end{bmatrix}, \begin{bmatrix} 1 \\ 2 \\ 4 \end{bmatrix}, \begin{bmatrix} 2 \\ 2 \\ 3 \end{bmatrix} \right\}$. Find the change of coordinates matrix Q from $\mathcal{B}$-coordinates to $\mathcal{S}$-coordinates. Find the change of coordinates matrix P from $\mathcal{S}$-coordinates to $\mathcal{B}$-coordinates. Verify that $PQ = I$.

PROBLEMS 4.4

Practice Problems

A1 In each case, check that the given vectors in $\mathcal{B}$ are linearly independent (and therefore form a basis for the subspace they span). Then determine the coordinates of $\mathbf{x}$ and $\mathbf{y}$ with respect to $\mathcal{B}$.

(a) $\mathcal{B} = \left\{ \begin{bmatrix} 1 \\ 0 \\ 1 \\ 1 \end{bmatrix}, \begin{bmatrix} -1 \\ 1 \\ -1 \\ 0 \end{bmatrix} \right\}, \mathbf{x} = \begin{bmatrix} 5 \\ -2 \\ 5 \\ 3 \end{bmatrix}, \mathbf{y} = \begin{bmatrix} -1 \\ 3 \\ -1 \\ 2 \end{bmatrix}$

(b) $\mathcal{B} = \{1 + x + x^2, 1 + 3x + 2x^2, 4 + x^2\}$, $\mathbf{x} = -2 + 8x + 5x^2$, $\mathbf{y} = -4 + 8x + 4x^2$

(c) $\mathcal{B} = \left\{ \begin{bmatrix} 1 & 1 \\ 1 & 0 \end{bmatrix}, \begin{bmatrix} 0 & 1 \\ 1 & 1 \end{bmatrix}, \begin{bmatrix} 2 & 0 \\ 0 & -1 \end{bmatrix} \right\}, \mathbf{x} = \begin{bmatrix} 0 & 1 \\ 1 & 2 \end{bmatrix}$, $\mathbf{y} = \begin{bmatrix} -4 & 1 \\ 1 & 4 \end{bmatrix}$

(d) $\mathcal{B} = \left\{ \begin{bmatrix} 1 & 1 & 0 \\ 0 & 1 & 1 \end{bmatrix}, \begin{bmatrix} 0 & 2 & -1 \\ 1 & 3 & -1 \end{bmatrix} \right\}$, $\mathbf{x} = \begin{bmatrix} 1 & 3 & -1 \\ 1 & 4 & 0 \end{bmatrix}, \mathbf{y} = \begin{bmatrix} 3 & -1 & 2 \\ -2 & -3 & 5 \end{bmatrix}$

(e) $\mathcal{B} = \{1 + x^2 + x^4, 1 + x + 2x^2 + x^3 + x^4, x - x^2 + x^3 - 2x^4\}$, $\mathbf{x} = 2 + x - 5x^2 + x^3 - 6x^4$, $\mathbf{y} = 1 + x + 4x^2 + x^3 + 3x^4$

A2 (a) Verify that $\mathcal{B} = \left\{ \begin{bmatrix} 1 \\ 2 \\ 0 \end{bmatrix}, \begin{bmatrix} 0 \\ 2 \\ -1 \end{bmatrix} \right\}$ is a basis for the plane $2x_1 - x_2 - 2x_3 = 0$.

(b) For each of the following vectors, determine whether it lies in the plane of part (a). If it does, find the vector's $\mathcal{B}$-coordinates.

(i) $\begin{bmatrix} 3 \\ 2 \\ 1 \end{bmatrix}$ (ii) $\begin{bmatrix} 3 \\ 2 \\ 2 \end{bmatrix}$ (iii) $\begin{bmatrix} 5 \\ 2 \\ 3 \end{bmatrix}$

A3 (a) Verify that $\mathcal{B} = \{1 + x^2, 1 - x + 2x^2, -1 - x + x^2\}$ is a basis for P_2.

(b) Determine the coordinates relative to $\mathcal{B}$ of the following polynomials.

(i) $p(x) = 1$

(ii) $q(x) = 4 - 2x + 7x^2$

(iii) $r(x) = -2 - 2x + 3x^2$

(c) Determine $[2 - 4x + 10x^2]_{\mathcal{B}}$ and use your answers to part (b) to check that

$$[4 - 2x + 7x^2]_{\mathcal{B}} + [-2 - 2x + 3x^2]_{\mathcal{B}} = [(4 - 2) + (-2 - 2)x + (7 + 3)x^2]_{\mathcal{B}}$$

A4 In each case, do the following.

(i) Determine whether the given matrix A belongs to Span $\mathcal{B}$.

(ii) Use your row reduction from (i) to explain whether $\mathcal{B}$ is a basis for Span $\mathcal{B}$.

(iii) If $\mathcal{B}$ is a basis for Span $\mathcal{B}$ and $A \in$ Span $\mathcal{B}$, determine $[A]_{\mathcal{B}}$.

(a) $\mathcal{B} = \left\{\begin{bmatrix} 1 & 2 \\ 1 & 3 \end{bmatrix}, \begin{bmatrix} 2 & 1 \\ -1 & 2 \end{bmatrix}, \begin{bmatrix} -2 & 2 \\ 4 & 10 \end{bmatrix}\right\}$, $A = \begin{bmatrix} -1 & 1 \\ 2 & 7 \end{bmatrix}$

(b) $\mathcal{B} = \left\{\begin{bmatrix} 1 & 3 \\ 2 & 3 \end{bmatrix}, \begin{bmatrix} 1 & -2 \\ 1 & 2 \end{bmatrix}, \begin{bmatrix} 0 & 1 \\ -1 & 1 \end{bmatrix}\right\}$, $A = \begin{bmatrix} 4 & -1 \\ 9 & 3 \end{bmatrix}$

A5 For the given basis $\mathcal{B}$, find the change of coordinates matrix to and from the standard basis of each vector space.

(a) $\mathcal{B} = \left\{\begin{bmatrix} 3 \\ 4 \\ 1 \end{bmatrix}, \begin{bmatrix} 0 \\ 1 \\ 0 \end{bmatrix}, \begin{bmatrix} -2 \\ -3 \\ 3 \end{bmatrix}\right\}$ for $\mathbb{R}^3$

(b) $\mathcal{B} = \{1 - 2x + 5x^2, 1 - 2x^2, x + x^2\}$ for P_2

(c) $\mathcal{B} = \left\{\begin{bmatrix} 1 & -1 \\ 0 & -1 \end{bmatrix}, \begin{bmatrix} 0 & -4 \\ 0 & -1 \end{bmatrix}, \begin{bmatrix} 2 & 1 \\ 0 & 1 \end{bmatrix}\right\}$ for the vector space of 2×2 upper-triangular matrices

Homework Problems

B1 In each case, check that the given vectors in $\mathcal{B}$ are linearly independent (and therefore form a basis for the subspace they span). Then determine the coordinates of $\mathbf{x}$ and $\mathbf{y}$ with respect to $\mathcal{B}$.

(a) $\mathcal{B} = \left\{\begin{bmatrix} 1 \\ 1 \\ 2 \\ 1 \end{bmatrix}, \begin{bmatrix} 2 \\ 1 \\ 1 \\ 1 \end{bmatrix}\right\}$, $\mathbf{x} = \begin{bmatrix} 5 \\ 1 \\ -2 \\ 1 \end{bmatrix}$, $\mathbf{y} = \begin{bmatrix} 0 \\ 1 \\ 3 \\ 1 \end{bmatrix}$

(b) $\mathcal{B} = \left\{\begin{bmatrix} 1 & 1 \\ 0 & -1 \end{bmatrix}, \begin{bmatrix} 1 & 0 \\ 1 & 1 \end{bmatrix}, \begin{bmatrix} 0 & 1 \\ 1 & 2 \end{bmatrix}\right\}$, $\mathbf{x} = \begin{bmatrix} 1 & -3 \\ 2 & 3 \end{bmatrix}$, $\mathbf{y} = \begin{bmatrix} -1 & 0 \\ 3 & 7 \end{bmatrix}$

(c) $\mathcal{B} = \{x + x^2, -x + 3x^2, 1 + x - x^2\}$, $\mathbf{x} = -1 + 3x - x^2$, $\mathbf{y} = 3 + 2x^2$

(d) $\mathcal{B} = \{1 + 2x + 2x^2, -3x - 3x^2, -3 - 3x\}$, $\mathbf{x} = 3 - 3x^2$, $\mathbf{y} = 1 + x^2$

(e) $\mathcal{B} = \{1 + x + x^3, 1 + 2x^2 + x^3 + x^4, x^2 + x^3 + 3x^4\}$, $\mathbf{x} = 2 - 2x + 5x^2 - x^3 - 5x^4$, $\mathbf{y} = -1 - 3x + 3x^2 - 2x^3 - x^4$

B2 (a) Verify that $\mathcal{B} = \left\{\begin{bmatrix} 3 \\ 2 \\ 0 \end{bmatrix}, \begin{bmatrix} 0 \\ 2 \\ 3 \end{bmatrix}\right\}$ is a basis for the plane $2x_1 - 3x_2 + 2x_3 = 0$.

(b) For each of the following vectors, determine whether it lies in the plane of part (a). If it does, find the vector's $\mathcal{B}$-coordinates.

(i) $\begin{bmatrix} 1 \\ 2 \\ 2 \end{bmatrix}$ (ii) $\begin{bmatrix} 4 \\ 2 \\ 2 \end{bmatrix}$ (iii) $\begin{bmatrix} 5 \\ 4 \\ 1 \end{bmatrix}$

B3 (a) Verify that $\mathcal{B} = \left\{\begin{bmatrix} 1 \\ 1 \\ -1 \end{bmatrix}, \begin{bmatrix} 1 \\ 0 \\ 2 \end{bmatrix}, \begin{bmatrix} 2 \\ 2 \\ 1 \end{bmatrix}\right\}$ is a basis for $\mathbb{R}^3$.

(b) Determine the coordinates relative to $\mathcal{B}$ of the following vectors.

(i) $\begin{bmatrix} 6 \\ 5 \\ 3 \end{bmatrix}$ (ii) $\begin{bmatrix} 4 \\ 5 \\ -4 \end{bmatrix}$ (iii) $\begin{bmatrix} 5 \\ 4 \\ 1 \end{bmatrix}$

(c) Determine $\begin{bmatrix} 9 \\ 9 \\ -3 \end{bmatrix}_{\mathcal{B}}$ and use your answers to part (b) to check that

$$\begin{bmatrix} 4 \\ 5 \\ -4 \end{bmatrix}_{\mathcal{B}} + \begin{bmatrix} 5 \\ 4 \\ 1 \end{bmatrix}_{\mathcal{B}} = \begin{bmatrix} 9 \\ 9 \\ -3 \end{bmatrix}_{\mathcal{B}}$$

B4 (a) Verify that $\mathcal{B} = \{1 + x^2, 1 + x, x + x^2\}$ is a basis for P_2.

(b) Determine the coordinates relative to $\mathcal{B}$ of the following polynomials.

(i) $p(x) = 3 + 4x + 5x^2$

(ii) $q(x) = 4 + 5x - 7x^2$

(iii) $r(x) = 1 + x + x^2$

(c) Determine $[4 + 5x + 6x^2]_{\mathcal{B}}$ and use your answers to part (b) to check that

$$\begin{aligned}&[3 + 4x + 5x^2]_{\mathcal{B}} + [1 + x + x^2]_{\mathcal{B}} \\ &= [(3 + 1) + (4 + 1)x + (5 + 1)x^2]_{\mathcal{B}}\end{aligned}$$

B5 In each case, do the following.

(i) Determine whether the given matrix A belongs to Span $\mathcal{B}$.

(ii) Use your row reduction from (i) to explain whether $\mathcal{B}$ is a basis for Span $\mathcal{B}$.

(iii) If $\mathcal{B}$ is a basis for Span $\mathcal{B}$ and $A \in$ Span $\mathcal{B}$, determine $[A]_{\mathcal{B}}$.

(a) $\mathcal{B} = \left\{ \begin{bmatrix} 1 & 1 \\ 3 & 2 \end{bmatrix}, \begin{bmatrix} -1 & 2 \\ 1 & -4 \end{bmatrix}, \begin{bmatrix} 0 & 3 \\ 4 & -2 \end{bmatrix} \right\}$,
$A = \begin{bmatrix} -1 & 5 \\ 13 & -9 \end{bmatrix}$

(b) $\mathcal{B} = \left\{ \begin{bmatrix} 1 & 1 \\ 1 & 0 \end{bmatrix}, \begin{bmatrix} 0 & 1 \\ 1 & 1 \end{bmatrix}, \begin{bmatrix} 1 & 0 \\ 1 & 1 \end{bmatrix} \right\}, A = \begin{bmatrix} 1 & 1 \\ -1 & -4 \end{bmatrix}$

B6 For the given basis $\mathcal{B}$, find the change of coordinates matrix to and from the standard basis of each vector space.

(a) $\mathcal{B} = \left\{ \begin{bmatrix} 1 \\ 1 \\ 0 \end{bmatrix}, \begin{bmatrix} 1 \\ 1 \\ -1 \end{bmatrix}, \begin{bmatrix} 5 \\ 1 \\ -3 \end{bmatrix} \right\}$ for $\mathbb{R}^3$

(b) $\mathcal{B} = \{-1 + 2x^2, 1 + x + x^2, 1 - x - 3x^2\}$ for P_2

(c) $\mathcal{B} = \left\{ \begin{bmatrix} 3 & 0 \\ 0 & -2 \end{bmatrix}, \begin{bmatrix} 5 & 0 \\ 0 & 3 \end{bmatrix} \right\}$ for the vector space of 2×2 diagonal matrices

Conceptual Problems

D1 Suppose that $\mathcal{B} = \{\mathbf{v}_1, \dots, \mathbf{v}_k\}$ is a basis for a vector space $\mathbb{V}$ and that $C = \{\mathbf{w}_1, \dots, \mathbf{w}_k\}$ is another basis for $\mathbb{V}$ and that for every $\mathbf{x} \in \mathbb{V}$, $[\mathbf{x}]_{\mathcal{B}} = [\mathbf{x}]_C$. Must it be true that $\mathbf{v}_i = \mathbf{w}_i$ for each $1 \leq i \leq k$? Explain or prove your conclusion.

D2 Suppose $\mathbb{V}$ is a vector space with basis $\mathcal{B} = \{\mathbf{v}_1, \mathbf{v}_2, \mathbf{v}_3, \mathbf{v}_4\}$. Then $C = \{\mathbf{v}_3, \mathbf{v}_2, \mathbf{v}_4, \mathbf{v}_1\}$ is also a basis of $\mathbb{V}$. Find a matrix P such that $P[\vec{x}]_{\mathcal{B}} = [\vec{x}]_C$.

D3 Let $B = \left\{ \begin{bmatrix} 2 \\ 3 \end{bmatrix}, \begin{bmatrix} 1 \\ 2 \end{bmatrix} \right\}$ and $C = \left\{ \begin{bmatrix} 2 \\ 1 \end{bmatrix}, \begin{bmatrix} 1 \\ 1 \end{bmatrix} \right\}$ and let $L : \mathbb{R}^2 \to \mathbb{R}^2$ be the linear mapping such that $[\vec{x}]_B = [L(\vec{x})]_C$.

(a) Find $L\left(\begin{bmatrix} 1 \\ 1 \end{bmatrix}\right)$.

(b) Find $L\left(\begin{bmatrix} x_1 \\ x_2 \end{bmatrix}\right)$.

D4 Prove Theorem 2.

4.5 General Linear Mappings

In Chapter 3 we looked at linear mappings $\mathrm{L}:\mathbb{R}^n\to\mathbb{R}^m$ *and found that they can be useful in solving some problems. Since vector spaces encompass the essential properties of* $\mathbb{R}^n$*, it makes sense that we can also define linear mappings whose domain and codomain are other vector spaces. This also turns out to be extremely useful and important in many real-world applications.*

Definition
Linear Mapping

If $\mathbb{V}$ and $\mathbb{W}$ are vector spaces over $\mathbb{R}$, a function $L:\mathbb{V}\to\mathbb{W}$ is a **linear mapping** if it satisfies the linearity properties

L1 $L(\mathbf{x}+\mathbf{y}) = L(\mathbf{x}) + L(\mathbf{y})$

L2 $L(t\mathbf{x}) = tL(\mathbf{x})$

for all $\mathbf{x},\mathbf{y}\in\mathbb{V}$ and $t\in\mathbb{R}$. If $\mathbb{W}=\mathbb{V}$, then L may be called a **linear operator**.

As before, the two properties can be combined into one statement:

$$L(t\mathbf{x}+\mathbf{y}) = tL(\mathbf{x}) + L(\mathbf{y}) \qquad \text{for all } \mathbf{x},\mathbf{y}\in\mathbb{V}, t\in\mathbb{R}$$

Moreover, we have that

$$L(t_1\mathbf{v}_1+\cdots+t_n\mathbf{v}_n) = t_1L(\mathbf{v}_1)+\cdots+t_nL(\mathbf{v}_n)$$

EXAMPLE 1

Let $L: M(2,2)\to P_2$ be defined by $L\left(\begin{bmatrix} a & b\\ c & d\end{bmatrix}\right) = d+(b+d)x+ax^2$. Prove that L is a linear mapping.

Solution: For any $\begin{bmatrix} a_1 & b_1\\ c_1 & d_1\end{bmatrix}, \begin{bmatrix} a_2 & b_2\\ c_2 & d_2\end{bmatrix} \in M(2,2)$ and $t\in\mathbb{R}$, we have

$$\begin{aligned}
L\left(t\begin{bmatrix} a_1 & b_1\\ c_1 & d_1\end{bmatrix} + \begin{bmatrix} a_2 & b_2\\ c_2 & d_2\end{bmatrix}\right) &= L\left(\begin{bmatrix} ta_1+a_2 & tb_1+b_2\\ tc_1+c_2 & td_1+d_2\end{bmatrix}\right)\\
&= (td_1+d_2) + (tb_1+b_2+td_1+d_2)x\\
&\quad + (ta_1+a_2)x^2\\
&= t(d_1+(b_1+d_1)x+a_1x^2)\\
&\quad + (d_2+(b_2+d_2)x+a_2x^2)\\
&= tL\left(\begin{bmatrix} a_1 & b_1\\ c_1 & d_1\end{bmatrix}\right) + L\left(\begin{bmatrix} a_2 & b_2\\ c_2 & d_2\end{bmatrix}\right)
\end{aligned}$$

So, L is linear.

EXAMPLE 2

Let $M: P_3\to P_3$ be defined by $M(a_0+a_1x+a_2x^2) = a_1+2a_2x$. Prove that M is a linear operator.

Solution: Let $p(x) = a_0+a_1x+a_2x^2$, $q(x) = b_0+b_1x+b_2x^2$, and $t\in\mathbb{R}$. Then,

$$\begin{aligned}
M(tp(x)+q(x)) &= M((ta_0+b_0)+(ta_1+b_1)x+(ta_2+b_2)x^2)\\
&= (ta_1+b_1)+2(ta_2+b_2)x\\
&= t(a_1+2a_2x)+(b_1+2b_2x)\\
&= tM(p(x))+M(q(x))
\end{aligned}$$

Hence, M is linear.

EXERCISE 1

Let $L : \mathbb{R}^3 \to M(2,2)$ be defined by $L\left(\begin{bmatrix} x_1 \\ x_2 \\ x_3 \end{bmatrix}\right) = \begin{bmatrix} x_1 & x_1 + x_2 + x_3 \\ 0 & x_2 \end{bmatrix}$. Prove that L is linear.

Remark

Since a linear mapping $L : \mathbb{V} \to \mathbb{W}$ is just a function, we define operations (addition, scalar multiplication, and composition) and the concept of invertibility on these more general linear mappings in the same way as we did for linear mappings $L : \mathbb{R}^n \to \mathbb{R}^m$.

As we did with linear mappings in Chapter 3, it is important to consider the range and nullspace of general linear mappings.

Definition

Range

Nullspace

Let $\mathbb{V}$ and $\mathbb{W}$ be vector spaces over $\mathbb{R}$. The **range** of a linear mapping $L : \mathbb{V} \to \mathbb{W}$ is defined to be the set

$$\text{Range}(L) = \{L(\mathbf{x}) \in \mathbb{W} \mid \mathbf{x} \in \mathbb{V}\}$$

The **nullspace** of L is the set of all vectors in $\mathbb{V}$ whose image under L is the zero vector $\mathbf{0}_{\mathbb{W}}$. We write

$$\text{Null}(L) = \{\mathbf{x} \in \mathbb{V} \mid L(\mathbf{x}) = \mathbf{0}_{\mathbb{W}}\}$$

EXAMPLE 3

Let $L : M(2,2) \to P_2$ be the linear mapping defined by $L\left(\begin{bmatrix} a & b \\ c & d \end{bmatrix}\right) = c + (b+d)x + ax^2$. Determine whether $1 + x + x^2$ is in the range of L, and if it is, determine a matrix A such that $L(A) = 1 + x + x^2$.

Solution: We want to find a, b, c and d such that

$$1 + x + x^2 = L\left(\begin{bmatrix} a & b \\ c & d \end{bmatrix}\right) = c + (b + d)x + ax^2$$

Hence, we must have $c = 1$, $b+d = 1$, and $a = 1$. Observe that we get a system of linear equations that is consistent with infinitely many solutions. Thus, $1 + x + x^2 \in \text{Range}(L)$ and one choice of A such that $L(A) = 1 + x + x^2$ is $A = \begin{bmatrix} 1 & 1 \\ 1 & 0 \end{bmatrix}$.

EXAMPLE 4

Let $L : P_2 \to \mathbb{R}^3$ be the linear mapping defined by $L(a + bx + cx^2) = \begin{bmatrix} a - b \\ b - c \\ c - a \end{bmatrix}$.

Determine whether $\begin{bmatrix} 1 \\ 1 \\ 1 \end{bmatrix}$ is in the range of L.

EXAMPLE 4 (continued)

Solution: We want to find a, b and c such that

$$\begin{bmatrix}1\\1\\1\end{bmatrix} = L(a + bx + cx^2) = \begin{bmatrix}a-b\\b-c\\c-a\end{bmatrix}$$

This gives us the system of linear equations $a - b = 1$, $b - c = 1$, and $c - a = 1$. Row reducing the corresponding augmented matrix gives

$$\left[\begin{array}{ccc|c}1 & -1 & 0 & 1\\0 & 1 & -1 & 1\\-1 & 0 & 1 & 1\end{array}\right] \sim \left[\begin{array}{ccc|c}1 & -1 & 0 & 1\\0 & 1 & -1 & 1\\0 & 0 & 0 & 3\end{array}\right]$$

Hence, the system is inconsistent, so $\begin{bmatrix}1\\1\\1\end{bmatrix}$ is not in the span of L.

Theorem 1

Let $\mathbb{V}$ and $\mathbb{W}$ be vector spaces and let $L : \mathbb{V} \to \mathbb{W}$ be a linear mapping. Then
(1) $L(\mathbf{0}_{\mathbb{V}}) = \mathbf{0}_{\mathbb{W}}$
(2) Null(L) is a subspace of $\mathbb{V}$
(3) Range(L) is a subspace of $\mathbb{W}$

The proof is left as Problem D1.

As with any other subspace, it is often useful to find a basis for the nullspace and range of a linear mapping.

EXAMPLE 5

Determine a basis for the range and nullspace of the linear mapping $L : P_1 \to \mathbb{R}^3$ defined by $L(a + bx) = \begin{bmatrix}a\\0\\a-2b\end{bmatrix}$.

Solution: If $a + bx \in \text{Null}(L)$, then we have $\begin{bmatrix}a\\0\\a-2b\end{bmatrix} = L(a + bx) = \begin{bmatrix}0\\0\\0\end{bmatrix}$. Hence, $a = 0$ and $a - 2b = 0$, which implies that $b = 0$. Thus, the only polynomial in the nullspace of L is the zero polynomial. That is, Null(L) = $\{0\}$, and so a basis for Null(L) is the empty set.

Any vector $\vec{y}$ in the range of L has the form

$$\vec{y} = \begin{bmatrix}a\\0\\a-2b\end{bmatrix} = a\begin{bmatrix}1\\0\\1\end{bmatrix} + b\begin{bmatrix}0\\0\\-2\end{bmatrix}$$

Thus, Range(L) = Span C, where $C = \left\{\begin{bmatrix}1\\0\\1\end{bmatrix}, \begin{bmatrix}0\\0\\-2\end{bmatrix}\right\}$. Moreover, C is clearly linearly independent. Hence C is a basis for the range of L.

EXAMPLE 6

Determine a basis for the range and nullspace of the linear mapping $L : M(2,2) \to P_2$ defined by $L\left(\begin{bmatrix} a & b \\ c & d \end{bmatrix}\right) = (b+c) + (c-d)x^2$.

Solution: If $\begin{bmatrix} a & b \\ c & d \end{bmatrix} \in \text{Null}(L)$, then $0 = L\left(\begin{bmatrix} a & b \\ c & d \end{bmatrix}\right) = (b+c) + (c-d)x^2$, so $b + c = 0$ and $c - d = 0$. Thus, $b = -c$ and $d = c$, so every matrix in the nullspace of L has the form

$$\begin{bmatrix} a & b \\ c & d \end{bmatrix} = \begin{bmatrix} a & -c \\ c & c \end{bmatrix} = a\begin{bmatrix} 1 & 0 \\ 0 & 0 \end{bmatrix} + c\begin{bmatrix} 0 & -1 \\ 1 & 1 \end{bmatrix}$$

Thus, $\mathcal{B} = \text{Span}\left\{\begin{bmatrix} 1 & 0 \\ 0 & 0 \end{bmatrix}, \begin{bmatrix} 0 & -1 \\ 1 & 1 \end{bmatrix}\right\}$ is a linearly independent spanning set for $\text{Null}(L)$ and hence is a basis for $\text{Null}(L)$.

Any polynomial in the range of L has the form $(b+c)+(c-d)x^2$. Hence $\text{Range}(L) = \text{Span}\{1, x^2\}$ since we can get any polynomial of the form $a_0 + a_2x^2$ by taking $b = a_0$, $c = 0$, and $d = -a_2$. Also, $\{1, x^2\}$ is clearly linearly independent and hence a basis for $\text{Range}(L)$.

EXERCISE 2

Determine a basis for the range and nullspace of the linear mapping $L : \mathbb{R}^3 \to M(2,2)$ defined by $L\left(\begin{bmatrix} x_1 \\ x_2 \\ x_3 \end{bmatrix}\right) = \begin{bmatrix} x_1 & x_2 + x_3 \\ x_2 + x_3 & x_1 \end{bmatrix}$.

Observe that in each of these examples, the dimension of the range of L plus the dimension of the nullspace of L equals the dimension of the domain of L. This result reminds us of the Rank Theorem (Theorem 3.4.8). Before we prove the similar result for general linear mappings, we make some definitions to make this look even more like the Rank Theorem.

Definition
Rank of a Linear Mapping

Let $\mathbb{V}$ and $\mathbb{W}$ be vector spaces over $\mathbb{R}$. The **rank of a linear mapping** $L : \mathbb{V} \to \mathbb{W}$ is the dimension of the range of L:

$$\text{rank}(L) = \dim(\text{Range}(L))$$

Definition
Nullity of a Linear Mapping

Let $\mathbb{V}$ and $\mathbb{W}$ be vector spaces over $\mathbb{R}$. The **nullity of a linear mapping** $L : \mathbb{V} \to \mathbb{W}$ is the dimension of the nullspace of L:

$$\text{nullity}(L) = \dim(\text{Null}(L))$$

Theorem 2 [Rank-Nullity Theorem]
Let $\mathbb{V}$ and $\mathbb{W}$ be vector spaces over $\mathbb{R}$ with $\dim \mathbb{V} = n$, and let $L : \mathbb{V} \to \mathbb{W}$ be a linear mapping. Then,

$$\text{rank}(L) + \text{nullity}(L) = n$$

Proof: The idea of the proof is to assume that a basis for the nullspace of L contains k vectors and show that we can then construct a basis for the range of L that contains $n - k$ vectors.

Let $\mathcal{B} = \{\mathbf{v}_1, \dots, \mathbf{v}_k\}$ be a basis for Null(L), so that nullity(L) $= k$. Since $\mathbb{V}$ is n-dimensional, we can use the procedure in Section 4.3. There exist vectors $\mathbf{u}_{k+1}, \dots, \mathbf{u}_n$ such that $\{\mathbf{v}_1, \dots, \mathbf{v}_k, \mathbf{u}_{k+1}, \dots, \mathbf{u}_n\}$ is a basis for $\mathbb{V}$.

Now consider any vector $\mathbf{w}$ in the range of L. Then $\mathbf{w} = L(\mathbf{x})$ for some $\mathbf{x} \in \mathbb{V}$. But any $\mathbf{x} \in \mathbb{V}$ can be written as a linear combination of the vectors in the basis $\{\mathbf{v}_1, \dots, \mathbf{v}_k, \mathbf{u}_{k+1}, \dots, \mathbf{u}_n\}$, so there exists $t_1, \dots, t_n$ such that $\mathbf{x} = t_1\mathbf{v}_1 + \cdots + t_k\mathbf{v}_k + t_{k+1}\mathbf{u}_{k+1} + \cdots + t_n\mathbf{u}_n$. Then,

$$\begin{aligned}\mathbf{w} &= L(t_1\mathbf{v}_1 + \cdots + t_k\mathbf{v}_k + t_{k+1}\mathbf{u}_{k+1} + \cdots + t_n\mathbf{u}_n) \\ &= t_1L(\mathbf{v}_1) + \cdots + t_kL(\mathbf{v}_k) + t_{k+1}L(\mathbf{u}_{k+1}) + \cdots + t_nL(\mathbf{u}_n)\end{aligned}$$

But each $\mathbf{v}_i$ is in the nullspace of L, so $L(\mathbf{v}_i) = \mathbf{0}$, and so we have

$$\mathbf{w} = t_{k+1}L(\mathbf{u}_{k+1}) + \cdots + t_nL(\mathbf{u}_n)$$

Therefore, any $\mathbf{w} \in$ Range(L) can be expressed as a linear combination of the vectors in the set $C = \{L(\mathbf{u}_{k+1}), \dots, L(\mathbf{u}_n)\}$. Thus, C is a spanning set for Range(L). Is it linearly independent? We consider

$$t_{k+1}L(\mathbf{u}_{k+1}) + \cdots + t_nL(\mathbf{u}_n) = \mathbf{0}_{\mathbb{W}}$$

By the linearity of L, this is equivalent to

$$L(t_{k+1}\mathbf{u}_{k+1} + \cdots + t_n\mathbf{u}_n) = \mathbf{0}_{\mathbb{W}}$$

If this is true, then $t_{k+1}\mathbf{u}_{k+1} + \cdots + t_n\mathbf{u}_n$ is a vector in the nullspace of L. Hence, for some $d_1, \dots, d_k$, we have

$$t_{k+1}\mathbf{u}_{k+1} + \cdots + t_n\mathbf{u}_n = d_1\mathbf{v}_1 + \cdots + d_k\mathbf{v}_k$$

But this is impossible unless all t_i and d_i are zero, because $\{\mathbf{v}_1, \dots, \mathbf{v}_k, \mathbf{u}_{k+1}, \dots, \mathbf{u}_n\}$ is a basis for $\mathbb{V}$ and hence linearly independent.

It follows that C is a linearly independent spanning set for Range(L). Hence it is a basis for Range(L) containing $n - k$ vectors. Thus, rank(L) $= n - k$ and

$$\text{rank}(L) + \text{nullity}(L) = (n - k) + k = n$$

as required. ■

EXAMPLE 7

Determine the rank and nullity of $L : M(2,2) \to P_3$ defined by

$$L\left(\begin{bmatrix} a & b \\ c & d \end{bmatrix}\right) = cx + (a+b)x^3$$

and verify the Rank-Nullity Theorem.

Solution: If $\begin{bmatrix} a & b \\ c & d \end{bmatrix} \in \text{Null}(L)$, then

$$0 = L\left(\begin{bmatrix} a & b \\ c & d \end{bmatrix}\right) = cx + (a+b)x^3$$

Hence, we have $a + b = 0$ and $c = 0$, and so every matrix in Null(L) has the form

$$\begin{bmatrix} a & -a \\ 0 & d \end{bmatrix} = a\begin{bmatrix} 1 & -1 \\ 0 & 0 \end{bmatrix} + d\begin{bmatrix} 0 & 0 \\ 0 & 1 \end{bmatrix}$$

Therefore, $\text{Null}(L) = \text{Span}\left\{\begin{bmatrix} 1 & -1 \\ 0 & 0 \end{bmatrix}, \begin{bmatrix} 0 & 0 \\ 0 & 1 \end{bmatrix}\right\}$. Moreover, we see that the set $\mathcal{B} = \left\{\begin{bmatrix} 1 & -1 \\ 0 & 0 \end{bmatrix}, \begin{bmatrix} 0 & 0 \\ 0 & 1 \end{bmatrix}\right\}$ is linearly independent, so $\mathcal{B}$ is a basis for Null(L) and hence nullity(L) = 2.

Clearly a basis for the range of L is $\{x, x^3\}$, since this set is linearly independent and spans the range. Thus rank(L) = 2.

Then, as predicted by the Rank-Nullity Theorem, we have

$$\text{rank}(L) + \text{nullity}(L) = 2 + 2 = 4 = \dim M(2,2)$$

PROBLEMS 4.5

Practice Problems

A1 Prove that the following mappings are linear.

(a) $L : \mathbb{R}^3 \to \mathbb{R}^2$ defined by $L(x_1, x_2, x_3) = (x_1 + x_2, x_1 + x_2 + x_3)$

(b) $L : \mathbb{R}^3 \to P_1$ defined by $L\left(\begin{bmatrix} a \\ b \\ c \end{bmatrix}\right) = (a+b) + (a+b+c)x$

(c) $\text{tr} : M(2,2) \to \mathbb{R}$ defined by $\text{tr}\left(\begin{bmatrix} a & b \\ c & d \end{bmatrix}\right) = a + d$ (Taking the trace of a matrix is a linear operation.)

(d) $T : P_3 \to M(2,2)$ defined by $T(a + bx + cx^2 + dx^3) = \begin{bmatrix} a & b \\ c & d \end{bmatrix}$

A2 Determine which of the following mappings are linear.

(a) $\det : M(2,2) \to \mathbb{R}$ defined by $\det\left(\begin{bmatrix} a & b \\ c & d \end{bmatrix}\right) = ad - bc$

(b) $L : P_2 \to P_2$ defined by $L(a + bx + cx^2) = (a - b) + (b + c)x^2$

(c) $T : \mathbb{R}^2 \to M(2,2)$ defined by $T\left(\begin{bmatrix} x_1 \\ x_2 \end{bmatrix}\right) = \begin{bmatrix} x_1 & 1 \\ 1 & x_2 \end{bmatrix}$

(d) $M : M(2,2) \to M(2,2)$ defined by $M\left(\begin{bmatrix} a & b \\ c & d \end{bmatrix}\right) = \begin{bmatrix} 0 & 0 \\ 0 & 0 \end{bmatrix}$

A3 For each of the following, determine whether the given vector $\mathbf{y}$ is in the range of the given linear mapping $L : \mathbb{V} \to \mathbb{W}$. If it is, find a vector $\mathbf{x} \in \mathbb{V}$ such that $L(\mathbf{x}) = \mathbf{y}$.

(a) $L : \mathbb{R}^3 \to \mathbb{R}^4$ defined by $L\left(\begin{bmatrix} x_1 \\ x_2 \\ x_3 \end{bmatrix}\right) = \begin{bmatrix} x_1 + x_3 \\ 0 \\ 0 \\ x_2 + x_3 \end{bmatrix}$, $\mathbf{y} = \begin{bmatrix} 2 \\ 0 \\ 0 \\ 3 \end{bmatrix}$

(b) $L : P_2 \to M(2,2)$ defined by $L(a + bx + cx^2) = \begin{bmatrix} a+c & 0 \\ 0 & b+c \end{bmatrix}$, $\mathbf{y} = \begin{bmatrix} 2 & 0 \\ 0 & 3 \end{bmatrix}$

(c) $L : P_2 \to P_1$ defined by $L(a + bx + cx^2) = (b + c) + (-b - c)x$, $\mathbf{y} = 1 + x$

(d) $L : \mathbb{R}^4 \to M(2,2)$ defined by $L\left(\begin{bmatrix} x_1 \\ x_2 \\ x_3 \\ x_4 \end{bmatrix}\right) = \begin{bmatrix} -2x_2 - 2x_3 - 2x_4 & x_1 + x_4 \\ -2x_1 - x_2 - x_4 & 2x_1 - 2x_2 - x_3 + 2x_4 \end{bmatrix}$, $\mathbf{y} = \begin{bmatrix} -1 & -1 \\ -2 & 2 \end{bmatrix}$

A4 Find a basis for the range and nullspace of the following linear mappings and verify the Rank-Nullity Theorem.

(a) $L : \mathbb{R}^3 \to \mathbb{R}^2$ defined by $L(x_1, x_2, x_3) = (x_1 + x_2, x_1 + x_2 + x_3)$

(b) $L : \mathbb{R}^3 \to P_1$ defined by $L\left(\begin{bmatrix} a \\ b \\ c \end{bmatrix}\right) = (a + b) + (a + b + c)x$

(c) $\operatorname{tr} : M(2,2) \to \mathbb{R}$ defined by $\operatorname{tr}\left(\begin{bmatrix} a & b \\ c & d \end{bmatrix}\right) = a + d$

(d) $T : P_3 \to M(2,2)$ defined by $T(a + bx + cx^2 + dx^3) = \begin{bmatrix} a & b \\ c & d \end{bmatrix}$

Homework Problems

B1 Prove that the following mappings are linear.

(a) $L : P_2 \to \mathbb{R}^3$ defined by $L(a + bx + cx^2) = \begin{bmatrix} a \\ b \\ c \end{bmatrix}$

(b) $L : P_2 \to P_1$ defined by $L\left(a + bx + cx^2\right) = b + 2cx$

(c) $L : M(2,2) \to \mathbb{R}$ defined by $L\left(\begin{bmatrix} a & b \\ c & d \end{bmatrix}\right) = 0$

(d) Let $\mathbb{D}$ be the subspace of $M(2,2)$ of diagonal matrices; $L : \mathbb{D} \to P_2$ defined by $L\left(\begin{bmatrix} a & 0 \\ 0 & b \end{bmatrix}\right) = a + (a + b)x + bx^2$

(e) $L : M(2,2) \to M(2,2)$ defined by $L\left(\begin{bmatrix} a & b \\ c & d \end{bmatrix}\right) = \begin{bmatrix} a-b & b-c \\ c-d & d-a \end{bmatrix}$

(f) $L : M(2,2) \to P_2$ defined by $L\left(\begin{bmatrix} a & b \\ c & d \end{bmatrix}\right) = (a + b + c + d)x^2$

(g) $T : M(2,2) \to M(2,2)$ defined by $T(A) = A^T$

B2 Determine which of the following mappings are linear.

(a) $L : P_3 \to \mathbb{R}$ defined by $L(a + bx + cx^2 + dx^3) = \left\|\begin{bmatrix} a \\ b \\ c \\ d \end{bmatrix}\right\|$

(b) $M : P_2 \to \mathbb{R}$ defined by $M(a + bx + cx^2) = b^2 - 4ac$

(c) $N : \mathbb{R}^3 \to M(2,2)$ defined by $N\left(\begin{bmatrix} x_1 \\ x_2 \\ x_3 \end{bmatrix}\right) = \begin{bmatrix} x_1 - x_3 & 0 \\ x_2 & x_2 \end{bmatrix}$

(d) $L : M(2,2) \to M(2,2)$ defined by $L(A) = \begin{bmatrix} 1 & 2 \\ 3 & 4 \end{bmatrix} A$

(e) $T : M(2,2) \to P_2$ defined by $T\left(\begin{bmatrix} a & b \\ c & d \end{bmatrix}\right) = a + (ab)x + (abc)x^2$

B3 For each of the following, determine whether the given vector $\mathbf{y}$ is in the range of the given linear mapping $L : \mathbb{V} \to \mathbb{W}$. If it is, find a vector $\mathbf{x} \in \mathbb{V}$ such that $L(\mathbf{x}) = \mathbf{y}$.

(a) $L : \mathbb{R}^3 \to \mathbb{R}^3$ defined by $L\left(\begin{bmatrix} x_1 \\ x_2 \\ x_3 \end{bmatrix}\right) = \begin{bmatrix} -x_1 - 2x_2 \\ 2x_1 + x_3 \\ -2x_1 + x_2 - 2x_3 \end{bmatrix}$, $\mathbf{y} = \begin{bmatrix} 1 \\ 1 \\ -1 \end{bmatrix}$

(b) $L : P_2 \to P_2$ defined by $L(a + bx + cx^2) = (-a - 2b) + (2a + c)x + (-2a + b - 2c)x^2$, $\mathbf{y} = 1 + x - x^2$

(c) $L : P_2 \to \mathbb{R}^3$ defined by $L(a + bx + cx^2) = \begin{bmatrix} a \\ b \\ a + b + c \end{bmatrix}$, $\mathbf{y} = \begin{bmatrix} 2 \\ -1 \\ 3 \end{bmatrix}$

(d) $L : \mathbb{R}^2 \to M(2,2)$ defined by $L\left(\begin{bmatrix} x_1 \\ x_2 \end{bmatrix}\right) = \begin{bmatrix} x_1 & x_2 \\ 0 & x_2 \end{bmatrix}$, $\mathbf{y} = \begin{bmatrix} 1 & 1 \\ 0 & 2 \end{bmatrix}$

(e) Let $\mathbb{T}$ denote the subspace of 2×2 upper-triangular matrices; $L : \mathbb{T} \to P_2$ defined by $L\left(\begin{bmatrix} a & b \\ 0 & c \end{bmatrix}\right) = (-a - c) + (a - 2b)x + (-2a + b + c)x^2$, $\mathbf{y} = 2 + x - x^2$

(f) $L : P_2 \to M(2,2)$ defined by $L(a + bx + cx^2) = \begin{bmatrix} -a - 2c & 2b - c \\ -2a + 2c & -2b - c \end{bmatrix}$, $\mathbf{y} = \begin{bmatrix} -2 & 2 \\ 0 & -2 \end{bmatrix}$

B4 Find a basis for the range and nullspace of the following linear mappings and verify the Rank-Nullity Theorem.

(a) $L : P_2 \to \mathbb{R}^3$ defined by $L(a + bx + cx^2) = \begin{bmatrix} a \\ b \\ c \end{bmatrix}$

(b) $L : P_2 \to P_1$ defined by $L\left(a + bx + cx^2\right) = b + 2cx$

(c) $L : M(2,2) \to \mathbb{R}$ defined by $L\left(\begin{bmatrix} a & b \\ c & d \end{bmatrix}\right) = 0$

(d) Let $\mathbb{D}$ be the subspace of $M(2,2)$ of diagonal matrices; $L : \mathbb{D} \to P_2$ defined by $L\left(\begin{bmatrix} a & 0 \\ 0 & b \end{bmatrix}\right) = a + (a + b)x + bx^2$

(e) $L : M(2,2) \to M(2,2)$ defined by $L\left(\begin{bmatrix} a & b \\ c & d \end{bmatrix}\right) = \begin{bmatrix} a - b & b - c \\ c - d & d - a \end{bmatrix}$

(f) $L : M(2,2) \to P_2$ defined by $L\left(\begin{bmatrix} a & b \\ c & d \end{bmatrix}\right) = (a + b + c + d)x^2$

(g) $T : M(2,2) \to M(2,2)$ defined by $T(A) = A^T$

(h) $L : P_2 \to M(2,2)$ defined by $L(a + bx + cx^2) = \begin{bmatrix} -a - 2c & 2b - c \\ -2a + 2c & -2b - c \end{bmatrix}$

Conceptual Problems

D1 Prove Theorem 1.

D2 For the given vector spaces $\mathbb{V}$ and $\mathbb{W}$, invent a linear mapping $L : \mathbb{V} \to \mathbb{W}$ that satisfies the given properties.

(a) $\mathbb{V} = \mathbb{R}^3$, $\mathbb{W} = P_2$; $L\left(\begin{bmatrix} 1 \\ 0 \\ 0 \end{bmatrix}\right) = x^2$, $L\left(\begin{bmatrix} 0 \\ 1 \\ 0 \end{bmatrix}\right) = 2x$, $L\left(\begin{bmatrix} 0 \\ 0 \\ 1 \end{bmatrix}\right) = 1 + x + x^2$

(b) $\mathbb{V} = P_2$, $\mathbb{W} = M(2,2)$; $\text{Null}(L) = \{0\}$ and $\text{Range}(L) = \text{Span}\left\{\begin{bmatrix} 1 & 0 \\ 0 & 0 \end{bmatrix}, \begin{bmatrix} 0 & 1 \\ 0 & 0 \end{bmatrix}, \begin{bmatrix} 0 & 0 \\ 0 & 1 \end{bmatrix}\right\}$

(c) $\mathbb{V} = M(2,2)$, $\mathbb{W} = \mathbb{R}^4$; $\text{nullity}(L) = 2$, $\text{rank}(L) = 2$, and $L\left(\begin{bmatrix} 1 & 0 \\ 0 & 1 \end{bmatrix}\right) = \begin{bmatrix} 0 \\ 1 \\ 1 \\ 0 \end{bmatrix}$

D3 (a) Let $\mathbb{V}$ and $\mathbb{W}$ be vector spaces and $L : \mathbb{V} \to \mathbb{W}$ be a linear mapping. Prove that if $\{L(\mathbf{v}_1), \dots, L(\mathbf{v}_k)\}$ is a linearly independent set in $\mathbb{W}$, then $\{\mathbf{v}_1, \dots, \mathbf{v}_k\}$ is a linearly independent set in $\mathbb{V}$.

(b) Give an example of a linear mapping $L : \mathbb{V} \to \mathbb{W}$, where $\{\mathbf{v}_1, \dots, \mathbf{v}_k\}$ is linearly independent in $\mathbb{V}$ but $\{L(\mathbf{v}_1), \dots, L(\mathbf{v}_k)\}$ is linearly dependent in $\mathbb{W}$.

D4 Let $\mathbb{V}$ and $\mathbb{W}$ be n-dimensional vector spaces and let $L : \mathbb{V} \to \mathbb{W}$ be a linear mapping. Prove that $\text{Range}(L) = \mathbb{W}$ if and only if $\text{Null}(L) = \{\mathbf{0}\}$.

D5 Let $\mathbb{U}$, $\mathbb{V}$ and $\mathbb{W}$ be finite-dimensional vector spaces over $\mathbb{R}$ and let $L : \mathbb{V} \to \mathbb{U}$ and $M : \mathbb{U} \to \mathbb{W}$ be linear mappings.

(a) Prove that $\text{rank}(M \circ L) \leq \text{rank}(M)$.

(b) Prove that $\text{rank}(M \circ L) \leq \text{rank}(L)$.

(c) Construct an example such that the rank of the composition is strictly less than the maximum of the ranks.

D6 Let $\mathbb{U}$ and $\mathbb{V}$ be finite-dimensional vector spaces and let $L : \mathbb{V} \to \mathbb{U}$ be a linear mapping and $M : \mathbb{U} \to \mathbb{U}$ be a linear operator such that $\text{Null}(M) = \{\mathbf{0}_{\mathbb{U}}\}$. Prove that $\text{rank}(M \circ L) = \text{rank}(L)$.

D7 Let S denote the set of all infinite sequences of real numbers. A typical element of S is $\mathbf{x} = (x_1, x_2, \dots, x_n, \dots)$. Define addition $\mathbf{x} + \mathbf{y}$ and scalar multiplication $t\mathbf{x}$ in the obvious way. Then S is a vector space. Define the left shift $L : S \to S$ by $L(x_1, x_2, x_3, \dots) = (x_2, x_3, x_4, \dots)$ and the right shift $R : S \to S$ by $R(x_1, x_2, x_3, \dots) = (0, x_1, x_2, x_3, \dots)$. Then it is easy to verify that L and R are linear. Check that $(L \circ R)(\mathbf{x}) = \mathbf{x}$ but that $(R \circ L)(\mathbf{x}) \neq \mathbf{x}$. L has a right inverse, but it does not have a left inverse. It is important in this example that S is infinite-dimensional.

4.6 Matrix of a Linear Mapping

The Matrix of L with Respect to the Basis B

In Section 3.2, we defined the standard matrix of a linear transformation $L : \mathbb{R}^n \to \mathbb{R}^m$ to be the matrix whose columns are the images of the standard basis vectors $\mathcal{S} = \{\vec{e}_1, \dots, \vec{e}_n\}$ of $\mathbb{R}^n$ under L. We now look at the matrix of L with respect to other bases. We shall use the subscript $\mathcal{S}$ to distinguish the standard matrix of L:

$$[L]_{\mathcal{S}} = \begin{bmatrix} [L(\vec{e}_1)]_{\mathcal{S}} & \cdots & [L(\vec{e}_n)]_{\mathcal{S}} \end{bmatrix}$$

The standard coordinates of the image under L of a vector $\vec{x} \in \mathbb{R}^n$ are given by the equation

$$[L(\vec{x})]_{\mathcal{S}} = [L]_{\mathcal{S}}[\vec{x}]_{\mathcal{S}} \tag{4.2}$$

These equations are exactly the same as those in Theorem 3.2.3 except that the notation is fancier so that we can compare this standard description with a description with respect to some other basis. We follow the same method as in the proof of Theorem 3.2.3 in defining the matrix of L with respect to another basis $\mathcal{B}$ of $\mathbb{R}^n$.

Let $\mathcal{B} = \{\vec{v}_1, \dots, \vec{v}_n\}$ be a basis for $\mathbb{R}^n$ and let $L : \mathbb{R}^n \to \mathbb{R}^n$ be a linear operator. Then, for any $\vec{x} \in \mathbb{R}^n$, we can write $\vec{x} = b_1\vec{v}_1 + \cdots + b_n\vec{v}_n$. Therefore,

$$L(\vec{x}) = L(b_1\vec{v}_1 + \cdots + b_n\vec{v}_n) = b_1L(\vec{v}_1) + \cdots + b_nL(\vec{v}_n)$$

Taking $\mathcal{B}$-coordinates of both sides gives

$$\begin{aligned} [L(\vec{x})]_{\mathcal{B}} &= [b_1L(\vec{v}_1) + \cdots + b_nL(\vec{v}_n)]_{\mathcal{B}} \\ &= b_1[L(\vec{v}_1)]_{\mathcal{B}} + \cdots + b_n[L(\vec{v}_n)]_{\mathcal{B}} \\ &= \begin{bmatrix} [L(\vec{v}_1)]_{\mathcal{B}} & \cdots & [L(\vec{v}_n)]_{\mathcal{B}} \end{bmatrix} \begin{bmatrix} b_1 \\ \vdots \\ b_n \end{bmatrix} \end{aligned}$$

Observe that $\begin{bmatrix} b_1 \\ \vdots \\ b_n \end{bmatrix} = [\vec{x}]_{\mathcal{B}}$, so this equation is the $\mathcal{B}$-coordinates version of equation (4.2) where the matrix $\begin{bmatrix} [L(\vec{v}_1)]_{\mathcal{B}} & \cdots & [L(\vec{v}_n)]_{\mathcal{B}} \end{bmatrix}$ is taking the place of the standard matrix of L.

Definition
Matrix of a Linear Operator

Suppose that $\mathcal{B} = \{\vec{v}_1, \dots, \vec{v}_n\}$ is any basis for $\mathbb{R}^n$ and that $L : \mathbb{R}^n \to \mathbb{R}^n$ is a linear operator. Define the **matrix of the linear operator** L with respect to the basis $\mathcal{B}$ to be the matrix

$$[L]_{\mathcal{B}} = \begin{bmatrix} [L(\vec{v}_1)]_{\mathcal{B}} & \cdots & [L(\vec{v}_n)]_{\mathcal{B}} \end{bmatrix}.$$

We then have that for any $\vec{x} \in \mathbb{R}^n$,

$$[L(\vec{x})]_{\mathcal{B}} = [L]_{\mathcal{B}}[\vec{x}]_{\mathcal{B}}$$

Note that the columns of $[L]_{\mathcal{B}}$ are the $\mathcal{B}$-coordinate vectors of the images of the $\mathcal{B}$-basis vectors under L. The pattern is exactly the same as before, except that everything is done in terms of the basis $\mathcal{B}$. It is important to emphasize again that by "basis," we always mean *ordered basis*; the order of the basis elements determines the order of the columns of the matrix $[L]_{\mathcal{B}}$.

EXAMPLE 1

Let $L : \mathbb{R}^3 \to \mathbb{R}^3$ be defined by $L(x_1, x_2, x_3) = (x_1 + 2x_2 - 2x_3, -x_2 + 2x_3, x_1 + 2x_2)$ and let $\mathcal{B} = \left\{ \begin{bmatrix} 2 \\ -1 \\ -1 \end{bmatrix}, \begin{bmatrix} 1 \\ 1 \\ 1 \end{bmatrix}, \begin{bmatrix} 0 \\ 0 \\ -1 \end{bmatrix} \right\}$. Find the $\mathcal{B}$-matrix of L and use it to determine $[L(\vec{x})]_{\mathcal{B}}$, where $[\vec{x}]_{\mathcal{B}} = \begin{bmatrix} 1 \\ 2 \\ 3 \end{bmatrix}$.

Solution: By definition, the columns of $[L]_{\mathcal{B}}$ are the $\mathcal{B}$-coordinates of the images of the vectors in $\mathcal{B}$ under L. So, we find these images and write them as a linear combination of the vectors in $\mathcal{B}$:

$$L(2, -1, -1) = \begin{bmatrix} 2 \\ -1 \\ 0 \end{bmatrix} = (1)\begin{bmatrix} 2 \\ -1 \\ -1 \end{bmatrix} + (0)\begin{bmatrix} 1 \\ 1 \\ 1 \end{bmatrix} + (-1)\begin{bmatrix} 0 \\ 0 \\ -1 \end{bmatrix}$$

$$L(1, 1, 1) = \begin{bmatrix} 1 \\ 1 \\ 3 \end{bmatrix} = (0)\begin{bmatrix} 2 \\ -1 \\ -1 \end{bmatrix} + (1)\begin{bmatrix} 1 \\ 1 \\ 1 \end{bmatrix} + (-2)\begin{bmatrix} 0 \\ 0 \\ -1 \end{bmatrix}$$

$$L(0, 0, -1) = \begin{bmatrix} 2 \\ -2 \\ 0 \end{bmatrix} = (4/3)\begin{bmatrix} 2 \\ -1 \\ -1 \end{bmatrix} + (-2/3)\begin{bmatrix} 1 \\ 1 \\ 1 \end{bmatrix} + (-2)\begin{bmatrix} 0 \\ 0 \\ -1 \end{bmatrix}$$

Hence,

$$[L]_{\mathcal{B}} = \begin{bmatrix} [L(2, -1, -1)]_{\mathcal{B}} & [L(1, 1, 1)]_{\mathcal{B}} & [L(0, 0, -1)]_{\mathcal{B}} \end{bmatrix} = \begin{bmatrix} 1 & 0 & 4/3 \\ 0 & 1 & -2/3 \\ -1 & -2 & -2 \end{bmatrix}$$

Thus,

$$[L(\vec{x})]_{\mathcal{B}} = \begin{bmatrix} 1 & 0 & 4/3 \\ 0 & 1 & -2/3 \\ -1 & -2 & -2 \end{bmatrix}\begin{bmatrix} 1 \\ 2 \\ 3 \end{bmatrix} = \begin{bmatrix} 5 \\ 0 \\ -11 \end{bmatrix}$$

We can verify that this answer is correct by calculating $L(\vec{x})$ in two ways. First, if $[\vec{x}]_{\mathcal{B}} = \begin{bmatrix} 1 \\ 2 \\ 3 \end{bmatrix}$, then

$$\vec{x} = 1\begin{bmatrix} 2 \\ -1 \\ -1 \end{bmatrix} + 2\begin{bmatrix} 1 \\ 1 \\ 1 \end{bmatrix} + 3\begin{bmatrix} 0 \\ 0 \\ -1 \end{bmatrix} = \begin{bmatrix} 4 \\ 1 \\ -2 \end{bmatrix}$$

EXAMPLE 1
(continued)

and by definition of the mapping, we have $L(\vec{x}) = L(4, 1, -2) = (10, -5, 6)$. Second, if $[L(\vec{x})]_{\mathcal{B}} = \begin{bmatrix} 5 \\ 0 \\ -11 \end{bmatrix}$, then

$$L(\vec{x}) = 5\begin{bmatrix} 2 \\ -1 \\ -1 \end{bmatrix} + 0\begin{bmatrix} 1 \\ 1 \\ 1 \end{bmatrix} - 11\begin{bmatrix} 0 \\ 0 \\ -1 \end{bmatrix} = \begin{bmatrix} 10 \\ -5 \\ 6 \end{bmatrix}$$

EXAMPLE 2

Let $\vec{v} = \begin{bmatrix} 3 \\ 4 \end{bmatrix}$. In Example 3.2.5, the standard matrix of the linear operator $\text{proj}_{\vec{v}} : \mathbb{R}^2 \to \mathbb{R}^2$ was found to be

$$[\text{proj}_{\vec{v}}]_S = \begin{bmatrix} 9/25 & 12/25 \\ 12/25 & 16/25 \end{bmatrix}$$

Find the matrix of $\text{proj}_{\vec{v}}$ with respect to a basis that shows the geometry of the transformation more clearly.

Solution: For this linear transformation, it is natural to use a basis for $\mathbb{R}^2$ consisting of the vector $\vec{v}$, which is the direction vector for the projection, and a second vector orthogonal to $\vec{v}$, say $\vec{w} = \begin{bmatrix} -4 \\ 3 \end{bmatrix}$. Then, with $\mathcal{B} = \{\vec{v}, \vec{w}\}$, by geometry,

$$\text{proj}_{\vec{v}}\, \vec{v} = \text{proj}_{\vec{v}} \begin{bmatrix} 3 \\ 4 \end{bmatrix} = \begin{bmatrix} 3 \\ 4 \end{bmatrix} = 1\vec{v} + 0\vec{w}$$

$$\text{proj}_{\vec{v}}\, \vec{w} = \text{proj}_{\vec{v}} \begin{bmatrix} -4 \\ 3 \end{bmatrix} = \begin{bmatrix} 0 \\ 0 \end{bmatrix} = 0\vec{v} + 0\vec{w}$$

Hence, $[\text{proj}_{\vec{v}}\, \vec{v}]_{\mathcal{B}} = \begin{bmatrix} 1 \\ 0 \end{bmatrix}$ and $[\text{proj}_{\vec{v}}\, \vec{w}]_{\mathcal{B}} = \begin{bmatrix} 0 \\ 0 \end{bmatrix}$. Thus,

$$[\text{proj}_{\vec{v}}]_{\mathcal{B}} = \begin{bmatrix} [\text{proj}_{\vec{v}}\, \vec{v}]_{\mathcal{B}} & [\text{proj}_{\vec{v}}\, \vec{w}]_{\mathcal{B}} \end{bmatrix} = \begin{bmatrix} 1 & 0 \\ 0 & 0 \end{bmatrix}$$

We now consider $\text{proj}_{\vec{v}}\, \vec{x}$ for any $\vec{x} \in \mathbb{R}^2$. We have

$$\begin{aligned} [\text{proj}_{\vec{v}}\, \vec{x}]_{\mathcal{B}} &= [\text{proj}_{\vec{v}}]_{\mathcal{B}}[\vec{x}]_{\mathcal{B}} \\ &= \begin{bmatrix} 1 & 0 \\ 0 & 0 \end{bmatrix}\begin{bmatrix} b_1 \\ b_2 \end{bmatrix} = \begin{bmatrix} b_1 \\ 0 \end{bmatrix} \end{aligned}$$

In terms of $\mathcal{B}$-coordinates, $\text{proj}_{\vec{v}}$ is described as the linear mapping that sends $\begin{bmatrix} b_1 \\ b_2 \end{bmatrix}$ to $\begin{bmatrix} b_1 \\ 0 \end{bmatrix}$.

This simple geometrical description is obtained when we use a basis $\mathcal{B}$ that is adapted to the geometry of the transformation (see Figure 4.6.4). This example will be discussed further below.

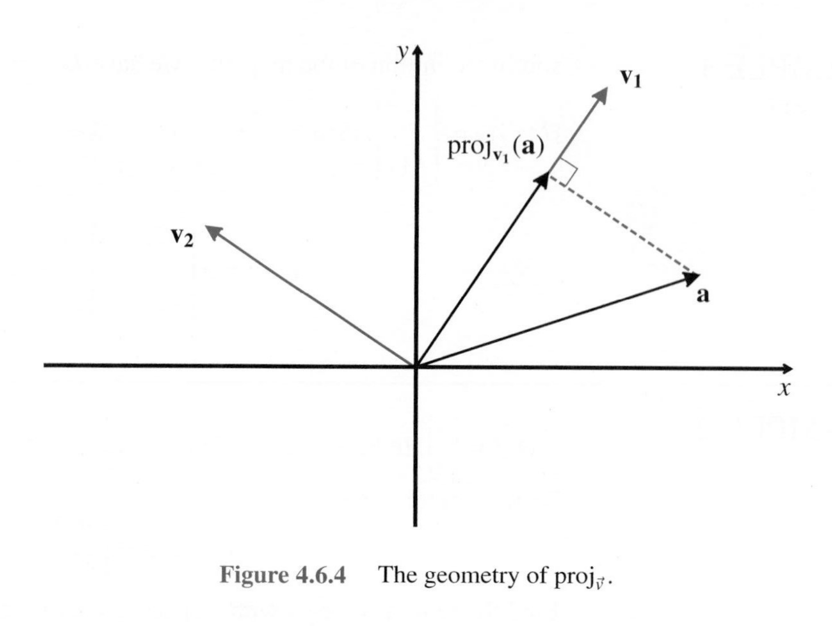

Figure 4.6.4 The geometry of $\text{proj}_{\vec{v}}$.

Of course, we want to generalize this to any linear operator $L : \mathbb{V} \to \mathbb{V}$ on a vector space $\mathbb{V}$ with respect to any basis $\mathcal{B}$ for $\mathbb{V}$. To do this, we repeat our argument above exactly.

Let $\mathcal{B} = \{\mathbf{v}_1, \dots, \mathbf{v}_n\}$ be a basis for a vector space $\mathbb{V}$ and let $L : \mathbb{V} \to \mathbb{V}$ be a linear operator. Then, for any $\mathbf{x} \in \mathbb{V}$, we can write $\mathbf{x} = b_1\mathbf{v}_1 + \cdots + b_n\mathbf{v}_n$. Therefore,

$$L(\mathbf{x}) = L(b_1\mathbf{v}_1 + \cdots + b_n\mathbf{v}_n) = b_1L(\mathbf{v}_1) + \cdots + b_nL(\mathbf{v}_n)$$

Taking $\mathcal{B}$-coordinates of both sides gives

$$\begin{aligned} [L(\mathbf{x})]_{\mathcal{B}} &= [b_1L(\mathbf{v}_1) + \cdots + b_nL(\mathbf{v}_n)]_{\mathcal{B}} \\ &= b_1[L(\mathbf{v}_1)]_{\mathcal{B}} + \cdots + b_n[L(\mathbf{v}_n)]_{\mathcal{B}} \\ &= \begin{bmatrix} [L(\mathbf{v}_1)]_{\mathcal{B}} & \cdots & [L(\mathbf{v}_n)]_{\mathcal{B}} \end{bmatrix} \begin{bmatrix} b_1 \\ \vdots \\ b_n \end{bmatrix} \end{aligned}$$

We make the following definition.

Definition
Matrix of a Linear Operator

Suppose that $\mathcal{B} = \{\mathbf{v}_1, \dots, \mathbf{v}_n\}$ is any basis for a vector space $\mathbb{V}$ and that $L : \mathbb{V} \to \mathbb{V}$ is a linear operator. Define the **matrix of the linear operator** L with respect to the basis $\mathcal{B}$ to be the matrix

$$[L]_{\mathcal{B}} = \begin{bmatrix} [L(\mathbf{v}_1)]_{\mathcal{B}} & \cdots & [L(\mathbf{v}_n)]_{\mathcal{B}} \end{bmatrix}$$

Then for any $\mathbf{x} \in \mathbb{V}$,

$$[L(\mathbf{x})]_{\mathcal{B}} = [L]_{\mathcal{B}}[\mathbf{x}]_{\mathcal{B}}$$

EXAMPLE 3

Let $L : P_2 \to P_2$ be defined by $L(a + bx + cx^2) = (a + b) + bx + (a + b + c)x^2$. Find the matrix of L with respect to the basis $\mathcal{B} = \{1, x, x^2\}$.

EXAMPLE 3
(continued)

Solution: We have

$$L(1) = 1 + x^2$$
$$L(x) = 1 + x + x^2$$
$$L(x^2) = x^2$$

Hence,

$$[L]_{\mathcal{B}} = \begin{bmatrix}[L(1)]_{\mathcal{B}} & [L(x)]_{\mathcal{B}} & [L(x^2)]_{\mathcal{B}}\end{bmatrix} = \begin{bmatrix}1 & 1 & 0\\ 0 & 1 & 0\\ 1 & 1 & 1\end{bmatrix}$$

We can check our answer by observing that $[L(a + bx + cx^2)]_{\mathcal{B}} = \begin{bmatrix}a+b\\ b\\ a+b+c\end{bmatrix}$ and

$$[L(a + bx + cx^2)]_{\mathcal{B}} = [L]_{\mathcal{B}}[a + bx + cx^2]_{\mathcal{B}} = \begin{bmatrix}1 & 1 & 0\\ 0 & 1 & 0\\ 1 & 1 & 1\end{bmatrix}\begin{bmatrix}a\\ b\\ c\end{bmatrix} = \begin{bmatrix}a+b\\ b\\ a+b+c\end{bmatrix}$$

EXAMPLE 4

Let $L : P_2 \to P_2$ be defined by $L(a + bx + cx^2) = (a + b) + bx + (a + b + c)x^2$. Find the matrix of L with respect to the basis $\mathcal{B} = \{1 + x + x^2, 1 - x, 2\}$.
Solution: We have

$$L(1 + x + x^2) = 2 + x + 3x^2 = (3)(1 + x + x^2) + (2)(1 - x) + (-3/2)(2)$$
$$L(1 - x) = 0 + (-1)x + 0x^2 = (0)(1 + x + x^2) + (1)(1 - x) + (-1/2)(2)$$
$$L(2) = 2 + 2x^2 = (2)(1 + x + x^2) + (2)(1 - x) + (-1)(2)$$

Hence,

$$[L]_{\mathcal{B}} = \begin{bmatrix}[L(1)]_{\mathcal{B}} & [L(x)]_{\mathcal{B}} & [L(x^2)]_{\mathcal{B}}\end{bmatrix} = \begin{bmatrix}3 & 0 & 2\\ 2 & 1 & 2\\ -3/2 & -1/2 & -1\end{bmatrix}$$

EXERCISE 1

Let $L : M(2, 2) \to M(2, 2)$ be defined by $L\left(\begin{bmatrix}a & b\\ c & d\end{bmatrix}\right) = \begin{bmatrix}a+b & a-b\\ c & a+b+d\end{bmatrix}$. Find the matrix of L with respect to the basis $\mathcal{B} = \left\{\begin{bmatrix}1 & 1\\ 1 & 1\end{bmatrix}, \begin{bmatrix}0 & 1\\ 1 & 1\end{bmatrix}, \begin{bmatrix}0 & 0\\ 1 & 1\end{bmatrix}, \begin{bmatrix}0 & 0\\ 0 & 1\end{bmatrix}\right\}$.

Observe that in each of the cases above, we have used the same basis for the domain and codomain of the linear mapping L. To make this as general as possible, we would like to define the matrix ${}_C[L]_{\mathcal{B}}$ of a linear mapping $L : \mathbb{V} \to \mathbb{W}$, where $\mathcal{B}$ is a basis for the vector space $\mathbb{V}$ and C is a basis for the vector space $\mathbb{W}$. This is left as Problem D3.

Change of Coordinates and Linear Mappings

In Example 2, we used special geometrical properties of the linear transformation and of the chosen basis $\mathcal{B}$ to determine the $\mathcal{B}$-coordinate vectors that make up the $\mathcal{B}$-matrix $[L]_{\mathcal{B}}$ of the linear transformation $L : \mathbb{R}^2 \to \mathbb{R}^2$. In some applications of these ideas, the geometry does not provide such a simple way of determining $[L]_{\mathcal{B}}$. We need a general method for determining $[L]_{\mathcal{B}}$, given the standard matrix $[L]_{\mathcal{S}}$ of a linear operator $L : \mathbb{R}^n \to \mathbb{R}^n$ and a new basis $\mathcal{B}$.

Let $L : \mathbb{R}^n \to \mathbb{R}^n$ be a linear operator. Let $\mathcal{S}$ denote the standard basis for $\mathbb{R}^n$ and let $\mathcal{B}$ be any other basis for $\mathbb{R}^n$. Denote the change of coordinates matrix from $\mathcal{B}$ to $\mathcal{S}$ by P so that we have

$$P[\vec{x}]_{\mathcal{B}} = [\vec{x}]_{\mathcal{S}} \qquad \text{and} \qquad [\vec{x}]_{\mathcal{B}} = P^{-1}[\vec{x}]_{\mathcal{S}}$$

If we apply the change of coordinates equation to the vector $L(\vec{x})$ (which is in $\mathbb{R}^n$), we get

$$[L(\vec{x})]_{\mathcal{B}} = P^{-1}[L(\vec{x})]_{\mathcal{S}}$$

Substitute for $[L(\vec{x})]_{\mathcal{B}}$ and $[L(\vec{x})]_{\mathcal{S}}$ to get

$$[L]_{\mathcal{B}}[\vec{x}]_{\mathcal{B}} = P^{-1}[L]_{\mathcal{S}}[\vec{x}]_{\mathcal{S}}$$

But, $P[\vec{x}]_{\mathcal{B}} = [\vec{x}]_{\mathcal{S}}$, so we have

$$[L]_{\mathcal{B}}[\vec{x}]_{\mathcal{B}} = P^{-1}[L]_{\mathcal{S}}P[\vec{x}]_{\mathcal{B}}$$

Since this is true for every $[\vec{x}]_{\mathcal{B}}$, we get, by Theorem 3.1.4, that

$$[L]_{\mathcal{B}} = P^{-1}[L]_{\mathcal{S}}P$$

Thus, we now have a method of determining the $\mathcal{B}$-matrix of L, given the standard matrix of L and a new basis $\mathcal{B}$.

We shall first apply this change of basis method to Example 2 to make sure that things work out as we expect.

EXAMPLE 5

Let $\vec{v} = \begin{bmatrix} 3 \\ 4 \end{bmatrix}$. In Example 2, we determined the matrix of $\text{proj}_{\vec{v}}$ with respect to a geometrically adapted basis $\mathcal{B}$. Let us verify that the change of basis method just described does transform the standard matrix $[\text{proj}_{\vec{v}}]_{\mathcal{S}}$ to the $\mathcal{B}$-matrix $[\text{proj}_{\vec{v}}]_{\mathcal{B}}$.

The matrix $[\text{proj}_{\vec{v}}]_{\mathcal{S}} = \begin{bmatrix} 9/25 & 12/25 \\ 12/25 & 16/25 \end{bmatrix}$. The basis $\mathcal{B} = \left\{ \begin{bmatrix} 3 \\ 4 \end{bmatrix}, \begin{bmatrix} -4 \\ 3 \end{bmatrix} \right\}$, so the change of coordinates matrix from $\mathcal{B}$ to $\mathcal{S}$ is $P = \begin{bmatrix} 3 & -4 \\ 4 & 3 \end{bmatrix}$. The inverse is found to be $P^{-1} = \begin{bmatrix} 3/25 & 4/25 \\ -4/25 & 3/25 \end{bmatrix}$. Hence, the $\mathcal{B}$-matrix of $\text{proj}_{\vec{v}}$ is given by

$$P^{-1}[\text{proj}_{\vec{v}}]_{\mathcal{S}}P = \begin{bmatrix} 3/25 & 4/25 \\ -4/25 & 3/25 \end{bmatrix} \begin{bmatrix} 9/25 & 12/25 \\ 12/25 & 16/25 \end{bmatrix} \begin{bmatrix} 3 & -4 \\ 4 & 3 \end{bmatrix} = \begin{bmatrix} 1 & 0 \\ 0 & 0 \end{bmatrix}$$

Thus, we obtain exactly the same $\mathcal{B}$-matrix $[\text{proj}_{\vec{v}}]_{\mathcal{B}}$ as we obtained by the earlier geometric argument.

EXAMPLE 5
(continued)

To make sure we understand exactly what this means, let us calculate the $\mathcal{B}$-coordinates of the image of the vector $\vec{x} = \begin{bmatrix} 5 \\ 2 \end{bmatrix}$ under $\text{proj}_{\vec{v}}$. We can do this in two ways.

Method 1. Use the fact that $[\text{proj}_{\vec{v}}\,\vec{x}]_{\mathcal{B}} = [\text{proj}_{\vec{v}}]_{\mathcal{B}}[\vec{x}]_{\mathcal{B}}$. We need the $\mathcal{B}$-coordinates of $\vec{x}$:

$$\begin{bmatrix} 5 \\ 2 \end{bmatrix}_{\mathcal{B}} = P^{-1}\begin{bmatrix} 5 \\ 2 \end{bmatrix}_{\mathcal{S}} = \begin{bmatrix} 3/25 & 4/25 \\ -4/25 & 3/25 \end{bmatrix}\begin{bmatrix} 5 \\ 2 \end{bmatrix} = \begin{bmatrix} 23/25 \\ -14/25 \end{bmatrix}$$

Hence,

$$[\text{proj}_{\vec{v}}\,\vec{x}]_{\mathcal{B}} = [\text{proj}_{\vec{v}}]_{\mathcal{B}}[\vec{x}]_{\mathcal{B}} = \begin{bmatrix} 1 & 0 \\ 0 & 0 \end{bmatrix}\begin{bmatrix} 23/25 \\ -14/25 \end{bmatrix} = \begin{bmatrix} 23/25 \\ 0 \end{bmatrix}$$

Method 2. Use the fact that $[\text{proj}_{\vec{v}}\,\vec{x}]_{\mathcal{B}} = P^{-1}[\text{proj}_{\vec{v}}\,\vec{x}]_{\mathcal{S}}$:

$$[\text{proj}_{\vec{v}}\,\vec{x}]_{\mathcal{S}} = [\text{proj}_{\vec{v}}]_{\mathcal{S}}\begin{bmatrix} 5 \\ 2 \end{bmatrix} = \begin{bmatrix} 9/25 & 12/25 \\ 12/25 & 16/25 \end{bmatrix}\begin{bmatrix} 5 \\ 2 \end{bmatrix} = \begin{bmatrix} 69/25 \\ 92/25 \end{bmatrix}$$

Therefore,

$$[\text{proj}_{\vec{v}}\,\vec{x}]_{\mathcal{B}} = \begin{bmatrix} 3/25 & 4/25 \\ -4/25 & 3/25 \end{bmatrix}\begin{bmatrix} 69/25 \\ 92/25 \end{bmatrix} = \begin{bmatrix} 23/25 \\ 0 \end{bmatrix}$$

The calculation is probably slightly easier if we use the first method, but that really is not the point. What is really important is that it is easy to get a geometrical understanding of what happens to vectors if you multiply by $\begin{bmatrix} 1 & 0 \\ 0 & 0 \end{bmatrix}$ (the $\mathcal{B}$-matrix); it is much more difficult to understand what happens if you multiply by $\begin{bmatrix} 9/25 & 12/25 \\ 12/25 & 16/25 \end{bmatrix}$ (the standard matrix). Using a non-standard basis may make it much easier to understand the geometry of a linear transformation.

EXAMPLE 6

Let L be the linear mapping with standard matrix $A = \begin{bmatrix} 2 & 3 \\ 4 & 5 \end{bmatrix}$. Let $\mathcal{B} = \left\{\begin{bmatrix} 3 \\ 1 \end{bmatrix}, \begin{bmatrix} -1 \\ 1 \end{bmatrix}\right\}$ be a basis for $\mathbb{R}^2$. Find the matrix of L with respect to the basis $\mathcal{B}$.

Solution: The change of coordinates matrix P is $P = \begin{bmatrix} 3 & -1 \\ 1 & 1 \end{bmatrix}$, and we have $P^{-1} = \frac{1}{4}\begin{bmatrix} 1 & 1 \\ -1 & 3 \end{bmatrix}$. It follows that the $\mathcal{B}$-matrix of L is

$$[L]_{\mathcal{B}} = P^{-1}AP = \frac{1}{4}\begin{bmatrix} 1 & 1 \\ -1 & 3 \end{bmatrix}\begin{bmatrix} 2 & 3 \\ 4 & 5 \end{bmatrix}\begin{bmatrix} 3 & -1 \\ 1 & 1 \end{bmatrix} = \begin{bmatrix} 13/2 & 1/2 \\ 21/2 & 1/2 \end{bmatrix}$$

EXAMPLE 7

Let L be the linear mapping with standard matrix $A = \begin{bmatrix} -3 & 5 & -5 \\ -7 & 9 & -5 \\ -7 & 7 & -3 \end{bmatrix}$. Let $\mathcal{B}$ be the basis $\mathcal{B} = \left\{\begin{bmatrix} 1 \\ 1 \\ 0 \end{bmatrix}, \begin{bmatrix} 1 \\ 1 \\ 1 \end{bmatrix}, \begin{bmatrix} 0 \\ 1 \\ 1 \end{bmatrix}\right\}$. Determine the matrix of L with respect to the basis $\mathcal{B}$.

EXAMPLE 7
(continued)

Solution: The change of coordinates matrix P is $P = \begin{bmatrix} 1 & 1 & 0 \\ 1 & 1 & 1 \\ 0 & 1 & 1 \end{bmatrix}$, and we have $P^{-1} = \begin{bmatrix} 0 & 1 & -1 \\ 1 & -1 & 1 \\ -1 & 1 & 0 \end{bmatrix}$. Thus, the $\mathcal{B}$-matrix of the mapping L is

$$[L]_{\mathcal{B}} = P^{-1}AP = \begin{bmatrix} 2 & 0 & 0 \\ 0 & -3 & 0 \\ 0 & 0 & 4 \end{bmatrix}$$

Observe that the resulting matrix is diagonal. What does this mean in terms of the geometry of the linear transformation? From the definition of $[L]_{\mathcal{B}}$ and the definition of $\mathcal{B}$-coordinates of a vector, we see that the linear transformation stretches the first basis vector $\begin{bmatrix} 1 \\ 1 \\ 0 \end{bmatrix}$ by a factor of 2, it reflects (because of the minus sign) the second basis vector $\begin{bmatrix} 1 \\ 1 \\ 1 \end{bmatrix}$ in the origin and stretches it by a factor of 3, and it stretches the third basis vector $\begin{bmatrix} 0 \\ 1 \\ 1 \end{bmatrix}$ by a factor of 4. This gives a very clear geometrical picture of how the linear transformation maps vectors. This picture is not obvious from looking at the standard matrix A.

At this point it is natural to ask whether for any linear mapping $L : \mathbb{R}^n \to \mathbb{R}^n$ there exists a basis $\mathcal{B}$ of $\mathbb{R}^n$ such that the $\mathcal{B}$-matrix of L is in diagonal form, and how can we find such a basis if it exists?

The answers to these questions are found in Chapter 6. However, in order to deal with these questions, one more computational tool is needed, the determinant, which is discussed in Chapter 5.

PROBLEMS 4.6

Practice Problems

A1 Determine the matrix of the linear mapping L with respect to the basis $\mathcal{B}$ in the following cases. Determine $[L(\vec{x})]_{\mathcal{B}}$ for the given $[\vec{x}]_{\mathcal{B}}$.

(a) In $\mathbb{R}^2$, $\mathcal{B} = \{\vec{v}_1, \vec{v}_2\}$ and $L(\vec{v}_1) = \vec{v}_2$, $L(\vec{v}_2) = 2\vec{v}_1 - \vec{v}_2$; $[\vec{x}]_{\mathcal{B}} = \begin{bmatrix} 4 \\ 3 \end{bmatrix}$

(b) In $\mathbb{R}^3$, $\mathcal{B} = \{\vec{v}_1, \vec{v}_2, \vec{v}_3\}$ and $L(\vec{v}_1) = 2\vec{v}_1 - \vec{v}_3$, $L(\vec{v}_2) = 2\vec{v}_1 - \vec{v}_3$, $L(\vec{v}_3) = 4\vec{v}_2 + 5\vec{v}_3$; $[\vec{x}]_{\mathcal{B}} = \begin{bmatrix} 3 \\ 3 \\ -1 \end{bmatrix}$

A2 Consider the basis $\mathcal{B} = \left\{ \begin{bmatrix} 1 \\ 1 \end{bmatrix}, \begin{bmatrix} -1 \\ 2 \end{bmatrix} \right\}$ of $\mathbb{R}^2$. In each of the following cases, assume that L is a linear mapping and determine $[L]_{\mathcal{B}}$.

(a) $L(1,1) = (-3,-3)$ and $L(-1,2) = (-4,8)$

(b) $L(1,1) = (-1,2)$ and $L(-1,2) = (2,2)$

A3 Consider the basis $\mathcal{B} = \{\vec{v}_1, \vec{v}_2, \vec{v}_3\} = \left\{ \begin{bmatrix} 1 \\ 1 \\ 1 \end{bmatrix}, \begin{bmatrix} -1 \\ 2 \\ 0 \end{bmatrix}, \begin{bmatrix} 0 \\ -1 \\ 4 \end{bmatrix} \right\}$ of $\mathbb{R}^3$. In each of the following cases, assume that L is a linear mapping.

Determine $[L(\vec{v}_1)]_{\mathcal{B}}$, $[L(\vec{v}_2)]_{\mathcal{B}}$, and $[L(\vec{v}_3)]_{\mathcal{B}}$ and hence determine $[L]_{\mathcal{B}}$.

(a) $L\left(\begin{bmatrix}1\\1\\1\end{bmatrix}\right) = \begin{bmatrix}-1\\2\\0\end{bmatrix}$, $L\left(\begin{bmatrix}-1\\2\\0\end{bmatrix}\right) = \begin{bmatrix}1\\1\\1\end{bmatrix}$, $L\left(\begin{bmatrix}0\\-1\\4\end{bmatrix}\right) = \begin{bmatrix}5\\5\\5\end{bmatrix}$

(b) $L\left(\begin{bmatrix}1\\1\\1\end{bmatrix}\right) = \begin{bmatrix}0\\-1\\4\end{bmatrix}$, $L\left(\begin{bmatrix}-1\\2\\0\end{bmatrix}\right) = \begin{bmatrix}1\\1\\1\end{bmatrix}$, $L\left(\begin{bmatrix}0\\-1\\4\end{bmatrix}\right) = \begin{bmatrix}-1\\2\\0\end{bmatrix}$

(c) $L\left(\begin{bmatrix}1\\1\\1\end{bmatrix}\right) = \begin{bmatrix}1\\1\\1\end{bmatrix}$, $L\left(\begin{bmatrix}-1\\2\\0\end{bmatrix}\right) = \begin{bmatrix}0\\2\\5\end{bmatrix}$, $L\left(\begin{bmatrix}0\\-1\\4\end{bmatrix}\right) = \begin{bmatrix}5\\0\\-1\end{bmatrix}$

A4 For each of the following linear transformations, determine a geometrically natural basis $\mathcal{B}$ (as in Examples 2 and 3) and determine the $\mathcal{B}$-matrix of the transformation.

(a) $\text{refl}_{(1,-2)}$

(b) $\text{proj}_{(2,1,-1)}$

(c) $\text{refl}_{(-1,-1,1)}$

A5 (a) Find the coordinates of $\begin{bmatrix}1\\2\\4\end{bmatrix}$ with respect to the basis $\mathcal{B} = \left\{\begin{bmatrix}1\\0\\1\end{bmatrix}, \begin{bmatrix}1\\-1\\0\end{bmatrix}, \begin{bmatrix}0\\1\\2\end{bmatrix}\right\}$ in $\mathbb{R}^3$.

(b) Suppose that $L : \mathbb{R}^3 \to \mathbb{R}^3$ is a linear transformation such that $L(1,0,1) = (1,2,4)$, $L(1,-1,0) = (0,1,2)$, and $L(0,1,2) = (2,-2,0)$. Determine the $\mathcal{B}$-matrix of L.

(c) Use parts (a) and (b) to determine $L(1,2,4)$.

A6 (a) Find the coordinates of $\begin{bmatrix}5\\3\\-5\end{bmatrix}$ with respect to the basis $\mathcal{B} = \left\{\begin{bmatrix}1\\0\\-1\end{bmatrix}, \begin{bmatrix}1\\2\\0\end{bmatrix}, \begin{bmatrix}0\\1\\1\end{bmatrix}\right\}$ in $\mathbb{R}^3$.

(b) Suppose that $L : \mathbb{R}^3 \to \mathbb{R}^3$ is a linear transformation such that $L(1,0,-1) = (0,1,1)$, $L(1,2,0) = (-2,0,2)$, and $L(0,1,1) = (5,3,-5)$. Determine the $\mathcal{B}$-matrix of L.

(c) Use parts (a) and (b) to determine $L(5,3,-5)$.

A7 Assume that each of the following matrices is the standard matrix of a linear mapping $L : \mathbb{R}^n \to \mathbb{R}^n$. Determine the matrix of L with respect to the given basis $\mathcal{B}$. You may find it helpful to use a computer to find inverses and to multiply matrices.

(a) $\begin{bmatrix}1 & 3\\-8 & 7\end{bmatrix}$, $\mathcal{B} = \left\{\begin{bmatrix}1\\2\end{bmatrix}, \begin{bmatrix}1\\4\end{bmatrix}\right\}$

(b) $\begin{bmatrix}1 & -6\\-4 & -1\end{bmatrix}$, $\mathcal{B} = \left\{\begin{bmatrix}3\\-2\end{bmatrix}, \begin{bmatrix}1\\1\end{bmatrix}\right\}$

(c) $\begin{bmatrix}4 & -6\\2 & 8\end{bmatrix}$, $\mathcal{B} = \left\{\begin{bmatrix}3\\1\end{bmatrix}, \begin{bmatrix}7\\3\end{bmatrix}\right\}$

(d) $\begin{bmatrix}16 & -20\\6 & -6\end{bmatrix}$, $\mathcal{B} = \left\{\begin{bmatrix}5\\3\end{bmatrix}, \begin{bmatrix}4\\2\end{bmatrix}\right\}$

(e) $\begin{bmatrix}3 & 1 & 1\\0 & 4 & 2\\1 & -1 & 5\end{bmatrix}$, $\mathcal{B} = \left\{\begin{bmatrix}1\\1\\0\end{bmatrix}, \begin{bmatrix}0\\1\\1\end{bmatrix}, \begin{bmatrix}1\\0\\1\end{bmatrix}\right\}$

(f) $\begin{bmatrix}4 & 1 & -3\\16 & 4 & -18\\6 & 1 & -5\end{bmatrix}$, $\mathcal{B} = \left\{\begin{bmatrix}1\\1\\1\end{bmatrix}, \begin{bmatrix}0\\3\\1\end{bmatrix}, \begin{bmatrix}1\\2\\1\end{bmatrix}\right\}$

A8 Find the $\mathcal{B}$-matrix of each of the following linear mappings.

(a) $L : \mathbb{R}^3 \to \mathbb{R}^3$ defined by $L(x_1, x_2, x_3) = (x_1 + x_2, x_2 + x_3, x_1 - x_3)$, $\mathcal{B} = \left\{\begin{bmatrix}1\\1\\1\end{bmatrix}, \begin{bmatrix}0\\1\\1\end{bmatrix}, \begin{bmatrix}0\\0\\1\end{bmatrix}\right\}$

(b) $L : P_2 \to P_2$ defined by $L(a + bx + cx^2) = a + (b + c)x^2$, $\mathcal{B} = \{1 + x^2, -1 + x, 1 - x + x^2\}$

(c) $D : P_2 \to P_2$ defined by $D(a + bx + cx^2) = b + 2cx$, $\mathcal{B} = \{1, x, x^2\}$

(d) $T : \mathbb{U} \to \mathbb{U}$, where $\mathbb{U}$ is the subspace of upper-triangular matrices in $M(2,2)$, defined by $T\left(\begin{bmatrix}a & b\\0 & c\end{bmatrix}\right) = \begin{bmatrix}a & b+c\\0 & a+b+c\end{bmatrix}$, $\mathcal{B} = \left\{\begin{bmatrix}1 & 1\\0 & 0\end{bmatrix}, \begin{bmatrix}1 & 0\\0 & 1\end{bmatrix}, \begin{bmatrix}1 & 1\\0 & 1\end{bmatrix}\right\}$

Homework Problems

B1 Determine the matrix of the linear mapping L with respect to the basis $\mathcal{B}$ in the following cases. Determine $[L(\vec{x})]_{\mathcal{B}}$ for the given $[\vec{x}]_{\mathcal{B}}$.

(a) In $\mathbb{R}^2$, $\mathcal{B} = \{\vec{v}_1, \vec{v}_2\}$ and $L(\vec{v}_1) = \vec{v}_1 + 3\vec{v}_2$, $L(\vec{v}_2) = 5\vec{v}_1 - 7\vec{v}_2$; $[\vec{x}]_{\mathcal{B}} = \begin{bmatrix}4\\-2\end{bmatrix}$

(b) In $\mathbb{R}^3$, $\mathcal{B} = \{\vec{v}_1, \vec{v}_2, \vec{v}_3\}$ and $L(\vec{v}_1) = 2\vec{v}_1 - 3\vec{v}_2$, $L(\vec{v}_2) = 3\vec{v}_1 + 4\vec{v}_2 - \vec{v}_3$, $L(\vec{v}_3) = -\vec{v}_1 + 2\vec{v}_2 + 6\vec{v}_3$; $[\vec{x}]_{\mathcal{B}} = \begin{bmatrix}5\\-3\\1\end{bmatrix}$

B2 Consider the basis $\mathcal{B} = \left\{ \begin{bmatrix} 1 \\ 2 \end{bmatrix}, \begin{bmatrix} 1 \\ -2 \end{bmatrix} \right\}$ of $\mathbb{R}^2$. In each of the following cases, assume that L is a linear mapping and determine $[L]_{\mathcal{B}}$.

(a) $L(1,2) = (1,-2)$ and $L(1,-2) = (4,8)$

(b) $L(1,2) = (5,10)$ and $L(1,-2) = (-3,6)$

(c) $L(1,2) = (0,0)$ and $L(1,-2) = (1,2)$

B3 Consider the basis $\mathcal{B} = \{\vec{v}_1, \vec{v}_2, \vec{v}_3\} = \left\{ \begin{bmatrix} 1 \\ 1 \\ 0 \end{bmatrix}, \begin{bmatrix} 1 \\ -1 \\ 1 \end{bmatrix}, \begin{bmatrix} 3 \\ 0 \\ 1 \end{bmatrix} \right\}$ of $\mathbb{R}^3$. In each of the following cases, assume that L is a linear mapping. Determine $[L(\vec{v}_1)]_{\mathcal{B}}$, $[L(\vec{v}_2)]_{\mathcal{B}}$, and $[L(\vec{v}_3)]_{\mathcal{B}}$ and hence determine $[L]_{\mathcal{B}}$.

(a) $L\left(\begin{bmatrix} 1 \\ 1 \\ 0 \end{bmatrix}\right) = \begin{bmatrix} 3 \\ 0 \\ 1 \end{bmatrix}, L\left(\begin{bmatrix} 1 \\ -1 \\ 1 \end{bmatrix}\right) = \begin{bmatrix} 1 \\ 1 \\ 0 \end{bmatrix}, L\left(\begin{bmatrix} 3 \\ 0 \\ 1 \end{bmatrix}\right) = \begin{bmatrix} 2 \\ -2 \\ 2 \end{bmatrix}$

(b) $L\left(\begin{bmatrix} 1 \\ 1 \\ 0 \end{bmatrix}\right) = \begin{bmatrix} 4 \\ 4 \\ 0 \end{bmatrix}, L\left(\begin{bmatrix} 1 \\ -1 \\ 1 \end{bmatrix}\right) = \begin{bmatrix} 6 \\ 0 \\ 2 \end{bmatrix}, L\left(\begin{bmatrix} 3 \\ 0 \\ 1 \end{bmatrix}\right) = \begin{bmatrix} -1 \\ 1 \\ -1 \end{bmatrix}$

(c) $L\left(\begin{bmatrix} 1 \\ 1 \\ 0 \end{bmatrix}\right) = \begin{bmatrix} 1 \\ 1 \\ 0 \end{bmatrix}, L\left(\begin{bmatrix} 1 \\ -1 \\ 1 \end{bmatrix}\right) = \begin{bmatrix} 1 \\ 1 \\ 0 \end{bmatrix}, L\left(\begin{bmatrix} 3 \\ 0 \\ 1 \end{bmatrix}\right) = \begin{bmatrix} 0 \\ 0 \\ 0 \end{bmatrix}$

(d) $L\left(\begin{bmatrix} 1 \\ 1 \\ 0 \end{bmatrix}\right) = \begin{bmatrix} -2 \\ 0 \\ 1 \end{bmatrix}, L\left(\begin{bmatrix} 1 \\ -1 \\ 1 \end{bmatrix}\right) = \begin{bmatrix} 5 \\ 0 \\ 2 \end{bmatrix}, L\left(\begin{bmatrix} 3 \\ 0 \\ 1 \end{bmatrix}\right) = \begin{bmatrix} 3 \\ 0 \\ 1 \end{bmatrix}$

B4 For each of the following linear transformations, determine a geometrically natural basis $\mathcal{B}$ (as in Examples 2 and 3) and determine the $\mathcal{B}$-matrix of the transformation.

(a) $\text{perp}_{(3,2)}$

(b) $\text{perp}_{(2,1,-2)}$

(c) $\text{refl}_{(1,2,3)}$

B5 (a) Find the coordinates of $\begin{bmatrix} 5 \\ 2 \\ 1 \end{bmatrix}$ with respect to the basis $\mathcal{B} = \left\{ \begin{bmatrix} 1 \\ 1 \\ 0 \end{bmatrix}, \begin{bmatrix} 0 \\ 1 \\ 1 \end{bmatrix}, \begin{bmatrix} 1 \\ 0 \\ 1 \end{bmatrix} \right\}$ in $\mathbb{R}^3$.

(b) Suppose that $L : \mathbb{R}^3 \to \mathbb{R}^3$ is a linear transformation such that $L(1,1,0) = (0,5,5)$, $L(0,1,1) = (2,0,2)$, and $L(1,0,1) = (5,2,1)$. Determine the $\mathcal{B}$-matrix of L.

(c) Use parts (a) and (b) to determine $L(5,2,1)$.

B6 (a) Find the coordinates of $\begin{bmatrix} 1 \\ 4 \\ 4 \end{bmatrix}$ with respect to the basis $\mathcal{B} = \left\{ \begin{bmatrix} 2 \\ 1 \\ 0 \end{bmatrix}, \begin{bmatrix} -1 \\ 0 \\ 1 \end{bmatrix}, \begin{bmatrix} 1 \\ 1 \\ 0 \end{bmatrix} \right\}$ in $\mathbb{R}^3$.

(b) Suppose that $L : \mathbb{R}^3 \to \mathbb{R}^3$ is a linear transformation such that $L(2,1,0) = (3,3,0)$, $L(-1,0,1) = (1,4,4)$, and $L(1,1,0) = (-2,0,2)$. Determine the $\mathcal{B}$-matrix of L.

(c) Use parts (a) and (b) to determine $L(1,4,4)$.

B7 Assume that each of the following matrices is the standard matrix of a linear mapping $L : \mathbb{R}^n \to \mathbb{R}^n$. Determine the matrix of L with respect to the given basis $\mathcal{B}$. You may find it helpful to use a computer to find inverses and to multiply matrices.

(a) $\begin{bmatrix} -3 & 1 \\ -16 & 5 \end{bmatrix}, \mathcal{B} = \left\{ \begin{bmatrix} 1 \\ 4 \end{bmatrix}, \begin{bmatrix} 1 \\ 5 \end{bmatrix} \right\}$

(b) $\begin{bmatrix} 6 & -10 \\ 2 & -6 \end{bmatrix}, \mathcal{B} = \left\{ \begin{bmatrix} 5 \\ 1 \end{bmatrix}, \begin{bmatrix} 1 \\ 1 \end{bmatrix} \right\}$

(c) $\begin{bmatrix} -6 & -2 & 9 \\ -5 & -1 & 7 \\ -7 & -2 & 10 \end{bmatrix}, \mathcal{B} = \left\{ \begin{bmatrix} 1 \\ 1 \\ 1 \end{bmatrix}, \begin{bmatrix} 1 \\ 0 \\ 1 \end{bmatrix}, \begin{bmatrix} 1 \\ 3 \\ 2 \end{bmatrix} \right\}$

(d) $\begin{bmatrix} -7 & -3 & 3 \\ 2 & 2 & -1 \\ -16 & -6 & 7 \end{bmatrix}, \mathcal{B} = \left\{ \begin{bmatrix} 1 \\ 0 \\ 2 \end{bmatrix}, \begin{bmatrix} 0 \\ 1 \\ 1 \end{bmatrix}, \begin{bmatrix} 1 \\ -1 \\ 2 \end{bmatrix} \right\}$

B8 Assume that each of the following matrices is the standard matrix of a linear mapping $L : \mathbb{R}^n \to \mathbb{R}^n$. Determine the matrix of L with respect to the given basis $\mathcal{B}$ and use it to determine $[L(\vec{x})]_{\mathcal{B}}$ for the given vector $\vec{x}$. You may find it helpful to use a computer to find inverses and to multiply matrices.

(a) $\begin{bmatrix} 12 & -15 \\ -16 & -7 \end{bmatrix}, \mathcal{B} = \left\{ \begin{bmatrix} 5 \\ 3 \end{bmatrix}, \begin{bmatrix} 3 \\ 2 \end{bmatrix} \right\}, [\vec{x}]_{\mathcal{B}} = \begin{bmatrix} 1 \\ 2 \end{bmatrix}$

(b) $\begin{bmatrix} 6 & -2 \\ 36 & -7 \end{bmatrix}, \mathcal{B} = \left\{ \begin{bmatrix} 1 \\ 5 \end{bmatrix}, \begin{bmatrix} 1 \\ 2 \end{bmatrix} \right\}, [\vec{x}]_{\mathcal{B}} = \begin{bmatrix} -2 \\ 1 \end{bmatrix}$

(c) $\begin{bmatrix} 10 & -20 & -24 \\ 5 & -10 & -15 \\ -2 & 4 & 8 \end{bmatrix}, \mathcal{B} = \left\{ \begin{bmatrix} 2 \\ 1 \\ 0 \end{bmatrix}, \begin{bmatrix} 1 \\ -1 \\ 1 \end{bmatrix}, \begin{bmatrix} 1 \\ 1 \\ -1 \end{bmatrix} \right\},$

$[\vec{x}]_{\mathcal{B}} = \begin{bmatrix} 1 \\ 2 \\ 3 \end{bmatrix}$

(d) $\begin{bmatrix} 3 & 6 & 1 \\ 5 & -4 & 5 \\ 3 & -6 & 5 \end{bmatrix}$, $\mathcal{B} = \left\{ \begin{bmatrix} 1 \\ 0 \\ -1 \end{bmatrix}, \begin{bmatrix} -1 \\ 1 \\ 1 \end{bmatrix}, \begin{bmatrix} 2 \\ 1 \\ 0 \end{bmatrix} \right\}$, $[\vec{x}]_{\mathcal{B}} = \begin{bmatrix} -1 \\ 0 \\ 1 \end{bmatrix}$

B9 Find the $\mathcal{B}$-matrix of each of the following linear mappings.

(a) $L : \mathbb{R}^3 \to \mathbb{R}^3$ defined by $L(x_1, x_2, x_3) = (x_1 + x_2 + x_3, x_1 + 2x_2, x_1 + x_3)$, $\mathcal{B} = \left\{ \begin{bmatrix} 1 \\ 1 \\ 0 \end{bmatrix}, \begin{bmatrix} 0 \\ 0 \\ 1 \end{bmatrix}, \begin{bmatrix} 1 \\ 0 \\ 1 \end{bmatrix} \right\}$

(b) $L : P_2 \to P_2$ defined by $L(a + bx + cx^2) = (a+b+c) + (a+2b)x + (a+c)x^2$, $\mathcal{B} = \{1, x, x^2\}$

(c) $L : M(2,2) \to M(2,2)$ defined by $L\left(\begin{bmatrix} a & b \\ c & d \end{bmatrix}\right) = \begin{bmatrix} a & b+c \\ 0 & d \end{bmatrix}$, $\mathcal{B} = \left\{ \begin{bmatrix} 1 & 1 \\ 0 & 0 \end{bmatrix}, \begin{bmatrix} 1 & 0 \\ 0 & 1 \end{bmatrix}, \begin{bmatrix} 1 & 1 \\ 0 & 1 \end{bmatrix}, \begin{bmatrix} 0 & 0 \\ 1 & 0 \end{bmatrix} \right\}$

(d) $D : P_2 \to P_2$ defined by $D(a + bx + cx^2) = b + 2ax$, $\mathcal{B} = \{1 + 2x + 3x^2, -2x + x^2, 1 + x + x^2\}$

(e) $T : \mathbb{D} \to \mathbb{D}$, where $\mathbb{D}$ is the subspace of diagonal matrices in $M(2,2)$, defined by $T\left(\begin{bmatrix} a & 0 \\ 0 & b \end{bmatrix}\right) = \begin{bmatrix} a+b & 0 \\ 0 & 2a+b \end{bmatrix}$, $\mathcal{B} = \left\{ \begin{bmatrix} 1 & 0 \\ 0 & 1 \end{bmatrix}, \begin{bmatrix} 2 & 0 \\ 0 & 3 \end{bmatrix} \right\}$

Conceptual Problems

D1 Suppose that $\mathcal{B}$ and $\mathcal{C}$ are bases for $\mathbb{R}^n$ and $\mathcal{S}$ is the standard basis of $\mathbb{R}^n$. Suppose that P is the change of coordinates matrix from $\mathcal{B}$ to $\mathcal{S}$ and that Q is the change of coordinates matrix from $\mathcal{C}$ to $\mathcal{S}$. Let $L : \mathbb{R}^n \to \mathbb{R}^n$ be a linear mapping. Express the matrix $[L]_{\mathcal{C}}$ in terms of $[L]_{\mathcal{B}}$, P, and Q.

D2 Suppose that a 2×2 matrix A is the standard matrix of a linear mapping $L : \mathbb{R}^2 \to \mathbb{R}^2$. Let $\mathcal{B} = \{\vec{v}_1, \vec{v}_2\}$ be a basis for $\mathbb{R}^2$ and let P denote the change of coordinates matrix from $\mathcal{B}$ to the standard basis. What conditions will have to be satisfied by the vectors $\vec{v}_1$ and $\vec{v}_2$ in order for $P^{-1}AP = \begin{bmatrix} d_1 & 0 \\ 0 & d_2 \end{bmatrix} = D$ for some $d_1, d_2 \in \mathbb{R}$? (Hint: Consider the equation $AP = PD$, or $A\begin{bmatrix} \vec{v}_1 & \vec{v}_2 \end{bmatrix} = \begin{bmatrix} \vec{v}_1 & \vec{v}_2 \end{bmatrix} D$.)

D3 Let $\mathbb{V}$ be a vector space with basis $\mathcal{B} = \{\mathbf{v}_1, \ldots, \mathbf{v}_n\}$, let $\mathbb{W}$ be a vector space with basis $\mathcal{C}$, and let $L : \mathbb{V} \to \mathbb{W}$ be a linear mapping. Prove that the matrix ${}_{\mathcal{C}}[L]_{\mathcal{B}}$ defined by

$$ {}_{\mathcal{C}}[L]_{\mathcal{B}} = \begin{bmatrix} [L(\mathbf{v}_1)]_{\mathcal{C}} & \cdots & [L(\mathbf{v}_n)]_{\mathcal{C}} \end{bmatrix} $$

satisfies $[L(\mathbf{x})]_{\mathcal{C}} = {}_{\mathcal{C}}[L]_{\mathcal{B}}[\mathbf{x}]_{\mathcal{B}}$ and hence is the **matrix of L with respect to basis $\mathcal{B}$ and $\mathcal{C}$**.

D4 Determine the matrix of the following linear mappings with respect to the given bases $\mathcal{B}$ and $\mathcal{C}$.

(a) $D : P_2 \to P_1$ defined by $D(a + bx + cx^2) = b + 2cx$, $\mathcal{B} = \{1, x, x^2\}$, $\mathcal{C} = \{1, x\}$

(b) $L : \mathbb{R}^2 \to P_2$ defined by $L(a_1, a_2) = (a_1 + a_2) + a_1x^2$, $\mathcal{B} = \left\{ \begin{bmatrix} 1 \\ -1 \end{bmatrix}, \begin{bmatrix} 1 \\ 2 \end{bmatrix} \right\}$, $\mathcal{C} = \{1 + x^2, 1 + x, -1 - x + x^2\}$

(c) $T : \mathbb{R}^2 \to M(2,2)$ defined by $T\left(\begin{bmatrix} a \\ b \end{bmatrix}\right) = \begin{bmatrix} a+b & 0 \\ 0 & a-b \end{bmatrix}$, $\mathcal{B} = \left\{ \begin{bmatrix} 2 \\ -1 \end{bmatrix}, \begin{bmatrix} 1 \\ 2 \end{bmatrix} \right\}$, $\mathcal{C} = \left\{ \begin{bmatrix} 1 & 1 \\ 0 & 0 \end{bmatrix}, \begin{bmatrix} 1 & 0 \\ 0 & 1 \end{bmatrix}, \begin{bmatrix} 1 & 1 \\ 0 & 1 \end{bmatrix}, \begin{bmatrix} 0 & 0 \\ 1 & 0 \end{bmatrix} \right\}$

(d) $L : P_2 \to \mathbb{R}^2$ defined by $L(a + bx + cx^2) = \begin{bmatrix} a+c \\ b-a \end{bmatrix}$, $\mathcal{B} = \{1 + x^2, 1 + x, -1 + x + x^2\}$, $\mathcal{C} = \left\{ \begin{bmatrix} 1 \\ 0 \end{bmatrix}, \begin{bmatrix} 1 \\ 1 \end{bmatrix} \right\}$

4.7 Isomorphisms of Vector Spaces

Some of the ideas discussed in Chapters 3 and 4 lead to generalizations that are important in the further development of linear algebra (and also in abstract algebra). Some of these generalizations are outlined in this section. Most of the proofs are easy or simple variations on proofs given earlier, so they will be left as exercises. Throughout this section, it is assumed that $\mathbb{U}$, $\mathbb{V}$, and $\mathbb{W}$ are vector spaces over $\mathbb{R}$ and that $L : \mathbb{U} \to \mathbb{V}$ and $M : \mathbb{V} \to \mathbb{W}$ are linear mappings.

Definition
One-to-One

L is said to be **one-to-one** if $L(\mathbf{u}_1) = L(\mathbf{u}_2)$ implies that $\mathbf{u}_1 = \mathbf{u}_2$.

Lemma 1

L is one-to-one if and only if $\text{Null}(L) = \{\mathbf{0}\}$. (Compare this to Theorem 3.5.6.)

You are asked to prove Lemma 1 as Problem D1.

EXAMPLE 1

Every invertible linear transformation $L : \mathbb{R}^n \to \mathbb{R}^n$ is one-to-one. The fact that such a transformation is one-to-one allows the definition of the inverse. The mapping inj : $\mathbb{R}^3 \to \mathbb{R}^4$ of Example 3.5.8 is a one-to-one mapping that is not invertible. The mapping $P : \mathbb{R}^4 \to \mathbb{R}^3$ of Example 3.5.8 is not one-to-one. For any $\vec{\mathbf{n}}$, $\text{proj}_{\vec{\mathbf{n}}} : \mathbb{R}^3 \to \mathbb{R}^3$ is *not* one-to-one, since many elements in the domain are mapped to the same vector in the range.

EXAMPLE 2

Prove that $L : \mathbb{R}^2 \to \mathbb{R}^3$ defined by $L(x_1, x_2) = (x_1, x_1 + x_2, x_2)$ is one-to-one.

Solution: Assume that $L(x_1, x_2) = L(y_1, y_2)$. Then we have $(x_1, x_1 + x_2, x_2) = (y_1, y_1 + y_2, y_2)$, and so $x_1 = y_1$ and $x_2 = y_2$. Thus, L is one-to-one.

EXAMPLE 3

Determine if $M : P_2 \to M(2, 2)$ defined by $L(a + bx + cx^2) = \begin{bmatrix} a & 0 \\ 0 & b \end{bmatrix}$ is one-to-one.

Solution: Let $p(x) = 1 + x$ and $q(x) = 1 + x + x^2$. Observe that $M(p) = \begin{bmatrix} 1 & 0 \\ 0 & 1 \end{bmatrix} = M(q)$, but $p(x) \neq q(x)$, so M is not one-to-one.

EXERCISE 1

Suppose that $\{\mathbf{u}_1, \dots, \mathbf{u}_k\}$ is a linearly independent set in $\mathbb{U}$ and L is one-to-one. Prove that $\{L(\mathbf{u}_1), \dots, L(\mathbf{u}_k)\}$ is linearly independent.

Definition
Onto

$L : \mathbb{U} \to \mathbb{V}$ is said to be **onto** if for every $\mathbf{v} \in \mathbb{V}$ there exists some $\mathbf{u} \in \mathbb{U}$ such that $L(\mathbf{u}) = \mathbf{v}$. That is, $\text{Range}(L) = \mathbb{V}$.

EXAMPLE 4

Invertible linear transformations of $\mathbb{R}^n$ are all onto mappings. The mapping $P : \mathbb{R}^4 \to \mathbb{R}^3$ of Example 3.5.8 is onto, but the mapping inj : $\mathbb{R}^3 \to \mathbb{R}^4$ of Example 3.5.8 is not onto.

EXAMPLE 5

Prove that $L : \mathbb{R}^2 \to P_1$ defined by $L(y_1, y_2) = y_1 + (y_1 + y_2)x$ is onto.

Solution: Let $a + bx$ be any polynomial in P_1. We need to find a vector $\vec{y} = \begin{bmatrix} y_1 \\ y_2 \end{bmatrix} \in \mathbb{R}^2$, such that $L(\vec{y}) = a + bx$. For this to be true, we require that $y_1 + (y_1 + y_2)x = L(y_1, y_2) = a + bx$, so $y_1 = a$ and $b = y_1 + y_2$, which gives $y_2 = b - y_1 = b - a$. Therefore, we have $L(a, b - a) = a + bx$, and so L is onto.

EXAMPLE 6

Determine if $M : \mathbb{R}^2 \to \mathbb{R}^3$ defined by $M(x_1, x_2) = (x_1, x_1 + x_2, x_2)$ is onto.

Solution: If M is onto, then for every vector $\vec{y} = \begin{bmatrix} y_1 \\ y_2 \\ y_3 \end{bmatrix} \in \mathbb{R}^3$, there exists $\vec{x} = \begin{bmatrix} x_1 \\ x_2 \end{bmatrix} \in \mathbb{R}^2$ such that $L(\vec{x}) = \vec{y}$. But, if $\vec{y} = \begin{bmatrix} 1 \\ 1 \\ 1 \end{bmatrix}$, then we have

$$\begin{bmatrix} 1 \\ 1 \\ 1 \end{bmatrix} = M(x_1, x_2) = \begin{bmatrix} x_1 \\ x_1 + x_2 \\ x_2 \end{bmatrix}$$

which implies that $x_1 = 1$, $x_2 = 1$, and $x_1 + x_2 = 1$, which is clearly inconsistent. Hence, M is not onto.

EXERCISE 2

Suppose that $\{\mathbf{u}_1, \dots, \mathbf{u}_k\}$ is a spanning set for $\mathbb{U}$ and L is onto. Prove that a spanning set for $\mathbb{V}$ is $\{L(\mathbf{u}_1), \dots, L(\mathbf{u}_k)\}$.

Theorem 2

The linear mapping $L : \mathbb{U} \to \mathbb{V}$ has an inverse linear mapping $L^{-1} : \mathbb{V} \to \mathbb{U}$ if and only if L is one-to-one and onto.

You are asked to prove Theorem 2 as problem D4.

Definition
Isomorphism
Isomorphic

If $\mathbb{U}$ and $\mathbb{V}$ are vector spaces over $\mathbb{R}$, and if $L : \mathbb{U} \to \mathbb{V}$ is a linear, one-to-one, and onto mapping, then L is called an **isomorphism** (or a vector space isomorphism), and $\mathbb{U}$ and $\mathbb{V}$ are said to be **isomorphic**.

The word *isomorphism* comes from Greek words meaning "same form." The concept of an isomorphism is a very powerful and important one. It implies that the essential structure of the isomorphic vector spaces is the same, so that a vector space statement that is true in one space is immediately true in any isomorphic space. Of course, some vector spaces such as $M(m, n)$ or P_n have some features that are not purely vector space properties (such as matrix decomposition and polynomial factorization), and these particular features cannot automatically be transferred from these spaces to spaces that are isomorphic as vector spaces.

EXAMPLE 7

Prove that P_2 and $\mathbb{R}^3$ are isomorphic by constructing an explicit isomorphism L.

Solution: We define $L : P_2 \to \mathbb{R}^3$ by $L(a_0 + a_1x + a_2x^2) = \begin{bmatrix} a_0 \\ a_1 \\ a_2 \end{bmatrix}$.

To prove that it is an isomorphism, we must prove that it is linear, one-to-one, and onto.

Linear: Let any two elements of P_2 be $p(x) = a_0 + a_1x + a_2x^2$ and $q(x) = b_0 + b_1x + b_2x^2$ and let $t \in \mathbb{R}$. Then,

$$\begin{aligned} L(tp + q) &= L(t(a_0 + a_1x + a_2x^2) + (b_0 + b_1x + b_2x^2)) \\ &= L(ta_0 + b_0 + (ta_1 + b_1)x + (ta_2 + b_2)x^2) \\ &= \begin{bmatrix} ta_0 + b_0 \\ ta_1 + b_1 \\ ta_2 + b_2 \end{bmatrix} \\ &= t\begin{bmatrix} a_0 \\ a_1 \\ a_2 \end{bmatrix} + \begin{bmatrix} b_0 \\ b_1 \\ b_2 \end{bmatrix} \\ &= tL(p) + L(q) \end{aligned}$$

Therefore, L is linear.

One-to-one: Let $a_0 + a_1x + a_2x^2 \in \text{Null}(L)$. Then, $\begin{bmatrix} 0 \\ 0 \\ 0 \end{bmatrix} = L(a_0 + a_1x + a_2x^2) = \begin{bmatrix} a_0 \\ a_1 \\ a_2 \end{bmatrix}$.

Hence, $a_0 = a_1 = a_2$, so $\text{Null}(L) = \{0\}$ and thus L is one-to-one by Lemma 1.

Onto: For any $\begin{bmatrix} a_0 \\ a_1 \\ a_2 \end{bmatrix} \in \mathbb{R}^3$, we have $L(a_0 + a_1x + a_2x^2) = \begin{bmatrix} a_0 \\ a_1 \\ a_2 \end{bmatrix}$. Hence, L is onto.

Thus, L is an isomorphism from P_2 to $\mathbb{R}^3$.

EXERCISE 3

Use Exercise 1 and Exercise 2 to prove that if $L : \mathbb{U} \to \mathbb{V}$ is an isomorphism and $\{\mathbf{u}_1, \dots, \mathbf{u}_n\}$ is a basis for $\mathbb{U}$, then $\{L(\mathbf{u}_1), \dots, L(\mathbf{u}_n)\}$ is a basis for $\mathbb{V}$.

Theorem 3

Suppose that $\mathbb{U}$ and $\mathbb{V}$ are finite-dimensional vector spaces over $\mathbb{R}$. Then $\mathbb{U}$ and $\mathbb{V}$ are isomorphic if and only if they are of the same dimension.

You are asked to prove Theorem 3 as Problem D5.

EXAMPLE 8

1. The vector space $M(m, n)$ is isomorphic to $\mathbb{R}^{mn}$.
2. The vector space P_n is isomorphic to $\mathbb{R}^{n+1}$.
3. Every k-dimensional subspace of $\mathbb{R}^n$ is isomorphic to every k-dimensional subspace of $M(m, n)$.

If we know that two vector spaces over $\mathbb{R}$ have the same dimension, then Theorem 3 says that they are isomorphic. However, even if we already know that two vector spaces are isomorphic, we may need to construct an explicit isomorphism between the two vector spaces. The following theorem shows that if we have two isomorphic vector spaces $\mathbb{U}$ and $\mathbb{V}$, then we only have to check if a linear mapping $L : \mathbb{U} \to \mathbb{V}$ is one-to-one or onto to prove that it is an isomorphism between these two spaces.

Theorem 4 If $\mathbb{U}$ and $\mathbb{V}$ are n-dimensional vector spaces over $\mathbb{R}$, then a linear mapping $L : \mathbb{U} \to \mathbb{V}$ is one-to-one if and only if it is onto.

You are asked to prove Theorem 4 as Problem D6.

PROBLEMS 4.7

Practice Problems

A1 For each of the following pairs of vector spaces, define an explicit isomorphism to establish that the spaces are isomorphic. Prove that your map is an isomorphism.

(a) P_3 and $\mathbb{R}^4$

(b) $M(2,2)$ and $\mathbb{R}^4$

(c) P_3 and $M(2,2)$

(d) $\mathbb{P} = \{p(x) \in P_2 \mid p(2) = 0\}$ and the vector space $\mathbb{U} = \left\{ \begin{bmatrix} a_1 & 0 \\ 0 & a_2 \end{bmatrix} \mid a_1, a_2 \in \mathbb{R} \right\}$ of 2×2 diagonal matrices

Homework Problems

B1 For each of the following pairs of vector spaces, define an explicit isomorphism to establish that the spaces are isomorphic. Prove that your map is an isomorphism.

(a) P_4 and $\mathbb{R}^5$

(b) $M(2,3)$ and $\mathbb{R}^6$

(c) $\mathbb{R}^2$ and the vector space $\mathcal{S} = \operatorname{Span}\left\{ \begin{bmatrix} 1 \\ 0 \\ 1 \end{bmatrix}, \begin{bmatrix} 1 \\ 2 \\ 1 \end{bmatrix} \right\}$

(d) $\mathbb{P} = \{p(x) \in P_3 \mid p(1) = 0\}$ and the vector space $\mathbb{T} = \left\{ \begin{bmatrix} a_1 & a_2 \\ 0 & a_3 \end{bmatrix} \mid a_1, a_2, a_3 \in \mathbb{R} \right\}$ of 2×2 upper-triangular matrices

Conceptual Problems

D1 Prove Lemma 1. (Hint: Suppose that L is one-to-one. What is the unique $\mathbf{u} \in \mathbb{U}$ such that $L(\mathbf{u}) = \mathbf{0}$? Conversely, suppose that $\operatorname{Null}(L) = \{\mathbf{0}\}$. If $L(\mathbf{u}_1) = L(\mathbf{u}_2)$, then what is $L(\mathbf{u}_1 - \mathbf{u}_2)$?)

D2 (a) Prove that if L and M are one-to-one, then $M \circ L$ is one-to-one.

(b) Give an example where M is not one-to-one but $M \circ L$ is one-to-one.

(c) Is it possible to give an example where L is not one-to-one but $M \circ L$ is one-to-one? Explain.

D3 Prove that if L and M are onto, then $M \circ L$ is onto.

D4 Prove Theorem 2.

D5 Prove Theorem 3. To prove "isomorphic $\Rightarrow$ same dimension," use Exercise 3. To prove "same dimension $\Rightarrow$ isomorphic," take a basis $\{\mathbf{u}_1, \dots, \mathbf{u}_n\}$ for $\mathbb{U}$, and a basis $\{\mathbf{v}_1, \dots, \mathbf{v}_n\}$ for $\mathbb{V}$. Define an isomorphism by taking $L(\mathbf{u}_i) = \mathbf{v}_i$ for $1 \le i \le n$, requiring that L be linear. (You must prove that this is an isomorphism.)

D6 Prove Theorem 4.

D7 Prove that any plane through the origin in $\mathbb{R}^3$ is isomorphic to $\mathbb{R}^2$.

D8 Recall the definition of the Cartesian product from Problem 4.1.D4. Prove that $\mathbb{U} \times \{\mathbf{0}_{\mathbb{V}}\}$ is a subspace of $\mathbb{U} \times \mathbb{V}$ that is isomorphic to $\mathbb{U}$.

D9 (a) Prove that $\mathbb{R}^2 \times \mathbb{R}$ is isomorphic to $\mathbb{R}^3$.
(b) Prove that $\mathbb{R}^n \times \mathbb{R}^m$ is isomorphic to $\mathbb{R}^{n+m}$.

D10 Suppose that $L : \mathbb{U} \to \mathbb{V}$ is a vector space isomorphism and that $M : \mathbb{V} \to \mathbb{V}$ is a linear mapping. Prove that $L^{-1} \circ M \circ L$ is a linear mapping from $\mathbb{U}$ to $\mathbb{U}$. Describe the nullspace and range of $L^{-1} \circ M \circ L$ in terms of the nullspace and range of M.

CHAPTER REVIEW

Suggestions for Student Review

Remember that if you have understood the ideas of Chapter 4, you should be able to give answers to these questions without looking them up. Try hard to answer them from your own understanding.

1 State the essential properties of a vector space over $\mathbb{R}$. Why is the empty set not a vector space? Describe two or three examples of vector spaces that are not subspaces of $\mathbb{R}^n$. (Section 4.2)

2 What is a basis? What are the important properties of a basis? (Section 4.3)

3 (a) Explain the concept of dimension. What theorem is required to ensure that the concept of dimension is well defined. (Section 4.3)
(b) Explain how knowing the dimension of a vector space is helpful when you have to find a basis for the vector space. (Section 4.3)

4 Why is linear independence of a spanning set important when we define coordinates with respect to the spanning set? (Section 4.4)

5 Invent and analyze an example as follows.
(a) Give a basis $\mathcal{B}$ for a three-dimensional subspace in $\mathbb{R}^5$. (Don't make it too easy by choosing any standard basis vectors, but don't make it too hard by choosing completely random components.) (Section 4.3)
(b) Determine the standard coordinates in $\mathbb{R}^5$ of the vector that has coordinate vector $\begin{bmatrix} 2 \\ -3 \\ 4 \end{bmatrix}$ with respect to your basis $\mathcal{B}$. (Section 4.4)
(c) Take the vector you found in (b) and carry out the standard procedure to determine its coordinates with respect to $\mathcal{B}$. Did you get the right answer, $\begin{bmatrix} 2 \\ -3 \\ 4 \end{bmatrix}$? (Section 4.4)
(d) Pick any two vectors in $\mathbb{R}^5$ and determine whether they lie in your subspace. Determine the coordinates of any vector that is in the subspace. (Section 4.4)

6 Write a short explanation of how you use information about consistency of systems and uniqueness of solutions in testing for linear independence and in determining whether a vector belongs to a given subspace. (Sections 4.3 and 4.4)

7 Give the definition of a linear mapping $L : \mathbb{V} \to \mathbb{W}$ and show how this implies that L preserves linear combinations. Explain the procedure for determining if a vector $\mathbf{y}$ is in the range of L. Describe how to find a basis for the nullspace and a basis for the range of L. (Section 4.5)

8 State how to determine the standard matrix and the $\mathcal{B}$-matrix of a linear mapping $L : \mathbb{V} \to \mathbb{V}$. Explain how $[L(\vec{x})]_{\mathcal{B}}$ is determined in terms of $[L]_{\mathcal{B}}$. (Section 4.6)

9 State the definition of an isomorphism of vector spaces and give some examples. Explain why a finite-dimensional vector space cannot be isomorphic to a proper subspace of itself. (Section 4.7)

Chapter Quiz

E1 Determine whether the following sets are vector spaces; explain briefly.

(a) The set of 4×3 matrices such that the sum of the entries in the first row is zero ($a_{11} + a_{12} + a_{13} = 0$) under standard addition and scalar multiplication of matrices.

(b) The set of polynomials $p(x)$ such that $p(1) = 0$ and $p(2) = 0$ under standard addition and scalar multiplication of polynomials.

(c) The set of 2×2 matrices such that all entries are integers under standard addition and scalar multiplication of matrices.

(d) The set of all vectors $\begin{bmatrix} x_1 \\ x_2 \\ x_3 \end{bmatrix}$ such that $x_1 + x_2 + x_3 = 0$ under standard addition and scalar multiplication of vectors.

E2 In each of the following cases, determine whether the given set of vectors is a basis for $M(2, 2)$.

(a) $\left\{ \begin{bmatrix} 1 & 1 \\ 2 & 1 \end{bmatrix}, \begin{bmatrix} 0 & 1 \\ 1 & -1 \end{bmatrix}, \begin{bmatrix} 0 & 1 \\ 1 & 3 \end{bmatrix}, \begin{bmatrix} 2 & 2 \\ 4 & -2 \end{bmatrix}, \begin{bmatrix} 0 & 2 \\ 3 & 0 \end{bmatrix} \right\}$

(b) $\left\{ \begin{bmatrix} 1 & 1 \\ 2 & 1 \end{bmatrix}, \begin{bmatrix} 0 & 1 \\ 1 & -1 \end{bmatrix}, \begin{bmatrix} 0 & 1 \\ 1 & 3 \end{bmatrix}, \begin{bmatrix} 2 & 2 \\ 4 & -2 \end{bmatrix} \right\}$

(c) $\left\{ \begin{bmatrix} 1 & 3 \\ -1 & 4 \end{bmatrix}, \begin{bmatrix} 2 & 2 \\ 0 & 3 \end{bmatrix}, \begin{bmatrix} 1 & 0 \\ 2 & 0 \end{bmatrix} \right\}$

E3 (a) Let $\mathbb{S}$ be the subspace spanned by $\vec{v}_1 = \begin{bmatrix} 1 \\ 0 \\ 1 \\ 1 \\ 3 \end{bmatrix}$, $\vec{v}_2 = \begin{bmatrix} 1 \\ 1 \\ 0 \\ 1 \\ 1 \end{bmatrix}$, $\vec{v}_3 = \begin{bmatrix} 3 \\ 3 \\ 1 \\ 0 \\ 2 \end{bmatrix}$, and $\vec{v}_4 = \begin{bmatrix} 1 \\ 1 \\ 1 \\ -2 \\ 0 \end{bmatrix}$. Pick a subset of the given vectors that forms a basis $\mathcal{B}$ for $\mathbb{S}$. Determine the dimension of $\mathbb{S}$.

(b) Determine the coordinates of $\vec{x} = \begin{bmatrix} 0 \\ 2 \\ -1 \\ -3 \\ -5 \end{bmatrix}$ with respect to $\mathcal{B}$.

E4 (a) Find a basis for the plane in $\mathbb{R}^3$ with equation $x_1 - x_3 = 0$.

(b) Extend the basis you found in (a) to a basis $\mathcal{B}$ for $\mathbb{R}^3$.

(c) Let $L : \mathbb{R}^3 \to \mathbb{R}^3$ be a reflection in the plane from part (a). Determine $[L]_{\mathcal{B}}$.

(d) Using your result from part (c), determine the standard matrix $[L]_S$ of the reflection.

E5 Let $L : \mathbb{R}^3 \to \mathbb{R}^3$ be a linear mapping with standard matrix $\begin{bmatrix} 1 & -1 & 2 \\ -1 & 0 & 1 \\ -2 & 1 & 0 \end{bmatrix}$ and let $\mathcal{B} = \left\{ \begin{bmatrix} 1 \\ 1 \\ 0 \end{bmatrix}, \begin{bmatrix} 0 \\ 1 \\ 1 \end{bmatrix}, \begin{bmatrix} 1 \\ -1 \\ 1 \end{bmatrix} \right\}$. Determine the matrix $[L]_{\mathcal{B}}$.

E6 Suppose that $L : \mathbb{V} \to \mathbb{W}$ is a linear mapping with $\text{Null}(L) = \{\mathbf{0}\}$. Suppose that $\{\mathbf{v}_1, \ldots, \mathbf{v}_k\}$ is a linearly independent set in $\mathbb{V}$. Prove that $\{L(\mathbf{v}_1), \ldots, L(\mathbf{v}_k)\}$ is a linearly independent set in $\mathbb{W}$.

E7 Decide whether each of the following statements is true or false. If it is true, explain briefly; if it is false, give an example to show that it is false.

(a) A subspace of $\mathbb{R}^n$ must have dimension less than n.

(b) A set of four polynomials in P_2 cannot be a basis for P_2.

(c) If $\mathcal{B}$ is a basis for a subspace of $\mathbb{R}^5$, the $\mathcal{B}$-coordinate vector of some vector $\vec{x} \in \mathbb{R}^5$ has five components.

(d) For any linear mapping $L : \mathbb{R}^n \to \mathbb{R}^n$ and any basis $\mathcal{B}$ of $\mathbb{R}^n$, the rank of the matrix $[L]_{\mathcal{B}}$ is the same as the rank of the matrix $[L]_S$.

(e) For any linear mapping $L : \mathbb{V} \to \mathbb{V}$ and any basis $\mathcal{B}$ of $\mathbb{V}$, the column space of $[L]_{\mathcal{B}}$ equals the range of L.

(f) If $L : \mathbb{V} \to \mathbb{W}$ is one-to-one, then $\dim \mathbb{V} = \dim \mathbb{W}$.

Further Problems

These problems are intended to be challenging. They may not be of interest to all students.

F1 Let $\mathbb{S}$ be a subspace of an n-dimensional vector space $\mathbb{V}$. Prove that there exists a linear operator $L : \mathbb{V} \to \mathbb{V}$ such that $\text{Null}(L) = \mathbb{S}$.

F2 Use the ideas of this chapter to prove the uniqueness of the reduced row echelon form for a given matrix A. (Hint: Begin by assuming that there are two reduced row echelon forms R and S. What can you say about the columns with leading 1s in the two matrices?)

F3 Magic Squares—An Exploration of Their Vector Space Properties

We say that any matrix $A \in M(3,3)$ is a 3×3 **magic square** if the three **row sums** (where each row sum is the sum of the entries in one row of A) of A, the three **column sums** of A, and the two **diagonal sums** of A ($a_{11} + a_{22} + a_{33}$ and $a_{13} + a_{22} + a_{31}$) all have the same value k. The common sum k is called the **weight** of the magic square A and is denoted by $wt(A) = k$.

For example, if $A = \begin{bmatrix} 2 & 2 & -1 \\ -2 & 1 & 4 \\ 3 & 0 & 0 \end{bmatrix}$, A is a magic square with $wt(A) = 3$.

The aim of this exploration is to find all 3×3 magic squares. The subset of $M(3,3)$ consisting of magic squares is denoted MS_3.

(a) Show that MS_3 is a subspace of $M(3,3)$.

(b) Observe that weight determines a map $wt : MS_3 \to \mathbb{R}$. Show that wt is linear.

(c) Compute the nullspace of wt. Suppose that

$$\underline{X}_1 = \begin{bmatrix} 1 & 0 & a \\ b & c & d \\ e & f & g \end{bmatrix}, \quad \underline{X}_2 = \begin{bmatrix} 0 & 1 & h \\ i & j & k \\ l & m & n \end{bmatrix}$$

and

$$\underline{0} = \begin{bmatrix} 0 & o & p \\ q & r & s \\ t & u & v \end{bmatrix}$$

are all in the nullspace, where $a, b, c, \dots, v$ denote unknown entries. Determine these unknown entries and prove that $\underline{X}_1$ and $\underline{X}_2$ form a basis for $\text{Null}(wt)$. (Hint: If $A \in \text{Null}(wt)$, consider $A - a_{11}\underline{X}_1 - a_{12}\underline{X}_2$.)

(d) Let $\underline{J} = \begin{bmatrix} 1 & 1 & 1 \\ 1 & 1 & 1 \\ 1 & 1 & 1 \end{bmatrix}$. Observe that $\underline{J}$ is a magic square with $wt(\underline{J}) = 3$. Show that all A in MS_3 that have weight k are of the form

$$(k/3)\underline{J} + p\underline{X}_1 + q\underline{X}_2$$

for some $p, q \in \mathbb{R}$.

(e) Show that $\mathcal{B} = \{\underline{J}, \underline{X}_1, \underline{X}_2\}$ is a basis for MS_3.

(f) As an example, find all 3×3 magic squares of weight 1.

(g) Find the coordinates of $A = \begin{bmatrix} 3 & 1 & 2 \\ 1 & 2 & 3 \\ 2 & 3 & 1 \end{bmatrix}$ with respect to the basis $\mathcal{B}$.

Conclusion: MS_3 is a three-dimensional subspace of $M(3,3)$.

Exercises F4–F7 require the following definitions.

If $\mathbb{S}$ and $\mathbb{T}$ are subspaces of the vector space $\mathbb{V}$, we define

$$\mathbb{S} + \mathbb{T} = \{p\mathbf{s} + q\mathbf{t} \mid p, q \in \mathbb{R}, \mathbf{s} \in \mathbb{S}, \mathbf{t} \in \mathbb{T}\}$$

If $\mathbb{S}$ and $\mathbb{T}$ are subspaces of $\mathbb{V}$ such that $\mathbb{S} + \mathbb{T} = \mathbb{V}$ and $\mathbb{S} \cap \mathbb{T} = \{\mathbf{0}\}$, we say that $\mathbb{S}$ is the **complement** of $\mathbb{T}$ (and $\mathbb{T}$ is the complement of $\mathbb{S}$). In general, given a subspace $\mathbb{S}$ of $\mathbb{V}$, one can choose a complement in many ways; the complement of $\mathbb{S}$ is not unique. For example, in $\mathbb{R}^2$, we may take a complement of $\text{Span}\left\{\begin{bmatrix} 1 \\ 0 \end{bmatrix}\right\}$ to be $\text{Span}\left\{\begin{bmatrix} 0 \\ 1 \end{bmatrix}\right\}$ or $\text{Span}\left\{\begin{bmatrix} 1 \\ 1 \end{bmatrix}\right\}$.

F4 In the vector space of continuous real-valued functions of a real variable, show that the even functions and the odd functions form subspaces such that each is the complement of the other.

F5 (a) If $\mathbb{S}$ is a k-dimensional subspace of $\mathbb{R}^n$, show that any complement of $\mathbb{S}$ must be of dimension $n - k$.

(b) Suppose that $\mathbb{S}$ is a subspace of $\mathbb{R}^n$ that has a unique complement. Must it be true that $\mathbb{S}$ is either $\{\vec{0}\}$ or $\mathbb{R}^n$?

F6 Suppose that $\mathbf{v}$ and $\mathbf{w}$ are vectors in a vector space $\mathbb{V}$. Suppose also that $\mathbb{S}$ is a subspace of $\mathbb{V}$. Let $\mathbb{T}$ be the subspace spanned by $\mathbf{v}$ and $\mathbb{S}$. Let $\mathbb{U}$ be the subspace spanned by $\mathbf{w}$ and $\mathbb{S}$. Prove that if $\mathbf{w}$ is in $\mathbb{T}$ but not in $\mathbb{S}$, then $\mathbf{v}$ is in $\mathbb{U}$.

F7 Show that if $\mathbb{S}$ and $\mathbb{T}$ are finite-dimensional subspaces of $\mathbb{V}$, then

$$\dim \mathbb{S} + \dim \mathbb{T} = \dim(\mathbb{S} + \mathbb{T}) + \dim(\mathbb{S} \cap \mathbb{T})$$

Visit the text's website at www.pearsoncanada.ca/norman for practice quizzes, additional applications, and an essay on linearity and superposition in physics.

CHAPTER 5

Determinants

CHAPTER OUTLINE

For each square matrix A, *we define a number called the determinant of* A. *Originally, the determinant was used to "determine" whether a system of* n *linear equations in* n *variables was consistent. Now the determinant also provides a second method for finding the inverse of a matrix. It also plays an important role in the discussion of volume. Finally, it is an important tool for finding eigenvalues in Chapter 6.*

5.1 Determinants in Terms of Cofactors

Consider the system of two linear equations in two variables:

$$\begin{aligned} a_{11}x_1 + a_{12}x_2 &= b_1 \\ a_{21}x_1 + a_{22}x_2 &= b_2 \end{aligned}$$

By the standard procedure of elimination, the system is found to be consistent if and only if $a_{11}a_{22} - a_{12}a_{21} \neq 0$. If it is consistent, then the solution is found to be

$$x_1 = \frac{a_{22}b_1 - a_{12}b_2}{a_{11}a_{22} - a_{12}a_{21}}, \quad x_2 = \frac{a_{11}b_2 - a_{21}b_1}{a_{11}a_{22} - a_{12}a_{21}}$$

This fact prompts the following definition.

Definition
Determinant of a 2×2 Matrix

The **determinant of a 2×2 matrix** $A = \begin{bmatrix} a_{11} & a_{12} \\ a_{21} & a_{22} \end{bmatrix}$ is defined by

$$\det A = \det \begin{bmatrix} a_{11} & a_{12} \\ a_{21} & a_{22} \end{bmatrix} = a_{11}a_{22} - a_{12}a_{21}$$

EXAMPLE 1

Find the determinant of $\begin{bmatrix} 1 & 3 \\ 2 & 4 \end{bmatrix}$, $\begin{bmatrix} 2 & 7 \\ 8 & -5 \end{bmatrix}$, and $\begin{bmatrix} 2 & 2 \\ 4 & 2 \end{bmatrix}$.

Solution: We have

$$\det\begin{bmatrix} 1 & 3 \\ 2 & 4 \end{bmatrix} = 1(4) - 3(2) = -2$$

$$\det\begin{bmatrix} 2 & 7 \\ 8 & -5 \end{bmatrix} = 2(-5) - 7(8) = -10 - 56 = -66$$

$$\det\begin{bmatrix} 2 & 2 \\ 4 & 4 \end{bmatrix} = 2(4) - 2(4) = 0$$

An Alternate Notation: The determinant is often denoted by vertical straight lines:

$$\begin{vmatrix} a_{11} & a_{12} \\ a_{21} & a_{22} \end{vmatrix} = \det\begin{bmatrix} a_{11} & a_{12} \\ a_{21} & a_{22} \end{bmatrix} = a_{11}a_{22} - a_{12}a_{21}$$

One risk with this notation is that one may fail to distinguish between a matrix and the determinant of the matrix. This is a rather gross error.

EXERCISE 1

Calculate the following determinants.

(a) $\begin{vmatrix} 3 & 2 \\ 2 & 1 \end{vmatrix}$ (b) $\begin{vmatrix} 1 & 3 \\ 0 & -2 \end{vmatrix}$ (c) $\begin{vmatrix} 2 & 4 \\ 1 & 2 \end{vmatrix}$

The 3×3 Case

Let A be a 3×3 matrix. We can show through elimination (with some effort) that the system is consistent with a unique solution if and only if

$$D = a_{11}a_{22}a_{33} - a_{11}a_{23}a_{32} - a_{12}a_{21}a_{33} + a_{12}a_{23}a_{31} + a_{13}a_{21}a_{32} - a_{13}a_{22}a_{31} \neq 0$$

We would like to simplify or reorganize this expression so that we can remember it more easily and so that we can determine how to generalize it to the $n \times n$ case.

Notice that a_{11} is a common factor in the first pair of terms in D, a_{12} is a common factor in the second pair, and a_{13} is a common factor in the third pair. Thus, D can be rewritten as

$$\begin{aligned} D &= a_{11}(a_{22}a_{33} - a_{23}a_{32}) - a_{12}(a_{21}a_{33} - a_{23}a_{31}) + a_{13}(a_{21}a_{32} - a_{22}a_{31}) \\ &= a_{11}\begin{vmatrix} a_{22} & a_{23} \\ a_{32} & a_{33} \end{vmatrix} + a_{12}(-1)\begin{vmatrix} a_{21} & a_{23} \\ a_{31} & a_{33} \end{vmatrix} + a_{13}\begin{vmatrix} a_{21} & a_{22} \\ a_{31} & a_{32} \end{vmatrix} \end{aligned}$$

Observe that the determinant being multiplied by a_{11} is the determinant of the 2×2 matrix formed by removing the first row and first column of A. Similarly, a_{12} is being multiplied by (-1) times the determinant of the matrix formed by removing the first row and second column of A, and a_{13} is being multiplied by the determinant of the matrix formed by removing the first row and third column of A. Hence, we make the following definitions.

Definition
Cofactors of a 3×3 Matrix

Let A be a 3×3 matrix. Let $A(i, j)$ denote the 2×2 submatrix obtained from A by deleting the i-th row and j-th column. Define the **cofactor** of **a 3×3 matrix** a_{ij} to be

$$C_{ij} = (-1)^{(i+j)} \det A(i, j)$$

Definition
Determinant of a 3×3 Matrix

The **determinant of a 3×3 matrix** A is defined by

$$\det A = a_{11}C_{11} + a_{12}C_{12} + a_{13}C_{13}$$

Remarks

1. This definition of the determinant of a 3×3 matrix is called the **expansion of the determinant along the first row**. As we shall see below, a determinant can be expanded along any row or column.

2. The signs attached to cofactors can cause trouble if you are not careful. One helpful way to remember which sign to attach to which cofactor is to take a blank matrix and put a + in the top-left corner and then alternate − and + both across and down: $\begin{bmatrix} + & - & + \\ - & + & - \\ + & - & + \end{bmatrix}$. This is shown for a 3×3 matrix, but it works for a square matrix of any size.

3. C_{ij} is called the cofactor of a_{ij} because it is the "factor with a_{ij}" in the expansion of the determinant. Note that each C_{ij} is a number not a matrix.

EXAMPLE 2

Let $A = \begin{bmatrix} 4 & -1 & 1 \\ 2 & 3 & 5 \\ 1 & 0 & 6 \end{bmatrix}$. Calculate the cofactors of the first row of A and use them to find the determinant of A.

Solution: By definition, the cofactor C_{11} is $(-1)^{1+1}$ times the determinant of the matrix obtained from A by deleting the first row and first column. Thus,

$$C_{11} = (-1)^{1+1} \det \begin{bmatrix} 3 & 5 \\ 0 & 6 \end{bmatrix} = 3(6) - 5(0) = 18$$

EXAMPLE 2 (continued)

The cofactor C_{12} is $(-1)^{1+2}$ times the determinant of the matrix obtained from A by deleting the first row and second column. So,

$$C_{12} = (-1)^{1+2} \det \begin{bmatrix} 2 & 5 \\ 1 & 6 \end{bmatrix} = -[2(6) - 5(1)] = -7$$

Finally, the cofactor C_{13} is

$$C_{13} = (-1)^{1+3} \det \begin{bmatrix} 2 & 3 \\ 1 & 0 \end{bmatrix} = 2(0) - 3(1) = -3$$

Hence,

$$\det A = a_{11}C_{11} + a_{12}C_{12} + a_{13}C_{13} = 4(18) + (-1)(-7) + 1(-3) = 76$$

EXERCISE 2

Let $A = \begin{bmatrix} 1 & 2 & 3 \\ 0 & -1 & -2 \\ 4 & 0 & -3 \end{bmatrix}$. Calculate the cofactors of the first row of A and use them to find the determinant of A.

Generally, when expanding a determinant, we write the steps more compactly, as in the next example.

EXAMPLE 3

Calculate $\det \begin{bmatrix} 1 & 2 & 3 \\ -2 & 2 & 1 \\ 5 & 0 & -1 \end{bmatrix}$.

Solution: By definition, we have

$$\det \begin{bmatrix} 1 & 2 & 3 \\ -2 & 2 & 1 \\ 5 & 0 & -1 \end{bmatrix} = a_{11}C_{11} + a_{12}C_{12} + a_{13}C_{13}$$

$$= 1(-1)^{1+1} \begin{vmatrix} 2 & 1 \\ 0 & -1 \end{vmatrix} + 2(-1)^{1+2} \begin{vmatrix} -2 & 1 \\ 5 & -1 \end{vmatrix} + 3(-1)^{1+3} \begin{vmatrix} -2 & 2 \\ 5 & 0 \end{vmatrix}$$

$$= 1[2(-1) - 1(0)] - 2[(-2)(-1) - 1(5)] + 3[(-2)(0) - 2(5)]$$

$$= -2 + 6 - 30 = -26$$

We now define the determinant of an $n \times n$ matrix by following the pattern of the definition for the 3×3 case.

Definition

Cofactors of an $n \times n$ Matrix

Let A be a $n \times n$ matrix. Let $A(i, j)$ denote the $(n-1) \times (n-1)$ submatrix obtained from A by deleting the i-th row and j-th column. The **cofactor of an $n \times n$ matrix** of a_{ij} is defined to be

$$C_{ij} = (-1)^{(i+j)} \det A(i, j)$$

Definition

Determinant of a $n \times n$ Matrix

The **determinant of an $n \times n$ matrix** A is defined by

$$\det A = a_{11}C_{11} + a_{12}C_{12} + \cdots + a_{1n}C_{1n}$$

Remark

This is a recursive definition. The result for the $n \times n$ case is defined in terms of the $(n-1) \times (n-1)$ case, which in turn must be calculated in terms of the $(n-2) \times (n-2)$ case, and so on, until we get back to the 2×2 case, for which the result is given explicitly.

EXAMPLE 4

We calculate the following determinant by using the definition of the determinant. Note that * and ** represent cofactors whose values are irrelevant because they are multiplied by 0.

$$\det \begin{bmatrix} 0 & 2 & 3 & 0 \\ 1 & 5 & 6 & 7 \\ -2 & 3 & 0 & 4 \\ -5 & 1 & 2 & 3 \end{bmatrix} = a_{11}C_{11} + a_{12}C_{12} + a_{13}C_{13} + a_{14}C_{14}$$

$$= 0(*) + 2(-1)^{1+2} \det \begin{vmatrix} 1 & 6 & 7 \\ -2 & 0 & 4 \\ -5 & 2 & 3 \end{vmatrix} + 3(-1)^{1+3} \begin{vmatrix} 1 & 5 & 7 \\ -2 & 3 & 4 \\ -5 & 1 & 3 \end{vmatrix} + 0(**)$$

$$= -2\left(1(-1)^{1+1} \begin{vmatrix} 0 & 4 \\ 2 & 3 \end{vmatrix} + 6(-1)^{1+2} \begin{vmatrix} -2 & 4 \\ -5 & 3 \end{vmatrix} + 7(-1)^{1+3} \begin{vmatrix} -2 & 0 \\ -5 & 2 \end{vmatrix}\right)$$

$$+3\left(1(-1)^{1+1} \begin{vmatrix} 3 & 4 \\ 1 & 3 \end{vmatrix} + 5(-1)^{1+2} \begin{vmatrix} -2 & 4 \\ -5 & 3 \end{vmatrix} + 7(-1)^{1+3} \begin{vmatrix} -2 & 3 \\ -5 & 1 \end{vmatrix}\right)$$

$$= -2((0-8) - 6(-6+20) + 7(-4-0))$$

$$+3((9-4) - 5(-6+20) + 7(-2+15))$$

$$= -2(-8-84-28) + 3(5-70+91)$$

$$= -2(-120) + 3(26) = 318$$

It is apparent that evaluating the determinant of a 4×4 matrix is a fairly lengthy calculation, and things will get worse for larger matrices. In applications it is not uncommon to have a matrix with thousands (or even millions) of columns, so it is very important to have results that simplify the evaluation of the determinant. We look at a few useful theorems here and some very helpful theorems in the next section.

Theorem 1

Suppose that A is an $n \times n$ matrix. Then the determinant of A may be obtained by a **cofactor expansion** along any row or any column. In particular, the expansion of the determinant along the i-th row of A is

$$\det A = a_{i1}C_{i1} + a_{i2}C_{i2} + \cdots + a_{in}C_{in}$$

The expansion of the determinant along the j-th column of A is

$$\det A = a_{1j}C_{1j} + a_{2j}C_{2j} + \cdots + a_{nj}C_{nj}$$

We omit a proof here since there is no conceptually helpful proof, and it would be a bit grim to verify the result in the general case.

Theorem 1 is a very practical result. It allows us to *choose* from A the row or column along which we are going to expand. If one row or column has many zeros, it is sensible to expand along it since we shall then not have to evaluate the cofactors of the zero entries. This was demonstrated in Example 4, where we had to compute only two cofactors.

EXAMPLE 5

Calculate the determinant of $A = \begin{bmatrix} 1 & 2 & -1 \\ 3 & 1 & 0 \\ -1 & 5 & 0 \end{bmatrix}$, $B = \begin{bmatrix} 3 & 2 & 0 & -1 \\ 0 & 6 & 0 & 0 \\ 4 & 1 & 2 & 1 \\ 3 & -1 & 0 & 1 \end{bmatrix}$, and

$$C = \begin{bmatrix} 4 & 2 & 1 & -1 \\ 0 & 2 & 2 & 2 \\ 0 & 0 & -1 & 3 \\ 0 & 0 & 0 & 4 \end{bmatrix}.$$

Solution: For A, we expand along the third column to get

$$\begin{aligned} \det A &= a_{13}C_{13} + a_{23}C_{23} + a_{33}C_{33} \\ &= (-1)(-1)^{1+3}\begin{vmatrix} 3 & 1 \\ -1 & 5 \end{vmatrix} + 0 + 0 \\ &= -1(15 - (-1)) = -16 \end{aligned}$$

For B, we expand along the second row to get

$$\begin{aligned} \det B &= a_{21}C_{21} + a_{22}C_{22} + a_{23}C_{23} + a_{24}C_{24} \\ &= 0 + (6)(-1)^{2+2}\begin{vmatrix} 3 & 0 & -1 \\ 4 & 2 & 1 \\ 3 & 0 & 1 \end{vmatrix} + 0 + 0 \\ &= 0 + 6\left(2(-1)^{2+2}\begin{vmatrix} 3 & -1 \\ 3 & 1 \end{vmatrix}\right) + 0 \\ &= 12(3 - (-3)) = 72 \end{aligned}$$

We expanded the 3×3 submatrix along the second column. For C we continuously expand along the bottom row to get

$$\begin{aligned} \det C &= a_{41}C_{41} + a_{42}C_{42} + a_{43}C_{43} + a_{44}C_{44} \\ &= 0 + 0 + 0 + 4(-1)^{4+4}\begin{vmatrix} 4 & 2 & 1 \\ 0 & 2 & 2 \\ 0 & 0 & -1 \end{vmatrix} \\ &= 4\left((-1)(-1)^{3+3}\begin{vmatrix} 4 & 2 \\ 0 & 2 \end{vmatrix}\right) \\ &= 4(-1)(4(2) - 0) = 4(-1)(4)(2) = -32 \end{aligned}$$

EXERCISE 3

Calculate the determinant of $A = \begin{bmatrix} 1 & 3 & 2 & 0 \\ 0 & 0 & -1 & 2 \\ 3 & 5 & -1 & 0 \\ -2 & 2 & -4 & 0 \end{bmatrix}$ by

1. Expanding along the first column
2. Expanding along the second row
3. Expanding along the fourth column

Exercise 3 demonstrates the usefulness of expanding along the row or column with the most zeros. Of course, if one row or column contains only zeros, then this is even easier.

Theorem 2 If one row (or column) of an $n \times n$ matrix A contains only zeros, then $\det A = 0$.

Proof: If the i-th row of A contains only zeros, then expanding the determinant along the i-th row of A gives

$$\det A = a_{i1}C_{i1} + a_{i2}C_{i2} + \cdots + a_{in}C_{in} = 0 + 0 + \cdots + 0 = 0$$ ■

As we saw with the matrix C in Example 5, another useful special case is when the matrix is upper or lower triangular.

Theorem 3 If A is an $n \times n$ upper- or lower-triangular matrix, then the determinant of A is the product of the diagonal entries of A. That is,

$$\det A = a_{11}a_{22} \cdots a_{nn}$$

The proof is left as Problem D1.

Finally, recall that taking the transpose of a matrix A turns rows into columns and vice versa. That is, the columns of A^T are identical to the rows of A. Thus, if we expand the determinant of A^T along its first column, we will get the same cofactors and coefficients we would get by expanding the determinant of A along its first row. So, we get the following theorem.

Theorem 4 If A is an $n \times n$ matrix, then $\det A = \det A^T$.

With the tools we have so far, evaluation of determinants is still a very tedious business. Properties of the determinant with respect to elementary row operations make the evaluation much easier. These properties are discussed in the next section.

PROBLEMS 5.1

Practice Problems

A1 Evaluate the following determinants.

(a) $\begin{vmatrix} 2 & -4 \\ 7 & 5 \end{vmatrix}$ (b) $\begin{vmatrix} -3 & 1 \\ 2 & 1 \end{vmatrix}$

(c) $\begin{vmatrix} 1 & 1 \\ 1 & 1 \end{vmatrix}$ (d) $\begin{vmatrix} -3 & 4 & 0 & -7 \\ 3 & -4 & 0 & 2 \\ 1 & 5 & 0 & -5 \\ 1 & 2 & 0 & 0 \end{vmatrix}$

(e) $\begin{vmatrix} 1 & 3 & -4 \\ 0 & 0 & 2 \\ 0 & 0 & 3 \end{vmatrix}$ (f) $\begin{vmatrix} 1 & 5 & -7 & 8 \\ 2 & -1 & 3 & 0 \\ -4 & 2 & 0 & 0 \\ 1 & 0 & 0 & 0 \end{vmatrix}$

A2 Evaluate the determinants of the following matrices by expanding along the first row.

(a) $\begin{bmatrix} 3 & 4 & 0 \\ 2 & 1 & -1 \\ -4 & -1 & 2 \end{bmatrix}$

(b) $\begin{bmatrix} 3 & 2 & 1 \\ -1 & 4 & 5 \\ 3 & 2 & 1 \end{bmatrix}$

(c) $\begin{bmatrix} 2 & 1 & 0 & -1 \\ 0 & 3 & 2 & 1 \\ -4 & 0 & 2 & -2 \\ 3 & -5 & 2 & 1 \end{bmatrix}$

(d) $\begin{bmatrix} 1 & 0 & 4 & 0 \\ 2 & -3 & 4 & 1 \\ -1 & 3 & 2 & 4 \\ 1 & 1 & -2 & 4 \end{bmatrix}$

A3 Evaluate the determinants of the following matrices by expanding along the row or column of your choice.

(a) $\begin{bmatrix} 3 & 5 & 0 \\ -2 & 6 & 0 \\ 4 & 1 & 0 \end{bmatrix}$

(b) $\begin{bmatrix} -5 & 2 & -4 \\ 2 & -4 & 6 \\ -6 & 2 & -3 \end{bmatrix}$

(c) $\begin{bmatrix} 2 & 1 & 5 \\ 4 & 3 & -1 \\ 0 & 1 & -2 \end{bmatrix}$

(d) $\begin{bmatrix} -3 & 4 & 0 & 1 \\ 4 & -1 & 0 & -6 \\ 1 & -1 & 0 & -3 \\ 4 & -2 & 3 & 6 \end{bmatrix}$

(e) $\begin{bmatrix} 0 & 6 & 1 & 2 \\ 0 & 5 & -1 & 1 \\ 3 & -5 & -3 & -5 \\ 5 & 6 & -3 & -6 \end{bmatrix}$

(f) $\begin{bmatrix} 1 & 3 & 4 & -5 & 7 \\ 0 & -3 & 1 & 2 & 3 \\ 0 & 0 & 4 & 1 & 0 \\ 0 & 0 & 0 & -1 & 8 \\ 0 & 0 & 0 & 4 & 3 \end{bmatrix}$

A4 Show that the determinant of each matrix is equal to the determinant of its transpose by expanding along the second column of A and the second row of A^T.

(a) $A = \begin{bmatrix} 1 & 3 & -1 \\ 2 & 1 & 0 \\ -1 & 0 & 5 \end{bmatrix}$

(b) $A = \begin{bmatrix} 1 & 2 & 3 & 4 \\ -2 & 0 & 2 & 5 \\ 3 & 0 & 1 & 4 \\ 4 & 5 & 1 & -2 \end{bmatrix}$

A5 Calculate the determinant of each of the following elementary matrices.

(a) $E_1 = \begin{bmatrix} 1 & 0 & 0 \\ 0 & 0 & 1 \\ 0 & 1 & 0 \end{bmatrix}$

(b) $E_2 = \begin{bmatrix} 1 & 0 & 3 \\ 0 & 1 & 0 \\ 0 & 0 & 1 \end{bmatrix}$

(c) $E_3 = \begin{bmatrix} -3 & 0 & 0 \\ 0 & 1 & 0 \\ 0 & 0 & 1 \end{bmatrix}$

Homework Problems

B1 Evaluate the following determinants.

(a) $\begin{vmatrix} 3 & 5 \\ 6 & 10 \end{vmatrix}$ (b) $\begin{vmatrix} 4 & -2 \\ -2 & 4 \end{vmatrix}$

(c) $\begin{vmatrix} 1 & -1 \\ 1 & 1 \end{vmatrix}$ (d) $\begin{vmatrix} 3 & -5 & 2 & 1 \\ 0 & 0 & 0 & 0 \\ 3 & -3 & 9 & 1 \\ 1 & 1 & -1 & 1 \end{vmatrix}$

(e) $\begin{vmatrix} 3 & 0 & 0 \\ 1 & 2 & 0 \\ -1 & 0 & 1 \end{vmatrix}$ (f) $\begin{vmatrix} 4 & 2 & 2 & -1 \\ 0 & 3 & 9 & -4 \\ 0 & 0 & 2 & -2 \\ 0 & 0 & 0 & 1 \end{vmatrix}$

B2 Evaluate the determinants of the following matrices by expanding along the first row.

(a) $\begin{bmatrix} 1 & 5 & 1 \\ 1 & 6 & 1 \\ 7 & 1 & 7 \end{bmatrix}$

(b) $\begin{bmatrix} 3 & -6 & 0 \\ -3 & 2 & 1 \\ 2 & 5 & 4 \end{bmatrix}$

(c) $\begin{bmatrix} -2 & 0 & 0 & -2 \\ 0 & 2 & 2 & 0 \\ 4 & 0 & 0 & -4 \\ 0 & -4 & 4 & 0 \end{bmatrix}$

(d) $\begin{bmatrix} -1 & -1 & 0 & 0 \\ 0 & 2 & 5 & -6 \\ 1 & 0 & 1 & -1 \\ 2 & 2 & 0 & -3 \end{bmatrix}$

B3 Evaluate the determinants of the following matrices by expanding along the row or column of your choice.

(a) $\begin{bmatrix} 2 & 5 & 4 \\ -3 & 2 & 1 \\ 3 & -6 & 0 \end{bmatrix}$

(b) $\begin{bmatrix} 3 & -2 & 4 \\ 3 & 0 & 0 \\ 0 & 8 & -3 \end{bmatrix}$

(c) $\begin{bmatrix} 0 & 1 & -2 \\ 2 & 0 & 1 \\ 0 & 1 & 0 \end{bmatrix}$

(d) $\begin{bmatrix} 2 & 3 & -3 & 1 \\ -3 & 1 & 0 & 5 \\ 1 & 2 & 1 & -2 \\ 3 & 0 & 1 & 0 \end{bmatrix}$

(e) $\begin{bmatrix} 4 & 2 & 5 & -4 \\ -1 & 0 & 2 & 3 \\ -2 & -1 & -4 & 3 \\ 3 & 0 & -3 & 2 \end{bmatrix}$

(f) $F = \begin{bmatrix} 1 & 3 & 4 & -5 & 7 \\ 3 & 3 & 1 & 2 & 0 \\ 2 & -1 & 4 & 1 & 0 \\ 5 & 3 & 0 & 0 & 0 \\ -2 & 0 & 0 & 0 & 0 \end{bmatrix}$

B4 Show that the determinant of each matrix is equal to the determinant of its transpose by expanding along the third row of A and the third column of A^T.

(a) $A = \begin{bmatrix} 2 & -1 & -2 \\ 1 & 0 & 3 \\ 3 & -1 & 0 \end{bmatrix}$

(b) $A = \begin{bmatrix} 3 & 1 & 6 & 2 \\ 1 & 6 & 3 & 5 \\ -5 & 6 & 0 & 0 \\ 4 & -3 & 4 & 0 \end{bmatrix}$

B5 Calculate the determinant of each of the following elementary matrices.

(a) $E_1 = \begin{bmatrix} 1 & 0 & 0 \\ 0 & 5 & 0 \\ 0 & 0 & 1 \end{bmatrix}$

(b) $E_2 = \begin{bmatrix} 0 & 0 & 1 \\ 0 & 1 & 0 \\ 1 & 0 & 0 \end{bmatrix}$

(c) $E_3 = \begin{bmatrix} 1 & 0 & 0 \\ 0 & 1 & 0 \\ 0 & -17 & 1 \end{bmatrix}$

Computer Problems

C1 Use a computer to evaluate the determinants of the following matrices.

(a) $\begin{bmatrix} 1.09 & 4.83 & 2.95 \\ 2.13 & -3.25 & 1.57 \\ 1.72 & 2.15 & -0.89 \end{bmatrix}$

(b) $\begin{bmatrix} 1.23 & 2.35 & 4.19 & -1.28 \\ -2.09 & 0.17 & 3.89 & 22.1 \\ 0.78 & 2.15 & -3.55 & 4.15 \\ 1.58 & -2.59 & 1.01 & 0.00 \end{bmatrix}$

(c) $\begin{bmatrix} 0.5 & 0.5 & 0.5 & 0.5 \\ 0.5 & -0.5 & 0.5 & -0.5 \\ 0.5 & 0.5 & -0.5 & -0.5 \\ -0.5 & 0.5 & -0.5 & 0.5 \end{bmatrix}$

Conceptual Problems

D1 Prove Theorem 3.

D2 (a) Consider the points (a_1, a_2) and (b_1, b_2) in $\mathbb{R}^2$. Show that the **determinantal equation** $\det \begin{bmatrix} x_1 & x_2 & 1 \\ a_1 & a_2 & 1 \\ b_1 & b_2 & 1 \end{bmatrix} = 0$ is the equation of the line containing the two points.

(b) Write a determinantal equation for a plane in $\mathbb{R}^3$ that contains the points (a_1, a_2, a_3), (b_1, b_2, b_3), and (c_1, c_2, c_3).

5.2 Elementary Row Operations and the Determinant

Calculating the determinant of a matrix can be lengthy. This calculation can often be simplified by applying elementary row operations to the matrix, provided that suitable adjustments are made to the determinant.

To see what happens to the determinant of a matrix A when we multiply a row of A by a constant, we first consider a 3×3 example. Following the example, we state and prove the general result.

EXAMPLE 1

Let $A = \begin{bmatrix} a_{11} & a_{12} & a_{13} \\ a_{21} & a_{22} & a_{23} \\ a_{31} & a_{32} & a_{33} \end{bmatrix}$ and let B be the matrix obtained from A by multiplying the third row of A by the real number r. Show that $\det B = r \det A$.

Solution: We have $B = \begin{bmatrix} a_{11} & a_{12} & a_{13} \\ a_{21} & a_{22} & a_{23} \\ ra_{31} & ra_{32} & ra_{33} \end{bmatrix}$. We wish to expand the determinant of B along its third row. The cofactors for this row are

$$C_{31} = (-1)^{3+1} \begin{vmatrix} a_{12} & a_{13} \\ a_{22} & a_{23} \end{vmatrix}, \quad C_{32} = (-1)^{3+2} \begin{vmatrix} a_{11} & a_{13} \\ a_{21} & a_{23} \end{vmatrix}, \quad C_{33} = (-1)^{3+3} \begin{vmatrix} a_{11} & a_{12} \\ a_{21} & a_{22} \end{vmatrix}$$

EXAMPLE 1 (continued)

Observe that these are also the cofactors for the third row of A. Hence,

$$\det B = ra_{31}C_{31} + ra_{32}C_{32} + ra_{33}C_{33} = r(a_{31}C_{31} + a_{32}C_{32} + a_{33}C_{33}) = r \det A$$

Theorem 1

Let A be an $n \times n$ matrix and let B be the matrix obtained from A by multiplying the i-th row of A by the real number r. Then, $\det B = r \det A$.

Proof: As in the example, we expand the determinant of B along the i-th row. Notice that the cofactors of the elements in this row are exactly the cofactors of the i-th row of A since all the other rows of B are identical to the corresponding rows in A. Therefore,

$$\det B = ra_{i1}C_{i1} + \cdots + ra_{in}C_{in} = r\left(a_{i1}C_{i1} + \cdots + a_{in}C_{in}\right) = r \det A$$ ■

Remark

It is important to be careful when using this theorem as it is not uncommon to incorrectly use the reciprocal $(1/r)$ of the factor. One way to counter this error is to think of "factoring out" the value of r from a row of the matrix. Keep this in mind when reading the following example.

EXAMPLE 2

By Theorem 1,

$$\begin{vmatrix} 1 & 3 & 4 \\ 5 & 10 & 15 \\ 2 & -1 & 0 \end{vmatrix} = 5 \begin{vmatrix} 1 & 3 & 4 \\ 1 & 2 & 3 \\ 2 & -1 & 0 \end{vmatrix}$$

and

$$\det \begin{bmatrix} -2 & -2 & 4 \\ 1 & 2 & 1 \\ 0 & 3 & 1 \end{bmatrix} = (-2) \det \begin{bmatrix} 1 & 1 & -2 \\ 1 & 2 & 1 \\ 0 & 3 & 1 \end{bmatrix}$$

EXERCISE 1

Let A be a 3×3 matrix and let $r \in \mathbb{R}$. Use Theorem 1 to show that $\det(rA) = r^3 \det A$.

Next, we consider the effect of swapping two rows.

EXAMPLE 3 The following calculations illustrate that swapping rows causes a change of sign of the determinant. We have

$$\det\begin{bmatrix} c & d \\ a & b \end{bmatrix} = cb - da = -(ad - bc) = -\det\begin{bmatrix} a & b \\ c & d \end{bmatrix}$$

Let $A = \begin{bmatrix} a_{11} & a_{12} & a_{13} \\ a_{21} & a_{22} & a_{23} \\ a_{31} & a_{32} & a_{33} \end{bmatrix}$ and let B be the matrix obtained from A by swapping the first and third rows of A. We expand the determinant of B along the second row (the row that has not been swapped):

$$\begin{aligned}
\det B &= \det\begin{bmatrix} a_{31} & a_{32} & a_{33} \\ a_{21} & a_{22} & a_{23} \\ a_{11} & a_{12} & a_{13} \end{bmatrix} \\
&= a_{21}(-1)^{2+1}\begin{vmatrix} a_{32} & a_{33} \\ a_{12} & a_{13} \end{vmatrix} + a_{22}(-1)^{2+2}\begin{vmatrix} a_{31} & a_{33} \\ a_{11} & a_{13} \end{vmatrix} + a_{23}(-1)^{2+3}\begin{vmatrix} a_{31} & a_{32} \\ a_{11} & a_{12} \end{vmatrix} \\
&= -a_{21}\left(-\begin{vmatrix} a_{12} & a_{13} \\ a_{32} & a_{33} \end{vmatrix}\right) + a_{22}\left(-\begin{vmatrix} a_{11} & a_{13} \\ a_{31} & a_{33} \end{vmatrix}\right) - a_{23}\left(-\begin{vmatrix} a_{11} & a_{12} \\ a_{31} & a_{32} \end{vmatrix}\right) \\
&= -\left(-a_{21}\begin{vmatrix} a_{12} & a_{13} \\ a_{32} & a_{33} \end{vmatrix} + a_{22}\begin{vmatrix} a_{11} & a_{13} \\ a_{31} & a_{33} \end{vmatrix} - a_{23}\begin{vmatrix} a_{11} & a_{12} \\ a_{31} & a_{32} \end{vmatrix}\right) \\
&= -\det\begin{bmatrix} a_{11} & a_{12} & a_{13} \\ a_{21} & a_{22} & a_{23} \\ a_{31} & a_{32} & a_{33} \end{bmatrix}
\end{aligned}$$

Theorem 2 Suppose that A is an $n \times n$ matrix and that B is the matrix obtained from A by swapping two rows. Then $\det B = -\det A$.

Proof: We use induction. We verified the case where $n = 2$ in Example 3.

Assume that the result holds for any $(n-1) \times (n-1)$ matrix and suppose that B is an $n \times n$ matrix obtained from A by swapping two rows. If the i-th row of A was *not* swapped, then the cofactors of the i-th row of B are $(n-1) \times (n-1)$ matrices. We can obtain these matrices from the cofactors of the i-th row of A by swapping the same two rows. Hence, by the inductive hypothesis, the cofactors C^*_{ij} of B and C_{ij} of A satisfy $C^*_{ij} = -C_{ij}$. Hence,

$$\begin{aligned}
\det B &= a_{i1}C^*_{i1} + \cdots + a_{in}C^*_{in} = a_{i1}(-C_{i1}) + \cdots + a_{in}(-C_{in}) \\
&= -(a_{i1}C_{i1} + \cdots + a_{in}C_{in}) = -\det A
\end{aligned}$$

■

Corollary 3

If two rows of A are equal, then $\det A = 0$.

Proof: Let B be the matrix obtained from A by interchanging the two equal rows. Obviously $B = A$, so $\det B = \det A$. But, by Theorem 2, $\det B = -\det A$, so $\det A = -\det A$. This implies that $\det A = 0$. ■

Finally, we show that the third type of elementary row operation is particularly useful as it does not change the determinant.

EXAMPLE 4

For any $r \in \mathbb{R}$ we have

$$\det\begin{bmatrix} a & b \\ c+ra & d+rb \end{bmatrix} = a(d+rb)-b(c+ra) = ad+arb-bc-arb = ad-bc = \det\begin{bmatrix} a & b \\ c & d \end{bmatrix}$$

Let $A = \begin{bmatrix} a_{11} & a_{12} & a_{13} \\ a_{21} & a_{22} & a_{23} \\ a_{31} & a_{32} & a_{33} \end{bmatrix}$ and let B be the matrix obtained from A by adding r times row 2 to row 3. Expanding the determinant of B along the first row and using the result above gives

$$\begin{aligned}
\det B &= \det\begin{bmatrix} a_{11} & a_{12} & a_{13} \\ a_{21} & a_{22} & a_{23} \\ a_{31}+ra_{21} & a_{32}+ra_{22} & a_{33}+ra_{23} \end{bmatrix} \\
&= a_{11}(-1)^{1+1}\begin{vmatrix} a_{22} & a_{23} \\ a_{32}+ra_{22} & a_{33}+ra_{23} \end{vmatrix} + a_{12}(-1)^{1+2}\begin{vmatrix} a_{21} & a_{23} \\ a_{31}+ra_{21} & a_{33}+ra_{23} \end{vmatrix} \\
&\quad + a_{13}(-1)^{1+3}\begin{vmatrix} a_{21} & a_{22} \\ a_{31}+ra_{21} & a_{32}+ra_{22} \end{vmatrix} \\
&= a_{11}\begin{vmatrix} a_{22} & a_{23} \\ a_{32} & a_{33} \end{vmatrix} - a_{12}\begin{vmatrix} a_{21} & a_{23} \\ a_{31} & a_{33} \end{vmatrix} + a_{13}\begin{vmatrix} a_{21} & a_{22} \\ a_{31} & a_{32} \end{vmatrix} \\
&= \det\begin{bmatrix} a_{11} & a_{12} & a_{13} \\ a_{21} & a_{22} & a_{23} \\ a_{31} & a_{32} & a_{33} \end{bmatrix}
\end{aligned}$$

Theorem 4

Suppose that A is an $n \times n$ matrix and that B is obtained from A by adding r times the i-th row of A to the k-th row. Then $\det B = \det A$.

Proof: We use induction. We verified the case where $n = 2$ in Example 4.

Assume that the result holds for any $(n-1) \times (n-1)$ matrix and suppose that B is the $n \times n$ matrix obtained from A by adding r times the i-th row of A to the k-th row. Then the cofactors of any other row, say the ℓ-th row, of B are $(n-1) \times (n-1)$ matrices. We can obtain these matrices from the cofactors of the ℓ-th row of A by adding r times the i-th row to the k-th row. Hence, by the inductive hypothesis, the cofactors C_{ij}^* of B and C_{ij} of A are equal. Hence,

$$\det B = a_{\ell 1}C_{\ell 1}^* + \cdots + a_{\ell n}C_{\ell n}^* = a_{\ell 1}C_{\ell 1} + \cdots + a_{\ell n}C_{\ell n} = \det A$$ ■

Theorem 1, Theorem 2, Theorem 4, and Theorem 5.1.3 suggest that an effective strategy for evaluating the determinant of a matrix is to row reduce the matrix to upper-triangular form. For $n > 3$, it can be shown that in general, this strategy will require fewer arithmetic operations than straight cofactor expansion. The following example illustrates this strategy.

EXAMPLE 5

Let $A = \begin{bmatrix} 1 & 3 & 1 & 5 \\ 1 & 3 & -3 & -3 \\ 0 & 3 & 1 & 0 \\ 1 & 6 & 2 & 11 \end{bmatrix}$. Find $\det A$.

Solution: By Theorem 4, performing the row operations $R_2 - R_1$ and $R_4 - R_1$ does not change the determinant, so

$$\det A = \begin{vmatrix} 1 & 3 & 1 & 5 \\ 0 & 0 & -4 & -8 \\ 0 & 3 & 1 & 0 \\ 0 & 3 & 1 & 6 \end{vmatrix}$$

To get the matrix into upper-triangular form, we now swap row 2 and row 3. By Theorem 2, this gives

$$\det A = (-1)\begin{vmatrix} 1 & 3 & 1 & 5 \\ 0 & 3 & 1 & 0 \\ 0 & 0 & -4 & -8 \\ 0 & 3 & 1 & 6 \end{vmatrix}$$

By Theorem 1, factoring out the common factor of (-4) from the third row gives

$$\det A = (-1)(-4)\begin{vmatrix} 1 & 3 & 1 & 5 \\ 0 & 3 & 1 & 0 \\ 0 & 0 & 1 & 2 \\ 0 & 3 & 1 & 6 \end{vmatrix}$$

Finally, by Theorem 4, we perform the row operation $R_4 - R_2$ to get

$$\det A = 4\begin{vmatrix} 1 & 3 & 1 & 5 \\ 0 & 3 & 1 & 0 \\ 0 & 0 & 1 & 2 \\ 0 & 0 & 0 & 6 \end{vmatrix} = 4(1)(3)(1)(6) = 72$$

EXERCISE 2

Let $A = \begin{bmatrix} 2 & 4 & -2 & 6 \\ -6 & -6 & -2 & 5 \\ 1 & 1 & 3 & -1 \\ 4 & 6 & -2 & 5 \end{bmatrix}$. Find $\det A$.

In some cases, it may be appropriate to use some combination of row operations and cofactor expansion. We demonstrate this in the following example.

EXAMPLE 6

Find the determinant of $A = \begin{bmatrix} 1 & 5 & 6 & 7 \\ 1 & 8 & 7 & 9 \\ 1 & 5 & 6 & 10 \\ 0 & 1 & 4 & -2 \end{bmatrix}$.

Solution: By Theorem 4,

$$\det A = \begin{vmatrix} 1 & 5 & 6 & 7 \\ 0 & 3 & 1 & 2 \\ 0 & 0 & 0 & 3 \\ 0 & 1 & 4 & -2 \end{vmatrix}$$

Expanding along the first column gives

$$\det A = (-1)^{1+1} \begin{vmatrix} 3 & 1 & 2 \\ 0 & 0 & 3 \\ 1 & 4 & -2 \end{vmatrix} + 0$$

Expanding along the second row gives

$$\det A = (1)(3)(-1)^{2+3} \begin{vmatrix} 3 & 1 \\ 1 & 4 \end{vmatrix} = (-3)(12 - 1) = -33$$

EXERCISE 3

Find the determinant of $A = \begin{bmatrix} -6 & -2 & 4 & -5 \\ 3 & 2 & -4 & 3 \\ -6 & 4 & 0 & 0 \\ -3 & 2 & -3 & -4 \end{bmatrix}$.

The Determinant and Invertibility

It follows from Theorem 1, Theorem 2, and Theorem 4 that there is a connection between the determinant of a square matrix, its rank, and whether it is invertible.

Theorem 5 If A is an $n \times n$ matrix, then the following are equivalent:

(1) $\det A \neq 0$
(2) $\text{rank}(A) = n$
(3) A is invertible

Proof: In Theorem 3.5.4, we proved that (2) if and only if (3), so it is only necessary to prove that (1) if and only if (2).

Notice that Theorem 1, Theorem 2, and Theorem 4 indicate that if $\det A \neq 0$, then the matrices obtained from A by elementary row operations will also have non-zero determinants. Every matrix is row equivalent to a matrix in reduced row echelon form; this reduced row echelon form has a leading 1 in every entry on the main diagonal if and only if the rank of the matrix is n. Hence, a given matrix A is of rank n if and only if $\det A \neq 0$. ■

Remark

Theorem 5 shows that $\det A \neq 0$ is equivalent to all of the statements in Theorem 3.5.4 and Theorem 3.5.6.

We shall see how to use the determinant in calculating the inverse in the next section. It is worth noting that Theorem 5 implies that "almost all" square matrices are invertible; a square matrix fails to be invertible only if it satisfies the special condition $\det A = 0$.

Determinant of a Product

Often it is necessary to calculate the determinant of the product of two square matrices A and B. When you remember that each entry of AB is the dot product of a row from A and a column from B, and that the rule for calculating determinants is quite complicated, you might expect a very complicated rule. The following theorem should be a welcome surprise. But first we prove a useful lemma.

Lemma 6 If E is an $n \times n$ elementary matrix and C is any $n \times n$ matrix, then

$$\det(EC) = (\det E)(\det C)$$

Proof: Observe that if E is an elementary matrix, then since E is obtained by performing a single row operation on the identity matrix, we get by Theorem 1, Theorem 2, and Theorem 4 that $\det E = \alpha$, where α is 1, -1, or r, depending on the elementary row operation used. Moreover, since EC is the matrix obtained by performing that row operation on C, we get

$$\det(EC) = \alpha \det C = \det E \det C$$

■

Theorem 7 If A and B are $n \times n$ matrices, then $\det(AB) = (\det A)(\det B)$.

Proof: If $\det A = 0$, then $A\vec{y} = \vec{0}$ has infinitely many solutions by Theorem 5 and Theorem 3.5.4. If B is invertible, then $\vec{y} = B\vec{x}$ has a solution for every $\vec{y} \in \mathbb{R}^n$, and hence there are infinitely many $\vec{x}$ such that $AB\vec{x} = \vec{0}$, and so AB is not invertible. If B is not invertible, then there are infinitely many $\vec{x}$ such that $B\vec{x} = \vec{0}$. Then for all such $\vec{x}$ we get $AB\vec{x} = A\vec{0} = \vec{0}$. So, AB is not invertible. Thus, if $\det A = 0$, then AB is not invertible and hence $\det(AB) = 0$. Therefore, if $\det A = 0$, then $\det(AB) = (\det A)(\det B)$.

On the other hand, if $\det A \neq 0$, then by Theorem 3.5.4 the RREF of A is I. Thus, by Theorem 3.6.3, there exists a sequence of elementary matrices $E_1, \dots, E_k$ such that $A = E_1 \cdots E_k$. Hence $AB = E_1 \cdots E_k B$ and, by repeated use of Lemma 6, we get

$$\det(AB) = \det(E_1 \cdots E_k B) = (\det E_1)(\det E_2) \cdots (\det E_k)(\det B) = (\det A)(\det B) \quad \blacksquare$$

EXAMPLE 7 Verify Theorem 7 for $A = \begin{bmatrix} 3 & 0 & 1 \\ 2 & -1 & 4 \\ 5 & 2 & 0 \end{bmatrix}$ and $B = \begin{bmatrix} -1 & 2 & 4 \\ 7 & 1 & 0 \\ 1 & -2 & 3 \end{bmatrix}$.

Solution: By Theorem 4 we get

$$\det A = \begin{vmatrix} 3 & 0 & 1 \\ -10 & -1 & 0 \\ 5 & 2 & 0 \end{vmatrix} = \begin{vmatrix} -10 & -1 \\ 5 & 2 \end{vmatrix} = -15$$

$$\det B = \begin{vmatrix} -1 & 2 & 4 \\ 0 & 15 & 28 \\ 0 & 0 & 7 \end{vmatrix} = -105$$

So $(\det A)(\det B) = (-15)(-105) = 1575$. On the other hand, using Theorem 1 and Theorem 4 gives

$$\det AB = \det \begin{bmatrix} -2 & 4 & 15 \\ -5 & -5 & 20 \\ 9 & 12 & 20 \end{bmatrix} = (-5) \begin{vmatrix} -2 & 4 & 15 \\ 1 & 1 & -4 \\ 9 & 12 & 20 \end{vmatrix}$$

$$= (-5) \begin{vmatrix} 0 & 6 & 7 \\ 1 & 1 & -4 \\ 0 & 3 & 56 \end{vmatrix} = (-5)(-1)^{2+1} \begin{vmatrix} 6 & 7 \\ 3 & 56 \end{vmatrix}$$

$$= (5)(315) = 1575$$

PROBLEMS 5.2

Practice Problems

A1 Use row operations and triangular form to compute the determinants of each of the following matrices. Show your work clearly. Decide whether each matrix is invertible.

(a) $A = \begin{bmatrix} 1 & 2 & 4 \\ 3 & 1 & 0 \\ -1 & 3 & 2 \end{bmatrix}$

(b) $A = \begin{bmatrix} 3 & 2 & 2 \\ 2 & 2 & 1 \\ 1 & 1 & 1 \end{bmatrix}$

(c) $A = \begin{bmatrix} 5 & 2 & -1 & 1 \\ 1 & 2 & -1 & 1 \\ 3 & 2 & 1 & 4 \\ -2 & 0 & 3 & 5 \end{bmatrix}$

(d) $A = \begin{bmatrix} 1 & 1 & 3 & 1 \\ -2 & -2 & -4 & -1 \\ 2 & 2 & 8 & 3 \\ 1 & 1 & 7 & 3 \end{bmatrix}$

(e) $A = \begin{bmatrix} 5 & 10 & 5 & -5 \\ 1 & 3 & 5 & 7 \\ 1 & 2 & 6 & 3 \\ -1 & 7 & 1 & 1 \end{bmatrix}$

A2 Use a combination of row operations and cofactor expansion to evaluate the following determinants.

(a) $\begin{vmatrix} 1 & -1 & 2 \\ 1 & 1 & -2 \\ 1 & 2 & 3 \end{vmatrix}$

(b) $\begin{vmatrix} 2 & 4 & 2 \\ 4 & 2 & 1 \\ -2 & 2 & 2 \end{vmatrix}$

(c) $\begin{vmatrix} 1 & 2 & 1 & 2 \\ 2 & 4 & 1 & 5 \\ 3 & 6 & 5 & 9 \\ 1 & 3 & 4 & 3 \end{vmatrix}$

(d) $\begin{vmatrix} 2 & 0 & -2 & -6 \\ 2 & -6 & -4 & -1 \\ -3 & -4 & 5 & 3 \\ -2 & -1 & -3 & 2 \end{vmatrix}$

A3 Use row operations to compute the determinant of each of the following matrices. In each case, determine all values of p such that the matrix is invertible.

(a) $A = \begin{bmatrix} 2 & 3 & 1 \\ 1 & 1 & -1 \\ 4 & p & -2 \end{bmatrix}$

(b) $A = \begin{bmatrix} 2 & 3 & 1 & p \\ 0 & 1 & 2 & 1 \\ 0 & 1 & 7 & 6 \\ 1 & 0 & 1 & 0 \end{bmatrix}$

(c) $A = \begin{bmatrix} 1 & 1 & 1 & 1 \\ 1 & 2 & 3 & 4 \\ 1 & 4 & 9 & 16 \\ 1 & 8 & 27 & p \end{bmatrix}$

A4 Verify Theorem 7 if

(a) $A = \begin{bmatrix} 2 & -1 \\ 3 & 5 \end{bmatrix}$, $B = \begin{bmatrix} 3 & 2 \\ -1 & 4 \end{bmatrix}$

(b) $A = \begin{bmatrix} 2 & 4 & 1 \\ -1 & 2 & -1 \\ 3 & 0 & 2 \end{bmatrix}$, $B = \begin{bmatrix} 1 & 3 & 2 \\ -1 & 0 & 5 \\ 4 & 1 & 1 \end{bmatrix}$

A5 Suppose that A is an $n \times n$ matrix.

(a) Determine $\det(rA)$.

(b) If A is invertible, show that $\det A^{-1} = \frac{1}{\det A}$.

(c) If $A^3 = I$, prove that A is invertible.

Homework Problems

B1 Use row operations and triangular form to compute the determinants of each of the following matrices. Show your work clearly. Decide whether each matrix is invertible.

(a) $A = \begin{bmatrix} 2 & 3 & 2 \\ 2 & -1 & 1 \\ 1 & 1 & 4 \end{bmatrix}$

(b) $A = \begin{bmatrix} 1 & 4 & -4 \\ 2 & -3 & 3 \\ -3 & -4 & 4 \end{bmatrix}$

(c) $A = \begin{bmatrix} 1 & 2 & 3 \\ 2 & -4 & 1 \\ 3 & 5 & -6 \end{bmatrix}$

(d) $A = \begin{bmatrix} 7 & 1 & -1 & 1 \\ 3 & 3 & -4 & 5 \\ 3 & 2 & 1 & 4 \\ 1 & 1 & -1 & 1 \end{bmatrix}$

(e) $A = \begin{bmatrix} 1 & 10 & 7 & -9 \\ 7 & -7 & 7 & 7 \\ 2 & -2 & 6 & 2 \\ -3 & -3 & 4 & 1 \end{bmatrix}$

(f) $A = \begin{bmatrix} 1 & 3 & 1 & 1 \\ -2 & 1 & 2 & 1 \\ 1 & 2 & 1 & -1 \\ -4 & 5 & 8 & -3 \end{bmatrix}$

B2 Use a combination of row operations and cofactor expansion to evaluate the following determinants.

(a) $\begin{vmatrix} 1 & 3 & 3 \\ 1 & -2 & 2 \\ -1 & 4 & 5 \end{vmatrix}$

(b) $\begin{vmatrix} 6 & 1 & 2 \\ -4 & -1 & 3 \\ -3 & 1 & 5 \end{vmatrix}$

(c) $\begin{vmatrix} 1 & 1 & -2 & 1 \\ 1 & 3 & -1 & -1 \\ 2 & 2 & -2 & 7 \\ 1 & 1 & 0 & 2 \end{vmatrix}$

(d) $\begin{vmatrix} 2 & 3 & 2 & -2 \\ 3 & 1 & 4 & 1 \\ 1 & 2 & 4 & 4 \\ 2 & -1 & 5 & 6 \end{vmatrix}$

B3 Use row operations to compute the determinant of each of the following matrices. In each case, determine all values of p such that the matrix is invertible.

(a) $A = \begin{bmatrix} 1 & 0 & -1 \\ 2 & 1 & 2 \\ p & 1 & -2 \end{bmatrix}$

(b) $A = \begin{bmatrix} 1 & 1 & 1 \\ 2 & p & p \\ 2 & 2 & 1 \end{bmatrix}$

(c) $A = \begin{bmatrix} 2 & 5 & 2 & 4 \\ 0 & 1 & -1 & 1 \\ 0 & 1 & 4 & 2 \\ 1 & 0 & 1 & p \end{bmatrix}$

(d) $A = \begin{bmatrix} 1 & 1 & 1 & 1 \\ 1 & 2 & 4 & 8 \\ 1 & 3 & 9 & 27 \\ 1 & 4 & 16 & p \end{bmatrix}$

B4 Verify Theorem 7 if

(a) $A = \begin{bmatrix} 4 & -1 \\ -3 & 2 \end{bmatrix}$, $B = \begin{bmatrix} -5 & 1 \\ 1 & 2 \end{bmatrix}$

(b) $A = \begin{bmatrix} 3 & 1 & -1 \\ 1 & -3 & 4 \\ 0 & 1 & 1 \end{bmatrix}$, $B = \begin{bmatrix} 2 & 1 & 3 \\ 3 & 1 & 0 \\ 1 & 1 & 4 \end{bmatrix}$

Conceptual Problems

D1 A square matrix A is **skew-symmetric** if $A^T = -A$. If A is an $n \times n$ skew-symmetric matrix, with n odd, prove that $\det A = 0$.

D2 Assume that A and B are $n \times n$ matrices. If $\det AB = 0$, is it necessarily true that $\det A = 0$ or $\det B = 0$? Prove it or give a counterexample.

D3 Suppose that A is a 3×3 matrix and that $\det A = 0$. What can you say about the rows of A? Argue that the range of the matrix mapping $L(\vec{x}) = A\vec{x}$ cannot be all of $\mathbb{R}^3$ and that its nullspace cannot be trivial.

D4 Let A be an $n \times n$ matrix. Prove that if P is any $n \times n$ invertible matrix, then $\det P^{-1}AP = \det A$.

D5 (a) Prove that $\det \begin{bmatrix} a+p & b+q & c+r \\ d & e & f \\ g & h & k \end{bmatrix}$

$= \det \begin{bmatrix} a & b & c \\ d & e & f \\ g & h & k \end{bmatrix} + \det \begin{bmatrix} p & q & r \\ d & e & f \\ g & h & k \end{bmatrix}$.

(b) Use part (a) to express $\det\begin{bmatrix} a+p & b+q & c+r \\ d+x & e+y & f+z \\ g & h & k \end{bmatrix}$ as the sum of determinants of matrices whose entries are not sums.

D6 Prove that

$$\det\begin{bmatrix} a+b & p+q & u+v \\ b+c & q+r & v+w \\ c+a & r+p & w+u \end{bmatrix} = 2\det\begin{bmatrix} a & p & u \\ b & q & v \\ c & r & w \end{bmatrix}.$$

D7 Prove that $\det\begin{bmatrix} 1 & 1 & 1 \\ 1 & 1+a & 1+2a \\ 1 & (1+a)^2 & (1+2a)^2 \end{bmatrix} = 2a^3.$

D8 (a) Prove that

$$\det\begin{bmatrix} a_{11} & a_{12} & ra_{13} \\ a_{21} & a_{22} & ra_{23} \\ a_{31} & a_{32} & ra_{33} \end{bmatrix} = r\det\begin{bmatrix} a_{11} & a_{12} & a_{13} \\ a_{21} & a_{22} & a_{23} \\ a_{31} & a_{32} & a_{33} \end{bmatrix}.$$

(b) Let A be an $n \times n$ matrix and B be the matrix obtained from A by multiplying the i-th *column* of A by a factor of r. Prove that $\det B = r \det A$. (Hint: How are the matrices A^T and B^T related?)

(c) Let A be an $n \times n$ matrix and B be the matrix obtained from A by adding a multiple of one column to another. Prove that $\det A = \det B$.

5.3 Matrix Inverse by Cofactors and Cramer's Rule

We will now see that the determinant provides an alternative method of calculating the inverse of a square matrix. Determinant calculations are generally much longer than row reduction. Therefore, this method of finding the inverse is not as efficient as the method based on row reduction. However, it is useful in some theoretical applications because it provides a *formula* for A^{-1} in terms of the entries of A.

This method is based on a simple calculation that makes use of the following theorem.

Theorem 1 [False Expansion Theorem]
If A is an $n \times n$ matrix and $i \neq k$, then

$$a_{i1}C_{k1} + \cdots + a_{in}C_{kn} = 0$$

Proof: Let B be the matrix obtained from A by replacing (not swapping) the k-th row of A by the i-th row of A. Then the i-th row of B is identical to the k-th row of B, hence $\det B = 0$ by Corollary 5.2.3. Since the cofactors C^*_{kj} of B are equal to the cofactors C_{kj} of A, and the coefficients b_{kj} of the k-th row of B are equal to the coefficients a_{ij} of the i-th row of A, we get

$$0 = \det B = b_{k1}C^*_{k1} + \cdots + b_{kn}C^*_{kn} = a_{i1}C_{k1} + \cdots + a_{in}C_{kn}$$

as required. ■

Definition
Cofactor Matrix

Let A be an $n \times n$ matrix. We define the **cofactor matrix** of A, denoted $\operatorname{cof} A$, by

$$(\operatorname{cof} A)_{ij} = C_{ij}$$

EXAMPLE 1

Let $A = \begin{bmatrix} 2 & 4 & -1 \\ 0 & 3 & 1 \\ 6 & -2 & 5 \end{bmatrix}$. Determine cof A.

Solution: The nine cofactors of A are

$$C_{11} = (1)\begin{vmatrix} 3 & 1 \\ -2 & 5 \end{vmatrix} = 17 \quad C_{12} = (-1)\begin{vmatrix} 0 & 1 \\ 6 & 5 \end{vmatrix} = 6 \quad C_{13} = (1)\begin{vmatrix} 0 & 3 \\ 6 & -2 \end{vmatrix} = -18$$

$$C_{21} = (-1)\begin{vmatrix} 4 & -1 \\ -2 & 5 \end{vmatrix} = -18 \quad C_{22} = (1)\begin{vmatrix} 2 & -1 \\ 6 & 5 \end{vmatrix} = 16 \quad C_{23} = (-1)\begin{vmatrix} 2 & 4 \\ 6 & -2 \end{vmatrix} = 28$$

$$C_{31} = (1)\begin{vmatrix} 4 & -1 \\ 3 & 1 \end{vmatrix} = 7 \quad C_{32} = (-1)\begin{vmatrix} 2 & -1 \\ 0 & 1 \end{vmatrix} = -2 \quad C_{33} = (1)\begin{vmatrix} 2 & 4 \\ 0 & 3 \end{vmatrix} = 6$$

Hence,

$$\text{cof}\,A = \begin{bmatrix} C_{11} & C_{12} & C_{13} \\ C_{21} & C_{22} & C_{23} \\ C_{31} & C_{32} & C_{33} \end{bmatrix} = \begin{bmatrix} 17 & 6 & -18 \\ -18 & 16 & 28 \\ 7 & -2 & 6 \end{bmatrix}$$

EXERCISE 1

Calculate the cofactor matrix of $A = \begin{bmatrix} 1 & 2 & 0 \\ -1 & 0 & 1 \\ 0 & 3 & 1 \end{bmatrix}$.

Observe that the cofactors of the i-th row of A form the i-th row of cof A, so they form the i-th column of $(\text{cof}\,A)^T$. Thus, the dot product of the i-th row of A and the i-th column of $(\text{cof}\,A)^T$ equals the determinant of A. Moreover, by the False Expansion Theorem, the dot product of the i-th row of A and the j-th column of cof A equals 0 if $i \neq j$. Hence, if $\vec{a}_i^T$ represents the i-th row of A and $\vec{c}_j$ represents the j-th column of $(\text{cof}\,A)^T$, then

$$\begin{aligned} A(\text{cof}\,A)^T &= \begin{bmatrix} \vec{a}_1^T \\ \vdots \\ \vec{a}_n^T \end{bmatrix} \begin{bmatrix} \vec{c}_1 & \cdots & \vec{c}_n \end{bmatrix} \\ &= \begin{bmatrix} \vec{a}_1 \cdot \vec{c}_1 & \cdots & \vec{a}_1 \cdot \vec{c}_n \\ \vdots & \ddots & \vdots \\ \vec{a}_n \cdot \vec{c}_1 & \cdots & \vec{a}_n \cdot \vec{c}_n \end{bmatrix} \\ &= \begin{bmatrix} \det A & 0 & \cdots & 0 \\ 0 & \det A & \ddots & \vdots \\ \vdots & \ddots & \ddots & 0 \\ 0 & \cdots & 0 & \det A \end{bmatrix} = (\det A)I \end{aligned}$$

where I is the identity matrix.

If $\det A \neq 0$, it follows that $A\left(\frac{1}{\det A}\right)(\text{cof}\,A)^T = I$, and, therefore,

$$A^{-1} = \left(\frac{1}{\det A}\right)(\text{cof}\,A)^T$$

If $\det A = 0$, then, by Theorem 5.2.4, A is not invertible. We shall refer to this method of finding the inverse as the *cofactor method*. (Some people refer to the transpose of the cofactor matrix as the **adjugate matrix** and therefore call this the *adjugate method*).

EXAMPLE 2

Find the inverse of the matrix $A = \begin{bmatrix} 2 & 4 & -1 \\ 0 & 3 & 1 \\ 6 & -2 & 5 \end{bmatrix}$ by the cofactor method.

Solution: We first find that

$$\det A = \begin{vmatrix} 2 & 4 & -1 \\ 0 & 3 & 1 \\ 6 & -2 & 5 \end{vmatrix} = \begin{vmatrix} 2 & 4 & -1 \\ 0 & 3 & 1 \\ 0 & -14 & 8 \end{vmatrix} = 2(24 + 14) = 76$$

Thus, A is invertible. Using the result of Example 1 gives

$$\begin{aligned} A^{-1} &= \frac{1}{\det A}(\operatorname{cof} A)^T \\ &= \frac{1}{76}\begin{bmatrix} 17 & -18 & 7 \\ 6 & 16 & -2 \\ -18 & 28 & 6 \end{bmatrix} \end{aligned}$$

EXERCISE 2

Use the cofactor method to find the inverse of $A = \begin{bmatrix} 1 & 2 & 0 \\ -1 & 0 & 1 \\ 0 & 3 & 1 \end{bmatrix}$.

For 3×3 matrices, the cofactor method requires the evaluation of nine 2×2 determinants. This is manageable, but it is more work than would be required by the row reduction method. Finding the inverse of a 4×4 matrix by using the cofactor method would require the evaluation of sixteen 3×3 determinants; this method becomes extremely unattractive.

Cramer's Rule

Consider the system of n linear equations in n variables, $A\vec{x} = \vec{b}$. If $\det A \neq 0$ so that A is invertible, then the solution may be written in the form

$$\vec{x} = A^{-1}\vec{b} = \frac{1}{\det A}(\operatorname{cof} A)^T\vec{b}$$

$$\begin{bmatrix} x_1 \\ \vdots \\ x_i \\ \vdots \\ x_n \end{bmatrix} = \frac{1}{\det A}\begin{bmatrix} C_{11} & C_{21} & \cdots & C_{n1} \\ \vdots & \vdots & \vdots & \vdots \\ C_{1i} & C_{2i} & \cdots & C_{ni} \\ \vdots & \vdots & \ddots & \vdots \\ C_{1n} & C_{2n} & \cdots & C_{nn} \end{bmatrix}\begin{bmatrix} b_1 \\ \vdots \\ b_i \\ \vdots \\ b_n \end{bmatrix}$$

Consider the value of the i-th component of $\vec{x}$:

$$x_i = \frac{1}{\det A}(b_1C_{1i} + b_2C_{2i} + \cdots + b_nC_{ni})$$

This is $\frac{1}{\det A}$ multiplied by the dot product of the vector $\vec{b}$ with the i-th row of $(\operatorname{cof} A)^T$. But the i-th row of $(\operatorname{cof} A)^T$ is the i-th column of the original cofactor matrix $\operatorname{cof} A$. So x_i is the dot product of the vector $\vec{b}$ with the i-th column of $\operatorname{cof} A$ divided by $\det A$.

Now, let N_i be the matrix obtained from A by replacing the i-th column of A by $\vec{b}$. Then the cofactors of the i-th column of N_i will equal the cofactors of the i-th column of A, and hence we get

$$\det N_i = b_1C_{1i} + b_2C_{2i} + \cdots + b_nC_{ni}$$

Therefore, the i-th component of $\vec{x}$ in the solution of $A\vec{x} = \vec{b}$ is

$$x_i = \frac{\det N_i}{\det A}$$

This is called Cramer's Rule (or Method). We now demonstrate Cramer's Rule with a couple of examples.

EXAMPLE 3

Use Cramer's Rule to solve the system of equations.

$$\begin{aligned} x_1 + x_2 - x_3 &= b_1 \\ 2x_1 + 4x_2 + 5x_3 &= b_2 \\ x_1 + x_2 + 2x_3 &= b_3 \end{aligned}$$

Solution: The coefficient matrix is $A = \begin{bmatrix} 1 & 1 & -1 \\ 2 & 4 & 5 \\ 1 & 1 & 2 \end{bmatrix}$, so

$$\det A = \begin{vmatrix} 1 & 1 & -1 \\ 2 & 4 & 5 \\ 1 & 1 & 2 \end{vmatrix} = \begin{vmatrix} 1 & 1 & -1 \\ 0 & 2 & 7 \\ 0 & 0 & 3 \end{vmatrix} = (1)(2)(3) = 6$$

Hence,

$$x_1 = \frac{\det N_1}{\det A} = \frac{1}{6}\begin{vmatrix} b_1 & 1 & -1 \\ b_2 & 4 & 5 \\ b_3 & 1 & 2 \end{vmatrix} = \frac{1}{6}\begin{vmatrix} b_1 & 1 & -1 \\ b_2 + 5b_1 & 9 & 0 \\ b_3 + 2b_1 & 3 & 0 \end{vmatrix} = \frac{3b_1 - 3b_2 + 9b_3}{6}$$

$$x_2 = \frac{\det N_2}{\det A} = \frac{1}{6}\begin{vmatrix} 1 & b_1 & -1 \\ 2 & b_2 & 5 \\ 1 & b_3 & 2 \end{vmatrix} = \frac{1}{6}\begin{vmatrix} 1 & b_1 & -1 \\ 7 & b_2 + 5b_1 & 0 \\ 3 & b_3 + 2b_1 & 0 \end{vmatrix} = \frac{b_1 + 3b_2 - 7b_3}{6}$$

$$x_3 = \frac{\det N_3}{\det A} = \frac{1}{6}\begin{vmatrix} 1 & 1 & b_1 \\ 2 & 4 & b_2 \\ 1 & 1 & b_3 \end{vmatrix} = \frac{1}{6}\begin{vmatrix} 1 & 1 & b_1 \\ 0 & 2 & b_2 - 2b_1 \\ 0 & 0 & b_3 - b_1 \end{vmatrix} = \frac{-2b_1 + 2b_3}{6}$$

EXAMPLE 4

Use Cramer's Rule to solve the system of equations

$$\begin{aligned}-\frac{2}{3}x_1 + \frac{3}{5}x_2 &= \frac{1}{5}\\ \frac{2}{5}x_1 - \frac{1}{3}x_2 &= \frac{1}{2}\end{aligned}$$

Solution: The coefficient matrix is $A = \begin{bmatrix} -2/3 & 3/5 \\ 2/5 & -1/3 \end{bmatrix}$, so

$$\det A = \frac{-2}{3}\cdot\frac{-1}{3} - \frac{3}{5}\cdot\frac{2}{5} = \frac{-4}{225}$$

Hence,

$$x_1 = \frac{1}{-4/225}\begin{vmatrix} 1/5 & 3/5 \\ 1/2 & -1/3 \end{vmatrix} = \frac{-225}{4}\cdot\frac{-11}{30} = \frac{165}{8}$$

$$x_2 = \frac{1}{-4/225}\begin{vmatrix} -2/3 & 1/5 \\ 2/5 & 1/2 \end{vmatrix} = \frac{-225}{4}\cdot\frac{93}{4} = \frac{93}{4}$$

To solve a system of n equations in n variables by using Cramer's Rule would require the evaluation of the determinant of $n+1$ matrices, each of which is $n \times n$. Thus, solving a system by using Cramer's Rule requires far more calculation than elimination, so Cramer's Rule is not considered a computationally useful solution method. However, like the cofactor method above, Cramer's Rule is sometimes used to write a formula for the solution of a problem.

PROBLEMS 5.3

Practice Problems

A1 Determine the inverse of each of the following matrices by using the cofactor method. Verify your answer by using multiplication.

(a) $\begin{bmatrix} 1 & 3 \\ 4 & 10 \end{bmatrix}$

(b) $\begin{bmatrix} 3 & -5 \\ 2 & -1 \end{bmatrix}$

(c) $\begin{bmatrix} 4 & 1 & 7 \\ 2 & -3 & 1 \\ -2 & 6 & 0 \end{bmatrix}$

(d) $\begin{bmatrix} 4 & 0 & -4 \\ 0 & -1 & 1 \\ -2 & 2 & -1 \end{bmatrix}$

A2 Let $A = \begin{bmatrix} 2 & -3 & t \\ -2 & 0 & 1 \\ 3 & 1 & 1 \end{bmatrix}$.

(a) Determine the cofactor matrix $\operatorname{cof} A$.

(b) Calculate $A(\operatorname{cof} A)^T$ and determine $\det A$ and A^{-1}.

A3 Use Cramer's Rule to solve the following systems.

(a) $2x_1 - 3x_2 = 6$
$3x_1 + 5x_2 = 7$

(b) $3x_1 + 3x_2 = 2$
$2x_1 - 3x_2 = 5$

(c) $7x_1 + x_2 - 4x_3 = 3$
$-6x_1 - 4x_2 + x_3 = 0$
$4x_1 - x_2 - 2x_3 = 6$

(d) $2x_1 + 3x_2 - 5x_3 = 2$
$3x_1 - x_2 + 2x_3 = 1$
$5x_1 + 4x_2 - 6x_3 = 3$

Homework Problems

B1 Determine the inverse of each of the following matrices by using the cofactor method. Verify your answer by using multiplication.

(a) $\begin{bmatrix} -3 & 4 \\ 7 & 2 \end{bmatrix}$

(b) $\begin{bmatrix} 5 & 4 \\ -2 & 3 \end{bmatrix}$

(c) $\begin{bmatrix} 0 & 2 & -3 \\ -3 & 3 & 0 \\ 2 & 5 & 4 \end{bmatrix}$

(d) $\begin{bmatrix} 1 & 0 & 2 \\ -2 & 1 & 0 \\ 3 & -1 & 1 \end{bmatrix}$

(e) $\begin{bmatrix} 6 & 2 & 2 \\ 0 & 4 & -2 \\ 0 & -7 & 2 \end{bmatrix}$

(f) $\begin{bmatrix} -2 & -2 & -4 & -4 \\ 3 & 1 & 0 & 2 \\ -3 & 0 & -3 & -3 \\ 2 & -2 & 0 & 4 \end{bmatrix}$

B2 Let $A = \begin{bmatrix} 2 & 1 & 3 \\ 3 & t & -1 \\ -2 & 1 & 4 \end{bmatrix}$.

(a) Determine the cofactor matrix $\operatorname{cof} A$.

(b) Calculate $A(\operatorname{cof} A)^T$ and determine $\det A$ and A^{-1}.

B3 Use Cramer's Rule to solve the following systems.

(a) $2x_1 - 7x_2 = 3$
$5x_1 + 4x_2 = -17$

(b) $3x_1 + 5x_2 = -2$
$x_1 + 3x_2 = -3$

(c) $x_1 - 5x_2 - 2x_3 = -2$
$2x_1 + 3x_3 = 3$
$4x_1 + x_2 - x_3 = 1$

(d) $x_1 + 2x_3 = -2$
$3x_1 - x_2 + 3x_3 = 5$
$-2x_1 + x_2 = 0$

Conceptual Problems

D1 Suppose that $A = \begin{bmatrix} \vec{a}_1 & \cdots & \vec{a}_n \end{bmatrix}$ is an invertible $n \times n$ matrix.

(a) Verify by Cramer's Rule that the system of equations $A\vec{x} = \vec{a}_j$ has the unique solution $\vec{x} = \vec{e}_j$ (the j-th standard basis vector).

(b) Explain the result of part (a) in terms of linear transformations and/or matrix multiplication.

D2 Let $A = \begin{bmatrix} 2 & -1 & 0 & 1 \\ 0 & -1 & 3 & 2 \\ 0 & 1 & 0 & 0 \\ 0 & 2 & 0 & 3 \end{bmatrix}$. Use the cofactor method to calculate $(A^{-1})_{23}$ and $(A^{-1})_{42}$. (If you calculate more than these two entries of A^{-1}, you have missed the point.)

5.4 Area, Volume, and the Determinant

So far, we have been using the determinant of a matrix only to determine if an $n \times n$ matrix is invertible. In particular, we have only been interested in whether the determinant of a matrix is zero or non-zero. We will now see that we are sometimes interested in the specific value of the determinant, as it can have an important geometric interpretation.

Area and the Determinant

Let $\vec{u} = \begin{bmatrix} u_1 \\ u_2 \end{bmatrix}$ and $\vec{v} = \begin{bmatrix} v_1 \\ v_2 \end{bmatrix}$. In Chapter 1, we saw that we could construct a parallelogram from these two vectors by making the vectors $\vec{u}$ and $\vec{v}$ as adjacent sides and having $\vec{u} + \vec{v}$ as the vertex of the parallelogram, opposite the origin, as in Figure 5.4.1. This is called the **parallelogram induced by $\vec{u}$ and $\vec{v}$**.

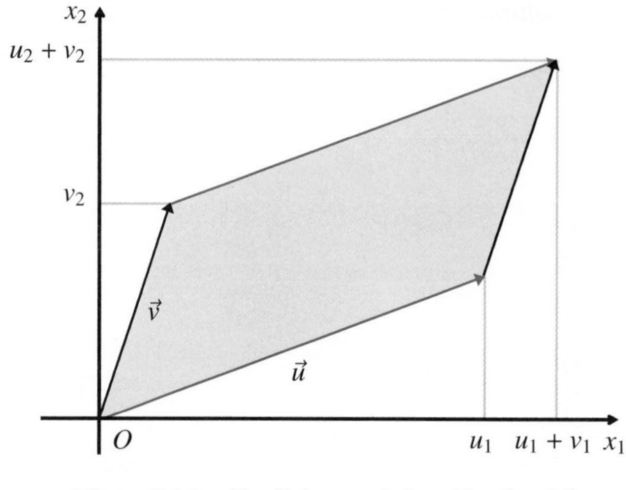

Figure 5.4.1 Parallelogram induced by $\vec{u}$ and $\vec{v}$.

We will calculate the area of this parallelogram by calculating the area of the rectangle with sides of length $u_1 + v_1$ and $u_2 + v_2$ and subtracting the area inside the rectangle and outside the parallelogram, as indicated in Figure 5.4.2.

This gives

$$\begin{aligned} \text{Area}(\vec{u}, \vec{v}) &= \text{Area of Square} - \text{Area 1} - \text{Area 2} - \text{Area 3} - \text{Area 4} - \text{Area 5} - \text{Area 6} \\ &= (u_1 + v_1)(u_2 + v_2) - \frac{1}{2}v_1v_2 - u_2v_1 - \frac{1}{2}u_1u_2 - \frac{1}{2}v_1v_2 - u_2v_1 - \frac{1}{2}u_1u_2 \\ &= u_1u_2 + u_1v_2 + u_2v_1 + v_1v_2 - v_1v_2 - 2u_2v_1 - u_1u_2 \\ &= u_1v_2 - u_2v_1 \end{aligned}$$

We immediately recognize this as the determinant of the matrix $\begin{bmatrix} u_1 & v_1 \\ u_2 & v_2 \end{bmatrix} = \begin{bmatrix} \vec{u} & \vec{v} \end{bmatrix}$.

Remark

At this time, you might be tempted to say that the area of the parallelogram induced by any two vectors $\vec{u}$ and $\vec{v}$ equals the determinant of $\begin{bmatrix} \vec{u} & \vec{v} \end{bmatrix}$. However, this would be slightly incorrect as we have made a hidden assumption in our calculation above. In particular, in our diagram we have drawn $\vec{u}$ and $\vec{v}$ as a right-handed system.

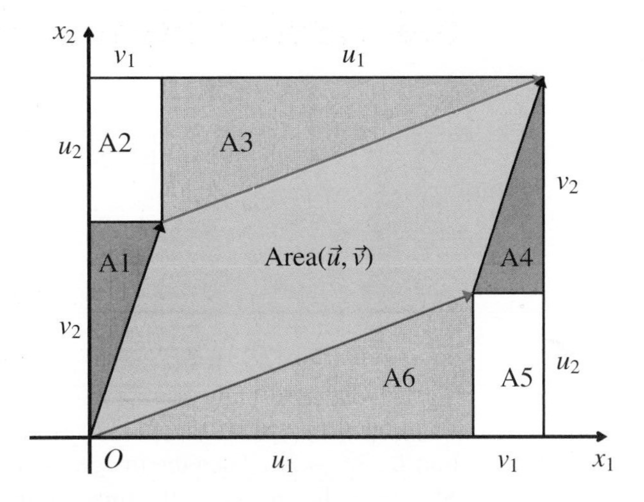

Figure 5.4.2 Area of the parallelogram induced by $\vec{u}$ and $\vec{v}$.

EXERCISE 1

Show that if $\vec{u} = \begin{bmatrix} u_1 \\ u_2 \end{bmatrix}$ and $\vec{v} = \begin{bmatrix} v_1 \\ v_2 \end{bmatrix}$ form a left-handed system, then the area of the parallelogram induced by $\vec{u}$ and $\vec{v}$ equals $\left|\det \begin{bmatrix} u_1 & v_1 \\ u_2 & v_2 \end{bmatrix}\right|$.

We have now shown that the area of the parallelogram induced by $\vec{u}$ and $\vec{v}$ is

$$\text{Area}(\vec{u}, \vec{v}) = \left|\det \begin{bmatrix} u_1 & v_1 \\ u_2 & v_2 \end{bmatrix}\right|$$

EXAMPLE 1

Draw the parallelogram induced by the following vectors and determine its area.

(a) $\vec{u} = \begin{bmatrix} 2 \\ 3 \end{bmatrix}, \vec{v} = \begin{bmatrix} -2 \\ 2 \end{bmatrix}$

(b) $\vec{u} = \begin{bmatrix} 1 \\ 1 \end{bmatrix}, \vec{v} = \begin{bmatrix} 1 \\ -2 \end{bmatrix}$.

Solution: For (a), we have

$$\text{Area}\,(\vec{u}, \vec{v}) = \left|\det \begin{bmatrix} 2 & 3 \\ -2 & 2 \end{bmatrix}\right| = 2(2) - 3(-2) = 10$$

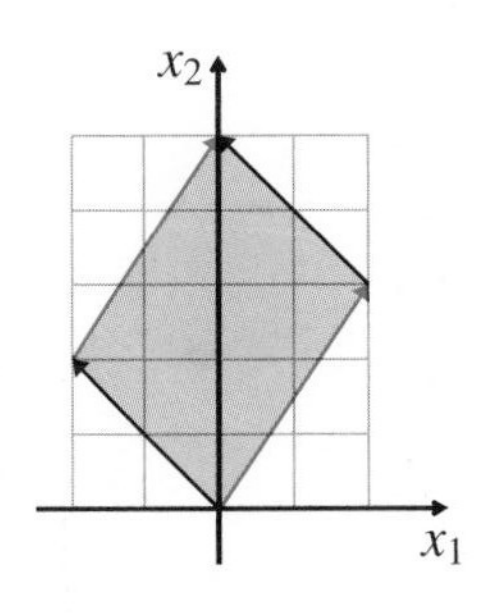

EXAMPLE 1 (continued)

For (b), we have

$$\text{Area}(\vec{u}, \vec{v}) = \left|\det\begin{bmatrix} 1 & 1 \\ 1 & -2 \end{bmatrix}\right| = 1(-2) - 1(1) = -3$$

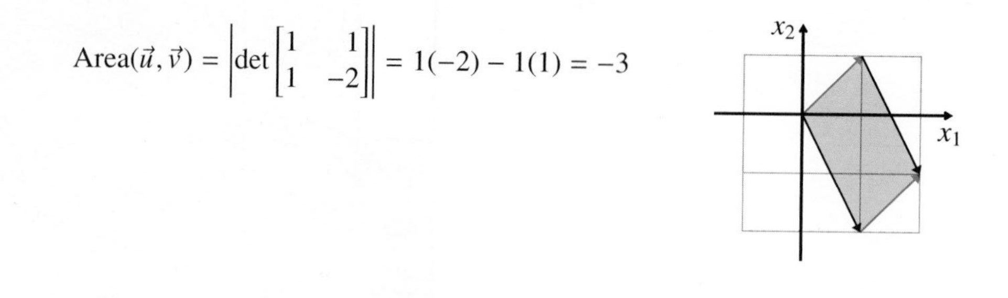

Now suppose that the 2×2 matrix A is the standard matrix of a linear transformation $L : \mathbb{R}^2 \to \mathbb{R}^2$. Then the images of $\vec{u}$ and $\vec{v}$ under L are $L(\vec{u}) = A\vec{u}$ and $L(\vec{v}) = A\vec{v}$. Moreover, the volume of the **image parallelogram** is

$$\text{Area}\,(A\vec{u}, A\vec{v}) = \left|\det\begin{bmatrix} A\vec{u} & A\vec{v} \end{bmatrix}\right| = \left|\det\left(A\begin{bmatrix} \vec{u} & \vec{v} \end{bmatrix}\right)\right|$$

Hence, we get

$$\text{Area}\,(A\vec{u}, A\vec{v}) = \left|\det\left(A\begin{bmatrix} \vec{u} & \vec{v} \end{bmatrix}\right)\right| = |\det A|\left|\det\begin{bmatrix} \vec{u} & \vec{v} \end{bmatrix}\right| = |\det A|\text{Area}(\vec{u}, \vec{v}) \qquad (5.1)$$

In words: the absolute value of the determinant of the standard matrix A of a linear transformation is the factor by which area is changed under the linear transformation L. The result is illustrated in Figure 5.4.3.

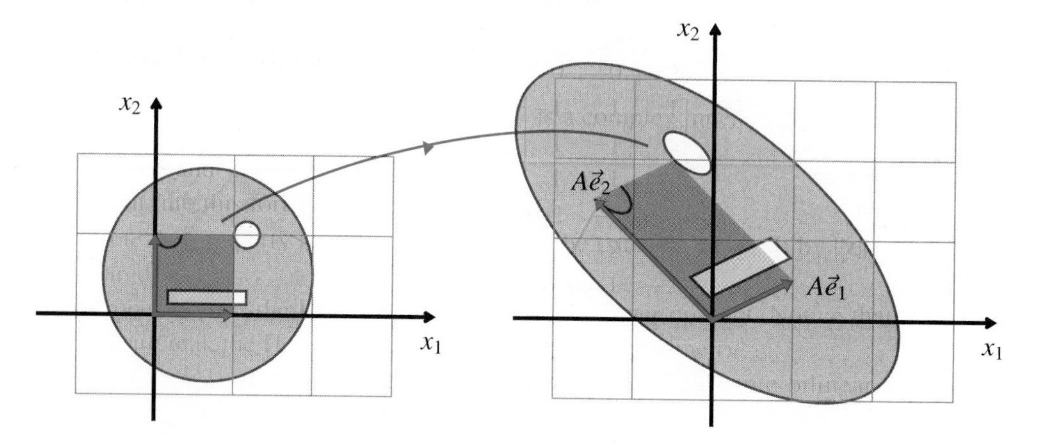

Figure 5.4.3 Under a linear transformation with matrix A, the area of a figure is changed by factor $|\det A|$.

EXAMPLE 2

Let $A = \begin{bmatrix} 1 & 3 \\ 0 & 2 \end{bmatrix}$ and L be the linear mapping $L(\vec{x}) = A\vec{x}$. Determine the image of $\vec{u} = \begin{bmatrix} 1 \\ 1 \end{bmatrix}$ and $\vec{v} = \begin{bmatrix} -1 \\ 1 \end{bmatrix}$ under L and compute the area determined by the image vectors in two ways.

EXAMPLE 2 (continued)

Solution: The image of each vector under L is

$$L(\vec{u}) = \begin{bmatrix} 1 & 3 \\ 0 & 2 \end{bmatrix}\begin{bmatrix} 1 \\ 1 \end{bmatrix} = \begin{bmatrix} 4 \\ 2 \end{bmatrix}, \qquad L(\vec{v}) = \begin{bmatrix} 1 & 3 \\ 0 & 2 \end{bmatrix}\begin{bmatrix} -1 \\ 1 \end{bmatrix} = \begin{bmatrix} 2 \\ 2 \end{bmatrix}$$

Hence, the area determined by the image vectors is

$$\text{Area}\,(L(\vec{u}), L(\vec{v})) = \left|\det \begin{bmatrix} 4 & 2 \\ 2 & 2 \end{bmatrix}\right| = |8 - 4| = 4$$

Or, using (5.1) gives

$$\text{Area}\,(L(\vec{u}), L(\vec{v})) = |\det A|\,\text{Area}\,(\vec{u}, \vec{v}) = \left|\det \begin{bmatrix} 1 & 3 \\ 0 & 2 \end{bmatrix}\right| \left|\det \begin{bmatrix} 1 & 1 \\ -1 & 1 \end{bmatrix}\right| = 2(2) = 4$$

EXERCISE 2

Let $A = \begin{bmatrix} t & 0 \\ 0 & 1 \end{bmatrix}$ be the standard matrix of the stretch $S : \mathbb{R}^2 \to \mathbb{R}^2$ in the x_1 direction by a factor of t. Determine the image of the standard basis vectors $\vec{e}_1$ and $\vec{e}_2$ under S and compute the area determined by the image vectors in two ways. Illustrate with a sketch.

The Determinant and Volume

Recall from Chapter 1 that if $\vec{u}$, $\vec{v}$, and $\vec{w}$ are vectors in $\mathbb{R}^3$, then the volume of the parallelepiped induced by $\vec{u}$, $\vec{v}$, and $\vec{w}$ is

$$\text{Volume}(\vec{u}, \vec{v}, \vec{w}) = |\vec{w} \cdot (\vec{u} \times \vec{v})|$$

Now observe that if $\vec{u} = \begin{bmatrix} u_1 \\ u_2 \\ u_3 \end{bmatrix}$, $\vec{v} = \begin{bmatrix} v_1 \\ v_2 \\ v_3 \end{bmatrix}$, and $\vec{w} = \begin{bmatrix} w_1 \\ w_2 \\ w_3 \end{bmatrix}$, then

$$|\vec{w} \cdot (\vec{u} \times \vec{v})| = |w_1(u_2v_3 - u_3v_2) - w_2(u_1v_3 - u_3v_1) + w_3(u_1v_2 - u_2v_1)| = \left|\det \begin{bmatrix} u_1 & v_1 & w_1 \\ u_2 & v_2 & w_2 \\ u_3 & v_3 & w_3 \end{bmatrix}\right|$$

Hence, the volume of the parallelepiped induced by $\vec{u}$, $\vec{v}$, and $\vec{w}$ is

$$\text{Volume}(\vec{u}, \vec{v}, \vec{w}) = \left|\det \begin{bmatrix} \vec{u} & \vec{v} & \vec{w} \end{bmatrix}\right|$$

EXAMPLE 3

Let $A = \begin{bmatrix} 4 & 1 & -1 \\ 2 & 4 & 1 \\ 1 & 1 & 4 \end{bmatrix}$, $\vec{u} = \begin{bmatrix} 1 \\ -1 \\ 0 \end{bmatrix}$, $\vec{v} = \begin{bmatrix} 0 \\ 1 \\ 2 \end{bmatrix}$, and $\vec{w} = \begin{bmatrix} -1 \\ 5 \\ 1 \end{bmatrix}$. Calculate the volume of the parallelepiped induced by $\vec{u}$, $\vec{v}$, and $\vec{w}$ and the volume of the parallelepiped induced by $A\vec{u}$, $A\vec{v}$, and $A\vec{w}$.

EXAMPLE 3
(continued)

Solution: The volume determined by $\vec{u}$, $\vec{v}$, and $\vec{w}$ is

$$\text{Volume}(\vec{u}, \vec{v}, \vec{w}) = \left|\det \begin{bmatrix} 1 & 0 & -1 \\ -1 & 1 & 5 \\ 0 & 2 & 1 \end{bmatrix}\right| = |-7| = 7$$

The volume determined by $A\vec{u}$, $A\vec{v}$, and $A\vec{w}$ is

$$\text{Volume}(A\vec{u}, A\vec{v}, A\vec{w}) = \left|\det \begin{bmatrix} A\vec{u} & A\vec{v} & A\vec{w} \end{bmatrix}\right| = \left|\det \begin{bmatrix} 3 & -1 & 0 \\ -2 & 6 & 19 \\ 0 & 9 & 8 \end{bmatrix}\right| = |-385| = 385$$

Moreover, $\det A = -55$, so

$$\text{Volume}(A\vec{u}, A\vec{v}, A\vec{w}) = |\det A| \text{Volume}(\vec{u}, \vec{v}, \vec{w})$$

which coincides with the result for the 2×2 case.

In general, if $\vec{v}_1, \dots, \vec{v}_n$ are n vectors in $\mathbb{R}^n$, then we say that they induce an **n-dimensional parallelotope** (the n-dimensional version of a parallelogram or parallelepiped). The **n-volume** of the parallelotope is

$$n\text{-Volume}(\vec{v}_1, \dots, \vec{v}_n) = \left|\det \begin{bmatrix} \vec{v}_1 & \cdots & \vec{v}_n \end{bmatrix}\right|$$

and if A is the standard matrix of a linear mapping $L : \mathbb{R}^n \to \mathbb{R}^n$, then

$$n\text{-Volume}(A\vec{v}_1, \dots, A\vec{v}_n) = |\det A| n\text{-Volume}(\vec{v}_1, \dots, \vec{v}_n)$$

PROBLEMS 5.4

Practice Problems

A1 (a) Calculate the area of the parallelogram induced by $\vec{u} = \begin{bmatrix} 3 \\ 5 \end{bmatrix}$ and $\vec{v} = \begin{bmatrix} 2 \\ 7 \end{bmatrix}$ in $\mathbb{R}^2$.

(b) Determine the image of $\vec{u}$ and $\vec{v}$ under the linear mapping with standard matrix $A = \begin{bmatrix} -4 & 6 \\ 3 & 2 \end{bmatrix}$.

(c) Compute the determinant of A.

(d) Compute the area of the parallelogram induced by the image vectors in two ways.

A2 Let $A = \begin{bmatrix} 0 & 1 \\ 1 & 0 \end{bmatrix}$ be the standard matrix of the reflection $R : \mathbb{R}^2 \to \mathbb{R}^2$ over the line $x_2 = x_1$. Determine the image of $\vec{u} = \begin{bmatrix} 1 \\ 3 \end{bmatrix}$ and $\vec{v} = \begin{bmatrix} -2 \\ 2 \end{bmatrix}$ under R and compute the area determined by the image vectors.

A3 (a) Compute the volume of the parallelepiped induced by $\vec{u} = \begin{bmatrix} 2 \\ 3 \\ 4 \end{bmatrix}$, $\vec{v} = \begin{bmatrix} 2 \\ -1 \\ -5 \end{bmatrix}$, and $\vec{w} = \begin{bmatrix} 1 \\ 5 \\ 2 \end{bmatrix}$.

(b) Compute the determinant of $A = \begin{bmatrix} 1 & -1 & 3 \\ 4 & 0 & 1 \\ 0 & 2 & 5 \end{bmatrix}$.

(c) What is the volume of the image of the parallelepiped of part (a) under the linear mapping with standard matrix A?

A4 Repeat Problem A3 with vectors $\vec{u} = \begin{bmatrix} 0 \\ 2 \\ 3 \end{bmatrix}$, $\vec{v} = \begin{bmatrix} 1 \\ -5 \\ 5 \end{bmatrix}$, $\vec{w} = \begin{bmatrix} 2 \\ 1 \\ 6 \end{bmatrix}$, and $A = \begin{bmatrix} 3 & -1 & 2 \\ 3 & 2 & -1 \\ 1 & 4 & 5 \end{bmatrix}$.

A5 (a) Calculate the 4-volume of the 4-dimensional parallelotope determined by $\vec{v}_1 = \begin{bmatrix} 1 \\ 0 \\ 2 \\ 0 \end{bmatrix}$, $\vec{v}_2 = \begin{bmatrix} 0 \\ 1 \\ 1 \\ 3 \end{bmatrix}$, $\vec{v}_3 = \begin{bmatrix} 0 \\ 2 \\ 3 \\ 0 \end{bmatrix}$, and $\vec{v}_4 = \begin{bmatrix} 1 \\ 0 \\ 2 \\ 5 \end{bmatrix}$.

(b) Calculate the 4-volume of the image of this parallelotope under the linear mapping with standard matrix $A = \begin{bmatrix} 2 & 3 & 1 & 1 \\ 5 & 4 & 3 & 0 \\ 0 & 0 & 7 & 3 \\ 0 & 0 & 0 & 1 \end{bmatrix}$.

A6 Let $\{\vec{v}_1, \dots, \vec{v}_n\}$ be vectors in $\mathbb{R}^n$. Prove that the n-volume of the parallelotope induced by $\vec{v}_1, \dots, \vec{v}_n$ is the same as the volume of the parallelotope induced by $\vec{v}_1, \dots, \vec{v}_{n-1}, \vec{v}_n + t\vec{v}_1$.

Homework Problems

B1 (a) Calculate the area of the parallelogram induced by $\vec{u} = \begin{bmatrix} 3 \\ 7 \end{bmatrix}$ and $\vec{v} = \begin{bmatrix} 2 \\ 4 \end{bmatrix}$ in $\mathbb{R}^2$.

(b) Determine the image of $\vec{u}$ and $\vec{v}$ under the linear mapping with standard matrix $A = \begin{bmatrix} 2 & 8 \\ -1 & 5 \end{bmatrix}$.

(c) Compute the determinant of A.

(d) Compute the area of the parallelogram induced by the image vectors in two ways.

B2 Let $A = \begin{bmatrix} 1 & t \\ 0 & 1 \end{bmatrix}$ be the standard matrix of the shear $H : \mathbb{R}^2 \to \mathbb{R}^2$ in the x_1 direction by a factor of t. Determine the image of the standard basis vectors $\vec{e}_1$ and $\vec{e}_2$ under H and compute the area determined by the image vectors.

B3 (a) Compute the volume of the parallelepiped induced by $\vec{u} = \begin{bmatrix} 2 \\ -1 \\ 0 \end{bmatrix}$, $\vec{v} = \begin{bmatrix} 1 \\ -3 \\ -1 \end{bmatrix}$, and $\vec{w} = \begin{bmatrix} 0 \\ 5 \\ 3 \end{bmatrix}$.

(b) Compute the determinant of $A = \begin{bmatrix} 2 & -1 & 3 \\ 4 & 5 & 0 \\ -1 & 0 & 0 \end{bmatrix}$.

(c) What is the volume of the image of the parallelepiped of part (a) under the linear mapping with standard matrix A?

B4 Repeat Problem B3 with vectors $\vec{u} = \begin{bmatrix} 1 \\ 1 \\ -4 \end{bmatrix}$, $\vec{v} = \begin{bmatrix} 2 \\ -1 \\ 3 \end{bmatrix}$, $\vec{w} = \begin{bmatrix} -1 \\ 3 \\ 5 \end{bmatrix}$, and $A = \begin{bmatrix} 4 & -2 & 1 \\ 0 & 1 & 3 \\ -1 & 4 & 2 \end{bmatrix}$.

B5 (a) Calculate the 4-volume of the 4-dimensional parallelotope determined by $\vec{v}_1 = \begin{bmatrix} 1 \\ 1 \\ 2 \\ 1 \end{bmatrix}$, $\vec{v}_2 = \begin{bmatrix} 1 \\ 1 \\ 1 \\ 3 \end{bmatrix}$, $\vec{v}_3 = \begin{bmatrix} 1 \\ 2 \\ 3 \\ 0 \end{bmatrix}$, and $\vec{v}_4 = \begin{bmatrix} 1 \\ 0 \\ 5 \\ 7 \end{bmatrix}$.

(b) Calculate the 4-volume of the image of this parallelotope under the linear mapping with standard matrix $A = \begin{bmatrix} 2 & 3 & 1 & 1 \\ 2 & 3 & 1 & 0 \\ -1 & 3 & 7 & 0 \\ 0 & 2 & 1 & 1 \end{bmatrix}$.

Conceptual Problems

D1 Let $\{\vec{v}_1, \dots, \vec{v}_n\}$ be vectors in $\mathbb{R}^n$. Prove that the n-volume of the parallelotope induced by $\vec{v}_1, \dots, \vec{v}_n$ is half the volume of the parallelotope induced by $2\vec{v}_1, \vec{v}_2, \dots, \vec{v}_n$.

D2 Suppose that $L, M : \mathbb{R}^3 \to \mathbb{R}^3$ are linear mappings with standard matrices A and B, respectively. Prove that the factor by which a volume is multiplied under the composite map $M \circ L$ is $|\det BA|$.

CHAPTER REVIEW

Suggestions for Student Review

Try to answer all of these questions before checking answers at the suggested locations. In particular, try to invent your own examples. These review suggestions are intended to help you carry out your review. They may not cover every idea you need to master. Working in small groups may improve your efficiency.

1 Define *cofactor* and explain cofactor expansion. Be especially careful about signs. (Section 5.1)

2 State as many facts as you can that simplify the evaluation of determinants. For each fact, explain why it is true. (Sections 5.1 and 5.2)

3 Explain and justify the cofactor method for finding a matrix inverse. Write down a 3×3 matrix A and calculate $A(\operatorname{cof} A)^T$. (Section 5.3)

4 How and why are determinants connected to volumes? (Section 5.4)

Chapter Quiz

E1 By cofactor expansion along some column, evaluate

$$\det \begin{bmatrix} -2 & 4 & 0 & 0 \\ 1 & -2 & 2 & 9 \\ -3 & 6 & 0 & 3 \\ 1 & -1 & 0 & 0 \end{bmatrix}.$$

E2 By row reducing to upper-triangular form, evaluate

$$\det \begin{bmatrix} 3 & 2 & 7 & -8 \\ -6 & -1 & -9 & 20 \\ 3 & 8 & 21 & -17 \\ 3 & 5 & 12 & 1 \end{bmatrix}.$$

E3 Evaluate $\det \begin{bmatrix} 0 & 2 & 0 & 0 & 0 \\ 0 & 0 & 0 & 3 & 0 \\ 0 & 0 & 0 & 0 & 1 \\ 0 & 0 & 4 & 0 & 0 \\ 5 & 0 & 0 & 0 & 6 \end{bmatrix}$.

E4 Determine all values of k such that the matrix $\begin{bmatrix} k & 2 & 1 \\ 0 & 3 & k \\ 2 & -4 & 1 \end{bmatrix}$ is invertible.

E5 Suppose that A is a 5×5 matrix and $\det A = 7$.

(a) If B is obtained from A by multiplying the fourth row of A by 3, what is $\det B$?

(b) If C is obtained from A by moving the first row to the bottom and moving all other rows up, what is $\det C$? Justify.

(c) What is $\det(2A)$?

(d) What is $\det(A^{-1})$?

(e) What is $\det(A^T A)$?

E6 Let $A = \begin{bmatrix} 2 & 3 & 1 \\ 1 & 1 & 1 \\ -2 & 0 & 2 \end{bmatrix}$. Determine $(A^{-1})_{31}$ by using the cofactor method.

E7 Determine x_2 by using Cramer's Rule if

$$\begin{aligned} 2x_1 + 3x_2 + x_3 &= 1 \\ x_1 + x_2 - x_3 &= -1 \\ -2x_1 \qquad + 2x_3 &= 1 \end{aligned}$$

E8 (a) What is the volume of the parallelepiped induced by $\vec{u} = \begin{bmatrix} 1 \\ 1 \\ -2 \end{bmatrix}$, $\vec{v} = \begin{bmatrix} 2 \\ -1 \\ 3 \end{bmatrix}$, and $\vec{w} = \begin{bmatrix} 0 \\ 3 \\ 4 \end{bmatrix}$?

(b) If $A = \begin{bmatrix} 2 & 1 & 5 \\ 0 & 3 & -2 \\ 0 & 0 & -4 \end{bmatrix}$, what is the volume of the parallelepiped induced by $A\vec{u}$, $A\vec{v}$, and $A\vec{w}$?

Further Problems

These exercises are intended to be challenging. They may not be of interest to all students.

F1 Suppose that A is an $n \times n$ matrix with all row sums equal to zero. (That is, $\sum_{j=1}^{n} a_{ij} = 0$ for $1 \leq i \leq n$.) Prove that $\det A = 0$.

F2 Suppose that A and A^{-1} both have all integer entries. Prove that $\det A = \pm 1$.

F3 Consider a triangle in the plane with side lengths a, b, and c. Let the angles opposite the sides with lengths a, b, and c be denoted by A, B, and C, respectively. By using trigonometry, show that

$$c = b\cos A + a\cos B$$

Write similar equations for the other two sides. Use Cramer's Rule to show that

$$\cos A = \frac{b^2 + c^2 - a^2}{2bc}$$

F4 (a) Let $V_3(a, b, c) = \det \begin{bmatrix} 1 & a & a^2 \\ 1 & b & b^2 \\ 1 & c & c^2 \end{bmatrix}$. Without expanding, argue that $(a - b)$, $(b - c)$, and $(c - a)$ are all factors of $V_3(a, b, c)$. By considering the cofactor of c^2, argue that

$$V_3(a, b, c) = (c - a)(c - b)(b - a)$$

(b) Let $V_4(a, b, c, d) = \det \begin{bmatrix} 1 & a & a^2 & a^3 \\ 1 & b & b^2 & b^3 \\ 1 & c & c^2 & c^3 \\ 1 & d & d^2 & d^3 \end{bmatrix}$. By using arguments similar to those in part (a) (and without expanding the determinant), argue that

$$V_4(a, b, c, d) = (d - a)(d - b)(d - c)V_3(a, b, c)$$

F5 Suppose that A is a 4×4 matrix partitioned into 2×2 blocks:

$$A = \left[\begin{array}{c|c} A_1 & A_2 \\ \hline A_3 & A_4 \end{array}\right]$$

(a) If $A_3 = O_{2,2}$ (the 2×2 zero matrix), show that $\det A = \det A_1 \det A_4$.

(b) Give an example to show that, in general,

$$\det A \neq \det A_1 \det A_4 - \det A_2 \det A_3$$

F6 Suppose that A is a 3×3 matrix and B is a 2×2 matrix.

(a) Show that $\det \left[\begin{array}{c|c} A & O_{3,2} \\ \hline O_{2,3} & B \end{array}\right] = \det A \det B$.

(b) What is $\det \left[\begin{array}{c|c} O_{2,3} & B \\ \hline A & O_{3,2} \end{array}\right]$?

Visit the text's website at www.pearsoncanada.ca/norman for practice quizzes, additional applications, and an essay on linearity and superposition in physics.

CHAPTER 6

Eigenvectors and Diagonalization

CHAPTER OUTLINE

An eigenvector *is a special or preferred vector of a linear transformation that is mapped by the linear transformation to a multiple of itself. Eigenvectors play an important role in many applications in the natural and physical sciences.*

6.1 Eigenvalues and Eigenvectors

Eigenvalues and Eigenvectors of a Mapping

Definition
Eigenvector
Eigenvalue

Suppose that $L : \mathbb{R}^n \to \mathbb{R}^n$ is a linear transformation. A non-zero vector $\vec{v} \in \mathbb{R}^n$ such that $L(\vec{v}) = \lambda\vec{v}$ is called an **eigenvector** of L; the scalar λ is called an **eigenvalue** of L. The pair $\lambda, \vec{v}$ is called an **eigenpair**.

Remarks

1. The pairing of eigenvalues and eigenvectors is not one-to-one. In particular, we will see that each eigenvector of L will have a distinct eigenvalue, while each eigenvalue will correspond to infinitely many eigenvectors.

2. We have restricted our definition of eigenvectors (and hence eigenvalues) to be real. In Chapter 9 we will consider the case where we allow eigenvalues and eigenvectors to be complex.

The restriction that an eigenvector $\vec{v}$ be non-zero is natural and important. It is natural because $L(\vec{0}) = \vec{0}$ for every linear transformation, so it is completely uninteresting to consider $\vec{0}$ as an eigenvector. It is important because many of the applications involving eigenvectors make sense only for non-zero vectors. In particular, we will see that we often want to look for a basis of $\mathbb{R}^n$ that contains eigenvectors of a linear transformation.

EXAMPLE 1

Let $A = \begin{bmatrix} 17 & -15 \\ 20 & -18 \end{bmatrix}$ and let $L : \mathbb{R}^2 \to \mathbb{R}^2$ be the linear transformation defined by $L(\vec{x}) = A\vec{x}$. Determine which of the following vectors are eigenvectors of L and give the corresponding eigenvalues: $\begin{bmatrix} 1 \\ 1 \end{bmatrix}, \begin{bmatrix} -2 \\ -2 \end{bmatrix}, \begin{bmatrix} 1 \\ 3 \end{bmatrix}$, and $\begin{bmatrix} 3 \\ 4 \end{bmatrix}$.

Solution: To test whether $\begin{bmatrix} 1 \\ 1 \end{bmatrix}$ is an eigenvector, we calculate $L(1, 1)$:

$$L(1,1) = \begin{bmatrix} 17 & -15 \\ 20 & -18 \end{bmatrix}\begin{bmatrix} 1 \\ 1 \end{bmatrix} = \begin{bmatrix} 2 \\ 2 \end{bmatrix} = 2\begin{bmatrix} 1 \\ 1 \end{bmatrix}$$

So, $\begin{bmatrix} 1 \\ 1 \end{bmatrix}$ is an eigenvector of L with eigenvalue 2.

Since $\begin{bmatrix} -2 \\ -2 \end{bmatrix} = (-2)\begin{bmatrix} 1 \\ 1 \end{bmatrix}$ and L is linear,

$$L(-2,-2) = (-2)L(1,1) = (-2)(2)\begin{bmatrix} 1 \\ 1 \end{bmatrix} = 2\begin{bmatrix} -2 \\ -2 \end{bmatrix}$$

So, $\begin{bmatrix} -2 \\ -2 \end{bmatrix}$ is also an eigenvector of L with eigenvalue 2. In fact, by a similar argument, any non-zero multiple of $\begin{bmatrix} 1 \\ 1 \end{bmatrix}$ is an eigenvector of L with eigenvalue 2.

$$L(1,3) = \begin{bmatrix} 17 & -15 \\ 20 & -18 \end{bmatrix}\begin{bmatrix} 1 \\ 3 \end{bmatrix} = \begin{bmatrix} -28 \\ -34 \end{bmatrix} \neq \lambda\begin{bmatrix} 1 \\ 3 \end{bmatrix}$$

for any real number λ, so $\begin{bmatrix} 1 \\ 3 \end{bmatrix}$ is not an eigenvector of L.

$$L(3,4) = \begin{bmatrix} 17 & -15 \\ 20 & -18 \end{bmatrix}\begin{bmatrix} 3 \\ 4 \end{bmatrix} = \begin{bmatrix} -9 \\ -12 \end{bmatrix} = -3\begin{bmatrix} 3 \\ 4 \end{bmatrix}$$

So, $\begin{bmatrix} 3 \\ 4 \end{bmatrix}$ (or any non-zero multiple of it) is an eigenvector of L with eigenvalue -3.

EXAMPLE 2

Eigenvectors and Eigenvalues of Projections and Reflections in $\mathbb{R}^3$

1. Since $\text{proj}_{\vec{n}}(\vec{n}) = 1\vec{n}$, $\vec{n}$ is an eigenvector of $\text{proj}_{\vec{n}}$ with corresponding eigenvalue 1. If $\vec{v}$ is orthogonal to $\vec{n}$, then $\text{proj}_{\vec{n}}(\vec{v}) = \vec{0} = 0\vec{v}$, so $\vec{v}$ is an eigenvector of $\text{proj}_{\vec{n}}$ with corresponding eigenvalue 0. Observe that this means there is a whole plane of eigenvectors corresponding to the eigenvalue 0 as the set of vectors orthogonal to $\vec{n}$ is a plane in $\mathbb{R}^3$. For an arbitrary vector $\vec{u} \in \mathbb{R}^3$, $\text{proj}_{\vec{n}}(\vec{u})$ is a multiple of $\vec{n}$, so that $\vec{u}$ is definitely *not* an eigenvector of $\text{proj}_{\vec{n}}$ unless it is a multiple of $\vec{n}$ or orthogonal to $\vec{n}$.

EXAMPLE 2
(continued)

2. On the other hand, $\text{perp}_{\vec{n}}(\vec{n}) = \vec{0} = 0\vec{n}$, so $\vec{n}$ is an eigenvector of $\text{perp}_{\vec{n}}$ with eigenvalue 0. Since $\text{perp}_{\vec{n}}(\vec{v}) = 1\vec{v}$ for any $\vec{v}$ orthogonal to $\vec{n}$, such a $\vec{v}$ is an eigenvector of $\text{perp}_{\vec{n}}$ with eigenvalue 1.

3. We have $\text{refl}_{\vec{n}}\,\vec{n} = -1\vec{n}$, so $\vec{n}$ is an eigenvector of $\text{refl}_{\vec{n}}$ with eigenvalue -1. For $\vec{v}$ orthogonal to $\vec{n}$, $\text{refl}_{\vec{n}}(\vec{v}) = 1\vec{v}$. Hence, such a $\vec{v}$ is an eigenvector of $\text{refl}_{\vec{n}}$ with eigenvalue 1.

EXAMPLE 3

Eigenvectors and Eigenvalues of Rotations in $\mathbb{R}^2$

Consider the rotation $R_\theta : \mathbb{R}^2 \to \mathbb{R}^2$ with matrix $\begin{bmatrix} \cos\theta & -\sin\theta \\ \sin\theta & \cos\theta \end{bmatrix}$, where θ is not an integer multiple of π. By geometry, it is clear that there is no non-zero vector $\vec{v}$ in $\mathbb{R}^2$ such that $R_\theta(\vec{v}) = \lambda\vec{v}$ for some real number λ. This linear transformation has no real eigenvalues or real eigenvectors. In Chapter 9 we will see that it does have complex eigenvalues and complex eigenvectors.

EXERCISE 1

Let $R_\theta : \mathbb{R}^3 \to \mathbb{R}^3$ denote the rotation in $\mathbb{R}^3$ with matrix $\begin{bmatrix} \cos\theta & -\sin\theta & 0 \\ \sin\theta & \cos\theta & 0 \\ 0 & 0 & 1 \end{bmatrix}$. Determine any real eigenvectors of R_θ and the corresponding eigenvalues.

Eigenvalues and Eigenvectors of a Matrix

The geometric meaning of eigenvectors is much clearer when we think of them as belonging to linear transformations. However, in many applications of these ideas, it is a matrix A that is given. Thus, we also speak of the eigenvalues and eigenvectors of the matrix A.

Definition
Eigenvector
Eigenvalue

Suppose that A is an $n \times n$ matrix. A non-zero vector $\vec{v} \in \mathbb{R}^n$ such that $A\vec{v} = \lambda\vec{v}$ is called an **eigenvector** of A; the scalar λ is called an **eigenvalue** of A. The pair $\lambda, \vec{v}$ is called an **eigenpair**.

In Example 1, we saw that $\begin{bmatrix} 1 \\ 1 \end{bmatrix}$ and $\begin{bmatrix} 3 \\ 4 \end{bmatrix}$ are eigenvectors of the matrix $A = \begin{bmatrix} 17 & -15 \\ 20 & -18 \end{bmatrix}$ with eigenvalues 2 and -3, respectively.

Finding Eigenvectors and Eigenvalues

If eigenvectors and eigenvalues are going to be of any use, we need a systematic method for finding them. Suppose that a square matrix A is given; then a non-zero vector $\vec{v} \in \mathbb{R}^n$ is an eigenvector if and only if $A\vec{v} = \lambda\vec{v}$. This condition can be rewritten as

$$A\vec{v} - \lambda\vec{v} = \vec{0}$$

It is tempting to write this as $(A - \lambda)\vec{v} = \vec{0}$, but this would be incorrect because A is a matrix and λ is a number, so their difference is not defined. To get around this, we use the fact that $\vec{v} = I\vec{v}$, where I is the appropriately sized identity matrix. Then the eigenvector condition can be rewritten

$$(A - \lambda I)\vec{v} = \vec{0}$$

The eigenvector $\vec{v}$ is thus any non-trivial solution (since it cannot be the zero vector) of the homogeneous system of linear equations with coefficient matrix $(A - \lambda I)$. By the Invertible Matrix Theorem, we know that a homogeneous system of n equations in n variables has non-trivial solutions if and only if it has a non-zero determinant. Hence, for λ to be an eigenvalue, we must have $\det(A - \lambda I) = 0$. This is the key result in the procedure for finding the eigenvalues and eigenvectors, so it is worth summarizing as a theorem.

Theorem 1

Suppose that A is an $n \times n$ matrix. A real number λ is an eigenvalue of A if and only if λ satisfies the equation

$$\det(A - \lambda I) = 0$$

If λ is an eigenvalue of A, then all non-trivial solutions of the homogeneous system

$$(A - \lambda I)\vec{v} = \vec{0}$$

are eigenvectors of A that correspond to λ.

Observe that the set of all eigenvectors corresponding to an eigenvalue λ is just the nullspace of $A - \lambda I$, excluding the zero vector. In particular, the set containing all eigenvectors corresponding to λ and the zero vector is a subspace of $\mathbb{R}^n$. We make the following definition.

Definition
Eigenspace

Let λ be an eigenvalue of an $n \times n$ matrix A. Then the set containing the zero vector and all eigenvectors of A corresponding to λ is called the **eigenspace** of λ.

Remark

From our work preceding the theorem, we see that the eigenspace of any eigenvalue λ must contain at least one non-zero vector. Hence, the dimension of the eigenspace must be at least 1.

EXAMPLE 4

Find the eigenvalues and eigenvectors of the matrix $A = \begin{bmatrix} 17 & -15 \\ 20 & -18 \end{bmatrix}$ of Example 1.

Solution: We have

$$A - \lambda I = \begin{bmatrix} 17 & -15 \\ 20 & -18 \end{bmatrix} - \lambda \begin{bmatrix} 1 & 0 \\ 0 & 1 \end{bmatrix} = \begin{bmatrix} 17 - \lambda & -15 \\ 20 & -18 - \lambda \end{bmatrix}$$

(You should set up your calculations like this: you will need $A - \lambda I$ *later when you find the eigenvectors.)* Then

$$\begin{aligned} \det(A - \lambda I) &= \begin{vmatrix} 17 - \lambda & -15 \\ 20 & -18 - \lambda \end{vmatrix} \\ &= (17 - \lambda)(-18 - \lambda) - (-15)20 \\ &= \lambda^2 + \lambda - 6 = (\lambda + 3)(\lambda - 2) \end{aligned}$$

so $\det(A - \lambda I) = 0$ when $\lambda = -3$ or $\lambda = 2$. These are all of the eigenvalues of A.

EXAMPLE 4
(continued)

To find all the eigenvectors of $\lambda = -3$, we solve the homogeneous system $(A - \lambda I)\vec{v} = \vec{0}$. Writing $A - \lambda I$ and row reducing gives

$$A - (-3)I = \begin{bmatrix} 20 & -15 \\ 20 & -15 \end{bmatrix} \sim \begin{bmatrix} 1 & -3/4 \\ 0 & 0 \end{bmatrix}$$

so that the general solution of $(A-\lambda I)\vec{v} = \vec{0}$ is $\vec{v} = t\begin{bmatrix} 3/4 \\ 1 \end{bmatrix}$, $t \in \mathbb{R}$. Thus, all eigenvectors of A corresponding to $\lambda = -3$ are $\vec{v} = t\begin{bmatrix} 3/4 \\ 1 \end{bmatrix}$ for any *non-zero* value of t, and the eigenspace for $\lambda = -3$ is Span $\left\{\begin{bmatrix} 3/4 \\ 1 \end{bmatrix}\right\}$.

We repeat the process for the eigenvalue $\lambda = 2$:

$$A - 2I = \begin{bmatrix} 15 & -15 \\ 20 & -20 \end{bmatrix} \sim \begin{bmatrix} 1 & -1 \\ 0 & 0 \end{bmatrix}$$

The general solution of $(A - \lambda I)\vec{v} = \vec{0}$ is $\vec{v} = t\begin{bmatrix} 1 \\ 1 \end{bmatrix}$, $t \in \mathbb{R}$, so the eigenspace for $\lambda = 2$ is Span $\left\{\begin{bmatrix} 1 \\ 1 \end{bmatrix}\right\}$. In particular, all eigenvectors of A corresponding to $\lambda = 2$ are all non-zero multiples of $\begin{bmatrix} 1 \\ 1 \end{bmatrix}$.

Observe in Example 4 that $\det(A - \lambda I)$ gave us a degree 2 polynomial. This motivates the following definition.

Definition
Characteristic Polynomial

Let A be an $n \times n$ matrix. Then $C(\lambda) = \det(A - \lambda I)$ is called the **characteristic polynomial** of A.

For an $n \times n$ matrix A, the characteristic polynomial $C(\lambda)$ is of degree n, and the roots of $C(\lambda)$ are the eigenvalues of A. Note that the term of highest degree λ^n has coefficient $(-1)^n$; some other books prefer to work with the polynomial $\det(\lambda I - A)$ so that the coefficient of λ^n is always 1. In our notation, the constant term in the characteristic polynomial is $\det A$ (see Problem 6.2.D7).

It is relevant here to recall some facts about the roots of an n-th degree polynomial:

(1) λ_1 is a root of $C(\lambda)$ if and only if $(\lambda - \lambda_1)$ is a factor of $C(\lambda)$.

(2) The total number of roots (real and complex, counting repetitions) is n.

(3) Complex roots of the equation occur in "conjugate pairs," so that the total number of complex roots must be even.

(4) If n is odd, there must be at least one real root.

(5) If the entries of A are integers, since the leading coefficient of the characteristic polynomial is ± 1, any rational root must in fact be an integer.

EXAMPLE 5

Find the eigenvalues and eigenvectors of $A = \begin{bmatrix} 1 & 1 \\ 0 & 1 \end{bmatrix}$.

Solution: The characteristic polynomial is

$$C(\lambda) = \det(A - \lambda I) = \begin{vmatrix} 1-\lambda & 1 \\ 0 & 1-\lambda \end{vmatrix} = (1-\lambda)(1-\lambda)$$

So, $\lambda = 1$ is a double root (that is, $(\lambda - 1)$ appears as a factor of $C(\lambda)$ twice) so $\lambda = 1$ is the only eigenvalue of A.

For $\lambda = 1$, we have

$$A - \lambda I = \begin{bmatrix} 0 & 1 \\ 0 & 0 \end{bmatrix}$$

which has the general solution $\vec{v} = t\begin{bmatrix} 1 \\ 0 \end{bmatrix}$, $t \in \mathbb{R}$. Thus, the eigenspace for $\lambda = 1$ is $\operatorname{Span}\left\{\begin{bmatrix} 1 \\ 0 \end{bmatrix}\right\}$.

EXERCISE 2

Find the eigenvalues and eigenvectors of $A = \begin{bmatrix} 1 & 2 \\ 2 & 4 \end{bmatrix}$.

EXAMPLE 6

Find the eigenvalues and eigenvectors of $A = \begin{bmatrix} -3 & 5 & -5 \\ -7 & 9 & -5 \\ -7 & 7 & -3 \end{bmatrix}$.

Solution: We have

$$C(\lambda) = \det(A - \lambda I) = \begin{vmatrix} -3-\lambda & 5 & -5 \\ -7 & 9-\lambda & -5 \\ -7 & 7 & -3-\lambda \end{vmatrix}$$

Expanding this determinant along some row or column will involve a fair number of calculations. Also, we will end up with a degree 3 polynomial, which may not be easy to factor. But this is just a determinant, so we can use properties of determinants to make it easier. Since adding a multiple of one row to another does not change the determinant, we get by subtracting row 2 from row 3,

$$C(\lambda) = \begin{vmatrix} -3-\lambda & 5 & -5 \\ -7 & 9-\lambda & -5 \\ 0 & -2+\lambda & 2-\lambda \end{vmatrix}$$

Expanding this along the bottom row gives

$$\begin{aligned} C(\lambda) &= (-2+\lambda)(-1)((-3-\lambda)(-5) - (-5)(-7)) \\ &\quad +(2-\lambda)((-3-\lambda)(9-\lambda) - 5(-7)) \\ &= (2-\lambda)((5\lambda + 15 - 35) + (\lambda^2 - 6\lambda - 27 + 35)) \\ &= -(\lambda - 2)(\lambda^2 - \lambda - 12) = -(\lambda - 2)(\lambda - 4)(\lambda + 3) \end{aligned}$$

EXAMPLE 6
(continued)

Hence, the eigenvalues of A are $\lambda_1 = 2$, $\lambda_2 = 4$, and $\lambda_3 = -3$.
For $\lambda_1 = 2$,

$$A - \lambda_1 I = \begin{bmatrix} -5 & 5 & -5 \\ -7 & 7 & -5 \\ -7 & 7 & -5 \end{bmatrix} \sim \begin{bmatrix} 1 & -1 & 0 \\ 0 & 0 & 1 \\ 0 & 0 & 0 \end{bmatrix}$$

Hence, the general solution of $(A - \lambda_1 I)\vec{v} = \vec{0}$ is $\vec{v} = t\begin{bmatrix} 1 \\ 1 \\ 0 \end{bmatrix}$. Thus, a basis for the eigenspace of λ_1 is $\left\{\begin{bmatrix} 1 \\ 1 \\ 0 \end{bmatrix}\right\}$.

For $\lambda_2 = 4$,

$$A - \lambda_2 I = \begin{bmatrix} -7 & 5 & -5 \\ -7 & 5 & -5 \\ -7 & 7 & -7 \end{bmatrix} \sim \begin{bmatrix} 1 & 0 & 0 \\ 0 & 1 & -1 \\ 0 & 0 & 0 \end{bmatrix}$$

Hence, the general solution of $(A - \lambda_2 I)\vec{v} = \vec{0}$ is $\vec{v} = t\begin{bmatrix} 0 \\ 1 \\ 1 \end{bmatrix}$. Thus, a basis for the eigenspace of λ_2 is $\left\{\begin{bmatrix} 0 \\ 1 \\ 1 \end{bmatrix}\right\}$.

For $\lambda_3 = -3$,

$$A - \lambda_3 I = \begin{bmatrix} 0 & 5 & -5 \\ -7 & 12 & -5 \\ -7 & 7 & 0 \end{bmatrix} \sim \begin{bmatrix} 1 & 0 & -1 \\ 0 & 1 & -1 \\ 0 & 0 & 0 \end{bmatrix}$$

Hence, the general solution of $(A - \lambda_3 I)\vec{v} = \vec{0}$ is $\vec{v} = t\begin{bmatrix} 1 \\ 1 \\ 1 \end{bmatrix}$. Thus, a basis for the eigenspace of λ_3 is $\left\{\begin{bmatrix} 1 \\ 1 \\ 1 \end{bmatrix}\right\}$.

EXAMPLE 7

Find the eigenvalues and eigenvectors of $A = \begin{bmatrix} 1 & 1 & 1 \\ 1 & 1 & 1 \\ 1 & 1 & 1 \end{bmatrix}$.

Solution: We have

$$\begin{aligned} C(\lambda) = \begin{vmatrix} 1-\lambda & 1 & 1 \\ 1 & 1-\lambda & 1 \\ 1 & 1 & 1-\lambda \end{vmatrix} &= \begin{vmatrix} 1-\lambda & 1 & 1 \\ 1 & 1-\lambda & 1 \\ 0 & \lambda & -\lambda \end{vmatrix} \\ &= \lambda(-1)((1-\lambda) - 1(1)) + (-\lambda)((1-\lambda)(1-\lambda) - 1(1)) \\ &= -\lambda(-\lambda + \lambda^2 - 2\lambda) = -\lambda^2(\lambda - 3) \end{aligned}$$

Therefore, the eigenvalues of A are $\lambda_1 = 0$ (which occurs twice) and $\lambda_2 = 3$.

For $\lambda_1 = 0$,

$$A - \lambda_1 I = \begin{bmatrix} 1 & 1 & 1 \\ 1 & 1 & 1 \\ 1 & 1 & 1 \end{bmatrix} \sim \begin{bmatrix} 1 & 1 & 1 \\ 0 & 0 & 0 \\ 0 & 0 & 0 \end{bmatrix}$$

EXAMPLE 7
(continued)

Hence, a basis for the eigenspace of $\lambda_1 = 0$ is $\left\{ \begin{bmatrix} -1 \\ 1 \\ 0 \end{bmatrix}, \begin{bmatrix} -1 \\ 0 \\ 1 \end{bmatrix} \right\}$.

For $\lambda_2 = 3$,

$$A - \lambda_2 I = \begin{bmatrix} -2 & 1 & 1 \\ 1 & -2 & 1 \\ 1 & 1 & -2 \end{bmatrix} \sim \begin{bmatrix} 1 & 0 & -1 \\ 0 & 1 & -1 \\ 0 & 0 & 0 \end{bmatrix}$$

Thus, a basis for the eigenspace of $\lambda_2 = 3$ is $\left\{ \begin{bmatrix} 1 \\ 1 \\ 1 \end{bmatrix} \right\}$.

These examples motivate the following definitions.

Definition
Algebraic Multiplicity
Geometric Multiplicity

Let A be an $n \times n$ matrix with eigenvalue λ. The **algebraic multiplicity** of λ is the number of times λ is repeated as a root of the characteristic polynomial. The **geometric multiplicity** of λ is the dimension of the eigenspace of λ.

EXAMPLE 8

In Example 5, the eigenvalue $\lambda = 1$ has algebraic multiplicity 2 since the characteristic polynomial is $(\lambda - 1)(\lambda - 1)$, and $\lambda = 1$ has geometric multiplicity 1 since a basis for its eigenspace is $\left\{ \begin{bmatrix} 1 \\ 0 \end{bmatrix} \right\}$.

In Example 6, each eigenvalue has algebraic multiplicity and geometric multiplicity 1. In Example 7, the eigenvalue 0 has algebraic and geometric multiplicity 2, and the eigenvalue 3 has algebraic and geometric multiplicity 1.

EXERCISE 3

Let $A = \begin{bmatrix} 5 & -3 & 2 \\ 0 & 0 & 2 \\ 0 & -2 & -4 \end{bmatrix}$. Show that $\lambda_1 = 5$ and $\lambda_2 = -2$ are both eigenvalues of A and determine the algebraic and geometric multiplicity of both of these eigenvalues.

These definitions lead to some theorems that will be very important in the next section.

Theorem 2

Let λ be an eigenvalue of an $n \times n$ matrix A. Then

$$1 \leq \text{geometric multiplicity} \leq \text{algebraic multiplicity}$$

If the geometric multiplicity of an eigenvalue is less than its algebraic multiplicity, then we say that the eigenvalue is **deficient**. However, if A is an $n \times n$ matrix with distinct eigenvalues $\lambda_1, \dots, \lambda_k$, which all have the property that their geometric multiplicity equals their algebraic multiplicity, then the sum of the geometric multiplicities of all eigenvalues equals the sum of the algebraic multiplicities, which equals n (since an n-th degree polynomials has exactly n roots). Hence, if we collect the basis vectors from the eigenspaces of all k eigenvalues, we will end up with n vectors in $\mathbb{R}^n$. We now prove that eigenvectors from eigenspaces of different eigenvalues are necessarily linearly independent, and hence this collection of n eigenvectors will form a basis for $\mathbb{R}^n$.

Theorem 3

Suppose that $\lambda_1, \dots, \lambda_k$ are distinct ($\lambda_i \neq \lambda_j$) eigenvalues of an $n \times n$ matrix A, with corresponding eigenvectors $\vec{v}_1, \dots, \vec{v}_k$, respectively. Then $\{\vec{v}_1, \dots, \vec{v}_k\}$ is linearly independent.

Proof: We will prove this theorem by induction. If $k = 1$, then the result is trivial, since by definition of an eigenvector, $\vec{v}_1 \neq \vec{0}$. Assume that the result is true for some $k \geq 1$. To show $\{\vec{v}_1, \dots, \vec{v}_k, \vec{v}_{k+1}\}$ is linearly independent, we consider

$$c_1\vec{v}_1 + \cdots + c_k\vec{v}_k + c_{k+1}\vec{v}_{k+1} = \vec{0} \tag{6.1}$$

Observe that since $A\vec{v}_i = \lambda_i\vec{v}_i$, we have $(A - \lambda_i I)\vec{v}_i = \vec{0}$ and

$$(A - \lambda_i I)\vec{v}_j = A\vec{v}_j - \lambda_i\vec{v}_j = \lambda_j\vec{v}_j - \lambda_i\vec{v}_j = (\lambda_j - \lambda_i)\vec{v}_j$$

Thus, multiplying both sides of (6.1) by $A - \lambda_{k+1}I$ gives

$$c_1(\lambda_1 - \lambda_{k+1})\vec{v}_1 + \cdots + c_k(\lambda_k - \lambda_{k+1})\vec{v}_k + \vec{0} = \vec{0}$$

By our induction hypothesis, $\{\vec{v}_1, \dots, \vec{v}_k\}$ is linearly independent; thus, all the coefficients must be 0. But $\lambda_i \neq \lambda_j$; hence, we must have $c_1 = \cdots = c_k = 0$. Thus (6.1) becomes

$$0 + c_{k+1}\vec{v}_{k+1} = \vec{0}$$

But $\vec{v}_{k+1} \neq \vec{0}$ since it is an eigenvector; hence, $c_{k+1} = 0$, and the set is linearly independent. ■

Remark

In this book, most eigenvalues turn out to be integers. This is somewhat unrealistic; in real world applications, eigenvalues are often not rational numbers. Effective computer methods for finding eigenvalues depend on the theory of eigenvectors and eigenvalues.

PROBLEMS 6.1

Practice Problems

A1 Let $A = \begin{bmatrix} -10 & 9 & 5 \\ -10 & 9 & 5 \\ 2 & 3 & -1 \end{bmatrix}$. Determine whether the following vectors are eigenvectors of A. If they are, determine the corresponding eigenvalues. Answer without calculating the characteristic polynomial.

$$\vec{v}_1 = \begin{bmatrix} 1 \\ 0 \\ 1 \end{bmatrix}, \vec{v}_2 = \begin{bmatrix} 1 \\ 0 \\ 2 \end{bmatrix}, \vec{v}_3 = \begin{bmatrix} 1 \\ 1 \\ -1 \end{bmatrix}, \vec{v}_4 = \begin{bmatrix} 1 \\ -1 \\ 1 \end{bmatrix}, \vec{v}_5 = \begin{bmatrix} 1 \\ 1 \\ 1 \end{bmatrix}$$

A2 Find the eigenvalues and corresponding eigenspaces of the following matrices.

(a) $\begin{bmatrix} 0 & 1 \\ -6 & 5 \end{bmatrix}$ (b) $\begin{bmatrix} 1 & 3 \\ 0 & 1 \end{bmatrix}$

(c) $\begin{bmatrix} 2 & 0 \\ 0 & 3 \end{bmatrix}$ (d) $\begin{bmatrix} -26 & 10 \\ -75 & 29 \end{bmatrix}$

(e) $\begin{bmatrix} 1 & 3 \\ 4 & 2 \end{bmatrix}$ (f) $\begin{bmatrix} 3 & -3 \\ 6 & -6 \end{bmatrix}$

A3 For each of the following matrices, determine the algebraic multiplicity of each eigenvalue and determine the geometric multiplicity of each eigenvalue by writing a basis for its eigenspace.

(a) $\begin{bmatrix} 2 & -2 \\ 0 & 3 \end{bmatrix}$ (b) $\begin{bmatrix} 2 & -2 \\ 0 & 2 \end{bmatrix}$

(c) $\begin{bmatrix} 1 & 1 \\ -1 & 3 \end{bmatrix}$ (d) $\begin{bmatrix} 0 & -5 & 3 \\ -2 & -6 & 6 \\ -2 & -7 & 7 \end{bmatrix}$

(e) $\begin{bmatrix} 2 & 2 & 2 \\ 2 & 2 & 2 \\ 2 & 2 & 2 \end{bmatrix}$ (f) $\begin{bmatrix} 3 & 1 & 1 \\ 1 & 3 & 1 \\ 1 & 1 & 3 \end{bmatrix}$

Homework Problems

B1 Let $A = \begin{bmatrix} -3 & 1 & -1 \\ 8 & -3 & 8 \\ 8 & -1 & 6 \end{bmatrix}$. Determine whether the following vectors are eigenvectors of A. If they are, determine the corresponding eigenvalues. Answer without calculating the characteristic polynomial.

$$\vec{v}_1 = \begin{bmatrix} 1 \\ 0 \\ 1 \end{bmatrix}, \vec{v}_2 = \begin{bmatrix} 1 \\ 0 \\ -1 \end{bmatrix}, \vec{v}_3 = \begin{bmatrix} 1 \\ 1 \\ 0 \end{bmatrix}, \vec{v}_4 = \begin{bmatrix} 1 \\ -1 \\ -1 \end{bmatrix}, \vec{v}_5 = \begin{bmatrix} 1 \\ 1 \\ 2 \end{bmatrix}$$

B2 Find the eigenvalues and corresponding eigenspaces of the following matrices.

(a) $\begin{bmatrix} 4 & -1 \\ -2 & 5 \end{bmatrix}$ (b) $\begin{bmatrix} 2 & 1 \\ -1 & 4 \end{bmatrix}$

(c) $\begin{bmatrix} -2 & 2 \\ -3 & 5 \end{bmatrix}$ (d) $\begin{bmatrix} 2 & 2 \\ -3 & -5 \end{bmatrix}$

(e) $\begin{bmatrix} 1 & 3 & 5 \\ 0 & 2 & 7 \\ 0 & 0 & 3 \end{bmatrix}$ (f) $\begin{bmatrix} -4 & 6 & 6 \\ -2 & 2 & 4 \\ -1 & 3 & 1 \end{bmatrix}$

B3 For each of the following matrices, determine the algebraic multiplicity of each eigenvalue and determine the geometric multiplicity of each eigenvalue by writing a basis for its eigenspace.

(a) $\begin{bmatrix} 3 & 0 \\ 0 & 7 \end{bmatrix}$ (b) $\begin{bmatrix} -3 & -3 \\ 0 & -3 \end{bmatrix}$

(c) $\begin{bmatrix} 7 & 3 \\ 2 & 2 \end{bmatrix}$ (d) $\begin{bmatrix} -4 & 0 & 0 \\ 2 & -8 & 4 \\ -4 & 5 & 0 \end{bmatrix}$

(e) $\begin{bmatrix} 4 & 2 & 2 \\ 2 & 4 & 2 \\ 2 & 2 & 4 \end{bmatrix}$ (f) $\begin{bmatrix} -9 & -7 & 7 \\ -9 & -7 & 2 \\ 8 & -8 & -3 \end{bmatrix}$

Computer Problems

C1 Use a computer to determine the eigenvalues and corresponding eigenspaces of the following matrices.

(a) $\begin{bmatrix} 9 & -9 & -8 \\ 0 & -5 & 3 \\ 0 & -2 & 5 \end{bmatrix}$

(b) $\begin{bmatrix} 5 & -5 & 4 \\ 3 & -2 & -1 \\ 1 & -2 & 6 \end{bmatrix}$

(c) $\begin{bmatrix} 2 & 1 & 0 & 3 \\ 2 & 3 & 1 & -4 \\ 4 & 2 & 2 & 4 \\ 4 & 2 & 2 & 4 \end{bmatrix}$

C2 Let $A = \begin{bmatrix} 2.89316 & -1.28185 & 2.42918 \\ -0.70562 & 0.76414 & -0.67401 \\ 1.67682 & -0.83198 & 2.34270 \end{bmatrix}$.

Verify that $\begin{bmatrix} 1.21 \\ -0.34 \\ 0.87 \end{bmatrix}$, $\begin{bmatrix} 1.31 \\ 2.15 \\ -0.21 \end{bmatrix}$, and $\begin{bmatrix} -1.85 \\ 0.67 \\ 2.10 \end{bmatrix}$ are (approximately) eigenvectors of A. Determine the corresponding eigenvalues.

Conceptual Problems

D1 Suppose that $\vec{v}$ is an eigenvector of both the matrix A and the matrix B, with corresponding eigenvalue λ for A and corresponding eigenvalue μ for B. Show that $\vec{v}$ is an eigenvector of $(A + B)$ and of AB. Determine the corresponding eigenvalues.

D2 (a) Show that if λ is an eigenvalue of a matrix A, then λ^n is an eigenvalue of A^n. How are the corresponding eigenvectors related?

(b) Give an example of a 2×2 matrix A such that A has no real eigenvalues, but A^3 does have real eigenvalues. (Hint: See Problem 3.3.D4.)

D3 Show that if A is invertible and $\vec{v}$ is an eigenvector of A, then $\vec{v}$ is also an eigenvector of A^{-1}. How are the corresponding eigenvalues related?

D4 (a) Let A be an $n \times n$ matrix with $\text{rank}(A) = r < n$. Prove that 0 is an eigenvalue of A and determine its geometric multiplicity.

(b) Give an example of a 3×3 matrix with $\text{rank}(A) = r < n$ such that the algebraic multiplicity of the eigenvalue 0 is greater than its geometric multiplicity.

D5 Suppose that A is an $n \times n$ matrix such that the sum of the entries in each row is the same. That is, $\sum_{k=1}^{n} a_{ik} = c$ for all $1 \leq i \leq n$. Show that $\vec{v} = \begin{bmatrix} 1 \\ \vdots \\ 1 \end{bmatrix}$ is an eigenvector of A. (Such matrices arise in probability theory.)

6.2 Diagonalization

At the end of the last section, we showed that if the k distinct eigenvalues $\lambda_1, \dots, \lambda_k$ of an $n \times n$ matrix A all had the property that their geometric multiplicity equalled their algebraic multiplicity, then we could find a basis for $\mathbb{R}^n$ of eigenvectors of A by collecting the basis vectors from the eigenspaces of each of the k eigenvalues. We now see that this basis of eigenvectors is extremely useful.

Suppose that A is an $n \times n$ matrix for which there is a basis $\{\vec{v}_1, \dots, \vec{v}_n\}$ of eigenvectors of A. Let the corresponding eigenvalues be denoted $\lambda_1, \dots, \lambda_n$, respectively. If we let $P = \begin{bmatrix} \vec{v}_1 & \cdots & \vec{v}_n \end{bmatrix}$, then we get

$$\begin{aligned} AP &= A\begin{bmatrix} \vec{v}_1 & \cdots & \vec{v}_n \end{bmatrix} \\ &= \begin{bmatrix} A\vec{v}_1 & \cdots & A\vec{v}_n \end{bmatrix} \\ &= \begin{bmatrix} \lambda_1\vec{v}_1 & \cdots & \lambda_n\vec{v}_n \end{bmatrix} \\ &= \begin{bmatrix} \vec{v}_1 & \cdots & \vec{v}_n \end{bmatrix} \begin{bmatrix} \lambda_1 & 0 & \cdots & 0 \\ 0 & \lambda_2 & \ddots & \vdots \\ \vdots & \ddots & \ddots & 0 \\ 0 & \cdots & 0 & \lambda_n \end{bmatrix} = PD \end{aligned}$$

Recall that a square matrix D such that $d_{ij} = 0$ for $i \neq j$ is said to be **diagonal** and can be denoted by $\text{diag}(d_{11}, \ldots, d_{nn})$. Thus, using the fact that P is invertible since the columns of P form a basis for $\mathbb{R}^n$, we can write $AP = PD$ as

$$P^{-1}AP = D = \text{diag}(\lambda_1, \ldots, \lambda_n)$$

Definition
Diagonalizable

If there exists an invertible matrix P and diagonal matrix D such that $P^{-1}AP = D$, then we say that A is **diagonalizable** (some people prefer "diagonable") and that the matrix P **diagonalizes** A to its **diagonal form** D.

It may be tempting to think that $P^{-1}AP = D$ implies that $A = D$ since P and P^{-1} are inverses. However, this is *not* true in general since matrix multiplication is not commutative. Not surprisingly, though, if A and B are matrices such that $P^{-1}AP = B$ for some invertible matrix P, then A and B have many similarities.

Theorem 1

If A and B are $n \times n$ matrices such that $P^{-1}AP = B$ for some invertible matrix P, then A and B have

(1) The same determinant
(2) The same eigenvalues
(3) The same rank
(4) The same **trace**, where the trace of a matrix A is defined by $\text{tr}\, A = \sum_{i=1}^{n} a_{ii}$

(1) was proved as Problem D4 in Section 5.2. The proofs of (2), (3), and (4) are left as Problems D1, D2, and D3, respectively.

This theorem motivates the following definition.

Definition
Similar Matrices

If A and B are $n \times n$ matrices such that $P^{-1}AP = B$ for some invertible matrix P, then A and B are said to be **similar**.

Thus, from our work above, if there is a basis of $\mathbb{R}^n$ consisting of eigenvectors of A, then A is similar to a diagonal matrix D and so A is diagonalizable. On the other hand, if at least one of the eigenvalues of A is deficient, then A will not have n linearly independent eigenvectors. Hence we will not be able to construct an invertible matrix P whose columns are eigenvectors of A. In this case, we say that A is not diagonalizable.

We get the following theorem.

Theorem 2 [Diagonalization Theorem]
An $n \times n$ matrix A can be diagonalized if and only if there exists a basis for $\mathbb{R}^n$ of eigenvectors of A. If such a basis $\{\vec{v}_1, \ldots, \vec{v}_n\}$ exists, the matrix $P = \begin{bmatrix} \vec{v}_1 & \cdots & \vec{v}_n \end{bmatrix}$ diagonalizes A to a diagonal matrix $D = \text{diag}(\lambda_1, \ldots, \lambda_n)$, where λ_i is an eigenvalue of A corresponding to $\vec{v}_i$ for $1 \leq i \leq n$.

From the Diagonalization Theorem and our work above, we immediately get the following two useful corollaries.

Corollary 3 A matrix A is diagonalizable if and only if every eigenvalue of a matrix A has its geometric multiplicity equal to its algebraic multiplicity.

Corollary 4 If an $n \times n$ matrix A has n distinct eigenvalues, then A is diagonalizable.

Remark

Observe that it is possible for a matrix A with real entries to have non-real eigenvalues, which will lead to non-real eigenvectors. In this case, there cannot exist a basis for $\mathbb{R}^n$ of eigenvectors of A, and so we will say that A is not diagonalizable over $\mathbb{R}$. In Chapter 9, we will examine the case where complex eigenvalues and eigenvectors are allowed.

EXAMPLE 1 Find an invertible matrix P and a diagonal matrix D such that $P^{-1}AP = D$, where $A = \begin{bmatrix} 2 & 3 \\ 3 & 2 \end{bmatrix}$.

Solution: We need to find a basis for $\mathbb{R}^2$ of eigenvectors of A. Hence, we need to find a basis for the eigenspace of each eigenvalue of A. The characteristic polynomial of A is

$$C(\lambda) = \det(A - \lambda I) = \begin{vmatrix} 2-\lambda & 3 \\ 3 & 2-\lambda \end{vmatrix} = \lambda^2 - 4\lambda - 5 = (\lambda - 5)(\lambda + 1)$$

Hence, the eigenvalues of A are $\lambda_1 = 5$ and $\lambda_2 = -1$.

For $\lambda_1 = 5$, we get

$$A - \lambda_1 I = \begin{bmatrix} -3 & 3 \\ 3 & -3 \end{bmatrix} \sim \begin{bmatrix} 1 & -1 \\ 0 & 0 \end{bmatrix}$$

So, $\vec{v}_1 = \begin{bmatrix} 1 \\ 1 \end{bmatrix}$ is an eigenvector for $\lambda_1 = 5$ and $\{\vec{v}_1\}$ is a basis for its eigenspace.

For $\lambda_2 = -1$, we get

$$A - \lambda_2 I = \begin{bmatrix} 3 & 3 \\ 3 & 3 \end{bmatrix} \sim \begin{bmatrix} 1 & 1 \\ 0 & 0 \end{bmatrix}$$

So, $\vec{v}_2 = \begin{bmatrix} -1 \\ 1 \end{bmatrix}$ is an eigenvector for $\lambda_2 = -1$ and $\{\vec{v}_2\}$ is a basis for its eigenspace.

Thus, $\{\vec{v}_1, \vec{v}_2\}$ is a basis for $\mathbb{R}^2$, and so if we let $P = \begin{bmatrix} \vec{v}_1 & \vec{v}_2 \end{bmatrix} = \begin{bmatrix} 1 & -1 \\ 1 & 1 \end{bmatrix}$, we get

$$P^{-1}AP = \text{diag}(\lambda_1, \lambda_2) = \begin{bmatrix} 5 & 0 \\ 0 & -1 \end{bmatrix} = D$$

EXAMPLE 1 (continued)

Note that we could have instead taken $P = \begin{bmatrix} \vec{v}_2 & \vec{v}_1 \end{bmatrix} = \begin{bmatrix} -1 & 1 \\ 1 & 1 \end{bmatrix}$, which would have given

$$P^{-1}AP = \text{diag}(\lambda_2, \lambda_1) = \begin{bmatrix} -1 & 0 \\ 0 & 5 \end{bmatrix}$$

EXAMPLE 2

Determine whether $A = \begin{bmatrix} 0 & 3 & -2 \\ -2 & 5 & -2 \\ -2 & 3 & 0 \end{bmatrix}$ is diagonalizable. If it is, find an invertible matrix P and a diagonal matrix D such that $P^{-1}AP = D$.

Solution: The characteristic polynomial of A is

$$\begin{aligned} C(\lambda) = \det(A - \lambda I) &= \begin{vmatrix} 0-\lambda & 3 & -2 \\ -2 & 5-\lambda & -2 \\ -2 & 3 & 0-\lambda \end{vmatrix} = \begin{vmatrix} 0-\lambda & 3 & -2 \\ -2 & 5-\lambda & -2 \\ 0 & -2+\lambda & 2-\lambda \end{vmatrix} \\ &= (-2+\lambda)(-1)(2\lambda - 4) + (2-\lambda)(\lambda^2 - 5\lambda + 6) \\ &= -(\lambda-2)(\lambda^2 - 3\lambda + 2) = -(\lambda-2)(\lambda-2)(\lambda-1) \end{aligned}$$

Hence, $\lambda_1 = 2$ is an eigenvalue with algebraic multiplicity 2 and $\lambda_2 = 1$ is an eigenvalue with algebraic multiplicity 1. By Theorem 6.1.2, the geometric multiplicity of $\lambda_2 = 1$ must equal 1. Thus, A is diagonalizable if and only if the geometric multiplicity of $\lambda_1 = 2$ is 2.

For $\lambda_1 = 2$, we get $A - \lambda_1 I = \begin{bmatrix} -2 & 3 & -2 \\ -2 & 3 & -2 \\ -2 & 3 & -2 \end{bmatrix} \sim \begin{bmatrix} 1 & -3/2 & 1 \\ 0 & 0 & 0 \\ 0 & 0 & 0 \end{bmatrix}$. Thus, a basis for the eigenspace is $\left\{ \begin{bmatrix} 3/2 \\ 1 \\ 0 \end{bmatrix}, \begin{bmatrix} -1 \\ 0 \\ 1 \end{bmatrix} \right\}$. Hence, the geometric multiplicity of $\lambda_1 = 2$ equals its algebraic multiplicity.

By Corollary 3, we see that A is diagonalizable. So, we also need to find a basis for the eigenspace of $\lambda_2 = 1$.

For $\lambda_2 = 1$, we get $A - \lambda_2 I = \begin{bmatrix} -1 & 3 & -2 \\ -2 & 4 & -2 \\ -2 & 3 & -1 \end{bmatrix} \sim \begin{bmatrix} 1 & 0 & -1 \\ 0 & 1 & -1 \\ 0 & 0 & 0 \end{bmatrix}$. Therefore, $\left\{ \begin{bmatrix} 1 \\ 1 \\ 1 \end{bmatrix} \right\}$ is a basis for the eigenspace.

So, we can take $P = \begin{bmatrix} 3/2 & -1 & 1 \\ 1 & 0 & 1 \\ 0 & 1 & 1 \end{bmatrix}$ and get $P^{-1}AP = \begin{bmatrix} 2 & 0 & 0 \\ 0 & 2 & 0 \\ 0 & 0 & 1 \end{bmatrix}$.

EXERCISE 1

Diagonalize $A = \begin{bmatrix} 1 & 0 & -1 \\ 11 & -4 & -7 \\ -7 & 3 & 4 \end{bmatrix}$.

EXAMPLE 3

Is the matrix $A = \begin{bmatrix} -1 & 7 & -5 \\ -4 & 11 & -6 \\ -4 & 8 & -3 \end{bmatrix}$ diagonalizable?

Solution: The characteristic polynomial is

$$C(\lambda) = \det(A - \lambda I) = \begin{vmatrix} -1-\lambda & 7 & -5 \\ -4 & 11-\lambda & -6 \\ -4 & 8 & -3-\lambda \end{vmatrix} = \begin{vmatrix} -1-\lambda & 7 & -5 \\ -4 & 11-\lambda & -6 \\ 0 & -3+\lambda & 3-\lambda \end{vmatrix}$$

$$= (-3+\lambda)(-1)(6\lambda + 6 - 20) + (3-\lambda)(\lambda^2 - 11\lambda + \lambda - 11 + 28)$$

$$= -(\lambda - 3)(\lambda^2 - 4\lambda + 3) = -(\lambda - 3)(\lambda - 3)(\lambda - 1)$$

Thus, $\lambda_1 = 3$ is an eigenvalue with algebraic multiplicity 2, and $\lambda_2 = 1$ is an eigenvalue with algebraic multiplicity 1.

For $\lambda_1 = 3$, we get $A - \lambda_1 I = \begin{bmatrix} -4 & 7 & -5 \\ -4 & 8 & -6 \\ -4 & 8 & -6 \end{bmatrix} \sim \begin{bmatrix} 1 & 0 & -1/2 \\ 0 & 1 & -1 \\ 0 & 0 & 0 \end{bmatrix}$. Thus, a basis for the eigenspace is $\left\{ \begin{bmatrix} 1/2 \\ 1 \\ 1 \end{bmatrix} \right\}$. Hence, the geometric multiplicity of $\lambda_1 = 1$ is less than its algebraic multiplicity, and so A is not diagonalizable by Corollary 3.

EXERCISE 2

Show that $A = \begin{bmatrix} 2 & 1 \\ 0 & 2 \end{bmatrix}$ is not diagonalizable.

EXAMPLE 4

Show that matrix $A = \begin{bmatrix} 0 & -1 \\ 1 & 0 \end{bmatrix}$ is not diagonalizable over $\mathbb{R}$.

Solution: The characteristic polynomial is

$$C(\lambda) = \det(A - \lambda I) = \begin{vmatrix} -\lambda & -1 \\ 1 & -\lambda \end{vmatrix} = \lambda^2 + 1$$

Since $\lambda^2 + 1 = 0$ has no real solutions, the matrix A has no real eigenvalues and hence is not diagonalizable over $\mathbb{R}$.

Some Applications of Diagonalization

A geometrical application of diagonalization occurs when we try to picture the graph of a quadratic equation in two variables, such as $ax^2 + 2bxy + cy^2 = d$. It turns out that we should consider the associated matrix $\begin{bmatrix} a & b \\ b & c \end{bmatrix}$. By diagonalizing this matrix, we can easily recognize the graph as an ellipse, a hyperbola, or perhaps some degenerate case. This problem will be discussed in Section 8.3.

A physical application related to these geometrical applications is the analysis of the deformation of a solid. Imagine, for example, a small steel block that experiences a small deformation when some forces are applied. The change of shape in the block can be described in terms of a 3×3 *strain matrix*. This matrix can always be diagonalized, so it turns out that we can identify the change of shape as the composition of three stretches along mutually orthogonal directions. This application is discussed in Section 8.4.

Diagonalization is also an important tool for studying systems of linear difference equations, which arise in many settings. Consider, for example, a population that is divided into two groups; we count these two groups at regular intervals (say, once a month) so that at every time n, we have a vector $\vec{p} = \begin{bmatrix} p_1(n) \\ p_2(n) \end{bmatrix}$ that tells us how many are in each group. For some situations, the change from month to month can be described by saying that the vector $\vec{p}$ changes according to the rule

$$\vec{p}(n+1) = A\vec{p}(n)$$

where A is some known 2×2 matrix. It follows that $p(n) = A^n p(0)$. We are often interested in understanding what happens to the population "in the long run." This requires us to calculate A^n for n large. This problem is easy to deal with if we can diagonalize A. Particular examples of this kind are Markov processes, which are discussed in Section 6.3.

One very important application of diagonalization and the related idea of eigenvectors is the solution of systems of linear differential equations. This application is discussed in Section 6.4.

In Section 4.6 we saw that if $L : \mathbb{R}^n \to \mathbb{R}^n$ is a linear transformation, then its matrix with respect to the basis $\mathcal{B}$ is determined from its standard matrix by the equation

$$[L]_{\mathcal{B}} = P^{-1}[L]_S P$$

where $P = \begin{bmatrix} \vec{v}_1 & \cdots & \vec{v}_n \end{bmatrix}$ is the change of basis matrix. Examples 5 and 7 in Section 4.6 show that we can more easily give a geometrical interpretation of a linear mapping L if there is a basis $\mathcal{B}$ such that $[L]_{\mathcal{B}}$ is in diagonal form. Hence, our diagonalization process is a method for finding such a geometrically natural basis. In particular, if the standard matrix of L is diagonalizable, then the basis for $\mathbb{R}^n$ of eigenvectors forms the geometrically natural basis.

PROBLEMS 6.2

Practice Problems

A1 By checking whether columns of P are eigenvectors of A, determine whether P diagonalizes A. If it does, determine P^{-1} and check that $P^{-1}AP$ is diagonal.

(a) $A = \begin{bmatrix} 11 & 6 \\ 9 & -4 \end{bmatrix}$, $P = \begin{bmatrix} 2 & -1 \\ 1 & 3 \end{bmatrix}$

(b) $A = \begin{bmatrix} 6 & 5 \\ 3 & -7 \end{bmatrix}$, $P = \begin{bmatrix} 1 & 2 \\ 1 & 1 \end{bmatrix}$

(c) $A = \begin{bmatrix} 5 & -8 \\ 4 & -7 \end{bmatrix}$, $P = \begin{bmatrix} 2 & 1 \\ 1 & 1 \end{bmatrix}$

(d) $A = \begin{bmatrix} 2 & 4 & 4 \\ 4 & 2 & 4 \\ 4 & 4 & 2 \end{bmatrix}$, $P = \begin{bmatrix} -1 & -1 & 1 \\ 1 & 0 & 1 \\ 0 & 1 & 1 \end{bmatrix}$

A2 For the following matrices, determine the eigenvalues and corresponding eigenvectors and determine whether each matrix is diagonalizable over $\mathbb{R}$. If it is diagonalizable, give a matrix P and diagonal matrix D such that $P^{-1}AP = D$.

(a) $A = \begin{bmatrix} 3 & 2 \\ 5 & 6 \end{bmatrix}$

(b) $A = \begin{bmatrix} -2 & 3 \\ 4 & -3 \end{bmatrix}$

(c) $A = \begin{bmatrix} 3 & 6 \\ -5 & -3 \end{bmatrix}$

(d) $A = \begin{bmatrix} 0 & 1 & 0 \\ 1 & 0 & 1 \\ 1 & 1 & 1 \end{bmatrix}$

(e) $A = \begin{bmatrix} 6 & -9 & -5 \\ -4 & 9 & 4 \\ 9 & -17 & -8 \end{bmatrix}$

(f) $A = \begin{bmatrix} -2 & 7 & 3 \\ -1 & 2 & 1 \\ 0 & 2 & 1 \end{bmatrix}$

(g) $A = \begin{bmatrix} -1 & 6 & 3 \\ 3 & -4 & -3 \\ -6 & 12 & 8 \end{bmatrix}$

A3 Follow the same instructions as for Problem A2.

(a) $A = \begin{bmatrix} 3 & 0 \\ -3 & 3 \end{bmatrix}$

(b) $A = \begin{bmatrix} 4 & 4 \\ 4 & 4 \end{bmatrix}$

(c) $A = \begin{bmatrix} -2 & 5 \\ 5 & -2 \end{bmatrix}$

(d) $A = \begin{bmatrix} 0 & 6 & -8 \\ -2 & 4 & -4 \\ -2 & 2 & -2 \end{bmatrix}$

(e) $A = \begin{bmatrix} 0 & 2 & 2 \\ 2 & 0 & 2 \\ 2 & 2 & 0 \end{bmatrix}$

(f) $A = \begin{bmatrix} 2 & 0 & 0 \\ -1 & 0 & 1 \\ -1 & -2 & 3 \end{bmatrix}$

(g) $A = \begin{bmatrix} -3 & -3 & 5 \\ 13 & 10 & -13 \\ 3 & 2 & -1 \end{bmatrix}$

Homework Problems

B1 By checking whether columns of P are eigenvectors of A, determine whether P diagonalizes A. If it does, determine P^{-1} and check that $P^{-1}AP$ is diagonal.

(a) $A = \begin{bmatrix} 4 & 2 \\ -5 & 3 \end{bmatrix}$, $P = \begin{bmatrix} 1 & 3 \\ -1 & 1 \end{bmatrix}$

(b) $A = \begin{bmatrix} 1 & 3 \\ 3 & 1 \end{bmatrix}$, $P = \begin{bmatrix} 1 & 1 \\ 1 & -1 \end{bmatrix}$

(c) $A = \begin{bmatrix} 1 & 2 \\ 3 & 2 \end{bmatrix}$, $P = \begin{bmatrix} 2 & 1 \\ 3 & -1 \end{bmatrix}$

(d) $A = \begin{bmatrix} -7 & 2 & -4 \\ 8 & -1 & 4 \\ 18 & -6 & 11 \end{bmatrix}$, $P = \begin{bmatrix} -1 & -1 & 1 \\ 1 & 2 & -1 \\ 3 & 3 & -2 \end{bmatrix}$

B2 For the following matrices, determine the eigenvalues and corresponding eigenvectors and determine whether each matrix is diagonalizable over $\mathbb{R}$. If it is diagonalizable, give a matrix P and diagonal matrix D such that $P^{-1}AP = D$.

(a) $A = \begin{bmatrix} -4 & 3 \\ 2 & 1 \end{bmatrix}$

(b) $A = \begin{bmatrix} 5 & 2 \\ 0 & 3 \end{bmatrix}$

(c) $A = \begin{bmatrix} 5 & 2 \\ 0 & 5 \end{bmatrix}$

(d) $A = \begin{bmatrix} 1 & 6 & 3 \\ 0 & -2 & 0 \\ 3 & 6 & 1 \end{bmatrix}$

(e) $A = \begin{bmatrix} -3 & 2 & 1 \\ 4 & -2 & -4 \\ -9 & 2 & 7 \end{bmatrix}$

(f) $A = \begin{bmatrix} 2 & -1 & -2 \\ -3 & 2 & 3 \\ 4 & -3 & -4 \end{bmatrix}$

(g) $A = \begin{bmatrix} 1 & -2 & -1 \\ -3 & 3 & 3 \\ 3 & 4 & 1 \end{bmatrix}$

B3 Follow the same instructions as for Problem B2.

(a) $A = \begin{bmatrix} 4 & 3 \\ 5 & 6 \end{bmatrix}$

(b) $A = \begin{bmatrix} 3 & -6 \\ -4 & 8 \end{bmatrix}$

(c) $A = \begin{bmatrix} 1 & -4 \\ -4 & 1 \end{bmatrix}$

(d) $A = \begin{bmatrix} -7 & 2 & 12 \\ -3 & 0 & 6 \\ -3 & 1 & 5 \end{bmatrix}$

(e) $A = \begin{bmatrix} 3 & 0 & -4 \\ 1 & 1 & -2 \\ 1 & 0 & -1 \end{bmatrix}$

(f) $A = \begin{bmatrix} 1 & 2 & 0 \\ 2 & 4 & 0 \\ 0 & 0 & -2 \end{bmatrix}$

(g) $A = \begin{bmatrix} -1 & -2 & 2 \\ 10 & 5 & -10 \\ 2 & 0 & -1 \end{bmatrix}$

Conceptual Problems

D1 Prove that if A and B are similar, then A and B have the same eigenvalues.

D2 Prove that if A and B are similar, then A and B have the same rank.

D3 (a) Let A and B be $n \times n$ matrices. Prove that $\operatorname{tr} AB = \operatorname{tr} BA$.

(b) Use the result of part (a) to prove that if A and B are similar, then $\operatorname{tr} A = \operatorname{tr} B$.

D4 (a) Suppose that P diagonalizes A and that the diagonal form is D. Show that $A = PDP^{-1}$.

(b) Use the result of part (a) and properties of eigenvectors to calculate a matrix that has eigenvalues 2 and 3 with corresponding eigenvectors $\begin{bmatrix} 1 \\ 2 \end{bmatrix}$ and $\begin{bmatrix} 1 \\ 3 \end{bmatrix}$, respectively.

(c) Determine a matrix that has eigenvalues 2, −2, and 3, with corresponding eigenvectors $\begin{bmatrix} 1 \\ 0 \\ 1 \end{bmatrix}$, $\begin{bmatrix} 1 \\ 1 \\ -1 \end{bmatrix}$, and $\begin{bmatrix} 1 \\ -1 \\ 2 \end{bmatrix}$, respectively.

D5 (a) Suppose that P diagonalizes A and that the diagonal form is D. Show that $A^k = PD^kP^{-1}$.

(b) Use the result of part (a) to calculate A^5, where $A = \begin{bmatrix} -1 & 6 & 3 \\ 3 & -4 & -3 \\ -6 & 12 & 8 \end{bmatrix}$ is the matrix from Problem A2 (g).

D6 (a) Suppose that A is diagonalizable. Prove that $\operatorname{tr} A$ is equal to the sum of the eigenvalues of A (including repeated eigenvalues) by using Theorem 1.

(b) Use the result of part (a) to determine, by inspection, the algebraic and geometric multiplicities of all of the eigenvalues of

$$A = \begin{bmatrix} a+b & a & a \\ a & a+b & a \\ a & a & a+b \end{bmatrix}.$$

D7 (a) Suppose that A is diagonalizable. Prove that $\det A$ is equal to the product of the eigenvalues of A (repeated according to their multiplicity) by considering $P^{-1}AP$.

(b) Show that the constant term in the characteristic polynomial is $\det A$. (Hint: How do you find the constant term in any polynomial $p(\lambda)$?)

(c) Without assuming that A is diagonalizable, show that $\det A$ is equal to the product of the roots of the characteristic equation of A (including any repeated roots and complex roots). (Hint: Consider the constant term in the characteristic equation and the factored version of that equation.)

D8 Let A be an $n \times n$ matrix. Prove that A is invertible if and only if A does not have 0 as an eigenvalue. (Hint: See Problem D7.)

D9 Suppose that A is diagonalized by the matrix P and that the eigenvalues of A are $\lambda_1, \dots, \lambda_n$. Show that the eigenvalues of $(A - \lambda_1 I)$ are $0, \lambda_2 - \lambda_1, \lambda_3 - \lambda_1, \dots, \lambda_n - \lambda_1$. (Hint: $A - \lambda_1 I$ is diagonalized by P.)

6.3 Powers of Matrices and the Markov Process

In some applications of linear algebra, it is necessary to calculate powers of a matrix. If the matrix A is diagonalized by P to the diagonal matrix D, it follows from $D = P^{-1}AP$ that $A = PDP^{-1}$, and then for any positive integer m we get

$$\begin{aligned} A^m &= (PDP^{-1})^m = (PDP^{-1})(PDP^{-1})\cdots(PDP^{-1}) \\ &= PD(P^{-1}P)D(P^{-1}P)D\cdots(P^{-1}P)DP = PD^mP^{-1} \end{aligned}$$

Thus, knowledge of the eigenvalues of A and the theory of diagonalization should be valuable tools in these applications. One such application is the study of Markov processes. After discussing Markov processes, we turn the question around and show how the "power method" uses powers of a matrix A to determine an eigenvalue of A. (This is an important step in the Google PageRank algorithm.) We begin with an example of a Markov process.

EXAMPLE 1

Smith and Jones are the only competing suppliers of communication services in their community. At present, they each have a 50% share of the market. However, Smith has recently upgraded his service, and a survey indicates that from one month to the next, 90% of Smith's customers remain loyal, while 10% switch to Jones. On the other hand, 70% of Jones's customers remain loyal and 30% switch to Smith. If this goes on for six months, how large are their market shares? If this goes on for a long time, how big will Smith's share become?

Solution: Let S_m be Smith's market share (as a decimal) at the end of the m-th month and let J_m be Jones's share. Then $S_m + J_m = 1$, since between them they have 100% of the market. At the end of the $(m+1)$-st month, Smith has 90% of his previous customers and 30% of Jones's previous customers, so

$$S_{m+1} = 0.90S_m + 0.3J_m$$

Similarly,

$$J_{m+1} = 0.1S_m + 0.7J_m$$

We can rewrite these equations in matrix-vector form:

$$\begin{bmatrix} S_{m+1} \\ J_{m+1} \end{bmatrix} = \begin{bmatrix} 0.9 & 0.3 \\ 0.1 & 0.7 \end{bmatrix} \begin{bmatrix} S_m \\ J_m \end{bmatrix}$$

The matrix $T = \begin{bmatrix} 0.9 & 0.3 \\ 0.1 & 0.7 \end{bmatrix}$ is called the **transition matrix** for this problem: it describes the transition (change) from the **state** $\begin{bmatrix} S_m \\ J_m \end{bmatrix}$ at time m to the state $\begin{bmatrix} S_{m+1} \\ J_{m+1} \end{bmatrix}$ at time $m+1$. Then we have answers to the questions if we can determine $T^6 \begin{bmatrix} 0.5 \\ 0.5 \end{bmatrix}$ and $T^m \begin{bmatrix} 0.5 \\ 0.5 \end{bmatrix}$ for m large.

To answer the first question, we might compute T^6 directly. For the second question, this approach is not reasonable, and instead we diagonalize. We find that $\lambda_1 = 1$

EXAMPLE 1
(continued)

is an eigenvalue of T with eigenvector $\vec{v}_1 = \begin{bmatrix} 3 \\ 1 \end{bmatrix}$, and $\lambda_2 = 0.6$ is the other eigenvalue, with eigenvector $\vec{v}_2 = \begin{bmatrix} 1 \\ -1 \end{bmatrix}$. Thus,

$$P = \begin{bmatrix} 3 & 1 \\ 1 & -1 \end{bmatrix} \quad \text{and} \quad P^{-1} = \frac{1}{4}\begin{bmatrix} 1 & 1 \\ 1 & -3 \end{bmatrix}$$

It follows that

$$T^m = PD^mP^{-1} = \begin{bmatrix} 3 & 1 \\ 1 & -1 \end{bmatrix}\begin{bmatrix} 1^m & 0 \\ 0 & (0.6)^m \end{bmatrix}\frac{1}{4}\begin{bmatrix} 1 & 1 \\ 1 & -3 \end{bmatrix}$$

We could now answer our question directly, but we get a simpler calculation if we observe that the eigenvectors form a basis, so we can write

$$\begin{bmatrix} S_0 \\ J_0 \end{bmatrix} = c_1\begin{bmatrix} 3 \\ 1 \end{bmatrix} + c_2\begin{bmatrix} 1 \\ -1 \end{bmatrix} = P\begin{bmatrix} c_1 \\ c_2 \end{bmatrix}$$

Then,

$$\begin{bmatrix} c_1 \\ c_2 \end{bmatrix} = P^{-1}\begin{bmatrix} S_0 \\ J_0 \end{bmatrix} = \frac{1}{4}\begin{bmatrix} S_0 + J_0 \\ S_0 - 3J_0 \end{bmatrix}$$

Then, by linearity,

$$\begin{aligned} T^m\begin{bmatrix} S_0 \\ J_0 \end{bmatrix} &= c_1T^m\vec{v}_1 + c_2T^m\vec{v}_2 \\ &= c_1\lambda_1^m\vec{v}_1 + c_2\lambda_2^m\vec{v}_2 \\ &= \frac{1}{4}(S_0 + J_0)\begin{bmatrix} 3 \\ 1 \end{bmatrix} + \frac{1}{4}(S_0 - 3J_0)(0.6)^m\begin{bmatrix} 1 \\ -1 \end{bmatrix} \end{aligned}$$

Now $S_0 = J_0 = 0.5$. When $m = 6$,

$$\begin{aligned} \begin{bmatrix} S_6 \\ J_6 \end{bmatrix} &= \frac{1}{4}\begin{bmatrix} 3 \\ 1 \end{bmatrix} - \frac{1}{4}(0.6)^6\begin{bmatrix} 1 \\ -1 \end{bmatrix} \\ &\approx \frac{1}{4}\begin{bmatrix} 3 - 0.0117 \\ 1 + 0.0117 \end{bmatrix} \\ &\approx \begin{bmatrix} 0.747 \\ 0.253 \end{bmatrix} \end{aligned}$$

Thus, after six months, Smith has approximately 74.7% of the market.

When m is very large, $(0.6)^m$ is nearly zero, so for m large enough ($m \to \infty$), we have $S_\infty = 0.75$ and $J_\infty = 0.25$.

Thus, in this problem, Smith's share approaches 75% as m gets large, but it never gets larger than 75%. Now look carefully: we get the same answer in the long run, no matter what the initial value of S_0 and J_0 are because $(0.6)^m \to 0$ and $S_0 + J_0 = 1$.

By emphasizing some features of Example 1, we will be led to an important definition and several general properties:

(1) Each column of T has sum 1. This means that all of Smith's customers show up a month later as customers of Smith or Jones; the same is true for Jones's customers. No customers are lost from the system and none are added after the process begins.

(2) It is natural to interpret the entries t_{ij} as **probabilities**. For example, $t_{11} = 0.9$ is the probability that a Smith customer remains a Smith customer, with $t_{21} = 0.1$ as the probability that a Smith customer becomes a Jones customer. If we consider "Smith customer" as "state 1" and "Jones customer" as "state 2," then t_{ij} is the probability of **transition** from state j to state i between time m and time $m+1$.

(3) The "initial state vector" is $\begin{bmatrix} S_0 \\ J_0 \end{bmatrix}$; $T^m \begin{bmatrix} S_0 \\ J_0 \end{bmatrix}$ is the state vector at time m.

(4) Note that

$$\begin{bmatrix} S_1 \\ J_1 \end{bmatrix} = T \begin{bmatrix} S_0 \\ J_0 \end{bmatrix} = S_0 \begin{bmatrix} t_{11} \\ t_{21} \end{bmatrix} + J_0 \begin{bmatrix} t_{12} \\ t_{22} \end{bmatrix}$$

Since $t_{11} + t_{21} = 1$ and $t_{12} + t_{22} = 1$, it follows that

$$S_1 + J_1 = S_0 + J_0$$

Thus, it follows from (1) that each state vector has the same column sum. In our example, S_0 and J_0 are decimal fractions, so $S_0 + J_0 = 1$, but we could consider a process whose states have some other constant column sum.

(5) Note that 1 is an eigenvalue of T with eigenvector $\begin{bmatrix} 3 \\ 1 \end{bmatrix}$. To get a state vector with the appropriate sum, we take the eigenvector to be $\begin{bmatrix} 3/4 \\ 1/4 \end{bmatrix}$. Thus,

$$T \begin{bmatrix} 3/4 \\ 1/4 \end{bmatrix} = \begin{bmatrix} 3/4 \\ 1/4 \end{bmatrix}$$

and the state vector $\begin{bmatrix} 3/4 \\ 1/4 \end{bmatrix}$ is **fixed** or **invariant** under the transformation with matrix T. Moreover, this fixed vector is the limiting state approached by $T^m \begin{bmatrix} S_0 \\ J_0 \end{bmatrix}$ for any $\begin{bmatrix} S_0 \\ J_0 \end{bmatrix}$.

The following definition captures the essential properties of this example.

Definition
Markov Matrix
Markov Process

An $n \times n$ matrix T is the **Markov matrix** (or transition matrix) of an n-state **Markov process** if

(1) $t_{ij} \geq 0$, for each i and j.

(2) Each column sum is 1: $\sum_{i=1}^{n} t_{ij} = 1$ for each j.

We take possible states of the process to be the vectors $S = \begin{bmatrix} s_1 \\ \vdots \\ s_n \end{bmatrix}$ such that $s_i \geq 0$ for each i, and $s_1 + \cdots + s_n = 1$.

Remark

With minor changes, we could develop the theory with $s_1 + \cdots + s_n =$ constant.

EXAMPLE 2

The matrix $\begin{bmatrix} 0.1 & 0.3 \\ 0.9 & 0.8 \end{bmatrix}$ is not a Markov matrix since the sum of the entries in the second column does not equal 1.

EXAMPLE 3

Find the fixed-state vector for the Markov matrix $A = \begin{bmatrix} 0.1 & 0.3 \\ 0.9 & 0.7 \end{bmatrix}$.

Solution: We know the fixed-state vector is an eigenvector for the eigenvalue $\lambda = 1$. We have

$$A - I = \begin{bmatrix} -0.9 & 0.3 \\ 0.9 & -0.3 \end{bmatrix} \sim \begin{bmatrix} 1 & -1/3 \\ 0 & 0 \end{bmatrix}$$

Therefore, an eigenvector corresponding to $\lambda = 1$ is $\begin{bmatrix} 1 \\ 3 \end{bmatrix}$. The components in the state vector must sum to 1, so the invariant state is $\begin{bmatrix} 1/4 \\ 3/4 \end{bmatrix}$.

It is easy to verify that

$$\begin{bmatrix} 0.1 & 0.3 \\ 0.9 & 0.7 \end{bmatrix}\begin{bmatrix} 1/4 \\ 3/4 \end{bmatrix} = \begin{bmatrix} 1/4 \\ 3/4 \end{bmatrix}$$

EXERCISE 1

Determine which of the following matrices is a Markov matrix. Find the fixed-state vector of the Markov matrix.

(a) $A = \begin{bmatrix} 0.4 & 0.6 \\ 0.5 & 0.5 \end{bmatrix}$ (b) $B = \begin{bmatrix} 0.4 & 0.6 \\ 0.6 & 0.4 \end{bmatrix}$

The goal with the Markov process is to establish the behaviour of a sequence with states $\vec{s}, T\vec{s}, T^2\vec{s}, \ldots, T^m\vec{s}$. If possible, we want to say something about the limit of $T^m\vec{s}$ as $m \to \infty$. As we saw in Example 1, diagonalization of T is a key to solving the problem. It is beyond the scope of this book to establish all the properties of the Markov process, but some of the properties are easy to prove, and others are easy to illustrate if we make extra assumptions.

PROPERTY 1. One eigenvalue of a Markov matrix is $\lambda_1 = 1$.

Proof: Since each column of T has sum 1, each column of $(T - 1I)$ has sum 0. Hence, the sum of the rows of $(T - 1I)$ is the zero vector. Thus the rows are linearly dependent, and $(T - 1I)$ has rank less than n, so $\det(T - 1I) = 0$. Therefore, 1 is an eigenvalue of T. ■

PROPERTY 2. The eigenvector $\vec{s}^*$ for $\lambda_1 = 1$ has $s_j^* \geq 0$ for $1 \leq j \leq n$.
This property is important because it means that the eigenvector $\vec{s}^*$ is a real state of the process. In fact, it is a fixed or invariant state:

$$T\vec{s}^* = \lambda_1 \vec{s}^* = \vec{s}^*$$

PROPERTY 3. All other eigenvalues satisfy $|\lambda_i| \leq 1$.
To see why we expect this, let us assume that T is diagonalizable, with distinct eigenvalues $1, \lambda_2, \dots, \lambda_n$ and corresponding eigenvectors $\vec{s}^*, \vec{s}_2, \dots, \vec{s}_n$. Then any initial state $\vec{s}$ can be written

$$\vec{s} = c_1\vec{s}^* + c_2\vec{s}_2 + \cdots + c_n\vec{s}_n$$

It follows that

$$T^m\vec{s} = c_1 1^m\vec{s}^* + c_2\lambda_2^m\vec{s}_2 + \cdots + c_n\lambda_n^m\vec{s}_n$$

If any $|\lambda_i| > 1$, the term $|\lambda_i^m|$ would become much larger than the other terms when m is large; it would follow that $T^m\vec{s}$ has some coordinates with magnitude greater than 1. This is impossible because state coordinates satisfy $0 \leq s_i \leq 1$, so we must have $|\lambda_i| \leq 1$.

PROPERTY 4. Suppose that for some m all the entries in T^m are not zero. Then all the eigenvalues of T except for $\lambda_1 = 1$ satisfy $|\lambda_i| < 1$. In this case, for any initial state $\vec{s}$, $T^m\vec{s} \to \vec{s}^*$ as $m \to \infty$: all states tend to the invariant state $\vec{s}^*$ under the process. Notice that in the diagonalizable case, the fact that $T^m\vec{s} \to \vec{s}^*$ follows from the expression for $T^m\vec{s}$ given under Property 3.

EXERCISE 2

The Markov matrix $T = \begin{bmatrix} 0 & 1 \\ 1 & 0 \end{bmatrix}$ has eigenvalues 1 and -1; it does not satisfy the conclusion of Property 4. However, it also does not satisfy the extra assumption of Property 4. It is worthwhile to explore this "bad" case.

Let $\vec{s} = \begin{bmatrix} s_1 \\ s_2 \end{bmatrix}$. Determine the behaviour of the sequence $\vec{s}, T\vec{s}, T^2\vec{s}, \dots$. What is the fixed-state vector for T?

Systems of Linear Difference Equations

If A is an $n \times n$ matrix and $\vec{s}(m)$ is a vector for each positive integer m, then the matrix vector equation

$$\vec{s}(m+1) = A\vec{s}(m)$$

may be regarded as a system of n linear first-order difference equations, describing the coordinates $s_1, s_2, \dots, s_n$ at times $m+1$ in terms of those at time m. They are "first-order difference" equations because they involve only one time difference from m to $m+1$; the Fibonacci equation $s(m+1) = s(m) + s(m-1)$ is a second-order difference equation.

Markov processes form a special class of this large class of systems of linear difference equations, but there are applications that do not fit the Markov assumptions. For example, in population models, we might wish to consider deaths (so that some column sums of A would be less than 1) or births, or even multiple births (so that some entries in A would be greater than 1). Similar considerations apply to some economic models, which are represented by matrix models. A proper discussion of such models requires more theory than is developed in this book.

The Power Method of Determining Eigenvalues

Practical applications of eigenvalues often involve larger matrices with non-integer entries. Such problems often require efficient computer methods for determining eigenvalues. A thorough discussion of such methods is beyond the scope of this book, but we can indicate how powers of matrices provide one tool for finding eigenvalues.

Let A be an $n \times n$ matrix. To simplify the discussion, we suppose that A has n distinct real eigenvalues $\lambda_1, \dots, \lambda_n$, with corresponding eigenvectors $\vec{v}_1, \dots, \vec{v}_n$. We suppose that $|\lambda_1| > |\lambda_i|$ for $2 \leq i \leq n$. We call λ_1 the **dominant** eigenvalue. Since $\{\vec{v}_1, \dots, \vec{v}_n\}$ will form a basis for $\mathbb{R}^n$, any vector $\vec{x} \in \mathbb{R}^n$ can be written

$$\vec{x} = c_1\vec{v}_1 + \cdots + c_n\vec{v}_n$$

Then

$$A\vec{x} = c_1\lambda_1\vec{v}_1 + \cdots + c_n\lambda_n\vec{v}_n$$

and

$$A^m\vec{x} = c_1\lambda_1^m\vec{v}_1 + \cdots + c_n\lambda_n^m\vec{v}_n$$

For m large, $|\lambda_1^m|$ is much greater than all other terms. If we divide by $c_1\lambda_1^m$, then all terms on the right-hand side will be negligibly small except for $\vec{v}_1$, so we will be able to identify $\vec{v}_1$. By calculating $A\vec{v}_1$, we determine λ_1.

To make this into an effective procedure, we must control the size of the vectors: if $\lambda_1 > 1$, then $\lambda_1^m \to \infty$ as m gets large, and the procedure would break down. Similarly, if all eigenvalues are between 0 and 1, then $A^m\vec{x} \to 0$, and the procedure would fail. To avoid these problems, we normalize the vector at each step (that is, convert it to a vector of length 1).

The procedure is as follows.

Algorithm 1

Guess $\vec{x}_0$; normalize $\vec{y}_0 = \vec{x}_0/\|\vec{x}_0\|$
$\vec{x}_1 = A\vec{y}_0$; normalize $\vec{y}_1 = \vec{x}_1/\|\vec{x}_1\|$
$\vec{x}_2 = A\vec{y}_1$; normalize $\vec{y}_2 = \vec{x}_2/\|\vec{x}_2\|$
and so on.

We seek convergence of $\vec{y}_m$ to some limiting vector; if such a vector exists, it must be $\vec{v}_1$, the eigenvector for the largest eigenvalue λ_1.

This procedure is illustrated in the following example, which is simple enough that you can check the calculations.

EXAMPLE 4

Determine the eigenvalue of largest absolute value for the matrix $A = \begin{bmatrix} 13 & 6 \\ -12 & -5 \end{bmatrix}$ by using the power method.

Solution: Choose any starting vector and let $\vec{x}_0 = \begin{bmatrix} 1 \\ 1 \end{bmatrix}$. Then

$$\vec{y}_0 = \frac{1}{\sqrt{2}}\begin{bmatrix} 1 \\ 1 \end{bmatrix} \approx \begin{bmatrix} 0.707 \\ 0.707 \end{bmatrix}$$

$$\vec{x}_1 = A\vec{y}_0 \approx \begin{bmatrix} 13 & 6 \\ -12 & -5 \end{bmatrix}\begin{bmatrix} 0.707 \\ 0.707 \end{bmatrix} \approx \begin{bmatrix} 13.44 \\ -12.02 \end{bmatrix}$$

$$\vec{y}_1 = \frac{\vec{x}_1}{\|\vec{x}_1\|} \approx \begin{bmatrix} 0.745 \\ -0.667 \end{bmatrix}$$

$$\vec{x}_2 = A\vec{y}_1 \approx \begin{bmatrix} 5.683 \\ -5.605 \end{bmatrix}, \qquad \vec{y}_2 \approx \begin{bmatrix} 0.712 \\ -0.702 \end{bmatrix}$$

$$\vec{x}_3 = A\vec{y}_2 \approx \begin{bmatrix} 5.044 \\ -5.034 \end{bmatrix}, \qquad \vec{y}_3 \approx \begin{bmatrix} 0.7078 \\ -0.7063 \end{bmatrix}$$

$$\vec{x}_4 = A\vec{y}_3 \approx \begin{bmatrix} 4.9636 \\ -4.9621 \end{bmatrix}, \qquad \vec{y}_4 \approx \begin{bmatrix} 0.7072 \\ -0.7070 \end{bmatrix}$$

At this point, we judge that $\vec{y}_m \to \begin{bmatrix} 0.707 \\ -0.707 \end{bmatrix}$, so $\vec{v}_1 = \begin{bmatrix} 0.707 \\ -0.707 \end{bmatrix}$ is an eigenvector of A, and the corresponding dominant eigenvalue is $\lambda_1 = 7$. (The answer is easy to check by using standard methods.)

Many questions arise with the power method. What if we poorly choose the initial vector? If we choose $\vec{x}_0$ in the subspace spanned by all eigenvectors of A *except* $\vec{v}_1$, the method will fail to give $\vec{v}_1$. How do we decide when to stop repeating the steps of the procedure? For a computer version of the algorithm, it would be important to have tests to decide that the procedure has converged—or that it will never converge.

Once we have determined the dominant eigenvalue of A, how can we determine other eigenvalues? If A is invertible, the dominant eigenvalue of A^{-1} would give the reciprocal of the eigenvalue of A with the smallest absolute value. Another approach is to observe that if one eigenvalue λ_1 is known, then eigenvalues of $A - \lambda_1 I$ will give us information about eigenvalues of A. (See Problem 6.2.D9.)

PROBLEMS 6.3

Practice Problems

A1 Determine which of the following matrices are Markov matrices. For each Markov matrix, determine the invariant or fixed state (corresponding to the eigenvalue $\lambda = 1$).

(a) $\begin{bmatrix} 0.2 & 0.6 \\ 0.8 & 0.3 \end{bmatrix}$

(b) $\begin{bmatrix} 0.3 & 0.6 \\ 0.7 & 0.4 \end{bmatrix}$

(c) $\begin{bmatrix} 0.7 & 0.3 & 0.0 \\ 0.1 & 0.6 & 0.1 \\ 0.2 & 0.2 & 0.9 \end{bmatrix}$

(d) $\begin{bmatrix} 0.9 & 0.1 & 0.0 \\ 0.0 & 0.9 & 0.1 \\ 0.1 & 0.0 & 0.9 \end{bmatrix}$

A2 Suppose that census data show that every decade, 15% of people dwelling in rural areas move into towns and cities, while 5% of urban dwellers move into rural areas.

(a) What would be the eventual steady-state population distribution?

(b) If the population were 50% urban, 50% rural at some census, what would be the distribution after 50 years?

A3 A car rental company serving one city has three locations: the airport, the train station, and the city centre. Of the cars rented at the airport, 8/10 are returned to the airport, 1/10 are left at the train station, and 1/10 are left at the city centre. Of cars rented at the train station, 3/10 are left at the airport, 6/10 are returned to the train station, and 1/10 are left at the city centre. Of cars rented at the city centre, 3/10 go to the airport, 1/10 go to the train station, and 6/10 are returned to the city centre. Model this as a Markov process and determine the steady-state distribution for the cars.

A4 To see how the power method works, use it to determine the largest eigenvalue of the given matrix, starting with the given initial vector. (You will need a calculator or computer.)

(a) $\begin{bmatrix} 5 & 0 \\ 0 & -2 \end{bmatrix}, \vec{x}_0 = \begin{bmatrix} 1 \\ 1 \end{bmatrix}$

(b) $\begin{bmatrix} 27 & 84 \\ -7 & -22 \end{bmatrix}, \vec{x}_0 = \begin{bmatrix} 1 \\ 0 \end{bmatrix}$

Homework Problems

B1 Determine which of the following matrices are Markov matrices. For each Markov matrix, determine the invariant or fixed state (corresponding to the eigenvalue $\lambda = 1$).

(a) $\begin{bmatrix} 0.4 & 0.7 \\ 0.5 & 0.3 \end{bmatrix}$

(b) $\begin{bmatrix} 0.5 & 0.6 \\ 0.5 & 0.4 \end{bmatrix}$

(c) $\begin{bmatrix} 0.8 & 0.3 & 0.2 \\ 0.0 & 0.6 & 0.2 \\ 0.2 & 0.1 & 0.6 \end{bmatrix}$

(d) $\begin{bmatrix} 0.8 & 0.1 & 0.2 \\ 0.1 & 0.9 & 0.6 \\ 0.1 & 0.1 & 0.2 \end{bmatrix}$

B2 The town of Markov Centre has only two suppliers of widgets—Johnson and Thomson. All inhabitants buy their supply on the first day of each month. Neither supplier is very successful at keeping customers. 70% of the customers who deal with Johnson decide that they will "try the other guy" next time. Thomson does even worse: only 20% of his customers come back the next month, and the rest go to Johnson.

(a) Model this as a Markov process and determine the steady-state distribution of customers.

(b) Determine a general expression for Johnson and Thomson's shares of the customers, given an initial state where Johnson has 25% and Thomson has 75%.

B3 A student society at a large university campus decides to create a pool of bicycles that can be used by the members of the society. Bicycles can be borrowed or returned at the residence, the library, or the athletic centre. The first day, 200 marked bicycles are left at each location. At the end of the day, at the residence, there are 160 bicycles that started at the residence, 40 that started at the library, and 60 that started at the athletic centre. At the library, there are 20 that started at the residence, 140 that started at the library, and 40 that started at the athletic centre. At the athletic centre, there are 20 that started at the residence, 20 that started at the library, and 100 that started at the athletic centre. If this pattern is repeated every day, what is the steady-state distribution of bicycles?

B4 Use the power method with initial vector $\begin{bmatrix} 1 \\ 0 \end{bmatrix}$ to determine the dominant eigenvalue of $\begin{bmatrix} 3.5 & 4.5 \\ 4.5 & 3.5 \end{bmatrix}$. Show your calculations clearly.

Computer Problems

C1 Use the power method with initial vector $\begin{bmatrix} 1 \\ 0 \\ 0 \end{bmatrix}$ to determine the dominant eigenvalue of the matrix $\begin{bmatrix} 2.89316 & -1.28185 & 2.42918 \\ -0.70562 & 0.76414 & -0.67401 \\ 1.67682 & -0.83198 & 2.34270 \end{bmatrix}$. You may do this by using software that includes matrix operations or by writing a program to carry out the procedure.

Conceptual Problems

D1 (a) Let T be the transition matrix for a two-state Markov process. Show that the eigenvalue that is not 1 is $\lambda_2 = t_{11} + t_{22} - 1$.

(b) For a two-state Markov process with $t_{21} = a$ and $t_{12} = b$, show that the fixed state (eigenvector for $\lambda = 1$) is $\frac{1}{a+b}\begin{bmatrix} b \\ a \end{bmatrix}$.

D2 Suppose that T is a Markov matrix.

(a) Show that for any state $\vec{x}$, $\sum_{k=1}^{n} (T\vec{x})_k = \sum_{k=1}^{n} x_k$.

(b) Show that if $\vec{v}$ is an eigenvector of T with eigenvalue $\lambda \neq 1$, then $\sum_{k=1}^{n} v_k = 0$.

6.4 Diagonalization and Differential Equations

This section requires knowledge of the exponential function, its derivative, and first-order linear differential equations. The ideas are not used elsewhere in this book.

Consider two tanks, Y and Z, each containing 1000 litres of a salt solution. At a initial time, $t = 0$ (in hours), the concentration of salt in tank Y is different from the concentration in tank Z. In each tank the solution is well stirred, so that the concentration is constant throughout the tank. The two tanks are joined by pipes; through one pipe, solution is pumped from Y to Z at a rate of 20 L/h; through the other, solution is pumped from Z to Y at the same rate. The problem is to determine the amount of salt in each tank at time t.

Let $y(t)$ be the amount of salt (in kilograms) in tank Y at time t, and let $z(t)$ be the amount in the tank Z at time t. Then the concentration in Y at time t is $(y/1000)$ kg/L. Similarly, $(z/1000)$ kg/L is the concentration in Z. Then for tank Y, salt is flowing out through one pipe at a rate of $(20)(y/1000)$ kg/h and in through the other pipe at a rate of $(20)(z/1000)$ kg/h. Since the rate of change is measured by the derivative, we have $\frac{dy}{dt} = -0.02y + 0.02z$. By consideration of Z, we get a second differential equation, so y and z are the solutions of the **system of linear ordinary differential equations**:

$$\begin{aligned}\frac{dy}{dt} &= -0.02y + 0.02z\\ \frac{dz}{dt} &= 0.02y - 0.02z\end{aligned}$$

It is convenient to rewrite this system in the form $\frac{d}{dt}\begin{bmatrix} y \\ z \end{bmatrix} = \begin{bmatrix} -0.02 & 0.02 \\ 0.02 & -0.02 \end{bmatrix}\begin{bmatrix} y \\ z \end{bmatrix}$.

How can we solve this system? Well, it might be easier if we could change variables so that the 2×2 matrix is diagonalized. By standard methods, one eigenvalue of $A = \begin{bmatrix} -0.02 & 0.02 \\ 0.02 & -0.02 \end{bmatrix}$ is $\lambda_1 = 0$, with corresponding eigenvector $\begin{bmatrix} 1 \\ 1 \end{bmatrix}$. The other eigenvalue is $\lambda_2 = -0.04$, with corresponding eigenvector $\begin{bmatrix} -1 \\ 1 \end{bmatrix}$. Hence, A is diagonalized by $P = \begin{bmatrix} 1 & -1 \\ 1 & 1 \end{bmatrix}$, with $P^{-1} = \frac{1}{2}\begin{bmatrix} 1 & 1 \\ -1 & 1 \end{bmatrix}$.

Introduce new coordinates $\begin{bmatrix} y^* \\ z^* \end{bmatrix}$ by the change of coordinates equation, as in Section 4.6: $\begin{bmatrix} y \\ z \end{bmatrix} = P\begin{bmatrix} y^* \\ z^* \end{bmatrix}$. Substitute this for $\begin{bmatrix} y \\ z \end{bmatrix}$ on both sides of the system to obtain

$$\frac{d}{dt}P\begin{bmatrix} y^* \\ z^* \end{bmatrix} = AP\begin{bmatrix} y^* \\ z^* \end{bmatrix}$$

Since the entries in P are constants, it is easy to check that

$$\frac{d}{dt}P\begin{bmatrix} y^* \\ z^* \end{bmatrix} = P\frac{d}{dt}\begin{bmatrix} y^* \\ z^* \end{bmatrix}$$

Multiply both sides of the system of equations (on the left) by P^{-1}. Since P diagonalizes A, we get

$$\frac{d}{dt}\begin{bmatrix} y^* \\ z^* \end{bmatrix} = P^{-1}AP\begin{bmatrix} y^* \\ z^* \end{bmatrix} = \begin{bmatrix} 0 & 0 \\ 0 & -0.04 \end{bmatrix}\begin{bmatrix} y^* \\ z^* \end{bmatrix}$$

Now write the pair of equations:

$$\frac{dy^*}{dt} = 0 \text{ and } \frac{dz^*}{dt} = -0.04z^*$$

These equations are "decoupled," and we can easily solve each of them by using simple one-variable calculus.

The only functions satisfying $\frac{dy^*}{dt} = 0$ are constants: we write $y^*(t) = a$. The only functions satisfying an equation of the form $\frac{dx}{dt} = kx$ are exponentials of the form $x(t) = ce^{kt}$ for a constant c. So, from $\frac{dz^*}{dt} = -0.04z^*$, we obtain $z^*(t) = be^{-0.04t}$, where b is a constant.

Now we need to express the solution in terms of the original variables y and z:

$$\begin{bmatrix} y \\ z \end{bmatrix} = P\begin{bmatrix} y^* \\ z^* \end{bmatrix} = \begin{bmatrix} 1 & -1 \\ 1 & 1 \end{bmatrix}\begin{bmatrix} y^* \\ z^* \end{bmatrix} = \begin{bmatrix} y^* - z^* \\ y^* + z^* \end{bmatrix} = \begin{bmatrix} a - be^{-0.04t} \\ a + be^{-0.04t} \end{bmatrix}$$

For later use, it is helpful to rewrite this as $\begin{bmatrix} y \\ z \end{bmatrix} = a\begin{bmatrix} 1 \\ 1 \end{bmatrix} + be^{-0.04t}\begin{bmatrix} -1 \\ 1 \end{bmatrix}$. This is the general solution of the problem. To determine the constants a and b, we would need to know the amounts $y(0)$ and $z(0)$ at the initial time $t = 0$. Then we would know y and z for all t. Note that as $t \to \infty$, y and z tend to a common value a, as we might expect.

A Practical Solution Procedure

The usual solution procedure takes advantage of the understanding obtained from this diagonalization argument, but it takes a major shortcut. Now that the expected form of the solution is known, we simply look for a solution of the form $\begin{bmatrix} y \\ z \end{bmatrix} = ce^{\lambda t}\begin{bmatrix} a \\ b \end{bmatrix}$. Substitute this into the original system and use the fact that $\frac{d}{dt}ce^{\lambda t}\begin{bmatrix} a \\ b \end{bmatrix} = \lambda ce^{\lambda t}\begin{bmatrix} a \\ b \end{bmatrix}$:

$$\lambda ce^{\lambda t}\begin{bmatrix} a \\ b \end{bmatrix} = Ace^{\lambda t}\begin{bmatrix} a \\ b \end{bmatrix}$$

After the common factor $ce^{\lambda t}$ is cancelled, this tells us that $\begin{bmatrix} a \\ b \end{bmatrix}$ is an eigenvector of A, with eigenvalue λ. We find the two eigenvalues λ_1 and λ_2 and the corresponding eigenvectors $\vec{v}_1$ and $\vec{v}_2$, as above. Observe that since our problem is a linear homogeneous problem, the general solution will be an arbitrary linear combination of the two solutions $e^{\lambda_1 t}\vec{v}_1$ and $e^{\lambda_2 t}\vec{v}_2$. This matches the general solution we found above.

General Discussion

There are many other problems that give rise to systems of linear homogeneous ordinary differential equations (for example, electrical circuits or a mechanical system consisting of springs). Many of these systems are much larger than the example we considered. Methods for solving these systems make extensive use of eigenvectors and eigenvalues, and they require methods for dealing with cases where the characteristic equation has complex roots.

PROBLEMS 6.4

Practice Problems

A1 Find the general solution of each of the following systems of linear differential equations.

(a) $\frac{d}{dt}\begin{bmatrix} y \\ z \end{bmatrix} = \begin{bmatrix} 3 & 2 \\ 4 & -4 \end{bmatrix}\begin{bmatrix} y \\ z \end{bmatrix}$

(b) $\frac{d}{dt}\begin{bmatrix} y \\ z \end{bmatrix} = \begin{bmatrix} 0.2 & 0.7 \\ 0.1 & -0.4 \end{bmatrix}\begin{bmatrix} y \\ z \end{bmatrix}$

Homework Problems

B1 Find the general solution of each of the following systems of linear differential equations.

(a) $\frac{d}{dt}\begin{bmatrix} y \\ z \end{bmatrix} = \begin{bmatrix} -0.5 & 0.3 \\ 0.1 & -0.7 \end{bmatrix}\begin{bmatrix} y \\ z \end{bmatrix}$

(b) $\frac{d}{dt}\begin{bmatrix} y \\ z \end{bmatrix} = \begin{bmatrix} -1 & 4 \\ 8 & -5 \end{bmatrix}\begin{bmatrix} y \\ z \end{bmatrix}$

(c) $\frac{d}{dt}\begin{bmatrix} x \\ y \\ z \end{bmatrix} = \begin{bmatrix} -1 & -1 & 0 \\ -13 & 3 & 8 \\ 11 & -5 & -8 \end{bmatrix}\begin{bmatrix} x \\ y \\ z \end{bmatrix}$

CHAPTER REVIEW

Suggestions for Student Review

1 Define eigenvectors and eigenvalues of a matrix A. Explain the connection between the statement that λ is an eigenvalue of A with eigenvector $\vec{v}$ and the condition $\det(A - \lambda I) = 0$. (Section 6.1)

2 What does it mean to say that matrices A and B are similar? Explain why this is an important question. (Section 6.2)

3 Suppose you are told that the $n \times n$ matrix A has eigenvalues $\lambda_1, \dots, \lambda_n$ (repeated according to multiplicity).

(a) What conditions on these eigenvalues guarantees that A is diagonalizable over $\mathbb{R}$? (Section 6.2)

(b) Is there any case where you can tell from the eigenvalues that A is not diagonalizable over $\mathbb{R}$? (Section 6.2)

4 Use the idea suggested in Problem 6.2.D4 to create matrices for your classmates to diagonalize. (Section 6.2)

5 Suppose that $P^{-1}AP = D$, where D is a diagonal matrix with distinct diagonal entries $\lambda_1, \dots, \lambda_n$. How can we use this information to solve the system of linear differential equations $\frac{d}{dt}\vec{x} = A\vec{x}$? (Section 6.4)

Chapter Quiz

E1 Let $A = \begin{bmatrix} 5 & -16 & -4 \\ 2 & -7 & -2 \\ -2 & 8 & 3 \end{bmatrix}$. Determine whether the following vectors are eigenvectors of A. If any is an eigenvector of A, state the corresponding eigenvalue.

(a) $\begin{bmatrix} 3 \\ 1 \\ 0 \end{bmatrix}$ (b) $\begin{bmatrix} 1 \\ 0 \\ 1 \end{bmatrix}$ (c) $\begin{bmatrix} 4 \\ 1 \\ 0 \end{bmatrix}$ (d) $\begin{bmatrix} 2 \\ 1 \\ -1 \end{bmatrix}$

E2 Determine whether the matrix $A = \begin{bmatrix} -3 & 1 & 0 \\ 13 & -7 & -8 \\ -11 & 5 & 4 \end{bmatrix}$ is diagonalizable. If it is, give an invertible matrix P and a diagonal matrix D such that $P^{-1}AP = D$.

E3 Determine the algebraic and geometric multiplicity of each eigenvalue of $A = \begin{bmatrix} 4 & 2 & 2 \\ -1 & 1 & -1 \\ 1 & 1 & 3 \end{bmatrix}$. Is A diagonalizable?

E4 If λ is an eigenvalue of the invertible matrix A, prove that λ^{-1} is an eigenvalue of A^{-1}.

E5 Suppose that A is a 3×3 matrix such that

$$\det A = 0, \quad \det(A + 2I) = 0, \quad \det(A - 3I) = 0$$

Answer the following questions and give a brief explanation in each case.

(a) What is the dimension of the solution space of $A\vec{x} = \vec{0}$?

(b) What is the dimension of the nullspace of the matrix $B = A - 2I$?

(c) What is the rank of A?

E6 Let $A = \begin{bmatrix} 0.9 & 0.1 & 0.0 \\ 0.0 & 0.8 & 0.1 \\ 0.1 & 0.1 & 0.9 \end{bmatrix}$. Verify that A is a Markov matrix and determine its invariant state $\vec{x}$ such that $\sum_{i=1}^{3} x_i = 1$.

E7 Find the general solution of the system of differential equations

$$\frac{d}{dt}\begin{bmatrix} y \\ z \end{bmatrix} = \begin{bmatrix} 0.1 & 0.2 \\ 0.3 & 0.2 \end{bmatrix}\begin{bmatrix} y \\ z \end{bmatrix}$$

Further Problems

F1 (a) Suppose that A and B are square matrices such that $AB = BA$. Suppose that the eigenvalues of A all have algebraic multiplicity 1. Prove that any eigenvector of A is also an eigenvector of B.

(b) Give an example to illustrate that the result in part (a) may not be true if A has eigenvalues with algebraic multiplicity greater than 1.

F2 If $\det B \neq 0$, prove that AB and BA have the same eigenvalues.

F3 Suppose that A is an $n \times n$ matrix with n distinct eigenvalues $\lambda_1, \dots, \lambda_n$ with corresponding eigenvectors $\vec{v}_1, \dots, \vec{v}_n$, respectively. By representing $\vec{x}$ with respect to the basis of eigenvectors, show that $(A - \lambda_1 I)(A - \lambda_2 I)\cdots(A - \lambda_n I)\vec{x} = \vec{0}$ for every $\vec{x} \in \mathbb{R}^n$, and hence conclude that "A is a root of its characteristic polynomial." That is, if the characteristic polynomial is

$$C(\lambda) = (-1)^n\lambda^n + c_{n-1}\lambda^{n-1} + \cdots + c_1\lambda + c_0$$

then

$$(-1)^n A^n + c_{n-1}A^{n-1} + \cdots + c_1 A + c_0 I = O_{n,n}$$

(Hint: Write the characteristic polynomial in factored form.) This result is called the Cayley–Hamilton Theorem and is true for any square matrix A.

F4 For an invertible $n \times n$ matrix, use the Cayley–Hamilton Theorem to show that A^{-1} can be written as a polynomial of degree less than or equal to $n - 1$ in A (that is, a linear combination of $\{A^{n-1}, \dots, A^2, A, I\}$.

Visit the text's website at www.pearsoncanada.ca/norman for practice quizzes, additional applications, and an essay on linearity and superposition in physics.

CHAPTER 7

Orthonormal Bases

CHAPTER OUTLINE

7.1 Orthonormal Bases and Orthogonal Matrices
7.2 Projections and the Gram-Schmidt Procedure
7.3 Method of Least Squares
7.4 Inner Product Spaces
7.5 Fourier Series

In Section 1.4 we saw that we can use a projection to find a point P *in a plane that is closest to some other point* Q *that is not in the plane. We can view this as finding the point in the plane that best approximates* Q. *In many applications, we want to find a best approximation. Thus, it is very useful to generalize our work with projections from Section 1.4 to not only general subspaces of* $\mathbb{R}^n$ *but also to general vector spaces. You may find it helpful to review Sections 1.3 and 1.4 carefully before proceeding with this chapter.*

7.1 Orthonormal Bases and Orthogonal Matrices

Most of our intuition about coordinate geometry is based on experience with the standard basis for $\mathbb{R}^n$. *It is therefore a little uncomfortable for many beginners to deal with the arbitrary bases that arise in Chapter 4. Fortunately, for many problems, it is possible to work with bases that have the most essential properties of the standard basis: the basis vectors are* **mutually orthogonal** *(that is, the dot product of any two vectors is 0), and each basis vector is a* **unit vector** *(a vector with length 1).*

Orthonormal Bases

Definition
Orthogonal

A set of vectors $\{\vec{v}_1, \dots, \vec{v}_k\}$ in $\mathbb{R}^n$ is **orthogonal** if $\vec{v}_i \cdot \vec{v}_j = 0$ whenever $i \neq j$.

EXAMPLE 1

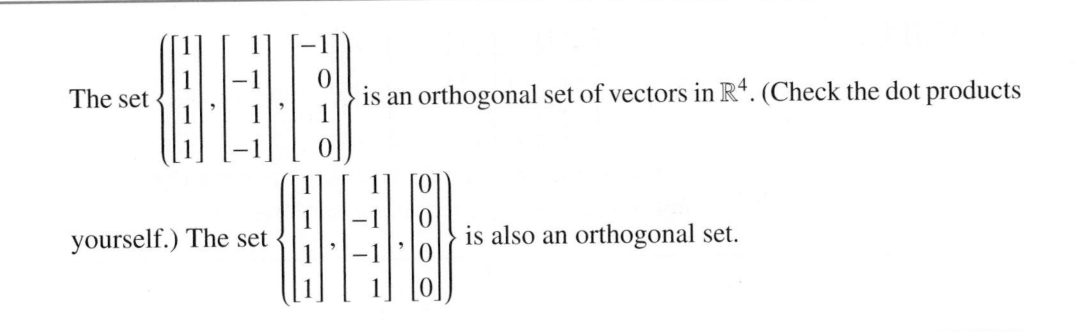

The set $\left\{ \begin{bmatrix} 1 \\ 1 \\ 1 \\ 1 \end{bmatrix}, \begin{bmatrix} 1 \\ -1 \\ 1 \\ -1 \end{bmatrix}, \begin{bmatrix} -1 \\ 0 \\ 1 \\ 0 \end{bmatrix} \right\}$ is an orthogonal set of vectors in $\mathbb{R}^4$. (Check the dot products yourself.) The set $\left\{ \begin{bmatrix} 1 \\ 1 \\ 1 \\ 1 \end{bmatrix}, \begin{bmatrix} 1 \\ -1 \\ -1 \\ 1 \end{bmatrix}, \begin{bmatrix} 0 \\ 0 \\ 0 \\ 0 \end{bmatrix} \right\}$ is also an orthogonal set.

If the zero vector is excluded, orthogonal sets have one very nice property.

Theorem 1 If $\{\vec{v}_1, \dots, \vec{v}_k\}$ is an orthogonal set of non-zero vectors in $\mathbb{R}^n$, it is linearly independent.

Proof: Consider the equation $c_1\vec{v}_1 + \cdots + c_k\vec{v}_k = \vec{0}$. Take the dot product of $\vec{v}_i$ with each side to get

$$\begin{aligned}(c_1\vec{v}_1 + \cdots + c_k\vec{v}_k) \cdot \vec{v}_i &= \vec{0} \cdot \vec{v}_i \\ c_1(\vec{v}_1 \cdot \vec{v}_i) + \cdots + c_i(\vec{v}_i \cdot \vec{v}_i) + \cdots + c_k(\vec{v}_k \cdot \vec{v}_i) &= 0 \\ 0 + \cdots + 0 + c_i\|\vec{v}_i\|^2 + 0 + \cdots + 0 &= 0\end{aligned}$$

since $\vec{v}_i \cdot \vec{v}_j = 0$ unless $i = j$. Moreover, $\vec{v}_i \neq \vec{0}$, so $\|\vec{v}_i\| \neq 0$ and hence $c_i = 0$. Since this is true for all $1 \leq i \leq k$, it follows that $\{\vec{v}_1, \dots, \vec{v}_k\}$ is linearly independent. ■

Remark

The trick used in this proof of taking the dot product of each side with one of the vectors $\vec{v}_i$ is an amazingly useful trick. Many of the things we do with orthogonal sets depend on it.

In addition to being mutually orthogonal, we want the vectors to be unit vectors.

Definition
Orthonormal

A set $\{\vec{v}_1, \dots, \vec{v}_k\}$ of vectors in $\mathbb{R}^n$ is **orthonormal** if it is orthogonal and each vector $\vec{v}_i$ is a unit vector (that is, each vector is normalized).

Notice that an orthonormal set of vectors does not contain the zero vector, since all vectors have length 1. It follows from Theorem 1 that orthonormal sets are necessarily linearly independent.

EXAMPLE 2

Any subset of the standard basis vectors in $\mathbb{R}^n$ is an orthonormal set. For example, in $\mathbb{R}^6$, $\{\vec{e}_1, \vec{e}_2, \vec{e}_5, \vec{e}_6\}$ is an orthonormal set of four vectors (where, as usual, $\vec{e}_i$ is the i-th standard basis vector).

EXAMPLE 3

The set $\left\{ \frac{1}{2}\begin{bmatrix}1\\1\\1\\1\end{bmatrix}, \frac{1}{2}\begin{bmatrix}1\\-1\\1\\-1\end{bmatrix}, \frac{1}{\sqrt{2}}\begin{bmatrix}-1\\0\\1\\0\end{bmatrix} \right\}$ is an orthonormal set in $\mathbb{R}^4$. The vectors are multiples of the vectors in Example 1, so they are certainly mutually orthogonal. They have been normalized so that each vector has length 1.

EXERCISE 1

Verify that the set $\left\{ \begin{bmatrix} 1 \\ 1 \\ 2 \end{bmatrix}, \begin{bmatrix} -2 \\ 0 \\ 1 \end{bmatrix}, \begin{bmatrix} 1 \\ -5 \\ 2 \end{bmatrix} \right\}$ is orthogonal and then normalize the vectors to produce the corresponding orthonormal set.

Many arguments based on orthonormal sets could be given for orthogonal sets of non-zero vectors. However, the general arguments are slightly simpler for orthonormal sets since $\|\vec{v}_i\| = 1$ in this case. In specific examples, it may be simpler to use orthogonal sets and postpone the normalization that often introduces square roots until the end. (Compare Examples 1 and 3.) The arguments here will usually be given for orthonormal sets.

Coordinates with Respect to an Orthonormal Basis

An orthonormal set of n vectors in $\mathbb{R}^n$ is necessarily a basis for $\mathbb{R}^n$ since it is automatically linearly independent and a set of n linearly independent vectors in $\mathbb{R}^n$ is a basis for $\mathbb{R}^n$ by Theorem 4.3.4. We will now see that there are several advantages to an orthonormal basis over an arbitrary basis.

The first of these advantages is that it is very easy to find the coordinates of a vector with respect to an orthonormal basis. Suppose that $\mathcal{B} = \{\vec{v}_1, \dots, \vec{v}_n\}$ is an orthonormal basis for $\mathbb{R}^n$ and that $\vec{x}$ is any vector in $\mathbb{R}^n$. To find the $\mathcal{B}$-coordinates of $\vec{x}$, we must find $b_1, \dots, b_n$ such that

$$\vec{x} = b_1\vec{v}_1 + \cdots + b_n\vec{v}_n \tag{7.1}$$

If $\mathcal{B}$ were an arbitrary basis, the procedure would be to solve the resulting system of n equations in n variables. However, since $\mathcal{B}$ is an orthonormal basis, we can use our amazingly useful trick: take the dot product of (7.1) with $\vec{v}_i$ to get

$$\vec{x} \cdot \vec{v}_i = 0 + \cdots + 0 + b_i(\vec{v}_i \cdot \vec{v}_i) + 0 + \cdots + 0 = b_i$$

because $\vec{v}_i \cdot \vec{v}_j = 0$ for $i \neq j$ and $\vec{v}_i \cdot \vec{v}_i = 1$. The result of this argument is important enough to summarize as a theorem.

Theorem 2

If $\mathcal{B} = \{\vec{v}_1, \dots, \vec{v}_n\}$ is an orthonormal basis for $\mathbb{R}^n$, then the i-th coordinate of a vector $\vec{x} \in \mathbb{R}^n$ with respect to $\mathcal{B}$ is

$$b_i = \vec{x} \cdot \vec{v}_i$$

It follows that $\vec{x}$ can be written as

$$\vec{x} = (\vec{x} \cdot \vec{v}_1)\vec{v}_1 + (\vec{x} \cdot \vec{v}_2)\vec{v}_2 + \cdots + (\vec{x} \cdot \vec{v}_n)\vec{v}_n$$

EXAMPLE 4

Find the coordinates of $\vec{x} = \begin{bmatrix} 1 \\ 2 \\ 3 \\ 4 \end{bmatrix}$ with respect to the orthonormal basis

$$\mathcal{B} = \left\{ \frac{1}{2}\begin{bmatrix} 1 \\ 1 \\ 1 \\ 1 \end{bmatrix}, \frac{1}{2}\begin{bmatrix} 1 \\ -1 \\ 1 \\ -1 \end{bmatrix}, \frac{1}{\sqrt{2}}\begin{bmatrix} -1 \\ 0 \\ 1 \\ 0 \end{bmatrix}, \frac{1}{\sqrt{2}}\begin{bmatrix} 0 \\ 1 \\ 0 \\ -1 \end{bmatrix} \right\}$$

Solution: By Theorem 2, the coordinates b_1, b_2, b_3, and b_4 of $\vec{x}$ are given by

$$\begin{aligned} b_1 &= \vec{x} \cdot \vec{v}_1 = \frac{1}{2}(1 + 2 + 3 + 4) = 5 \\ b_2 &= \vec{x} \cdot \vec{v}_2 = \frac{1}{2}(1 - 2 + 3 - 4) = -1 \\ b_3 &= \vec{x} \cdot \vec{v}_3 = \frac{1}{\sqrt{2}}(-1 + 0 + 3 + 0) = \sqrt{2} \\ b_4 &= \vec{x} \cdot \vec{v}_4 = \frac{1}{\sqrt{2}}(0 + 2 + 0 - 4) = -\sqrt{2} \end{aligned}$$

Thus the $\mathcal{B}$-coordinate vector of $\vec{x}$ is $[\vec{x}]_{\mathcal{B}} = \begin{bmatrix} 5 \\ -1 \\ \sqrt{2} \\ -\sqrt{2} \end{bmatrix}$. (It is easy to check that $\vec{x} = b_1\vec{v}_1 + b_2\vec{v}_2 + b_3\vec{v}_3 + b_4\vec{v}_4$.)

EXERCISE 2

Let $\mathcal{B} = \left\{ \frac{1}{\sqrt{3}}\begin{bmatrix} 1 \\ 1 \\ 1 \end{bmatrix}, \frac{1}{\sqrt{2}}\begin{bmatrix} 1 \\ 0 \\ -1 \end{bmatrix}, \frac{1}{\sqrt{6}}\begin{bmatrix} 1 \\ -2 \\ 1 \end{bmatrix} \right\}$. Verify that $\mathcal{B}$ is an orthonormal basis for $\mathbb{R}^3$ and then find the coordinates of $\vec{x} = \begin{bmatrix} 3 \\ 1 \\ 4 \end{bmatrix}$ with respect to $\mathcal{B}$.

Another technical advantage of using orthonormal bases is related to the first one. Often it is necessary to calculate the lengths and dot products of vectors whose coordinates are given with respect to some basis other than the standard basis. If the basis is not orthonormal, the calculations are a little ugly, but they are quite simple when the basis is orthonormal. Let $\{\vec{v}_1, \dots, \vec{v}_n\}$ be an orthonormal basis for $\mathbb{R}^n$ and let $\vec{x} = x_1\vec{v}_1 + \cdots + x_n\vec{v}_n$ and $\vec{y} = y_1\vec{v}_1 + \cdots + y_n\vec{v}_n$ be any vectors in $\mathbb{R}^n$. Using the fact that $\vec{v}_i \cdot \vec{v}_j = 0$ for $i \neq j$ and $\vec{v}_i \cdot \vec{v}_i = 1$ gives

$$\begin{aligned}\vec{x} \cdot \vec{y} &= (x_1\vec{v}_1 + \cdots + x_n\vec{v}_n) \cdot (y_1\vec{v}_1 + \cdots + y_n\vec{v}_n)\\ &= x_1y_1(\vec{v}_1 \cdot \vec{v}_1) + x_1y_2(\vec{v}_1 \cdot \vec{v}_2) + \cdots + x_1y_n(\vec{v}_1 \cdot \vec{v}_n) + x_2y_1(\vec{v}_2 \cdot \vec{v}_1)\\ &\quad + \cdots + x_2y_n(\vec{v}_2 \cdot \vec{v}_n) + \cdots + x_ny_n(\vec{v}_n \cdot \vec{v}_n)\\ &= x_1y_1 + x_2y_2 + \cdots + x_ny_n\end{aligned}$$

and

$$\|\vec{x}\|^2 = \vec{x} \cdot \vec{x} = x_1^2 + \cdots + x_n^2$$

Thus, the formulas in the new coordinates look exactly like the formulas in standard coordinates. This fact will be used in Section 7.2.

EXAMPLE 5

Let $\mathcal{B} = \left\{ \frac{1}{\sqrt{3}}\begin{bmatrix}1\\1\\1\end{bmatrix}, \frac{1}{\sqrt{2}}\begin{bmatrix}1\\0\\-1\end{bmatrix}, \frac{1}{\sqrt{6}}\begin{bmatrix}1\\-2\\1\end{bmatrix} \right\}$, and let $\vec{x}, \vec{y} \in \mathbb{R}^3$ such that $[\vec{x}]_{\mathcal{B}} = \begin{bmatrix}1\\2\\-1\end{bmatrix}$ and $[\vec{y}]_{\mathcal{B}} = \begin{bmatrix}3\\-2\\1\end{bmatrix}$. Determine $\|\vec{x}\|^2$ and $\vec{x} \cdot \vec{y}$.

Solution: Using our work above, we get

$$\|\vec{x}\|^2 = [\vec{x}]_{\mathcal{B}} \cdot [\vec{x}]_{\mathcal{B}} = \begin{bmatrix}1\\2\\-1\end{bmatrix} \cdot \begin{bmatrix}1\\2\\-1\end{bmatrix} = 6$$

and

$$\vec{x} \cdot \vec{y} = [\vec{x}]_{\mathcal{B}} \cdot [\vec{y}]_{\mathcal{B}} = \begin{bmatrix}1\\2\\-1\end{bmatrix} \cdot \begin{bmatrix}3\\-2\\1\end{bmatrix} = -2$$

EXERCISE 3

Verify the result of Example 5 by finding $\vec{x}$ and $\vec{y}$ explicitly and computing $\|\vec{x}\|^2$ and $\vec{x} \cdot \vec{y}$ directly. This will demonstrate the usefulness of coordinates with respect to an orthonormal basis.

Dot products and orthonormal bases in $\mathbb{R}^n$ have important generalizations to **inner products** and orthonormal bases in general vector spaces. These will be considered in Section 7.4.

Change of Coordinates and Orthogonal Matrices

A third technical advantage of using orthonormal bases is that it is very easy to invert a change of coordinates matrix between the standard basis and an orthonormal basis. To keep the writing short, we give the argument in $\mathbb{R}^3$, but the corresponding argument works in any dimension.

Let $\mathcal{B} = \{\vec{v}_1, \vec{v}_2, \vec{v}_3\}$ be an orthonormal basis for $\mathbb{R}^3$ and let $P = \begin{bmatrix} \vec{v}_1 & \vec{v}_2 & \vec{v}_3 \end{bmatrix}$. From Section 4.4, P is the change of coordinates matrix from $\mathcal{B}$-coordinates to coordinates with respect to the standard basis $\mathcal{S}$. Now, consider the product of the transpose of P with P: the rows of P^T are $\vec{v}_1^T, \vec{v}_2^T, \vec{v}_3^T$, so

$$\begin{aligned} P^T P &= \begin{bmatrix} \vec{v}_1^T \\ \vec{v}_2^T \\ \vec{v}_3^T \end{bmatrix} \begin{bmatrix} \vec{v}_1 & \vec{v}_2 & \vec{v}_3 \end{bmatrix} \\ &= \begin{bmatrix} \vec{v}_1^T\vec{v}_1 & \vec{v}_1^T\vec{v}_2 & \vec{v}_1^T\vec{v}_3 \\ \vec{v}_2^T\vec{v}_1 & \vec{v}_2^T\vec{v}_2 & \vec{v}_2^T\vec{v}_3 \\ \vec{v}_3^T\vec{v}_1 & \vec{v}_3^T\vec{v}_2 & \vec{v}_3^T\vec{v}_3 \end{bmatrix} \\ &= \begin{bmatrix} \vec{v}_1\cdot\vec{v}_1 & \vec{v}_1\cdot\vec{v}_2 & \vec{v}_1\cdot\vec{v}_3 \\ \vec{v}_2\cdot\vec{v}_1 & \vec{v}_2\cdot\vec{v}_2 & \vec{v}_2\cdot\vec{v}_3 \\ \vec{v}_3\cdot\vec{v}_1 & \vec{v}_3\cdot\vec{v}_2 & \vec{v}_3\cdot\vec{v}_3 \end{bmatrix} \\ &= \begin{bmatrix} 1 & 0 & 0 \\ 0 & 1 & 0 \\ 0 & 0 & 1 \end{bmatrix} \end{aligned}$$

Remark

We have used the fact that the matrix multiplication $\vec{x}^T\vec{y}$ equals the dot product $\vec{x}\cdot\vec{y}$. We will use this very important fact many times throughout the rest of the book.

It follows that if P is a square matrix whose columns are orthonormal, then P is invertible and $P^{-1} = P^T$. Matrices with this property are given a special name.

Definition
Orthogonal Matrix

An $n \times n$ matrix P such that $P^T P = I$ is called an **orthogonal matrix**. It follows that $P^{-1} = P^T$ and that $PP^T = I = P^T P$.

It is important to observe that the definition of an orthogonal matrix is equivalent to the orthonormality of either the columns or the rows of the matrix.

Theorem 3

The following are equivalent for an $n \times n$ matrix P:

(1) P is orthogonal.
(2) The columns of P form an orthonormal set.
(3) The rows of P form an orthonormal set.

Proof: Let $P = \begin{bmatrix} \vec{v}_1 & \cdots & \vec{v}_n \end{bmatrix}$. By the usual rule for matrix multiplication,

$$(P^T P)_{ij} = \vec{v}_i^T\vec{v}_j = \vec{v}_i\cdot\vec{v}_j$$

Hence, $PP^T = I$ if and only if $\vec{v}_i\cdot\vec{v}_i = 1$ and $\vec{v}_i\cdot\vec{v}_j = 0$ for all $i \neq j$. But this is true if and only if the columns of P form an orthonormal set.

The result for the rows of P follows from consideration of the product $PP^T = I$. You are asked to show this in Problem D4. ■

Remark

Observe that such matrices should probably be called orthonormal matrices, but the name *orthogonal matrix* is the name everybody uses. Be sure that you remember that an orthogonal matrix has *orthonormal* columns and rows.

EXAMPLE 6

The set $\left\{ \begin{bmatrix} \cos\theta \\ \sin\theta \end{bmatrix}, \begin{bmatrix} -\sin\theta \\ \cos\theta \end{bmatrix} \right\}$ is orthonormal for any θ (verify). So, the matrix $P = \begin{bmatrix} \cos\theta & -\sin\theta \\ \sin\theta & \cos\theta \end{bmatrix}$ is an orthogonal matrix. Hence, $P^{-1} = P^T = \begin{bmatrix} \cos\theta & \sin\theta \\ -\sin\theta & \cos\theta \end{bmatrix}$.

EXAMPLE 7

The set $\left\{ \begin{bmatrix} 1 \\ 1 \\ 0 \end{bmatrix}, \begin{bmatrix} 1 \\ -1 \\ -1 \end{bmatrix}, \begin{bmatrix} 1 \\ -1 \\ 2 \end{bmatrix} \right\}$ is orthogonal. (Verify this.) If the vectors are normalized, the resulting set is orthonormal, so the following matrix P is orthogonal:

$$P = \begin{bmatrix} 1/\sqrt{2} & 1/\sqrt{3} & 1/\sqrt{6} \\ 1/\sqrt{2} & -1/\sqrt{3} & -1/\sqrt{6} \\ 0 & -1/\sqrt{3} & 2/\sqrt{6} \end{bmatrix}$$

Thus, P is invertible and

$$P^{-1} = P^T = \begin{bmatrix} 1/\sqrt{2} & 1/\sqrt{2} & 0 \\ 1/\sqrt{3} & -1/\sqrt{3} & -1/\sqrt{3} \\ 1/\sqrt{6} & -1/\sqrt{6} & 2/\sqrt{6} \end{bmatrix}$$

Moreover, observe that P^{-1} is also orthogonal.

EXAMPLE 8

The vectors of Example 4 are orthonormal, so the following matrix is orthogonal:

$$\begin{bmatrix} 1/2 & 1/2 & -1/\sqrt{2} & 0 \\ 1/2 & -1/2 & 0 & 1/\sqrt{2} \\ 1/2 & 1/2 & 1/\sqrt{2} & 0 \\ 1/2 & -1/2 & 0 & -1/\sqrt{2} \end{bmatrix}$$

EXERCISE 4

Verify that $P = \begin{bmatrix} 1/\sqrt{2} & 1/\sqrt{2} & 0 \\ 1/\sqrt{3} & -1/\sqrt{3} & 1/\sqrt{3} \\ -1/\sqrt{6} & 1/\sqrt{6} & 2/\sqrt{6} \end{bmatrix}$ is orthogonal by showing that $PP^T = I$.

The most important application of orthogonal matrices considered in this book is the diagonalization of symmetric matrices in Chapter 8, but there are many other geometrical applications as well. In the next example, an orthogonal change of coordinates matrix is used to find the standard matrix of a rotation transformation about an axis that is not a coordinate axis. (This was one question we could not answer in Chapter 3.)

EXAMPLE 9

Find the standard matrix of the linear transformation $L : \mathbb{R}^3 \to \mathbb{R}^3$ that rotates vectors about the axis defined by the vector $\vec{v} = \begin{bmatrix} 1 \\ 1 \\ 1 \end{bmatrix}$ counterclockwise through an angle $\frac{\pi}{3}$.

Solution: If the rotation were about the standard x_1-axis (that is, the axis defined by $\vec{e}_1$), the matrix of the rotation would be

$$R_1 = \begin{bmatrix} 1 & 0 & 0 \\ 0 & \cos \pi/3 & -\sin \pi/3 \\ 0 & \sin \pi/3 & \cos \pi/3 \end{bmatrix} = \begin{bmatrix} 1 & 0 & 0 \\ 0 & 1/2 & -\sqrt{3}/2 \\ 0 & \sqrt{3}/2 & 1/2 \end{bmatrix}$$

This will also be the $\mathcal{B}$-matrix of the rotation in this problem if there exists a basis $\mathcal{B} = \{\vec{f}_1, \vec{f}_2, \vec{f}_3\}$ such that

(1) $\vec{f}_1$ is a unit vector in the direction of the axis $\vec{v}$.
(2) $\mathcal{B}$ is orthonormal.
(3) $\mathcal{B}$ is right-handed (so that we can correctly include the counterclockwise sense of the rotation with respect to a right-handed basis).

Let us find such a basis.

To start, let $\vec{f}_1 = \frac{\vec{v}}{\|\vec{v}\|} = \frac{1}{\sqrt{3}} \begin{bmatrix} 1 \\ 1 \\ 1 \end{bmatrix}$. We must find two vectors that are orthogonal to $\vec{f}_1$ and to each other. Solving the equation

$$\vec{0} = \vec{f}_1 \cdot \vec{x} = \frac{1}{\sqrt{3}}(x_1 + x_2 + x_3)$$

by inspection, we find that the vector $\begin{bmatrix} 1 \\ -1 \\ 0 \end{bmatrix}$ is orthogonal to $\vec{f}_1$. (There are infinitely many other choices for this vector; this is just one simple choice.) To form a right-handed system, we can now take the third vector to be

$$\vec{f}_1 \times \vec{f}_2 = \begin{bmatrix} 1 \\ 1 \\ 1 \end{bmatrix} \times \begin{bmatrix} 1 \\ -1 \\ 0 \end{bmatrix} = \begin{bmatrix} 1 \\ 1 \\ -2 \end{bmatrix}$$

Normalizing these vectors, we get

$$\vec{f}_2 = \frac{1}{\sqrt{2}} \begin{bmatrix} 1 \\ -1 \\ 0 \end{bmatrix} \quad \text{and} \quad \vec{f}_3 = \frac{1}{\sqrt{6}} \begin{bmatrix} 1 \\ 1 \\ -2 \end{bmatrix}$$

The required right-handed orthonormal basis is thus $\mathcal{B} = \{\vec{f}_1, \vec{f}_2, \vec{f}_3\}$, and the orthogonal change of coordinates matrix from this basis to the standard basis is

$$P = \begin{bmatrix} \vec{f}_1 & \vec{f}_2 & \vec{f}_3 \end{bmatrix} = \begin{bmatrix} 1/\sqrt{3} & 1/\sqrt{2} & 1/\sqrt{6} \\ 1/\sqrt{3} & -1/\sqrt{2} & 1/\sqrt{6} \\ 1/\sqrt{3} & 0 & -2/\sqrt{6} \end{bmatrix}$$

EXAMPLE 9
(continued)

Since $[L]_{\mathcal{B}} = R_1$ (above), the standard matrix is given by

$$[L]_S = P[L]_{\mathcal{B}}P^{-1} = \begin{bmatrix} 2/3 & -1/3 & 2/3 \\ 2/3 & 2/3 & -1/3 \\ -1/3 & 2/3 & 2/3 \end{bmatrix}$$

It is easy to check that $\begin{bmatrix} 1 \\ 1 \\ 1 \end{bmatrix}$ is an eigenvector of this matrix with eigenvalue 1, and it should be since it defines the axis of the rotation represented by the matrix. Notice also that the matrix $[L]_S$ is itself an orthogonal matrix. Since a rotation transformation always maps the standard basis to a new orthonormal basis, its matrix can always be taken as a change of coordinates matrix, and it must be orthogonal.

A Note on Rotation Transformations and Rotation of Axes in $\mathbb{R}^2$

The matrix $P = \begin{bmatrix} \cos\theta & -\sin\theta \\ \sin\theta & \cos\theta \end{bmatrix}$ is the change of coordinates matrix from the basis $\mathcal{B} = \left\{ \begin{bmatrix} \cos\theta \\ \sin\theta \end{bmatrix}, \begin{bmatrix} -\sin\theta \\ \cos\theta \end{bmatrix} \right\}$ to the standard basis $\mathcal{S}$ of $\mathbb{R}^2$. This change of basis is often described as a "rotation of axes through angle θ" because each of the basis vectors in $\mathcal{B}$ is obtained from the corresponding standard basis vector by a rotation through angle θ.

Treatments of rotation of axes often emphasize the change of coordinates equation. Recall Theorem 4.4.2, which tells us that if P is the change of coordinates matrix from $\mathcal{B}$ to $\mathcal{S}$, then P^{-1} is the change of coordinates matrix from $\mathcal{S}$ to $\mathcal{B}$. Then, for any $\vec{x} \in \mathbb{R}^2$ with $[\vec{x}]_{\mathcal{B}} = \begin{bmatrix} b_1 \\ b_2 \end{bmatrix}$, the change of coordinates equation can be written in the form $[\vec{x}]_{\mathcal{B}} = P^{-1}[\vec{x}]_{\mathcal{S}}$. If the change of basis is a rotation of axes, then P is an orthogonal matrix, so $P^{-1} = P^T$. Thus, the change of coordinates equation for this rotation of axes can be written as

$$\begin{bmatrix} b_1 \\ b_2 \end{bmatrix} = \begin{bmatrix} \cos\theta & \sin\theta \\ -\sin\theta & \cos\theta \end{bmatrix} \begin{bmatrix} x_1 \\ x_2 \end{bmatrix}$$

Or it can be written as two equations for the "new" coordinates in terms of the "old":

$$b_1 = x_1\cos\theta + x_2\sin\theta$$
$$b_2 = -x_1\sin\theta + x_2\cos\theta$$

These equations could also be derived using a fairly simple trigonometric argument.

The matrix $\begin{bmatrix} \cos\theta & -\sin\theta \\ \sin\theta & \cos\theta \end{bmatrix}$ also appeared in Section 3.3 as the standard matrix $[R_\theta]$ of the linear transformation of $\mathbb{R}^2$ that rotates vectors counterclockwise through angle θ. Conceptually, this is quite different from a rotation of axes.

It can be confusing that the matrix for a rotation through θ as a linear transformation is the transpose of the change of coordinates matrix for a rotation of axes through θ. In fact, what may seem even more confusing is the fact that if you replace θ with $(-\theta)$, one matrix turns into the other (because $\cos(-\theta) = \cos\theta$ and $\sin(-\theta) = -\sin\theta$). One way to understand this is to imagine what happens to the vector $\vec{e}_1$ under the two different scenarios. First, consider R to be the transformation that rotates each vector by angle θ, with $0 < \theta < \frac{\pi}{2}$; then $R(\vec{e}_1)$ is a vector in the first quadrant that makes an angle θ with the x_1-axis. Next, consider a rotation of axes through $(-\theta)$; denote the new axes by y_1 and y_2, respectively. Then $\vec{e}_1$ has not moved but the axes have, and with respect to the new axes, $\vec{e}_1$ is in the new first quadrant and makes an angle of θ with respect to the y_1-axis. Therefore, the new coordinates of $\vec{e}_1$ relative to the rotated axes are exactly the same as the standard coordinates of $R(\vec{e}_1)$. Compare Figures 7.1.1 and 7.1.2.

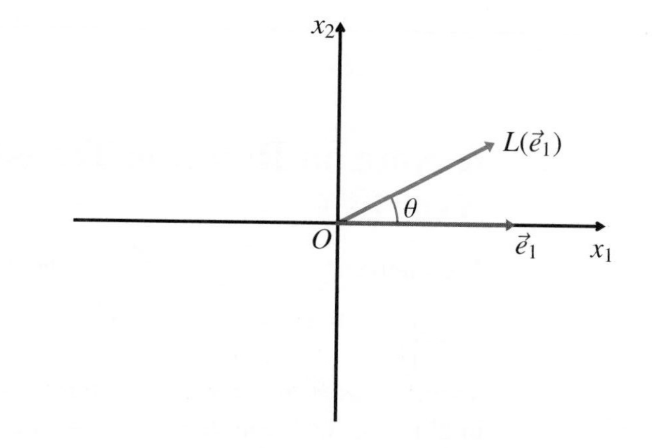

Figure 7.1.1 The transformation $L : \mathbb{R}^2 \to \mathbb{R}^2$ rotates vectors by angle θ.

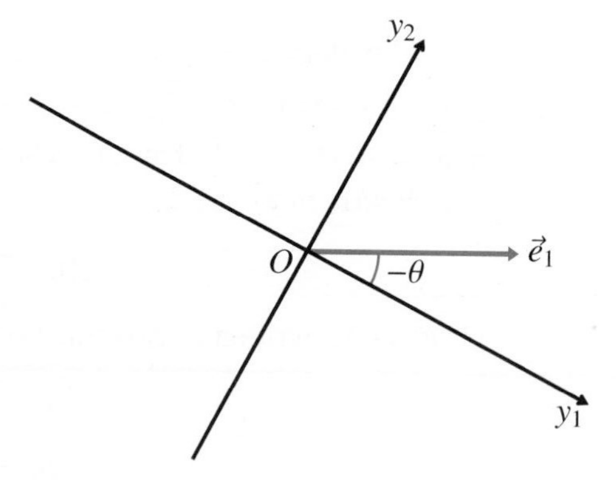

Figure 7.1.2 The standard basis vector $\vec{e}_1$ is shown relative to axes obtained from the standard axis by rotation through $-\theta$.

PROBLEMS 7.1

Practice Problems

A1 Determine which of the following sets are orthogonal. For each orthogonal set, produce the corresponding orthonormal set and the orthogonal change of coordinates matrix P.

(a) $\left\{\begin{bmatrix}1\\2\end{bmatrix}, \begin{bmatrix}2\\-1\end{bmatrix}\right\}$

(b) $\left\{\begin{bmatrix}1\\-1\\1\end{bmatrix}, \begin{bmatrix}1\\2\\1\end{bmatrix}, \begin{bmatrix}2\\-1\\1\end{bmatrix}\right\}$

(c) $\left\{\begin{bmatrix}1\\1\\3\end{bmatrix}, \begin{bmatrix}3\\0\\-1\end{bmatrix}, \begin{bmatrix}1\\-10\\3\end{bmatrix}\right\}$

(d) $\left\{\begin{bmatrix}1\\0\\1\\1\end{bmatrix}, \begin{bmatrix}2\\1\\-1\\-1\end{bmatrix}, \begin{bmatrix}1\\-3\\-1\\0\end{bmatrix}, \begin{bmatrix}-2\\1\\1\\0\end{bmatrix}\right\}$

A2 Let $\mathcal{B} = \left\{\frac{1}{3}\begin{bmatrix}1\\-2\\2\end{bmatrix}, \frac{1}{3}\begin{bmatrix}2\\2\\1\end{bmatrix}, \frac{1}{3}\begin{bmatrix}-2\\1\\2\end{bmatrix}\right\}$. Find the coordinates of each of the following vectors with respect to the orthonormal basis $\mathcal{B}$.

(a) $\vec{w} = \begin{bmatrix}4\\3\\5\end{bmatrix}$ (b) $\vec{x} = \begin{bmatrix}3\\-7\\2\end{bmatrix}$

(c) $\vec{y} = \begin{bmatrix}2\\-4\\6\end{bmatrix}$ (d) $\vec{z} = \begin{bmatrix}6\\6\\3\end{bmatrix}$

A3 Let $\mathcal{B} = \left\{\frac{1}{2}\begin{bmatrix}1\\1\\1\\1\end{bmatrix}, \frac{1}{2}\begin{bmatrix}1\\-1\\1\\-1\end{bmatrix}, \frac{1}{\sqrt{2}}\begin{bmatrix}-1\\0\\1\\0\end{bmatrix}, \frac{1}{\sqrt{2}}\begin{bmatrix}0\\1\\0\\-1\end{bmatrix}\right\}$.

Find the coordinates of each of the following vectors with respect to the orthonormal basis $\mathcal{B}$.

(a) $\vec{x} = \begin{bmatrix}2\\4\\-3\\5\end{bmatrix}$ (b) $\vec{y} = \begin{bmatrix}-4\\1\\3\\-5\end{bmatrix}$

(c) $\vec{w} = \begin{bmatrix}3\\1\\0\\1\end{bmatrix}$ (d) $\vec{z} = \begin{bmatrix}-1\\3\\2\\-3\end{bmatrix}$

A4 For each of the following matrices, decide whether A is orthogonal by calculating A^TA. If A is not orthogonal, indicate how the columns of A fail to form an orthonormal set (for example, "the second and third columns are not orthogonal").

(a) $A = \begin{bmatrix}5/13 & 12/13\\12/13 & -5/13\end{bmatrix}$

(b) $A = \begin{bmatrix}3/5 & 4/5\\-4/5 & -3/5\end{bmatrix}$

(c) $A = \begin{bmatrix}2/5 & -1/5\\1/5 & 2/5\end{bmatrix}$

(d) $A = \begin{bmatrix}1/3 & 2/3 & -2/3\\2/3 & -2/3 & 1/3\\2/3 & 1/3 & 2/3\end{bmatrix}$

(e) $A = \begin{bmatrix}1/3 & 2/3 & 2/3\\2/3 & -2/3 & 1/3\\2/3 & 1/3 & -2/3\end{bmatrix}$

A5 Let $L : \mathbb{R}^3 \to \mathbb{R}^3$ be the rotation through angle $\frac{\pi}{4}$ about the axis determined by $\vec{g}_3 = \begin{bmatrix}2\\2\\-1\end{bmatrix}$.

(a) Verify that $\vec{g}_1 = \begin{bmatrix}-1\\2\\2\end{bmatrix}$ is orthogonal to $\vec{g}_3$. Define $\vec{g}_2 = \vec{g}_3 \times \vec{g}_1$ and calculate the components of $\vec{g}_2$.

(b) Let $\vec{f}_i = \vec{g}_i/\|\vec{g}_i\|$ for $i = 1, 2, 3$, so that $\mathcal{B} = \{\vec{f}_1, \vec{f}_2, \vec{f}_3\}$ is an orthonormal basis. Write the change of coordinates matrix P for the change from $\mathcal{B}$ to the standard basis $\mathcal{S}$ in the form $P = \frac{1}{3}\begin{bmatrix}\ddots\end{bmatrix}$.

(c) Write the $\mathcal{B}$-matrix of L, $[L]_{\mathcal{B}}$. For part (d), it is probably easiest to write this in the form $[L]_{\mathcal{B}} = \frac{1}{\sqrt{2}}\begin{bmatrix}\ddots\end{bmatrix}$.

(d) Determine the standard matrix of $[L]_{\mathcal{S}}$.

A6 Given that $\mathcal{B} = \left\{ \begin{bmatrix} 1/\sqrt{6} \\ -2/\sqrt{6} \\ 1/\sqrt{6} \end{bmatrix}, \begin{bmatrix} 1/\sqrt{3} \\ 1/\sqrt{3} \\ 1/\sqrt{3} \end{bmatrix}, \begin{bmatrix} 1/\sqrt{2} \\ 0 \\ -1/\sqrt{2} \end{bmatrix} \right\}$ is an orthonormal basis for $\mathbb{R}^3$. Determine another orthonormal basis for $\mathbb{R}^3$ that includes the vector $\begin{bmatrix} 1/\sqrt{6} \\ 1/\sqrt{3} \\ 1/\sqrt{2} \end{bmatrix}$ and briefly explain why your basis is orthonormal.

Homework Problems

B1 Determine which of the following sets are orthogonal. For each orthogonal set, produce the corresponding orthonormal set and the orthogonal change of coordinates matrix P.

(a) $\left\{ \begin{bmatrix} 1 \\ 3 \end{bmatrix}, \begin{bmatrix} 3 \\ -1 \end{bmatrix} \right\}$

(b) $\left\{ \begin{bmatrix} 2 \\ -1 \\ 1 \end{bmatrix}, \begin{bmatrix} 1 \\ 0 \\ -2 \end{bmatrix}, \begin{bmatrix} 2 \\ 5 \\ 1 \end{bmatrix} \right\}$

(c) $\left\{ \begin{bmatrix} 1 \\ 1 \\ 1 \end{bmatrix}, \begin{bmatrix} -1 \\ 0 \\ -1 \end{bmatrix}, \begin{bmatrix} 1 \\ -1 \\ 0 \end{bmatrix} \right\}$

(d) $\left\{ \begin{bmatrix} 1 \\ -2 \\ 0 \\ 1 \end{bmatrix}, \begin{bmatrix} -1 \\ 0 \\ 1 \\ 1 \end{bmatrix}, \begin{bmatrix} 2 \\ 0 \\ 1 \\ 1 \end{bmatrix}, \begin{bmatrix} 2 \\ 1 \\ -2 \\ 0 \end{bmatrix} \right\}$

B2 Let $\mathcal{B} = \left\{ \frac{1}{\sqrt{2}}\begin{bmatrix} 1 \\ 0 \\ 1 \end{bmatrix}, \frac{1}{\sqrt{3}}\begin{bmatrix} -1 \\ 1 \\ 1 \end{bmatrix}, \frac{1}{\sqrt{6}}\begin{bmatrix} 1 \\ 2 \\ -1 \end{bmatrix} \right\}$. Find the coordinates of each of the following vectors with respect to the orthonormal basis $\mathcal{B}$.

(a) $\vec{w} = \begin{bmatrix} 6 \\ 2 \\ 1 \end{bmatrix}$ (b) $\vec{x} = \begin{bmatrix} -4 \\ 2 \\ 3 \end{bmatrix}$

(c) $\vec{y} = \begin{bmatrix} 3 \\ 3 \\ -5 \end{bmatrix}$ (d) $\vec{z} = \begin{bmatrix} -1 \\ 4 \\ 2 \end{bmatrix}$

B3 Let $\mathcal{B} = \left\{ \frac{1}{2}\begin{bmatrix} 1 \\ 1 \\ 1 \\ 1 \end{bmatrix}, \frac{1}{2}\begin{bmatrix} 1 \\ -1 \\ 1 \\ -1 \end{bmatrix}, \frac{1}{\sqrt{2}}\begin{bmatrix} -1 \\ 0 \\ 1 \\ 0 \end{bmatrix}, \frac{1}{\sqrt{2}}\begin{bmatrix} 0 \\ 1 \\ 0 \\ -1 \end{bmatrix} \right\}$. Find the coordinates of each of the following vectors with respect to the orthonormal basis $\mathcal{B}$.

(a) $\vec{w} = \begin{bmatrix} 3 \\ -2 \\ 6 \\ 1 \end{bmatrix}$ (b) $\vec{x} = \begin{bmatrix} 2 \\ -4 \\ 0 \\ 4 \end{bmatrix}$

(c) $\vec{y} = \begin{bmatrix} 5 \\ 0 \\ -2 \\ 2 \end{bmatrix}$ (d) $\vec{z} = \begin{bmatrix} 4 \\ 2 \\ -2 \\ 3 \end{bmatrix}$

B4 For each of the following matrices, decide whether A is orthogonal by calculating A^TA. If A is not orthogonal, indicate how the columns of A fail to form an orthonormal set (for example, "the second and third columns are not orthogonal").

(a) $A = \begin{bmatrix} 2/\sqrt{5} & -1/\sqrt{5} \\ 1/\sqrt{5} & -2/\sqrt{5} \end{bmatrix}$

(b) $A = \begin{bmatrix} 2/\sqrt{5} & -1/\sqrt{5} \\ -1/\sqrt{5} & -2/\sqrt{5} \end{bmatrix}$

(c) $A = \begin{bmatrix} 1/2 & 1/2 \\ -1/2 & 1/2 \end{bmatrix}$

(d) $A = \begin{bmatrix} 1/\sqrt{3} & 1/\sqrt{2} & -1/\sqrt{6} \\ -1/\sqrt{3} & 1/\sqrt{2} & 1/\sqrt{6} \\ 1/\sqrt{3} & 0 & 2/\sqrt{6} \end{bmatrix}$

(e) $A = \begin{bmatrix} 1/\sqrt{3} & 1/\sqrt{6} & 1/\sqrt{2} \\ 1/\sqrt{3} & 1/\sqrt{6} & 1/\sqrt{2} \\ 1/\sqrt{3} & -1/\sqrt{6} & 0 \end{bmatrix}$

B5 (a) Let $\vec{w}_1 = \begin{bmatrix} 1 \\ 1 \\ 2 \end{bmatrix}$ and $\vec{w}_2 = \begin{bmatrix} -2 \\ 0 \\ 1 \end{bmatrix}$. Determine a third vector $\vec{w}_3$ such that $\{\vec{w}_1, \vec{w}_2, \vec{w}_3\}$ forms a right-handed orthogonal set.

(b) Let $\vec{v}_i = \vec{w}_i/\|\vec{w}_i\|$ so that $\mathcal{B} = \{\vec{v}_1, \vec{v}_2, \vec{v}_3\}$ is an orthonormal basis. Find $\begin{bmatrix} 2 \\ 3 \\ -4 \end{bmatrix}_{\mathcal{B}}$ and $\begin{bmatrix} -3 \\ 1 \\ 3 \end{bmatrix}_{\mathcal{B}}$.

Conceptual Problems

D1 Verify that the product of two orthogonal matrices is an orthogonal matrix.

D2 (a) Prove that if P is an orthogonal matrix, then $\det P = \pm 1$.
(b) Give an example of a 2×2 matrix A such that $\det A = 1$, but A is not orthogonal.

D3 (a) Use the fact that $\vec{x} \cdot \vec{y} = \vec{x}^T \vec{y}$ to show that if an $n \times n$ matrix P is orthogonal, then $\|P\vec{x}\| = \|\vec{x}\|$ for every $\vec{x} \in \mathbb{R}^n$.
(b) Show that any real eigenvalue of an orthogonal matrix must be either 1 or -1.

D4 Prove that an $n \times n$ matrix P is orthogonal if and only if the rows of P form an orthonormal set.

7.2 Projections and the Gram-Schmidt Procedure

Projections onto a Subspace

The projection of a vector $\vec{y}$ onto another vector $\vec{x}$ was defined in Chapter 1 by finding a scalar multiple of $\vec{x}$, denoted $\text{proj}_{\vec{x}}\, \vec{y}$, and a vector perpendicular to $\vec{x}$, denoted $\text{perp}_{\vec{x}}\, \vec{y}$, such that

$$\vec{y} = \text{proj}_{\vec{x}}\, \vec{y} + \text{perp}_{\vec{x}}\, \vec{y}$$

Since we were just trying to find a scalar multiple of $\vec{x}$, the projection of $\vec{y}$ onto $\vec{x}$ could be viewed as projecting $\vec{y}$ onto the subspace spanned by $\vec{x}$. Similarly, we saw how to find the projection of $\vec{y}$ onto a plane, which is just a 2-dimensional subspace. It is natural and useful to define the projection of vectors onto more general subspaces.

Let $\vec{y} \in \mathbb{R}^n$ and let $\mathbb{S}$ be a subspace of $\mathbb{R}^n$. To match what we did in Chapter 1, we want to write $\vec{y}$ as

$$\vec{y} = \text{proj}_{\mathbb{S}}\, \vec{y} + \text{perp}_{\mathbb{S}}\, \vec{y}$$

where $\text{proj}_{\mathbb{S}}\, \vec{y}$ is a vector in $\mathbb{S}$ and $\text{perp}_{\mathbb{S}}\, \vec{y}$ is a vector orthogonal to $\mathbb{S}$. To do this, we first observe that we need to define precisely what we mean by a vector orthogonal to a subspace.

Definition
Orthogonal
Orthogonal Complement

Let $\mathbb{S}$ be a subspace of $\mathbb{R}^n$. We shall say that a vector $\vec{x}$ is **orthogonal** to $\mathbb{S}$ if

$$\vec{x} \cdot \vec{s} = 0 \qquad \text{for all } \vec{s} \in \mathbb{S}$$

We call the set of all vectors orthogonal to $\mathbb{S}$ the **orthogonal complement** of $\mathbb{S}$ and denote it $\mathbb{S}^\perp$. That is,

$$\mathbb{S}^\perp = \{\vec{x} \in \mathbb{R}^n \mid \vec{x} \cdot \vec{s} = 0 \text{ for all } \vec{s} \in \mathbb{S}\}$$

Remark

Note that if $\mathcal{B} = \{\vec{v}_1, \ldots, \vec{v}_k\}$ is a basis for $\mathbb{S}$, then

$$\mathbb{S}^\perp = \{\vec{x} \in \mathbb{R}^n \mid \vec{x} \cdot \vec{v}_i = 0 \text{ for } 1 \leq i \leq k\}$$

EXAMPLE 1

If $\mathbb{S}$ is a plane in $\mathbb{R}^3$ with normal vector $\vec{n}$, then by definition $\vec{n}$ is orthogonal to every vector in the plane, so we say that $\vec{n}$ is orthogonal to the plane. On the other hand, we saw in Chapter 1 that the plane is the set of all vectors orthogonal to $\vec{n}$ (or any scalar multiple of $\vec{n}$), so the orthogonal complement of the subspace Span$\{\vec{n}\}$ is the plane.

EXAMPLE 2

Let $\mathbb{W} = \text{Span}\left\{\begin{bmatrix}1\\0\\0\\1\end{bmatrix}, \begin{bmatrix}1\\0\\1\\0\end{bmatrix}\right\}$. Find $\mathbb{W}^\perp$ in $\mathbb{R}^4$.

Solution: We want to find all $\vec{v} = \begin{bmatrix}v_1\\v_2\\v_3\\v_4\end{bmatrix} \in \mathbb{R}^4$ such that $\vec{v} \cdot \begin{bmatrix}1\\0\\0\\1\end{bmatrix} = 0$ and $\vec{v} \cdot \begin{bmatrix}1\\0\\1\\0\end{bmatrix} = 0$. This gives the system of equations $v_1 + v_4 = 0$ and $v_1 + v_3 = 0$, which has solution space $\text{Span}\left\{\begin{bmatrix}0\\1\\0\\0\end{bmatrix}, \begin{bmatrix}-1\\0\\1\\1\end{bmatrix}\right\}$. Hence, $\mathbb{W}^\perp = \text{Span}\left\{\begin{bmatrix}0\\1\\0\\0\end{bmatrix}, \begin{bmatrix}-1\\0\\1\\1\end{bmatrix}\right\}$.

EXERCISE 1

Let $\mathbb{S} = \text{Span}\left\{\begin{bmatrix}1\\1\\1\\0\end{bmatrix}, \begin{bmatrix}-1\\0\\1\\1\end{bmatrix}, \begin{bmatrix}0\\1\\-1\\1\end{bmatrix}\right\}$. Find $\mathbb{S}^\perp$.

We get the following important facts about $\mathbb{S}$ and $\mathbb{S}^\perp$.

Theorem 1

Let $\mathbb{S}$ be a k-dimensional subspace of $\mathbb{R}^n$. Then

(1) $\mathbb{S} \cap \mathbb{S}^\perp = \{\vec{0}\}$

(2) $\dim(\mathbb{S}^\perp) = n - k$

(3) If $\{\vec{v}_1, \dots, \vec{v}_k\}$ is an orthonormal basis for $\mathbb{S}$ and $\{\vec{v}_{k+1}, \dots, \vec{v}_n\}$ is an orthonormal basis for $\mathbb{S}^\perp$, then $\{\vec{v}_1, \dots, \vec{v}_k, \vec{v}_{k+1}, \dots, \vec{v}_n\}$ is an orthonormal basis for $\mathbb{R}^n$.

You are asked to prove these facts in Problems D1, D2, and D4.

We are now able to return to our goal of defining the projection of a vector $\vec{x}$ onto a subspace $\mathbb{S}$ of $\mathbb{R}^n$. Assume that we have an orthonormal basis $\{\vec{v}_1, \dots, \vec{v}_k\}$ for $\mathbb{S}$ and an orthonormal basis $\{\vec{v}_{k+1}, \dots, \vec{v}_n\}$ for $\mathbb{S}^\perp$. Then, by (3) of Theorem 1, we know that $\{\vec{v}_1, \dots, \vec{v}_n\}$ is an orthonormal basis for $\mathbb{R}^n$. Therefore, from our work in

Section 7.1, we can find the coordinates of $\vec{x}$ with respect to this orthonormal basis. We get

$$\vec{x} = (\vec{x} \cdot \vec{v}_1)\vec{v}_1 + \cdots + (\vec{x} \cdot \vec{v}_k)\vec{v}_k + (\vec{x} \cdot \vec{v}_{k+1})\vec{v}_{k+1} + \cdots + (\vec{x} \cdot \vec{v}_n)\vec{v}_n$$

Observe that this is exactly what we have been looking for. In particular, we have written $\vec{x}$ as a sum of $(\vec{x} \cdot \vec{v}_1)\vec{v}_1 + \cdots + (\vec{x} \cdot \vec{v}_k)\vec{v}_k$, which is a vector in $\mathbb{S}$, and $(\vec{x} \cdot \vec{v}_{k+1})\vec{v}_{k+1} + \cdots + (\vec{x} \cdot \vec{v}_n)\vec{v}_n$, which is a vector in $\mathbb{S}^\perp$. Thus, we make the following definition.

Definition
Projection onto a Subspace

Let $\mathbb{S}$ be a k-dimensional subspace of $\mathbb{R}^n$ and let $\mathcal{B} = \{\vec{v}_1, \dots, \vec{v}_k\}$ be an *orthonormal basis* of $\mathbb{S}$. If $\vec{x}$ is any vector in $\mathbb{R}^n$, the **projection** of $\vec{x}$ onto $\mathbb{S}$ is defined to be

$$\text{proj}_{\mathbb{S}}\, \vec{x} = (\vec{x} \cdot \vec{v}_1)\vec{v}_1 + (\vec{x} \cdot \vec{v}_2)\vec{v}_2 + \cdots + (\vec{x} \cdot \vec{v}_k)\vec{v}_k$$

The **projection of $\vec{x}$ perpendicular to** $\mathbb{S}$ is defined to be

$$\text{perp}_{\mathbb{S}}\, \vec{x} = \vec{x} - \text{proj}_{\mathbb{S}}\, \vec{x}$$

Remark

Observe that a key component for this definition is that we have an orthonormal basis for the subspace $\mathbb{S}$. We could, of course, make a similar definition for the projection if we have only an orthogonal basis. See Problem D6.

We have defined $\text{perp}_{\mathbb{S}}\, \vec{x}$ so that we do not require an orthonormal basis for $\mathbb{S}^\perp$. However, we have to ensure that this is a valid equation by verifying that $\text{perp}_{\mathbb{S}}\, \vec{x} \in \mathbb{S}^\perp$. For any $1 \le i \le k$, we have

$$\begin{aligned}
\vec{v}_i \cdot \text{perp}_{\mathbb{S}}\, \vec{x} &= \vec{v}_i \cdot [\vec{x} - ((\vec{x} \cdot \vec{v}_1)\vec{v}_1 + \cdots + (\vec{x} \cdot \vec{v}_k)\vec{v}_k)] \\
&= \vec{v}_i \cdot \vec{x} - \vec{v}_i \cdot ((\vec{x} \cdot \vec{v}_1)\vec{v}_1 + \cdots + (\vec{x} \cdot \vec{v}_k)\vec{v}_k) \\
&= \vec{v}_i \cdot \vec{x} - (0 + \cdots + 0 + (\vec{x} \cdot \vec{v}_i)(\vec{v}_i \cdot \vec{v}_i) + 0 + \cdots + 0) \\
&= \vec{v}_i \cdot \vec{x} - \vec{v}_i \cdot \vec{x} \\
&= 0
\end{aligned}$$

since $\mathcal{B}$ is an orthonormal basis and the dot product is symmetric. Hence, $\text{perp}_{\mathbb{S}}\, \vec{x}$ is orthogonal to every vector in the orthonormal basis $\{\vec{v}_1, \dots, \vec{v}_k\}$ of $\mathbb{S}$. Hence, it is orthogonal to every vector in $\mathbb{S}$.

EXAMPLE 3

Let $\mathbb{S} = \text{Span}\left\{\frac{1}{2}\begin{bmatrix}1\\1\\1\\1\end{bmatrix}, \frac{1}{2}\begin{bmatrix}1\\-1\\1\\-1\end{bmatrix}\right\}$ and let $\vec{x} = \begin{bmatrix}2\\5\\-7\\3\end{bmatrix}$. Determine $\text{proj}_{\mathbb{S}}\, \vec{x}$ and $\text{perp}_{\mathbb{S}}\, \vec{x}$.

Solution: An orthonormal basis for $\mathbb{S}$ is $\mathcal{B} = \{\vec{v}_1, \vec{v}_2\} = \left\{\frac{1}{2}\begin{bmatrix}1\\1\\1\\1\end{bmatrix}, \frac{1}{2}\begin{bmatrix}1\\-1\\1\\-1\end{bmatrix}\right\}$. Thus,

EXAMPLE 3
(continued)

we get

$$\text{proj}_{\mathbb{S}}\, \vec{x} = (\vec{x}\cdot\vec{v}_1)\vec{v}_1 + (\vec{x}\cdot\vec{v}_2)\vec{v}_2 = \left(\frac{3}{2}\right)\frac{1}{2}\begin{bmatrix}1\\1\\1\\1\end{bmatrix} + \left(\frac{-13}{2}\right)\frac{1}{2}\begin{bmatrix}1\\-1\\1\\-1\end{bmatrix} = \begin{bmatrix}-5/2\\4\\-5/2\\4\end{bmatrix}$$

$$\text{perp}_{\mathbb{S}}\, \vec{x} = \vec{x} - \text{proj}_{\mathbb{S}}\, \vec{x} = \begin{bmatrix}2\\5\\-7\\3\end{bmatrix} - \begin{bmatrix}-5/2\\4\\-5/2\\4\end{bmatrix} = \begin{bmatrix}9/2\\1\\-9/2\\-1\end{bmatrix}$$

EXERCISE 2

Let $\mathcal{B} = \left\{\begin{bmatrix}1\\2\\1\end{bmatrix}, \begin{bmatrix}-1\\1\\-1\end{bmatrix}\right\}$ be an orthogonal basis for $\mathbb{S}$ and let $\vec{x} = \begin{bmatrix}2\\1\\3\end{bmatrix}$. Determine $\text{proj}_{\mathbb{S}}\, \vec{x}$ and $\text{perp}_{\mathbb{S}}\, \vec{x}$.

Recall that we showed in Chapter 1 that the projection of a vector $\vec{x} \in \mathbb{R}^3$ onto a plane in $\mathbb{R}^3$ is the vector in the plane that is closest to $\vec{x}$. We now prove that the projection of $\vec{x} \in \mathbb{R}^n$ onto a subspace $\mathbb{S}$ of $\mathbb{R}^n$ is the vector in $\mathbb{S}$ that is closest to $\vec{x}$.

Theorem 2 **Approximation Theorem**
Let $\mathbb{S}$ be a subspace of $\mathbb{R}^n$. Then, for any $\vec{x} \in \mathbb{R}^n$, the unique vector $\vec{s} \in \mathbb{S}$ that minimizes the distance $\|\vec{x} - \vec{s}\|$ is $\vec{s} = \text{proj}_{\mathbb{S}}\, \vec{x}$.

Proof: Let $\{\vec{v}_1, \dots, \vec{v}_k\}$ be an orthonormal basis for $\mathbb{S}$ and let $\{\vec{v}_{k+1}, \dots, \vec{v}_n\}$ be an orthonormal basis for $\mathbb{S}^\perp$. Then, for any $\vec{x} \in \mathbb{R}^n$, we can write

$$\vec{x} = x_1\vec{v}_1 + \cdots + x_k\vec{v}_k + x_{k+1}\vec{v}_{k+1} + \cdots + x_n\vec{v}_n$$

Any vector $\vec{s} \in \mathbb{S}$ can be expressed as $\vec{s} = s_1\vec{v}_1 + \cdots + s_k\vec{v}_k$, so that the square of the distance from $\vec{x}$ to $\vec{s}$ is given by

$$\|\vec{x} - \vec{s}\|^2 = (x_1 - s_1)^2 + \cdots + (x_k - s_k)^2 + x_{k+1}^2 + \cdots + x_n^2$$

To minimize the distance, we must choose $s_i = x_i$ for $1 \le i \le k$. But this means that

$$\vec{s} = x_1\vec{v}_1 + \cdots + x_k\vec{v}_k = \text{proj}_{\mathbb{S}}\, \vec{x}$$

■

The Gram-Schmidt Procedure

For many of the calculations in this chapter, we need an orthonormal (or orthogonal) basis for a subspace $\mathbb{S}$ of $\mathbb{R}^n$. If $\mathbb{S}$ is a k-dimensional subspace of $\mathbb{R}^n$, it is certainly possible to use the methods of Section 4.3 to produce some basis $\{\vec{w}_1, \dots, \vec{w}_k\}$ for $\mathbb{S}$. We will now show that we can convert any such basis for $\mathbb{S}$ into an orthonormal basis $\{\vec{v}_1, \dots, \vec{v}_k\}$ for $\mathbb{S}$. To simplify the description (and calculations), we first produce an orthogonal basis and then normalize each of the vectors to get an orthonormal basis.

The construction is inductive. That is, we first take $\vec{v}_1 = \vec{w}_1$ so that $\{\vec{v}_1\}$ is an orthogonal basis for $\operatorname{Span}\{\vec{w}_1\}$. Then, given an orthogonal basis $\{\vec{v}_1, \dots, \vec{v}_{i-1}\}$ for $\operatorname{Span}\{\vec{w}_1, \dots, \vec{w}_{i-1}\}$, we want to find a vector $\vec{v}_i$ such that $\{\vec{v}_1, \dots, \vec{v}_{i-1}, \vec{v}_i\}$ is an orthogonal basis for $\operatorname{Span}\{\vec{w}_1, \dots, \vec{w}_i\}$. We will repeat this procedure, called the **Gram-Schmidt Procedure**, until we have the desired orthogonal basis $\{\vec{v}_1, \dots, \vec{v}_k\}$ for $\operatorname{Span}\{\vec{w}_1, \dots, \vec{w}_k\}$. To do this, we will use the following theorem.

Theorem 3

Suppose that $\vec{v}_1, \dots, \vec{v}_k \in \mathbb{R}^n$. Then

$$\operatorname{Span}\{\vec{v}_1, \dots, \vec{v}_k\} = \operatorname{Span}\{\vec{v}_1, \dots, \vec{v}_{k-1}, \vec{v}_k + t_1\vec{v}_1 + \cdots + t_{k-1}\vec{v}_{k-1}\}$$

for any $t_1, \dots, t_{k-1} \in \mathbb{R}$.

You are asked to prove Theorem 3 in Problem D5.

Algorithm 1

The Gram-Schmidt Procedure is as follows.

First step: Let $\vec{v}_1 = \vec{w}_1$. Then the one-dimensional subspace spanned by $\vec{v}_1$ is obviously the same as the subspace spanned by $\vec{w}_1$. We will denote this subspace as $\mathbb{S}_1 = \operatorname{Span}\{\vec{v}_1\}$.

Second step: We want to find a vector $\vec{v}_2$ such that it is orthogonal to $\vec{v}_1$ and $\operatorname{Span}\{\vec{v}_1, \vec{v}_2\} = \operatorname{Span}\{\vec{w}_1, \vec{w}_2\}$. We know that the perpendicular of a projection onto a subspace is orthogonal to the subspace, so we take

$$\vec{v}_2 = \operatorname{perp}_{\mathbb{S}_1} \vec{w}_2 = \vec{w}_2 - \frac{\vec{w}_2 \cdot \vec{v}_1}{\|\vec{v}_1\|^2}\vec{v}_1$$

Then $\vec{v}_2$ is orthogonal to $\vec{v}_1$ and $\operatorname{Span}\{\vec{v}_1, \vec{v}_2\} = \operatorname{Span}\{\vec{w}_1, \vec{w}_2\}$ by Theorem 3. We denote the two-dimensional subspace by $\mathbb{S}_2 = \operatorname{Span}\{\vec{v}_1, \vec{v}_2\}$.

***i*-th step:** Suppose that $i-1$ steps have been carried out so that $\{\vec{v}_1, \dots, \vec{v}_{i-1}\}$ is orthogonal, and $\mathbb{S}_{i-1} = \operatorname{Span}\{\vec{v}_1, \dots, \vec{v}_{i-1}\} = \operatorname{Span}\{\vec{w}_1, \dots, \vec{w}_{i-1}\}$. Let

$$\vec{v}_i = \operatorname{perp}_{\mathbb{S}_{i-1}} \vec{w}_i = \vec{w}_i - \frac{\vec{w}_i \cdot \vec{v}_1}{\|\vec{v}_1\|^2}\vec{v}_1 - \cdots - \frac{\vec{w}_i \cdot \vec{v}_{i-1}}{\|\vec{v}_{i-1}\|^2}\vec{v}_{i-1}$$

By Theorem 3, $\{\vec{v}_1, \dots, \vec{v}_i\}$ is orthogonal and $\operatorname{Span}\{\vec{v}_1, \dots, \vec{v}_i\} = \operatorname{Span}\{\vec{w}_1, \dots, \vec{w}_i\}$. We continue this procedure until $i = k$, so that

$$\mathbb{S}_k = \operatorname{Span}\{\vec{v}_1, \dots, \vec{v}_k\} = \operatorname{Span}\{\vec{w}_1, \dots, \vec{w}_k\} = \mathbb{S}$$

and an orthogonal basis has been produced for the original subspace $\mathbb{S}$.

Remarks

1. It is an important feature of the construction that $\mathbb{S}_{i-1}$ is a subspace of the next $\mathbb{S}_i$.

2. Since it is really only the direction of $\vec{v}_i$ that is important in this procedure, we can rescale each $\vec{v}_i$ in any convenient fashion to simplify the calculations.

3. Notice that the order of the vectors in the original basis has an effect on the calculations because each step takes the perpendicular part of the next vector. If the original vectors were given in a different order, the procedure might produce a different orthonormal basis.

4. Observe that the procedure does not actually require that we start with a basis $\mathbb{S}$; only a spanning set is required. The procedure will actually detect a linearly dependent vector by returning the zero vector when we take the perpendicular part. This is demonstrated in Example 5.

EXAMPLE 4

Use the Gram-Schmidt Procedure to find an orthonormal basis for the subspace of $\mathbb{R}^5$ defined by $\mathbb{S} = \text{Span}\left\{\begin{bmatrix}1\\1\\0\\1\\1\end{bmatrix}, \begin{bmatrix}-1\\2\\1\\0\\1\end{bmatrix}, \begin{bmatrix}0\\1\\1\\1\\2\end{bmatrix}\right\}$.

Solution: Call the vectors in the basis $\vec{w}_1, \vec{w}_2$, and $\vec{w}_3$, respectively.

First step: Let $\vec{v}_1 = \vec{w}_1 = \begin{bmatrix}1\\1\\0\\1\\1\end{bmatrix}$ and $\mathbb{S}_1 = \text{Span}\{\vec{v}_1\}$.

Second step: Determine $\text{perp}_{\mathbb{S}_1} \vec{w}_2$:

$$\text{perp}_{\mathbb{S}_1} \vec{w}_2 = \vec{w}_2 - \frac{\vec{w}_2 \cdot \vec{v}_1}{\|\vec{v}_1\|^2}\vec{v}_1 = \begin{bmatrix}-3/2\\3/2\\1\\-1/2\\1/2\end{bmatrix}$$

(It is wise to check your arithmetic by verifying that $\vec{v}_1 \cdot \text{perp}_{\mathbb{S}_1} \vec{w}_2 = 0$.)

As mentioned above, we can take any scalar multiple of $\text{perp}_{\mathbb{S}_1} \vec{w}_2$, so we take

$\vec{v}_2 = \begin{bmatrix}-3\\3\\2\\-1\\1\end{bmatrix}$ and $\mathbb{S}_2 = \text{Span}\{\vec{v}_1, \vec{v}_2\}$.

EXAMPLE 4
(continued)

Third step: Determine $\text{perp}_{\mathbb{S}_2} \vec{w}_3$:

$$\text{perp}_{\mathbb{S}_2} \vec{w}_3 = \vec{w}_3 - \frac{\vec{w}_3 \cdot \vec{v}_1}{\|\vec{v}_1\|^2}\vec{v}_1 - \frac{\vec{w}_3 \cdot \vec{v}_2}{\|\vec{v}_2\|^2}\vec{v}_2 = \begin{bmatrix} -1/4 \\ -3/4 \\ 1/2 \\ 1/4 \\ 3/4 \end{bmatrix}$$

(Again, it is wise to check that $\text{perp}_{\mathbb{S}_2}\vec{w}_3$ is orthogonal to $\vec{v}_1$ and $\vec{v}_2$.)

We now see that $\{\vec{v}_1, \vec{v}_2, \vec{v}_3\}$ is an orthogonal basis for $\mathbb{S}$. To obtain an orthonormal basis for $\mathbb{S}$, we divide each vector in this basis by its length. Thus, we find that an orthonormal basis for $\mathbb{S}$ is

$$\left\{ \frac{1}{2}\begin{bmatrix} 1 \\ 1 \\ 0 \\ 1 \\ 1 \end{bmatrix}, \frac{1}{2\sqrt{6}}\begin{bmatrix} -3 \\ 3 \\ 2 \\ -1 \\ 1 \end{bmatrix}, \frac{1}{2\sqrt{6}}\begin{bmatrix} -1 \\ -3 \\ 2 \\ 1 \\ 3 \end{bmatrix} \right\}$$

EXAMPLE 5

Use the Gram-Schmidt Procedure to find an orthogonal basis for the subspace $\mathbb{S} = \text{Span}\left\{ \begin{bmatrix} 1 \\ 2 \\ 3 \end{bmatrix}, \begin{bmatrix} 4 \\ 5 \\ 6 \end{bmatrix}, \begin{bmatrix} 7 \\ 8 \\ 9 \end{bmatrix} \right\}$ of $\mathbb{R}^3$.

Solution: Call the vectors in the spanning set $\vec{w}_1$, $\vec{w}_2$, and $\vec{w}_3$, respectively.

First step: Let $\vec{v}_1 = \vec{w}_1 = \begin{bmatrix} 1 \\ 2 \\ 3 \end{bmatrix}$ and $\mathbb{S}_1 = \text{Span}\{\vec{v}_1\}$.

Second step: Determine $\text{perp}_{\mathbb{S}_1} \vec{w}_2$:

$$\text{perp}_{\mathbb{S}_1} \vec{w}_2 = \vec{w}_2 - \frac{\vec{w}_2 \cdot \vec{v}_1}{\|\vec{v}_1\|^2}\vec{v}_1 = \begin{bmatrix} 12/7 \\ 3/7 \\ -6/7 \end{bmatrix}$$

We take $\vec{v}_2 = \begin{bmatrix} 4 \\ 1 \\ -2 \end{bmatrix}$ and $\mathbb{S}_2 = \text{Span}\{\vec{v}_1, \vec{v}_2\}$.

Third step: Determine $\text{perp}_{\mathbb{S}_2} \vec{w}_3$:

$$\text{perp}_{\mathbb{S}_2} \vec{w}_3 = \vec{w}_3 - \frac{\vec{w}_3 \cdot \vec{v}_1}{\|\vec{v}_1\|^2}\vec{v}_1 - \frac{\vec{w}_3 \cdot \vec{v}_2}{\|\vec{v}_2\|^2}\vec{v}_2 = \begin{bmatrix} 0 \\ 0 \\ 0 \end{bmatrix}$$

Hence, $\vec{w}_3$ was in $\mathbb{S}_2$ and so $\vec{w}_3 \in \text{Span}\{\vec{v}_1, \vec{v}_2\} = \text{Span}\{\vec{w}_1, \vec{w}_2\}$. Therefore, $\{\vec{v}_1, \vec{v}_2\}$ is an orthogonal basis for $\mathbb{S}$.

PROBLEMS 7.2

Practice Problems

A1 Each of the following sets is orthogonal but not orthonormal. Determine the projection of $\vec{x} = \begin{bmatrix} 2 \\ 3 \\ 5 \\ 6 \end{bmatrix}$ onto the subspace spanned by each set.

(a) $\mathcal{A} = \left\{ \begin{bmatrix} 1 \\ -1 \\ -1 \\ 1 \end{bmatrix}, \begin{bmatrix} 1 \\ 2 \\ 1 \\ 2 \end{bmatrix} \right\}$

(b) $\mathcal{B} = \left\{ \begin{bmatrix} 1 \\ 0 \\ 1 \\ 0 \end{bmatrix}, \begin{bmatrix} 0 \\ 1 \\ 0 \\ 1 \end{bmatrix}, \begin{bmatrix} 1 \\ 0 \\ -1 \\ 0 \end{bmatrix} \right\}$

(c) $C = \left\{ \begin{bmatrix} 1 \\ 1 \\ 1 \\ 1 \end{bmatrix}, \begin{bmatrix} -1 \\ -1 \\ 1 \\ 1 \end{bmatrix}, \begin{bmatrix} -1 \\ 1 \\ -1 \\ 1 \end{bmatrix} \right\}$

A2 Find a basis for the orthogonal complement of each of the following subspaces.

(a) $\mathbb{S} = \text{Span}\left\{ \begin{bmatrix} 1 \\ 2 \\ -1 \end{bmatrix} \right\}$

(b) $\mathbb{S} = \text{Span}\left\{ \begin{bmatrix} 1 \\ 1 \\ 1 \end{bmatrix}, \begin{bmatrix} -1 \\ 1 \\ 3 \end{bmatrix} \right\}$

(c) $\mathbb{S} = \text{Span}\left\{ \begin{bmatrix} 1 \\ 0 \\ 1 \\ 0 \end{bmatrix}, \begin{bmatrix} 2 \\ -1 \\ 1 \\ 3 \end{bmatrix} \right\}$

A3 Use the Gram-Schmidt Procedure to produce an orthogonal basis for the subspace spanned by each set.

(a) $\left\{ \begin{bmatrix} 1 \\ 0 \\ 0 \end{bmatrix}, \begin{bmatrix} 1 \\ 1 \\ 0 \end{bmatrix}, \begin{bmatrix} 1 \\ 1 \\ 1 \end{bmatrix} \right\}$

(b) $\left\{ \begin{bmatrix} 1 \\ 1 \\ 1 \end{bmatrix}, \begin{bmatrix} 1 \\ 1 \\ 0 \end{bmatrix}, \begin{bmatrix} 1 \\ 0 \\ 0 \end{bmatrix} \right\}$

(c) $\left\{ \begin{bmatrix} 1 \\ 0 \\ 0 \\ 1 \end{bmatrix}, \begin{bmatrix} 2 \\ -1 \\ 1 \\ 2 \end{bmatrix}, \begin{bmatrix} -1 \\ 1 \\ 1 \\ 1 \end{bmatrix} \right\}$

(d) $\left\{ \begin{bmatrix} 1 \\ 1 \\ 0 \\ 1 \end{bmatrix}, \begin{bmatrix} 1 \\ 0 \\ 1 \\ 1 \end{bmatrix}, \begin{bmatrix} 1 \\ 3 \\ -2 \\ 1 \end{bmatrix}, \begin{bmatrix} 1 \\ 0 \\ 0 \\ 1 \end{bmatrix} \right\}$

A4 Use the Gram-Schmidt Procedure to produce an orthonormal basis for the subspace spanned by each set.

(a) $\left\{ \begin{bmatrix} 1 \\ 1 \\ 0 \end{bmatrix}, \begin{bmatrix} 1 \\ 1 \\ 1 \end{bmatrix}, \begin{bmatrix} 1 \\ 0 \\ 0 \end{bmatrix} \right\}$

(b) $\left\{ \begin{bmatrix} 2 \\ 1 \\ -2 \end{bmatrix}, \begin{bmatrix} 2 \\ 1 \\ 1 \end{bmatrix}, \begin{bmatrix} -1 \\ -1 \\ 1 \end{bmatrix} \right\}$

(c) $\left\{ \begin{bmatrix} 1 \\ 1 \\ 0 \\ 1 \end{bmatrix}, \begin{bmatrix} 0 \\ 1 \\ 1 \\ 1 \end{bmatrix}, \begin{bmatrix} 1 \\ -1 \\ -1 \\ -1 \end{bmatrix} \right\}$

(d) $\left\{ \begin{bmatrix} 1 \\ 0 \\ 1 \\ 0 \\ 1 \end{bmatrix}, \begin{bmatrix} 1 \\ 0 \\ -1 \\ 1 \\ 0 \end{bmatrix}, \begin{bmatrix} 1 \\ 1 \\ 1 \\ 1 \\ 1 \end{bmatrix} \right\}$

A5 Let $\mathbb{S}$ be a subspace of $\mathbb{R}^n$. Prove that $\text{perp}_{\mathbb{S}}\, \vec{x} = \text{proj}_{\mathbb{S}^\perp}\, \vec{x}$ for any $\vec{x} \in \mathbb{R}^n$.

Homework Problems

B1 Each of the following sets is orthogonal but not orthonormal. Determine the projection of $\vec{x} = \begin{bmatrix} 4 \\ 3 \\ -2 \\ 5 \end{bmatrix}$ onto the subspace spanned by each set.

(a) $\mathcal{A} = \left\{ \begin{bmatrix} 1 \\ 0 \\ -1 \\ 1 \end{bmatrix}, \begin{bmatrix} 1 \\ 1 \\ -1 \\ -2 \end{bmatrix} \right\}$

(b) $\mathcal{B} = \left\{ \begin{bmatrix} 2 \\ 1 \\ 0 \\ 1 \end{bmatrix}, \begin{bmatrix} -1 \\ 1 \\ 1 \\ 1 \end{bmatrix} \right\}$

(c) $C = \left\{ \begin{bmatrix} 1 \\ 0 \\ 1 \\ 1 \end{bmatrix}, \begin{bmatrix} 0 \\ 1 \\ -1 \\ 1 \end{bmatrix}, \begin{bmatrix} 1 \\ 1 \\ 0 \\ -1 \end{bmatrix} \right\}$

(d) $\mathcal{D} = \left\{ \begin{bmatrix} 1 \\ 1 \\ 1 \\ 0 \end{bmatrix}, \begin{bmatrix} 1 \\ 0 \\ -1 \\ 1 \end{bmatrix}, \begin{bmatrix} 0 \\ 1 \\ -1 \\ -1 \end{bmatrix} \right\}$

B2 Find a basis for the orthogonal complement of each of the following subspaces.

(a) $\mathbb{S} = \text{Span}\left\{ \begin{bmatrix} 3 \\ 2 \\ 1 \end{bmatrix} \right\}$

(b) $\mathbb{S} = \text{Span}\left\{ \begin{bmatrix} 1 \\ 2 \\ -4 \end{bmatrix}, \begin{bmatrix} -1 \\ 2 \\ 2 \end{bmatrix} \right\}$

(c) $\mathbb{S} = \text{Span}\left\{ \begin{bmatrix} 1 \\ 2 \\ -1 \end{bmatrix}, \begin{bmatrix} 2 \\ 1 \\ 1 \end{bmatrix}, \begin{bmatrix} -1 \\ 1 \\ 1 \end{bmatrix} \right\}$

(d) $\mathbb{S} = \text{Span}\left\{ \begin{bmatrix} 2 \\ 1 \\ -1 \\ 0 \end{bmatrix}, \begin{bmatrix} 1 \\ 2 \\ 1 \\ 1 \end{bmatrix}, \begin{bmatrix} 0 \\ -1 \\ 3 \\ 1 \end{bmatrix} \right\}$

B3 Use the Gram-Schmidt Procedure to produce an orthogonal basis for the subspace spanned by each set.

(a) $\left\{ \begin{bmatrix} 1 \\ 0 \\ 1 \end{bmatrix}, \begin{bmatrix} 1 \\ -1 \\ -1 \end{bmatrix}, \begin{bmatrix} 2 \\ 2 \\ 1 \end{bmatrix} \right\}$

(b) $\left\{ \begin{bmatrix} 0 \\ 1 \\ 1 \end{bmatrix}, \begin{bmatrix} 1 \\ 1 \\ 2 \end{bmatrix}, \begin{bmatrix} 3 \\ 0 \\ -2 \end{bmatrix} \right\}$

(c) $\left\{ \begin{bmatrix} 1 \\ 0 \\ -1 \\ 1 \end{bmatrix}, \begin{bmatrix} 1 \\ 1 \\ 1 \\ 1 \end{bmatrix}, \begin{bmatrix} 2 \\ 0 \\ 1 \\ 1 \end{bmatrix} \right\}$

(d) $\left\{ \begin{bmatrix} 1 \\ 0 \\ 0 \\ 1 \end{bmatrix}, \begin{bmatrix} 1 \\ 1 \\ 0 \\ 2 \end{bmatrix}, \begin{bmatrix} 0 \\ -1 \\ 1 \\ 1 \end{bmatrix} \right\}$

B4 Use the Gram-Schmidt Procedure to produce an orthonormal basis for the subspace spanned by each set.

(a) $\left\{ \begin{bmatrix} 1 \\ 0 \\ 1 \end{bmatrix}, \begin{bmatrix} 1 \\ -1 \\ -1 \end{bmatrix}, \begin{bmatrix} 2 \\ 2 \\ 1 \end{bmatrix} \right\}$

(b) $\left\{ \begin{bmatrix} 1 \\ 1 \\ 3 \end{bmatrix}, \begin{bmatrix} 2 \\ 2 \\ -3 \end{bmatrix}, \begin{bmatrix} 3 \\ 3 \\ -4 \end{bmatrix}, \begin{bmatrix} 1 \\ 0 \\ 1 \end{bmatrix} \right\}$

(c) $\left\{ \begin{bmatrix} 1 \\ 1 \\ 0 \\ 1 \end{bmatrix}, \begin{bmatrix} 0 \\ -1 \\ 1 \\ 1 \end{bmatrix}, \begin{bmatrix} 3 \\ 0 \\ 1 \\ 1 \end{bmatrix} \right\}$

(d) $\left\{ \begin{bmatrix} 1 \\ 1 \\ 0 \\ 1 \\ 0 \end{bmatrix}, \begin{bmatrix} -1 \\ 0 \\ 1 \\ 1 \\ 1 \end{bmatrix}, \begin{bmatrix} 1 \\ 1 \\ 0 \\ 2 \\ 1 \end{bmatrix} \right\}$

B5 Let $\mathbb{S} = \text{Span}\left\{ \begin{bmatrix} 1 \\ 2 \\ 1 \\ 0 \end{bmatrix}, \begin{bmatrix} 1 \\ 1 \\ 1 \\ 1 \end{bmatrix} \right\}$ and let $\vec{x} = \begin{bmatrix} 1 \\ 1 \\ -1 \\ 1 \end{bmatrix}$.

(a) Find an orthonormal basis $\mathcal{B}$ for $\mathbb{S}$.
(b) Calculate $\text{perp}_{\mathbb{S}}\, \vec{x}$.
(c) Find an orthonormal basis C for $\mathbb{S}^{\perp}$.
(d) Calculate $\text{proj}_{\mathbb{S}^{\perp}}\, \vec{x}$.

Conceptual Problems

D1 Prove that if $\mathbb{S}$ is a k-dimensional subspace of $\mathbb{R}^n$, then $\mathbb{S} \cap \mathbb{S}^{\perp} = \{\vec{0}\}$.

D2 Prove that if $\mathbb{S}$ is a k-dimensional subspace of $\mathbb{R}^n$, then $\mathbb{S}^{\perp}$ is an $(n-k)$-dimensional subspace.

D3 Prove that if $\mathbb{S}$ is a k-dimensional subspace of $\mathbb{R}^n$, then $(\mathbb{S}^{\perp})^{\perp} = \mathbb{S}$.

D4 Prove that if $\{\vec{v}_1, \dots, \vec{v}_k\}$ is an orthonormal basis for $\mathbb{S}$ and $\{\vec{v}_{k+1}, \dots, \vec{v}_n\}$ is an orthonormal basis for $\mathbb{S}^{\perp}$, then $\{\vec{v}_1, \dots, \vec{v}_n\}$ is an orthonormal basis for $\mathbb{R}^n$.

D5 Suppose that $\vec{v}_1, \dots, \vec{v}_k \in \mathbb{R}^n$ and $t_1, \dots, t_{k-1} \in \mathbb{R}$. Prove that

$$\operatorname{Span}\{\vec{v}_1, \dots, \vec{v}_k\} = \operatorname{Span}\{\vec{v}_1, \dots, \vec{v}_{k-1}, \vec{v}_k + t_1\vec{v}_1 + \cdots + t_{k-1}\vec{v}_{k-1}\}$$

D6 Suppose that $\mathbb{S}$ is a k-dimensional subspace of $\mathbb{R}^n$ and let $\mathcal{B} = \{\vec{v}_1, \dots, \vec{v}_k\}$ be an *orthogonal basis* of $\mathbb{S}$. For any $\vec{x} \in \mathbb{S}$, find the coordinates of $\operatorname{proj}_{\mathbb{S}} \vec{x}$ with respect to $\mathcal{B}$.

D7 If $\{\vec{v}_1, \dots, \vec{v}_k\}$ is an orthonormal basis for a subspace $\mathbb{S}$, verify that the standard matrix of $\operatorname{proj}_{\mathbb{S}}$ can be written

$$[\operatorname{proj}_{\mathbb{S}}] = \vec{v}_1\vec{v}_1^T + \vec{v}_2\vec{v}_2^T + \cdots + \vec{v}_k\vec{v}_k^T$$

7.3 Method of Least Squares

Suppose that an experimenter measures a variable y at times $t_1, t_2, \dots, t_n$ and obtains the values $y_1, y_2, \dots, y_n$. For example, y might be the position of a particle at time t or the temperature of some body fluid. Suppose that the experimenter believes that the data fit (more or less) a curve of the form $y = a + bt + ct^2$. How should she choose a, b, and c to get the best-fitting curve of this form?

Let us consider some particular a, b, and c. If the data fit the curve $y = a + bt + ct^2$ perfectly, then for each i, $y_i = a + bt_i + ct_i^2$. However, for arbitrary a, b, and c, we expect that at each t_i, there will be an error, denoted by e_i and measured by the vertical distance

$$e_i = y_i - (a + bt_i + ct_i^2)$$

as shown in Figure 7.3.3.

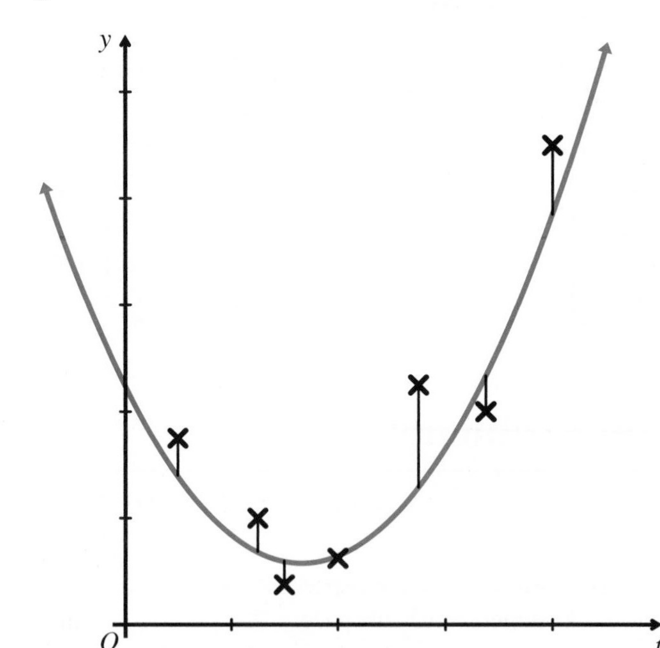

Figure 7.3.3 Some data points and a curve $y = a + bt + ct^2$. Vertical line segments measure the error in the fit at each t_i.

One approach to finding the best-fitting curve might be to try to minimize the *total error* $\sum_{i=1}^{n} e_i$. This would be unsatisfactory, however, because we might get a small total error by having large positive errors cancelled by large negative errors. Thus, we instead choose to minimize the sum of the squares of the errors

$$\sum_{i=1}^{n} e_i^2 = \sum_{i=1}^{n} (y_i - (a + bt_i + ct_i^2))^2$$

This method is called the **method of least squares**.

To find the parameters a, b, and c that minimize this expression for given values $t_1, \dots, t_n$ and $y_1, \dots, y_n$, one could use calculus, but we will proceed by using a projection. This requires us to set up the problem as follows.

Let $\vec{y} = \begin{bmatrix} y_1 \\ \vdots \\ y_n \end{bmatrix}$, $\vec{1} = \begin{bmatrix} 1 \\ \vdots \\ 1 \end{bmatrix}$, $\vec{t} = \begin{bmatrix} t_1 \\ \vdots \\ t_n \end{bmatrix}$, and $\vec{t}^2 = \begin{bmatrix} t_1^2 \\ \vdots \\ t_n^2 \end{bmatrix}$ be vectors in $\mathbb{R}^n$. Now consider the distance from $\vec{y}$ to $(a\vec{1} + b\vec{t} + c\vec{t}^2)$. *Observe that the square of this distance is exactly the sum of the squares of the errors*

$$\|\vec{y} - (a\vec{1} + b\vec{t} + c\vec{t}^2)\|^2 = \sum_{i=1}^{n} (y_i - (a + bt_i + ct_i^2))^2$$

Next, observe that $(a\vec{1} + b\vec{t} + c\vec{t}^2)$ is a vector in the subspace $\mathbb{S}$ of $\mathbb{R}^n$ spanned by $\mathcal{B} = \{\vec{1}, \vec{t}, \vec{t}^2\}$. If at least four of the t_i are distinct, then $\mathcal{B}$ is linearly independent (see Problem D2), so it is a basis for $\mathbb{S}$. Thus, the problem of finding the curve of best fit is reduced to finding a, b, and c such that $a\vec{1} + b\vec{t} + c\vec{t}^2$ is the vector in $\mathbb{S}$ that is closest to $\vec{y}$. By the Approximation Theorem, this vector is $\text{proj}_{\mathbb{S}}\, \vec{y}$ and the required a, b, and c are the $\mathcal{B}$-coordinates of $\text{proj}_{\mathbb{S}}\, \vec{y}$.

Given what we know so far, it might seem that we should proceed by transforming $\mathcal{B}$ into an orthonormal basis for $\mathbb{S}$ so that we can find the projection. However, we can use the theory of orthogonality and projections to simplify the problem. If a, b, and c have been chosen correctly, the **error vector** $\vec{e} = \vec{y} - a\vec{1} - b\vec{t} - c\vec{t}^2$ is equal to $\text{perp}_{\mathbb{S}}\, \vec{y}$. In particular, it must be orthogonal to every vector in $\mathbb{S}$, so it is orthogonal to $\vec{1}$, $\vec{t}$, and $\vec{t}^2$. Therefore,

$$\begin{aligned} \vec{1} \cdot \vec{e} &= \vec{1} \cdot (\vec{y} - a\vec{1} - b\vec{t} - c\vec{t}^2) = 0 \\ \vec{t} \cdot \vec{e} &= \vec{t} \cdot (\vec{y} - a\vec{1} - b\vec{t} - c\vec{t}^2) = 0 \\ \vec{t}^2 \cdot \vec{e} &= \vec{t}^2 \cdot (\vec{y} - a\vec{1} - b\vec{t} - c\vec{t}^2) = 0 \end{aligned}$$

The required a, b, and c are determined as the solutions of this homogeneous system of three equations in three variables.

It is helpful to rewrite these equations by introducing the matrix

$$X = \begin{bmatrix} \vec{1} & \vec{t} & \vec{t}^2 \end{bmatrix}$$

and the vector $\vec{a} = \begin{bmatrix} a \\ b \\ c \end{bmatrix}$ of parameters. Then the error vector can be written as $\vec{e} = \vec{y} - X\vec{a}$.

Since the three equations are obtained by taking dot products of $\vec{e}$ with the columns of X, the system of equations can be written in the form

$$X^T(\vec{y} - X\vec{a}) = \vec{0}$$

The equations in this form are called the **normal equations** for the least squares fit. Since the columns of X are linearly independent, the matrix $X^T X$ is a 3×3 invertible matrix (see Problem D2), and the normal equations can be rewritten as

$$\vec{a} = (X^T X)^{-1} X^T \vec{y}$$

This is consistent with a unique solution.

For a more general situation, we use a similar construction. The matrix X, called the **design matrix**, depends on the desired model curve and the way the data are collected. This will be demonstrated in Example 2 below.

EXAMPLE 1

Suppose that the experimenter's data are as follows:

t	1.0	2.1	3.1	4.0	4.9	6.0
y	6.1	12.6	21.1	30.2	40.9	55.5

Solution: As in the earlier discussion, the experimenter wants to find the curve $y = a + bt + ct^2$ that best fits the data, in the sense of minimizing the sum of the squares of the errors. We let

$$X^T = \begin{bmatrix} \vec{1} & \vec{t} & \vec{t^2} \end{bmatrix}^T = \begin{bmatrix} 1 & 1 & 1 & 1 & 1 & 1 \\ 1.0 & 2.1 & 3.1 & 4.0 & 4.9 & 6.0 \\ 1.0 & 4.41 & 9.61 & 16.0 & 24.01 & 36.0 \end{bmatrix}$$

$$\vec{y} = \begin{bmatrix} 6.1 \\ 12.6 \\ 21.1 \\ 30.2 \\ 40.9 \\ 55.5 \end{bmatrix}$$

Using a computer, we can find that the solution for the system $\vec{a} = (X^T X)^{-1} X^T \vec{y}$ is $\vec{a} = \begin{bmatrix} 1.63175 \\ 3.38382 \\ 0.93608 \end{bmatrix}$. The data do not justify retaining so many decimal places, so we take the best-fitting quadratic curve to be $y = 1.63 + 3.38t + 0.94t^2$. The results are shown in Figure 7.3.4.

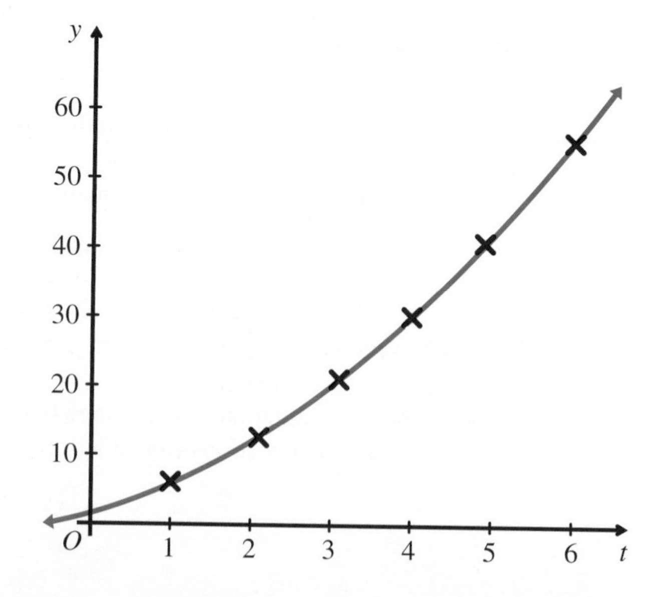

Figure 7.3.4 The data points and the best-fitting curve from Example 1.

EXAMPLE 2

Find a and b to obtain the best-fitting equation of the form $y = at^2 + bt$ for the following data:

t	-1	0	1
y	4	1	1

Solution: Using the method above, we observe that we want the error vector $\vec{e} = \vec{y} - a\vec{t}^2 - b\vec{t}$ to be equal to $\text{perp}_{\mathbb{S}}\,\vec{y}$. In particular, $\vec{e}$ must be orthogonal to $\vec{t}^2$ and $\vec{t}$. Therefore,

$$\vec{t}^2 \cdot \vec{e} = \vec{t}^2 \cdot (\vec{y} - a\vec{t}^2 - b\vec{t}) = 0$$
$$\vec{t} \cdot \vec{e} = \vec{t} \cdot (\vec{y} - a\vec{t}^2 - b\vec{t}) = 0$$

In this case, we want to pick X to be of the form

$$X = \begin{bmatrix} \vec{t}^2 & \vec{t} \end{bmatrix} = \begin{bmatrix} t_1^2 & t_1 \\ t_2^2 & t_2 \\ t_3^2 & t_3 \end{bmatrix} = \begin{bmatrix} 1 & -1 \\ 0 & 0 \\ 1 & 1 \end{bmatrix}$$

Taking $\vec{y} = \begin{bmatrix} 4 \\ 1 \\ 1 \end{bmatrix}$ then gives

$$\begin{bmatrix} a \\ b \end{bmatrix} = (X^T X)^{-1} X^T \vec{y} = \begin{bmatrix} 2 & 0 \\ 0 & 2 \end{bmatrix}^{-1} \begin{bmatrix} 5 \\ -3 \end{bmatrix} = \begin{bmatrix} 5/2 \\ -3/2 \end{bmatrix}$$

So, $y = \frac{5}{2}t^2 - \frac{3}{2}t$ is the equation of best fit for the given data.

Overdetermined Systems

The problem of finding the best-fitting curve can be viewed as a special case of the problem of "solving" an **overdetermined system**. Suppose that $A\vec{x} = \vec{b}$ is a system of p equations in q variables, where p is greater than q. With more equations than variables, we expect the system to be inconsistent unless $\vec{b}$ has some special properties. If the system is inconsistent, we say that the system is overdetermined—that is, there are too many equations to be satisfied.

Note that the problem in Example 1 of finding the best-fitting quadratic curve was of this form: we needed to solve $X\vec{a} = \vec{y}$ for the three variables a, b, and c, where there were n equations. Thus, for $n > 3$, this is an overdetermined system.

If there is no $\vec{x}$ such that $A\vec{x} = \vec{b}$, the next best "solution" is to find a vector $\vec{x}$ that minimizes the "error" $\|A\vec{x} - \vec{b}\|$. However, $A\vec{x} = x_1\vec{a}_1 + \cdots + x_q\vec{a}_q$, which is a vector in the columnspace of A. Therefore, our challenge is to find $\vec{x}$ such that $A\vec{x}$ is the point in the columnspace of A that is closest to $\vec{b}$. By the Approximation Theorem, we know that this vector is $\text{proj}_{\mathbb{S}}\,\vec{x}$. Thus, to find a vector $\vec{x}$ that minimizes the "error"

$\|A\vec{x} - \vec{b}\|$, we want to solve the consistent system $A\vec{x} = \text{proj}_{\text{Col}(A)}\,\vec{x}$. Using an argument analogous to that in the special case above, it can be shown that this vector $\vec{x}$ must also satisfy the normal system

$$A^T A\vec{x} = A^T\vec{b}$$

EXAMPLE 3

Verify that the following system $A\vec{x} = \vec{b}$ is inconsistent and then determine the vector $\vec{x}$ that minimizes $\|A\vec{x} - \vec{b}\|$:

$$\begin{aligned} 3x_1 - x_2 &= 4 \\ x_1 + 2x_2 &= 0 \\ 2x_1 + x_2 &= 1 \end{aligned}$$

Solution: Write the augmented matrix $\left[A \mid \vec{b}\right]$ and row reduce:

$$\left[\begin{array}{cc|c} 3 & -1 & 4 \\ 1 & 2 & 0 \\ 2 & 1 & 1 \end{array}\right] \sim \left[\begin{array}{cc|c} 1 & 2 & 0 \\ 0 & 1 & -2 \\ 0 & 0 & -5 \end{array}\right]$$

The last row indicates that the system is inconsistent. The $\vec{x}$ that minimizes $\|A\vec{x} - \vec{b}\|$ must satisfy $A^T A\vec{x} = A^T\vec{b}$. Solving for $\vec{x}$ in this system gives

$$\begin{aligned} \vec{x} &= (A^T A)^{-1} A^T\vec{b} \\ &= \begin{bmatrix} 14 & 1 \\ 1 & 6 \end{bmatrix}^{-1} \begin{bmatrix} 3 & 1 & 2 \\ -1 & 2 & 1 \end{bmatrix} \begin{bmatrix} 4 \\ 0 \\ 1 \end{bmatrix} \\ &= \frac{1}{83}\begin{bmatrix} 6 & -1 \\ -1 & 14 \end{bmatrix} \begin{bmatrix} 14 \\ -3 \end{bmatrix} \\ &= \begin{bmatrix} 87/83 \\ -56/83 \end{bmatrix} \end{aligned}$$

So, $\vec{x} = \begin{bmatrix} 87/83 \\ -56/83 \end{bmatrix}$ is the vector that minimizes $\|A\vec{x} - \vec{b}\|$.

PROBLEMS 7.3

Practice Problems

A1 Find a and b to obtain the best-fitting equation of the form $y = a + bt$ for the given data. Make a graph showing the data and the best-fitting line.

(a)

t	1	2	3	4	5
y	9	6	5	3	1

(b)

t	−2	−1	0	1	2
y	2	2	4	4	5

A2 Find the best-fitting equation of the given form for each set of data.

(a) $y = at + bt^2$ for the data

t	−1	0	1
y	4	1	1

(b) $y = a + bt^2$ for the data

t	−2	−1	0	1	2
y	1	1	2	3	−2

A3 Verify that the following systems $A\vec{x} = \vec{b}$ are inconsistent and then determine for each system the vector $\vec{x}$ that minimizes $\|A\vec{x} - \vec{b}\|$.

(a)
$$\begin{aligned} x_1 + 2x_2 &= 5 \\ 2x_1 - 3x_2 &= 6 \\ x_1 - 12x_2 &= -4 \end{aligned}$$

(b)
$$\begin{aligned} 2x_1 + 3x_2 &= -4 \\ 3x_1 - 2x_2 &= 4 \\ x_1 - 6x_2 &= 7 \end{aligned}$$

Homework Problems

B1 Find a and b to obtain the best-fitting equation of the form $y = a + bt$ for the given data. Make a graph showing the data and the best-fitting line.

(a)

t	−2	−1	0	1	2
y	9	8	5	3	1

(b)

t	1	2	3	4	5
y	4	3	4	5	5

B2 Find a, b, and c to obtain the best-fitting equation of the form $y = a + bt + ct^2$ for the given data. Make a graph showing the data and the best-fitting curve.

t	−2	−1	0	1	2
y	3	2	0	2	8

B3 Find the best-fitting equation of the given form for each set of data.

(a) $y = at + bt^2$ for the data

t	0	1	2
y	−1	1	−1

(b) $y = a + bt + ct^3$ for the data

t	−2	−1	0	1	2
y	−5	−2	−1	−1	2

B4 Verify that the following systems $A\vec{x} = \vec{b}$ are inconsistent and then determine for each system the vector $\vec{x}$ that minimizes $\|A\vec{x} - \vec{b}\|$.

(a)
$$\begin{aligned} x_1 - x_2 &= 4 \\ 3x_1 + 2x_2 &= 5 \\ x_1 - 6x_2 &= 10 \end{aligned}$$

(b)
$$\begin{aligned} x_1 + x_2 &= 7 \\ x_1 - x_2 &= 4 \\ x_1 + 3x_2 &= 14 \end{aligned}$$

Computer Problems

C1 Find a, b, and c to obtain the best-fitting equation of the form $y = a + bt + ct^2$ for the following data. Make a graph showing the data and the curve.

t	0.0	1.1	1.9	3.0	4.1	5.2
y	4.0	3.6	4.1	5.6	7.9	11.8

Conceptual Problems

D1 Let $X = \begin{bmatrix} \vec{1} & \vec{t} & \vec{t}^2 \end{bmatrix}$, where $\vec{t} = \begin{bmatrix} t_1 \\ \vdots \\ t_n \end{bmatrix}$ and $\vec{t}^2 = \begin{bmatrix} t_1^2 \\ \vdots \\ t_n^2 \end{bmatrix}$.

Then show that

$$X^T X = \begin{bmatrix} n & \sum_{i=1}^n t_i & \sum_{i=1}^n t_i^2 \\ \sum_{i=1}^n t_i & \sum_{i=1}^n t_i^2 & \sum_{i=1}^n t_i^3 \\ \sum_{i=1}^n t_i^2 & \sum_{i=1}^n t_i^3 & \sum_{i=1}^n t_i^4 \end{bmatrix}$$

D2 Let $X = \begin{bmatrix} \vec{1} & \vec{t} & \vec{t}^2 & \cdots & \vec{t}^m \end{bmatrix}$, where $\vec{t} = \begin{bmatrix} t_1 \\ \vdots \\ t_n \end{bmatrix}$ and $\vec{t}^i = \begin{bmatrix} t_1^i \\ \vdots \\ t_n^i \end{bmatrix}$ for $1 \leq i \leq n$. Assume that at least $m + 1$ of the numbers $t_1, \dots, t_n$ are distinct.

(a) Prove that the columns of X are linearly independent by showing that the only solution to $c_0\vec{1} + c_1\vec{t} + \cdots + c_m\vec{t}^m = \vec{0}$ is $c_0 = \cdots = c_m = 0$. (Hint: Let $p(t) = c_0 + c_1 t + \cdots + c_m t^m$ and show that if $c_0\vec{1} + c_1\vec{t} + \cdots + c_m\vec{t}^m = \vec{0}$, $p(t)$ must be the zero polynomial.)

(b) Use the result from part (a) to prove that $X^T X$ is invertible. (Hint: Show the only solution to $X^T X\vec{v} = \vec{0}$ is $\vec{v} = \vec{0}$ by considering $\|X\vec{v}\|^2$.)

7.4 Inner Product Spaces

In Sections 1.3, 1.4, and 7.2, we saw that the dot product plays an essential role in the discussion of lengths, distances, and projections in $\mathbb{R}^n$. In Chapter 4, we saw that the ideas of vector spaces and linear mappings apply to more general sets, including some function spaces. If ideas such as projections are going to be used in these more general spaces, it will be necessary to have a generalization of the dot product to general vector spaces.

Inner Product Spaces

Consideration of the most essential properties of the dot product in Section 1.3 leads to the following definition.

Definition
Inner Product
Inner Product Space

Let $\mathbb{V}$ be a vector space over $\mathbb{R}$. An **inner product** on $\mathbb{V}$ is a function $\langle\,,\rangle : \mathbb{V} \times \mathbb{V} \to \mathbb{R}$ such that

(1) $\langle \mathbf{v}, \mathbf{v} \rangle \geq 0$ for all $\mathbf{v} \in \mathbb{V}$ and $\langle \mathbf{v}, \mathbf{v} \rangle = 0$ if and only if $\mathbf{v} = \mathbf{0}$. (positive definite)

(2) $\langle \mathbf{v}, \mathbf{w} \rangle = \langle \mathbf{w}, \mathbf{v} \rangle$ for all $\mathbf{v}, \mathbf{w} \in \mathbb{V}$. (symmetric)

(3) $\langle \mathbf{v}, s\mathbf{w} + t\mathbf{z} \rangle = s\langle \mathbf{v}, \mathbf{w} \rangle + t\langle \mathbf{v}, \mathbf{z} \rangle$ for all $s, t \in \mathbb{R}$ and $\mathbf{v}, \mathbf{w}, \mathbf{z} \in \mathbb{V}$. (bilinear)

A vector space $\mathbb{V}$ with an inner product is called an **inner product space**.

Remark

Every non-trivial finite-dimensional vector space $\mathbb{V}$ has in fact infinitely many different inner products. When we talk about an inner product space, we mean the vector space and one particular inner product.

EXAMPLE 1

The dot product on $\mathbb{R}^n$ is an inner product on $\mathbb{R}^n$.

EXAMPLE 2

Show that the function $\langle\,,\rangle$ defined by

$$\langle\vec{x},\vec{y}\rangle = 2x_1y_1 + 3x_2y_2$$

is an inner product on $\mathbb{R}^2$.

Solution: We verify that $\langle\,,\rangle$ satisfies the three properties of an inner product:

1. $\langle\vec{x},\vec{x}\rangle = 2x_1^2 + 3x_2^2 \geq 0$ and $\langle\vec{x},\vec{x}\rangle = 0$ if and only if $\vec{x} = \vec{0}$. Thus, it is positive definite.

2. $\langle\vec{x},\vec{y}\rangle = 2x_1y_1 + 3x_2y_2 = 2y_1x_1 + 3y_2x_2 = \langle\vec{y},\vec{x}\rangle$. Thus, it is symmetric.

3. For any $\vec{x},\vec{y},\vec{z} \in \mathbb{R}^2$ and $s,t \in \mathbb{R}$,

$$\begin{aligned}\langle\vec{x}, s\vec{w} + t\vec{z}\rangle &= 2x_1(sw_1 + tz_1) + 3x_2(sw_2 + tz_2)\\ &= s(2x_1w_1 + 3x_2w_2) + t(2x_1z_1 + 3x_2z_2)\\ &= s\langle\vec{x},\vec{w}\rangle + t\langle\vec{x},\vec{z}\rangle\end{aligned}$$

So, $\langle\,,\rangle$ is bilinear. Thus, $\langle\,,\rangle$ is an inner product on $\mathbb{R}^2$.

Remark

Although there are infinitely many inner products on $\mathbb{R}^n$, it can be proven that for any inner product on $\mathbb{R}^n$ there exists an orthonormal basis such that the inner product is just the dot product on $\mathbb{R}^n$ with respect to this basis. See Problem D1.

EXAMPLE 3

Verify that $\langle p,q\rangle = p(0)q(0) + p(1)q(1) + p(2)q(2)$ defines an inner product on the vector space P_2 and determine $\langle 1 + x + x^2, 2 - 3x^2\rangle$.

Solution: We first verify that $\langle\,,\rangle$ satisfies the three properties of an inner product:

(1) $\langle p,p\rangle = (p(0))^2 + (p(1))^2 + (p(2))^2 \geq 0$ for all $p \in P_2$. Moreover, $\langle p,p\rangle = 0$ if and only if $p(0) = p(1) = p(2) = 0$, and the only $p \in P_2$ that is zero for three values of x is the zero polynomial, $p(x) = 0$. Thus $\langle\,,\rangle$ is positive definite.

(2) $\langle p,q\rangle = p(0)q(0)+p(1)q(1)+p(2)q(2) = q(0)p(0)+q(1)p(1)+q(2)p(2) = \langle q,p\rangle$. So, $\langle\,,\rangle$ is symmetric.

(3) For any $p,q,r \in P_2$ and $s,t \in \mathbb{R}$,

$$\begin{aligned}\langle p, sq + tr\rangle &= p(0)(sq(0) + tr(0)) + p(1)(sq(1) + tr(1)) + p(2)(sq(2) + tr(2))\\ &= s(p(0)q(0) + p(1)q(1) + p(2)q(2)) + t(p(0)r(0) + p(1)r(1)\\ &\quad + p(2)r(2)) = s\langle p,q\rangle + t\langle p,r\rangle\end{aligned}$$

So, $\langle\,,\rangle$ is bilinear. Thus, $\langle\,,\rangle$ is an inner product on P_2. That is, P_2 is an inner product space under the inner product $\langle\,,\rangle$.

In this inner product space, we have

$$\begin{aligned}\langle 1 + x + x^2, 2 - 3x^2\rangle &= (1 + 0 + 0)(2 - 0) + (1 + 1 + 1)(2 - 3) + (1 + 2 + 4)(2 - 12)\\ &= 2 - 3 - 70 = -71\end{aligned}$$

EXAMPLE 4

Let $\operatorname{tr}(C)$ represent the trace of a matrix (the sum of the diagonal entries). Then, $M(2,2)$ is an inner product space under the inner product defined by $\langle A, B\rangle = \operatorname{tr}(B^T A)$. If $A = \begin{bmatrix} 1 & 2 \\ 3 & -1 \end{bmatrix}$ and $B = \begin{bmatrix} 4 & 5 \\ 0 & 6 \end{bmatrix}$, then under this inner product, we have

$$\left\langle \begin{bmatrix} 1 & 2 \\ 3 & -1 \end{bmatrix}, \begin{bmatrix} 4 & 5 \\ 0 & 6 \end{bmatrix} \right\rangle = \operatorname{tr}\left(\begin{bmatrix} 4 & 0 \\ 5 & 6 \end{bmatrix} \begin{bmatrix} 1 & 2 \\ 3 & -1 \end{bmatrix} \right)$$
$$= \operatorname{tr}\left(\begin{bmatrix} 4 & 8 \\ 23 & 4 \end{bmatrix} \right) = 4 + 4 = 8$$

EXERCISE 1

Verify that $\langle A, B\rangle = \operatorname{tr}(B^T A)$ is an inner product for $M(2,2)$. Do you notice a relationship between this inner product and the dot product on $\mathbb{R}^4$?

Since these properties of the inner product mimic the properties of the dot product, it makes sense to define the norm or length of a vector and the distance between vectors in terms of the inner product.

Definition

Norm

Distance

Let $\mathbb{V}$ be an inner product space. Then, for any $\mathbf{v} \in \mathbb{V}$, we define the **norm** (or **length**) of $\mathbf{v}$ to be

$$\|\mathbf{v}\| = \sqrt{\langle \mathbf{v}, \mathbf{v}\rangle}$$

For any vectors $\mathbf{v}, \mathbf{w} \in \mathbb{V}$, the **distance** between $\mathbf{v}$ and $\mathbf{w}$ is

$$\|\mathbf{v} - \mathbf{w}\|$$

Definition

Unit Vector

A vector $\mathbf{v}$ in an inner product space $\mathbb{V}$ is called a **unit vector** if $\|\mathbf{v}\| = 1$.

EXAMPLE 5

Find the norm of $A = \begin{bmatrix} 1 & 0 \\ 2 & 1 \end{bmatrix}$ in $M(2,2)$ under the inner product $\langle A, B\rangle = \operatorname{tr}(B^T A)$.

Solution: We have

$$\|A\| = \sqrt{\langle A, A\rangle} = \sqrt{\operatorname{tr}(A^T A)} = \sqrt{5+1} = \sqrt{6}$$

EXAMPLE 6

Find the norm of $p(x) = 1 - 2x - x^2$ in P_2 under the inner product $\langle p, q\rangle = p(0)q(0) + p(1)q(1) + p(2)q(2)$.

Solution: We have

$$\begin{aligned}\|1 - 2x - x^2\| &= \sqrt{\langle p, p\rangle}\\ &= \sqrt{(p(0))^2 + (p(1))^2 + (p(2))^2}\\ &= \sqrt{1^2 + (1-2-1)^2 + (1-4-4)^2}\\ &= \sqrt{54}\end{aligned}$$

EXERCISE 2

Find the norm of $p(x) = 1$ and $q(x) = x$ in P_2 under the inner product $\langle p, q\rangle = p(0)q(0) + p(1)q(1) + p(2)q(2)$.

EXERCISE 3

Find the norm of $q(x) = x$ in P_2 under the inner product $\langle p, q\rangle = p(-1)q(-1) + p(0)q(0) + p(1)q(1)$.

In Sections 7.1 and 7.2 we saw that the concept of orthogonality is very useful. Hence, we extend this concept to general inner product spaces.

Definition
Orthogonal
Orthonormal

Let $\mathbb{V}$ be an inner product space with inner product $\langle\ ,\ \rangle$. Then two vectors $\mathbf{v}, \mathbf{w} \in \mathbb{V}$ are said to be **orthogonal** if $\langle \mathbf{v}, \mathbf{w}\rangle = 0$. The set of vectors $\{\mathbf{v}_1, \dots, \mathbf{v}_k\}$ in $\mathbb{V}$ is said to be **orthogonal** if $\langle \mathbf{v}_i, \mathbf{v}_j\rangle = 0$ for all $i \neq j$. The set is said to be **orthonormal** if we also have $\langle \mathbf{v}_i, \mathbf{v}_i\rangle = 1$ for all i.

With this definition, we can now repeat our arguments from Sections 7.1 and 7.2 for coordinates with respect to an orthonormal basis and projections. In particular, we get that if $\mathcal{B} = \{\mathbf{v}_1, \dots, \mathbf{v}_k\}$ is an orthonormal basis for a subspace $\mathbb{S}$ of an inner product space $\mathbb{V}$ with inner product $\langle\ ,\ \rangle$, then for any $\mathbf{x} \in \mathbb{V}$ we have

$$\operatorname{proj}_{\mathbb{S}} \mathbf{x} = \langle \mathbf{x}, \mathbf{v}_1\rangle \mathbf{v}_1 + \cdots + \langle \mathbf{x}, \mathbf{v}_k\rangle \mathbf{v}_k$$

Additionally, the Gram-Schmidt Procedure is also identical. If we have a basis $\{\mathbf{w}_1, \dots, \mathbf{w}_n\}$ for an inner product space $\mathbb{V}$ with inner product $\langle , \rangle$, then the set $\{\mathbf{v}_1, \dots, \mathbf{v}_n\}$ defined by

$$\begin{aligned}\mathbf{v}_1 &= \mathbf{v}_1\\ \mathbf{v}_2 &= \mathbf{w}_2 - \frac{\langle \mathbf{w}_2, \mathbf{v}_1\rangle}{\|\mathbf{v}_1\|^2}\mathbf{v}_1\\ &\vdots\\ \mathbf{v}_n &= \mathbf{w}_n - \frac{\langle \mathbf{w}_n, \mathbf{v}_1\rangle}{\|\mathbf{v}_1\|^2}\mathbf{v}_1 - \cdots - \frac{\langle \mathbf{w}_n, \mathbf{v}_{n-1}\rangle}{\|\mathbf{v}_{n-1}\|^2}\mathbf{v}_{n-1}\end{aligned}$$

is an orthogonal basis for $\mathbb{V}$.

EXAMPLE 7

Use the Gram-Schmidt Procedure to determine an orthonormal basis for $\mathbb{S} = \text{Span}\{1, x\}$ of P_2 under the inner product $\langle p, q\rangle = p(0)q(0) + p(1)q(1) + p(2)q(2)$. Use this basis to determine $\text{proj}_{\mathbb{S}}\, x^2$.

Solution: Denote the basis vectors of $\mathbb{S}$ by $p_1(x) = 1$ and $p_2(x) = x$. We want to find an orthogonal basis $\{q_1(x), q_2(x)\}$ for $\mathbb{S}$. By using the Gram-Schmidt Procedure, we take $q_1(x) = p_1(x) = 1$ and then let

$$q_2 = p_2 - \frac{\langle p_2, q_1\rangle}{\|q_1\|^2}q_1 = x - \frac{0(1) + 1(1) + 2(1)}{1^2 + 1^2 + 1^2}1 = x - 1$$

Therefore, our orthogonal basis is $\{q_1, q_2\} = \{1, x - 1\}$.

Hence, we have

$$\begin{aligned}\text{proj}_{\mathbb{S}}\, x^2 &= \frac{\langle x^2, 1\rangle}{\|1\|^2}1 + \frac{\langle x^2, x-1\rangle}{\|x-1\|^2}(x-1)\\ &= \frac{0(1) + 1(1) + 4(1)}{1^2 + 1^2 + 1^2}1 + \frac{0(-1) + 1(0) + 4(1)}{(-1)^2 + 0^2 + 1^2}(x-1)\\ &= \frac{5}{3}1 + 2(x-1) = 2x - \frac{1}{3}\end{aligned}$$

PROBLEMS 7.4

Practice Problems

A1 On P_2, define the inner product $\langle p, q\rangle = p(0)q(0) + p(1)q(1) + p(2)q(2)$. Calculate the following.

(a) $\langle x - 2x^2, 1 + 3x\rangle$ (b) $\langle 2 - x + 3x^2, 4 - 3x^2\rangle$

(c) $\|3 - 2x + x^2\|$ (d) $\|9 + 9x + 9x^2\|$

A2 In each of the following cases, determine whether $\langle\, ,\rangle$ defines an inner product on P_2.

(a) $\langle p, q\rangle = p(0)q(0) + p(1)q(1)$

(b) $\langle p, q\rangle = |p(0)q(0)| + |p(1)q(1)| + |p(2)q(2)|$

(c) $\langle p, q\rangle = p(-1)q(-1) + 2p(0)q(0) + p(1)q(1)$

(d) $\langle p, q\rangle = p(-1)q(1) + 2p(0)q(0) + p(1)q(-1)$

A3 On $M(2, 2)$ define the inner product $\langle A, B\rangle = \text{tr}(B^T A)$.

(i) Use the Gram-Schmidt Procedure to determine an orthonormal basis for the following subspaces of $M(2, 2)$.

(ii) Use the orthonormal basis you found in part (i) to determine $\text{proj}_{\mathbb{S}} \begin{bmatrix} 4 & 3 \\ -2 & 1 \end{bmatrix}$.

(a) $\mathbb{S} = \text{Span}\left\{\begin{bmatrix} 1 & 0 \\ -1 & 1 \end{bmatrix}, \begin{bmatrix} 1 & 1 \\ 0 & 1 \end{bmatrix}, \begin{bmatrix} 2 & 0 \\ 1 & -1 \end{bmatrix}\right\}$

(b) $\mathbb{S} = \text{Span}\left\{\begin{bmatrix} 1 & 1 \\ 0 & 1 \end{bmatrix}, \begin{bmatrix} 0 & -1 \\ 1 & 1 \end{bmatrix}, \begin{bmatrix} 2 & 0 \\ -1 & 1 \end{bmatrix}\right\}$

A4 Define the inner product $\langle \vec{x}, \vec{y}\rangle = 2x_1y_1 + x_2y_2 + 3x_3y_3$ on $\mathbb{R}^3$.

(a) Use the Gram-Schmidt Procedure to determine an orthogonal basis for

$$\mathbb{S} = \text{Span}\left\{\begin{bmatrix} 1 \\ 1 \\ 0 \end{bmatrix}, \begin{bmatrix} -1 \\ 1 \\ 0 \end{bmatrix}, \begin{bmatrix} -1 \\ 1 \\ -2 \end{bmatrix}\right\}$$

(b) Determine the coordinates of $\vec{x} = \begin{bmatrix}1\\0\\0\end{bmatrix}$ with respect to the orthogonal basis you found in (a).

A5 Let $\{\mathbf{v}_1, \dots, \mathbf{v}_k\}$ be an orthogonal set in an inner product space $\mathbb{V}$. Prove that

$$\|\mathbf{v}_1 + \cdots + \mathbf{v}_k\|^2 = \|\mathbf{v}_1\|^2 + \cdots + \|\mathbf{v}_k\|^2$$

Homework Problems

B1 On P_2, define the inner product $\langle p, q\rangle = p(0)q(0) + p(1)q(1) + p(2)q(2)$. Calculate the following.

(a) $\langle 1-3x^2, 1+x+2x^2\rangle$
(b) $\langle 3-x, -2-x-x^2\rangle$
(c) $\|1-5x+2x^2\|$
(d) $\|73x+73x^2\|$

B2 In each of the following cases, determine whether $\langle\ ,\ \rangle$ defines an inner product on P_3.

(a) $\langle p, q\rangle = p(-1)q(-1) + p(0)q(0) + p(1)q(1)$
(b) $\langle p, q\rangle = p(0)q(0) + p(1)q(1) + p(3)q(3) + p(4)q(4)$
(c) $\langle p, q\rangle = p(-1)q(0) + p(1)q(1) + p(2)q(2) + p(3)q(3)$

B3 On P_2 define the inner product $\langle p, q\rangle = p(-1)q(-1) + p(0)q(0) + p(1)q(1)$.

(i) Use the Gram-Schmidt Procedure to determine an orthogonal basis for the following subspaces of P_2.

(ii) Use the orthogonal basis you found in part (i) to determine $\text{proj}_{\mathbb{S}}(1 + x + x^2)$.

(a) $\mathbb{S} = \text{Span}\left\{1, x - x^2\right\}$
(b) $\mathbb{S} = \text{Span}\left\{1 + x^2, x - x^2\right\}$

B4 Define the inner product $\langle\vec{x}, \vec{y}\rangle = x_1y_1 + 3x_2y_2 + 2x_3y_3$ on $\mathbb{R}^3$.

(a) Use the Gram-Schmidt Procedure to determine an orthogonal basis for

$$\mathbb{S} = \text{Span}\left\{\begin{bmatrix}1\\1\\0\end{bmatrix}, \begin{bmatrix}-1\\1\\0\end{bmatrix}, \begin{bmatrix}-1\\1\\-2\end{bmatrix}\right\}$$

(b) Determine the coordinates of $\vec{x} = \begin{bmatrix}1\\0\\0\end{bmatrix}$ with respect to the orthogonal basis you found in (a).

Conceptual Problems

D1 (a) Let $\{\vec{e}_1, \vec{e}_2\}$ be the standard basis for $\mathbb{R}^2$ and suppose that $\langle\ ,\ \rangle$ is an inner product on $\mathbb{R}^2$. Show that if $\vec{x}, \vec{y} \in \mathbb{R}^2$,

$$\begin{aligned}\langle\vec{x}, \vec{y}\rangle = {} & x_1y_1\langle\vec{e}_1, \vec{e}_1\rangle + x_1y_2\langle\vec{e}_1, \vec{e}_2\rangle \\ & + x_2y_1\langle\vec{e}_2, \vec{e}_1\rangle + x_2y_2\langle\vec{e}_2, \vec{e}_2\rangle\end{aligned}$$

(b) For the inner product in part (a), define a matrix G, called the **standard matrix of the inner product** $\langle\ ,\ \rangle$, by $g_{ij} = \langle\vec{e}_i, \vec{e}_j\rangle$ for $i, j = 1, 2$. Show that G is symmetric and that

$$\langle\vec{x}, \vec{y}\rangle = \sum_{i,j=1}^{2} g_{ij}x_iy_j = \vec{x}^T G\vec{y}$$

(c) Apply the Gram-Schmidt Procedure, using the inner product $\langle\ ,\ \rangle$ and the corresponding norm, to produce an orthonormal basis $\mathcal{B} = \{\vec{v}_1, \vec{v}_2\}$ for $\mathbb{R}^2$.

(d) Define $\tilde{G}$, the $\mathcal{B}$-matrix of the inner product $\langle\ ,\ \rangle$, by $\tilde{g}_{ij} = \langle\vec{v}_i, \vec{v}_j\rangle$ for $i, j = 1, 2$. Show that $\tilde{G} = I$ and that for $\vec{x} = \tilde{x}_1\vec{v}_1 + \tilde{x}_2\vec{v}_2$ and $\vec{y} = \tilde{y}_1\vec{v}_1 + \tilde{y}_2\vec{v}_2$,

$$\langle\vec{x}, \vec{y}\rangle = \tilde{x}_1\tilde{y}_1 + \tilde{x}_2\tilde{y}_2$$

Conclusion. For an arbitrary inner product $\langle\ ,\ \rangle$ on $\mathbb{R}^2$, there exists a basis for $\mathbb{R}^2$ that is orthonormal with respect to this inner product. Moreover, when $\vec{x}$ and $\vec{y}$ are expressed in terms of this basis, $\langle\vec{x}, \vec{y}\rangle$ looks just like the standard inner product in $\mathbb{R}^2$.

This argument generalizes in a straightforward way to $\mathbb{R}^n$; see Problem 8.2.D6.

D2 Suppose that $\{\mathbf{v}_1, \mathbf{v}_2, \mathbf{v}_3\}$ is a basis for an inner product space $\mathbb{V}$ with inner product $\langle\ ,\ \rangle$. Define a matrix G by

$$g_{ij} = \langle \mathbf{v}_i, \mathbf{v}_j \rangle, \qquad \text{for } i, j = 1, 2, 3$$

(a) Prove that G is symmetric ($G^T = G$).

(b) Show that if $[\vec{x}]_B = \begin{bmatrix} x_1 \\ x_2 \\ x_3 \end{bmatrix}$ and $[\vec{y}]_B = \begin{bmatrix} y_1 \\ y_2 \\ y_3 \end{bmatrix}$, then

$$\langle \vec{x}, \vec{y} \rangle = \begin{bmatrix} x_1 & x_2 & x_3 \end{bmatrix} G \begin{bmatrix} y_1 \\ y_2 \\ y_3 \end{bmatrix}$$

(c) Determine the matrix G of the inner product $\langle p, q \rangle = p(0)q(0) + p(1)q(1) + p(2)q(2)$ for P_2 with respect to the basis $\{1, x, x^2\}$.

7.5 Fourier Series

The Inner Product $\int_a^b f(x)g(x)\,dx$

Let $C[a, b]$ be the space of functions $f : \mathbb{R} \to \mathbb{R}$ that are continuous on the interval $[a, b]$. Then, for any $f, g \in C[a, b]$ we have that the product fg is also continuous on $[a, b]$ and hence integrable on $[a, b]$. Therefore, it makes sense to define an inner product as follows.

The inner product $\langle\ ,\ \rangle$ is defined on $C[a, b]$ by

$$\langle f, g \rangle = \int_a^b f(x)g(x)\,dx$$

The three properties of an inner product are satisfied because

(1) $\langle f, f \rangle = \int_a^b f(x)f(x)\,dx \geq 0$ for all $f \in C[a, b]$ and $\langle f, f \rangle = \int_a^b f(x)f(x)\,dx = 0$ if and only if $f(x) = 0$ for all $x \in [a, b]$.

(2) $\langle f, g \rangle = \int_a^b f(x)g(x)\,dx = \int_a^b g(x)f(x)\,dx = \langle g, f \rangle$

(3) $\langle f, sg + th \rangle = \int_a^b f(x)(sg(x) + th(x))\,dx = s\int_a^b f(x)g(x)\,dx + t\int_a^b f(x)h(x)\,dx = s\langle f, g \rangle + t\langle f, h \rangle$ for any $s, t \in \mathbb{R}$.

Since an integral is the limit of sums, this inner product defined as the *integral of the product of the values of* f *and* g *at each* x is a fairly natural generalization of the dot product in $\mathbb{R}^n$ defined as a *sum of the product of the* i*-th components of* $\vec{x}$ *and* $\vec{y}$ *for each* i.

One interesting consequence is that the norm of a function f with respect to this inner product is

$$\|f\| = \left(\int_a^b f^2(x)\,dx\right)^{1/2}$$

Intuitively, this is quite satisfactory as a measure of how far the function is from the zero function.

One of the most interesting and important applications of this inner product involves Fourier series.

Fourier Series

Let $CP_{2\pi}$ denote the space of continuous real-valued functions of a real variable that are periodic with period 2π. Such functions satisfy $f(x+2\pi) = f(x)$ for all x. Examples of such functions are $f(x) = c$ for any constant c, $\cos x$, $\sin x$, $\cos 2x$, $\sin 3x$, etc. (Note that the function $\cos 2x$ is periodic with period 2π because $\cos(2(x + 2\pi)) = \cos 2x$. However, its "fundamental (smallest) period" is π.) In some electrical engineering applications, it is of interest to consider a signal described by functions such as the function

$$f(x) = \begin{cases} -\pi - x & \text{if } -\pi \leq x \leq -\pi/2 \\ x & \text{if } -\pi/2 < x \leq \pi/2 \\ \pi - x & \text{if } \pi/2 < x \leq \pi \end{cases}$$

This function is shown in Figure 7.5.5.

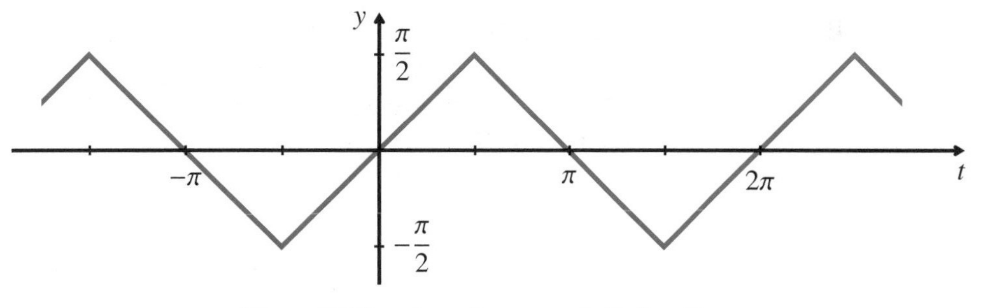

Figure 7.5.5 A continuous periodic function.

In the early nineteenth century, while studying the problem of the conduction of heat, Fourier had the brilliant idea of trying to represent an arbitrary function in $CP_{2\pi}$ as a linear combination of the set of functions

$$\{1, \cos x, \sin x, \cos 2x, \sin 2x, \ldots, \cos nx, \sin nx, \ldots\}$$

This idea developed into Fourier analysis, which is now one of the essential tools in quantum physics, communication engineering, and many other areas.

We formulate the questions and ideas as follows. (The proofs of the statements are discussed below.)

(i) For any n, the set of functions $\{1, \cos x, \sin x, \cos 2x, \sin 2x, \ldots, \cos nx, \sin nx\}$ is an orthogonal set with respect to the inner product

$$\langle f, g\rangle = \int_{-\pi}^{\pi} f(x)g(x)\,dx$$

The set is therefore an orthogonal basis for the subspace of $CP_{2\pi}$ that it spans. This subspace will be denoted $CP_{2\pi,n}$.

(ii) Given an arbitrary function f in $CP_{2\pi}$, how well can it be approximated by a function in $CP_{2\pi,n}$? We expect from our experience with distance and subspaces that the closest approximation to f in $CP_{2\pi,n}$ is $\text{proj}_{CP_{2\pi,n}} f$. The coefficients for Fourier's representation of f by a linear combination of $\{1, \cos x, \sin x, \dots, \cos nx, \sin nx, \dots\}$, called **Fourier coefficients**, are found by considering this projection.

(iii) We hope that the approximation improves as n gets larger. Since the distance from f to the n-th approximation $\text{proj}_{CP_{2\pi,n}} f$ is $\| \text{perp}_{CP_{2\pi,n}} f\|$, to test if the approximation improves, we must examine whether $\| \text{perp}_{CP_{2\pi,n}} f\| \to 0$ as $n \to \infty$.

Let us consider these statements in more detail.

(i) **The orthogonality of constants, sines, and cosines with respect to the inner product** $\langle f, g\rangle = \int_{-\pi}^{\pi} f(x)g(x)\, dx$

These results follow by standard trigonometric integrals and trigonometric identities:

$$\int_{-\pi}^{\pi} \sin nx\, dx = -\frac{1}{n}\cos nx\Big|_{-\pi}^{\pi} = 0$$

$$\int_{-\pi}^{\pi} \cos nx\, dx = \frac{1}{n}\sin nx\Big|_{-\pi}^{\pi} = 0$$

$$\int_{-\pi}^{\pi} \cos mx \sin nx\, dx = \int_{-\pi}^{\pi} \frac{1}{2}(\sin(m+n)x - \sin(m-n)x)\, dx = 0$$

and for $m \neq n$,

$$\int_{-\pi}^{\pi} \cos mx \cos nx\, dx = \int_{-\pi}^{\pi} \frac{1}{2}(\cos(m+n)x + \cos(m-n)x)\, dx = 0$$

$$\int_{-\pi}^{\pi} \sin mx \sin nx\, dx = \int_{-\pi}^{\pi} \frac{1}{2}(\cos(m-n)x - \cos(m+n)x)\, dx = 0$$

Hence, the set $\{1, \cos x, \sin x, \dots, \cos nx, \sin nx\}$ is orthogonal. To use this as a basis for projection arguments, it is necessary to calculate $\|1\|^2$, $\|\cos mx\|^2$, and $\|\sin mx\|^2$:

$$\|1\|^2 = \int_{-\pi}^{\pi} 1\, dx = 2\pi$$

$$\|\cos mx\|^2 = \int_{-\pi}^{\pi} \cos^2 mx\, dx = \int_{-\pi}^{\pi} \frac{1}{2}(1 + \cos 2mx)\, dx = \pi$$

$$\|\sin mx\|^2 = \int_{-\pi}^{\pi} \sin^2 mx\, dx = \int_{-\pi}^{\pi} \frac{1}{2}(1 - \cos 2mx)\, dx = \pi$$

(ii) **The Fourier coefficients of f as coordinates of a projection with respect to the orthogonal basis for $CP_{2\pi,n}$**

The procedure for finding the closest approximation $\text{proj}_{CP_{2\pi,n}} f$ in $CP_{2\pi,n}$ to an arbitrary function f in $CP_{2\pi}$ is parallel to the procedure in Sections 7.2 and 7.4. That is, we use the projection formula, given an orthogonal basis $\{\vec{v}_1, \dots, \vec{v}_k\}$ for a subspace $\mathbb{S}$:

$$\text{proj}_{\mathbb{S}}\, \vec{x} = \frac{\langle \vec{x}, \vec{v}_1\rangle}{\|\vec{v}_1\|^2}\vec{v}_1 + \cdots + \frac{\langle \vec{x}, \vec{v}_k\rangle}{\|\vec{v}_k\|^2}\vec{v}_k$$

There is a standard way to label the coefficients of this linear combination:

$$\begin{aligned}\operatorname{proj}_{CP_{2\pi,n}} f = &\frac{a_0}{2}1 + a_1 \cos x + a_2 \cos 2x + \cdots + a_n \cos nx \\ &+ b_1 \sin x + b_2 \sin 2x + \cdots + b_n \sin nx\end{aligned}$$

The factor $\frac{1}{2}$ in the coefficient of 1 appears here because $\|1\|^2$ is equal to 2π, while the other basis vectors have length squared equal to π. Thus, we have

$$\begin{aligned}a_0 &= \frac{\langle f, 1\rangle}{\|1\|^2} = \frac{1}{\pi}\int_{-\pi}^{\pi} f(x)\,dx \\ a_m &= \frac{\langle f, \cos mx\rangle}{\|\cos mx\|^2} = \frac{1}{\pi}\int_{-\pi}^{\pi} f(x)\cos mx\,dx \\ b_m &= \frac{\langle f, \sin mx\rangle}{\|\sin mx\|^2} = \frac{1}{\pi}\int_{-\pi}^{\pi} f(x)\sin mx\,dx\end{aligned}$$

(iii) **Is** $\operatorname{proj}_{CP_{2\pi,n}} f$ **equal to** f **in the limit as** $n \to \infty$**?**

As $n \to \infty$, the sum becomes an infinite series called the **Fourier series** for f. The question being asked is a question about the convergence of series—and in fact, about series of functions. Such questions are raised in calculus (or analysis) and are beyond the scope of this book. (The short answer is "yes, the series converges to f provided that f is continuous." The problem becomes more complicated if f is allowed to be piecewise continuous.) Questions about convergence are important in physical and engineering applications.

EXAMPLE 1

Determine $\operatorname{proj}_{CP_{2\pi,3}} f$ for the function $f(x)$ defined by $f(x) = |x|$ if $-\pi \le x \le \pi$ and $f(x + 2\pi) = f(x)$ for all x.

Solution: We have

$$\begin{aligned}a_0 &= \frac{1}{\pi}\int_{-\pi}^{\pi} |x|\,dx = \pi \\ a_1 &= \frac{1}{\pi}\int_{-\pi}^{\pi} |x|\cos x\,dx = -\frac{4}{\pi} \\ a_2 &= \frac{1}{\pi}\int_{-\pi}^{\pi} |x|\cos 2x\,dx = 0 \\ a_3 &= \frac{1}{\pi}\int_{-\pi}^{\pi} |x|\cos 3x\,dx = -\frac{4}{9\pi} \\ b_1 &= \frac{1}{\pi}\int_{-\pi}^{\pi} |x|\sin x\,dx = 0 \\ b_2 &= \frac{1}{\pi}\int_{-\pi}^{\pi} |x|\sin 2x\,dx = 0 \\ b_3 &= \frac{1}{\pi}\int_{-\pi}^{\pi} |x|\sin 3x\,dx = 0\end{aligned}$$

Hence, $\operatorname{proj}_{CP_{2\pi,3}} f = \frac{\pi}{2} - \frac{4}{\pi}\cos x - \frac{4}{9\pi}\cos 3x$. The results are shown in Figure 7.5.6.

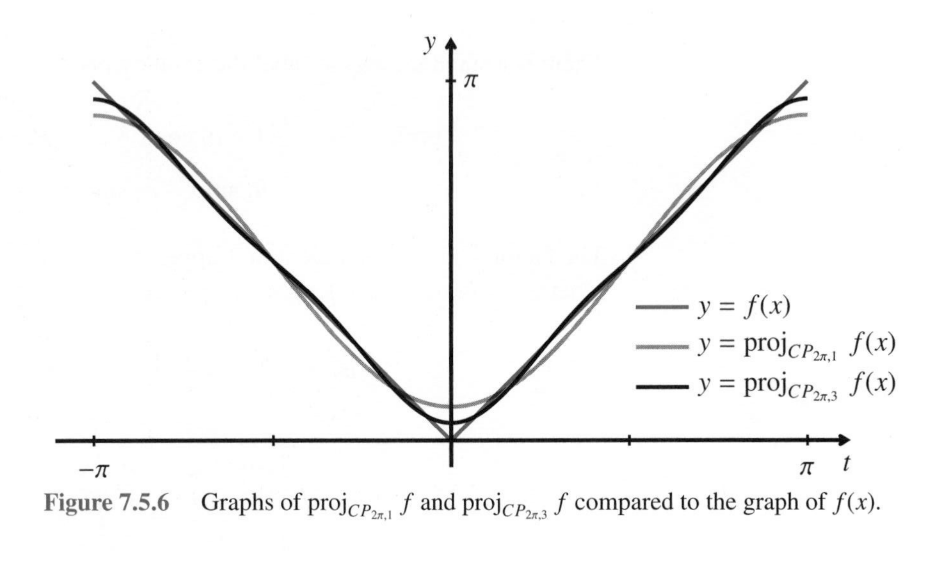

Figure 7.5.6 Graphs of $\text{proj}_{CP_{2\pi,1}} f$ and $\text{proj}_{CP_{2\pi,3}} f$ compared to the graph of $f(x)$.

EXAMPLE 2

Determine $\text{proj}_{CP_{2\pi,3}} f$ for the function $f(x)$ defined by

$$f(x) = \begin{cases} -\pi - x & \text{if } -\pi \leq x \leq -\pi/2 \\ x & \text{if } -\pi/2 < x \leq \pi/2 \\ \pi - x & \text{if } \pi/2 < x \leq \pi \end{cases}$$

Solution: We have

$$a_0 = \frac{1}{\pi}\int_{-\pi}^{\pi} f\, dx = 0$$
$$a_1 = \frac{1}{\pi}\int_{-\pi}^{\pi} f \cos x\, dx = 0$$
$$a_2 = \frac{1}{\pi}\int_{-\pi}^{\pi} f \cos 2x\, dx = 0$$
$$a_3 = \frac{1}{\pi}\int_{-\pi}^{\pi} f \cos 3x\, dx = 0$$
$$b_1 = \frac{1}{\pi}\int_{-\pi}^{\pi} f \sin x\, dx = \frac{4}{\pi}$$
$$b_2 = \frac{1}{\pi}\int_{-\pi}^{\pi} f \sin 2x\, dx = 0$$
$$b_3 = \frac{1}{\pi}\int_{-\pi}^{\pi} f \sin 3x\, dx = -\frac{4}{9\pi}$$

Hence, $\text{proj}_{CP_{2\pi,3}} f = \frac{4}{\pi}\sin x - \frac{4}{9\pi}\sin 3x$. The results are shown in Figure 7.5.7.

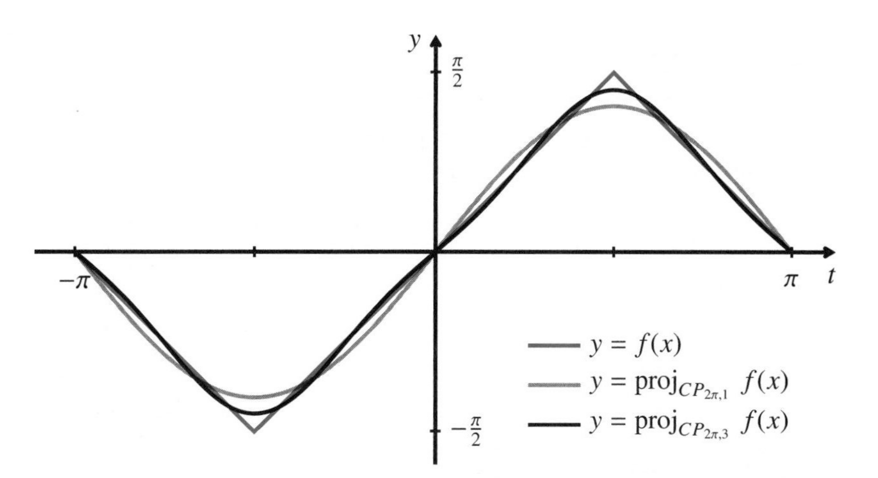

Figure 7.5.7 Graphs of $\text{proj}_{CP_{2\pi,1}} f$ and $\text{proj}_{CP_{2\pi,3}} f$ compared to the graph of $f(x)$.

PROBLEMS 7.5
Computer Problems

C1 Use a computer to calculate $\text{proj}_{CP_{2\pi,n}} f$ for n =3,7, and 11 for each of the following functions. Graph the function f and each of the projections on the same plot.

(a) $f(x) = x^2, -\pi \leq x \leq \pi$

(b) $f(x) = e^x, -\pi \leq x \leq \pi$

(c) $f(x) = \begin{cases} 0 & \text{if } -\pi \leq x \leq 0 \\ 1 & \text{if } 0 < x \leq \pi \end{cases}$

PROBLEMS 7.5
Suggestions for Student Review

1 What is meant by *an orthogonal set of vectors* in $\mathbb{R}^n$? What is the difference between an orthogonal basis and an orthonormal basis? (Section 7.1)

2 Why is it easier to determine coordinates with respect to an orthonormal basis than with respect to an arbitrary basis? What are some special features of the change of coordinates matrix from an orthonormal basis to the standard basis? What is an orthogonal matrix? (Section 7.1)

3 Does every subspace of $\mathbb{R}^n$ have an orthonormal basis? What about the zero subspace? How do you find an orthonormal basis? Describe the Gram-Schmidt Procedure. (Section 7.2)

4 What are the essential properties of a projection onto a subspace of $\mathbb{R}^n$? How do you calculate a projection onto a subspace? (Section 7.2)

5 Outline how to use the ideas of orthogonality to find the best-fitting *line* for a given set of data points $\{(t_i, y_i) \mid i = 1, \dots, n\}$. (Section 7.3)

6 What are the essential properties of an inner product? Give an example of an inner product on P_2. Give an example of an inner product on $M(2, 3)$. (Section 7.4)

Chapter Quiz

E1 Determine whether the following sets are orthogonal, and which are orthonormal. Show how you decide.

(a) $\left\{\frac{1}{3}\begin{bmatrix}1\\0\\1\\1\end{bmatrix}, \frac{1}{3}\begin{bmatrix}0\\1\\1\\-1\end{bmatrix}, \frac{1}{2}\begin{bmatrix}0\\1\\-1\\0\end{bmatrix}\right\}$

(b) $\left\{\frac{1}{\sqrt{3}}\begin{bmatrix}1\\0\\1\\1\end{bmatrix}, \frac{1}{\sqrt{5}}\begin{bmatrix}0\\0\\1\\-2\end{bmatrix}\right\}$

(c) $\left\{\frac{1}{\sqrt{3}}\begin{bmatrix}1\\0\\1\\1\end{bmatrix}, \frac{1}{\sqrt{3}}\begin{bmatrix}1\\1\\-1\\0\end{bmatrix}, \frac{1}{\sqrt{3}}\begin{bmatrix}0\\-1\\-1\\1\end{bmatrix}\right\}$

E2 Consider the orthonormal set

$$\mathcal{B} = \left\{\frac{1}{2}\begin{bmatrix}1\\0\\1\\1\\1\end{bmatrix}, \frac{1}{\sqrt{3}}\begin{bmatrix}1\\1\\0\\0\\-1\end{bmatrix}, \frac{1}{\sqrt{3}}\begin{bmatrix}-1\\1\\0\\1\\0\end{bmatrix}\right\}.$$

Let $\mathbb{S}$ be the subspace of $\mathbb{R}^n$ spanned by $\mathcal{B}$. Given that $\vec{x} = \begin{bmatrix}2\\5\\1\\3\\-2\end{bmatrix}$ is a vector in $\mathbb{S}$, use the orthonormality of $\mathcal{B}$ to determine the coordinates of $\vec{x}$ with respect to $\mathcal{B}$.

E3 (a) Prove that if P is an orthogonal matrix, $\det P = \pm 1$.

(b) Prove that if P and R are $n \times n$ orthogonal matrices, then so is PR.

E4 Let $\mathbb{S}$ be the subspace of $\mathbb{R}^4$ defined by $\mathbb{S} = \operatorname{Span}\left\{\begin{bmatrix}1\\0\\1\\0\end{bmatrix}, \begin{bmatrix}1\\1\\1\\1\end{bmatrix}, \begin{bmatrix}1\\3\\3\\1\end{bmatrix}\right\}$.

(a) Apply the Gram-Schmidt Procedure to the given spanning set to produce an orthonormal basis for $\mathbb{S}$.

(b) Determine the point in $\mathbb{S}$ closest to $\vec{x} = \begin{bmatrix}1\\-2\\-1\\0\end{bmatrix}$.

E5 Determine whether each of the following functions $\langle\ ,\ \rangle$ defines an inner product on $M(2,2)$. Explain how you decide in each case.

(a) $\langle A, B\rangle = \det(AB)$

(b) $\langle A, B\rangle = a_{11}b_{11} + 2a_{12}b_{12} + 2a_{21}b_{21} + a_{22}b_{22}$

Further Problems

F1 **(Isometries of $\mathbb{R}^3$)**

(a) A linear mapping is an **isometry** of $\mathbb{R}^3$ if $\|L(\vec{x})\| = \|\vec{x}\|$ for every $\vec{x} \in \mathbb{R}^3$. Prove that an isometry preserves dot products and angles as well as lengths.

(b) Show that L is an isometry if and only if the standard matrix of L is orthogonal. (Hint: See Problem 3.F5 and Problem 7.1.D3.)

(c) Explain why an isometry of $\mathbb{R}^3$ must have one or three real characteristic roots, counting multiplicity. Based on Problem 7.1.D3 (b), these must be ± 1.

(d) Let A be the standard matrix of L. Suppose that 1 is an eigenvalue of A with eigenvector $\vec{u}$. Let $\vec{v}$ and $\vec{w}$ be vectors such that $\{\vec{u}, \vec{v}, \vec{w}\}$ is an orthonormal basis for $\mathbb{R}^3$ and let $P = \begin{bmatrix}\vec{u} & \vec{v} & \vec{w}\end{bmatrix}$. Show that

$$P^T AP = \begin{bmatrix}1 & 0_{12}\\ 0_{21} & A^*\end{bmatrix}$$

where the right-hand side is a partitioned matrix, with 0_{ij} being the $i \times j$ zero matrix, and with A^* being a 2×2 orthogonal matrix. Moreover, show that the characteristic roots of A are 1 and the characteristic roots of A^*.

Note that an analogous form can be obtained for P^TAP *in the case where one eigenvalue is* -1.

(e) Use Problem 3.F5 to analyze the A^* of part (d) and explain why every isometry of $\mathbb{R}^3$ is the identity mapping, a reflection, a composition of reflections, a rotation, or a composition of a reflection and a rotation.

F2 A linear mapping $L : \mathbb{R}^n \to \mathbb{R}^n$ is called an **involution** if $L \circ L = \text{Id}$. In terms of its standard matrix, this means that $A^2 = I$. Prove that any two of the following imply the third.

(a) A is the matrix of an involution.
(b) A is symmetric.
(c) A is an isometry.

F3 The sum $\mathbb{S} + \mathbb{T}$ of subspaces of a finite dimensional vector space $\mathbb{V}$ is defined in the Chapter 4 Further Problems. Prove that $(\mathbb{S} + \mathbb{T})^\perp = \mathbb{S}^\perp \cap \mathbb{T}^\perp$.

F4 A problem of finding a sequence of approximations to some vector (or function) $\mathbf{v}$ in a possibly infinite-dimensional inner product space $\mathbb{V}$ can often be described by requiring the i-th approximation to be the closest vector $\mathbf{v}$ in some finite-dimensional subspace $\mathbb{S}_i$ of $\mathbb{V}$, where the subspaces are required to satisfy

$$\mathbb{S}_1 \subset \mathbb{S}_2 \subset \cdots \subset \mathbb{S}_i \subset \cdots \subset \mathbb{V}$$

The i-th approximation is then $\text{proj}_{\mathbb{S}_i} \mathbf{v}$. Prove that the approximations improve as i increases in the sense that

$$\|\mathbf{v} - \text{proj}_{\mathbb{S}_{i+1}} \mathbf{v}\| \leq \|\mathbf{v} - \text{proj}_{\mathbb{S}_i} \mathbf{v}\|$$

F5 ***QR*-factorization.** Suppose that A is an invertible $n \times n$ matrix. Prove that A can be written as the product of an orthogonal matrix Q and a upper-triangular matrix $R : A = QR$.

(Hint: Apply the Gram-Schmidt Procedure to the columns of A, starting at the first column.)

Note that this QR*-factorization is important in a numerical procedure for determining eigenvalues of symmetric matrices.*

Visit the text's website at www.pearsoncanada.ca/norman for practice quizzes, additional applications, and an essay on linearity and superposition in physics.

CHAPTER 8

Symmetric Matrices and Quadratic Forms

CHAPTER OUTLINE

Symmetric matrices and quadratic forms arise naturally in many physical applications. For example, the strain matrix describing the deformation of a solid and the inertia tensor of a rotating body are symmetric (Section 8.4). We have also seen that the matrix of a projection is symmetric since a real inner product is symmetric. We now use our work with diagonalization and inner products to explore the theory of symmetric matrices and quadratic forms.

8.1 Diagonalization of Symmetric Matrices

Definition
Symmetric Matrix

A matrix A is symmetric if $A^T = A$ or, equivalently, if $a_{ij} = a_{ji}$ for all i and j.

In Chapter 6, we saw that diagonalization of a square matrix may not be possible if some of the roots of its characteristic polynomial are complex or if the geometric multiplicity of an eigenvalue is less than the algebraic multiplicity of that eigenvalue. As we will see later in this section, a symmetric matrix can always be diagonalized: all the roots of its characteristic polynomial are real, and we can always find a basis of eigenvectors. Before considering why this works, we give three examples.

EXAMPLE 1

Determine the eigenvalues and corresponding eigenvectors of the symmetric matrix $A = \begin{bmatrix} 0 & 1 \\ 1 & -2 \end{bmatrix}$. What is the diagonal matrix corresponding to A, and what is the matrix that diagonalizes A?

Solution: We have

$$C(\lambda) = \det(A - \lambda I) = \begin{vmatrix} 0-\lambda & 1 \\ 1 & -2-\lambda \end{vmatrix} = \lambda^2 + 2\lambda - 1$$

Using the quadratic formula, we find that the roots of the characteristic polynomial are $\lambda_1 = -1 + \sqrt{2}$ and $\lambda_2 = -1 - \sqrt{2}$. Thus, the resulting diagonal matrix is

$$D = \begin{bmatrix} -1+\sqrt{2} & 0 \\ 0 & -1-\sqrt{2} \end{bmatrix}$$

EXAMPLE 1
(continued)

For $\lambda_1 = -1 + \sqrt{2}$, we have

$$A - \lambda_1 I = \begin{bmatrix} 1-\sqrt{2} & 1 \\ 1 & -1-\sqrt{2} \end{bmatrix} \sim \begin{bmatrix} 1 & -1-\sqrt{2} \\ 0 & 0 \end{bmatrix}$$

Thus, a basis for the eigenspace is $\left\{ \begin{bmatrix} 1+\sqrt{2} \\ 1 \end{bmatrix} \right\}$.

Similarly, for $\lambda_2 = -1 - \sqrt{2}$, we have

$$A - \lambda_2 I = \begin{bmatrix} 1+\sqrt{2} & 1 \\ 1 & -1+\sqrt{2} \end{bmatrix} \sim \begin{bmatrix} 1 & -1+\sqrt{2} \\ 0 & 0 \end{bmatrix}$$

Thus, a basis for the eigenspace is $\left\{ \begin{bmatrix} 1-\sqrt{2} \\ 1 \end{bmatrix} \right\}$.

Hence, A is diagonalized by $P = \begin{bmatrix} 1+\sqrt{2} & 1-\sqrt{2} \\ 1 & 1 \end{bmatrix}$.

Observe in Example 1 that the columns of P are orthogonal. That is,

$$\begin{bmatrix} 1+\sqrt{2} \\ 1 \end{bmatrix} \cdot \begin{bmatrix} 1-\sqrt{2} \\ 1 \end{bmatrix} = (1+\sqrt{2})(1-\sqrt{2}) + 1(1) = 1 - 2 + 1 = 0$$

Hence, if we normalized the columns of P, we would find that A is diagonalized by an orthogonal matrix. (It is important to remember that an orthogonal matrix has *orthonormal* columns.)

EXAMPLE 2

Diagonalize the symmetric $A = \begin{bmatrix} 4 & 0 & 0 \\ 0 & 1 & -2 \\ 0 & -2 & 1 \end{bmatrix}$. Show that A can be diagonalized by an orthogonal matrix.

Solution: We have

$$C(\lambda) = \begin{vmatrix} 4-\lambda & 0 & 0 \\ 0 & 1-\lambda & -2 \\ 0 & -2 & 1-\lambda \end{vmatrix} = -(\lambda-4)(\lambda-3)(\lambda+1)$$

The eigenvalues are $\lambda_1 = 4$, $\lambda_2 = 3$, and $\lambda_3 = -1$, each with algebraic multiplicity 1.

For $\lambda_1 = 4$, we get

$$A - \lambda_1 I = \begin{bmatrix} 0 & 0 & 0 \\ 0 & -3 & -2 \\ 0 & -2 & -3 \end{bmatrix} \sim \begin{bmatrix} 0 & 1 & 0 \\ 0 & 0 & 1 \\ 0 & 0 & 0 \end{bmatrix}$$

Thus, a basis for the eigenspace is $\left\{ \begin{bmatrix} 1 \\ 0 \\ 0 \end{bmatrix} \right\}$.

For $\lambda_2 = 3$, we get

$$A - \lambda_2 I = \begin{bmatrix} 1 & 0 & 0 \\ 0 & -2 & -2 \\ 0 & -2 & -2 \end{bmatrix} \sim \begin{bmatrix} 1 & 0 & 0 \\ 0 & 1 & 1 \\ 0 & 0 & 0 \end{bmatrix}$$

EXAMPLE 2
(continued)

Thus, a basis for the eigenspace is $\left\{\begin{bmatrix}0\\-1\\1\end{bmatrix}\right\}$.

For $\lambda_3 = -1$, we get

$$A - \lambda_2 I = \begin{bmatrix}5 & 0 & 0\\0 & 2 & -2\\0 & -2 & 2\end{bmatrix} \sim \begin{bmatrix}1 & 0 & 0\\0 & 1 & -1\\0 & 0 & 0\end{bmatrix}$$

Thus, a basis for the eigenspace is $\left\{\begin{bmatrix}0\\1\\1\end{bmatrix}\right\}$.

Observe that the vectors $\vec{v}_1 = \begin{bmatrix}1\\0\\0\end{bmatrix}$, $\vec{v}_2 = \begin{bmatrix}0\\-1\\1\end{bmatrix}$, and $\vec{v}_3 = \begin{bmatrix}0\\1\\1\end{bmatrix}$ form an orthogonal set. Hence, if we normalize them, we find that A is diagonalized by the orthogonal matrix $P = \begin{bmatrix}1 & 0 & 0\\0 & -1/\sqrt{2} & 1/\sqrt{2}\\0 & 1/\sqrt{2} & 1/\sqrt{2}\end{bmatrix}$ to $D = \begin{bmatrix}4 & 0 & 0\\0 & 3 & 0\\0 & 0 & -1\end{bmatrix}$.

EXAMPLE 3

Diagonalize the symmetric $A = \begin{bmatrix}5 & -4 & -2\\-4 & 5 & -2\\-2 & -2 & 8\end{bmatrix}$. Show that A can be diagonalized by an orthogonal matrix.

Solution: We have

$$C(\lambda) = \begin{vmatrix}5-\lambda & -4 & -2\\-4 & 5-\lambda & -2\\-2 & -2 & 8-\lambda\end{vmatrix} = -\lambda(\lambda-9)^2$$

The eigenvalues are $\lambda_1 = 9$ with algebraic multiplicity 2 and $\lambda_2 = 0$ with algebraic multiplicity 1.

For $\lambda_1 = 9$, we get

$$A - \lambda_1 I = \begin{bmatrix}-4 & -4 & -2\\-4 & -4 & -2\\-2 & -2 & -1\end{bmatrix} \sim \begin{bmatrix}2 & 2 & 1\\0 & 0 & 0\\0 & 0 & 0\end{bmatrix}$$

Thus, a basis for the eigenspace of λ_1 is $\{\vec{w}_1, \vec{w}_2\} = \left\{\begin{bmatrix}-1\\1\\0\end{bmatrix}, \begin{bmatrix}-1\\0\\2\end{bmatrix}\right\}$. However, observe that these vectors are not orthogonal to each other. Since we require an orthonormal basis of eigenvectors of A, we need to find an orthonormal basis for the eigenspace of λ_1. We can do this by applying the Gram-Schmidt Procedure to this set.

Pick $\vec{v}_1 = \vec{w}_1$. Then $\mathbb{S}_1 = \operatorname{Span}\{\vec{v}_1\}$ and

$$\vec{v}_2 = \operatorname{perp}_{\mathbb{S}_1} \vec{w}_2 = \vec{w}_2 - \frac{\vec{w}_2 \cdot \vec{v}_1}{\|\vec{v}_1\|^2}\vec{v}_1 = \frac{1}{2}\begin{bmatrix}-1\\-1\\4\end{bmatrix}$$

EXAMPLE 3 (continued)

Then, $\{\vec{v}_1, \vec{v}_2\}$ is an orthogonal basis for the eigenspace of λ_1.

For $\lambda_2 = 0$, we get

$$A - \lambda_2 I = \begin{bmatrix} 5 & -4 & -2 \\ -4 & 5 & -2 \\ -2 & -2 & 8 \end{bmatrix} \sim \begin{bmatrix} 1 & 0 & -2 \\ 0 & 1 & -2 \\ 0 & 0 & 0 \end{bmatrix}$$

Thus, a basis for the eigenspace of λ_2 is $\{\vec{v}_3\} = \left\{ \begin{bmatrix} 2 \\ 2 \\ 1 \end{bmatrix} \right\}$.

Observe that $\vec{v}_3$ is orthogonal to $\vec{v}_1$ and $\vec{v}_2$. Hence, normalizing $\vec{v}_1$, $\vec{v}_2$, and $\vec{v}_3$, we find that A is diagonalized by the orthogonal matrix

$$\begin{bmatrix} 2/3 & -1/\sqrt{2} & -1/\sqrt{18} \\ 2/3 & 1/\sqrt{2} & -1/\sqrt{18} \\ 1/3 & 0 & 4/\sqrt{18} \end{bmatrix}$$

to $D = \begin{bmatrix} 0 & 0 & 0 \\ 0 & 9 & 0 \\ 0 & 0 & 9 \end{bmatrix}$.

The Principal Axis Theorem

We will now proceed to show that not only can every symmetric matrix be diagonalized, but that it can be diagonalized by an orthogonal matrix.

Definition Orthogonally Diagonalizable

A matrix A is said to be **orthogonally diagonalizable** if there exists an orthogonal matrix P and a diagonal matrix D such that

$$P^T AP = D$$

Remark

Two matrices A and B are said to be **orthogonally similar** if there exists an orthogonal matrix P such that $P^T AP = B$. Observe that orthogonally similar matrices are similar since an orthogonal matrix P satisfies $P^T = P^{-1}$. In particular, $P^T AP = D$ is equivalent to $P^{-1}AP = D$ for an orthogonal matrix P. Hence, all of our theory of similar matrices and diagonalization applies to orthogonal diagonalization.

Lemma 1

An $n \times n$ matrix A is symmetric if and only if

$$\vec{x} \cdot (A\vec{y}) = (A\vec{x}) \cdot \vec{y}$$

for all $\vec{x}, \vec{y} \in \mathbb{R}^n$.

Proof: Suppose that A is symmetric. For any $\vec{x}, \vec{y} \in \mathbb{R}^n$,

$$\vec{x} \cdot (A\vec{y}) = \vec{x}^T A\vec{y} = \vec{x}^T A^T \vec{y} = (A\vec{x})^T \vec{y} = (A\vec{x}) \cdot \vec{y}$$

Conversely, if $\vec{x} \cdot (A\vec{y}) = (A\vec{x}) \cdot \vec{y}$ for all $\vec{x}, \vec{y} \in \mathbb{R}^n$,

$$\vec{x}^T A\vec{y} = \vec{x} \cdot (A\vec{y}) = (A\vec{x}) \cdot \vec{y} = (A\vec{x})^T \vec{y}$$

Since $\vec{x}^T A\vec{y} = (A\vec{x})^T \vec{y}$ for all $\vec{y} \in \mathbb{R}^n$, based on Theorem 3.1.4, $\vec{x}^T A = (A\vec{x})^T$. Hence, $(\vec{x}^T A)^T = A\vec{x}$ or $A^T \vec{x} = A\vec{x}$ for all $\vec{x} \in \mathbb{R}^n$. Applying Theorem 3.1.4 again gives $A^T = A$, as required. ■

In Examples 1–3, we saw that the basis vectors of distinct eigenspaces are orthogonal. This implies that every eigenvector in the eigenspace of one eigenvalue is orthogonal to every eigenvector in the eigenspace of a different eigenvalue. We now use Lemma 1 to prove that is always true.

Theorem 2

If $\vec{v}_1$ and $\vec{v}_2$ are eigenvectors of a symmetric matrix A corresponding to distinct eigenvalues λ_1 and λ_2, then $\vec{v}_1$ is orthogonal to $\vec{v}_2$.

Proof: By definition of eigenvalues and eigenvectors, $A\vec{v}_1 = \lambda_1\vec{v}_1$ and $A\vec{v}_2 = \lambda_2\vec{v}_2$. Hence,

$$\lambda_1(\vec{v}_1 \cdot \vec{v}_2) = (\lambda_1\vec{v}_1) \cdot \vec{v}_2 = (A\vec{v}_1) \cdot \vec{v}_2$$

Using Lemma 1, we get

$$(A\vec{v}_1) \cdot \vec{v}_2 = \vec{v}_1 \cdot (A\vec{v}_2)$$

Hence,

$$\lambda_1(\vec{v}_1 \cdot \vec{v}_2) = \vec{v}_1 \cdot (A\vec{v}_2) = \vec{v}_1 \cdot (\lambda_2\vec{v}_2) = \lambda_2(\vec{v}_1 \cdot \vec{v}_2)$$

It follows that

$$(\lambda_1 - \lambda_2)(\vec{v}_1 \cdot \vec{v}_2) = 0$$

It was assumed that $\lambda_1 \neq \lambda_2$, so $\vec{v}_1 \cdot \vec{v}_2$ must be zero, and the eigenvectors corresponding to distinct eigenvalues are mutually orthogonal, as claimed. ■

Note that this theorem applies only to eigenvectors that correspond to different eigenvalues. As we saw in Example 3, eigenvectors that correspond to the same eigenvalue do not need to be orthogonal. Thus, as in Example 3, if an eigenvalue has algebraic multiplicity greater than 1, it may be necessary to apply the Gram-Schmidt Procedure to find an orthogonal basis for its eigenspace.

Theorem 3

If A is a symmetric matrix, then all eigenvalues of A are real.

The proof of this theorem requires properties of complex numbers and hence is postponed until Chapter 9. See Problem 9.5.D6.

We can now prove that every symmetric matrix is orthogonally diagonalizable. We begin with a lemma that will be the main step in the proof by induction in the theorem.

Lemma 4

Suppose that λ_1 is an eigenvalue of the $n \times n$ symmetric matrix A, with corresponding unit eigenvector $\vec{v}_1$. Then there is an orthogonal matrix P whose first column is $\vec{v}_1$, such that

$$P^T AP = \begin{bmatrix} \lambda_1 & O_{1,n-1} \\ O_{n-1,1} & A_1 \end{bmatrix}$$

where A_1 is an $(n-1)\times(n-1)$ symmetric matrix and $O_{m,n}$ is the $m\times n$ zero matrix.

Proof: By extending the set $\{\vec{v}_1\}$ to a basis for $\mathbb{R}^n$, applying the Gram-Schmidt Procedure, and normalizing the vectors, we can produce an orthonormal basis $\mathcal{B} = \{\vec{v}_1, \vec{w}_2, \dots, \vec{w}_n\}$ for $\mathbb{R}^n$. Let

$$P = \begin{bmatrix} \vec{v}_1 & \vec{w}_2 & \cdots & \vec{w}_n \end{bmatrix}$$

Then

$$\begin{aligned} P^T AP &= \begin{bmatrix} \vec{v}_1^T \\ \vec{w}_2^T \\ \vdots \\ \vec{w}_n^T \end{bmatrix} A \begin{bmatrix} \vec{v}_1 & \vec{w}_2 & \cdots & \vec{w}_n \end{bmatrix} \\ &= \begin{bmatrix} \vec{v}_1^T \\ \vec{w}_2^T \\ \vdots \\ \vec{w}_n^T \end{bmatrix} \begin{bmatrix} A\vec{v}_1 & A\vec{w}_2 & \cdots & A\vec{w}_n \end{bmatrix} \\ &= \begin{bmatrix} \vec{v}_1^T A\vec{v}_1 & \vec{v}_1^T A\vec{w}_2 & \cdots & \vec{v}_1^T A\vec{w}_n \\ \vec{w}_2^T A\vec{v}_1 & \vec{w}_2^T A\vec{w}_2 & \cdots & \vec{w}_2^T A\vec{w}_n \\ \vdots & \vdots & \ddots & \vdots \\ \vec{w}_n^T A\vec{v}_1 & \vec{w}_n^T A\vec{w}_2 & \cdots & \vec{w}_n^T A\vec{w}_n \end{bmatrix} \\ &= \begin{bmatrix} \vec{v}_1 \cdot A\vec{v}_1 & \vec{v}_1 \cdot A\vec{w}_2 & \cdots & \vec{v}_1 \cdot A\vec{w}_n \\ \vec{w}_2 \cdot A\vec{v}_1 & \vec{w}_2 \cdot A\vec{w}_2 & \cdots & \vec{w}_2 \cdot A\vec{w}_n \\ \vdots & \vdots & \ddots & \vdots \\ \vec{w}_n \cdot A\vec{v}_1 & \vec{w}_n \cdot A\vec{w}_2 & \cdots & \vec{w}_n \cdot A\vec{w}_n \end{bmatrix} \end{aligned}$$

First observe that $P^T AP$ is symmetric since

$$(P^T AP)^T = P^T A^T P^{TT} = P^T AP$$

Also, $\vec{v}_1$ is a unit eigenvector of λ_1, so we have

$$(P^T AP)_{11} = \vec{v}_1 \cdot (A\vec{v}_1) = \vec{v}_1 \cdot (\lambda_1 \vec{v}_1) = \lambda_1(\vec{v}_1 \cdot \vec{v}_1) = \lambda_1$$

Since $\mathcal{B}$ is orthonormal, we get

$$(P^T AP)_{i1} = \vec{w}_i \cdot (A\vec{v}_1) = \vec{v}_1 \cdot (\lambda_1 \vec{v}_1) = \lambda_1(\vec{w}_i \cdot \vec{v}_1) = 0$$

So, all other entries in the first column are 0. Hence, all other entries in the first row are also 0 since $P^T AP$ is symmetric. Moreover, the $(n-1) \times (n-1)$ block A_1 is also symmetric since $P^T AP$ is symmetric. ■

Theorem 5

Principal Axis Theorem

Suppose A is an $n \times n$ symmetric matrix. Then there exists an orthogonal matrix P and diagonal matrix D such that $P^TAP = D$. That is, every symmetric matrix is orthogonally diagonalizable.

Proof: The proof is by induction on n. If $n = 1$, then A is a diagonal matrix and hence is orthogonally diagonalizable with $P = [1]$. Now suppose the result is true for $(n-1) \times (n-1)$ symmetric matrices, and consider an $n \times n$ symmetric matrix A. Pick an eigenvalue λ_1 of A (note that λ_1 is real by Theorem 3) and find a corresponding unit eigenvector $\vec{v}_1$. Then, by Lemma 4, there exists an orthogonal matrix $R = \begin{bmatrix} \vec{v}_1 & \vec{w}_2 & \cdots & \vec{w}_n \end{bmatrix}$ such that

$$R^TAR = \begin{bmatrix} \lambda_1 & O_{1,n-1} \\ O_{n-1,1} & A_1 \end{bmatrix}$$

where A_1 is an $(n-1) \times (n-1)$ symmetric matrix. Then, by our hypothesis, there is an $(n-1) \times (n-1)$ orthogonal matrix P_1 such that

$$P_1^TA_1P_1 = D_1$$

where D_1 is an $(n-1) \times (n-1)$ diagonal matrix. Define

$$P_2 = \begin{bmatrix} 1 & O_{1,n-1} \\ O_{n-1,1} & P_1 \end{bmatrix}$$

The columns of P_2 form an orthonormal basis for $\mathbb{R}^n$. Hence, P_2 is orthogonal. Since a product of orthogonal matrices is orthogonal, we get that $P = P_2R$ is an $n \times n$ orthogonal matrix and, by block multiplication,

$$\begin{aligned} P^TAP &= (P_2R)^TA(P_2R) = P_2^TR^TARP_2 \\ &= P_2^T \begin{bmatrix} \lambda_1 & O_{1,n-1} \\ O_{n-1,1} & A_1 \end{bmatrix} P_2 \\ &= \begin{bmatrix} \lambda_1 & O_{1,n-1} \\ O_{n-1,1} & D_1 \end{bmatrix} = D \end{aligned}$$

This is diagonal, as required. ■

EXERCISE 1

Orthogonally diagonalize $A = \begin{bmatrix} 5 & -3 \\ -3 & 5 \end{bmatrix}$.

EXERCISE 2

Orthogonally diagonalize $A = \begin{bmatrix} 2 & -1 & -1 \\ -1 & 2 & -1 \\ -1 & -1 & 2 \end{bmatrix}$.

Remarks

1. The eigenvectors in an orthogonal matrix that diagonalizes a symmetric matrix A are called the **principal axes** for A. We will see why this definition makes sense in Section 8.3.

2. The converse of the Principal Axis Theorem is also true. That is, every orthogonally diagonalizable matrix is symmetric. You are asked to prove this in Problem D2. Hence, we can say that a matrix is orthogonally diagonalizable if and only if it is symmetric.

PROBLEMS 8.1

Practice Problems

A1 Determine which of the following matrices are symmetric.

(a) $A = \begin{bmatrix} 0 & 2 \\ 2 & -1 \end{bmatrix}$

(b) $B = \begin{bmatrix} 0 & 0 \\ 0 & 0 \end{bmatrix}$

(c) $C = \begin{bmatrix} 1 & 2 & 1 \\ -2 & 1 & 2 \\ -1 & -2 & 1 \end{bmatrix}$

(d) $D = \begin{bmatrix} 0 & -1 & 1 \\ -1 & 0 & -1 \\ 1 & -1 & 0 \end{bmatrix}$

A2 For each of the following symmetric matrices, find an orthogonal matrix P and diagonal matrix D such that $P^TAP = D$.

(a) $A = \begin{bmatrix} 1 & -2 \\ -2 & 1 \end{bmatrix}$

(b) $A = \begin{bmatrix} 5 & 3 \\ 3 & -3 \end{bmatrix}$

(c) $A = \begin{bmatrix} 0 & 1 & 1 \\ 1 & 0 & 1 \\ 1 & 1 & 0 \end{bmatrix}$

(d) $A = \begin{bmatrix} 1 & 0 & -2 \\ 0 & -1 & -2 \\ -2 & -2 & 0 \end{bmatrix}$

(e) $A = \begin{bmatrix} 1 & 8 & 4 \\ 8 & 1 & -4 \\ 4 & -4 & 7 \end{bmatrix}$

Homework Problems

B1 Determine which of the following matrices are symmetric.

(a) $A = \begin{bmatrix} 1 & 1 \\ 1 & 1 \end{bmatrix}$

(b) $B = \begin{bmatrix} 1 & 2 \\ 1 & 2 \end{bmatrix}$

(c) $C = \begin{bmatrix} 1 & 0 & 1 \\ 0 & 1 & 1 \\ 0 & 1 & 1 \end{bmatrix}$

(d) $D = \begin{bmatrix} 1 & 1 & 0 \\ 0 & 1 & 1 \\ 1 & 0 & 1 \end{bmatrix}$

(e) $E = \begin{bmatrix} 2 & 3 & 0 \\ 3 & 3 & 2 \\ 0 & 2 & -4 \end{bmatrix}$

B2 For each of the following symmetric matrices, find an orthogonal matrix P and diagonal matrix D such that $P^TAP = D$.

(a) $A = \begin{bmatrix} 5 & 2 \\ 2 & 2 \end{bmatrix}$

(b) $A = \begin{bmatrix} 4 & 2 \\ 2 & 1 \end{bmatrix}$

(c) $A = \begin{bmatrix} 7 & 3 \\ 3 & -1 \end{bmatrix}$

(d) $A = \begin{bmatrix} 1 & 2 & 1 \\ 2 & 1 & 1 \\ 1 & 1 & 2 \end{bmatrix}$

(e) $A = \begin{bmatrix} 0 & 1 & -1 \\ 1 & 0 & 1 \\ -1 & 1 & 0 \end{bmatrix}$

(f) $A = \begin{bmatrix} 1 & 0 & -1 \\ 0 & 1 & 1 \\ -1 & 1 & 2 \end{bmatrix}$

(g) $A = \begin{bmatrix} 1 & 2 & -4 \\ 2 & -2 & -2 \\ -4 & -2 & 1 \end{bmatrix}$

(h) $A = \begin{bmatrix} -2 & 2 & -1 \\ 2 & 1 & -2 \\ -1 & -2 & -2 \end{bmatrix}$

(i) $A = \begin{bmatrix} 2 & -2 & 2 \\ -2 & -1 & -5 \\ 2 & -5 & -1 \end{bmatrix}$

Computer Problems

C1 Use a computer to determine the eigenvalues and a basis of orthonormal eigenvectors of the following symmetric matrices.

(a) The matrix in Problem A2 (e).

(b) $\begin{bmatrix} 4.1 & 1.9 & 0.5 \\ 1.9 & 1.2 & 0.6 \\ 0.5 & 0.6 & -2.1 \end{bmatrix}$

(c) $\begin{bmatrix} 0.15 & 0.05 & 0.95 & 0.25 \\ 0.05 & 0.15 & 0.25 & 0.95 \\ 0.95 & 0.25 & 0.15 & 0.05 \\ 0.25 & 0.95 & 0.05 & 0.15 \end{bmatrix}$

C2 Let $S(t)$ denote the symmetric matrix

$$S(t) = \begin{bmatrix} 2+t & -2t & t \\ -2t & 2-t & -t \\ t & -t & 1+t \end{bmatrix}$$

By calculating the eigenvalues of $S(-0.1)$, $S(-0.05)$, $S(0)$, $S(0.05)$, and $S(0.1)$, explore how the eigenvalues of $S(t)$ change as t varies.

Conceptual Problems

D1 Let A and B be $n \times n$ symmetric matrices. Determine which of the following is symmetric. (a) $A + B$ (b) A^TA (c) AB (d) A^2

D2 Show that if A is orthogonally diagonalizable, then A is symmetric.

D3 Prove that if A is an invertible symmetric matrix, then A^{-1} is orthogonally diagonalizable.

8.2 Quadratic Forms

We saw earlier in the book the relationship between matrix mappings and linear mappings. We now explore the relationship between symmetric matrices and an important class of functions called **quadratic forms**, *which are not linear. Quadratic forms appear in geometry, statistics, calculus, topology, and many other areas. We shall see in Section 8.3 how quadratic forms and our special theory of diagonalization of symmetric matrices can be used to graph conic sections and quadric surfaces.*

Quadratic Forms

Consider the symmetric matrix $A = \begin{bmatrix} a & b/2 \\ b/2 & c \end{bmatrix}$. If $\vec{x} = \begin{bmatrix} x_1 \\ x_2 \end{bmatrix}$, then

$$\begin{aligned} \vec{x}^T A\vec{x} &= \begin{bmatrix} x_1 & x_2 \end{bmatrix} \begin{bmatrix} a & b/2 \\ b/2 & c \end{bmatrix} \begin{bmatrix} x_1 \\ x_2 \end{bmatrix} \\ &= \begin{bmatrix} x_1 & x_2 \end{bmatrix} \begin{bmatrix} ax_1 + bx_2/2 \\ bx_1/2 + cx_2 \end{bmatrix} \\ &= ax_1^2 + bx_1x_2 + cx_2^2 \end{aligned}$$

We call the expression $ax_1^2 + bx_1x_2 + cx_2^2$ a quadratic form on $\mathbb{R}^2$ (or in the variables x_1 and x_2). Thus, corresponding to every symmetric matrix A, there is a quadratic form

$$Q(\vec{x}) = \vec{x}^T A\vec{x} = ax_1^2 + bx_1x_2 + cx_2^2$$

On the other hand, given a quadratic form $Q(\vec{x}) = ax_1^2 + bx_1x_2 + cx_2^2$, we can reconstruct the symmetric matrix $A = \begin{bmatrix} a & b/2 \\ b/2 & c \end{bmatrix}$ by choosing $(A)_{11}$ to be the coefficient of x_1^2, $(A)_{12} = (A)_{21}$ to be half of the coefficient of x_1x_2, and $(A)_{22}$ to be the coefficient of x_2^2. We deal with the coefficient of x_1x_2 in this way to ensure that A is symmetric.

EXAMPLE 1

Determine the symmetric matrix corresponding to the quadratic form

$$Q(\vec{x}) = 2x_1^2 - 4x_1x_2 - x_2^2$$

Solution: The corresponding symmetric matrix A is

$$A = \begin{bmatrix} 2 & -2 \\ -2 & -1 \end{bmatrix}$$

Notice that we could have written $ax_1^2 + bx_1x_2 + cx_2^2$ in terms of other *asymmetric* matrices. For example,

$$ax_1^2 + bx_1x_2 + cx_2^2 = \vec{x}^T \begin{bmatrix} a & b \\ 0 & c \end{bmatrix} \vec{x} = \vec{x}^T \begin{bmatrix} a & 2b \\ -b & c \end{bmatrix} \vec{x}$$

Many choices are possible. However, we agree always to choose the symmetric matrix for two reasons. First, it gives us a unique (symmetric) matrix corresponding to a given quadratic form. Second, the choice of the symmetric matrix A allows us to apply the special theory available for symmetric matrices. We now use this to extend the definition of quadratic form to n variables.

Definition
Quadratic Form

A **quadratic form** on $\mathbb{R}^n$, with corresponding symmetric matrix A, is a function $Q : \mathbb{R}^n \to \mathbb{R}$ defined by

$$Q(\vec{x}) = \sum_{i,j=1}^{n} a_{ij}x_i x_j = \vec{x}^T A\vec{x}$$

As above, given a quadratic form $Q(\vec{x})$ on $\mathbb{R}^n$, we can easily construct the corresponding symmetric matrix A by taking $(A)_{ii}$ to be the coefficient of x_i^2 and $(A)_{ij}$ to be half of the coefficient of $x_i x_j$ for $i \neq j$.

EXAMPLE 2

$Q(\vec{x}) = 3x_1^2 + 5x_1x_2 + 2x_2^2$ is a quadratic form on $\mathbb{R}^2$ with corresponding symmetric matrix $A = \begin{bmatrix} 3 & 5/2 \\ 5/2 & 2 \end{bmatrix}$.

$Q(\vec{x}) = x_1^2 + 4x_2^2 + 2x_3^2 + 4x_1x_2 + x_1x_3 + 2x_2x_3$ is a quadratic form on $\mathbb{R}^3$ with corresponding symmetric matrix $A = \begin{bmatrix} 1 & 2 & 1/2 \\ 2 & 4 & 1 \\ 1/2 & 1 & 2 \end{bmatrix}$.

$Q(\vec{x}) = 2x_1^2 + 4x_1x_3 - x_2^2 + 2x_2x_3 + 6x_2x_4 + x_3^2 + x_4^2$ is a quadratic form on $\mathbb{R}^2$ with corresponding symmetric matrix $A = \begin{bmatrix} 2 & 0 & 2 & 0 \\ 0 & -1 & 1 & 3 \\ 2 & 1 & 1 & 0 \\ 0 & 3 & 0 & 1 \end{bmatrix}$.

EXERCISE 1

Find the quadratic form corresponding to each of the following symmetric matrices.

(a) $\begin{bmatrix} 4 & 1/2 \\ 1/2 & \sqrt{2} \end{bmatrix}$.

(b) $\begin{bmatrix} 1 & -1 & 0 \\ -1 & 2 & 3 \\ 0 & 3 & -1 \end{bmatrix}$

EXERCISE 2

Find the corresponding symmetric matrix for each of the following quadratic forms.

1. $Q(\vec{x}) = x_1^2 - 2x_1x_2 - 3x_2^2$
2. $Q(\vec{x}) = 2x_1^2 + 3x_1x_2 - x_1x_3 + 4x_2^2 + x_3^2$
3. $Q(\vec{x}) = x_1^2 + 2x_2^2 + 3x_3^2 + 4x_4^2$

Observe that the symmetric matrix corresponding to $Q(\vec{x}) = x_1^2 + 2x_2^2 + 3x_3^2 + 4x_4^2$ is in fact diagonal. This motivates the following definition.

Definition
Diagonal Form

A quadratic form $Q(\vec{x})$ is in **diagonal form** if all the coefficients a_{jk} with $j \neq k$ are equal to 0. Equivalently, $Q(\vec{x})$ is in diagonal form if its corresponding symmetric matrix is diagonal.

EXAMPLE 3

The quadratic form $Q(\vec{x}) = 3x_1^2 - 2x_2^2 + 4x_3^2$ is in diagonal form. The quadratic form $Q(\vec{x}) = 2x_1^2 - 4x_1x_2 + 3x_2^2$ is not in diagonal form.

Since each quadratic form has an associated symmetric matrix, we should expect that diagonalizing the symmetric matrix should also diagonalize the quadratic form. We first demonstrate this with an example and then prove the result in general.

EXAMPLE 4

Let $A = \begin{bmatrix} 2 & 1 \\ 1 & 2 \end{bmatrix}$ and let $Q(\vec{x}) = \vec{x}^T A\vec{x} = 2x_1^2 + 2x_1x_2 + 2x_2^2$. Let $\vec{x} = P\vec{y}$, where $P = \begin{bmatrix} 1/\sqrt{2} & 1/\sqrt{2} \\ -1/\sqrt{2} & 1/\sqrt{2} \end{bmatrix}$. Express $Q(\vec{x})$ in terms of $\vec{y} = \begin{bmatrix} y_1 \\ y_2 \end{bmatrix}$.

Solution: We first observe that P diagonalizes A. In particular, we have

$$P^T AP = \begin{bmatrix} 1/\sqrt{2} & -1/\sqrt{2} \\ 1/\sqrt{2} & 1/\sqrt{2} \end{bmatrix} \begin{bmatrix} 2 & 1 \\ 1 & 2 \end{bmatrix} \begin{bmatrix} 1/\sqrt{2} & 1/\sqrt{2} \\ -1/\sqrt{2} & 1/\sqrt{2} \end{bmatrix} = \begin{bmatrix} 1 & 0 \\ 0 & 3 \end{bmatrix}$$

We have

$$\begin{aligned} Q(\vec{x}) &= \vec{x}^T A\vec{x} \\ &= (P\vec{y})^T A(P\vec{y}) \\ &= \vec{y}^T P^T AP\vec{y} \\ &= \vec{y}^T \begin{bmatrix} 1 & 0 \\ 0 & 3 \end{bmatrix} \vec{y} \\ &= y_1^2 + 3y_2^2 \end{aligned}$$

Recall that if $P = \begin{bmatrix} \vec{v}_1 & \vec{v}_2 \end{bmatrix}$ is an orthogonal matrix, it is a change of coordinates matrix from standard coordinates to coordinates with respect to the basis $\mathcal{B} = \{\vec{v}_1, \vec{v}_2\}$. In particular, in Example 4, we put $Q(\vec{x})$ into diagonal form by writing it with respect to the orthonormal basis $\mathcal{B} = \left\{ \begin{bmatrix} 1/\sqrt{2} \\ 1/\sqrt{2} \end{bmatrix}, \begin{bmatrix} -1/\sqrt{2} \\ 1/\sqrt{2} \end{bmatrix} \right\}$. The vector $\vec{y}$ is just the $\mathcal{B}$-coordinates with respect to $\vec{x}$. That is, $[\vec{x}]_{\mathcal{B}} = \vec{y}$. We now prove this in general.

Theorem 1

Let $Q(\vec{x}) = \vec{x}^T A\vec{x}$ be a quadratic form on $\mathbb{R}^n$. Then there is an orthonormal basis $\mathcal{B}$ of $\mathbb{R}^n$ such that when $Q(\vec{x})$ is expressed in terms of $\mathcal{B}$-coordinates, it is in diagonal form.

Proof: Since A is symmetric, we can apply the Principal Axis Theorem to get an orthogonal matrix P such that $P^T AP = D = \text{diag}(\lambda_1, \dots, \lambda_n)$, where $\lambda_1, \dots, \lambda_n$ are the eigenvalues of A. Recall from Section 4.4 that the change of coordinates matrix P from $\mathcal{B}$-coordinates to standard coordinates satisfies $\vec{x} = P[\vec{x}]_{\mathcal{B}}$. Hence,

$$\begin{aligned} Q(\vec{x}) &= \vec{x}^T A\vec{x} \\ &= (P[\vec{x}]_{\mathcal{B}})^T A(P[\vec{x}]_{\mathcal{B}}) \\ &= [\vec{x}]_{\mathcal{B}}^T P^T AP[\vec{x}]_{\mathcal{B}} \\ &= [\vec{x}]_{\mathcal{B}}^T D[\vec{x}]_{\mathcal{B}} \end{aligned}$$

Let $[\vec{x}]_{\mathcal{B}} = \begin{bmatrix} y_1 \\ \vdots \\ y_n \end{bmatrix}$. Then we have

$$Q(\vec{x}) = \begin{bmatrix} y_1 & \cdots & y_n \end{bmatrix} \begin{bmatrix} \lambda_1 & 0 & \cdots & 0 \\ 0 & \lambda_2 & \ddots & \vdots \\ \vdots & \ddots & \ddots & 0 \\ 0 & \cdots & 0 & \lambda_n \end{bmatrix} \begin{bmatrix} y_1 \\ \vdots \\ y_n \end{bmatrix}$$
$$= \lambda_1 y_1^2 + \lambda_2 y_2^2 + \cdots + \lambda_n y_n^2$$

as required. ■

EXAMPLE 5

Let $Q(\vec{x}) = x_1^2 + 4x_1x_2 + x_2^2$. Find a diagonal form of $Q(\vec{x})$ and an orthogonal matrix P that brings it into this form.

Solution: The corresponding symmetric matrix is $A = \begin{bmatrix} 1 & 2 \\ 2 & 1 \end{bmatrix}$. We have

$$C(\lambda) = \begin{vmatrix} 1-\lambda & 2 \\ 2 & 1-\lambda \end{vmatrix} = (\lambda - 3)(\lambda + 1)$$

The eigenvalues are $\lambda_1 = 3$ with algebraic multiplicity 1 and $\lambda_2 = -1$ with algebraic multiplicity 1.

For $\lambda_1 = 3$, we get

$$A - \lambda_1 I = \begin{bmatrix} -2 & 2 \\ 2 & -2 \end{bmatrix} \sim \begin{bmatrix} 1 & -1 \\ 0 & 0 \end{bmatrix}$$

An eigenvector for λ_1 is $\vec{v}_1 = \begin{bmatrix} 1 \\ 1 \end{bmatrix}$, and a basis for the eigenspace is $\{\vec{v}_1\}$.

For $\lambda_2 = -1$, we get

$$A - \lambda_2 I = \begin{bmatrix} 2 & 2 \\ 2 & 2 \end{bmatrix} \sim \begin{bmatrix} 1 & 1 \\ 0 & 0 \end{bmatrix}$$

An eigenvector for λ_2 is $\vec{v}_2 = \begin{bmatrix} -1 \\ 1 \end{bmatrix}$, and a basis for the eigenspace is $\{\vec{v}_2\}$.

Therefore, we see that A is orthogonally diagonalized by $P = \frac{1}{\sqrt{2}} \begin{bmatrix} 1 & -1 \\ 1 & 1 \end{bmatrix}$ to $D = \begin{bmatrix} 3 & 0 \\ 0 & -1 \end{bmatrix}$. Let $\vec{x} = P\vec{y}$. Then, we get

$$Q(\vec{x}) = \vec{x}^T A\vec{x} = (P\vec{y})^T A(P\vec{y}) = \vec{y}^T (P^T AP)\vec{y} = \vec{y}^T D\vec{y} = 3y_1^2 - y_2^2$$

Hence, $Q(\vec{x})$ is brought into diagonal form by P.

EXERCISE 3

Let $Q(\vec{x}) = 4x_1x_2 - 3x_2^2$. Find a diagonal form of $Q(\vec{x})$ and an orthogonal matrix P that brings it into this form.

Classifications of Quadratic Forms

Definition

Positive Definite
Negative Definite
Indefinite
Semidefinite

A quadratic form $Q(\vec{x})$ on $\mathbb{R}^n$ is

1. **Positive definite** if $Q(\vec{x}) > 0$ for all $\vec{x} \neq \vec{0}$
2. **Negative definite** if $Q(\vec{x}) < 0$ for all $\vec{x} \neq \vec{0}$
3. **Indefinite** if $Q(\vec{x}) > 0$ for some $\vec{x}$ and $Q(\vec{x}) < 0$ for some $\vec{x}$
4. **Positive semidefinite** if $Q(\vec{x}) \geq 0$ for all $\vec{x}$
5. **Negative semidefinite** if $Q(\vec{x}) \leq 0$ for all $\vec{x}$

These concepts are useful in applications. For example, we shall see in Section 8.3 that the graph of $Q(\vec{x}) = 1$ in $\mathbb{R}^2$ is an ellipse if and only if $Q(\vec{x})$ is positive definite.

EXAMPLE 6

Classify the quadratic forms $Q_1(\vec{x}) = 3x_1^2 + 4x_2^2$, $Q_2(\vec{x}) = xy$, and $Q_3(\vec{x}) = 2x_1^2 + 4x_1x_2 + x_2^2$.

Solution: $Q_1(\vec{x})$ is positive definite since $Q(\vec{x}) = 3x_1^2 + 4x_2^2 > 0$ for all $\vec{x} \neq \vec{0}$.
$Q_2(\vec{x})$ is indefinite since $Q(1, 1) = 1 > 0$ and $Q(-1, 1) = -1 < 0$.
$Q_3(\vec{x})$ is indefinite since $Q(1, 1) = 8 > 0$ and $Q(1, -2) = -2 < 0$.

Observe that classifying $Q_3(\vec{x})$ was a little more difficult than classifying $Q_1(\vec{x})$ or $Q_2(\vec{x})$. A general quadratic form $Q(\vec{x})$ on $\mathbb{R}^n$ would be difficult to classify by inspection. The following theorem gives us an easier way to classify a quadratic form.

Theorem 2

Let $Q(\vec{x}) = \vec{x}^T A\vec{x}$, where A is a symmetric matrix. Then

1. $Q(\vec{x})$ is positive definite if and only if all eigenvalues of A are positive.
2. $Q(\vec{x})$ is negative definite if and only if all eigenvalues of A are negative.
3. $Q(\vec{x})$ is indefinite if and only if some of the eigenvalues of A are positive and some are negative.

Proof: We prove (1) and leave (2) and (3) as Problems D1 and D2.

By Theorem 1, there exists an orthogonal matrix P such that

$$Q(\vec{x}) = \lambda_1 y_1^2 + \lambda_2 y_2^2 + \cdots + \lambda_n y_n^2$$

where $\vec{x} = P\vec{y}$ and $\lambda_1, \ldots, \lambda_n$ are the eigenvalues of A. Clearly, $Q(\vec{x}) > 0$ for all $\vec{y} \neq \vec{0}$ if and only if the eigenvalues are all positive. Moreover, since P is orthogonal, it is invertible. Hence, $\vec{x} = \vec{0}$ if and only if $\vec{y} = \vec{0}$ since $\vec{x} = P\vec{y}$. Thus we have shown that $Q(\vec{x})$ is positive definite if and only if all eigenvalues of A are positive. ■

EXAMPLE 7

Classify the following quadratic forms.

(a) $Q(\vec{x}) = 4x_1^2 + 8x_1x_2 + 3x_2^2$

Solution: The symmetric matrix corresponding to $Q(\vec{x})$ is $A = \begin{bmatrix} 4 & 4 \\ 4 & 3 \end{bmatrix}$. The characteristic polynomial of A is $C(\lambda) = \lambda^2 - 7\lambda - 4$. Using the quadratic formula, we find that the eigenvalues of A are $\lambda = \frac{7 \pm \sqrt{33}}{2}$. These are both positive. Hence, $Q(\vec{x})$ is positive definite.

EXAMPLE 7
(continued)

(b) $Q(\vec{x}) = -2x_1^2 - 2x_1x_2 + 2x_1x_3 - 2x_2^2 + 2x_2x_3 - 2x_3^2$

Solution: The symmetric matrix corresponding to $Q(\vec{x})$ is $A = \begin{bmatrix} -2 & -1 & 1 \\ -1 & -2 & 1 \\ 1 & 1 & -2 \end{bmatrix}$. The characteristic polynomial of A is $C(\lambda) = -(\lambda + 1)^2(\lambda + 4)$. Thus, the eigenvalues of A are -1, -1, and -4. Therefore, $Q(\vec{x})$ is negative definite.

EXERCISE 4

Classify the following quadratic forms.

(a) $Q(\vec{x}) = 4x_1^2 + 16x_1x_2 + 4x_2^2$

(b) $Q(\vec{x}) = 2x_1^2 - 6x_1x_2 - 6x_1x_3 + 3x_2^2 + 4x_2x_3 + 3x_3^2$

Since every symmetric matrix corresponds uniquely to a quadratic form, it makes sense to classify a symmetric matrix by classifying its corresponding quadratic form. That is, for example, we will say a symmetric matrix A is positive definite if and only if the quadratic form $Q(\vec{x}) = \vec{x}^T A\vec{x}$ is positive definite. Observe that this implies that we can use Theorem 2 to classify symmetric matrices as well.

EXAMPLE 8

Classify the following symmetric matrices.

(a) $A = \begin{bmatrix} 3 & 2 \\ 2 & 3 \end{bmatrix}$

Solution: We have $C(\lambda) = \lambda^2 - 6\lambda + 5$. Thus, the eigenvalues of A are 5 and 1, so A is positive definite.

(b) $A = \begin{bmatrix} -2 & 2 & -4 \\ 2 & -4 & -2 \\ -4 & -2 & 2 \end{bmatrix}$

Solution: We have $C(\lambda) = (\lambda + 4)(\lambda^2 - 28)$. Thus, the eigenvalues of A are -4, $2\sqrt{7}$, and $-2\sqrt{7}$, so A is indefinite.

PROBLEMS 8.2

Practice Problems

A1 Determine the quadratic form corresponding to the given symmetric matrix.

(a) $A = \begin{bmatrix} 1 & 3 \\ 3 & -1 \end{bmatrix}$

(b) $A = \begin{bmatrix} 1 & 0 & 0 \\ 0 & -2 & 3 \\ 0 & 3 & -1 \end{bmatrix}$

(c) $A = \begin{bmatrix} -2 & 1 & 1 \\ 1 & 1 & -1 \\ 1 & -1 & 0 \end{bmatrix}$

A2 For each of the following quadratic forms $Q(\vec{x})$,

(i) Determine the corresponding symmetric matrix A.

(ii) Express $Q(\vec{x})$ in diagonal form and give the orthogonal matrix that brings it into this form.

(iii) Classify $Q(\vec{x})$.

(a) $Q(\vec{x}) = x_1^2 - 3x_1x_2 + x_2^2$

(b) $Q(\vec{x}) = 5x_1 - 4x_1x_2 + 2x_2^2$

(c) $Q(\vec{x}) = -2x_1^2 + 12x_1x_2 + 7x_2^2$

(d) $Q(\vec{x}) = x_1^2 - 2x_1x_2 + 6x_1x_3 + x_2^2 + 6x_2x_3 - 3x_3^2$

(e) $Q(\vec{x}) = -4x_1^2 + 2x_1x_2 - 5x_2^2 - 2x_2x_3 - 4x_3^2$

A3 Classify each of the following symmetric matrices.

(a) $A = \begin{bmatrix} 4 & -2 \\ -2 & 4 \end{bmatrix}$

(b) $A = \begin{bmatrix} 1 & 0 & 0 \\ 0 & 2 & 0 \\ 0 & 0 & 3 \end{bmatrix}$

(c) $A = \begin{bmatrix} 1 & 0 & 0 \\ 0 & -2 & 6 \\ 0 & 6 & 7 \end{bmatrix}$

(d) $A = \begin{bmatrix} -3 & 1 & -1 \\ 1 & -3 & 1 \\ -1 & 1 & -3 \end{bmatrix}$

(e) $A = \begin{bmatrix} 7 & 2 & -1 \\ 2 & 10 & -2 \\ -1 & -2 & 7 \end{bmatrix}$

(f) $A = \begin{bmatrix} -4 & -5 & 5 \\ -5 & 2 & 1 \\ 5 & 1 & 2 \end{bmatrix}$

Homework Problems

B1 Determine the quadratic form corresponding to the given symmetric matrix.

(a) $A = \begin{bmatrix} 2 & 4 \\ 4 & 3 \end{bmatrix}$

(b) $A = \begin{bmatrix} -1 & 1 & 2 \\ 1 & 3 & 1 \\ 2 & 1 & -2 \end{bmatrix}$

(c) $A = \begin{bmatrix} 0 & 2 & 3 \\ 2 & 0 & -1 \\ 3 & -1 & 0 \end{bmatrix}$

B2 For each of the following quadratic forms $Q(\vec{x})$,

(i) Determine the corresponding symmetric matrix A.

(ii) Express $Q(\vec{x})$ in diagonal form and give the orthogonal matrix that brings it into this form.

(iii) Classify $Q(\vec{x})$.

(a) $Q(\vec{x}) = 7x_1^2 + 4x_1x_2 + 4x_2^2$

(b) $Q(\vec{x}) = 2x_1 + 6x_1x_2 + 2x_2^2$

(c) $Q(\vec{x}) = x_1^2 + 3x_1x_2 + x_2^2$

(d) $Q(\vec{x}) = 3x_1^2 - 2x_1x_2 - 2x_1x_3 + 5x_2^2 + 2x_2x_3 + 3x_3^2$

(e) $Q(\vec{x}) = 2x_1^2 - 4x_1x_2 + 6x_1x_3 + 2x_2^2 + 6x_2x_3 - 3x_3^2$

B3 Classify each of the following symmetric matrices.

(a) $A = \begin{bmatrix} 4 & -1 \\ -1 & 4 \end{bmatrix}$

(b) $A = \begin{bmatrix} 1 & -5 \\ -5 & 1 \end{bmatrix}$

(c) $A = \begin{bmatrix} -1 & 0 & 0 \\ 0 & -2 & 0 \\ 0 & 0 & -3 \end{bmatrix}$

(d) $A = \begin{bmatrix} -1 & 1 & 1 \\ 1 & -1 & 1 \\ 1 & 1 & -1 \end{bmatrix}$

(e) $A = \begin{bmatrix} -2 & -1 & 0 \\ -1 & -2 & -1 \\ 0 & -1 & -2 \end{bmatrix}$

Computer Problems

C1 Classify each of the following quadratic forms with the help of a computer.

(a) $Q(\vec{x}) = -9x_1^2 + 8x_1x_2 + 8x_1x_3 - 5x_2^2 - 5x_3^2$

(b) $Q(\vec{x}) = -0.1x_1^2 - 0.8x_1x_2 + 1.2x_1x_4 + 2.1x_2^2 + 1.6x_2x_3 + 1.3x_3^2 + 4.2x_3x_4 + 1.1x_4^2$

(c) $Q(\vec{x}) = 0.85(x_1^2 + x_2^2 + x_3^2 + x_4^2) - 0.1x_1x_2 + 0.6x_1x_3 + 0.2x_1x_4 + 0.2x_2x_3 + 0.6x_2x_4 - 0.1x_3x_4$

Conceptual Problems

D1 Let $Q(\vec{x}) = \vec{x}^T A\vec{x}$, where A is a symmetric matrix. Prove that $Q(\vec{x})$ is negative definite if and only if all eigenvalues of A are negative.

D2 Let $Q(\vec{x}) = \vec{x}^T A\vec{x}$, where A is a symmetric matrix. Prove that $Q(\vec{x})$ is indefinite if and only if some of the eigenvalues of A are positive and some are negative.

D3 (a) Let $Q(\vec{x}) = \vec{x}^T A\vec{x}$, where A is a symmetric matrix. Prove that $Q(\vec{x})$ is positive semidefinite if and only if all of the eigenvalues of A are non-negative.

(b) Let A be an $m \times n$ matrix. Prove that $A^T A$ is positive semidefinite.

D4 Let A be a positive definite symmetric matrix. Prove that

(a) The diagonal entries of A are all positive.

(b) A is invertible.

(c) A^{-1} is positive definite.

(d) $P^T AP$ is positive definite for any orthogonal matrix P.

D5 A matrix B is called **skew-symmetric** if $B^T = -B$. Given a square matrix A, define the **symmetric part** of A to be

$$A^+ = \frac{1}{2}(A + A^T)$$

and the **skew-symmetric** part of A to be

$$A^- = \frac{1}{2}(A - A^T)$$

(a) Verify that A^+ is symmetric, A^- is skew-symmetric, and $A = A^+ + A^-$.

(b) Prove that the diagonal entries of A^- are 0.

(c) Determine expressions for typical entries $(A^+)_{ij}$ and $(A^-)_{ij}$ in terms of the entries of A.

(d) Prove that for every $\vec{x} \in \mathbb{R}^n$,

$$\vec{x}^T A\vec{x} = \vec{x}^T A^+\vec{x}$$

(Hint: Use the fact that $A = A^+ + A^-$ and prove that $\vec{x}^T A^-\vec{x} = \vec{0}$.)

D6 In this problem, we show that general inner products on $\mathbb{R}^n$ are not different in interesting ways from the standard inner product.

Let $\langle , \rangle$ be an inner product on $\mathbb{R}^n$ and let $\mathcal{S} = \{\vec{e}_1, \dots, \vec{e}_n\}$ be the standard basis.

(a) Verify that for any $\vec{x}, \vec{y} \in \mathbb{R}^n$,

$$\langle \vec{x}, \vec{y} \rangle = \sum_{i=1}^{n} \sum_{j=1}^{n} x_i y_j \langle \vec{e}_i, \vec{e}_j \rangle$$

(b) Let G be the $n \times n$ matrix defined by $g_{ij} = \langle \vec{e}_i, \vec{e}_j \rangle$. Verify that

$$\langle \vec{x}, \vec{y} \rangle = \vec{x}^T G\vec{y}$$

(c) Use the properties of an inner product to verify that G is symmetric and positive definite.

(d) By adapting the proof of Theorem 1, show that there is a basis $\mathcal{B} = \{\vec{v}_1, \dots, \vec{v}_n\}$ such that in $\mathcal{B}$-coordinates,

$$\langle \vec{x}, \vec{y} \rangle = \lambda_1 \tilde{x}_1 \tilde{y}_1 + \cdots + \lambda_n \tilde{x}_n \tilde{y}_n$$

where $\lambda_1, \dots, \lambda_n$ are the eigenvalues of G. In particular,

$$\langle \vec{x}, \vec{y} \rangle = \|\vec{x}\|^2 = \sum_{i=1}^{n} \lambda_i x_i^2$$

(e) Introduce a new basis $C = \{\vec{w}_1, \dots, \vec{w}_n\}$ by defining $\vec{w}_i = \vec{v}_i / \sqrt{\lambda_i}$. Use an asterisk to denote C-coordinates, so that $\vec{x} = x_1^* \vec{w}_1 + \cdots + x_n^* \vec{w}_n$. Verify that

$$\langle \vec{w}_i, \vec{w}_j \rangle = \begin{cases} 1 & \text{if } i = k \\ 0 & \text{if } i \neq k \end{cases}$$

and that

$$\langle \vec{x}, \vec{y} \rangle = x_1^* y_1^* + \cdots + x_n^* y_n^*$$

Thus, with respect to the inner product $\langle , \rangle$, C is an orthonormal basis, and in C-coordinates, the inner product of two vectors looks just like the standard dot product.

8.3 Graphs of Quadratic Forms

In $\mathbb{R}^2$, it is often of interest to know the graph of an equation of the form $Q(\vec{x}) = k$, where $Q(\vec{x})$ is a quadratic form on $\mathbb{R}^2$ and k is a constant. If we were interested in only one or two particular graphs, it might be sensible to simply use a computer to produce these graphs. However, by applying diagonalization to the problem of determining these graphs, we see a very clear interpretation of eigenvectors. We also consider a concrete useful application of a change of coordinates. Moreover, this approach to these graphs leads to a *classification* of the various possibilities; all of the graphs of the form $Q(\vec{x}) = k$ in $\mathbb{R}^2$ can be divided into a few standard cases. Classification is a useful process because it allows us to say "I really only need to understand these few standard cases." A classification of these graphs is given later in this section.

Observe that in general it is difficult to identify the shape of the graph of $ax_1^2 + bx_1x_2 + cx_2^2 = k$. It is even more difficult to try to sketch the graph. However, it is easy to sketch the graph of $ax_1^2 + cx_2^2 = k$. Thus, our strategy to sketch the graph of a quadratic form $Q(\vec{x}) = k$ is to first bring it into diagonal form. Of course, we first need to determine how diagonalizing the quadratic form will affect the graph.

Theorem 1 Let $Q(\vec{x}) = ax_1^2 + bx_1x_2 + cx_2^2$, where $a, b,$ and c are not all zero. Then an orthogonal matrix P with $\det P = 1$, which diagonalizes $Q(\vec{x})$, corresponds to a rotation in $\mathbb{R}^2$.

Proof: Let $A = \begin{bmatrix} a & b/2 \\ b/2 & c \end{bmatrix}$. Since A is symmetric, by the Principal Axis Theorem, there exists an orthonormal basis $\{\vec{v}, \vec{w}\}$ of $\mathbb{R}^2$ of eigenvectors of A. Let $\vec{v} = \begin{bmatrix} v_1 \\ v_2 \end{bmatrix}$ and $\vec{w} = \begin{bmatrix} w_1 \\ w_2 \end{bmatrix}$. Since $\vec{v}$ is a unit vector, we must have

$$1 = \|\vec{v}\|^2 = v_1^2 + v_2^2$$

Hence, the entries v_1 and v_2 lie on the unit circle. Therefore, there exists an angle θ such that $v_1 = \cos\theta$ and $v_2 = \sin\theta$. Moreover, since $\vec{w}$ is a unit vector orthogonal to $\vec{w}$, we must have $\vec{w} = \pm \begin{bmatrix} -\sin\theta \\ \cos\theta \end{bmatrix}$. We choose $\vec{w} = + \begin{bmatrix} -\sin\theta \\ \cos\theta \end{bmatrix}$ so that $\det P = 1$. Hence we have

$$P = \begin{bmatrix} \cos\theta & -\sin\theta \\ \sin\theta & \cos\theta \end{bmatrix}$$

This corresponds to a rotation by θ. Finally, from our work in Section 8.2, we know that this change of coordinates matrix brings Q into diagonal form. ■

Remark

If we picked $\vec{w} = -\begin{bmatrix} -\sin\theta \\ \cos\theta \end{bmatrix}$, we would find that P corresponds to a rotation and a reflection.

In practice, we do not need to calculate the angle of rotation. When we orthogonally diagonalize $Q(\vec{x})$ with $P = \begin{bmatrix} \vec{v}_1 & \vec{v}_2 \end{bmatrix}$, the change of coordinates $\vec{x} = P\vec{y}$ causes a rotation of the new y_1- and y_2-axes. In particular, taking $\vec{y} = \begin{bmatrix} 1 \\ 0 \end{bmatrix}$, we get

$$\vec{x} = P\vec{y} = \begin{bmatrix} \vec{v}_1 & \vec{v}_2 \end{bmatrix} \begin{bmatrix} 1 \\ 0 \end{bmatrix} = \vec{v}_1$$

and taking $\vec{y} = \begin{bmatrix} 0 \\ 1 \end{bmatrix}$ gives

$$\vec{x} = P\vec{y} = \begin{bmatrix} \vec{v}_1 & \vec{v}_2 \end{bmatrix} \begin{bmatrix} 0 \\ 1 \end{bmatrix} = \vec{v}_2$$

That is, the new y_1-axis corresponds to the vector $\vec{v}_1$ in the x_1x_2-plane, and the y_2-axis corresponds the vector $\vec{v}_2$ in the x_1x_2-plane.

We demonstrate this with two examples.

EXAMPLE 1

Sketch the graph of the equation $3x_1^2 + 4x_1x_2 = 16$.

Solution: The quadratic form $Q(\vec{x}) = 3x_1^2 + 4x_1x_2$ corresponds to the symmetric matrix $A = \begin{bmatrix} 3 & 2 \\ 2 & 0 \end{bmatrix}$, so the characteristic polynomial is

$$C(\lambda) = \det(A - \lambda I) = \lambda^2 - 3\lambda - 4 = (\lambda - 4)(\lambda + 1)$$

Thus, the eigenvalues of A are $\lambda_1 = 4$ and $\lambda_2 = -1$. Thus, by an orthogonal change of coordinates, the equation can be brought into the diagonal form:

$$4y_1^2 - y_2^2 = 16$$

This is an equation of a hyperbola, and we can sketch the graph in the y_1y_2-plane. We observe that the y_1-intercepts are $(2, 0)$ and $(-2, 0)$, and there are no intercepts on the y_2-axis. The asymptotes of the hyperbola are determined by the equation $4y_1^2 - y_2^2 = 0$. By factoring, we determine that the asymptotes are lines with equations $2y_1 - y_2 = 0$ and $2y_1 + y_2 = 0$. With this information, we obtain the graph in Figure 8.3.1.

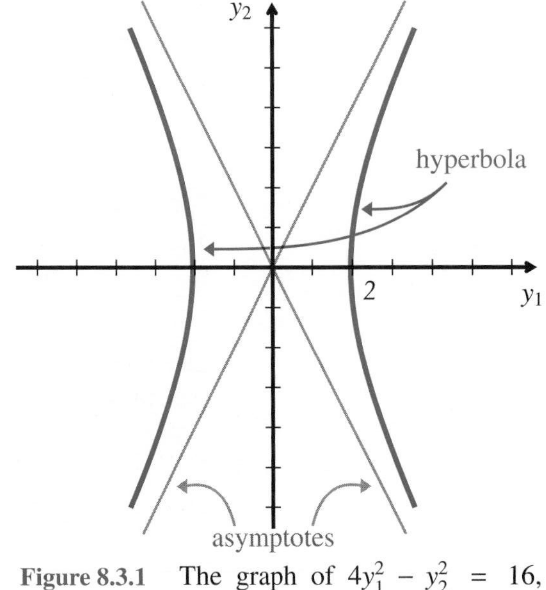

Figure 8.3.1 The graph of $4y_1^2 - y_2^2 = 16$, shown with horizontal y_1-axis and asymptotes.

However, we want a picture of the graph $3x_1^2 + 4x_1x_2 = 16$ relative to the original x_1-axis and x_2-axis—that is, in the x_1x_2-plane. Hence, we need to find the eigenvectors of A.

For $\lambda_1 = 4$,

$$A - \lambda_1 I = \begin{bmatrix} -1 & 2 \\ 2 & -4 \end{bmatrix} \sim \begin{bmatrix} 1 & -2 \\ 0 & 0 \end{bmatrix}$$

Thus, a basis for the eigenspace is $\{\vec{v}_1\}$, where $\vec{v}_1 = \begin{bmatrix} 2 \\ 1 \end{bmatrix}$.

For $\lambda_2 = -1$,

$$A - \lambda_2 I = \begin{bmatrix} 4 & 2 \\ 2 & 1 \end{bmatrix} \sim \begin{bmatrix} 2 & 1 \\ 0 & 0 \end{bmatrix}$$

EXAMPLE 1
(continued)

Thus, a basis for the eigenspace is $\{\vec{v}_2\}$, where $\vec{v}_2 = \begin{bmatrix} -1 \\ 2 \end{bmatrix}$. (We could have chosen $\begin{bmatrix} 1 \\ -2 \end{bmatrix}$, but $\begin{bmatrix} -1 \\ 2 \end{bmatrix}$ is better because $\{\vec{v}_1, \vec{v}_2\}$ is right-handed.)

Now we sketch the graph of $3x_1^2 + 4x_1x_2 = 16$. In the x_1x_2-plane, we draw the new y_1-axis in the direction of $\vec{v}_1$. (For clarity, in Figure 8.3.2 we have shown the vector $\begin{bmatrix} 4 \\ 2 \end{bmatrix}$ instead of $\begin{bmatrix} 2 \\ 1 \end{bmatrix}$.) We also draw the new y_2-axis in the direction of $\vec{v}_2$. Then, relative to these new axes, we sketch the graph of the hyperbola $4y_1^2 - y_2^2 = 16$. The graph in Figure 8.3.2 is also the graph of the original equation $3x_1^2 + 4x_1x_2 = 16$.

In order to include the asymptotes in the sketch, we rewrite their equations in standard coordinates. The orthogonal change of coordinates matrix in this case is given by $P = \frac{1}{\sqrt{5}}\begin{bmatrix} 2 & -1 \\ 1 & 2 \end{bmatrix}$. (This is a rotation of the axes through angle $\theta \approx 0.46$ radians.) Thus, the change of coordinates equation can be written

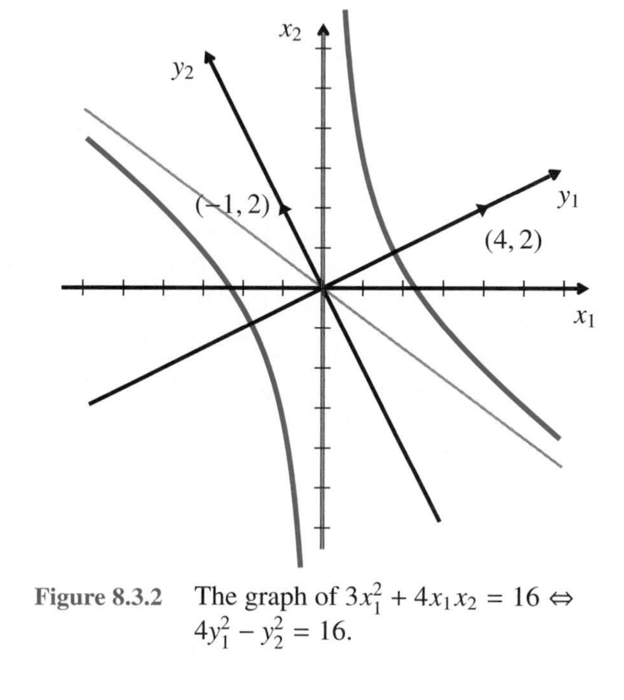

Figure 8.3.2 The graph of $3x_1^2 + 4x_1x_2 = 16 \Leftrightarrow 4y_1^2 - y_2^2 = 16$.

$$\begin{bmatrix} y_1 \\ y_2 \end{bmatrix} = \frac{1}{\sqrt{5}}\begin{bmatrix} 2 & 1 \\ -1 & 2 \end{bmatrix}\begin{bmatrix} x_1 \\ x_2 \end{bmatrix}$$

since $P^T = P^{-1}$ as P is orthogonal. This gives

$$y_1 = \frac{1}{\sqrt{5}}(2x_1 + x_2) \qquad \text{and} \qquad y_2 = \frac{1}{\sqrt{5}}(-x_1 + 2x_2)$$

Then one asymptote is

$$0 = 2y_1 + y_2 = \frac{2}{\sqrt{5}}(2x_1 + x_2) + \frac{1}{\sqrt{5}}(-x_1 + 2x_2) = \frac{1}{\sqrt{5}}(3x_1 + 4x_2)$$

The other asymptote is

$$0 = 2y_1 - y_2 = \frac{2}{\sqrt{5}}(2x_1 + x_2) - \frac{1}{\sqrt{5}}(-x_1 + 2x_2) = \frac{1}{\sqrt{5}}(5x_1)$$

Thus, in standard coordinates, the asymptotes are $3x_1 + 4x_2 = 0$ and $x_1 = 0$.

EXAMPLE 2

Sketch the graph of the equation $6x_1^2 + 4x_1x_2 + 3x_2^2 = 14$.

Solution: The corresponding symmetric matrix is $\begin{bmatrix} 6 & 2 \\ 2 & 3 \end{bmatrix}$. The eigenvalues are $\lambda_1 = 2$ and $\lambda_2 = 7$. A basis for the eigenspace of λ_1 is $\{\vec{v}_1\}$, where $\vec{v}_1 = \begin{bmatrix} 1 \\ -2 \end{bmatrix}$. A basis for the eigenspace of λ_2 is $\{\vec{v}_2\}$, where $\vec{v}_2 = \begin{bmatrix} 2 \\ 1 \end{bmatrix}$. If $\vec{v}_1$ is taken to define the y_1-axis and $\vec{v}_2$ is taken to define the y_2-axis, then the original equation is equivalent to

$$2y_1^2 + 7y_2^2 = 14$$

This is the equation of an ellipse with y_1-intercepts $(\sqrt{7}, 0)$ and $(-\sqrt{7}, 0)$ and y_2-intercepts $(0, \sqrt{2})$ and $(0, -\sqrt{2})$.

In Figure 8.3.3, the ellipse is shown relative to the y_1- and y_2-axes. In Figure 8.3.4, the new y_1- and y_2-axes determined by the eigenvectors are shown relative to the standard axes, and the ellipse from Figure 8.3.3 is rotated into place. The resulting ellipse is the graph of $6x_1^2 + 4x_1x_2 + 3x_2^2 = 14$.

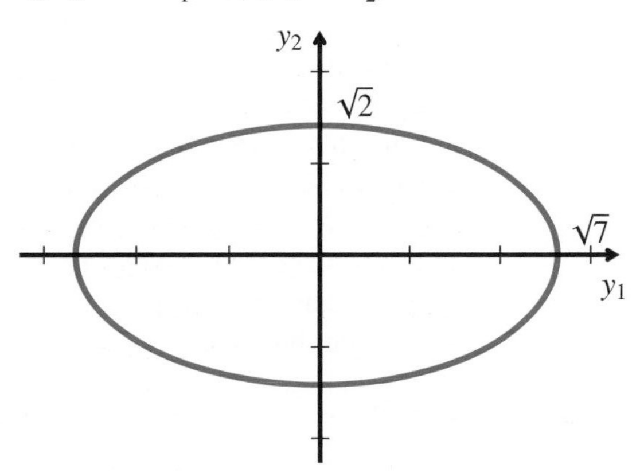

Figure 8.3.3 The graph of $2y_1^2 + 7y_2^2 = 14$ in the y_1y_2-plane.

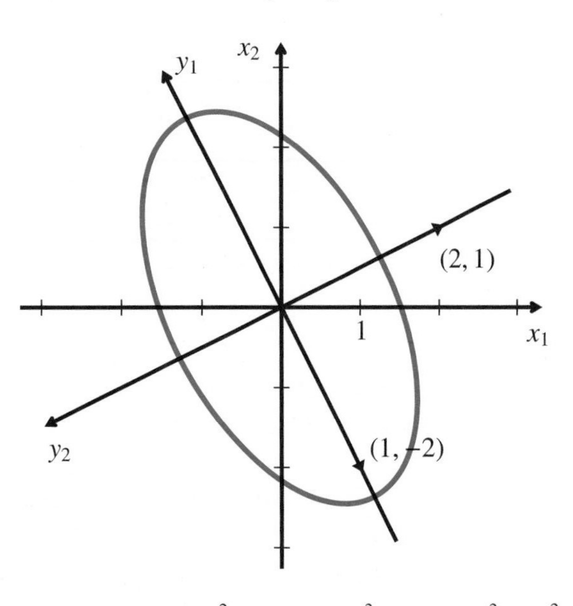

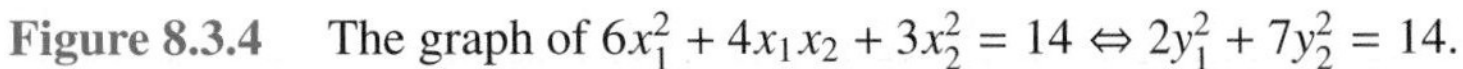

Figure 8.3.4 The graph of $6x_1^2 + 4x_1x_2 + 3x_2^2 = 14 \Leftrightarrow 2y_1^2 + 7y_2^2 = 14$.

Since diagonalizing a quadratic form corresponds to a rotation, to classify all the graphs of equations of the form $Q(\vec{x}) = k$, we diagonalize and rewrite the equation in the form $\lambda_1 y_1^2 + \lambda_2 y_2^2 = k$. Here, λ_1 and λ_2 are the eigenvalues of the corresponding symmetric matrix. The distinct possibilities are displayed in Table 8.3.1.

Table 8.3.1 Graphs of $\lambda_1 x_1^2 + \lambda_2 x_2^2 = k$

	$k > 0$	$k = 0$	$k < 0$
$\lambda_1 > 0, \lambda_2 > 0$	ellipse	point (0,0)	empty set
$\lambda_1 > 0, \lambda_2 = 0$	parallel lines	line $x = 0$	empty set
$\lambda_1 > 0, \lambda_2 < 0$	hyperbola	intersecting lines	hyperbola
$\lambda_1 = 0, \lambda_2 < 0$	empty set	line $y = 0$	parallel lines
$\lambda_1 < 0, \lambda_2 < 0$	empty set	point (0,0)	ellipse

The cases where $k = 0$ or one eigenvalue is zero may be regarded as **degenerate cases** (not general cases). The **nondegenerate cases** are the ellipses and hyperbolas, which are **conic sections**. (A conic section is a curve obtained in $\mathbb{R}^3$ as the intersection of a cone and a plane.) Notice that the cases of a single point, a single line, and intersecting lines can also be obtained as the intersection of a cone and a plane passing through the vertex of the cone. However, the cases of parallel lines (in Table 8.3.1.) are not obtained as the intersection of a cone and a plane.

It is also important to realize that one class of conic sections, parabolas, does not appear in Table 8.3.1. In $\mathbb{R}^2$, the equation of a parabola is a quadratic equation, but it contains first-degree terms. Since a quadratic form contains only second-degree terms, an equation of the form $Q(\vec{x}) = k$ cannot be a parabola.

The classification provided by Table 8.3.1 suggests that it might be interesting to consider how degenerate cases arise as limiting cases of nondegenerate cases. For example, Figure 8.3.5 shows that the case of parallel lines ($y = \pm$ constant) arises from the family of ellipses $\lambda x_1^2 + x_2^2 = 1$ as λ tends to 0.

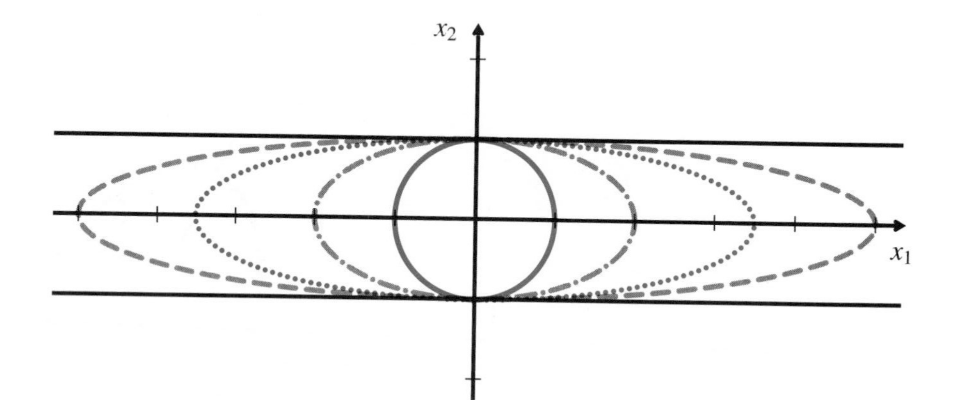

Figure 8.3.5 A family of ellipses $\lambda x_1^2 + x_2^2 = 1$. The circle occurs for $\lambda = 1$; as λ decreases, the ellipses get "fatter"; for $\lambda = 0$, the graph is a pair of lines.

Table 8.3.1 could have been constructed using only the cases $k > 0$ and $k = 0$. The graphs obtained for $k < 0$ are all obtained for $k > 0$, although they may be oriented differently. For example, the graph of $x_1^2 - x_2^2 = -1$ is the same as the graph of $-x_1^2 + x_2^2 = 1$, and this hyperbola may be obtained from the hyperbola $x_1^2 - x_2^2 = 1$ by reflecting over the line $x_1 = x_2$. However, for purposes of illustration, it is convenient to include both $k > 0$ and $k < 0$. Figure 8.3.6 shows that the case of intersecting lines ($k = 0$) separates the hyperbolas with intercepts on the x_1-axis ($x_1^2 - 2x_2^2 = k > 0$) from the hyperbolas with intercepts on the x_2-axis ($x_1^2 - 2x_2^2 = k < 0$).

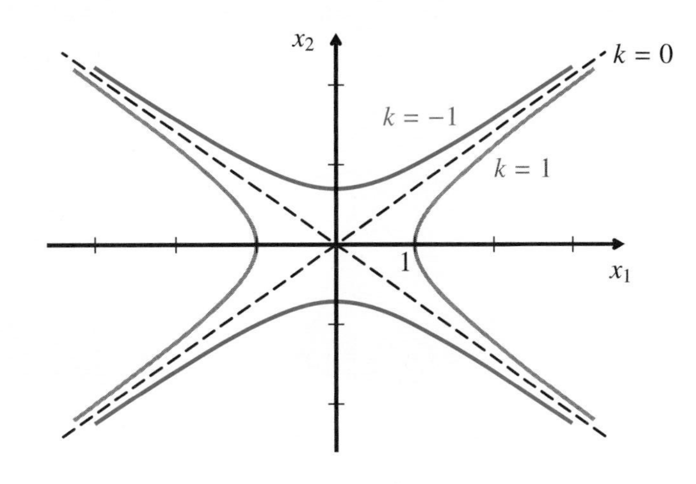

Figure 8.3.6 Graphs of $x_1^2 - 2x_2^2 = k$ for $k \in \{-1, 0, 1\}$.

EXERCISE 1

Diagonalize the quadratic form and sketch the graph of the equation $x_1^2 + 2x_1x_2 + x_2^2 = 2$. Show both the original axes and the new axes.

Graphs of $Q(\vec{x}) = k$ in $\mathbb{R}^3$

For a quadratic equation of the form $Q(\vec{x}) = k$ in $\mathbb{R}^3$, there are similar results to what we did above. However, because there are three variables instead of two, there are more possibilities. The nondegenerate cases give ellipsoids, hyperboloids of one sheet, and hyperboloids of two sheets. These graphs are called **quadric surfaces**.

The usual standard form for the equation of an ellipsoid is $\frac{x_1^2}{a^2} + \frac{x_2^2}{b^2} + \frac{x_3^2}{c^2} = 1$. This is the case obtained by diagonalizing $Q(\vec{x}) = k$ if the eigenvalues and k are all non-zero and have the same sign. In particular, if we write

$$a^2 = k/\lambda_1, \qquad b^2 = k/\lambda_2, \qquad c^2 = k/\lambda_3$$

An ellipsoid is shown in Figure 8.3.7. The positive intercepts on the coordinate axes are $(a, 0, 0)$, $(0, b, 0)$, and $(0, 0, c)$.

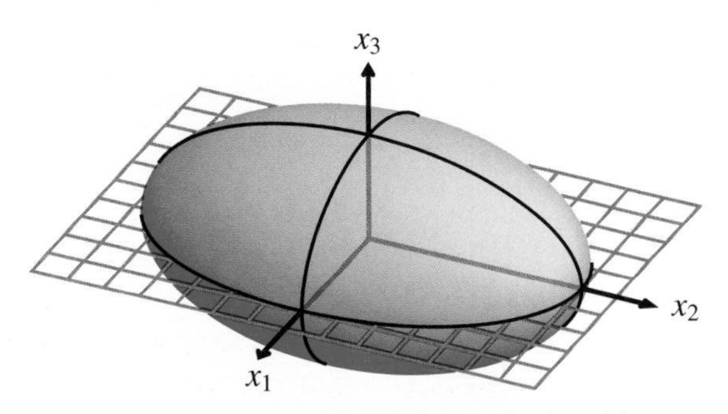

Figure 8.3.7 An ellipsoid in standard position.

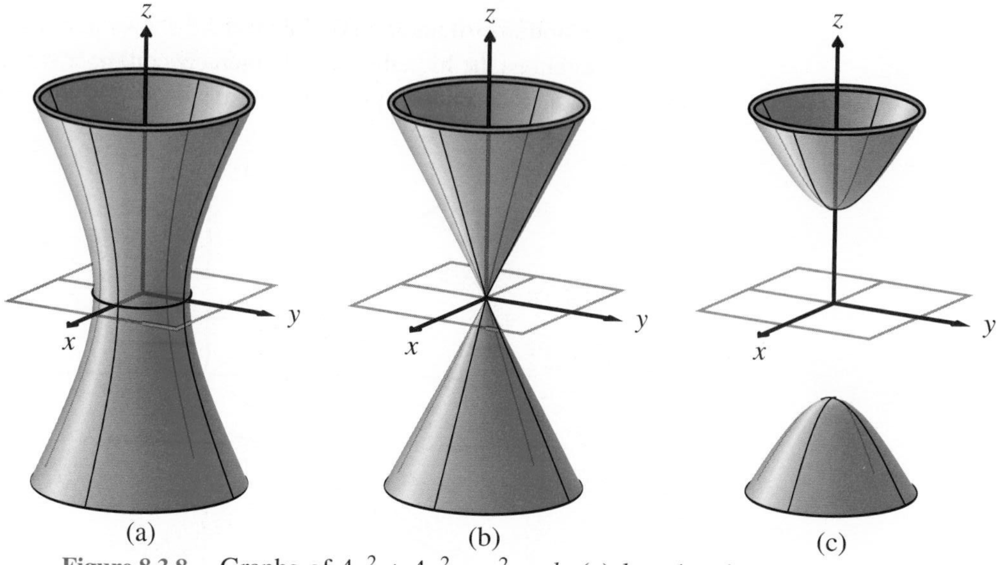

Figure 8.3.8 Graphs of $4x_1^2 + 4x_2^2 - x_3^2 = k$. (a) $k = 1$; a hyperboloid of one sheet. (b) $k = 0$; a cone. (c) $k = -1$; a hyperboloid of two sheets.

The standard form of the equation for a hyperboloid of one sheet is $\frac{x_1^2}{a^2} + \frac{x_2^2}{b^2} - \frac{x_3^2}{c^2} = 1$. This form is obtained when k and two eigenvalues of the matrix of Q are positive and the third eigenvalue is negative. It is also obtained when k and two eigenvalues are negative and the other eigenvalue is positive. Notice that if this is rewritten $\frac{x_1^2}{a^2} + \frac{x_2^2}{b^2} = 1 - \frac{x_3^2}{c^2}$, it is clear that for every z there are values of x_1 and x_2 that satisfy the equation, so that the surface is all one piece (or one sheet). A hyperboloid of one sheet is shown if Figure 8.3.8 (a).

The standard form of the equation for a hyperboloid of two sheets is $\frac{x_1^2}{a^2} + \frac{x_2^2}{b^2} - \frac{x_3^2}{c^2} = -1$. This form is obtained when k and one eigenvalue is negative and the other eigenvalues are positive, or when k and one eigenvalue are positive and the other eigenvalues are negative. Notice that if this is rewritten $\frac{x_1^2}{a^2} + \frac{x_2^2}{b^2} = -1 - \frac{x_3^2}{c^2}$, it is clear that for every $|x_3| < c$, there are no values of x_1 and x_2 that satisfy the equation. Therefore, the graph consists of two pieces (or two sheets), one with $x_3 \geq c$ and the other with $x_3 \leq -c$. A hyperboloid of two sheets is shown in Figure 8.3.8 (c).

It is interesting to consider the family of surfaces obtained by varying k in the equation $\frac{x_1^2}{a^2} + \frac{x_2^2}{b^2} - \frac{x_3^2}{c^2} = k$, as in Figure 8.3.8. When $k = 1$, the surface is a hyperboloid of one sheet; as k decreases toward 0, the "waist" of the hyperboloid shrinks until at $k = 0$ it has "pinched in" to a single point and the hyperboloid of one sheet becomes a cone. As k decreases towards -1, the waist has disappeared, and the graph is now a hyperboloid of two sheets.

Table 8.3.2 Graphs of $\lambda_1 x_1^2 + \lambda_2 x_2^2 + \lambda_3 x_3^2 = k$

	$k > 0$	$k = 0$
$\lambda_1, \lambda_2, \lambda_3 > 0$	ellipsoid	point $(0, 0, 0)$
$\lambda_1, \lambda_2 > 0, \lambda_3 = 0$	elliptic cylinder	x_3-axis
$\lambda_1, \lambda_2 > 0, \lambda_3 < 0$	hyperboloid of one sheet	cone
$\lambda_1 > 0, \lambda_2 = 0, \lambda_3 < 0$	hyperbolic cylinder	intersecting planes
$\lambda_1 > 0, \lambda_2, \lambda_3 < 0$	hyperboloid of two sheets	cone
$\lambda_1 = 0, \lambda_2, \lambda_3 < 0$	empty set	x_1-axis
$\lambda_1, \lambda_2, \lambda_3 < 0$	empty set	point $(0, 0, 0)$

Table 8.3.2 displays the possible cases for $Q(\vec{x}) = k$ in $\mathbb{R}^3$. The nondegenerate cases are the ellipsoids and hyperboloids. Note that the hyperboloid of two sheets appears in the form $\frac{x_1^2}{a^2} - \frac{x_2^2}{b^2} - \frac{x_3^2}{c^2} = k$, $k > 0$.

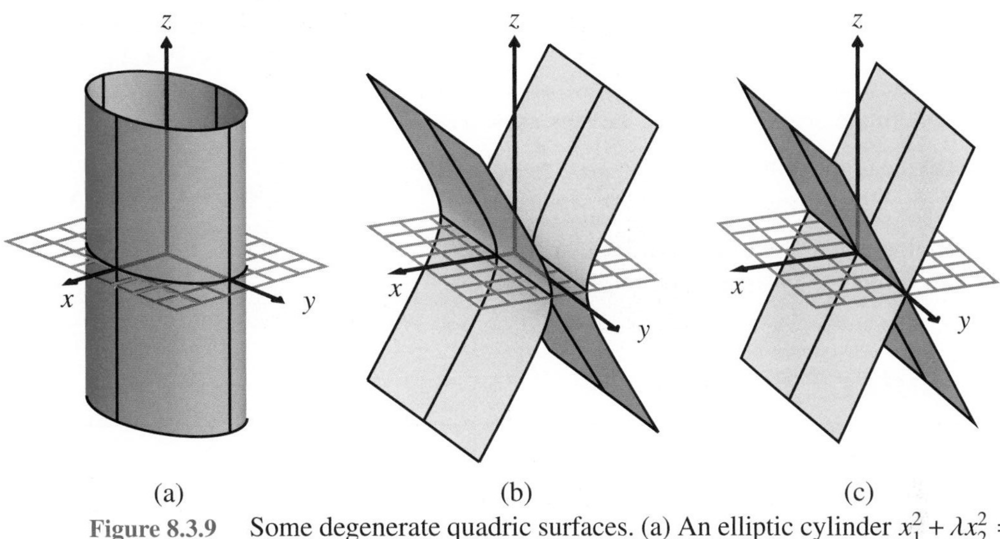

Figure 8.3.9 Some degenerate quadric surfaces. (a) An elliptic cylinder $x_1^2 + \lambda x_2^2 = 1$, parallel to the x_3-axis. (b) A hyperbolic cylinder $\lambda x_1^2 - x_3^2 = 1$, parallel to the x_2-axis. (c) Intersecting planes $\lambda x_1^2 - x_3^2 = 0$.

Figure 8.3.9 shows some degenerate quadric surfaces. Note that paraboloidal surfaces do not appear as graphs of the form $Q(\vec{x}) = k$ in $\mathbb{R}^3$ for the same reason that parabolas do not appear in Table 8.3.1 for $\mathbb{R}^2$: their equations contain first-degree terms.

PROBLEMS 8.3

Practice Problems

A1 Sketch the graph of the equation $2x_1^2 + 4x_1x_2 - x_2^2 = 6$. Show both the original axes and the new axes.

A2 Sketch the graph of the equation $2x_1^2 + 6x_1x_2 + 10x_2^2 = 11$. Show both the original axes and the new axes.

A3 Sketch the graph of the equation $4x_1^2 - 6x_1x_2 + 4x_2^2 = 12$. Show both the original axes and the new axes.

A4 Sketch the graph of the equation $5x_1^2 + 6x_1x_2 - 3x_2^2 = 15$. Show both the original axes and the new axes.

A5 For each of the following symmetric matrices, identify the shape of the graph $\vec{x}^T A\vec{x} = 1$ and the shape of the graph $\vec{x}^T A\vec{x} = -1$.

(a) $A = \begin{bmatrix} 4 & 2 \\ 2 & 1 \end{bmatrix}$

(b) $A = \begin{bmatrix} 5 & 3 \\ 3 & -3 \end{bmatrix}$

(c) $A = \begin{bmatrix} 0 & 1 & 1 \\ 1 & 0 & 1 \\ 1 & 1 & 0 \end{bmatrix}$

(d) $A = \begin{bmatrix} 1 & 0 & -2 \\ 0 & -1 & -2 \\ -2 & -2 & 0 \end{bmatrix}$

(e) $A = \begin{bmatrix} 1 & 8 & 4 \\ 8 & 1 & -4 \\ 4 & -4 & 7 \end{bmatrix}$

Homework Problems

B1 Sketch the graph of the equation $9x_1^2+4x_1x_2+6x_2^2 = 90$. Show both the original axes and the new axes.

B2 Sketch the graph of the equation $x_1^2+6x_1x_2-7x_2^2 = 32$. Show both the original axes and the new axes.

B3 Sketch the graph of the equation $x_1^2-4x_1x_2+x_2^2 = 8$. Show both the original axes and the new axes.

B4 Sketch the graph of the equation $x_1^2+4x_1x_2+x_2^2 = 8$. Show both the original axes and the new axes.

B5 Sketch the graph of the equation $3x_1^2-4x_1x_2+3x_2^2 = 32$. Show both the original axes and the new axes.

B6 In each of the following cases, diagonalize the quadratic form. Then determine the shape of the surface $Q(\vec{x}) = k$ for $k = 1, 0, -1$. Note that two of the quadratic forms are degenerate.

(a) $Q(\vec{x}) = x_1^2 + 4x_1x_2 + x_2^2$
(b) $Q(\vec{x}) = x_1^2 + 6x_1x_2 + 2x_1x_3 + x_2^2 + 2x_2x_3 + 5x_3^2$
(c) $Q(\vec{x}) = x_1^2 + 4x_1x_2 + 4x_1x_3 + 5x_2^2 + 6x_2x_3 + 5x_3^2$
(d) $Q(\vec{x}) = -x_1^2 + 2x_1x_2 - 6x_1x_3 + x_2^2 - 2x_2x_3 - x_3^2$
(e) $Q(\vec{x}) = 4x_1^2 + 2x_1x_2 + 5x_2^2 - 2x_2x_3 + 4x_3^2$

Computer Problems

C1 Identify the following surfaces by using a computer to find the eigenvalues of the corresponding symmetric matrix. Plot graphs of the original system and of the diagonalized system.

(a) $x_1^2 - 14x_1x_2 + 6x_1x_3 - x_2^2 + 8x_2x_3 = 10$
(b) $3x_1^2 + 10x_1x_2 + 4x_1x_3 + 16x_2x_3 - 6x_3^2 = 37$

8.4 Applications of Quadratic Forms

Applying quadratic forms requires some knowledge of calculus and physics. This section may be omitted with no loss of continuity.

Some may think of mathematics as only a set of rules for doing calculations. However, a theorem such as the Principal Axis Theorem is often important because it provides a simple way of thinking about complicated situations. The Principal Axis Theorem plays an important role in the two applications described here.

Small Deformations

A small deformation of a solid body may be understood as the composition of three stretches along the principal axes of a symmetric matrix together with a rigid rotation of the body.

Consider a body of material that can be deformed when it is subjected to some external forces. This might be, for example, a piece of steel under some load. Fix an origin of coordinates $\vec{0}$ in the body; to simplify the story, suppose that this origin is left unchanged by the deformation. Suppose that a material point in the body, which is at $\vec{x}$ before the forces are applied, is moved by the forces to the point $f(\vec{x}) = (f_1(\vec{x}), f_2(\vec{x}), f_3(\vec{x}))$; we have assumed that $f(\vec{0}) = \vec{0}$. The problem is to understand this deformation f so that it can be related to the properties of the body. (Note that f represents the displacement of the point initially at $\vec{x}$, not the force at $\vec{x}$.)

For many materials under reasonable forces, the deformation is small; this means that the point $f(\vec{x})$ is not far from $\vec{x}$. It is convenient to introduce a parameter β to describe how small the deformation is and a function $h(\vec{x})$ and write

$$f(\vec{x}) = \vec{x} + \beta h(\vec{x})$$

This equation is really the definition of $h(\vec{x})$ in terms of the point $\vec{x}$, the given function $f(\vec{x})$, and the parameter β.

For many materials, an arbitrary small deformation is well approximated by its "best linear approximation," the derivative. In this case, the map $f : \mathbb{R}^3 \to \mathbb{R}^3$ is approximated near the origin by the linear transformation with matrix $\left[\frac{\partial f_j}{\partial x_k}(\vec{0})\right]$, so that in this approximation, a point originally at $\vec{v}$ is moved (approximately) to $\left[\frac{\partial f_j}{\partial x_k}(\vec{0})\right]\vec{v}$. (This is a standard calculus approximation.)

In terms of the parameter β and the function h, this matrix can be written as

$$\left[\frac{\partial f_j}{\partial x_k}(\vec{0})\right] = I + \beta G$$

where $G = \left[\frac{\partial h_j}{\partial x_k}(\vec{0})\right]$. In this situation, it is useful to write G as $G = E + W$, where

$$E = \frac{1}{2}(G + G^T)$$

is its symmetric part, and

$$W = \frac{1}{2}(G - G^T)$$

is its skew-symmetric part, as in Problem 8.2.D5.

The next step is to observe that we can write

$$I + \beta G = I + \beta(E + W) = (I + \beta E)(I + \beta W) - \beta^2 EW$$

Since β is assumed to be small, β^2 is very small and may be ignored. (Such treatment of terms like β^2 can be justified by careful discussion of the limit at $\beta \to 0$.)

The small deformation we started with is now described as the composition of two linear transformations, one with matrix $I + \beta E$ and the other with matrix $I + \beta W$. It can be shown that $I + \beta W$ describes a small rigid rotation of the body; a rigid rotation does not alter the distance between any two points in the body. (The matrix βW is called an *infinitesimal rotation*.)

Finally, we have the linear transformation with matrix $I + \beta E$. This matrix is symmetric, so there exist principal axes such that the symmetric matrix is diagonalized to $\begin{bmatrix} 1+\epsilon_1 & 0 & 0 \\ 0 & 1+\epsilon_2 & 0 \\ 0 & 0 & 1+\epsilon_3 \end{bmatrix}$. (It is equivalent to diagonalize βE and add the result to I, because I is transformed to itself under any orthonormal change of coordinates.) Since β is small, it follows that the numbers ϵ_j are small in magnitude, and therefore $1 + \epsilon_j > 0$. This diagonalized matrix can be written as the product of the three matrices:

$$\begin{bmatrix} 1+\epsilon_1 & 0 & 0 \\ 0 & 1+\epsilon_2 & 0 \\ 0 & 0 & 1+\epsilon_3 \end{bmatrix} = \begin{bmatrix} 1+\epsilon_1 & 0 & 0 \\ 0 & 1 & 0 \\ 0 & 0 & 1 \end{bmatrix}\begin{bmatrix} 1 & 0 & 0 \\ 0 & 1+\epsilon_2 & 0 \\ 0 & 0 & 1 \end{bmatrix}\begin{bmatrix} 1 & 0 & 0 \\ 0 & 1 & 0 \\ 0 & 0 & 1+\epsilon_3 \end{bmatrix}$$

It is now apparent that, excluding rotation, the small deformation can be represented as the composition of three stretches along the principal axes of the matrix βE. The quantities ϵ_1, ϵ_2, and ϵ_3 are related to the external and internal forces in the material by elastic properties of the material. (βE is called the infinitesimal strain; this notation is not quite the standard notation. This will be important if you read further about this topic in a book on continuum mechanics.)

The Inertia Tensor

For the purpose of discussing the rotation motion of a rigid body, information about the mass distribution within the body is summarized in a symmetric matrix N called the **inertia tensor**. (1) The tensor is easiest to understand if principal axes are used so that the matrix is diagonal; in this case, the diagonal entries are simply the moments of inertia about the principal axes, and the moment of inertia about any other axis can be calculated in terms of these **principal moments of inertia**. (2) In general, the **angular momentum** vector $\vec{J}$ of the rotating body is equal to $N\vec{\omega}$, where $\vec{\omega}$ is the **instantaneous angular velocity** vector. The vector $\vec{J}$ is a scalar multiple of $\vec{\omega}$ if and only if $\vec{\omega}$ is an eigenvector of N—that is, if and only if the axis of rotation is one of the principal axes of the body. This is a beginning to an explanation of how the body wobbles during rotation ($\vec{\omega}$ need not be constant) even though $\vec{J}$ is a conserved quantity (that is, $\vec{J}$ is constant if no external force is applied).

Suppose that a rigid body is rotating about some point in the body that remains fixed in space throughout the rotation. Make this fixed point the origin $(0, 0, 0)$. Suppose that there are coordinate axes fixed in space and also three reference axes that are fixed in the body (so that they rotate with the body). At any time t, these body axes make certain angles with respect to the space axes; at a later time $t + \Delta t$, the body axes have moved to a new position. Since $(0, 0, 0)$ is fixed and the body is rigid, the body axes have moved only by a rotation, and it is a fact that any rotation in $\mathbb{R}^3$ is determined by its axis and an angle. Call the unit vector along this axis $\vec{u}(t + \Delta t)$ and denote the angle by $\Delta\theta$. Now let $\Delta t \to 0$; the unit vector $\vec{u}(t + \Delta t)$ must tend to a limit $\vec{u}(t)$, and this determines the **instantaneous axis of rotation at time** t. Also, as $\Delta t \to 0$, $\frac{\Delta\theta}{\Delta t} \to \frac{d\theta}{dt}$, the **instantaneous rate of rotation about the axis**. The **instantaneous angular velocity** is defined to be the vector $\vec{\omega} = \left(\frac{d\theta}{dt}\right)\vec{u}(t)$.

(It is a standard exercise to show that the instantaneous linear velocity $\vec{v}(t)$ at some point in the body whose space coordinates are given by $\vec{x}(t)$ is determined by $\vec{v} = \vec{\omega} \times \vec{x}$.)

To use concepts such as energy and momentum in the discussion of rotating motion, it is necessary to introduce moments of inertia.

For a single mass m at the point (x_1, x_2, x_3) the **moment of inertia about the** x_3**-axis** is defined to be $m(x_1^2 + x_2^2)$; this will be denoted by n_{33}. The factor $(x_1^2 + x_2^2)$ is simply the square of the distance of the mass from the x_3-axis. There are similar definitions of the moments of inertia about the x_1-axis (denoted by n_{11}) and about the x_2-axis (denoted by n_{22}).

For a general axis ℓ through the origin with unit direction vector $\vec{u}$, the moment of inertia of the mass about ℓ is defined to be m multiplied by the square of the distance of m from ℓ. Thus, if we let $\vec{x} = \begin{bmatrix} x_1 \\ x_2 \\ x_3 \end{bmatrix}$, the moment of inertia in this case is

$$m\|\operatorname{perp}_{\vec{u}} \vec{x}\|^2 = m[\vec{x} - (\vec{x} \cdot \vec{u})\vec{u}]^T[\vec{x} - (\vec{x} \cdot \vec{u})\vec{u}] = m(\|\vec{x}\|^2 - (\vec{x} \cdot \vec{u})^2)$$

With some manipulation, using $\vec{u}^T\vec{u} = 1$ and $\vec{x} \cdot \vec{u} = \vec{x}^T\vec{u}$, we can verify that this is equal to the expression

$$\vec{u}^T m(\|\vec{x}\|^2 I - \vec{x}\vec{x}^T)\vec{u}$$

Because of this, for the single point mass m at $\vec{x}$, we define the **inertia tensor** N to be the 3×3 matrix

$$N = m(\|\vec{x}\|^2 I - \vec{x}\vec{x}^T)$$

(Vectors and matrices are special kinds of "tensors"; for our present purposes, we simply treat N as a matrix.) With this definition, the moment of inertia about an axis with unit direction $\vec{u}$ is

$$\vec{u}^T N \vec{u}$$

It is easy to check that N is the matrix with components n_{11}, n_{22}, and n_{33} as given above, and for $i \neq j$, $n_{ij} = -mx_ix_j$. It is clear that this matrix N is symmetric because $\vec{x}\vec{x}^T$ is a symmetric 3×3 matrix. (The term mx_ix_j is called a *product of inertia*. This name has no special meaning; the term is simply a product that appears as an entry in the inertia tensor.)

It is easy to extend the definition of moments of inertia and the inertia tensor to bodies that are more complicated than a single point mass. Consider a rigid body that can be thought of as k masses joined to each other by weightless rigid rods. The moment of inertia of the body about the x_3-axis is determined by taking the moment of inertia about the x_3-axis of each mass and simply adding these moments; the moments about the x_1- and x_2-axes, and the products of the inertia are defined similarly. The inertia tensor of this body is just the sum of the inertia tensors of the k masses; since it is the sum of symmetric matrices, it is also symmetric. If the mass is distributed continuously, the various moments and products of inertia are determined by definite integrals. In any case, the inertia tensor N is still defined, and is still a symmetric matrix.

Since N is a symmetric matrix, it can be brought into diagonal form by the Principal Axis Theorem. The diagonal entries are then the moments of inertia with respect to the principal axes, and these are called the **principal moments of inertia**. Denote these by N_1, N_2, and N_3. Let $\mathcal{P}$ denote the orthonormal basis consisting of eigenvectors of N (which means these vectors are unit vectors along the principal axes). Suppose an arbitrary axis ℓ is determined by the unit vector $\vec{u}$ such that $[\vec{u}]_{\mathcal{P}} = \begin{bmatrix} p_1 \\ p_2 \\ p_3 \end{bmatrix}$. Then, from the discussion of quadratic forms in Section 8.2, the moment of inertia about this axis ℓ is simply

$$\begin{aligned} \vec{u}^T N \vec{u} &= \begin{bmatrix} p_1 & p_2 & p_3 \end{bmatrix} \begin{bmatrix} N_1 & 0 & 0 \\ 0 & N_2 & 0 \\ 0 & 0 & N_3 \end{bmatrix} \begin{bmatrix} p_1 \\ p_2 \\ p_3 \end{bmatrix} \\ &= p_1^2 N_1 + p_2^2 N_2 + p_3^2 N_3 \end{aligned}$$

This formula is greatly simplified because of the use of the principal axes.

It is important to get equations for rotating motion that corresponds to Newton's equation:

The rate of change of momentum equals the applied force.

The appropriate equation is

The rate of change of angular momentum is the applied torque.

It turns out that the right way to define the angular momentum vector $\vec{J}$ for a general body is

$$\vec{J} = N(t)\vec{\omega}(t)$$

Note that in general N is a function of t since it depends on the positions at time t of each of the masses making up the solid body. Understanding the possible motions of a rotating body depends on determining $\vec{\omega}(t)$, or at least saying something about it. In general, this is a very difficult problem, but there will often be important simplifications if N is diagonalized by the Principal Axis Theorem. Note that $\vec{J}(t)$ is parallel to $\vec{\omega}(t)$ if and only if $\vec{\omega}(t)$ is an eigenvector of $N(t)$.

PROBLEM 8.4

Conceptual Problem

D1 Show that if P is an orthogonal matrix that diagonalizes the symmetric matrix βE to a matrix with diagonal entries ϵ_1, ϵ_2, and ϵ_3, then P also diagonalizes $(I + \beta E)$ to a matrix with diagonal entries $1 + \epsilon_1$, $1 + \epsilon_2$, and $1 + \epsilon_3$.

CHAPTER REVIEW

Suggestions for Student Review

1. How does the theory of diagonalization of symmetric matrices differ from the theory for general square matrices? (Section 8.1)
2. Explain the connection between quadratic forms and symmetric matrices. How do you find the symmetric matrix corresponding to a quadratic form? How does diagonalization of the symmetric matrix enable us to diagonalize the quadratic form? (Section 8.2)
3. List the classifications of a quadratic form. How does diagonalizing the corresponding symmetric matrix help us classify a quadratic form? (Section 8.2)
4. What role do eigenvectors play in helping us understand the graphs of equations $Q(\vec{x}) = k$, where $Q(\vec{x})$ is a quadratic form? (Section 8.3)
5. Define the principal axes of a symmetric matrix A. How do the principal axes of A relate to the graph of $Q(\vec{x}) = \vec{x}^T A\vec{x} = k$? (Section 8.3)
6. When diagonalizing a symmetric matrix A, we know that we can choose the eigenvalues in any order. How would changing the order in which we pick the eigenvalues change the graph of $Q(\vec{x}) = \vec{x}^T A\vec{x} = k$? Explain. (Section 8.3)

Chapter Quiz

E1 Let $A = \begin{bmatrix} 2 & -3 & 2 \\ -3 & 3 & 3 \\ 2 & 3 & 2 \end{bmatrix}$. Find an orthogonal matrix P such that $P^T AP = D$ is diagonal.

E2 For each of the following quadratic forms $Q(\vec{x})$,
- (i) Determine the corresponding symmetric matrix A.
- (ii) Express $Q(\vec{x})$ in diagonal form and give the orthogonal matrix that brings it into this form.
- (iii) Classify $Q(\vec{x})$.
- (iv) Describe the shape of $Q(\vec{x}) = 1$ and $Q(\vec{x}) = 0$.

(a) $Q(\vec{x}) = 5x_1^2 + 4x_1x_2 + 5x_2^2$
(b) $Q(\vec{x}) = 2x_1^2 - 6x_1x_2 - 6x_1x_3 - 3x_2^2 + 4x_2x_3 - 3x_3^2$

E3 By diagonalizing the quadratic form, make a sketch of the graph of

$$5x_1^2 - 2x_1x_2 + 5x_2^2 = 12$$

in the x_1x_2-plane. Show the new and old coordinate axes.

E4 Prove that if A is a positive definite symmetric matrix, then $\langle \vec{x}, \vec{y} \rangle = \vec{x}^T A\vec{y}$ is an inner product on $\mathbb{R}^n$.

E5 Prove that if A is a 4×4 symmetric matrix with characteristic polynomial $C(\lambda) = (\lambda - 3)^4$, then $A = 3I$.

Further Problems

F1 In Problem 7.F5, we saw the QR-factorization: an invertible $n \times n$ matrix A can be expressed in the form $A = QR$, where Q is orthogonal and R is upper triangular. Let $A_1 = RQ$, and prove that A_1 is orthogonally similar to A and hence has the same eigenvalues as A. (By repeating this process, $A = Q_1R_1, A_1 = R_1Q_1, A_1 = Q_2R_2, A_2 = R_2Q_2, \ldots$, one obtains an effective numerical procedure for determining eigenvalues of a symmetric matrix.)

F2 Suppose that A is an $n \times n$ positive semidefinite symmetric matrix. Prove that A has a square root. That is, show that there is a positive semidefinite symmetric matrix B such that $B^2 = A$. (Hint: Suppose that Q diagonalizes A to D so that $Q^TAQ = D$. Define C to be a positive square root for D and let $B = QCQ^T$.)

F3 (a) If A is any $n \times n$ matrix, prove that A^TA is symmetric and positive semidefinite. (Hint: Consider $A\vec{x} \cdot A\vec{x}$.)

(b) If A is invertible, prove that A^TA is positive definite.

F4 (a) Suppose that A is an invertible $n \times n$ matrix. Prove that A can be expressed as a product of an orthogonal matrix Q and a positive definite symmetric matrix U, $A = QU$. This is known as a **polar decomposition** of A. (Hint: Use Problems F2 and F3, let U be the square root of A^TA, and let $Q = AU^{-1}$.)

(b) Let $V = QUQ^T$. Show that V is symmetric and that $A = VQ$. Moreover, show that $V^2 = AA^T$, so that V is a positive definite symmetric square root of AA^T.

(c) Suppose that the 3×3 matrix A is the matrix of an orientation-preserving linear mapping L. Show that L is the composition of a rotation following three stretches along mutually orthogonal axes. (This follows from part (a), facts about isometries of $\mathbb{R}^3$, and ideas in Section 8.4. In fact, this is a finite version of the result for infinitesimal strain in Section 8.4.)

Companion Website

Visit the text's website at www.pearsoncanada.ca/norman for practice quizzes, additional applications, and an essay on linearity and superposition in physics.

CHAPTER 9

Complex Vector Spaces

CHAPTER OUTLINE

When they first encounter imaginary numbers, many students wonder why we look at numbers that are not real. In fact, it was not until Rafael Bombelli showed in 1572 that numbers involving square roots of negative numbers can be used to solve real-world problems. Currently, complex numbers are used to solve problems in a wide variety of areas. Some examples are electronics, control theory, quantum mechanics, and fluid dynamics. Our goal in this chapter is to extend everything we did in Chapters 1–8 to allow for the use of complex numbers instead of just real numbers.

9.1 Complex Numbers

The first numbers we encounter as children are the natural numbers $1, 2, 3$, and so on. In school, we soon find that in order to perform certain subtractions, we must extend our concept of number to the integers, which include the natural numbers. Then, so that division can always be carried out, we extend the concept of number to the rational numbers, which include the integers. Next we have to extend our understanding to the real numbers, which include all the rationals and also include irrational numbers.

To solve the equation $x^2 + 1 = 0$, we have to extend our concept of number one more time. We define the number i to be a number such that $i^2 = -1$. The system of numbers of the form $x + yi$ where $x, y \in \mathbb{R}$ is called the *complex numbers*. Note that the real numbers are included as those complex numbers with $b = 0$. As in the case with all the previous extensions of our understanding of number, some people are initially uncertain about the meaning of the "new" numbers. However, the complex numbers have a consistent set of rules of arithmetic, and the extension to complex numbers is justified by the fact that they allow us to solve important mathematical and physical problems that we could not solve using only real numbers.

The Arithmetic of Complex Numbers

Definition
Complex Number

A **complex number** is a number of the form $z = x + yi$, where $x, y \in \mathbb{R}$, and i is an element such that $i^2 = -1$. The set of all complex numbers is denoted by $\mathbb{C}$.

Addition of complex numbers $z_1 = x_1 + y_1 i$ and $z_2 = x_2 + y_2 i$ is defined by

$$z_1 + z_2 = (x_1 + x_2) + (y_1 + y_2)i$$

Multiplication of complex numbers $z_1 = x_1 + y_1 i$ and $z_2 = x_2 + y_2 i$ is defined by

$$\begin{aligned} z_1 z_2 &= (x_1 + y_1 i)(x_2 + y_2 i) \\ &= x_1 x_2 + x_1 y_2 i + x_2 y_1 i + y_1 y_2 i^2 \\ &= (x_1 x_2 - y_1 y_2) + (x_1 y_2 + x_2 y_1)i \end{aligned}$$

EXAMPLE 1

Perform the following operations

(a) $(2 + 3i) + (5 - 4i)$

Solution: $(2 + 3i) + (5 - 4i) = 2 + 5 + (3 - 4)i = 7 - i$

(b) $(2 + 3i) - (5 - 4i)$

Solution: $(2 + 3i) - (5 - 4i) = 2 - 5 + (3 - (-4))i = -3 + 7i$

(c) $(3 - 2i)(-2 + 5i)$

Solution: $(3 - 2i)(-2 + 5i) = [3(-2) - (-2)(5)] + [3(5) + (-2)(-2)]i = 4 + 19i$

EXERCISE 1

Calculate the following:

(a) $(1 - 4\text{i}) + (2 + 5\text{i})$

(b) $(2 + 2\text{i})\text{i}$

(c) $(1 - 3\text{i})(2 + \text{i})$

(d) $(3 - 2\text{i})(3 + 2\text{i})$

Remarks

1. Notice that for a complex number $z = x + yi$, we have that $z = 0$ if and only if $x = 0$ and $y = 0$.
2. If $z = x + yi$, we say that the **real part** of z is x and write $\text{Re}(z) = x$. We say that the **imaginary part** of z is y (*not* yi), and we write $\text{Im}(z) = y$. It is important to remember that the imaginary part of z is a real number.
3. If $x = 0$, $z = yi$ is said to be "purely imaginary." If $y = 0$, $z = x$ is "purely real." Notice that the real numbers are the subset of $\mathbb{C}$ of purely real complex numbers.
4. It is best not to use $\sqrt{-1}$ as a notation for the number i; doing so can lead to confusion in some cases. In physics and engineering, it is common to use j in place of i since the letter i is often used to denote electric current.
5. It is sometimes convenient to write $x + iy$ instead of $x + yi$. This is particularly common with the polar form for complex numbers, which is discussed below.

The Complex Conjugate and Division

We have not yet discussed division of complex numbers. From the multiplication in Example 1, we can say that

$$\frac{4+19i}{3-2i} = -2+5i$$

In order to give a systematic method for expressing the quotient of two complex numbers as a complex number in standard form, it is useful to introduce the complex conjugate.

Definition
Complex Conjugate

The **complex conjugate** of the complex number $z = x + yi$ is $x - yi$ and is denoted

$$\overline{z} = x - yi$$

EXAMPLE 2

$$\overline{2+5i} = 2-5i$$
$$\overline{-3-2i} = -3+2i$$
$$\overline{x} = x, \quad \text{for any } x \in \mathbb{R}$$

Theorem 1

Properties of the Complex Conjugate

For complex numbers $z_1 = x + iy$ and z_2 with $x, y \in \mathbb{R}$ we have

(1) $\overline{\overline{z_1}} = z_1$
(2) z_1 is purely real if and only if $\overline{z_1} = z_1$
(3) z_1 is purely imaginary if and only if $\overline{z_1} = -z$
(4) $\overline{z_1 + z_2} = \overline{z_1} + \overline{z_2}$
(5) $\overline{z_1 z_2} = \overline{z_1}\,\overline{z_2}$
(6) $\overline{z_1^n} = \overline{z_1}^n$
(7) $z_1 + \overline{z_1} = 2\,\mathrm{Re}(z_1) = 2x_1$
(8) $z_1 - \overline{z_1} = i2\,\mathrm{Im}(z_1) = i2y_1$
(9) $z_1\overline{z_1} = x_1^2 + y_1^2$

EXERCISE 2

Prove properties (1), (2), and (4) in Theorem 1.

The proofs of the remaining properties are left as Problem D1.

The **quotient** of two complex numbers can now be displayed as a complex number in standard form by multiplying both the numerator and the denominator by the complex conjugate of the denominator and simplifying. If $z_1 = x_1 + y_1 i$ and $z_2 = x_2 + y_2 i \neq 0$, then

$$\begin{aligned}\frac{z_1}{z_2} = \frac{z_1\overline{z_2}}{z_2\overline{z_2}} &= \frac{(x_1+y_1 i)(x_2-y_2 i)}{(x_2+y_2 i)(x_2-y_2 i)} \\ &= \frac{(x_1x_2+y_1y_2)+(y_1x_2-x_1y_2)i}{x_2^2+y_2^2} \\ &= \frac{x_1x_2+y_1y_2}{x_2^2+y_2^2} + \frac{y_1x_2-x_1y_2}{x_2^2+y_2^2}i\end{aligned}$$

Notice that the quotient is defined for every pair of complex numbers z_1, z_2, provided that the denominator is not zero.

EXAMPLE 3

$$\frac{2+5i}{3-4i} = \frac{(2+5i)(3+4i)}{(3-4i)(3+4i)} = \frac{(6-20)+(8+15)i}{9+16} = -\frac{14}{25} + \frac{23}{25}i$$

EXERCISE 3

Calculate the following quotients.

(a) $\frac{1+i}{1-i}$ (b) $\frac{2i}{1+i}$ (c) $\frac{4-i}{1+5i}$

Roots of Polynomial Equations

Complex conjugates are not only used to determine quotients of complex numbers, but occur naturally as roots of polynomials with real coefficients.

Theorem 2

Let $p(x) = a_n x^n + \cdots + a_1 x + a_0$, where $a_i \in \mathbb{R}$ for $1 \leq i \leq n$. If z is a root of $p(x)$, then $\bar{z}$ is also a root of $p(x)$.

Proof: Suppose that z is a root, so that

$$a_n z^n + \cdots + a_1 z + a_0 = 0$$

Using Theorem 1, we get

$$0 = \overline{0} = \overline{a_n z^n + \cdots + a_1 z + a_0} = a_n \bar{z}^n + \cdots + a_1 \bar{z} + a_0$$

Thus, $\bar{z}$ is a root of $p(x)$. ■

EXAMPLE 4

Find the roots of $p(x) = x^3 + 1$.

Solution: By the Rational Roots Theorem (or by observation), we see that $x = -1$ is a root of $p(x)$. Therefore, by the Factor Theorem, $(x + 1)$ is a factor of $p(x)$. Thus,

$$x^3 + 1 = (x+1)(x^2 - x + 1)$$

Using the quadratic formula, we find that the other roots are

$$z = \frac{1 + \sqrt{3}i}{2} \quad \text{and} \quad \bar{z} = \frac{1 - \sqrt{3}i}{2}$$

The Complex Plane

For some purposes, it is convenient to represent the complex numbers as ordered pairs of real numbers: instead of $z = x + yi$, we write $z = (x, y)$. Then addition and multiplication appear as follows:

$$z_1 + z_2 = (x_1, y_1) + (x_2, y_2) = (x_1 + x_2, y_1 + y_2)$$
$$z_1 z_2 = (x_1, y_1)(x_2, y_2) = (x_1 x_2 - y_1 y_2, x_1 y_2 + x_2 y_1)$$

In terms of this ordered pair notation, it is natural to represent complex numbers as points in the plane, with the real part of z being the x-coordinate and the imaginary part being the y-coordinate. We will speak of the **real axis** and the **imaginary axis** in the **complex plane**. See Figure 9.1.1. A picture of this kind is sometimes called an **Argand diagram**.

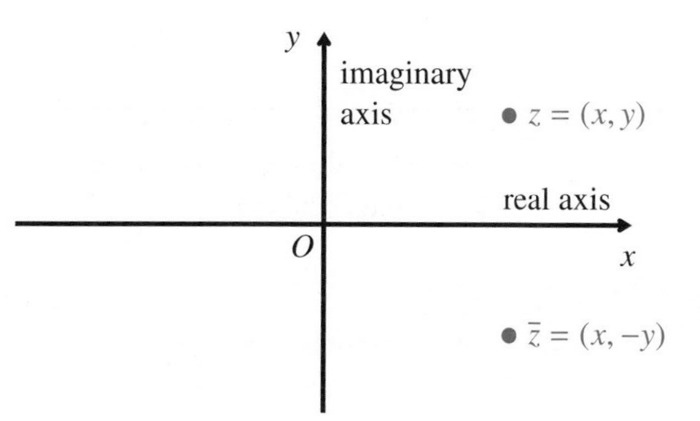

Figure 9.1.1 The complex plane.

If c is purely real, then $cz = c(x, y) = (cx, cy)$. Thus, with respect to addition and scalar multiplication by *real* scalars, the complex plane is just like the usual plane $\mathbb{R}^2$. However, in the complex plane, we can also multiply by complex scalars. This has a natural geometrical interpretation, which we shall see in the next section.

EXERCISE 4

Plot the following complex numbers in the complex plane.
(a) $2 + i$ (b) $\overline{2 + i}$ (c) $(2 + i)i$ (d) $(2 + i)(1 + i)$

Polar Form

Given a complex number $z = x + yi$, the real number

Definition
Modulus
Argument
Polar Form

$$|z| = r = \sqrt{x^2 + y^2}$$

is called the **modulus** of z. If $|z| \neq 0$, let θ be the angle measured counterclockwise from the positive x-axis such that

$$x = r\cos\theta \quad \text{and} \quad y = r\sin\theta$$

The angle θ is unique up to a multiple of 2π, and it is called an **argument** of z. As shown in Figure 9.1.2, a **polar form** of z is

$$z = r(\cos\theta + i\sin\theta)$$

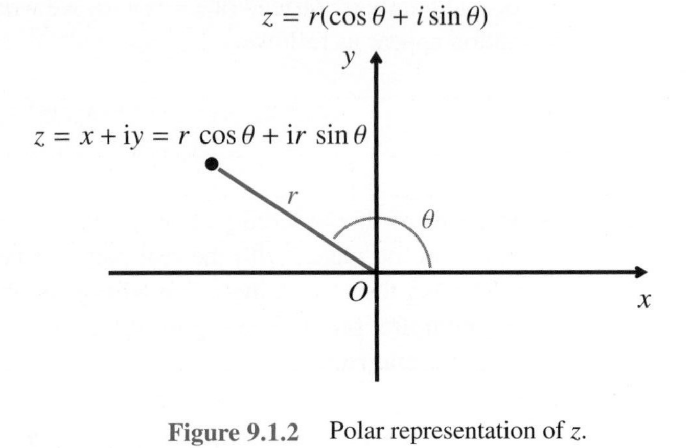

Figure 9.1.2 Polar representation of z.

EXAMPLE 5

Determine the modulus, an argument, and a polar form of $z_1 = 2-2i$ and $z_2 = -1+\sqrt{3}i$.
Solution: We have

$$|z_1| = |2-2i| = \sqrt{2^2+2^2} = 2\sqrt{2}$$

Any argument θ satisfies

$$2 = 2\sqrt{2}\cos\theta \quad \text{and} \quad -2 = 2\sqrt{2}\sin\theta$$

so $\cos\theta = \frac{1}{\sqrt{2}}$ and $\sin\theta = -\frac{1}{\sqrt{2}}$, which gives $\theta = -\frac{\pi}{4} + 2\pi k$, $k \in \mathbb{Z}$. Hence, a polar form of z_1 is

$$z_1 = 2\sqrt{2}\left(\cos\left(-\frac{\pi}{4}\right) + i\sin\left(-\frac{\pi}{4}\right)\right)$$

For z_2, we have

$$|z_2| = \left|-1+\sqrt{3}i\right| = \sqrt{(-1)^2+(\sqrt{3})^2} = 2$$

Since $-1 = 2\cos\theta$ and $\sqrt{3} = 2\sin\theta$, we get $\theta = \frac{2\pi}{3} + 2\pi k$, $k \in \mathbb{Z}$. Thus, a polar form of z_2 is

$$z_2 = 2\left(\cos\left(\frac{2\pi}{3}\right) + i\sin\left(\frac{2\pi}{3}\right)\right)$$

Remarks

1. An important consequence of the definition is

$$|z|^2 = x^2 + y^2 = z\bar{z}$$

2. The angles may be measured in radians or degrees. We will always use radians.

3. Notice that every complex number z has infinitely many arguments and hence infinitely many polar forms.

4. It is tempting but incorrect to write $\theta = \arctan(y/x)$. Remember that you need two trigonometric functions to locate the correct quadrant for z. Also note that y/x is not defined if $x = 0$.

EXERCISE 5

Determine the modulus, an argument, and a polar form of $z_1 = \sqrt{3} + i$ and $z_2 = -1 - i$.

EXERCISE 6

Let $z = r(\cos\theta + i\sin\theta)$. Prove that the modulus of $\bar{z}$ equals the modulus of z and an argument of $\bar{z}$ is $-\theta$.

The polar form is particularly convenient for multiplication and division because of the trigonometric identities

$$\begin{aligned}\cos(\theta_1 + \theta_2) &= \cos\theta_1\cos\theta_2 - \sin\theta_1\sin\theta_2\\ \sin(\theta_1 + \theta_2) &= \sin\theta_1\cos\theta_2 + \cos\theta_1\sin\theta_2\end{aligned}$$

It follows that

$$\begin{aligned}z_1z_2 &= r_1(\cos\theta_1 + i\sin\theta_1)r_2(\cos\theta_2 + i\sin\theta_2)\\ &= r_1r_2((\cos\theta_1\cos\theta_2 - \sin\theta_1\sin\theta_2) + i(\cos\theta_1\sin\theta_2 + \sin\theta_1\cos\theta_2))\\ &= r_1r_2(\cos(\theta_1 + \theta_2) + i\sin(\theta_1 + \theta_2))\end{aligned}$$

In words, the modulus of a product is the product of the moduli of the factors, while an argument of a product is the sum of the arguments.

Theorem 3

For any complex numbers $z_1 = r_1(\cos\theta_1 + i\sin\theta_1)$ and $z_2 = r_2(\cos\theta_2 + i\sin\theta_2)$, with $z_2 \neq 0$, we have

$$\frac{z_1}{z_2} = \frac{r_1}{r_2}(\cos(\theta_1 - \theta_2) + i\sin(\theta_1 - \theta_2))$$

The proof is left for you to complete in Problem D4.

Corollary 4

Let $z = r(\cos\theta + i\sin\theta)$ with $r \neq 0$. Then $z^{-1} = \frac{1}{r}(\cos(-\theta) + i\sin(-\theta))$.

EXERCISE 7

Describe Theorem 3 in words and use it to prove Corollary 4.

EXAMPLE 6

Calculate $(1 - i)(-\sqrt{3} + i)$ and $\dfrac{2 + 2i}{1 + \sqrt{3}i}$ using polar form.

Solution: We have

$$\begin{aligned}(1 - i)(-\sqrt{3} + i) &= \sqrt{2}\left(\cos\left(-\frac{\pi}{4}\right) + i\sin\left(-\frac{\pi}{4}\right)\right)2\left(\cos\left(\frac{5\pi}{6}\right) + i\sin\left(\frac{5\pi}{6}\right)\right)\\ &= 2\sqrt{2}\left(\cos\left(\frac{7\pi}{12}\right) + i\sin\left(\frac{7\pi}{12}\right)\right)\\ &\approx -0.732 + i(2.732)\\ \frac{2 + 2i}{1 + \sqrt{3}i} &= \frac{2\sqrt{2}\left(\cos\left(\frac{\pi}{4}\right) + i\sin\left(\frac{\pi}{4}\right)\right)}{2\left(\cos\left(\frac{\pi}{3}\right) + i\sin\left(\frac{\pi}{3}\right)\right)}\\ &= \sqrt{2}\left(\cos\left(-\frac{\pi}{12}\right) + i\sin\left(-\frac{\pi}{12}\right)\right)\\ &\approx 1.366 + i(0.366)\end{aligned}$$

EXERCISE 8

Calculate $(2 - 2i)(-1 + \sqrt{3}i)$ and $\dfrac{2 - 2i}{-1 + \sqrt{3}i}$ using polar form.

Powers and the Complex Exponential

From the rule for products, we find that

$$z^2 = r^2(\cos 2\theta + i\sin 2\theta)$$

Then

$$\begin{aligned}z^3 = z^2 z &= r^2 r(\cos(2\theta + \theta) + i\sin\theta(2\theta + \theta))\\ &= r^3(\cos 3\theta + i\sin 3\theta)\end{aligned}$$

Theorem 5

[de Moivre's Formula]
Let $z = r(\cos\theta + i\sin\theta)$ with $r \neq 0$. Then, for any integer n, we have

$$z^n = r^n(\cos n\theta + i\sin n\theta)$$

Proof: For $n = 0$, we have $z^0 = 1 = r^0(\cos 0 + i\sin 0)$. To prove that the theorem holds for positive integers, we proceed by induction. Assume that the result is true for some integer $k \geq 0$. Then

$$\begin{aligned}z^{k+1} = z^k z &= r^k r[\cos(k\theta + \theta) + i\sin(k\theta + \theta)]\\ &= r^{k+1}[\cos((k + 1)\theta) + i\sin((k + 1)\theta)]\end{aligned}$$

Therefore, the result is true for all non-negative integers n. Then, by Theorem 4, for any positive integer m, we have

$$\begin{aligned} z^{-m} &= (z^m)^{-1} = \left(r^m(\cos m\theta + i\sin m\theta)\right)^{-1} \\ &= r^{-m}(\cos(-m\theta) + i\sin(-m\theta)) \end{aligned}$$

Hence, the result also holds for all negative integers $n = -m$. ■

EXAMPLE 7

Calculate $(2 + 2i)^3$.

Solution:

$$\begin{aligned} (2+2i)^3 &= \left[2\sqrt{2}\left(\cos\left(\frac{\pi}{4}\right) + \sin\left(\frac{\pi}{4}\right)\right)\right]^3 \\ &= (2\sqrt{2})^3\left(\cos\left(\frac{3\pi}{4}\right) + i\sin\left(\frac{3\pi}{4}\right)\right) \\ &= 16\sqrt{2}\left(-\frac{1}{\sqrt{2}} + i\frac{1}{\sqrt{2}}\right) \\ &= -16 + 16i \end{aligned}$$

In the case where $r = 1$, de Moivre's Formula reduces to

$$(\cos\theta + i\sin\theta)^n = \cos n\theta + i\sin n\theta$$

This is formally just like one of the exponential laws, $(e^\theta)^n = e^{n\theta}$. We use this idea to define e^z for any $z \in \mathbb{C}$, where e is the usual natural base for exponentials ($e \approx 2.71828$). We begin with a useful formula of Euler.

Definition

Euler's Formula

$$e^{i\theta} = \cos\theta + i\sin\theta$$

Definition

e^{x+iy}

For any complex number $z = x + iy$, we define

$$e^z = e^{x+iy} = e^x e^{iy}$$

Remarks

1. One interesting consequence of Euler's Formula is that

$$e^{i\pi} + 1 = 0$$

In one formula, we have five of the most important numbers in mathematics: 0, 1, e, i, and π.

2. One area where Euler's Formula has important applications is ordinary differential equations. There, one often uses the fact that

$$e^{(a+bi)t} = e^{at}e^{ibt} = e^{at}(\cos bt + i\sin bt)$$

Observe that Euler's Formula allows us to write every complex number z in the form

$$z = re^{i\theta}$$

where $r = |z|$ and θ is any argument of z. Hence, in this form, de Moivre's Formula becomes

$$z^n = r^n e^{in\theta}$$

EXAMPLE 8

Calculate the following using the polar form.

(a) $(2 + 2i)^3$

Solution: $(2 + 2i)^3 = \left(2\sqrt{2}e^{i\pi/4}\right)^3 = (2\sqrt{2})^3 e^{i(3\pi/4)} = -16 + 16i$

(b) $(2i)^3$

Solution: $(2i)^3 = \left(2e^{i\pi/2}\right)^3 = 2^3 e^{i(3\pi/2)} = -8i$

(c) $(\sqrt{3} + i)^5$

Solution: $(\sqrt{3} + i)^5 = \left(2e^{i\pi/6}\right)^5 = 2^5 e^{i5\pi/6} = -16\sqrt{3} + 16i$

EXERCISE 9

Use polar form to calculate $(1 - i)^5$ and $(-1 - \sqrt{3}i)^5$.

---

n-th Roots

Using de Moivre's Formula for n-th powers is the key to finding n-th roots. Suppose that we need to find the n-th root of the non-zero complex number $z = re^{i\theta}$. That is, we need a number w such that $w^n = z$. Suppose that $w = Re^{i\phi}$. Then $w^n = z$ implies that

$$R^n e^{in\phi} = re^{i\theta}$$

Then R is the real n-th root of the positive real number r. However, because arguments of complex numbers are determined only up to the addition of $2\pi k$, all we can say about ϕ is that

$$n\phi = \theta + 2\pi k, \quad k \in \mathbb{Z}$$

or

$$\phi = \frac{\theta + 2\pi k}{n}, \quad k \in \mathbb{Z}$$

EXAMPLE 9

Find all the cube roots of 8.

Solution: We have $8 = 8e^{i(0+2\pi k)}$, $k \in \mathbb{Z}$. Thus, for any $k \in \mathbb{Z}$.

$$8^{1/3} = \left(8e^{i(0+2\pi k)}\right)^{1/3} = 8^{1/3} e^{i2k\pi/3}$$

EXAMPLE 9
(continued)

If $k = 0$, we have the root $w_0 = 2e^0 = 2$.
If $k = 1$, we have the root $w_1 = 2e^{i2\pi/3} = -1 + \sqrt{3}i$.
If $k = 2$, we have the root $w_2 = 2e^{i4\pi/3} = -1 - \sqrt{3}i$.
If $k = 3$, we have the root $2e^{i2\pi} = 2 = w_0$.

By increasing k further, we simply repeat the roots we have already found. Similarly, consideration of negative k gives us no further roots. The number 8 has three third roots, ω_0, ω_1, and ω_2. In particular, these are the roots of equation $w^3 - 8 = 0$.

Theorem 6 Let z be a non-zero complex number. Then the n distinct n-th roots of $z = re^{i\theta}$ are

$$w_k = r^{1/n} e^{i(\theta + 2\pi k)/n}, \quad k = 0, 1, \ldots, n-1$$

EXAMPLE 10

Find the fourth roots of -81.

Solution: We have $-81 = 81e^{i(\pi+2\pi k)}$. Thus, the fourth roots are

$$(81)^{1/4} e^{i(\pi+2\pi k)/4}, \quad k = 0, 1, 2, 3$$

This gives $3e^{i\pi/4}$, $3e^{i3\pi/4}$, $3e^{i5\pi/4}$, and $3e^{i7\pi/4}$.

In Examples 9 and 10, we took roots of numbers that were purely real: we were really solving $x^n - a = 0$, where $a \in \mathbb{R}$. By our earlier theorem, when the coefficients of the equations are real, the roots that are not real occur in complex conjugate pairs. As a contrast, let us consider roots of a number that is not real.

EXAMPLE 11

Find the third roots of $5i$ and illustrate in an Argand diagram.

Solution: $5i = 5e^{i(\frac{\pi}{2}+2k\pi)}$, so the cube roots are

$$\begin{aligned} w_0 &= 5^{1/3}\, e^{i\pi/6} = 5^{1/3}\left(\frac{\sqrt{3}}{2} + i\frac{1}{2}\right) \\ w_1 &= 5^{1/3}\, e^{i5\pi/6} = 5^{1/3}\left(\frac{-\sqrt{3}}{2} + i\frac{1}{2}\right) \\ w_2 &= 5^{1/3}\, e^{i5\pi/6} = 5^{1/3}(-i) \end{aligned}$$

Plotting these roots shows that all three are points on the circle of radius $5^{1/3}$ centred at the origin and that they are separated by equal angles of $\frac{2\pi}{3}$.

Examples 9, 10, and 11 all illustrate a general rule: the n-th roots of a complex number $z = re^{i\theta}$ all lie on the circle of radius $r^{1/n}$, and they are separated by equal angles of $2\pi/n$.

PROBLEMS 9.1

Practice Problems

A1 Determine the following sums or differences.
(a) $(2 + 5i) + (3 + 2i)$
(b) $(2 - 7i) + (-5 + 3i)$
(c) $(-3 + 5i) - (4 + 3i)$
(d) $(-5 - 6i) - (9 - 11i)$

A2 Express the following products in standard form.
(a) $(1 + 3i)(3 - 2i)$
(b) $(-2 - 4i)(3 - i)$
(c) $(1 - 6i)(-4 + i)$
(d) $(-1 - i)(1 - i)$

A3 Determine the complex conjugates of the following numbers.
(a) $3 - 5i$ (b) $2 + 7i$
(c) 3 (d) $-4i$

A4 Determine the real and imaginary parts of the following.
(a) $z = 3 - 6i$ (b) $z = (2 + 5i)(1 - 3i)$
(c) $z = \dfrac{4}{6 - i}$ (d) $z = \dfrac{-1}{i}$

A5 Express the following quotients in standard form.
(a) $\dfrac{1}{2 + 3i}$ (b) $\dfrac{3}{2 - 7i}$
(c) $\dfrac{2 - 5i}{3 + 2i}$ (d) $\dfrac{1 + 6i}{4 - i}$

A6 Use polar form to determine $z_1 z_2$ and $\dfrac{z_1}{z_2}$ if
(a) $z_1 = 1 + i$, $z_2 = 1 + \sqrt{3}i$
(b) $z_1 = -\sqrt{3} - i$, $z_2 = 1 - i$
(c) $z_1 = 1 + 2i$, $z_2 = -2 - 3i$
(d) $z_1 = -3 + i$, $z_2 = 6 - i$

A7 Use polar form to determine the following.
(a) $(1 + i)^4$ (b) $(3 - 3i)^3$
(c) $(-1 - \sqrt{3}i)^4$ (d) $(-2\sqrt{3} + 2i)^5$

A8 Use polar form to determine all the indicated roots.
(a) $(-1)^{1/5}$ (b) $(-16i)^{1/4}$
(c) $(-\sqrt{3} - i)^{1/3}$ (d) $(1 + 4i)^{1/3}$

Homework Problems

B1 Determine the following sums or differences.
(a) $(3 + 4i) + (1 + 5i)$
(b) $(3 - 2i) + (-7 + 6i)$
(c) $(-5 + 7i) - (2 + 6i)$
(d) $(-7 - 2i) - (-8 - 9i)$

B2 Express the following products in standard form.
(a) $(2 + i)(5 - 3i)$
(b) $(-3 - 2i)(5 - 2i)$
(c) $(3 - 5i)(-1 + 6i)$
(d) $(-3 - i)(3 - i)$

B3 Determine the complex conjugates of the following numbers.
(a) $2i$
(b) 17
(c) $4 - 8i$
(d) $5 + 11i$

B4 Determine the real and imaginary parts of the following.
(a) $4 - 7i$
(b) $(3 + 2i)(2 - 3i)$
(c) $\dfrac{5}{4 - i}$
(d) $\dfrac{1 - 2i}{1 + i}$

B5 Express the following quotients in standard form.
(a) $\dfrac{1}{3 + 4i}$
(b) $\dfrac{2}{3 - 5i}$
(c) $\dfrac{1 - 4i}{3 + 5i}$
(d) $\dfrac{1 + 4i}{4 - 5i}$

B6 Use polar form to determine $z_1 z_2$ and $\dfrac{z_1}{z_2}$ if
(a) $z_1 = 1 - \sqrt{3}i$, $z_2 = -1 + i$
(b) $z_1 = -\sqrt{3} + i$, $z_2 = -3 - 3i$
(c) $z_1 = 1 + 3i$, $z_2 = -1 - 2i$
(d) $z_1 = -2 + i$, $z_2 = 4 - i$

B7 Use polar form to determine the following.

(a) $(1+\sqrt{3}i)^4$
(b) $(-2-2i)^3$
(c) $(\sqrt{3}-i)^4$
(d) $(-2+2\sqrt{3}i)^5$

B8 Use polar form to determine all the indicated roots.

(a) $(32)^{1/5}$
(b) $(81i)^{1/5}$
(c) $(-\sqrt{3}+i)^{1/3}$
(d) $(4+i)^{1/3}$

Conceptual Problems

D1 Prove properties (3), (5), (6), (7), (8), and (9) of Theorem 1.

D2 If $z = r(\cos\theta + i\sin\theta)$, what is $|\overline{z}|$? What is an argument of $\overline{z}$?

D3 Use Euler's Formula to show that

(a) $\overline{e^{i\theta}} = e^{-i\theta}$
(b) $\cos\theta = \frac{1}{2}\left(e^{i\theta} + e^{-i\theta}\right)$
(c) $\sin\theta = \frac{1}{2i}\left(e^{i\theta} - e^{-i\theta}\right)$

D4 Prove Theorem 3.

9.2 Systems with Complex Numbers

In some applications, it is necessary to consider systems of linear equations with complex coefficients and complex right-hand sides. One physical application, discussed later in this section, is the problem of determining currents in electrical circuits with capacitors and inductive coils as well as resistance. We can solve systems with complex coefficients by using exactly the same elimination/row reduction procedures as for systems with real coefficients. Of course, our solutions will be complex, and any free variables will be allowed to take any complex value.

EXAMPLE 1

Solve the system of linear equations

$$\begin{aligned} z_1 + z_2 + z_3 &= 0 \\ (1-i)z_1 + z_2 + \quad &= i \\ (3-i)z_1 + 2z_2 + z_3 &= 1+2i \end{aligned}$$

Solution: The solution procedure is, as usual, to write the augmented matrix for the system and row reduce the coefficient matrix to row echelon form:

$$\left[\begin{array}{ccc|c} 1 & 1 & 1 & 0 \\ 1-i & 1 & 0 & i \\ 3-i & 2 & 1 & 1+2i \end{array}\right] \begin{array}{l} \\ R_2-(1-i)R_1 \\ R_3-(3-i)R_1 \end{array} \sim \left[\begin{array}{ccc|c} 1 & 1 & 1 & 0 \\ 0 & i & -1+i & i \\ 0 & -1+i & -2+i & 1+2i \end{array}\right] -iR_2 \sim$$

$$\left[\begin{array}{ccc|c} 1 & 1 & 1 & 0 \\ 0 & 1 & 1+i & 1 \\ 0 & -1+i & -2+i & 1+2i \end{array}\right] \begin{array}{l} R_1-R_2 \\ \\ R_3+(1-i)R_2 \end{array} \sim \left[\begin{array}{ccc|c} 1 & 0 & -i & -1 \\ 0 & 1 & 1+i & 1 \\ 0 & 0 & i & 2+i \end{array}\right] \begin{array}{l} \\ \sim \\ -iR_3 \end{array}$$

$$\left[\begin{array}{ccc|c} 1 & 0 & -i & -1 \\ 0 & 1 & 1+i & 1 \\ 0 & 0 & 1 & 1-2i \end{array}\right] \begin{array}{l} R_1+iR_3 \\ \\ R_2-(1+i)R_3 \end{array} \sim \left[\begin{array}{ccc|c} 1 & 0 & 0 & 1+i \\ 0 & 1 & 0 & -2+i \\ 0 & 0 & 1 & 1-2i \end{array}\right]$$

Hence, the solution is $\vec{z} = \begin{bmatrix} 1+i \\ -2+i \\ 1-2i \end{bmatrix}$.

EXAMPLE 2 Solve the system

$$\begin{aligned} (1+i)z_1 \quad + 2iz_2 + \quad &= 1 \\ (1+i)z_2 + z_3 &= \frac{1}{2} - \frac{1}{2}i \\ z_1 \qquad\quad - z_3 &= 0 \end{aligned}$$

Solution: Row reducing the augmented matrix gives

$$\left[\begin{array}{ccc|c} 1+i & 2i & 0 & 1 \\ 0 & 1+i & 1 & \frac{1}{2}-\frac{1}{2}i \\ 1 & 0 & -1 & 0 \end{array}\right] \sim \left[\begin{array}{ccc|c} 1 & 0 & -1 & 0 \\ 0 & 1+i & 1 & \frac{1}{2}-\frac{1}{2}i \\ 1+i & 2i & 0 & 1 \end{array}\right] \sim$$

$$\left[\begin{array}{ccc|c} 1 & 0 & -1 & 0 \\ 0 & 1+i & 1 & \frac{1}{2}-\frac{1}{2}i \\ 0 & 2i & 1+i & 1 \end{array}\right] \sim \left[\begin{array}{ccc|c} 1 & 0 & -1 & 0 \\ 0 & 1 & \frac{1-i}{2} & -\frac{1}{2}i \\ 0 & 2i & 1+i & 1 \end{array}\right] \sim \left[\begin{array}{ccc|c} 1 & 0 & -1 & 0 \\ 0 & 1 & \frac{1-i}{2} & -\frac{1}{2}i \\ 0 & 0 & 0 & 0 \end{array}\right]$$

Hence, z_3 is a free variable, so we let $z_3 = t \in \mathbb{C}$. Then $z_1 = z_3 = t$, $z_2 = -\frac{1}{2}i - \frac{1-i}{2}t$, and the general solution is

$$\begin{bmatrix} z_1 \\ z_2 \\ z_3 \end{bmatrix} = \begin{bmatrix} 0 \\ -i/2 \\ 0 \end{bmatrix} + t \begin{bmatrix} 1 \\ -(1-i)/2 \\ 1 \end{bmatrix}$$

EXERCISE 1 Solve the system

$$\begin{aligned} iz_1 + z_2 + 3z_3 &= -1 - 2i \\ iz_1 + iz_2 + (1+2i)z_3 &= 2 + i \\ 2z_1 + (1+i)z_2 + 2z_3 &= 5 - i \end{aligned}$$

Complex Numbers in Electrical Circuit Equations

This application requires some knowledge of calculus and physics. It can be omitted with no loss of continuity.

For purposes of the following discussion only, we switch to a notation commonly used by engineers and physicists and denote by j the complex number such that $j^2 = -1$, so that we can use i to denote current.

In Section 2.4, we discussed electrical circuits with resistors. We now also consider capacitors and inductors, as well as alternating current. A simple capacitor can be thought of as two conducting plates separated by a vacuum or some dielectric. Charge can be stored on these plates, and it is found that the voltage across a capacitor at time t is proportional to the charge stored at that time:

$$V(t) = \frac{Q(t)}{C}$$

where Q is the charge and the constant C is called the *capacitance* of the capacitor.

The usual model of an inductor is a coil; because of magnetic effects, it is found that with time-varying current $i(t)$, the voltage across an inductor is proportional to the rate of change of current:

$$V(t) = L\frac{d\,i(t)}{dt}$$

where the constant of proportionality L is called the *inductance*.

As in the case of the resistor circuits, Kirchhoff's Laws applies: the sum of the voltage drops across the circuit elements must be equal to the applied electromotive force (voltage). Thus, for a simple loop with inductance L, capacitance C, resistance R, and applied electromotive force $E(t)$ (Figure 9.2.3), the circuit equation is

$$L\frac{d\,i(t)}{dt} + R\,i(t) + \frac{1}{C}Q(t) = E(t)$$

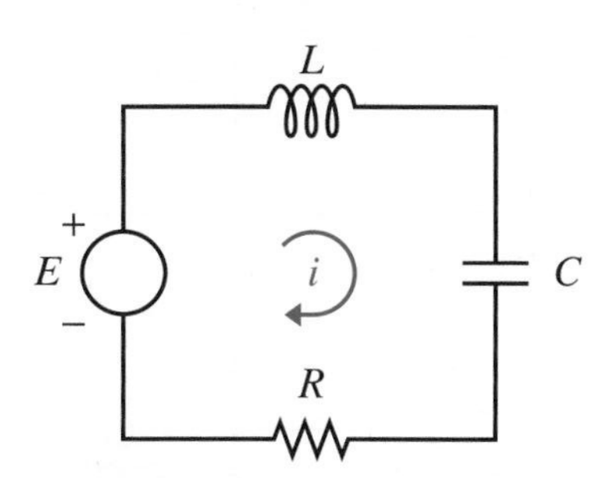

Figure 9.2.3 Kirchhoff's voltage law applied to an alternating current circuit.

For our purposes, it is easier to work with the derivative of this equation and use the fact that $\frac{dQ}{dt} = i$:

$$L\frac{d^2\,i(t)}{dt^2} + R\,\frac{d\,i(t)}{dt} + \frac{1}{C}i(t) = \frac{d\,E(t)}{dt}$$

In general, the solution to such an equation will involve the superposition (sum) of a steady-state solution and a transient solution. Here we will be looking only for the steady-state solution, in the special case where the applied electromotive force, and hence any current, is a single-frequency sinusoidal function. Thus, we can assume that

$$E(t) = Be^{j\omega t} \quad \text{and} \quad i(t) = Ae^{j\omega t}$$

where A and B are complex numbers that determine the amplitudes and phases of voltage and current, and ω is 2π multiplied by the frequency. Then

$$\frac{d\,i}{dt} = j\omega Ae^{j\omega t} = j\omega i$$
$$\frac{d^2 i}{dt^2} = (j\omega)^2 i = -\omega^2 i$$

and the circuit equation can be rewritten

$$-\omega^2 L\,i + j\omega R\,i + \frac{1}{C}i = \frac{dE}{dt}$$

Now consider a network of circuits with resistors, capacitors, inductors, electromotive force, and currents, as shown in Figure 9.2.4. As in Section 2.4, the currents are loops, so that the actual current across some circuit elements is the difference of two loop currents. (For example, across R_1, the actual current is $i_1 - i_2$.) From our assumption that we have only one single frequency source, we may conclude that the steady-state loop currents must be of the form

$$i_1(t) = A_1 e^{j\omega t}, \quad i_2(t) = A_2 e^{j\omega t}, \quad i_3(t) = A_3 e^{j\omega t}$$

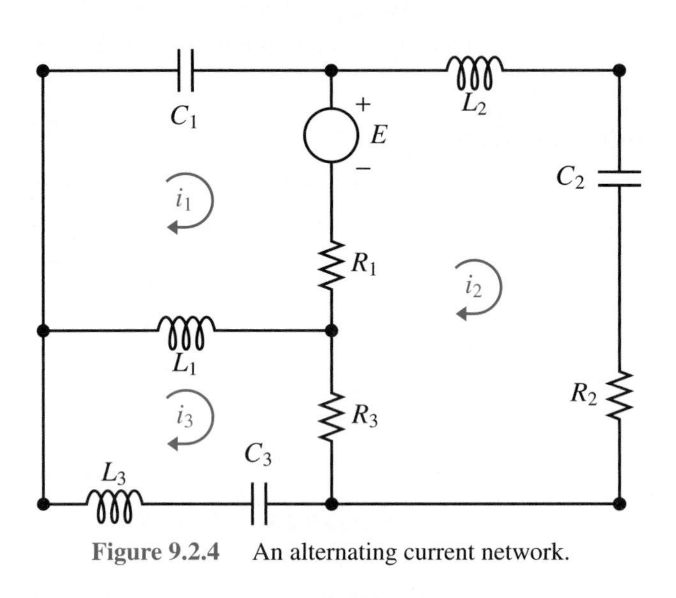

Figure 9.2.4 An alternating current network.

By applying Kirchhoff's laws to the top-left loop, we find that

$$\left[-\omega^2 L_1(A_1 - A_3) + j\omega R_1(A_1 - A_2) + \frac{1}{C}A_1\right] e^{j\omega t} = -j\omega B e^{j\omega t}$$

If we write the corresponding equations for the other two loops, reorganize each equation, and divide out the non-zero common factor $e^{j\omega t}$, we obtain the following system of linear equations for the three variables A_1, A_2, and A_3:

$$\left[-\omega^2 L_1 + \frac{1}{C_1} + j\omega R_1\right] A_1 - j\omega R_1 A_2 + \omega^2 L_1 A_3 = -j\omega B$$

$$-j\omega R_1 A_1 + \left[-\omega^2 L_2 + \frac{1}{C_2} + j\omega(R_1 + R_2 + R_3)\right] A_2 - j\omega R_3 A_3 = j\omega B$$

$$\omega^2 L_1 A_1 - j\omega R_3 A_2 + \left[-\omega^2(L_1 + L_3) + \frac{1}{C_3} + j\omega R_3\right] A_3 = 0$$

Thus, we have a system of three linear equations with complex coefficients for the three variables A_1, A_2, and A_3. We can solve this system by standard elimination. We emphasize that this example is for illustrative purposes only: we have constructed a completely arbitrary network and provided the solution method for only part of the problem, in a special case. A much more extensive discussion is required before a reader will be ready to start examining realistic circuits to discover what they can do. But even this limited example illustrates the general point that to analyze some electrical networks, we need to solve systems of linear equations with complex coefficients.

PROBLEMS 9.2

Practice Problems

A1 Determine whether each system is consistent, and if it is, determine the general solution.

(a)
$$\begin{aligned} z_1 + iz_2 + (1+i)z_3 &= 1 - i \\ -2z_1 + (1-2i)z_2 - 2z_3 &= 2i \\ 2iz_1 - 2z_2 - (2+3i)z_3 &= -1 + 3i \end{aligned}$$

(b)
$$\begin{aligned} z_1 + (1+i)z_2 + 2z_3 + z_4 &= 1 - i \\ 2z_1 + (2+i)z_2 + 5z_3 + (2+i)z_4 &= 4 - i \\ iz_1 + (-1+i)z_2 + (1+2i)z_3 + 2iz_4 &= 1 \end{aligned}$$

Homework Problems

B1 Determine whether each system is consistent, and if it is, determine the general solution.

(a)
$$\begin{aligned} z_2 - iz_3 &= 1 + 3i \\ iz_1 - z_2 + (-1+i)z_3 &= 1 + 2i \\ 2z_1 + 2iz_2 + (3+2i)z_3 &= 4 \end{aligned}$$

(b)
$$\begin{aligned} z_1 + (2+i)z_2 + iz_3 \qquad\qquad &= 1 + i \\ iz_1 + (-1+2i)z_2 \qquad\qquad + 2iz_4 &= -i \\ z_1 + (2+i)z_2 + (1+i)z_3 + 2iz_4 &= 2 - i \end{aligned}$$

(c)
$$\begin{aligned} iz_1 + 2z_2 - (3+i)z_3 &= 1 \\ (1+i)z_1 + (2-2i)z_2 - 4z_3 &= i \\ iz_1 + 2z_2 - (3+3i)z_3 &= 1 + 2i \end{aligned}$$

(d)
$$\begin{aligned} z_1 + z_2 + iz_3 + (1+i)z_4 &= 0 \\ iz_1 + iz_2 + (-1-i)z_3 + (-2+i)z_4 &= 0 \\ 2z_1 + 2z_2 + (2+2i)z_3 + 2z_4 &= 0 \end{aligned}$$

9.3 Vector Spaces over $\mathbb{C}$

The definition of a vector space in Section 4.2 is given in the case where the scalars are real numbers. In fact, the definition makes sense when the scalars are taken from any one system of numbers such that addition, subtraction, multiplication, and division are defined for any pairs of numbers (excluding division by 0) and satisfy the usual commutative, associative, and distributive rules for doing arithmetic. Thus, the vector space axioms make sense if we allow the scalars to be the set of complex numbers. In such cases, we say that we have a vector space over $\mathbb{C}$, or a complex vector space.

EXERCISE 1

Let $\mathbb{C}^2 = \left\{ \begin{bmatrix} z_1 \\ z_2 \end{bmatrix} \mid z_1, z_2 \in \mathbb{C} \right\}$, with addition defined by

$$\begin{bmatrix} z_1 \\ z_2 \end{bmatrix} + \begin{bmatrix} w_1 \\ w_2 \end{bmatrix} = \begin{bmatrix} z_1 + w_1 \\ z_2 + w_2 \end{bmatrix}$$

and scalar multiplication by $\alpha \in \mathbb{C}$ defined by

$$\alpha \begin{bmatrix} z_1 \\ z_2 \end{bmatrix} = \begin{bmatrix} \alpha z_1 \\ \alpha z_2 \end{bmatrix}$$

Show that $\mathbb{C}^2$ is a vector space over $\mathbb{C}$.

It is instructive to look carefully at the ideas of basis and dimension for complex vector spaces. We begin by considering the set of complex numbers $\mathbb{C}$ itself as a vector space.

As a vector space over the complex numbers, $\mathbb{C}$ has a basis consisting of a single element, $\{1\}$. That is, every complex number can be written in the form $\alpha 1$, where α is a complex number. Thus, with respect to this basis, the coordinate of the complex number z is z itself. Alternatively, we could choose to use the basis $\{i\}$. Then the coordinate of z would be $-iz$ since

$$z = (-iz)i$$

In either case, we see that $\mathbb{C}$ has a basis consisting of one element, so $\mathbb{C}$ is a one-dimensional complex vector space.

Another way of looking at this is to observe that when we use complex scalars, any two non-zero elements of the space $\mathbb{C}$ are linearly dependent. That is, given $z_1, z_2 \in \mathbb{C}$, there exist complex scalars, not both zero, such that

$$\alpha_1 z_1 + \alpha_2 z_2 = 0$$

For example, we may take $\alpha_1 = 1$ and $\alpha_2 = -\frac{z_1}{z_2}$, since we have assumed that $z_2 \neq 0$. It follows that with respect to complex scalars, a basis for $\mathbb{C}$ must have fewer than two dimensions.

However, we could also view $\mathbb{C}$ as a vector space over $\mathbb{R}$. Addition of complex numbers is defined as usual, and multiplication of $z = x + iy$ by a real scalar k gives

$$kz = kx + kyi$$

Observe that if we use real scalars, then the elements 1 and i in $\mathbb{C}$ are linearly independent. Hence, viewed as a vector space over $\mathbb{R}$, the set of complex numbers is two-dimensional, with "standard" basis $\{1, i\}$.

As we saw in Section 9.2, we sometimes write complex numbers in a way that exhibits the property that $\mathbb{C}$ is a two-dimensional real vector space: we write a complex number z in the form

$$z = x + iy = (x, y) = x(1, 0) + y(0, 1)$$

Note that $(1, 0)$ denotes the complex number 1 and that $(0, 1)$ denotes the complex number i. With this notation, we see that the set of complex numbers is isomorphic to $\mathbb{R}^2$, which justifies our work with the complex plane in Section 9.1. However, notice that this representation of the complex numbers as a real vector space does not include multiplication by complex scalars.

Using arguments similar to those above, we see that $\mathbb{C}^2$ is a two-dimensional complex vector space, but it can be viewed as a real vector space of dimension four.

Definition
$\mathbb{C}^n$

The vector space $\mathbb{C}^n$ is defined to be the set

$$\mathbb{C}^n = \left\{ \begin{bmatrix} z_1 \\ \vdots \\ z_n \end{bmatrix} \mid z_1, \dots, z_n \in \mathbb{C} \right\}$$

with addition of vectors and scalar multiplication defined as above.

Remark

These vector spaces play an important role in much of modern mathematics.

Since the complex conjugate is so useful in $\mathbb{C}$, we extend the definition of a complex conjugate to vectors in $\mathbb{C}^n$.

Definition
Complex Conjugate

The **complex conjugate** of $\vec{z} = \begin{bmatrix} z_1 \\ \vdots \\ z_n \end{bmatrix} \in \mathbb{C}^n$ is defined to be $\overline{\vec{z}} = \begin{bmatrix} \overline{z_1} \\ \vdots \\ \overline{z_n} \end{bmatrix}$.

EXAMPLE 1

Let $\vec{z} = \begin{bmatrix} 1+i \\ -2i \\ 3 \\ 1-2i \end{bmatrix}$. Then $\overline{\vec{z}} = \begin{bmatrix} \overline{1+i} \\ \overline{-2i} \\ \overline{3} \\ \overline{1-2i} \end{bmatrix} = \begin{bmatrix} 1-i \\ 2i \\ 3 \\ 1+2i \end{bmatrix}$.

Linear Mappings and Subspaces

We can now extend the definition of a linear mapping $L : \mathbb{V} \to \mathbb{W}$ to the case where $\mathbb{V}$ and $\mathbb{W}$ are both vector spaces over the complex numbers. We say that L is linear over the complex numbers if for any $\alpha \in \mathbb{C}$ and $\mathbf{v}_1, \mathbf{v}_2 \in \mathbb{V}$ we have

$$L(\alpha \mathbf{v}_1 + \mathbf{v}_2) = \alpha L(\mathbf{v}_1) + L(\mathbf{v}_2)$$

We can also define subspaces just as we did for real vector spaces, and the range and nullspace of a linear mapping will be subspaces of the appropriate vector spaces, as before.

EXAMPLE 2

Let $L : \mathbb{C}^3 \to \mathbb{C}^2$ be the linear mapping such that

$$L(1,0,0) = \begin{bmatrix} 1+i \\ 2 \end{bmatrix}, \quad L(0,1,0) = \begin{bmatrix} -2i \\ 1-i \end{bmatrix}, \quad L(0,0,1) = \begin{bmatrix} 1+2i \\ 3+i \end{bmatrix}$$

Then, the standard matrix of L is

$$[L] = \begin{bmatrix} 1+i & -2i & 1+2i \\ 2 & 1-i & 3+i \end{bmatrix}$$

The image of $\vec{z} = \begin{bmatrix} 1 \\ 2i \\ 1-i \end{bmatrix}$ under L is calculated by

$$L(\vec{x}) = [L]\vec{z} = \begin{bmatrix} 1+i & -2i & 1+2i \\ 2 & 1-i & 3+i \end{bmatrix} \begin{bmatrix} 1 \\ 2i \\ 1-i \end{bmatrix} = \begin{bmatrix} 8+2i \\ 8 \end{bmatrix}$$

The range of L is the subspace of the codomain $\mathbb{C}^2$ spanned by the columns of $[L]$, and the nullspace is the solution space of the system of linear equations $A\vec{z} = \vec{0}$. (Remember that $\vec{z}$ is a vector in $\mathbb{C}^3$.)

Similarly, the rowspace, columnspace, and nullspace of a matrix are also defined as in the real case.

EXAMPLE 3

Let $A = \begin{bmatrix} 1 & i & 1+i & -i \\ 1 & i & 2+i & -2i \\ i & -1 & -1+2i & 2 \end{bmatrix}$. Find a basis for the rowspace, columnspace, and nullspace of A.

Solution: We row reduce and find that the reduced row echelon form of A is

$$R = \begin{bmatrix} 1 & i & 0 & -1 \\ 0 & 0 & 1 & -i \\ 0 & 0 & 0 & 0 \end{bmatrix}$$

As in the real case, a basis for Row(A) is the non-zero rows of the reduced row echelon for of A. That is, a basis for Row(A) is $\left\{ \begin{bmatrix} 1 \\ i \\ 0 \\ -1 \end{bmatrix}, \begin{bmatrix} 0 \\ 0 \\ 1 \\ -i \end{bmatrix} \right\}$.

We next recall that the columns of A corresponding to the columns of R that contain leading 1s form a basis for Col(A). So, a basis for Col(A) is $\left\{ \begin{bmatrix} 1 \\ 1 \\ i \end{bmatrix}, \begin{bmatrix} 1+i \\ 2+i \\ -1+2i \end{bmatrix} \right\}$.

To find the nullspace, we solve the homogeneous systems $A\vec{x} = \vec{0}$. Using the reduced row echelon form of A, we find that the homogeneous system is equivalent to

$$\begin{aligned} z_1 + iz_2 - z_4 &= 0 \\ z_3 - iz_4 &= 0 \end{aligned}$$

The free variables are z_2 and z_4, so we let $z_2 = s \in \mathbb{C}$, and $z_4 = t \in \mathbb{C}$. Then we get $z_1 = -is + t$ and $z_3 = it$. Hence, the general solution to the homogeneous system is

$$\begin{bmatrix} z_1 \\ z_2 \\ z_3 \\ z_4 \end{bmatrix} = \begin{bmatrix} -is + t \\ s \\ it \\ t \end{bmatrix} = s \begin{bmatrix} -i \\ 1 \\ 0 \\ 0 \end{bmatrix} + t \begin{bmatrix} 1 \\ 0 \\ i \\ 1 \end{bmatrix}$$

Thus, a basis for Null(L) is $\left\{ \begin{bmatrix} -i \\ 1 \\ 0 \\ 0 \end{bmatrix}, \begin{bmatrix} 1 \\ 0 \\ i \\ 1 \end{bmatrix} \right\}$.

EXERCISE 2

Let $A = \begin{bmatrix} 1 & -1 & 1+i \\ 1 & 1 & 1-i \\ 1+i & 1 & 1 \\ -i & 1-2i & -2-2i \end{bmatrix}$. Find a basis for the rowspace, columnspace, and nullspace of A.

Complex Multiplication as a Matrix Mapping

We have seen that $\mathbb{C}$ can be regarded as a *real* two-dimensional vector space. We now want to represent multiplication by a complex number as a matrix mapping. We first consider a special case.

As before, to regard $\mathbb{C}$ as a real vector space we write

$$z = x + iy = (x, y)$$

Let us consider multiplication by i:

$$iz = i(x + iy) = ix - y = (-y, x)$$

It is easy to see that this corresponds to a linear mapping:

$$M_i : \mathbb{R}^2 \to \mathbb{R}^2 \quad \text{defined by} \quad M_i(x, y) = (-y, x)$$

The standard matrix is

$$[M_i] = \begin{bmatrix} 0 & -1 \\ 1 & 0 \end{bmatrix}$$

Observe that $[M_i] = [R_{\frac{\pi}{2}}]$. That is, multiplication by i corresponds to a rotation by an angle of $\frac{\pi}{2}$.

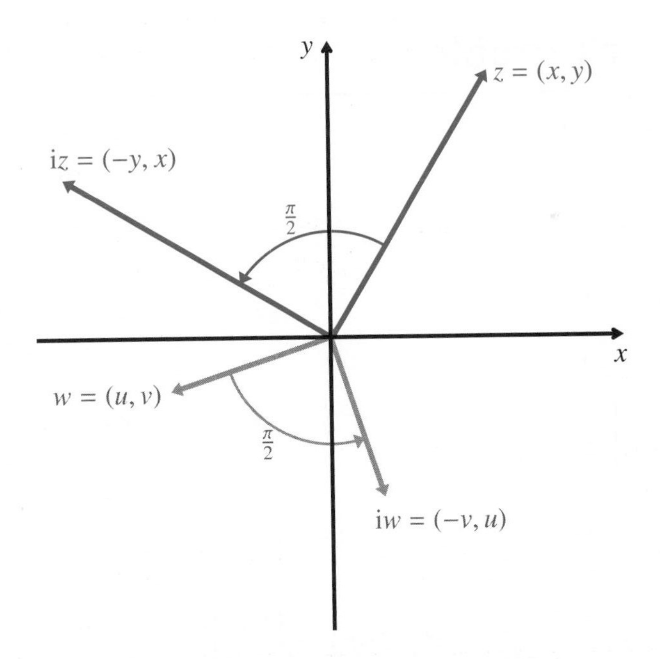

Figure 9.3.5 Multiplication by i corresponds to rotation by angle $\frac{\pi}{2}$.

More generally, we can consider multiplication of complex numbers by a complex number $\alpha = x + yi$. In Problem D1 you are asked to prove that multiplication by any complex number $\alpha = a + bi$ can be represented as a linear mapping M_α of $\mathbb{R}^2$ with standard matrix $\begin{bmatrix} a & -b \\ b & a \end{bmatrix}$.

PROBLEMS 9.3

Practice Problems

A1 Calculate the following.

(a) $\begin{bmatrix} -2+i \\ 1 \end{bmatrix} - \begin{bmatrix} 3+4i \\ 1-i \end{bmatrix}$

(b) $\begin{bmatrix} 2-i \\ 3+i \\ 2-5i \end{bmatrix} + \begin{bmatrix} 3-2i \\ 4+7i \\ -3-4i \end{bmatrix}$

(c) $2i\begin{bmatrix} 2+5i \\ 3-2i \end{bmatrix}$

(d) $(-1-2i)\begin{bmatrix} 2-i \\ 3+i \\ 2-5i \end{bmatrix}$

A2 (a) Write the standard matrix of the linear mapping $L : \mathbb{C}^2 \to \mathbb{C}^2$ such that

$$L(1,0) = \begin{bmatrix} 1+2i \\ 1 \end{bmatrix} \quad \text{and} \quad L(0,1) = \begin{bmatrix} 3+i \\ 1-i \end{bmatrix}$$

(b) Determine $L(2+3i, 1-4i)$.

(c) Find a basis for the range and nullspace of L.

A3 Find a basis for the rowspace, columnspace, and nullspace of the following matrices.

(a) $A = \begin{bmatrix} 1+i & 1 & i \\ -2i & 2i & 2+2i \end{bmatrix}$

(b) $B = \begin{bmatrix} 1 & i \\ 1+i & -1+i \\ -1 & i \end{bmatrix}$

(c) $C = \begin{bmatrix} 1 & i & -1+i & -1 \\ 2 & 1+2i & -2+3i & -2 \\ 1+i & i & -2+i & -1-i \end{bmatrix}$

Homework Problems

B1 Calculate the following.

(a) $\begin{bmatrix} 4-3i \\ -i \end{bmatrix} - \begin{bmatrix} -2-4i \\ 1-i \end{bmatrix}$

(b) $\begin{bmatrix} 3+2i \\ -2-i \\ 1+3i \end{bmatrix} - \begin{bmatrix} 2+i \\ 3+i \\ 1-i \end{bmatrix}$

(c) $-3i\begin{bmatrix} 1+3i \\ 5-3i \end{bmatrix}$

(d) $(-1-i)\begin{bmatrix} i \\ 2+3i \\ -2-7i \end{bmatrix}$

B2 (a) Write the standard matrix of the linear mapping $L : \mathbb{C}^2 \to \mathbb{C}^2$ such that

$$L(1,0) = \begin{bmatrix} -i \\ -1+i \end{bmatrix} \quad \text{and} \quad L(0,1) = \begin{bmatrix} 1+i \\ -2i \end{bmatrix}$$

(b) Determine $L(2-i, -4+i)$.

(c) Find a basis for the range and nullspace of L.

B3 Determine which of the following sets is a basis for $\mathbb{C}^3$.

(a) $\left\{ \begin{bmatrix} 1 \\ 0 \\ 0 \end{bmatrix}, \begin{bmatrix} 0 \\ i \\ 0 \end{bmatrix}, \begin{bmatrix} 0 \\ 0 \\ -i \end{bmatrix} \right\}$

(b) $\left\{ \begin{bmatrix} i \\ i \\ i \end{bmatrix}, \begin{bmatrix} 1 \\ 0 \\ i \end{bmatrix}, \begin{bmatrix} -1 \\ 2i \\ -1 \end{bmatrix} \right\}$

B4 Find a basis for the rowspace, columnspace, and nullspace of the following matrices.

(a) $A = \begin{bmatrix} 1 & 1 & i \\ i & i & 1 \\ 1+i & 1+i & -1-i \end{bmatrix}$

(b) $B = \begin{bmatrix} 1+i & 2-i \\ 1 & -i \\ -1+i & 2i \end{bmatrix}$

(c) $C = \begin{bmatrix} i & -1 & 2 & i \\ 2i & 6 & 8 & -4i \end{bmatrix}$

(d) $D = \begin{bmatrix} 0 & i & 2-i \\ 1+4i & i & 3 \\ 1 & 0 & i \end{bmatrix}$

Conceptual Problems

D1 (a) Prove that multiplication by any complex number $\alpha = a + bi$ can be represented as a linear mapping M_α of $\mathbb{R}^2$ with standard matrix $\begin{bmatrix} a & -b \\ b & a \end{bmatrix}$.

(b) Interpret multiplication by an arbitrary complex number as a composition of a contraction or dilation, and a rotation in the plane $\mathbb{R}^2$.

(c) Verify the result by calculating M_α for $\alpha = 3 - 4i$ and interpreting it as in part (b).

D2 Let $\mathbb{C}(2,2)$ denote the set of all 2×2 matrices with complex entries with standard addition and scalar multiplication of matrices.

(a) Prove that $\mathbb{C}(2,2)$ is a complex vector space.

(b) Write a basis for $\mathbb{C}(2,2)$ and determine its dimension.

D3 Define *isomorphisms for complex vector spaces* and check that the arguments and results of Section 4.7 are correct, provided that the scalars are always taken to be complex numbers.

D4 Let $\{\vec{v}_1, \vec{v}_2, \vec{v}_3\}$ be a basis for $\mathbb{R}^3$. Prove that $\{\vec{v}_1, \vec{v}_2, \vec{v}_3\}$ is also a basis for $\mathbb{C}^3$ (taken as a complex vector space).

9.4 Eigenvectors in Complex Vector Spaces

We now look at eigenvectors and diagonalization for linear mappings $L : \mathbb{C}^n \to \mathbb{C}^n$ or, equivalently, in the case of $n \times n$ matrices with complex entries.

Eigenvalues and eigenvectors are defined in the same way as before, except that the scalars and the coordinates of the vectors are complex numbers.

Definition
Eigenvalue
Eigenvector

Let $L : \mathbb{C}^n \to \mathbb{C}^n$ be a linear mapping. If for some $\lambda \in \mathbb{C}$ there exists a non-zero vector $\vec{z} \in \mathbb{C}^n$ such that $L(\vec{z}) = \lambda\vec{z}$, then λ is an **eigenvalue** of L and $\vec{z}$ is called an **eigenvector** of L that corresponds to λ. Similarly, a complex number λ is an eigenvalue of an $n \times n$ matrix A with complex entries with corresponding eigenvector $\vec{z} \in \mathbb{C}^n$, $\vec{z} \neq \vec{0}$, if $A\vec{z} = \lambda\vec{z}$.

Since the theory of solving systems of equations, inverting matrices, and finding coordinates with respect to a basis is exactly the same for complex vector spaces as the theory for real vector spaces, the basic results on diagonalization are unchanged except that the vector space is now $\mathbb{C}^n$. A complex $n \times n$ matrix A is diagonalized by a matrix P if and only if the columns of P form a basis for $\mathbb{C}^n$ consisting of eigenvectors of A. Since the Fundamental Theorem of Algebra guarantees that every n-th degree polynomial has exactly n roots over $\mathbb{C}$, the only way a matrix cannot be diagonalizable over $\mathbb{C}$ is if it has an eigenvalue with geometric multiplicity less than its algebraic multiplicity.

We do not often have to carry out the diagonalization procedure for complex matrices. However, a simple example is given to illustrate the theory.

EXAMPLE 1

Determine whether the matrix $A = \begin{bmatrix} 3-8i & -11+7i \\ -1-4i & -2+6i \end{bmatrix}$ is diagonalizable. If it is, determine the invertible matrix P that diagonalizes A.

EXAMPLE 1
(continued)

Solution: Consider

$$A - \lambda I = \begin{bmatrix} 3 - 8i - \lambda & -11 + 7i \\ -1 - 4i & -2 + 6i - \lambda \end{bmatrix}$$

The characteristic equation is

$$\det(A - \lambda I) = \lambda^2 - (1 - 2i)\lambda + (3 - 3i) = 0$$

Using the quadratic formula, we find that

$$\lambda = \frac{(1 - 2i) \pm [(1 - 2i)^2 - 4(1)(3 - 3i)]^{1/2}}{2}$$

Using the methods from Section 9.1, we find that the eigenvalues are $\lambda_1 = 1 + i$ and $\lambda_2 = -3i$. For $\lambda_1 = 1 + i$, we get

$$A - \lambda_1 I = \begin{bmatrix} 2 - 9i & -11 + 7i \\ -1 - 4i & -3 + 5i \end{bmatrix} \sim \begin{bmatrix} 1 & -1 - i \\ 0 & 0 \end{bmatrix}$$

Hence, the general solution is $\alpha \begin{bmatrix} 1 + i \\ 1 \end{bmatrix}$, $\alpha \in \mathbb{C}$. Thus, an eigenvector corresponding to $\lambda = 1 + i$ is $\begin{bmatrix} 1 + i \\ 1 \end{bmatrix}$. For $\lambda_2 = -3i$, we get

$$A - \lambda_2 I = \begin{bmatrix} 3 - 5i & -11 + 7i \\ -1 - 4i & -2 + 9i \end{bmatrix} \sim \begin{bmatrix} 1 & -2 - i \\ 0 & 0 \end{bmatrix}$$

Hence, the general solution is $\alpha \begin{bmatrix} 2 + i \\ 1 \end{bmatrix}$, $\alpha \in \mathbb{C}$. Thus, an eigenvector corresponding to $\lambda = -3i$ is $\begin{bmatrix} 2 + i \\ 1 \end{bmatrix}$.

Hence, $P = \begin{bmatrix} 1 + i & 2 + i \\ 1 & 1 \end{bmatrix}$. Using the formula $\begin{bmatrix} a & b \\ c & d \end{bmatrix}^{-1} = \frac{1}{ad-bc}\begin{bmatrix} d & -b \\ -c & a \end{bmatrix}$, we find that $P^{-1} = \begin{bmatrix} -1 & 2 + i \\ 1 & -1 - i \end{bmatrix}$ and

$$P^{-1}AP = \begin{bmatrix} 1 + i & 0 \\ 0 & -3i \end{bmatrix}$$

Complex Characteristic Roots of a Real Matrix and a Real Canonical Form

Does diagonalizing a complex matrix tell us anything useful about diagonalizing a real matrix? First, note that since the real numbers form a subset of the complex numbers, we may regard a matrix A with real entries as being a matrix with complex entries; all of the entries just happen to have zero imaginary part. In this context, if the real matrix A has a complex characteristic root λ, we speak of λ as a complex eigenvalue of A, with a corresponding complex eigenvector. We can then proceed to diagonalize A over $\mathbb{C}$.

EXAMPLE 2

Let $A = \begin{bmatrix} 5 & -6 \\ 3 & -1 \end{bmatrix}$. Find its eigenvectors and diagonalize over $\mathbb{C}$.

Solution: We have

$$A - \lambda I = \begin{bmatrix} 5-\lambda & -6 \\ 3 & -1-\lambda \end{bmatrix}$$

so $\det(A - \lambda I) = \lambda^2 - 4\lambda + 13$, and the roots of the characteristic equation are $\lambda_1 = 2+3i$ and $\lambda_2 = 2 - 3i = \overline{\lambda_1}$.

For $\lambda_1 = 2 + 3i$,

$$A - \lambda_1 I = \begin{bmatrix} 3-3i & -6 \\ 3 & -3-3i \end{bmatrix} \sim \begin{bmatrix} 1 & -(1+i) \\ 0 & 0 \end{bmatrix}$$

Hence, a complex eigenvector corresponding to $\lambda_1 = 2 + 3i$ is $\vec{z}_1 = \begin{bmatrix} 1+i \\ 1 \end{bmatrix}$.

For $\lambda_2 = 2 - 3i$,

$$A - \lambda_2 I = \begin{bmatrix} 3+3i & -6 \\ 3 & -3+3i \end{bmatrix} \sim \begin{bmatrix} 1 & -(1-i) \\ 0 & 0 \end{bmatrix}$$

Thus, an eigenvector corresponding to $\lambda_2 = 2 - 3i$ is $\vec{z}_2 = \begin{bmatrix} 1-i \\ 1 \end{bmatrix}$.

If follows that A is diagonalized to $\begin{bmatrix} 2+3i & 0 \\ 0 & 2-3i \end{bmatrix}$ by $P = \begin{bmatrix} 1+i & 1-i \\ 1 & 1 \end{bmatrix}$.

Observe in Example 2 that the eigenvalues of A were complex conjugates. This makes sense since we know that by Theorem 9.1.2, complex roots of real polynomials come in pairs of complex conjugates. Before proving this, we first extend the definition of complex conjugates to matrices.

Definition
Complex Conjugate

Let A be an $m \times n$ matrix A. We define the complex conjugate of A, $\overline{A}$ by

$$\left(\overline{A}\right)_{ij} = \overline{(A)_{ij}}$$

Theorem 1

Suppose that A is an $n \times n$ matrix with real entries and that $\lambda = a + bi$, $b \neq 0$ is an eigenvalue of A, with corresponding eigenvector $\vec{z}$. Then $\overline{\lambda}$ is also an eigenvalue, with corresponding eigenvector $\overline{\vec{z}}$.

Proof: Suppose that $A\vec{z} = \lambda\vec{z}$. Taking complex conjugates of both sides gives

$$\overline{A\vec{z}} = \overline{\lambda\vec{z}} \Rightarrow A\overline{\vec{z}} = \overline{\lambda}\overline{\vec{z}}$$

since $\left(\overline{A}\right)_{ij} = \overline{(A)_{ij}} = (A_{ij})$. Hence, $\overline{\lambda}$ is an eigenvalue of A with corresponding eigenvector $\overline{\vec{z}}$, as required. ■

Now we note that the solution to Example 2 was not completely satisfying. The point of diagonalization is to simplify the matrix of the linear mapping. However, in Example 2, we have changed from a real matrix to a complex matrix. Given a square matrix A with real entries, we would like to determine a similar matrix $P^{-1}AP$ that also has real entries and that reveals information about the eigenvalues of A, even if the eigenvalues are complex. The rest of this section is concerned with the problem of finding such a real matrix.

If the eigenvalues of A are all real, then by diagonalizing A, we have our desired similar matrix. Thus, we will just consider the case where A is an $n \times n$ real matrix with at least one eigenvalue $\lambda = a + bi$, where $a, b \in \mathbb{R}$ and $b \neq 0$.

Since we are splitting the eigenvalue into real and imaginary parts, it makes sense to also split the corresponding eigenvector $\vec{z}$ into real and imaginary parts:

$$\vec{z} = \begin{bmatrix} x_1 + y_1 i \\ \vdots \\ x_n + y_n i \end{bmatrix} = \begin{bmatrix} x_1 \\ \vdots \\ x_n \end{bmatrix} + i \begin{bmatrix} y_1 \\ \vdots \\ y_n \end{bmatrix} = \vec{x} + i\vec{y}, \qquad \vec{x}, \vec{y} \in \mathbb{R}^n$$

Thus, we have $A\vec{z} = \lambda\vec{z}$, or

$$A(\vec{x} + i\vec{y}) = (a + bi)(\vec{x} + i\vec{y}) = (a\vec{x} - b\vec{y}) + i(b\vec{x} + a\vec{y})$$

Upon considering real and imaginary parts, we get

$$A\vec{x} = a\vec{x} - b\vec{y} \qquad \text{and} \qquad A\vec{y} = b\vec{x} + a\vec{y} \tag{9.1}$$

Observe that equation (9.1) shows that the image of any linear combination of $\vec{x}$ and $\vec{y}$ under A will be a linear combination of $\vec{x}$ and $\vec{y}$. That is, if $\vec{v} \in \text{Span}\{\vec{x}, \vec{y}\}$, then $A\vec{v} \in \text{Span}\{\vec{x}, \vec{y}\}$.

Definition
Invariant Subspace

If $T : \mathbb{V} \to \mathbb{V}$ is a linear operator and $\mathbb{U}$ is a subspace of $\mathbb{V}$ such that $T(\mathbf{u}) \in \mathbb{U}$ for all $\mathbf{u} \in \mathbb{U}$, then $\mathbb{U}$ is called an **invariant subspace** of T.

Note that $\vec{x}$ and $\vec{y}$ must be linearly independent, for if $\vec{x} = k\vec{y}$, equation (9.1) would say that they are real eigenvectors of A with corresponding real eigenvalues, and this is impossible with our assumption that $b \neq 0$. (See Problem D2b.) Moreover, it can be shown that no vector in $\text{Span}\{\vec{x}, \vec{y}\}$ is a real eigenvector of A. (See Problem D2c.) This discussion together with Problem D2 is summarized in Theorem 2.

Theorem 2

Suppose that $\lambda = a + bi$, $b \neq 0$ is an eigenvalue of an $n \times n$ real matrix A with corresponding eigenvector $\vec{z} = \vec{x} + i\vec{y}$. Then $\text{Span}\{\vec{x}, \vec{y}\}$ is a two-dimensional subspace of $\mathbb{R}^n$ that is invariant under A and contains no real eigenvector of A.

The Case of a 2×2 Matrix

If A is a 2×2 real matrix, then it follows from Theorem 2 that $\mathcal{B} = \{\vec{x}, \vec{y}\}$ is a basis for $\mathbb{R}^2$. From (9.1) we get that the $\mathcal{B}$-matrix of the linear mapping L associated with A is $[L]_{\mathcal{B}} = \begin{bmatrix} a & b \\ -b & a \end{bmatrix}$. Moreover, from our work in Section 4.6, we know that $[L]_{\mathcal{B}} = P^{-1}AP$, where $P = \begin{bmatrix} \vec{x} & \vec{y} \end{bmatrix}$ is the change of coordinates matrix from $\mathcal{B}$-coordinates to standard coordinates.

Thus, we have a real matrix that is similar to A and gives information about the eigenvalues of A, as desired.

Definition
Real Canonical Form

Let A be a 2×2 real matrix with eigenvalue $\lambda = a + ib$, $b \neq 0$. The matrix $\begin{bmatrix} a & b \\ -b & a \end{bmatrix}$ is called a **real canonical form** for A.

EXAMPLE 3

Find a real canonical form of the matrix C of $A = \begin{bmatrix} 2 & -5 \\ 1 & -2 \end{bmatrix}$ and find a change of coordinates matrix P such that $P^{-1}AP = C$.

Solution: We have

$$\det(A - \lambda I) = \begin{vmatrix} 2-\lambda & -5 \\ 1 & -2-\lambda \end{vmatrix} = \lambda^2 + 1$$

So, the eigenvalues of A are $\lambda_1 = 0 + i$ and $\lambda_2 = 0 - i = \overline{\lambda_1}$. Thus, we have $a = 0$ and $b = 1$ and hence a real canonical form of A is $\begin{bmatrix} 0 & 1 \\ -1 & 0 \end{bmatrix}$.

For $\lambda_1 = i$, we have

$$A - \lambda_1 I = \begin{bmatrix} 2-i & -5 \\ 1 & -2-i \end{bmatrix} \sim \begin{bmatrix} 1 & -2-i \\ 0 & 0 \end{bmatrix}$$

so an eigenvector corresponding to λ is

$$\vec{z} = \begin{bmatrix} 2+i \\ 1 \end{bmatrix} = \begin{bmatrix} 2 \\ 1 \end{bmatrix} + \begin{bmatrix} 1 \\ 0 \end{bmatrix} i$$

Hence, a change of coordinates matrix P is

$$P = \begin{bmatrix} 2 & 1 \\ 1 & 0 \end{bmatrix}$$

EXAMPLE 4

Find a real canonical form of the matrix $A = \begin{bmatrix} 5 & -6 \\ 3 & -1 \end{bmatrix}$.

Solution: In Example 2, we saw that A has eigenvalues $\lambda = 2 + 3i$ and $\overline{\lambda} = 2 - 3i$. Thus, a real canonical form of A is $\begin{bmatrix} 2 & 3 \\ -3 & 2 \end{bmatrix}$.

EXERCISE 1

Find a change of coordinates matrix P for Example 4 and verify that $P^{-1}AP = \begin{bmatrix} 2 & 3 \\ -3 & 2 \end{bmatrix}$.

Remarks

1. A matrix of the form $\begin{bmatrix} a & b \\ -b & a \end{bmatrix}$ can be rewritten as

$$\sqrt{a^2+b^2}\begin{bmatrix} \cos\theta & -\sin\theta \\ \sin\theta & \cos\theta \end{bmatrix}$$

where $\cos\theta = a/\sqrt{a^2+b^2}$ and $\sin\theta = -b/\sqrt{a^2+b^2}$. But, since the new basis vectors $\vec{x}$ and $\vec{y}$ are not necessarily orthogonal, the matrix does not represent a true rotation of $\operatorname{Span}\{\vec{x}, \vec{y}\}$.

2. Observe in Example 3 that we could have taken $\lambda_1 = -i$ and $\lambda_2 = \overline{\lambda_1}$. Thus, $\begin{bmatrix} 0 & -1 \\ 1 & 0 \end{bmatrix}$ is also a real canonical form of A.

3. The complex eigenvector $\vec{z}$ is determined only up to multiplication by an arbitrary non-zero complex number. This means that the vectors $\vec{x}$ and $\vec{y}$ are not uniquely determined. For example, in Example 3, $\vec{z} = (1+i)\begin{bmatrix} 2+i \\ 1 \end{bmatrix} = \begin{bmatrix} 1+3i \\ 1+i \end{bmatrix}$ is also an eigenvector corresponding to λ. Thus, taking $P = \begin{bmatrix} 1 & 3 \\ 1 & 1 \end{bmatrix}$ would also give $P^{-1}AP = \begin{bmatrix} 0 & 1 \\ -1 & 0 \end{bmatrix}$.

EXERCISE 2

Find a real canonical form of the matrix C of $A = \begin{bmatrix} 3 & 2 \\ -1 & 3 \end{bmatrix}$ and find a change of coordinates matrix P such that $P^{-1}AP = C$.

The Case of a 3×3 Matrix

If A is a 3×3 real matrix with one real eigenvalue μ with corresponding eigenvector $\vec{v}$ and complex eigenvalues $\lambda = a + bi$, $b \neq 0$ and $\overline{\lambda}$ with corresponding eigenvectors $\vec{z} = \vec{x} + i\vec{y}$ and $\overline{\vec{z}}$, then we have the equations

$$A\vec{v} = \mu\vec{v}, \qquad A\vec{x} = a\vec{x} - b\vec{y}, \qquad A\vec{y} = b\vec{x} + a\vec{y}$$

Then, the matrix of A with respect to the basis $\mathcal{B} = \{\vec{v}, \vec{x}, \vec{y}\}$ is $\begin{bmatrix} \mu & 0 & 0 \\ 0 & a & b \\ 0 & -b & a \end{bmatrix}$.

The matrix can be brought into this real canonical form by a change of coordinates with matrix $P = \begin{bmatrix} \vec{v} & \vec{x} & \vec{y} \end{bmatrix}$.

EXAMPLE 5

Find a real canonical form of the matrix C of $A = \begin{bmatrix} -1 & -4 & -4 \\ 4 & 7 & 4 \\ 0 & -2 & -1 \end{bmatrix}$ and find a change of coordinates matrix P such that $P^{-1}AP = C$.

EXAMPLE 5
(continued)

Solution: We have

$$\det(A - \lambda I) = \begin{vmatrix} -1-\lambda & -4 & -4 \\ 4 & 7-\lambda & 4 \\ 0 & -2 & -1-\lambda \end{vmatrix} = -(\lambda - 3)(\lambda^2 - 2\lambda + 5)$$

Thus, the eigenvalues of A are $\mu = 3$, $\lambda_1 = 1 + 2i$, and $\lambda_2 = 1 - 2i = \overline{\lambda_1}$. Thus, a real canonical form of A is $\begin{bmatrix} 3 & 0 & 0 \\ 0 & 1 & 2 \\ 0 & -2 & 1 \end{bmatrix}$.

For $\mu = 3$, we have

$$A - \mu I = \begin{bmatrix} -4 & -4 & -4 \\ 4 & 4 & 4 \\ 0 & -2 & -4 \end{bmatrix} \sim \begin{bmatrix} 1 & 0 & -1 \\ 0 & 1 & 2 \\ 0 & 0 & 0 \end{bmatrix}$$

Thus, an eigenvector corresponding to μ is $\vec{v} = \begin{bmatrix} 1 \\ -2 \\ 1 \end{bmatrix}$.

For $\lambda_1 = 1 + 2i$, we have

$$A - \lambda_1 I = \begin{bmatrix} -2-2i & -4 & -4 \\ 4 & 6-2i & 4 \\ 0 & -2 & -2-2i \end{bmatrix} \sim \begin{bmatrix} 1 & 0 & -1-i \\ 0 & 1 & 1+i \\ 0 & 0 & 0 \end{bmatrix}$$

Thus, an eigenvector corresponding to λ_1 is

$$\vec{z} = \begin{bmatrix} 1+i \\ -1-i \\ 1 \end{bmatrix} = \begin{bmatrix} 1 \\ -1 \\ 1 \end{bmatrix} + \begin{bmatrix} 1 \\ -1 \\ 0 \end{bmatrix} i$$

Hence, a change of coordinates matrix P is

$$P = \begin{bmatrix} 1 & 1 & 1 \\ -2 & -1 & -1 \\ 1 & 1 & 0 \end{bmatrix}$$

If A is an $n \times n$ real matrix with complex eigenvalue $\lambda = a + bi$, $b \neq 0$ and corresponding eigenvector $\vec{z} = \vec{x} + i\vec{y}$, then Span$\{\vec{x}, \vec{y}\}$ is a two-dimensional invariant subspace of $\mathbb{R}^n$. If we use $\vec{x}$ and $\vec{y}$ as consecutive vectors in a basis for $\mathbb{R}^n$, with the other vectors being eigenvectors for other eigenvalues, then in a matrix similar to A, there is a block $\begin{bmatrix} a & b \\ -b & a \end{bmatrix}$, with the a's occurring on the diagonal in positions determined by the position of $\vec{x}$ and $\vec{y}$ in the basis.

For repeated complex eigenvalues, the situation is the same as for repeated real eigenvectors. In some cases, it is possible to find a basis of eigenvectors and diagonalize the matrix over $\mathbb{C}$. In other cases, further theory is required, leading to the Jordan normal form.

PROBLEMS 9.4

Practice Problems

A1 For each of the following matrices, determine a diagonal matrix D similar to the given matrix over $\mathbb{C}$. Also determine a real canonical form and give a change of coordinates matrix P that brings the matrix into this form.

(a) $\begin{bmatrix} 0 & 4 \\ -1 & 0 \end{bmatrix}$

(b) $\begin{bmatrix} -1 & 2 \\ -1 & -3 \end{bmatrix}$

(c) $\begin{bmatrix} 2 & 2 & -1 \\ -4 & 1 & 2 \\ 2 & 2 & -1 \end{bmatrix}$

(d) $\begin{bmatrix} 2 & 1 & -1 \\ 2 & 1 & 0 \\ 3 & -1 & 2 \end{bmatrix}$

Homework Problems

B1 For each of the following matrices, determine a diagonal matrix D similar to the given matrix over $\mathbb{C}$. Also determine a real canonical form and give a change of coordinates matrix P that brings the matrix into this form.

(a) $\begin{bmatrix} 1 & -5 \\ 1 & -3 \end{bmatrix}$

(b) $\begin{bmatrix} 1 & -5 \\ 1 & 3 \end{bmatrix}$

(c) $\begin{bmatrix} 0 & -2 & 1 \\ 2 & 2 & -1 \\ 0 & -2 & 2 \end{bmatrix}$

(d) $\begin{bmatrix} -1 & 2 & -2 \\ -2 & -1 & -1 \\ 4 & -2 & 5 \end{bmatrix}$

(e) $\begin{bmatrix} 6 & 0 & -4 \\ 0 & 1 & 1 \\ 8 & -1 & -5 \end{bmatrix}$

(f) $\begin{bmatrix} 2 & 2 & 2 \\ 1 & 1 & -2 \\ -2 & -1 & 2 \end{bmatrix}$

Conceptual Problems

D1 Verify that if $\vec{z}$ is an eigenvector of a matrix A with complex entries, then $\overline{\vec{z}}$ is an eigenvector of $\overline{A}$ (the matrix obtained from A by taking complex conjugates of each entry of A).

D2 Suppose that A is an $n \times n$ real matrix and that $\lambda = a + bi$ is a complex eigenvalue of A with $b \neq 0$. Let the corresponding eigenvector be $\vec{x} + i\vec{y}$.

(a) Prove that $\vec{x} \neq \vec{0}$ and $\vec{y} \neq \vec{0}$.

(b) Show that $\vec{x} \neq k\vec{y}$ for any real number k. (Hint: Suppose $\vec{x} = k\vec{y}$ for some k, and use equation (9.1) to show that this requires $\beta = 0$.)

(c) Prove that Span$\{\vec{x}, \vec{y}\}$ does not contain an eigenvector of A corresponding to a real eigenvalue of A.

9.5 Inner Products in Complex Vector Spaces

We would like to have an inner product defined for complex vector spaces because the concepts of length, orthogonality, and projection are powerful tools for solving certain problems.

Our first thought would be to determine if we can extend the dot product to $\mathbb{C}^n$. Does this define an inner product on $\mathbb{C}^n$? Let $\vec{z} = \vec{x} + i\vec{y}$, then we have

$$\begin{aligned}\vec{z} \cdot \vec{z} &= z_1^2 + \cdots + z_n^2 \\ &= (x_1^2 + \cdots + x_n^2 - y_1^2 - \cdots - y_n^2) + 2i(x_1y_1 + \cdots + x_ny_n)\end{aligned}$$

Observe that $\vec{z} \cdot \vec{z}$ does not even need to be a real number and so the condition $\vec{z} \cdot \vec{z} \geq 0$ does not even make sense. Thus *we cannot use the dot product as a rule for defining an inner product in* $\mathbb{C}^n$.

As in the real case, we want $\langle \vec{z}, \vec{z} \rangle$ to be a non-negative real number so that we can define a vector by $\|\vec{z}\| = \sqrt{\langle \vec{z}, \vec{z} \rangle}$. We recall that if $z \in \mathbb{C}$, then $z\overline{z} = |z|^2 \geq 0$. Hence, it makes sense to choose

$$\langle \vec{z}, \vec{w} \rangle = \vec{z} \cdot \overline{\vec{w}}$$

as this gives us

$$\langle \vec{z}, \vec{z} \rangle = \vec{z} \cdot \overline{\vec{z}} = z_1\overline{z_1} + \cdots + z_n\overline{z_n} = |z_1|^2 + \cdots + |z_n|^2 \geq 0$$

Definition
Standard Inner Product on $\mathbb{C}^n$

In $\mathbb{C}^n$ the **standard inner product** $\langle\, , \rangle$ is defined by

$$\langle \vec{z}, \vec{w} \rangle = \vec{z} \cdot \overline{\vec{w}} = z_1\overline{w_1} + \cdots + z_n\overline{w_n}, \qquad \text{for } \vec{w}, \vec{z} \in \mathbb{C}^n$$

EXAMPLE 1

Let $\vec{u} = \begin{bmatrix} 1+i \\ 2-i \end{bmatrix}$, $\vec{v} = \begin{bmatrix} -2+i \\ 3+2i \end{bmatrix}$. Determine $\langle \vec{v}, \vec{u} \rangle$, $\langle \vec{u}, \vec{v} \rangle$, and $\langle \vec{v}, (2-i)\vec{u} \rangle$.

Solution:

$$\begin{aligned}\langle \vec{v}, \vec{u} \rangle = \vec{v} \cdot \overline{\vec{u}} &= \begin{bmatrix} -2+i \\ 3+2i \end{bmatrix} \cdot \overline{\begin{bmatrix} 1+i \\ 2-i \end{bmatrix}} \\ &= \begin{bmatrix} -2+i \\ 3+2i \end{bmatrix} \cdot \begin{bmatrix} 1-i \\ 2+i \end{bmatrix} \\ &= (-2+i)(1-i) + (3+2i)(2+i) \\ &= -1 + 3i + 4 + 7i = 3 + 10i \\ \langle \vec{u}, \vec{v} \rangle &= \begin{bmatrix} 1+i \\ 2-i \end{bmatrix} \cdot \begin{bmatrix} -2-i \\ 3-2i \end{bmatrix} \\ &= -1 - 3i + 4 - 7i = 3 - 10i \\ \langle \vec{v}, (2-i)\vec{u} \rangle = \vec{v} \cdot \overline{(2-i)\vec{u}} &= \begin{bmatrix} -2+i \\ 3+2i \end{bmatrix} \cdot \overline{(2-i)\begin{bmatrix} 1-i \\ 2+i \end{bmatrix}} \\ &= \begin{bmatrix} -2+i \\ 3+2i \end{bmatrix} \cdot (2+i)\begin{bmatrix} 1+i \\ 2-i \end{bmatrix} \\ &= (2+i)(3+10i) \\ &= -4 + 23i\end{aligned}$$

Observe that this does not satisfy the properties of the real inner product. In particular, $\langle \vec{u}, \vec{v} \rangle \neq \langle \vec{v}, \vec{u} \rangle$ and $\langle \vec{v}, \alpha\vec{u} \rangle \neq \alpha\langle \vec{v}, \vec{u} \rangle$.

EXERCISE 1

Let $\vec{u} = \begin{bmatrix} i \\ 1+2i \end{bmatrix}$ and $\vec{v} = \begin{bmatrix} 2+2i \\ 1-3i \end{bmatrix}$. Determine $\langle \vec{u}, \vec{v} \rangle$, $\langle 2i\vec{u}, \vec{v} \rangle$, and $\langle \vec{u}, 2i\vec{v} \rangle$.

Properties of Complex Inner Products

Example 1 warns us that for complex vector spaces, we must modify the requirements of symmetry and bilinearity stated for real inner products.

Definition
Complex Inner Product

Let $\mathbb{V}$ be a vector space over $\mathbb{C}$. A **complex inner product** on $\mathbb{V}$ is a function $\langle\,,\rangle : \mathbb{V} \times \mathbb{V} \to \mathbb{C}$ such that

(1) $\langle \mathbf{z}, \mathbf{z} \rangle \geq 0$ for all $\mathbf{z} \in \mathbb{V}$ and $\langle \mathbf{z}, \mathbf{z} \rangle = 0$ if and only if $\mathbf{z} = \mathbf{0}$

(2) $\langle \mathbf{z}, \mathbf{w} \rangle = \overline{\langle \mathbf{w}, \mathbf{z} \rangle}$ for all $\mathbf{w}, \mathbf{z} \in \mathbb{V}$

(3) For all $\mathbf{u}, \mathbf{v}, \mathbf{w}, \mathbf{z} \in \mathbb{V}$ and $\alpha \in \mathbb{C}$

(i) $\langle \mathbf{v} + \mathbf{z}, \mathbf{w} \rangle = \langle \mathbf{v}, \mathbf{w} \rangle + \langle \mathbf{z}, \mathbf{w} \rangle$
(ii) $\langle \mathbf{z}, \mathbf{w} + \mathbf{u} \rangle = \langle \mathbf{z}, \mathbf{w} \rangle + \langle \mathbf{z}, \mathbf{u} \rangle$
(iii) $\langle \alpha\mathbf{z}, \mathbf{w} \rangle = \alpha\langle \mathbf{z}, \mathbf{w} \rangle$
(iv) $\langle \mathbf{z}, \alpha\mathbf{w} \rangle = \overline{\alpha}\langle \mathbf{z}, \mathbf{w} \rangle$

EXERCISE 2

Verify that the standard inner product on $\mathbb{C}^n$ is a complex inner product.

Note that property (1) allows us to define the (standard) **length** by $\|\mathbf{z}\| = \langle \mathbf{z}, \mathbf{z} \rangle^{1/2}$, as desired.

Property (2) is the **Hermitian** property of the inner product. Notice that if all the vectors are real, the Hermitian property simplifies to symmetry.

Property (3) says that the complex inner product is not quite bilinear. However, this property reduces to bilinearity when the scalars are all real.

The Cauchy-Schwarz and Triangle Inequalities

Complex inner products satisfy the Cauchy-Schwarz and triangle inequalities. However, new proofs are required.

Theorem 1

Let $\mathbb{V}$ be a complex inner product space with inner product $\langle\,,\rangle$. Then, for all $\mathbf{w}, \mathbf{z} \in \mathbb{V}$,

(4) $|\langle \mathbf{z}, \mathbf{w} \rangle| \leq \|\mathbf{z}\|\,\|\mathbf{w}\|$
(5) $\|\mathbf{z} + \mathbf{w}\| \leq \|\mathbf{z}\| + \|\mathbf{w}\|$

Proof: We prove (4) and leave the proof of (5) as Problem D1.

If $\mathbf{w} = \mathbf{0}$, then (4) is immediate, so assume that $\mathbf{w} \neq \mathbf{0}$, and let

$$\alpha = \frac{\langle \mathbf{z}, \mathbf{w} \rangle}{\langle \mathbf{w}, \mathbf{w} \rangle}$$

Then, we get

$$\begin{aligned}
0 &\leq \langle \mathbf{z} - \alpha\mathbf{w}, \mathbf{z} - \alpha\mathbf{w} \rangle \\
&= \langle \mathbf{z}, \mathbf{z} - \alpha\mathbf{w} \rangle - \alpha\langle \mathbf{w}, \mathbf{z} - \alpha\mathbf{w} \rangle \\
&= \langle \mathbf{z}, \mathbf{z} \rangle - \overline{\alpha}\langle \mathbf{z}, \mathbf{w} \rangle - \alpha\langle \mathbf{w}, \mathbf{z} \rangle + \alpha\overline{\alpha}\langle \mathbf{w}, \mathbf{w} \rangle \\
&= \langle \mathbf{z}, \mathbf{z} \rangle - \frac{|\langle \mathbf{z}, \mathbf{w} \rangle|^2}{\langle \mathbf{w}, \mathbf{w} \rangle} - \frac{|\langle \mathbf{z}, \mathbf{w} \rangle|^2}{\langle \mathbf{w}, \mathbf{w} \rangle} + \frac{|\langle \mathbf{z}, \mathbf{w} \rangle|^2}{\langle \mathbf{w}, \mathbf{w} \rangle} \\
&= ||\mathbf{z}||^2 - \frac{|\langle \mathbf{z}, \mathbf{w} \rangle|^2}{||\mathbf{w}||^2}
\end{aligned}$$

and (4) follows. ■

Let $\mathbb{C}(m, n)$ denote the complex vector space of $m \times n$ matrices with complex entries. How should we define the standard complex inner product on this vector space? We saw with real vector spaces that we defined $\langle A, B \rangle = \operatorname{tr}(B^T A)$ and found that this inner product was equivalent to the dot product on $\mathbb{R}^{mn}$. Of course, we want to define the standard inner product on $\mathbb{C}(m, n)$ in a similar way. However, because of the way the complex inner product is defined on $\mathbb{C}^n$, we see that we also need to take a complex conjugate. Thus, we define the inner product $\langle\, , \rangle$ on $C(m, n)$ by $\langle A, B \rangle = \operatorname{tr}(\overline{B}^T A)$.

In particular, let $A = \begin{bmatrix} a_{11} & \cdots & a_{1n} \\ \vdots & & \vdots \\ a_{m1} & \cdots & a_{mn} \end{bmatrix}$ and $B = \begin{bmatrix} b_{11} & \cdots & b_{1n} \\ \vdots & & \vdots \\ b_{m1} & \cdots & b_{mn} \end{bmatrix}$. Then

$$\overline{B}^T A = \begin{bmatrix} \overline{b_{11}} & \cdots & \overline{b_{m1}} \\ \vdots & & \vdots \\ \overline{b_{1n}} & \cdots & \overline{b_{mn}} \end{bmatrix} \begin{bmatrix} a_{11} & \cdots & a_{1n} \\ \vdots & & \vdots \\ a_{m1} & \cdots & a_{mn} \end{bmatrix}$$

Since we want the trace of this, we just consider the diagonal entries in the product and find that

$$\begin{aligned}
\operatorname{tr}(\overline{B}^T A) &= \sum_{i=1}^{m} a_{i1}\overline{b_{i1}} + \sum_{i=1}^{m} a_{i2}\overline{b_{i2}} + \cdots + \sum_{i=1}^{m} a_{in}\overline{b_{in}} \\
&= \sum_{j=1}^{n}\sum_{i=1}^{m} a_{ij}\overline{b_{ij}}
\end{aligned}$$

which corresponds to the standard inner product of the corresponding vectors under the obvious isomorphism with $\mathbb{C}^{mn}$.

EXAMPLE 2

Let $A = \begin{bmatrix} 2+i & 1 \\ i & 1-i \end{bmatrix}$ and $B = \begin{bmatrix} 3 & 2-3i \\ -2i & 1+2i \end{bmatrix}$. Find $\langle A, B\rangle$ and show that this corresponds to the standard inner product of $\vec{a} = \begin{bmatrix} 2+i \\ 1 \\ i \\ 1-i \end{bmatrix}$ and $\vec{b} = \begin{bmatrix} 3 \\ 2-3i \\ -2i \\ 1+2i \end{bmatrix}$.

Solution: We have

$$\begin{aligned}
\langle A, B\rangle = \operatorname{tr}(\overline{B}^T A) &= \operatorname{tr}\left(\begin{bmatrix} 3 & 2i \\ 2+3i & 1-2i \end{bmatrix}\begin{bmatrix} 2+i & 1 \\ i & 1-i \end{bmatrix}\right) \\
&= \operatorname{tr}\begin{bmatrix} 3(2+i)+2i(i) & 3(1)+2i(1-i) \\ (2+3i)(2+i)+(1-2i)(i) & (2+3i)(1)+(1-2i)(1-i) \end{bmatrix} \\
&= 3(2+i)+2i(i)+(2+3i)(1)+(1-2i)(1-i) \\
&= \begin{bmatrix} 2+i \\ 1 \\ i \\ 1-i \end{bmatrix} \cdot \begin{bmatrix} 3 \\ 2+3i \\ 2i \\ 1-2i \end{bmatrix} \\
&= \vec{a} \cdot \overline{\vec{b}} \\
&= \langle \vec{a}, \vec{b}\rangle
\end{aligned}$$

This matrix $\overline{B}^T$ is very important, so we make the following definition.

Definition

Conjugate Transpose

Let A be an $n \times n$ matrix with complex entries. We define the **conjugate transpose** A^* of A to be

$$A^* = \overline{A}^T$$

EXAMPLE 3

Let $A = \begin{bmatrix} 1+i & 1-2i & i \\ 2 & -i & 3+i \end{bmatrix}$. Then $A^* = \begin{bmatrix} 1-i & 2 \\ 1+2i & i \\ -i & 3-i \end{bmatrix}$.

Observe that if A is a real matrix, then $A^* = A^T$.

Theorem 2

Let A and B be complex matrices and let $\alpha \in \mathbb{C}$. Then

(1) $\langle A\vec{z}, \vec{w}\rangle = \langle \vec{z}, A^*\vec{w}\rangle$ for all $\vec{z}, \vec{w} \in \mathbb{C}^n$
(2) $A^{**} = A$
(3) $(A+B)^* = A^* + B^*$
(4) $(\alpha A)^* = \overline{\alpha} A^*$
(5) $(AB)^* = B^* A^*$

The proof is left as Problem D2.

Orthogonality in $\mathbb{C}^n$ and Unitary Matrices

With a complex inner product defined, we can proceed and introduce orthogonality, projections, and distance, as we did in the real case. No new approaches or ideas are required, so we omit a detailed discussion. However, we must be very careful that we have the vectors in the correct order when calculating an inner product since the complex inner product is not symmetric. For example, if $\mathcal{B} = \{\vec{v}_1, \dots, \vec{v}_k\}$ is an orthonormal basis for a subspace $\mathbb{S}$ of $\mathbb{C}^n$, then the projection of $\vec{z} \in \mathbb{C}^n$ onto $\mathbb{S}$ is given by

$$\text{proj}_{\mathbb{S}}\, \vec{z} = \langle \vec{z}, \vec{v}_1 \rangle \vec{v}_1 + \cdots + \langle \vec{z}, \vec{v}_k \rangle \vec{v}_k$$

If we were to calculate $\langle \vec{v}_1, \vec{z} \rangle$ instead of $\langle \vec{z}, \vec{v}_1 \rangle$, we would likely get the wrong answer.

Notice that since $\langle \vec{z}, \vec{w} \rangle$ may be complex, there is no obvious way to define the angle between vectors.

EXAMPLE 4

Let $\vec{v}_1 = \begin{bmatrix} i \\ 0 \\ 0 \\ 0 \end{bmatrix}$, $\vec{v}_2 = \begin{bmatrix} 1 \\ 1 \\ -i \\ i \end{bmatrix}$, and $\vec{v}_3 = \begin{bmatrix} 1 \\ i \\ 1 \\ 0 \end{bmatrix}$ and consider the subspace $\mathbb{S} = \text{Span}\{\vec{v}_1, \vec{v}_2, \vec{v}_3\}$ of $\mathbb{C}^4$.

(a) Use the Gram-Schmidt Procedure to find an orthogonal basis for $\mathbb{S}$.

Solution: Let $\vec{w}_1 = \begin{bmatrix} i \\ 0 \\ 0 \\ 0 \end{bmatrix}$. Then

$$\begin{aligned} \vec{w}_2 &= \vec{v}_2 - \frac{\langle \vec{v}_2, \vec{w}_1 \rangle}{\|\vec{w}_1\|^2}\vec{w}_1 = \begin{bmatrix} 0 \\ 1 \\ -i \\ i \end{bmatrix} \\ \vec{w}_3 &= \vec{v}_3 - \frac{\langle \vec{v}_3, \vec{w}_1 \rangle}{\|\vec{w}_1\|^2}\vec{w}_1 - \frac{\langle \vec{v}_3, \vec{w}_2 \rangle}{\|\vec{w}_2\|^2}\vec{w}_2 = \begin{bmatrix} 1 \\ i \\ 1 \\ 0 \end{bmatrix} + i \begin{bmatrix} i \\ 0 \\ 0 \\ 0 \end{bmatrix} - \frac{2i}{3} \begin{bmatrix} 0 \\ 1 \\ -i \\ i \end{bmatrix} = \begin{bmatrix} 0 \\ i/3 \\ 1/3 \\ 2/3 \end{bmatrix} \end{aligned}$$

Since we can take any scalar multiple of this, we instead take $\vec{w}_3 = \begin{bmatrix} 0 \\ i \\ 1 \\ 2 \end{bmatrix}$ and get that $\{\vec{w}_1, \vec{w}_2, \vec{w}_3\}$ is an orthogonal basis for $\mathbb{S}$.

(b) Let $\vec{z} = \begin{bmatrix} 1 \\ -i \\ 1 \\ -1 \end{bmatrix}$. Find $\text{proj}_{\mathbb{S}}\, \vec{z}$.

EXAMPLE 4
(continued)

Solution: Using the orthogonal basis we found in (a), we get

$$\operatorname{proj}_{\mathbb{S}} \vec{z} = \frac{\langle \vec{z}, \vec{w}_1 \rangle}{\|\vec{w}_1\|^2}\vec{w}_1 + \frac{\langle \vec{z}, \vec{w}_2 \rangle}{\|\vec{w}_2\|^2}\vec{w}_2 + \frac{\langle \vec{z}, \vec{w}_3 \rangle}{\|\vec{w}_3\|^2}\vec{w}_3$$
$$= -i\begin{bmatrix} i \\ 0 \\ 0 \\ 0 \end{bmatrix} + \frac{i}{3}\begin{bmatrix} 0 \\ 1 \\ -i \\ i \end{bmatrix} - \frac{2}{6}\begin{bmatrix} 0 \\ i \\ 1 \\ 2 \end{bmatrix} = \begin{bmatrix} 1 \\ 0 \\ 0 \\ -1 \end{bmatrix}$$

When working with real inner products, we saw that orthogonal matrices are very important. So, we now consider a complex version of these matrices.

Definition
Unitary Matrix

An $n \times n$ matrix with complex entries is said to be **unitary** if its columns form an orthonormal basis for $\mathbb{C}^n$.

For an orthogonal matrix P, we saw that the defining property is equivalent to the matrix condition $P^{-1} = P^T$. We get the associated result for unitary matrices.

Theorem 3

If U is an $n \times n$ matrix, then the following are equivalent.

(1) The columns of U form an orthonormal basis for $\mathbb{C}^n$
(2) The rows of U form an orthonormal basis for $\mathbb{C}^n$
(3) $U^{-1} = U^*$

Proof: Let $U = \begin{bmatrix} \vec{z}_1 & \cdots & \vec{z}_n \end{bmatrix}$. By definition, we have that $\{\vec{z}_1, \dots, \vec{z}_n\}$ is orthonormal if and only if $\langle \vec{z}_i, \vec{z}_i \rangle = 1$ and

$$0 = \langle \vec{z}_i, \vec{z}_j \rangle = \vec{z}_i \cdot \overline{\vec{z}_j} = \vec{z}_i^{\,T} \overline{\vec{z}_j}, \qquad \text{for } i \neq j$$

Thus, since

$$(U^*U)_{ij} = \overline{\vec{z}_i}^{\,T} \vec{z}_j = \overline{\vec{z}_i^{\,T} \overline{\vec{z}_j}} = \overline{\langle \vec{z}_i, \vec{z}_j \rangle}$$

we get that $U^*U = I$ if and only if $\{\vec{z}_1, \dots, \vec{z}_n\}$ is orthonormal. The proof that (2) is equivalent to (3) is similar. ■

Observe that if the entries of A are all real, then A is unitary if and only if it is orthogonal.

EXAMPLE 5

Are the matrices $U = \begin{bmatrix} 1 & -i \\ i & 1 \end{bmatrix}$ and $V = \begin{bmatrix} \frac{1}{\sqrt{3}}(1+i) & \frac{1}{\sqrt{6}}(1+i) \\ -\frac{1}{\sqrt{3}}i & \frac{2}{\sqrt{6}}i \end{bmatrix}$ unitary?

Solution: Observe that

$$\left\langle \begin{bmatrix} 1 \\ i \end{bmatrix}, \begin{bmatrix} 1 \\ i \end{bmatrix} \right\rangle = \begin{bmatrix} 1 \\ i \end{bmatrix} \cdot \overline{\begin{bmatrix} 1 \\ i \end{bmatrix}} = \begin{bmatrix} 1 \\ i \end{bmatrix} \cdot \begin{bmatrix} 1 \\ -i \end{bmatrix} = 1(1) + i(-i) = 2$$

Hence, $\begin{bmatrix} 1 \\ i \end{bmatrix}$ is not a unit vector. Thus, U is not unitary since its columns are not orthonormal.

EXAMPLE 5
(continued)

For V, we have $V^* = \begin{bmatrix} \frac{1}{\sqrt{3}}(1-i) & \frac{1}{\sqrt{3}}i \\ \frac{1}{\sqrt{6}}(1-i) & -\frac{2}{\sqrt{6}}i \end{bmatrix}$, so

$$V^*V = \begin{bmatrix} \frac{1}{3}(2+1) & \frac{1}{3\sqrt{2}}(2-2) \\ \frac{1}{3\sqrt{2}}(2-2) & \frac{1}{6}(2+4) \end{bmatrix} = \begin{bmatrix} 1 & 0 \\ 0 & 1 \end{bmatrix}$$

Thus, V is unitary.

EXERCISE 3

Determine if either of the following matrices is unitary.

$$A = \begin{bmatrix} \frac{2+i}{\sqrt{6}} & \frac{1}{\sqrt{6}} \\ -\frac{1}{\sqrt{6}} & \frac{2+i}{\sqrt{6}} \end{bmatrix}, \qquad B = \begin{bmatrix} \frac{1+i}{\sqrt{6}} & \frac{1+i}{\sqrt{3}} \\ \frac{2}{\sqrt{6}} & -\frac{1}{\sqrt{3}} \end{bmatrix}$$

PROBLEMS 9.5

Practice Problems

A1 Use the standard inner product in $\mathbb{C}^n$ to calculate $\langle \vec{u}, \vec{v} \rangle$, $\langle \vec{v}, \vec{u} \rangle$, $\|\vec{u}\|$, and $\|\vec{v}\|$.

(a) $\vec{u} = \begin{bmatrix} 2+3i \\ -1-2i \end{bmatrix}$, $\vec{v} = \begin{bmatrix} -2i \\ 2-5i \end{bmatrix}$

(b) $\vec{u} = \begin{bmatrix} -1+4i \\ 2-i \end{bmatrix}$, $\vec{v} = \begin{bmatrix} 3+i \\ 1+3i \end{bmatrix}$

(c) $\vec{u} = \begin{bmatrix} 1-i \\ 3 \end{bmatrix}$, $\vec{v} = \begin{bmatrix} 1+i \\ 1-i \end{bmatrix}$

(d) $\vec{u} = \begin{bmatrix} 1+2i \\ -1-3i \end{bmatrix}$, $\vec{v} = \begin{bmatrix} -i \\ -2i \end{bmatrix}$

A2 Are the following matrices unitary?

(a) $A = \frac{1}{\sqrt{3}}\begin{bmatrix} 1 & 1+i \\ 1+i & 1 \end{bmatrix}$

(b) $B = \begin{bmatrix} i & 0 \\ 0 & -1 \end{bmatrix}$

(c) $C = \frac{1}{\sqrt{2}}\begin{bmatrix} 1 & i \\ -1 & i \end{bmatrix}$

(d) $D = \begin{bmatrix} (-1+i)/\sqrt{3} & (1-i)/\sqrt{6} \\ 1/\sqrt{3} & 2/\sqrt{6} \end{bmatrix}$

A3 (a) Verify that $\vec{u}$ is orthogonal to $\vec{v}$ if $\vec{u} = \begin{bmatrix} 1 \\ 0 \\ i \end{bmatrix}$ and $\vec{v} = \begin{bmatrix} i \\ 1 \\ 1 \end{bmatrix}$.

(b) Determine the projection of $\vec{w} = \begin{bmatrix} 1+i \\ 2+i \\ 3+i \end{bmatrix}$ onto the subspace of $\mathbb{C}^3$ spanned by $\vec{u}$ and $\vec{v}$.

A4 (a) Prove that any unitary matrix U satisfies $|\det U| = 1$. (Hint: See Problem D3.)

(b) Give a 2×2 unitary matrix such that $\det U \neq \pm 1$.

Homework Problems

B1 Use the standard inner product in $\mathbb{C}^n$ to calculate $\langle \vec{u}, \vec{v} \rangle$, $\langle \vec{v}, \vec{u} \rangle$, $\|\vec{u}\|$, and $\|\vec{v}\|$.

(a) $\vec{u} = \begin{bmatrix} -2-3i \\ 2+i \end{bmatrix}$, $\vec{v} = \begin{bmatrix} 4-i \\ 4+i \end{bmatrix}$

(b) $\vec{u} = \begin{bmatrix} 3-i \\ 1+2i \end{bmatrix}$, $\vec{v} = \begin{bmatrix} 1+i \\ 2+i \end{bmatrix}$

(c) $\vec{u} = \begin{bmatrix} 4-3i \\ 2-i \end{bmatrix}$, $\vec{v} = \begin{bmatrix} 3 \\ -2i \end{bmatrix}$

(d) $\vec{u} = \begin{bmatrix} 1+i \\ -1+i \end{bmatrix}$, $\vec{v} = \begin{bmatrix} 0 \\ 0 \end{bmatrix}$

B2 Are the following matrices unitary?

(a) $A = \frac{1}{\sqrt{5}}\begin{bmatrix} 1 & 2 \\ -2 & 1 \end{bmatrix}$

(b) $B = \begin{bmatrix} 2 & 0 \\ 0 & 2 \end{bmatrix}$

(c) $C = \begin{bmatrix} (1+i)/\sqrt{7} & -5/\sqrt{35} \\ (1+2i)/\sqrt{7} & (3+i)/\sqrt{35} \end{bmatrix}$

(d) $D = \begin{bmatrix} (1+i)/\sqrt{6} & (1+i)/\sqrt{3} \\ 2i/\sqrt{6} & i/\sqrt{3} \end{bmatrix}$

B3 (a) Verify that $\vec{u}$ is orthogonal to $\vec{v}$ if $\vec{u} = \begin{bmatrix} 1+i \\ 1 \\ 2 \end{bmatrix}$ and $\vec{v} = \begin{bmatrix} 1-i \\ 2i \\ 0 \end{bmatrix}$.

(b) Determine the projection of $\vec{w} = \begin{bmatrix} 2+i \\ 2-i \\ 1-2i \end{bmatrix}$ onto the subspace of $\mathbb{C}^3$ spanned by $\vec{u}$ and $\vec{v}$.

B4 Consider $\mathbb{C}^3$ with its standard inner product. Let $\vec{v}_1 = \begin{bmatrix} 1+i \\ 2-i \\ -1+i \end{bmatrix}$ and $\vec{v}_2 = \begin{bmatrix} 1-i \\ -2-3i \\ -1 \end{bmatrix}$.

(a) Find an orthonormal basis for $\mathbb{S} = \text{Span}\{\vec{v}_1, \vec{v}_2\}$.

(b) Determine $\text{proj}_{\mathbb{S}}\,\vec{u}$ where $\vec{u} = \begin{bmatrix} 1 \\ 0 \\ i \end{bmatrix}$.

Conceptual Problems

D1 Prove property (5) of Theorem 1.

D2 Prove Theorem 2.

D3 Prove that for any $n \times n$ matrix,

$$\det \overline{A} = \overline{\det A}$$

D4 Prove that if A and B are unitary, then AB is unitary.

D5 (a) Show that if U is unitary, then $\|U\vec{z}\| = \|\vec{z}\|$ for all $\vec{z} \in \mathbb{C}^n$.

(b) Show that if U is unitary, all of its eigenvalues satisfy $|\lambda| = 1$.

(c) Give a 2×2 unitary matrix such that none of its eigenvalues are real.

D6 Prove that all eigenvalues of a real symmetric matrix are real.

9.6 Hermitian Matrices and Unitary Diagonalization

In Section 8.1, we saw that every real symmetric matrix is orthogonally diagonalizable. It is natural to ask if there is a comparable result in the case of matrices with complex entries.

First, we observe that if A is a real symmetric matrix, then the condition $A^T = A$ is equivalent to $A^* = A$. Hence, this condition should take the place of the condition that A be symmetric.

Definition
Hermitian Matrix

An $n \times n$ matrix A with complex entries is called **Hermitian** if $A^* = A$ or, equivalently, if $\overline{A} = A^T$.

EXAMPLE 1

Which of the following matrices are Hermitian?

$$A = \begin{bmatrix} 2 & 3-i \\ 3+i & 4 \end{bmatrix}, \quad B = \begin{bmatrix} 1 & 2i \\ -2i & 3-i \end{bmatrix}, \quad C = \begin{bmatrix} 0 & i & i \\ -i & 0 & i \\ -i & i & 0 \end{bmatrix}$$

Solution: We have $\overline{A} = \begin{bmatrix} 2 & 3+i \\ 3-i & 4 \end{bmatrix} = A^T$, so A is Hermitian.

$\overline{B} = \begin{bmatrix} 1 & -2i \\ 2i & 3+i \end{bmatrix} \neq B^T$, so B is not Hermitian.

$\overline{C} = \begin{bmatrix} 0 & -i & -i \\ i & 0 & -i \\ i & -i & 0 \end{bmatrix} \neq C^T$, so C is not Hermitian.

Observe that if A is Hermitian, then we have $\overline{(A)_{ij}} = A_{ji}$, so the diagonal entries of A must be real and for $i \neq j$ the ij-th entry must be the complex conjugate of the ji-th entry.

Theorem 1

An $n \times n$ matrix A is Hermitian if and only if for all $\vec{z}, \vec{w} \in \mathbb{C}^n$, we have

$$\langle \vec{z}, A\vec{w} \rangle = \langle A\vec{z}, \vec{w} \rangle$$

Proof: If A is Hermitian, then we get

$$\langle \vec{z}, A\vec{w} \rangle = \vec{z}^T \overline{A\vec{w}} = \vec{z}^T \overline{A}\,\overline{\vec{w}} = \vec{z}^T A^T \overline{\vec{w}} = (A\vec{z})^T \overline{\vec{w}} = \langle A\vec{z}, \vec{w} \rangle$$

If $\langle \vec{z}, A\vec{w} \rangle = \langle A\vec{z}, \vec{w} \rangle$ for all $\vec{z}, \vec{w} \in \mathbb{C}^n$, then we have

$$\vec{z}^T \overline{A}\,\overline{\vec{w}} = \vec{z}^T \overline{A\vec{w}} = \langle \vec{z}, A\vec{w} \rangle = \langle A\vec{z}, \vec{w} \rangle = (A\vec{z})^T \overline{\vec{w}} = \vec{z}^T A^T \overline{\vec{w}}$$

Since this is valid for all $\vec{z}, \vec{w} \in \mathbb{C}^n$, we have that $\overline{A} = A^T$. Thus, A is Hermitian. ■

Remark

A linear operator $L : \mathbb{V} \to \mathbb{V}$ is called Hermitian if $\langle \vec{x}, L(\vec{y}) \rangle = \langle L(\vec{x}), \vec{y} \rangle$ for all $\vec{x}, \vec{y} \in \mathbb{V}$. A linear operator is Hermitian if and only if its matrix with respect to any orthonormal basis of $\mathbb{V}$ is a Hermitian matrix. Hermitian linear operators play an important role in quantum mechanics.

Theorem 2

Suppose that A is an $n \times n$ Hermitian matrix. Then

(1) All eigenvalues of A are real.
(2) Eigenvectors corresponding to distinct eigenvalues are orthogonal to each other.

Proof: To prove (1), suppose that λ is an eigenvalue of A with corresponding unit eigenvector $\vec{z}$. Then,

$$\langle \vec{z}, A\vec{z} \rangle = \langle \vec{z}, \lambda\vec{z} \rangle = \overline{\lambda}\langle \vec{z}, \vec{z} \rangle = \overline{\lambda} \quad \text{and} \quad \langle A\vec{z}, \vec{z} \rangle = \langle \lambda\vec{z}, \vec{z} \rangle = \lambda$$

But, since A is Hermitian, we have $\langle \vec{z}, A\vec{z} \rangle = \langle A\vec{z}, \vec{z} \rangle$, so $\overline{\lambda} = \lambda$. Thus, λ must be real.

To prove (2), suppose that λ_1 and λ_2 are distinct eigenvalues of A with corresponding eigenvectors $\vec{z}_1, \vec{z}_2$. Then

$$\langle \vec{z}_1, A\vec{z}_2 \rangle = \langle \vec{z}_1, \lambda_2\vec{z}_2 \rangle = \overline{\lambda_2}\langle \vec{z}_1, \vec{z}_2 \rangle = \lambda_2\langle \vec{z}_1, \vec{z}_2 \rangle \quad \text{and} \quad \langle A\vec{z}_1, \vec{z}_2 \rangle = \lambda_1\langle \vec{z}_1, \vec{z}_2 \rangle$$

Since A is Hermitian, we get $\lambda_2\langle \vec{z}_1, \vec{z}_2 \rangle = \lambda_1\langle \vec{z}_1, \vec{z}_2 \rangle$. Thus, since $\lambda_1 \neq \lambda_2$, we must have $\langle \vec{z}_1, \vec{z}_2 \rangle = 0$, as required. ■

From this result, we expect to get something very similar to the principal axis theorem for Hermitian matrices. Let's consider an example.

EXAMPLE 2

Let $A = \begin{bmatrix} 2 & 1+i \\ 1-i & 3 \end{bmatrix}$. Verify that A is Hermitian and diagonalize A.

Solution: We have $A^* = A$, so A is Hermitian. Consider $A - \lambda I = \begin{bmatrix} 2-\lambda & 1+i \\ 1-i & 3-\lambda \end{bmatrix}$.

Then the characteristic polynomial is $C(\lambda) = \lambda^2 - 5\lambda + 4 = (\lambda - 4)(\lambda - 1)$, so $\lambda = 4$ or $\lambda = 1$.

For $\lambda = 4$,

$$A - \lambda I = \begin{bmatrix} -2 & 1+i \\ 1-i & -1 \end{bmatrix} \sim \begin{bmatrix} -2 & 1+i \\ 0 & 0 \end{bmatrix}.$$

Thus, a corresponding eigenvector is $\vec{z}_1 = \begin{bmatrix} 1+i \\ 2 \end{bmatrix}$. If $\lambda = 1$,

$$A - \lambda I = \begin{bmatrix} 1 & 1+i \\ 1-i & 2 \end{bmatrix} \sim \begin{bmatrix} 1 & 1+i \\ 0 & 0 \end{bmatrix}.$$

Thus, a corresponding eigenvector is $\vec{z}_1 = \begin{bmatrix} 1+i \\ -1 \end{bmatrix}$. Hence, A is diagonalized to $\begin{bmatrix} 4 & 0 \\ 0 & 1 \end{bmatrix}$ by $Q = \begin{bmatrix} 1+i & 1+i \\ 2 & -1 \end{bmatrix}$.

Observe in Example 2 that since the columns of Q are orthogonal, we can make Q unitary by normalizing the columns. Hence, we have that the Hermitian matrix A is diagonalized by a unitary matrix. We now prove that we can do this for any Hermitian matrix.

Theorem 3 **Spectral Theorem for Hermitian Matrices**
Suppose that A is an $n \times n$ Hermitian matrix. Then there exist a unitary matrix U and a diagonal matrix D such that $U^*AU = D$.

The proof is essentially the same as the proof of the Principal Axis Theorem, with appropriate changes to allow for complex numbers. You are asked to prove the theorem as Problem D5.

Remarks

1. If A and B are matrices such that $B = U^*AU$ for some U, we say that A and B are **unitarily similar**. If B is diagonal, then we say that A is **unitarily diagonalizable**.
2. Unlike, the Principal Axis Theorem, the converse of the Spectral Theorem for Hermitian Matrices is not true. That is, there exist matrices that are unitarily diagonalizable but not Hermitian.

EXAMPLE 3

The matrix $A = \begin{bmatrix} 2 & 1+i \\ 1-i & 3 \end{bmatrix}$ from Example 2 is Hermitian and hence unitarily diagonalizable. In particular, by normalizing the columns of Q, we find that A is unitarily diagonalized by $U = \begin{bmatrix} (1+i)/\sqrt{6} & (1+i)/\sqrt{3} \\ 2/\sqrt{6} & -1/\sqrt{3} \end{bmatrix}$.

PROBLEMS 9.6

Practice Problems

A1 For each of the following matrices,
(i) Determine whether it is Hermitian.
(ii) If it is Hermitian, unitarily diagonalize it.

(a) $A = \begin{bmatrix} 4 & \sqrt{2}+i \\ \sqrt{2}-i & 2 \end{bmatrix}$

(b) $B = \begin{bmatrix} 5 & \sqrt{2}-i \\ \sqrt{2}+i & \sqrt{3}+i \end{bmatrix}$

(c) $C = \begin{bmatrix} 6 & \sqrt{3}-i \\ \sqrt{3}+i & 3 \end{bmatrix}$

(d) $F = \begin{bmatrix} 1 & 1+i & 0 \\ 1-i & 0 & 1-i \\ 0 & 1+i & -1 \end{bmatrix}$

Homework Problems

B1 For each of the following matrices,

(i) Determine whether it is Hermitian.

(ii) If it is Hermitian, unitarily diagonalize it.

(a) $A = \begin{bmatrix} 2 & \sqrt{2}+i \\ \sqrt{2}+i & \sqrt{3} \end{bmatrix}$

(b) $B = \begin{bmatrix} 5 & \sqrt{2}-i \\ \sqrt{2}+i & 3 \end{bmatrix}$

(c) $C = \begin{bmatrix} 5 & \sqrt{3}+i \\ \sqrt{3}-i & 2 \end{bmatrix}$

(d) $F = \begin{bmatrix} 1 & i & -i \\ -i & -1 & i \\ i & -i & 0 \end{bmatrix}$

Conceptual Problems

D1 Suppose that A and B are $n \times n$ Hermitian matrices and that A is invertible. Determine which of the following are Hermitian.

(a) AB

(b) A^2

(c) A^{-1}

D2 Prove (without appealing to diagonalization) that if A is Hermitian, then $\det A$ is real.

D3 A general 2×2 Hermitian matrix can be written as $A = \begin{bmatrix} a & b+ci \\ b-ci & d \end{bmatrix}$, $a, b, c, d \in \mathbb{R}$.

(a) What can you say about $a, b, c,$ and d if A is unitary as well as Hermitian.

(b) What can you say about $a, b, c,$ and d if A is Hermitian, unitary, and diagonal.

(c) What can you say about the form of a 3×3 matrix that is Hermitian, unitary, and diagonal?

D4 Let $\mathbb{V}$ be a complex inner product space. Prove that a linear operator $L : \mathbb{V} \to \mathbb{V}$ is Hermitian ($\langle \vec{x}, L(\vec{y}) \rangle = \langle L(\vec{x}), \vec{y} \rangle$) if and only if its matrix with respect to any orthonormal basis of $\mathbb{V}$ is a Hermitian matrix.

D5 Prove the Spectral Theorem for Hermitian Matrices.

CHAPTER REVIEW

Suggestions for Student Review

1. What is the complex conjugate of a complex number? List some properties of the complex conjugate. How does the complex conjugate relate to division of complex numbers? How does it relate to the length of a complex number? (Section 9.1)

2. Define the polar form of a complex number. Explain how to convert a complex number from standard form to polar form. Is the polar form unique? How does the polar form relate to Euler's Formula? (Section 9.1)

3. List some of the similarities and some of the differences between complex vector spaces and real vector spaces. Discuss the differences between viewing $\mathbb{C}$ as a complex vector space and as a real vector space. (Section 9.3)

4. Explain how diagonalization of matrices over $\mathbb{C}$ differs from diagonalization over $\mathbb{R}$. (Section 9.4)

5. Define the real canonical form of a real matrix A. In what situations would we find the real canonical form of A instead of diagonalizing A? (Section 9.4)

6. Discuss the standard inner product in $\mathbb{C}^n$. How are the essential properties of an inner product modified in generalizing from the real case to the complex case? (Section 9.5)

7. Define the conjugate transpose of a matrix. List some similarities between the conjugate transpose of a complex matrix and the transpose of a real matrix. (Section 9.5)

8. What is a Hermitian matrix? State what you can about diagonalizing a Hermitian matrix. (Section 9.6)

Chapter Quiz

E1 Let $z_1 = 1 - \sqrt{3}i$ and $z_2 = 2 + 2i$. Use polar form to determine $z_1 z_2$ and $\frac{z_1}{z_2}$.

E2 Use polar form to determine all values of $(i)^{1/2}$.

E3 Let $\vec{u} = \begin{bmatrix} 3-i \\ i \\ 2 \end{bmatrix}$ and $\vec{v} = \begin{bmatrix} 1 \\ 3 \\ 4-i \end{bmatrix}$. Calculate the following.

(a) $2\vec{u} + (1+i)\vec{v}$ (b) $\overline{\vec{u}}$
(c) $\langle \vec{u}, \vec{v} \rangle$ (d) $\langle \vec{v}, \vec{u} \rangle$
(e) $\|\vec{v}\|$ (f) $\text{proj}_{\vec{u}}\, \vec{v}$

E4 Let $A = \begin{bmatrix} 0 & 13 \\ -1 & 4 \end{bmatrix}$.

(a) Determine a diagonal matrix similar to A over $\mathbb{C}$ and give a diagonalizing matrix P.
(b) Determine a real canonical form of A and the corresponding change of coordinates matrix P.

E5 Prove that $U = \frac{1}{\sqrt{3}} \begin{bmatrix} 1-i & -i \\ 1 & -1+i \end{bmatrix}$ is a unitary matrix.

E6 Let $A = \begin{bmatrix} 0 & 3+ki \\ 3+i & 3 \end{bmatrix}$.

(a) Determine k such that A is Hermitian.
(b) With the value of k as determined in part (a), find the eigenvalues of A and corresponding eigenvectors. Verify that the eigenvectors are orthogonal.

Further Problems

F1 Suppose that A is a Hermitian matrix with all non-negative eigenvalues. Prove that A has a square root. That is, show that there is a Hermitian matrix B such that $B^2 = A$. (Hint: Suppose that U diagonalizes A to D so that $U^*AU = D$. Define C to be a square root for D and let $B = UCU^*$.)

F2 (a) If A is any $n \times n$ matrix, prove that A^*A is Hermitian and has all non-negative real eigenvalues.
(b) If A is invertible, prove that A^*A has all positive real eigenvalues.

F3 A matrix A is said to be **unitarily triangularizable** if there exists a unitary matrix U and an upper triangular matrix T such that $U^*AU = T$. By adapting the proof of the Principal Axis Theorem (or the Spectral Theorem for Hermitian Matrices), prove that every $n \times n$ matrix A is unitarily triangularizable. (This is called Schur's Theorem.)

F4 A matrix is said to be **normal** if $A^*A = AA^*$.

(a) Show that every unitarily diagonalizable matrix is normal.
(b) Use (a) to show that if A is normal, then A is unitarily similar to an upper triangular matrix T and that T is normal.
(c) Prove that every upper triangular normal matrix is diagonal and hence conclude that every normal matrix is unitarily diagonalizable. (This is often called the Spectral Theorem for Normal Matrices.)

Visit the text's website at www.pearsoncanada.ca/norman for practice quizzes, additional applications, and an essay on linearity and superposition in physics.

APPENDIX A

Answers to Mid-Section Exercises

CHAPTER 1

Section 1.1

1. (a) $\begin{bmatrix}1\\0\end{bmatrix}$ (b) $\begin{bmatrix}-2\\-1\end{bmatrix}$ (c) $\begin{bmatrix}-1\\-1\end{bmatrix}$

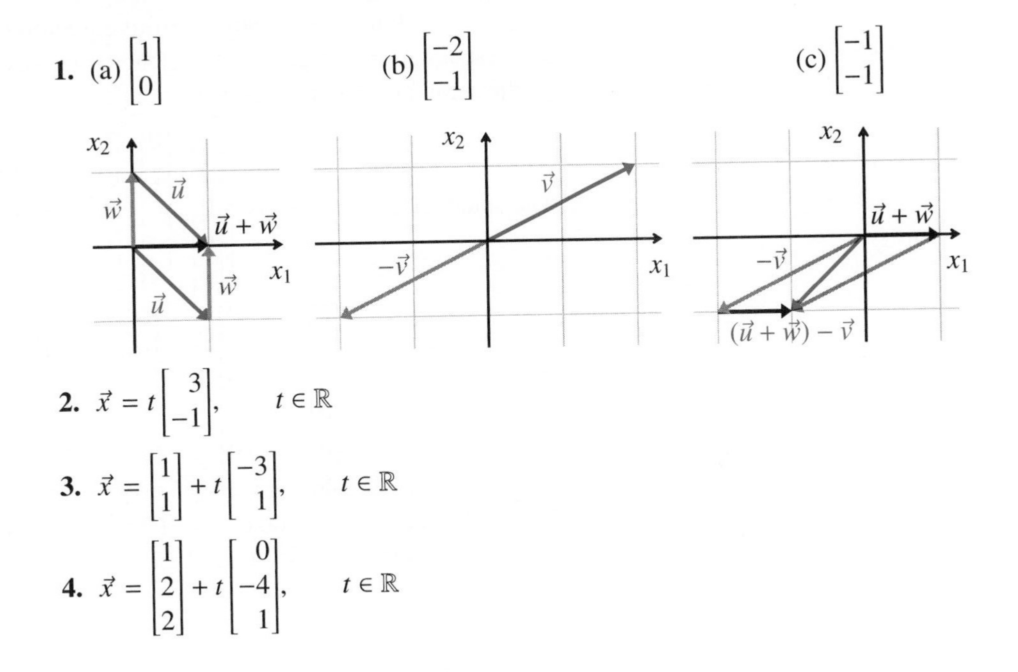

2. $\vec{x} = t\begin{bmatrix}3\\-1\end{bmatrix}, \quad t \in \mathbb{R}$

3. $\vec{x} = \begin{bmatrix}1\\1\end{bmatrix} + t\begin{bmatrix}-3\\1\end{bmatrix}, \quad t \in \mathbb{R}$

4. $\vec{x} = \begin{bmatrix}1\\2\\2\end{bmatrix} + t\begin{bmatrix}0\\-4\\1\end{bmatrix}, \quad t \in \mathbb{R}$

Section 1.2

1. (5) Let $\vec{x} = \begin{bmatrix}x_1\\\vdots\\x_n\end{bmatrix}$. Then $-\vec{x} = \begin{bmatrix}-x_1\\\vdots\\-x_n\end{bmatrix}$ since $\vec{x} + (-\vec{x}) = \begin{bmatrix}x_1 + (-x_1)\\\vdots\\x_n + (-x_n)\end{bmatrix} = \begin{bmatrix}0\\\vdots\\0\end{bmatrix} = \vec{0}$.

(6) $t\vec{x} = \begin{bmatrix}tx_1\\\vdots\\tx_n\end{bmatrix} \in \mathbb{R}^n$ since $tx_i \in \mathbb{R}$ for $1 \leq i \leq n$.

(7) $s(t\vec{x}) = s\begin{bmatrix}tx_1\\\vdots\\tx_n\end{bmatrix} = \begin{bmatrix}stx_1\\\vdots\\stx_n\end{bmatrix} = st\begin{bmatrix}x_1\\\vdots\\x_n\end{bmatrix} = (st)\vec{x}$.

2. By definition, S is a subset of $\mathbb{R}^2$. We have that $\begin{bmatrix} 0 \\ 0 \end{bmatrix} \in S$ since $2(0) = 0$. Let $\vec{x} = \begin{bmatrix} x_1 \\ x_2 \end{bmatrix}$ and $\vec{y} = \begin{bmatrix} y_1 \\ y_2 \end{bmatrix}$ be vectors in S. Then $2x_1 = x_2$ and $2y_1 = y_2$. We find that $\vec{x} + \vec{y} = \begin{bmatrix} x_1 + y_1 \\ x_2 + y_2 \end{bmatrix} \in S$ since $2(x_1 + y_1) = 2x_1 + 2y_1 = x_2 + y_2$. Similarly, $t\vec{x} = \begin{bmatrix} tx_1 \\ tx_2 \end{bmatrix} \in S$ since $2(tx_1) = t2x_1 = tx_2$. Hence, S is a subspace.

   T is not a subspace since $\begin{bmatrix} 0 \\ 0 \end{bmatrix}$ is not in T.

3. Consider $\begin{bmatrix} 0 \\ 0 \\ 0 \end{bmatrix} = c_1 \begin{bmatrix} 1 \\ 0 \\ 1 \end{bmatrix} + c_2 \begin{bmatrix} 0 \\ 1 \\ 1 \end{bmatrix} + c_3 \begin{bmatrix} 1 \\ 1 \\ 0 \end{bmatrix} = \begin{bmatrix} c_1 + c_3 \\ c_2 + c_3 \\ c_1 + c_2 \end{bmatrix}$.
   This gives $c_1 + c_3 = 0$, $c_2 + c_3 = 0$, and $c_1 + c_2 = 0$. The first two equations imply that $c_1 = c_2$. Putting this into the third equation gives $c_1 = c_2 = 0$. Then, from the first equation, we get $c_3 = 0$. Hence, the only solution is $c_1 = c_2 = c_3 = 0$, so the set is linearly independent.

4. The standard basis for $\mathbb{R}^5$ is $\left\{ \begin{bmatrix} 1 \\ 0 \\ 0 \\ 0 \\ 0 \end{bmatrix}, \begin{bmatrix} 0 \\ 1 \\ 0 \\ 0 \\ 0 \end{bmatrix}, \begin{bmatrix} 0 \\ 0 \\ 1 \\ 0 \\ 0 \end{bmatrix}, \begin{bmatrix} 0 \\ 0 \\ 0 \\ 1 \\ 0 \end{bmatrix}, \begin{bmatrix} 0 \\ 0 \\ 0 \\ 0 \\ 1 \end{bmatrix} \right\}$.

   Consider $\begin{bmatrix} x_1 \\ x_2 \\ x_3 \\ x_4 \\ x_5 \end{bmatrix} = t_1 \begin{bmatrix} 1 \\ 0 \\ 0 \\ 0 \\ 0 \end{bmatrix} + t_2 \begin{bmatrix} 0 \\ 1 \\ 0 \\ 0 \\ 0 \end{bmatrix} + t_3 \begin{bmatrix} 0 \\ 0 \\ 1 \\ 0 \\ 0 \end{bmatrix} + t_4 \begin{bmatrix} 0 \\ 0 \\ 0 \\ 1 \\ 0 \end{bmatrix} + t_5 \begin{bmatrix} 0 \\ 0 \\ 0 \\ 0 \\ 1 \end{bmatrix} = \begin{bmatrix} t_1 \\ t_2 \\ t_3 \\ t_4 \\ t_5 \end{bmatrix}$. Thus, for every $\vec{x} \in \mathbb{R}^5$ we have a solution $t_i = x_i$ for $1 \leq i \leq 5$. Hence, the set is a spanning set for $\mathbb{R}^5$. Moreover, if we take $\vec{x} = \vec{0}$, we get that the only solution is $t_i = 0$ for $1 \leq i \leq 5$, so the set is also linearly independent.

Section 1.3

1. $\cos\theta = \dfrac{\vec{v} \cdot \vec{w}}{\|\vec{v}\|\|\vec{w}\|} = 0$, so $\theta = \frac{\pi}{2}$ rads.

2. $\|\vec{x}\| = \sqrt{1 + 4 + 1} = \sqrt{6}$, $\|\vec{y}\| = \sqrt{\frac{1}{6} + \frac{4}{6} + \frac{1}{6}} = 1$

3. By definition, $\hat{x}$ is a scalar multiple of $\vec{x}$, so it is parallel. Using Theorem 1.3.2 (2), we get
$$\|\hat{x}\| = \left\| \frac{1}{\|\vec{x}\|}\vec{x} \right\| = \left| \frac{1}{\|\vec{x}\|} \right| \|\vec{x}\| = \frac{\|\vec{x}\|}{\|\vec{x}\|} = 1$$

4. $x_1 - 3x_2 - 2x_3 = 1(1) + (-3)(2) + (-2)(3) = -11$

Section 1.4

1. $\text{proj}_{\vec{v}}\, \vec{u} = \dfrac{\vec{u} \cdot \vec{v}}{\|\vec{v}\|^2}\vec{v} = \dfrac{8}{9}\begin{bmatrix} 1 \\ 2 \\ 2 \end{bmatrix}$, $\text{proj}_{\vec{u}}\, \vec{v} = \dfrac{\vec{v} \cdot \vec{u}}{\|\vec{u}\|^2}\vec{u} = \dfrac{8}{17}\begin{bmatrix} -2 \\ 3 \\ 2 \end{bmatrix}$

2. $\text{proj}_{\vec{v}}\,\vec{u} = \dfrac{\vec{u}\cdot\vec{v}}{\|\vec{v}\|^2}\vec{v} = \dfrac{1}{14}\begin{bmatrix}3\\1\\2\end{bmatrix}$, $\text{perp}_{\vec{v}}\,\vec{u} = \vec{u} - \text{proj}_{\vec{v}}\,\vec{u} = \begin{bmatrix}1\\-2\\0\end{bmatrix} - \begin{bmatrix}3/14\\1/14\\2/14\end{bmatrix} = \begin{bmatrix}11/14\\-29/14\\-1/7\end{bmatrix}$

3. $\text{proj}_{\vec{x}}(\vec{y}+\vec{z}) = \dfrac{(\vec{y}+\vec{z})\cdot\vec{x}}{\|\vec{x}\|^2}\vec{x} = \dfrac{\vec{y}\cdot\vec{x}+\vec{z}\cdot\vec{x}}{\|\vec{x}\|^2}\vec{x} = \dfrac{\vec{y}\cdot\vec{x}}{\|\vec{x}\|^2}\vec{x} + \dfrac{\vec{z}\cdot\vec{x}}{\|\vec{x}\|^2}\vec{x} = \text{proj}_{\vec{x}}\,\vec{y} + \text{proj}_{\vec{x}}\,\vec{z}$

$\text{proj}_{\vec{x}}(t\vec{y}) = \dfrac{(t\vec{y})\cdot\vec{x}}{\|\vec{x}\|^2}\vec{x} = t\dfrac{\vec{y}\cdot\vec{x}}{\|\vec{x}\|^2}\vec{x} = t\,\text{proj}_{\vec{x}}\,\vec{y}$

Section 1.5

1. $\vec{w}\cdot\vec{u} = (u_2v_3 - u_3v_2)(u_1) + (u_3v_1 - u_1v_3)(u_2) + (u_1v_2 - u_2v_1)(u_3) = 0$
$\vec{w}\cdot\vec{v} = (u_2v_3 - u_3v_2)(v_1) + (u_3v_1 - u_1v_3)(v_2) + (u_1v_2 - u_2v_1)(v_3) = 0$

2. $\begin{bmatrix}3\\-2\\1\end{bmatrix} \times \begin{bmatrix}2\\3\\7\end{bmatrix} = \begin{bmatrix}-17\\-19\\13\end{bmatrix}$

3. The six cross-products are easily checked.

4. Area $= \|\vec{u}\times\vec{v}\| = \left\|\begin{bmatrix}-2\\-1\\-2\end{bmatrix}\right\| = \sqrt{9} = 3$

5. The direction vector is $\vec{d} = \begin{bmatrix}-1\\-2\\1\end{bmatrix} \times \begin{bmatrix}2\\1\\-2\end{bmatrix} = \begin{bmatrix}3\\0\\3\end{bmatrix}$. Put $x_3 = 0$ and then solve $-x_1 - 2x_2 = -2$ and $2x_1 + x_2 = 1$. This gives $x_1 = 0$ and $x_2 = 1$. So, a vector equation of the line of intersection is $\vec{x} = \begin{bmatrix}0\\1\\0\end{bmatrix} + t\begin{bmatrix}3\\0\\3\end{bmatrix}$, $t \in \mathbb{R}$.

CHAPTER 2

Section 2.1

1. Add (–1/2) times the first equation to the second equation:

$$\begin{aligned} 2x_1 + 4x_2 + 0x_3 &= 12 \\ -x_3 &= -2 \end{aligned}$$

Since x_2 does not appear as a leading 1, it is a free variable. Thus, we let $x_2 = t \in \mathbb{R}$. Then we rewrite the first equation as

$$x_1 = \frac{1}{2}(12 - 4x_2) = 6 - 2t$$

Thus, the general solution is

$$\begin{bmatrix}x_1\\x_2\\x_3\end{bmatrix} = \begin{bmatrix}6-2t\\t\\2\end{bmatrix} = \begin{bmatrix}6\\0\\2\end{bmatrix} + t\begin{bmatrix}-2\\1\\0\end{bmatrix}, \quad t \in \mathbb{R}$$

2. $\left[\begin{array}{ccc|c}2 & 4 & 0 & 12\\1 & 2 & -1 & 4\end{array}\right] \begin{array}{c} \\ R_2 - \frac{1}{2}R_1\end{array} \sim \left[\begin{array}{ccc|c}2 & 4 & 0 & 12\\0 & 0 & -1 & -2\end{array}\right]$

Section 2.2

1. For Example 2.1.9, we get

$$\left[\begin{array}{cccc|c} 1 & 2 & -4 & 0 & 11 \\ 0 & 1 & -3 & 1 & -1 \\ 0 & 0 & 0 & -1 & 4 \end{array}\right] \begin{array}{c} \\ \\ (-1)R_3 \end{array} \sim \left[\begin{array}{cccc|c} 1 & 2 & -4 & 0 & 11 \\ 0 & 1 & -3 & 1 & -1 \\ 0 & 0 & 0 & 1 & -4 \end{array}\right] R_2 - R_3 \sim$$

$$\left[\begin{array}{cccc|c} 1 & 2 & -4 & 0 & 11 \\ 0 & 1 & -3 & 0 & 3 \\ 0 & 0 & 0 & 1 & -4 \end{array}\right] \begin{array}{c} R_1 - 2R_2 \\ \\ \\ \end{array} \sim \left[\begin{array}{cccc|c} 1 & 0 & 2 & 0 & 5 \\ 0 & 1 & -3 & 0 & 3 \\ 0 & 0 & 0 & 1 & -4 \end{array}\right]$$

For Example 2.1.12, we get

$$\left[\begin{array}{ccc|c} 1 & 1 & 1 & 180 \\ 0 & 1 & 0 & 60 \\ 0 & 0 & 4 & 80 \end{array}\right] \begin{array}{c} \\ \\ \frac{1}{4}R_3 \end{array} \sim \left[\begin{array}{ccc|c} 1 & 1 & 1 & 180 \\ 0 & 1 & 0 & 60 \\ 0 & 0 & 1 & 20 \end{array}\right] R_1 - R_3 \sim$$

$$\left[\begin{array}{ccc|c} 1 & 1 & 0 & 160 \\ 0 & 1 & 0 & 60 \\ 0 & 0 & 1 & 20 \end{array}\right] \begin{array}{c} R_1 - R_2 \\ \\ \\ \end{array} \sim \left[\begin{array}{ccc|c} 1 & 0 & 0 & 100 \\ 0 & 1 & 0 & 60 \\ 0 & 0 & 1 & 20 \end{array}\right]$$

2. (a) $\operatorname{rank}(A) = 2$ (b) $\operatorname{rank}(B) = 2$

3. We write the reduced row echelon form of the coefficient matrix as a homogeneous system. We get

$$\begin{aligned} x_1 + x_4 + 2x_5 &= 0 \\ x_2 + x_3 + x_5 &= 0 \end{aligned}$$

We see that x_3, x_4, and x_5 are free variables, so we let $x_3 = t_1$, $x_4 = t_2$, and $x_5 = t_3$. Then, we rewrite the equations as

$$\begin{aligned} x_1 &= -t_2 - 2t_3 \\ x_2 &= -t_1 - t_3 \end{aligned}$$

Hence, the general solution is

$$\begin{bmatrix} x_1 \\ x_2 \\ x_3 \\ x_4 \\ x_5 \end{bmatrix} = \begin{bmatrix} -t_2 - 2t_3 \\ -t_1 - t_3 \\ t_1 \\ t_2 \\ t_3 \end{bmatrix} = t_1 \begin{bmatrix} 0 \\ -1 \\ 1 \\ 0 \\ 0 \end{bmatrix} + t_2 \begin{bmatrix} -1 \\ 0 \\ 0 \\ 1 \\ 0 \end{bmatrix} + t_3 \begin{bmatrix} -2 \\ -1 \\ 0 \\ 0 \\ 1 \end{bmatrix}, \quad t_1, t_2, t_3 \in \mathbb{R}$$

Section 2.3

1. Consider $t_1 \begin{bmatrix} 1 \\ -3 \\ -3 \end{bmatrix} + t_2 \begin{bmatrix} 2 \\ -2 \\ 1 \end{bmatrix} + t_3 \begin{bmatrix} -2 \\ 2 \\ -3 \end{bmatrix} = \begin{bmatrix} 1 \\ 3 \\ 1 \end{bmatrix}$.
   Simplifying and comparing entries gives the system

$$\begin{aligned} t_1 + 2t_2 - 2t_3 &= 1 \\ -3t_1 - 2t_2 + 2t_3 &= 3 \\ -3t_1 \quad + t_2 - 3t_3 &= 1 \end{aligned}$$

Solving the system by row reducing the augmented matrix, we find that it is consistent so that $\vec{v}$ is in the span. In particular, we find that

$$(-2)\begin{bmatrix}1\\-3\\-3\end{bmatrix}+\frac{19}{4}\begin{bmatrix}2\\-2\\1\end{bmatrix}+\frac{13}{4}\begin{bmatrix}-2\\2\\-3\end{bmatrix}=\begin{bmatrix}1\\3\\1\end{bmatrix}$$

2. Consider

$$\begin{bmatrix}0\\0\\0\end{bmatrix}=t_1\begin{bmatrix}-1\\1\\-3\end{bmatrix}+t_2\begin{bmatrix}-2\\-3\\-3\end{bmatrix}+t_3\begin{bmatrix}1\\-1\\3\end{bmatrix}$$

Simplifying and comparing entries gives the homogeneous system

$$\begin{aligned}-t_1-2t_2+t_3&=0\\t_1-3t_2-t_3&=0\\-3t_1-3t_2+3t_3&=0\end{aligned}$$

Row reducing the coefficient matrix, we find that the rank of the coefficient matrix is less than the number of variables. Thus, there is at least one parameter. Hence, there are infinitely many solutions. Therefore, the set is linearly dependent.

CHAPTER 3

Section 3.1

1. Consider $\begin{bmatrix}0&0\\0&0\end{bmatrix}=t_1\begin{bmatrix}1&2\\1&1\end{bmatrix}+t_2\begin{bmatrix}1&1\\3&1\end{bmatrix}+t_3\begin{bmatrix}3&5\\5&3\end{bmatrix}+t_4\begin{bmatrix}0&-1\\-2&0\end{bmatrix}$. Simplifying the right-hand side, we get the homogeneous system

$$\begin{aligned}t_1+t_2+3t_3&=0\\2t_1+t_2+5t_3-t_4&=0\\t_1+3t_2+5t_3-2t_4&=0\\t_1+t_2+3t_3&=0\end{aligned}$$

Using the methods of Chapter 2, we find that this system has infinitely many solutions, and hence $\mathcal{B}$ is linearly dependent.

Similar to our work above, to determine whether $X\in\operatorname{Span}\mathcal{B}$, we solve the system

$$\begin{aligned}t_1+t_2+3t_3&=1\\2t_1+t_2+5t_3-t_4&=5\\t_1+3t_2+5t_3-2t_4&=-5\\t_1+t_2+3t_3&=1\end{aligned}$$

Using the methods of Chapter 2, we find that this system is consistent and hence X is in the span of $\mathcal{B}$.

2. For any 2×2 matrix $\begin{bmatrix}x_1&x_2\\x_3&x_4\end{bmatrix}$, we have

$$x_1\begin{bmatrix}1&0\\0&0\end{bmatrix}+x_2\begin{bmatrix}0&1\\0&0\end{bmatrix}+x_3\begin{bmatrix}0&0\\1&0\end{bmatrix}+x_4\begin{bmatrix}0&0\\0&1\end{bmatrix}=\begin{bmatrix}x_1&x_2\\x_3&x_4\end{bmatrix}$$

Hence, $\mathcal{B}$ spans all 2×2 matrices. Moreover, if we take $x_1=x_2=x_3=x_4=0$, then the only solution is the trivial solution, so the set is linearly independent.

3. $A^T = \begin{bmatrix} 2 & -1 \\ 3 & 0 \\ 1 & 5 \end{bmatrix}$ and $(A^T)^T = \begin{bmatrix} 2 & 3 & 1 \\ -1 & 0 & 5 \end{bmatrix} = A$

$$3A^T = \begin{bmatrix} 6 & -3 \\ 9 & 0 \\ 3 & 15 \end{bmatrix} = (3A)^T$$

4. (a) AB is not defined since A has three columns and B has two rows.

(b) $BA = \begin{bmatrix} 4 & 7 & -1 \\ 1 & 2 & -1 \end{bmatrix}$

(c) $A^TA = \begin{bmatrix} 5 & 8 & 1 \\ 8 & 13 & 1 \\ 1 & 1 & 2 \end{bmatrix}$

(d) $BB^T = \begin{bmatrix} 5 & 2 \\ 2 & 1 \end{bmatrix}$

Section 3.2

1. $f_A(1,0) = \begin{bmatrix} 1 \\ 3 \end{bmatrix}$, $f_A(0,1) = \begin{bmatrix} 2 \\ -1 \end{bmatrix}$, $f_A(2,3) = \begin{bmatrix} 8 \\ 3 \end{bmatrix}$

We have $f_A(2,3) = 2f_A(1,0) + 3f_A(0,1)$.

2. $f_A(-1,1,1,0) = \begin{bmatrix} 0 \\ 0 \\ 0 \end{bmatrix}$, $f_A(-3,1,0,1) = \begin{bmatrix} 0 \\ 0 \\ 0 \end{bmatrix}$

3. (a) f is not linear. $f(1,0) + f(2,0) \neq f(3,0)$.

(b) G is linear since for all $\vec{x}, \vec{y} \in \mathbb{R}^2$ and $t \in \mathbb{R}$, we have

$$g(t\vec{x} + \vec{y}) = g(tx_1 + y_1, tx_2 + y_2) = \begin{bmatrix} tx_2 + y_2 \\ (tx_1 + y_1) - (tx_2 + y_2) \end{bmatrix}$$
$$= t\begin{bmatrix} x_2 \\ x_1 - x_2 \end{bmatrix} + \begin{bmatrix} y_2 \\ y_1 - y_2 \end{bmatrix} = tg(\vec{x}) + g(\vec{y})$$

4. H is linear since for all $\vec{x}, \vec{y} \in \mathbb{R}^4$ and $t \in \mathbb{R}$, we have

$$H(t\vec{x} + \vec{y}) = H(tx_1 + y_1, tx_2 + y_2, tx_3 + y_3, tx_4 + y_4) = \begin{bmatrix} tx_3 + y_3 + tx_4 + y_4 \\ tx_1 + y_1 \end{bmatrix}$$
$$= t\begin{bmatrix} x_3 + x_4 \\ x_1 \end{bmatrix} + \begin{bmatrix} y_3 + y_4 \\ y_1 \end{bmatrix} = tH(\vec{x}) + H(\vec{y})$$

By definition of the standard matrix, we get

$$[H] = \begin{bmatrix} H(\vec{e}_1) & H(\vec{e}_2) & H(\vec{e}_3) & H(\vec{e}_4) \end{bmatrix} = \begin{bmatrix} 0 & 0 & 1 & 1 \\ 1 & 0 & 0 & 0 \end{bmatrix}$$

5. Let $\vec{x}, \vec{y} \in \mathbb{R}^n$ and $t \in \mathbb{R}$. Then, since L and M are linear, we get

$$(M \circ L)(t\vec{x} + \vec{y}) = M(L(t\vec{x} + \vec{y})) = M(tL(\vec{x}) + L(\vec{y}))$$
$$= tM(L(\vec{x})) + M(L(\vec{y})) = t(M \circ L)(\vec{x}) + (M \circ L)(\vec{y})$$

Hence $(M \circ L)$ is linear.

Section 3.3

1. $[R_{\pi/4}] = \begin{bmatrix} \sqrt{2}/2 & -\sqrt{2}/2 \\ \sqrt{2}/2 & \sqrt{2}/2 \end{bmatrix}$, $R_{\pi/4}(1,1) = \begin{bmatrix} \sqrt{2}/2 & -\sqrt{2}/2 \\ \sqrt{2}/2 & \sqrt{2}/2 \end{bmatrix}\begin{bmatrix} 1 \\ 1 \end{bmatrix} = \begin{bmatrix} 0 \\ \sqrt{2} \end{bmatrix}$

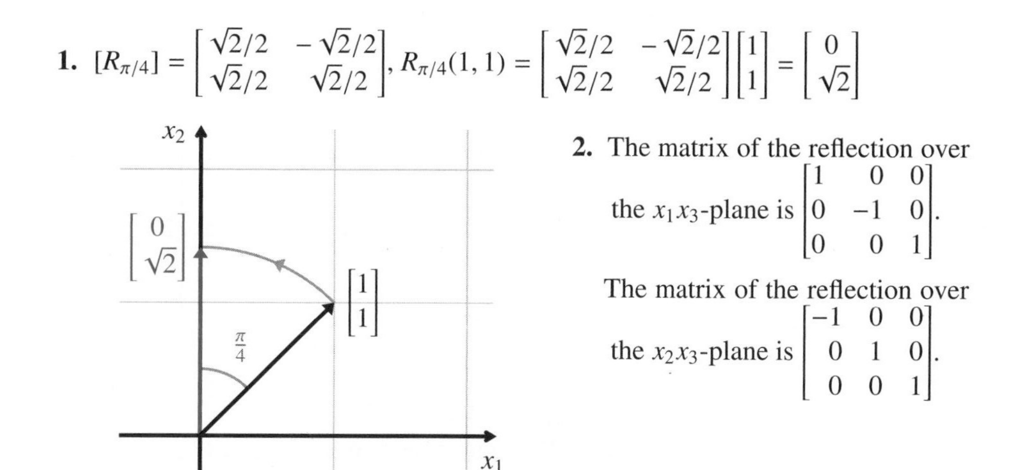

2. The matrix of the reflection over the x_1x_3-plane is $\begin{bmatrix} 1 & 0 & 0 \\ 0 & -1 & 0 \\ 0 & 0 & 1 \end{bmatrix}$.

The matrix of the reflection over the x_2x_3-plane is $\begin{bmatrix} -1 & 0 & 0 \\ 0 & 1 & 0 \\ 0 & 0 & 1 \end{bmatrix}$.

Section 3.4

1. The solution space is Span $\left\{ \begin{bmatrix} -2 \\ -3 \\ 1 \\ 0 \\ 0 \end{bmatrix}, \begin{bmatrix} -4 \\ -5 \\ 0 \\ -6 \\ 1 \end{bmatrix} \right\}$.

2. If $\vec{x} = \begin{bmatrix} x_1 \\ x_2 \\ x_3 \end{bmatrix} \in \text{Null}(L)$, then we must have $x_1 - x_2 = 0$ and $-2x_1 + 2x_3 + x_3 = 0$.

Solving this homogeneous system, we find that $\text{Null}(L) = \text{span}\left\{ \begin{bmatrix} 1 \\ 1 \\ 0 \end{bmatrix} \right\}$.

3. The general solution is $\vec{x} = \begin{bmatrix} 7 \\ 8 \\ 0 \\ 9 \\ 0 \end{bmatrix} + s\begin{bmatrix} -2 \\ -3 \\ 1 \\ 0 \\ 0 \end{bmatrix} + t\begin{bmatrix} -4 \\ -5 \\ 0 \\ -6 \\ 1 \end{bmatrix}, \quad s, t \in \mathbb{R}$.

4. We have

$$\text{Range}(L) = x_1\begin{bmatrix} 1 \\ -2 \end{bmatrix} + x_2\begin{bmatrix} -1 \\ 2 \end{bmatrix} + x_3\begin{bmatrix} 0 \\ 1 \end{bmatrix}$$

Thus,

$$\text{Range}(L) = \text{Span}\left\{ \begin{bmatrix} 1 \\ -2 \end{bmatrix}, \begin{bmatrix} -1 \\ 2 \end{bmatrix}, \begin{bmatrix} 0 \\ 1 \end{bmatrix} \right\} = \text{Span}\left\{ \begin{bmatrix} 1 \\ -2 \end{bmatrix}, \begin{bmatrix} 0 \\ 1 \end{bmatrix} \right\} = \mathbb{R}^2$$

5. $[L] = \begin{bmatrix} 1 & -1 & 0 \\ -2 & 2 & 1 \end{bmatrix}$. So $\text{Col}(L) = \text{Span}\left\{ \begin{bmatrix} 1 \\ -2 \end{bmatrix}, \begin{bmatrix} -1 \\ 2 \end{bmatrix}, \begin{bmatrix} 0 \\ 1 \end{bmatrix} \right\} = \text{Range}(L)$.

6. Row reducing A to RREF gives $\begin{bmatrix} 1 & 1 & 3 \\ 2 & 1 & 5 \\ 1 & 3 & 5 \\ 1 & 1 & 3 \end{bmatrix} \sim \begin{bmatrix} 1 & 0 & 2 \\ 0 & 1 & 1 \\ 0 & 0 & 0 \\ 0 & 0 & 0 \end{bmatrix}$.

Thus, a basis for Row(A) is $\left\{ \begin{bmatrix} 1 \\ 0 \\ 2 \end{bmatrix}, \begin{bmatrix} 0 \\ 1 \\ 1 \end{bmatrix} \right\}$.

7. Row reducing A to RREF gives $\begin{bmatrix} 1 & 1 & 2 & 0 & 3 \\ 1 & -1 & 0 & 2 & -3 \\ -1 & 2 & 1 & -3 & -2 \end{bmatrix} \sim \begin{bmatrix} 1 & 0 & 1 & 1 & 0 \\ 0 & 1 & 1 & -1 & 0 \\ 0 & 0 & 0 & 0 & 1 \end{bmatrix}$.

Thus, a basis for Col(A) is $\left\{ \begin{bmatrix} 1 \\ 1 \\ -1 \end{bmatrix}, \begin{bmatrix} 1 \\ -1 \\ 2 \end{bmatrix}, \begin{bmatrix} 3 \\ -3 \\ -2 \end{bmatrix} \right\}$.

8. Row reducing A^T to RREF gives $\begin{bmatrix} 1 & 1 & -1 \\ 1 & -1 & 2 \\ 2 & 0 & 1 \\ 0 & 2 & -3 \\ 3 & -3 & -2 \end{bmatrix} \sim \begin{bmatrix} 1 & 0 & 0 \\ 0 & 1 & 0 \\ 0 & 0 & 1 \\ 0 & 0 & 0 \\ 0 & 0 & 0 \end{bmatrix}$

Thus, a basis for Row(A^T) = Col(A) is $\left\{ \begin{bmatrix} 1 \\ 0 \\ 0 \end{bmatrix}, \begin{bmatrix} 0 \\ 1 \\ 0 \end{bmatrix}, \begin{bmatrix} 0 \\ 0 \\ 1 \end{bmatrix} \right\}$.

9. Row reducing A to RREF gives

$$\begin{bmatrix} 1 & 1 & -3 & 1 \\ 2 & 3 & -8 & 4 \\ 0 & 1 & -2 & 3 \end{bmatrix} \sim \begin{bmatrix} 1 & 0 & -1 & 0 \\ 0 & 1 & -2 & 0 \\ 0 & 0 & 0 & 1 \end{bmatrix}$$

Thus, $\left\{ \begin{bmatrix} 1 \\ 0 \\ -1 \\ 0 \end{bmatrix}, \begin{bmatrix} 0 \\ 1 \\ -2 \\ 0 \end{bmatrix}, \begin{bmatrix} 0 \\ 0 \\ 0 \\ 1 \end{bmatrix} \right\}, \left\{ \begin{bmatrix} 1 \\ 2 \\ 0 \end{bmatrix}, \begin{bmatrix} 1 \\ 3 \\ 1 \end{bmatrix}, \begin{bmatrix} 1 \\ 4 \\ 3 \end{bmatrix} \right\}$, and $\left\{ \begin{bmatrix} 1 \\ 2 \\ 1 \\ 0 \end{bmatrix} \right\}$ are bases for Row(A), Col(A), and Null(A), respectively. Thus, rank(A) = dim Row(A) = 3 and nullity(A) = dim Null(A) = 1, and so rank(A) + nullity(A) = 3 + 1 = 4, as required.

Section 3.5

1. We can row reduce A to the identity matrix so it is invertible. In particular, we have $\left[\begin{array}{cc|cc} 2 & 3 & 1 & 0 \\ 4 & 5 & 0 & 1 \end{array}\right] \sim \left[\begin{array}{cc|cc} 1 & 0 & -5/2 & 3/2 \\ 0 & 1 & 2 & -1 \end{array}\right]$, so $A^{-1} = \begin{bmatrix} -5/2 & 3/2 \\ 2 & -1 \end{bmatrix}$.

2. We have $\left[\begin{array}{cc|cc} 1 & 3 & 1 & 0 \\ -2 & 1 & 0 & 1 \end{array}\right] \sim \left[\begin{array}{cc|cc} 1 & 0 & 1/7 & -3/7 \\ 0 & 1 & 2/7 & 1/7 \end{array}\right]$, so $A^{-1} = \frac{1}{7}\begin{bmatrix} 1 & -3 \\ 2 & 1 \end{bmatrix}$.

Thus, the solution to $A\vec{x} = \vec{b}$ is $\vec{x} = A^{-1}\vec{b} = \begin{bmatrix} -5 \\ 4 \end{bmatrix}$.

3. (a) Let R denote the reflection over the lines $x_2 = x_1$. Then R is its own inverse. We have $[R] = \begin{bmatrix} 0 & 1 \\ 1 & 0 \end{bmatrix}$, and we find that $[R][R] = \begin{bmatrix} 1 & 0 \\ 0 & 1 \end{bmatrix}$, as required.

(b) Let S denote the shear in the x_1-direction by a factor of t. Then S^{-1} is a shear in the x_1-direction by a factor of $-t$. We get $[S][S^{-1}] = \begin{bmatrix} 1 & t \\ 0 & 1 \end{bmatrix}\begin{bmatrix} 1 & -t \\ 0 & 1 \end{bmatrix} = \begin{bmatrix} 1 & 0 \\ 0 & 1 \end{bmatrix}$, as required.

Section 3.6

1. Let $E = \begin{bmatrix} 1 & 0 & 0 \\ 0 & k & 0 \\ 0 & 0 & 0 \end{bmatrix}$. Then,

$$\begin{bmatrix} a_{11} & a_{12} & a_{13} \\ a_{21} & a_{22} & a_{23} \\ a_{31} & a_{32} & a_{33} \end{bmatrix} \; kR_2 \sim \begin{bmatrix} a_{11} & a_{12} & a_{13} \\ ka_{21} & ka_{22} & ka_{23} \\ a_{31} & a_{32} & a_{33} \end{bmatrix}$$

while $EA = \begin{bmatrix} 1 & 0 & 0 \\ 0 & k & 0 \\ 0 & 0 & 0 \end{bmatrix}\begin{bmatrix} a_{11} & a_{12} & a_{13} \\ a_{21} & a_{22} & a_{23} \\ a_{31} & a_{32} & a_{33} \end{bmatrix} = \begin{bmatrix} a_{11} & a_{12} & a_{13} \\ ka_{21} & ka_{22} & ka_{23} \\ a_{31} & a_{32} & a_{33} \end{bmatrix}$, as required.

2. We row reduce A to its reduced row echelon form $R = \begin{bmatrix} 1 & 0 \\ 0 & 1 \\ 0 & 0 \end{bmatrix}$. We find that $E_5E_4E_3E_2E_1A = R$, where

$$E_1 = \begin{bmatrix} 1 & 0 & 0 \\ -2 & 1 & 0 \\ 0 & 0 & 1 \end{bmatrix}, E_2 = \begin{bmatrix} 1 & 0 & 0 \\ 0 & 1 & 0 \\ -3 & 0 & 1 \end{bmatrix}, E_3 = \begin{bmatrix} 1 & 0 & 0 \\ 0 & 1 & 0 \\ 0 & 0 & -1 \end{bmatrix}, E_4 = \begin{bmatrix} 1 & 0 & 0 \\ 0 & 0 & 1 \\ 0 & 1 & 0 \end{bmatrix},$$

$$E_5 = \begin{bmatrix} 1 & -1 & 0 \\ 0 & 1 & 0 \\ 0 & 0 & 1 \end{bmatrix}$$

(Alternative solutions are possible.)

Section 3.7

1. We row reduce and get

$$\begin{bmatrix} -1 & 1 & 2 \\ 4 & -1 & -3 \\ -3 & -3 & 1 \end{bmatrix} \begin{matrix} \\ R_2 + 4R_1 \\ R_3 - 3R_1 \end{matrix} \sim \begin{bmatrix} -1 & 1 & 2 \\ 0 & 3 & 5 \\ 0 & -6 & -5 \end{bmatrix} \quad \Rightarrow \quad L = \begin{bmatrix} 1 & 0 & 0 \\ -4 & 1 & 0 \\ 3 & * & 1 \end{bmatrix}$$

$$\begin{bmatrix} -1 & 1 & 2 \\ 0 & 3 & 5 \\ 0 & -6 & -5 \end{bmatrix} \begin{matrix} \\ \\ R_3 + 2R_2 \end{matrix} \sim \begin{bmatrix} -1 & 1 & 2 \\ 0 & 3 & 5 \\ 0 & 0 & 5 \end{bmatrix} \quad \Rightarrow \quad L = \begin{bmatrix} 1 & 0 & 0 \\ -4 & 1 & 0 \\ 3 & -2 & 1 \end{bmatrix}$$

Therefore, we have

$$A = LU = \begin{bmatrix} 1 & 0 & 0 \\ -4 & 1 & 0 \\ 3 & -2 & 1 \end{bmatrix}\begin{bmatrix} -1 & 1 & 2 \\ 0 & 3 & 5 \\ 0 & 0 & 5 \end{bmatrix}$$

2. (a) We have $A = LU = \begin{bmatrix} 1 & 0 & 0 \\ -4 & 1 & 0 \\ 3 & -2 & 1 \end{bmatrix}\begin{bmatrix} -1 & 1 & 2 \\ 0 & 3 & 5 \\ 0 & 0 & 5 \end{bmatrix}$. Write $A\vec{x} = \vec{b}$ as $LU\vec{x} = \vec{b}$ and take $\vec{y} = U\vec{x}$. Writing out the system $L\vec{y} = \vec{b}$, we get

$$\begin{aligned} y_1 &= 3 \\ -4y_1 + y_2 &= 2 \\ 3y_1 - 2y_2 + y_3 &= 6 \end{aligned}$$

Using forward substitution, we find that $y_1 = 3$, so $y_2 = 2 + 4(3) = 14$, and $y_3 = 6 - 3(3) + 2(14) = 25$. Hence $\vec{y} = \begin{bmatrix} 3 \\ 14 \\ 25 \end{bmatrix}$.

Thus, our system $U\vec{x} = \vec{y}$ is

$$\begin{aligned} -x_1 + x_2 + 2x_3 &= 3 \\ 3x_2 + 5x_3 &= 14 \\ 5x_3 &= 25 \end{aligned}$$

Using back-substitution, we get $x_3 = 5$, $3x_2 = 14 - 5(5) \Rightarrow x_2 = -\frac{11}{3}$, and $-x_1 = 3 + \frac{11}{3} - 2(5) \Rightarrow x_1 = \frac{10}{3}$. Thus, the solution is $\vec{x} = \begin{bmatrix} 10/3 \\ -11/3 \\ 5 \end{bmatrix}$.

(b) Write $A\vec{x} = \vec{b}$ as $LU\vec{x} = \vec{b}$ and take $\vec{y} = U\vec{x}$. Writing out the system $L\vec{y} = \vec{b}$, we get

$$\begin{aligned} y_1 &= 8 \\ -4y_1 + y_2 &= 2 \\ 3y_1 - 2y_2 + y_3 &= -9 \end{aligned}$$

Using forward-substitution, we find that $y_1 = 8$, so $y_2 = 2 + 4(8) = 34$, and $y_3 = -9 - 3(8) + 2(34) = 35$. Hence $\vec{y} = \begin{bmatrix} 8 \\ 34 \\ 35 \end{bmatrix}$.

Thus, our system $U\vec{x} = \vec{y}$ is

$$\begin{aligned} -x_1 + x_2 + 2x_3 &= 8 \\ 3x_2 + 5x_3 &= 34 \\ 5x_3 &= 35 \end{aligned}$$

Using back-substitution, we get $x_3 = 7$, $3x_2 = 34 - 5(7) \Rightarrow x_2 = -\frac{1}{3}$, and $-x_1 = 8 + \frac{1}{3} - 2(7) \Rightarrow x_1 = \frac{17}{3}$. Thus, the solution is $\vec{x} = \begin{bmatrix} 17/3 \\ -1/3 \\ 7 \end{bmatrix}$.

CHAPTER 4

Section 4.1

1. Consider

$$0 = t_1(1+2x+x^2+x^3)+t_2(1+x+3x^2+x^3)+t_3(3+5x+5x^2+3x^3)+t_4(-x-2x^2)$$

Row reducing the coefficient matrix of the associated homogeneous system of linear equations gives

$$\begin{bmatrix} 1 & 1 & 3 & 0 \\ 2 & 1 & 5 & -1 \\ 1 & 3 & 5 & -2 \\ 1 & 1 & 3 & 0 \end{bmatrix} \sim \begin{bmatrix} 1 & 0 & 2 & 0 \\ 0 & 1 & 1 & 0 \\ 0 & 0 & 0 & 1 \\ 0 & 0 & 0 & 0 \end{bmatrix}$$

Hence, there are infinitely many solutions, so $\mathcal{B}$ is linearly independent.
To determine if $p(x) \in$ span $\mathcal{B}$ we need to find if there exists t_1, t_2, t_3, and t_4 such that

$$\begin{aligned} 1 + 5x - 5x^2 + x^3 &= t_1(1 + 2x + x^2 + x^3) + t_2(1 + x + 3x^2 + x^3) \\ &\quad + t_3(3 + 5x + 5x^2 + 3x^3) + t_4(-x - 2x^2) \end{aligned}$$

Row reducing the corresponding augmented matrix gives

$$\left[\begin{array}{cccc|c} 1 & 1 & 3 & 0 & 1 \\ 2 & 1 & 5 & -1 & 5 \\ 1 & 3 & 5 & -2 & -5 \\ 1 & 1 & 3 & 0 & 1 \end{array}\right] \sim \left[\begin{array}{cccc|c} 1 & 0 & 2 & 0 & 4 \\ 0 & 1 & 1 & 0 & -3 \\ 0 & 0 & 0 & 1 & 0 \\ 0 & 0 & 0 & 0 & 0 \end{array}\right]$$

Since the system is consistent, we have that $p(x) \in \mathcal{B}$.

2. Clearly, if we pick any polynomial $a + bx + cx^2 + dx^3$, then it is a linear combination of the vectors in $\mathcal{B}$. Moreover, if we consider

$$t_1(1) + t_2x + t_3x^2 + t_4x^3 = 0 = 0 + 0x + 0x^2 + 0x^3$$

we get $t_1 = t_2 = t_3 = t_4 = 0$. Thus, $\mathcal{B}$ is also linearly independent.

Section 4.2

1. The set $\mathbb{S}$ is not closed under scalar multiplication. For example, $\begin{bmatrix} 1 \\ 1 \end{bmatrix} \in \mathbb{S}$, but $\sqrt{2}\begin{bmatrix} 1 \\ 1 \end{bmatrix} = \begin{bmatrix} \sqrt{2} \\ \sqrt{2} \end{bmatrix} \notin \mathbb{S}$.

2. Observe that axioms V2, V5, V7, V8, V9, and V10 must hold since we are using the operations of the vector space $M(2, 2)$.
Let $A = \begin{bmatrix} a_1 & 0 \\ 0 & a_2 \end{bmatrix}$ and $B = \begin{bmatrix} b_1 & 0 \\ 0 & b_2 \end{bmatrix}$ be vectors in $\mathbb{S}$.

V1 We have

$$A + B = \begin{bmatrix} a_1 & 0 \\ 0 & a_2 \end{bmatrix} + \begin{bmatrix} b_1 & 0 \\ 0 & b_2 \end{bmatrix} = \begin{bmatrix} a_1 + b_1 & 0 \\ 0 & a_2 + b_2 \end{bmatrix} \in \mathbb{S}$$

Therefore, $\mathbb{S}$ is closed under addition.

V3 The matrix $O_{2,2} = \begin{bmatrix} 0 & 0 \\ 0 & 0 \end{bmatrix} \in \mathbb{S}$ (take $a_1 = a_2 = 0$) satisfies $A + O_{2,2} = A = O_{2,2} + A$. Hence, it is the zero vector of $\mathbb{S}$.

V4 The additive inverse of A is $(-A) = \begin{bmatrix} -a_1 & 0 \\ 0 & -a_2 \end{bmatrix}$, which is clearly in $\mathbb{S}$.

V6 $tA = t\begin{bmatrix} a_1 & 0 \\ 0 & a_2 \end{bmatrix} = \begin{bmatrix} ta_1 & 0 \\ 0 & ta_2 \end{bmatrix} \in \mathbb{S}$. Therefore, $\mathbb{S}$ is closed under scalar multiplication.

Thus, $\mathbb{S}$ with these operators is a vector space as it satisfies all 10 axioms.

3. By definition, $\mathbb{U}$ is a subset of P_2. Taking $a = b = c = 0$, we get $0 \in \mathbb{U}$, so $\mathbb{U}$ is non-empty. Let $p = a_1 + b_1x + c_1x^2, q = a_2 + b_2x + c_2x^2 \in \mathbb{U}$, and $s \in \mathbb{R}$. Then $b_1 + c_1 = a_1$ and $b_2 + c_2 = a_2$.

S1 $p + q = (a_1 + a_2) + (b_1 + b_2)x + (c_1 + c_2)x^2$ and $(b_1 + b_2) + (c_1 + c_2) = b_1 + c_1 + b_2 + c_2 = a_1 + a_2$ so $p + q \in \mathbb{U}$

S2 $sp = sa_1 + sb_1x + sc_1x^2$ and $sb_1 + sc_1 = s(b_1 + c_1) = sa_1$, so $sp \in \mathbb{U}$

Hence, $\mathbb{U}$ is a subspace of P_2.

4. By definition, $\{\mathbf{0}\}$ is a non-empty subset of $\mathbb{V}$. Let $\mathbf{x}, \mathbf{y} \in \{\mathbf{0}\}$ and $s \in \mathbb{R}$. Then, $\mathbf{x} = \mathbf{0}$ and $\mathbf{y} = \mathbf{0}$. Thus,

S1 $\mathbf{x} + \mathbf{y} = \mathbf{0} + \mathbf{0} = \mathbf{0}$ by V4. Hence, $\mathbf{x} + \mathbf{y} \in \{\mathbf{0}\}$.

S2 $s\mathbf{x} = s\mathbf{0} = \mathbf{0}$ by Theorem 4.2.1. Hence, $s\mathbf{x} \in \{\mathbf{0}\}$.

Hence, $\{\mathbf{0}\}$ is a subspace of $\mathbb{V}$ and therefore is a vector space under the same operations as $\mathbb{V}$.

Section 4.3

1. Consider the equation

$$a_0 + a_1x + a_2x^2 = t_1(1 + 2x + x^2) + t_2(1 + x^2) + t_3(1 + x)$$

Row reducing the coefficient matrix of the corresponding system gives

$$\begin{bmatrix} 1 & 1 & 1 \\ 2 & 0 & 1 \\ 1 & 1 & 0 \end{bmatrix} \sim \begin{bmatrix} 1 & 0 & 0 \\ 0 & 1 & 0 \\ 0 & 0 & 1 \end{bmatrix}$$

Observe that this implies that the system is consistent and has a unique solution for all $a_0 + a_1x + a_2x^2 \in P_2$. Thus, $\mathcal{B}$ is a basis for P_2.

2. Consider the equation

$$0 = t_1(1 - x) + t_2(2 + 2x + x^2) + t_3(x + x^2) + (1 + x^2)$$

Row reducing the coefficient matrix of the corresponding system gives

$$\begin{bmatrix} 1 & 2 & 0 & 1 \\ -1 & 2 & 1 & 0 \\ 0 & 1 & 1 & 1 \end{bmatrix} \sim \begin{bmatrix} 1 & 0 & 0 & 1 \\ 0 & 1 & 0 & 0 \\ 0 & 0 & 1 & 1 \end{bmatrix}$$

Thus, $1 + x^2$ can be written as a linear combination of the other vectors. Moreover, $\{1 - x, 2 + 2x + x^2, x + x^2\}$ is a linearly independent set and hence is a basis for Span $\mathcal{B}$.

3. Observe that every polynomial in $\mathbb{S}$ can be written in the form

$$a + bx + cx^2 + (-a - b - c)x^3 = a(1 - x^3) + b(x - x^3) + c(x^2 - x^3)$$

Hence, $\mathcal{B} = \{1 - x^3, x - x^3, x^2 - x^3\}$ spans $\mathbb{S}$. Verify that $\mathcal{B}$ is also linearly independent and hence a basis for $\mathbb{S}$. Thus, $\dim \mathbb{S} = 3$.

4. By the procedure, we first add a vector not in Span $\left\{\begin{bmatrix}1\\1\\1\end{bmatrix}\right\}$. We pick $\begin{bmatrix}1\\0\\0\end{bmatrix}$. Thus, $\left\{\begin{bmatrix}1\\1\\1\end{bmatrix}, \begin{bmatrix}1\\0\\0\end{bmatrix}\right\}$ is linearly independent. Consider

$$\begin{bmatrix}x_1\\x_2\\x_3\end{bmatrix} = t_1\begin{bmatrix}1\\1\\1\end{bmatrix} + t_2\begin{bmatrix}1\\0\\0\end{bmatrix}$$

Row reducing the corresponding augmented matrix gives

$$\left[\begin{array}{cc|c}1 & 1 & x_1\\1 & 0 & x_2\\1 & 0 & x_3\end{array}\right] \sim \left[\begin{array}{cc|c}1 & 1 & x_1\\0 & 1 & x_1 - x_2\\0 & 0 & x_3 - x_2\end{array}\right]$$

Thus, the vector $\vec{x} = \begin{bmatrix}0\\1\\0\end{bmatrix}$ is not in the span of the first two vectors. Hence, $\mathcal{B} = \left\{\begin{bmatrix}1\\1\\1\end{bmatrix}, \begin{bmatrix}1\\0\\0\end{bmatrix}, \begin{bmatrix}0\\1\\0\end{bmatrix}\right\}$ is linearly independent. Moreover, repeating the row reduction above with $\vec{x} = \begin{bmatrix}0\\1\\0\end{bmatrix}$, we find that the augmented matrix has a leading 1 in each row, and hence $\mathcal{B}$ also spans $\mathbb{R}^3$. Thus, $\mathcal{B}$ is a basis for $\mathbb{R}^3$. Note that there are many possible correct answers.

5. A hyperplane in $\mathbb{R}^4$ is three-dimensional; therefore, we need to pick three linearly independent vectors that satisfy the equation of the hyperplane. We pick $\mathcal{B} = \left\{\begin{bmatrix}1\\1\\0\\0\end{bmatrix}, \begin{bmatrix}0\\1\\1\\0\end{bmatrix}, \begin{bmatrix}2\\0\\0\\1\end{bmatrix}\right\}$. It is easy to verify that $\mathcal{B}$ is linearly independent and hence forms a basis for the hyperplane. To extend this to a basis for $\mathbb{R}^4$, we just need to add one vector that does not lie in the hyperplane. We observe that $\begin{bmatrix}1\\0\\0\\0\end{bmatrix}$ does not satisfy the equation of the hyperplane. Thus, $\left\{\begin{bmatrix}1\\1\\0\\0\end{bmatrix}, \begin{bmatrix}0\\1\\1\\0\end{bmatrix}, \begin{bmatrix}2\\0\\0\\1\end{bmatrix}, \begin{bmatrix}1\\0\\0\\0\end{bmatrix}\right\}$ is a basis for $\mathbb{R}^4$.

Section 4.4

1. Consider $t_1\begin{bmatrix}2\\1\\-1\end{bmatrix}+t_2\begin{bmatrix}-3\\-1\\2\end{bmatrix}=\begin{bmatrix}0\\2\\2\end{bmatrix}$.

Row reducing the corresponding augmented matrix gives

$$\left[\begin{array}{cc|c}2&-3&0\\1&-1&2\\-1&2&2\end{array}\right]\sim\left[\begin{array}{cc|c}1&0&6\\0&1&4\\0&0&0\end{array}\right]$$

Thus, $\begin{bmatrix}0\\2\\2\end{bmatrix}_{\mathcal{B}}=\begin{bmatrix}6\\4\end{bmatrix}$.

2. To find the change of coordinates matrix Q we need to find the coordinates of the vectors in $\mathcal{B}$ with respect to the standard basis $\mathcal{S}$. We get

$$Q=\begin{bmatrix}\begin{bmatrix}1\\1\\2\end{bmatrix}_{\mathcal{S}} & \begin{bmatrix}1\\2\\4\end{bmatrix}_{\mathcal{S}} & \begin{bmatrix}2\\3\\3\end{bmatrix}_{\mathcal{S}}\end{bmatrix}=\begin{bmatrix}1&1&2\\1&2&2\\2&4&3\end{bmatrix}$$

To find the change of coordinates matrix P, we need to find the coordinates of the standard basis vectors with respect to the basis $\mathcal{B}$. To do this, we solve the triple augmented systems

$$\left[\begin{array}{ccc|ccc}1&1&2&1&0&0\\1&2&2&0&1&0\\2&1&3&0&0&1\end{array}\right]\sim\left[\begin{array}{ccc|ccc}1&0&0&-4&1&2\\0&1&0&-1&1&0\\0&0&1&3&-1&-1\end{array}\right]$$

Thus,

$$P=\begin{bmatrix}-4&1&2\\-1&1&0\\3&-1&-1\end{bmatrix}$$

We now verify that

$$\begin{bmatrix}-4&1&2\\-1&1&0\\3&-1&-1\end{bmatrix}\begin{bmatrix}1&1&2\\1&2&2\\2&1&3\end{bmatrix}=\begin{bmatrix}1&0&0\\0&1&0\\0&0&1\end{bmatrix}$$

Section 4.5

1. Let $\vec{x}=\begin{bmatrix}x_1\\x_2\\x_3\end{bmatrix}, \vec{y}=\begin{bmatrix}y_1\\y_2\\y_3\end{bmatrix}\in\mathbb{R}^3$ and $t\in\mathbb{R}$. Then,

$$\begin{aligned}L(t\vec{x}+\vec{y})&=L\left(\begin{bmatrix}tx_1+y_1\\tx_2+y_2\\tx_3+y_3\end{bmatrix}\right)=\begin{bmatrix}tx_1+y_1 & tx_1+y_1+tx_2+y_2+tx_3+y_3\\0 & tx_2+y_2\end{bmatrix}\\&=t\begin{bmatrix}x_1&x_1+x_2+x_3\\0&x_2\end{bmatrix}+\begin{bmatrix}y_1&y_1+y_2+y_3\\0&y_2\end{bmatrix}=tL(\vec{x})+L(\vec{y})\end{aligned}$$

Hence, L is linear.

2. If $\vec{x} = \begin{bmatrix} x_1 \\ x_2 \\ x_3 \end{bmatrix} \in \text{Null}(L)$, then $\begin{bmatrix} 0 & 0 \\ 0 & 0 \end{bmatrix} = L(\vec{x}) = \begin{bmatrix} x_1 & x_2 + x_3 \\ x_2 + x_3 & x_1 \end{bmatrix}$.
Hence, $x_1 = 0$ and $x_2 + x_3 = 0$. Thus, every vector in Null(L) has the form $\vec{x} = x_2 \begin{bmatrix} 0 \\ 1 \\ -1 \end{bmatrix}$. Hence, a basis for Null(L) is $\left\{ \begin{bmatrix} 0 \\ 1 \\ -1 \end{bmatrix} \right\}$.
Every vector in the range of L has the form

$$\begin{bmatrix} x_1 & x_2 + x_3 \\ x_2 + x_3 & x_1 \end{bmatrix} = x_1 \begin{bmatrix} 1 & 0 \\ 0 & 1 \end{bmatrix} + (x_2 + x_3) \begin{bmatrix} 0 & 1 \\ 1 & 0 \end{bmatrix}$$

Since $\left\{ \begin{bmatrix} 1 & 0 \\ 0 & 1 \end{bmatrix}, \begin{bmatrix} 0 & 1 \\ 1 & 0 \end{bmatrix} \right\}$ is also clearly linearly independent, it is a basis for the range of L.

Section 4.6

1. We have

$$L\left(\begin{bmatrix} 1 & 1 \\ 1 & 1 \end{bmatrix}\right) = \begin{bmatrix} 2 & 0 \\ 1 & 3 \end{bmatrix} = 2\begin{bmatrix} 1 & 1 \\ 1 & 1 \end{bmatrix} - 2\begin{bmatrix} 0 & 1 \\ 1 & 1 \end{bmatrix} + 1\begin{bmatrix} 0 & 0 \\ 1 & 1 \end{bmatrix} + 2\begin{bmatrix} 0 & 0 \\ 0 & 1 \end{bmatrix}$$

$$L\left(\begin{bmatrix} 0 & 1 \\ 1 & 1 \end{bmatrix}\right) = \begin{bmatrix} 1 & -1 \\ 1 & 2 \end{bmatrix} = 1\begin{bmatrix} 1 & 1 \\ 1 & 1 \end{bmatrix} - 2\begin{bmatrix} 0 & 1 \\ 1 & 1 \end{bmatrix} + 2\begin{bmatrix} 0 & 0 \\ 1 & 1 \end{bmatrix} + 1\begin{bmatrix} 0 & 0 \\ 0 & 1 \end{bmatrix}$$

$$L\left(\begin{bmatrix} 0 & 0 \\ 1 & 1 \end{bmatrix}\right) = \begin{bmatrix} 0 & 0 \\ 1 & 1 \end{bmatrix} = 0\begin{bmatrix} 1 & 1 \\ 1 & 1 \end{bmatrix} + 0\begin{bmatrix} 0 & 1 \\ 1 & 1 \end{bmatrix} + 1\begin{bmatrix} 0 & 0 \\ 1 & 1 \end{bmatrix} + 0\begin{bmatrix} 0 & 0 \\ 0 & 1 \end{bmatrix}$$

$$L\left(\begin{bmatrix} 0 & 0 \\ 0 & 1 \end{bmatrix}\right) = \begin{bmatrix} 0 & 0 \\ 0 & 1 \end{bmatrix} = 0\begin{bmatrix} 1 & 1 \\ 1 & 1 \end{bmatrix} + 0\begin{bmatrix} 0 & 1 \\ 1 & 1 \end{bmatrix} + 0\begin{bmatrix} 0 & 0 \\ 1 & 1 \end{bmatrix} + 1\begin{bmatrix} 0 & 0 \\ 0 & 1 \end{bmatrix}$$

Hence,

$$[L]_{\mathcal{B}} = \begin{bmatrix} 2 & 1 & 0 & 0 \\ -2 & -2 & 0 & 0 \\ 1 & 2 & 1 & 0 \\ 2 & 1 & 0 & 1 \end{bmatrix}$$

Section 4.7

1. If $\mathbf{0} = t_1 L(\mathbf{u}_1) + \cdots + t_k L(\mathbf{u}_k) = L(t_1\mathbf{u}_1 + \cdots t_k\mathbf{u}_k)$, then $t_1\mathbf{u}_1 + \cdots + t\mathbf{u}_k \in \text{Null}(L)$. But, L is one-to-one, so $\text{Null}(L) = \{\mathbf{0}\}$. Thus, $t_1\mathbf{u}_1 + \cdots + t\mathbf{u}_k = \mathbf{0}$ and hence $t_1 = \cdots = t_k = 0$ since $\{\mathbf{u}_1, \dots, \mathbf{u}_k\}$ is linearly independent. Thus, $\{L(\mathbf{u}_1), \dots, L(\mathbf{u}_k)\}$ is linearly independent.
2. Let $\mathbf{v}$ be any vector in $\mathbb{V}$. Since L is onto, there exists a vector $\mathbf{x} \in \mathbb{U}$ such that $L(\mathbf{x}) = \mathbf{v}$. Since $\mathbf{x} \in \mathbb{U}$, we can write it as a linear combination of the vectors $\{\mathbf{u}_1, \dots, \mathbf{u}_k\}$. Hence, we have

$$\mathbf{v} = L(\mathbf{x}) = L(t_1\mathbf{u}_1 + \cdots + t_k\mathbf{u}_k) = t_1L(\mathbf{u}_1) + \cdots + t_kL(\mathbf{u}_k)$$

Therefore, $\text{Span}\{L(\mathbf{u}_1), \dots, L(\mathbf{u}_k)\} = \mathbb{V}$.
3. If L is an isomorphism, then it is one-to-one and onto. Since $\{\mathbf{u}_1, \dots, \mathbf{u}_n\}$ is a basis, it is a linearly independent spanning set. Thus, $\{L(\mathbf{u}_1), \dots, L(\mathbf{u}_n)\}$ is also linearly independent by Exercise 1, and it is a spanning set for $\mathbb{V}$ by Exercise 2. Thus, it is a basis for $\mathbb{V}$.

CHAPTER 5

Section 5.1

1. (a) $\begin{vmatrix} 3 & 2 \\ 2 & 1 \end{vmatrix} = 3(1) - 2(2) = -1$

(b) $\begin{vmatrix} 1 & 3 \\ 0 & -2 \end{vmatrix} = 1(-2) - 3(0) = -2$

(c) $\begin{vmatrix} 2 & 4 \\ 1 & 2 \end{vmatrix} = 2(2) - 4(1) = 0$

2. We have $C_{11} = (-1)^{1+1}\begin{vmatrix} -1 & -2 \\ 0 & -3 \end{vmatrix} = 3$, $C_{12} = (-1)^{1+2}\begin{vmatrix} 0 & -2 \\ 4 & -3 \end{vmatrix} = -8$, and $C_{13} = (-1)^{1+3}\begin{vmatrix} 0 & -1 \\ 4 & 0 \end{vmatrix} = 4$. So, $\det A = 1C_{11}+2C_{12}+3C_{13} = 1(3)+2(-8)+3(4) = -1$.

3. (a) $\det A = 1C_{11} + 0C_{21} + 3C_{31} + (-2)C_{41} = 1(-36) + 3(32) + (-2)(-26) = 112$

(b) $\det A = 0C_{21} + 0C_{22} + (-1)C_{23} + 2C_{24} = (-1)(0) + 2(56) = 112$

(c) $\det A = 0C_{14} + 2C_{24} + 0C_{34} + 0C_{44} = 2(56) = 112$

Section 5.2

1. rA is obtained by multiplying each row of A by r. Thus, by Theorem 5.2.1, $\det(rA) = (r)(r)(r)\det A = r^3 \det A$.

2.

$$\det A = \begin{vmatrix} 0 & 2 & -8 & 8 \\ 0 & 0 & 16 & -1 \\ 1 & 1 & 3 & -1 \\ 0 & 2 & -14 & 9 \end{vmatrix} = (-1)^2 \begin{vmatrix} 1 & 1 & 3 & -1 \\ 0 & 2 & -14 & 9 \\ 0 & 2 & -8 & 8 \\ 0 & 0 & 16 & -1 \end{vmatrix}$$

$$= (-1)^2 \begin{vmatrix} 1 & 1 & 3 & -1 \\ 0 & 2 & -14 & 9 \\ 0 & 0 & 6 & -1 \\ 0 & 0 & 16 & -1 \end{vmatrix} = (-1)^2 \begin{vmatrix} 1 & 1 & 3 & -1 \\ 0 & 2 & -14 & 9 \\ 0 & 0 & 6 & -1 \\ 0 & 0 & 0 & 5/3 \end{vmatrix} = 20$$

3.

$$\det A = (-6)(-1)^{3+1}\begin{vmatrix} -2 & 4 & -5 \\ 2 & -4 & 3 \\ 2 & -3 & -4 \end{vmatrix} + 4(-1)^{3+2}\begin{vmatrix} -6 & 4 & -5 \\ 3 & -4 & 3 \\ -3 & -3 & -4 \end{vmatrix}$$

$$= -6\begin{vmatrix} -2 & 4 & -5 \\ 0 & 0 & -2 \\ 0 & 1 & -9 \end{vmatrix} - 4\begin{vmatrix} 0 & -4 & 1 \\ 3 & -4 & 3 \\ 0 & -7 & -1 \end{vmatrix}$$

$$= (-6)(-1)\begin{vmatrix} -2 & 4 & -5 \\ 0 & 1 & -9 \\ 0 & 0 & -2 \end{vmatrix} - 4(3)(-1)^{2+1}\begin{vmatrix} -4 & 1 \\ -7 & -1 \end{vmatrix}$$

$$= 6(4) + 12(11) = 156$$

Section 5.3

1. $\operatorname{cof} A = \begin{bmatrix} -3 & 1 & -3 \\ -2 & 1 & -3 \\ 2 & -1 & 2 \end{bmatrix}$

2. $A^{-1} = \frac{1}{\det A}(\operatorname{cof} A)^T = \begin{bmatrix} 3 & 2 & -2 \\ -1 & -1 & 1 \\ 3 & 3 & -2 \end{bmatrix}$

Section 5.4

1. We draw $\vec{u}$ and $\vec{v}$ to form a left-handed system and repeat the calculations for the right-handed system:

$$\begin{aligned} \text{Area}(\vec{u}, \vec{v}) &= \text{Area of Square} - \text{Area 1} - \text{Area 2} - \text{Area 3} - \text{Area 4} - \text{Area 5} - \text{Area 6} \\ &= (v_1 + u_1)(v_2 + u_2) - \frac{1}{2}u_1u_2 - v_2u_1 - \frac{1}{2}v_1v_2 - \frac{1}{2}u_1u_2 - v_2u_1 - \frac{1}{2}v_1v_2 \\ &= v_1v_2 + v_1u_2 + v_2u_2 + v_2u_2 - u_1u_2 - 2v_2u_1 - v_2v_1 \\ &= v_1u_2 - v_2u_1 = -(u_1v_2 - u_2v_1) \end{aligned}$$

So, Area $= \left|\det \begin{bmatrix} u_1 & v_1 \\ u_2 & v_2 \end{bmatrix}\right|$.

2. We have $L(\vec{e}_1) = \begin{bmatrix} t \\ 0 \end{bmatrix}$ and $L(\vec{e}_2) = \begin{bmatrix} 0 \\ 1 \end{bmatrix}$. Hence, the area determined by the image vectors is

$$\text{Area}(L(\vec{e}_1), L(\vec{e}_2)) = \left|\det \begin{bmatrix} t & 0 \\ 0 & 1 \end{bmatrix}\right| = |t| = t$$

Alternatively,

$$\text{Area}(L(\vec{e}_1), L(\vec{e}_2)) = |\det A|\text{Area}(\vec{e}_1, \vec{e}_2) = \left|\det \begin{bmatrix} t & 0 \\ 0 & 1 \end{bmatrix}\right| \left|\det \begin{bmatrix} 1 & 0 \\ 0 & 1 \end{bmatrix}\right| = t(1) = t$$

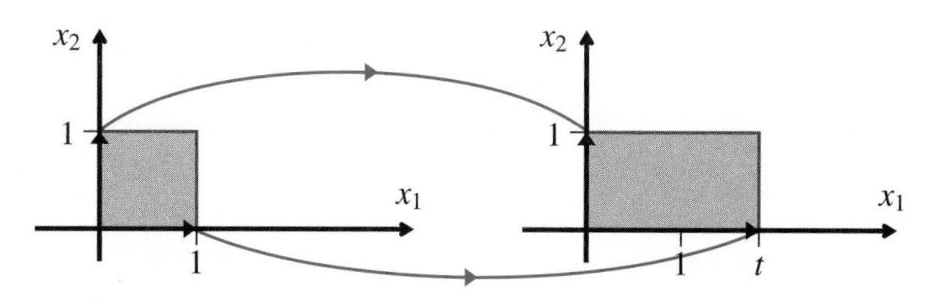

CHAPTER 6

Section 6.1

1. The only eigenvectors are multiples of $\vec{e}_3$ with eigenvalue 1.
2. $C(\lambda) = \det(A - \lambda I) = \lambda^2 - 5\lambda = \lambda(\lambda - 5)$, so the eigenvalues are $\lambda_1 = 0$ and $\lambda_2 = 5$. For $\lambda_1 = 0$, we have

$$A - 0I = \begin{bmatrix} 1 & 2 \\ 2 & 4 \end{bmatrix} \sim \begin{bmatrix} 1 & 2 \\ 0 & 0 \end{bmatrix}$$

Thus, the eigenspace of $\lambda_1 = 0$ is Span $\left\{ \begin{bmatrix} -2 \\ 1 \end{bmatrix} \right\}$. For $\lambda_2 = 5$, we have

$$A - 5I = \begin{bmatrix} -4 & 2 \\ 2 & -1 \end{bmatrix} \sim \begin{bmatrix} 1 & -1/2 \\ 0 & 0 \end{bmatrix}$$

Thus, the eigenspace of $\lambda_1 = 0$ is Span $\left\{ \begin{bmatrix} 1 \\ 2 \end{bmatrix} \right\}$.

3. We have

$$C(\lambda) = \det(A - \lambda I) = \begin{vmatrix} 5-\lambda & -3 & 2 \\ 0 & -\lambda & 2 \\ 0 & -2 & -4-\lambda \end{vmatrix}$$

$$= (5-\lambda)\begin{vmatrix} -\lambda & 2 \\ -2 & -4-\lambda \end{vmatrix} = -(\lambda - 5)(\lambda + 2)^2$$

So $\lambda_1 = 5$ has algebraic multiplicity 1, and $\lambda_2 = -2$ has algebraic multiplicity 2. For $\lambda_1 = 5$, we have

$$A - 5I = \begin{bmatrix} 0 & -3 & 2 \\ 0 & -5 & 2 \\ 0 & -2 & -9 \end{bmatrix} \sim \begin{bmatrix} 0 & 1 & 0 \\ 0 & 0 & 1 \\ 0 & 0 & 0 \end{bmatrix}$$

Thus, a basis for the eigenspace of $\lambda_1 = 5$ is $\left\{ \begin{bmatrix} 1 \\ 0 \\ 0 \end{bmatrix} \right\}$, so it has geometric multiplicity 1. For $\lambda_2 = -2$, we have

$$A - 5I = \begin{bmatrix} 7 & -3 & 2 \\ 0 & 2 & 2 \\ 0 & -2 & -2 \end{bmatrix} \sim \begin{bmatrix} 1 & 0 & 5/7 \\ 0 & 1 & 1 \\ 0 & 0 & 0 \end{bmatrix}$$

A basis for the eigenspace of $\lambda_2 = 2$ is $\left\{ \begin{bmatrix} -5/7 \\ -1 \\ 1 \end{bmatrix} \right\}$, so it also has geometric multiplicity 1.

Section 6.2

1. We have $C(\lambda) = \det(A - \lambda I) = -\lambda(\lambda - 2)(\lambda + 1)$. Hence, the eigenvalues are $\lambda_1 = 0$, $\lambda_2 = 2$, and $\lambda_3 = -1$. For $\lambda_1 = 0$, we have

$$A - 0I = \begin{bmatrix} 1 & 0 & -1 \\ 11 & -4 & -7 \\ -7 & 3 & 4 \end{bmatrix} \sim \begin{bmatrix} 1 & 0 & -1 \\ 0 & 1 & -1 \\ 0 & 0 & 0 \end{bmatrix}$$

Thus, a basis for the eigenspace of λ_1 is $\left\{ \begin{bmatrix} 1 \\ 1 \\ 1 \end{bmatrix} \right\}$.

For $\lambda_2 = 2$, we have

$$A - 2I = \begin{bmatrix} -1 & 0 & -1 \\ 11 & -6 & -7 \\ -7 & 3 & 2 \end{bmatrix} \sim \begin{bmatrix} 1 & 0 & 1 \\ 0 & 1 & 3 \\ 0 & 0 & 0 \end{bmatrix}$$

Thus, a basis for the eigenspace of λ_1 is $\left\{ \begin{bmatrix} -1 \\ -3 \\ 1 \end{bmatrix} \right\}$.

For $\lambda_3 = -1$, we have

$$A - (-1)I = \begin{bmatrix} 2 & 0 & -1 \\ 11 & -3 & -7 \\ -7 & 3 & 5 \end{bmatrix} \sim \begin{bmatrix} 1 & 0 & -1/2 \\ 0 & 1 & -1/2 \\ 0 & 0 & 0 \end{bmatrix}$$

Thus, a basis for the eigenspace of λ_1 is $\left\{ \begin{bmatrix} 1 \\ 1 \\ 2 \end{bmatrix} \right\}$. So, we can take $P = \begin{bmatrix} 1 & -1 & 1 \\ 1 & -3 & 1 \\ 1 & 1 & 2 \end{bmatrix}$ and get $P^{-1}AP = D = \begin{bmatrix} 0 & 0 & 0 \\ 0 & 2 & 0 \\ 0 & 0 & -1 \end{bmatrix}$.

2. We have $C(\lambda) = \det(A - \lambda I) = (\lambda - 2)^2$. Thus, $\lambda = 2$ has algebraic multiplicity 2. We have $A - 2I = \begin{bmatrix} 0 & 1 \\ 0 & 0 \end{bmatrix}$, so a basis for its eigenspace is $\left\{ \begin{bmatrix} 1 \\ 0 \end{bmatrix} \right\}$. Hence, the geometric multiplicity is less than the algebraic multiplicity. Thus, A is not diagonalizable.

Section 6.3

1. (a) A is not a Markov matrix.

(b) B is a Markov matrix. We have $B - I = \begin{bmatrix} -0.6 & 0.6 \\ 0.6 & -0.6 \end{bmatrix} \sim \begin{bmatrix} 1 & -1 \\ 0 & 0 \end{bmatrix}$. Thus, its fixed-state vector is $\begin{bmatrix} 1/2 \\ 1/2 \end{bmatrix}$.

2. We find that $\mathbb{T} = \begin{bmatrix} s_2 \\ s_1 \end{bmatrix}$, $\mathbb{T}^2 = \begin{bmatrix} s_1 \\ s_2 \end{bmatrix}$, $\mathbb{T}^3 = \begin{bmatrix} s_2 \\ s_1 \end{bmatrix}$, etc. On the other hand, the fixed-state vector is the eigenvector corresponding to $\lambda = 1$, which is $\begin{bmatrix} 1 \\ 1 \end{bmatrix}$.

CHAPTER 7

Section 7.1

1. We have $\begin{bmatrix}1\\1\\2\end{bmatrix} \cdot \begin{bmatrix}-2\\0\\1\end{bmatrix} = 0$, $\begin{bmatrix}1\\1\\2\end{bmatrix} \cdot \begin{bmatrix}1\\-5\\2\end{bmatrix} = 0$, and $\begin{bmatrix}-2\\0\\1\end{bmatrix} \cdot \begin{bmatrix}1\\-5\\2\end{bmatrix} = 0$. Thus, the set is orthogonal. Dividing each vector by its length gives the orthonormal set $\left\{\begin{bmatrix}1/\sqrt{6}\\1/\sqrt{6}\\2/\sqrt{6}\end{bmatrix}, \begin{bmatrix}-2/\sqrt{5}\\0\\1/\sqrt{5}\end{bmatrix}, \begin{bmatrix}1/\sqrt{30}\\-5\sqrt{30}\\2\sqrt{30}\end{bmatrix}\right\}$.

2. It is easy to verify that $\mathcal{B}$ is orthonormal. We have $b_1 = \vec{x} \cdot \vec{v}_1 = 8/\sqrt{3}$, $b_2 = \vec{x} \cdot \vec{v}_2 = -1/\sqrt{2}$, and $b_3 = \vec{x} \cdot \vec{v}_3 = 5/\sqrt{6}$. Hence, $[\vec{x}]_{\mathcal{B}} = \begin{bmatrix}8/\sqrt{3}\\-1/\sqrt{2}\\5/\sqrt{6}\end{bmatrix}$.

3. We have $\vec{x} = \begin{bmatrix}(\sqrt{2}+2\sqrt{3}-1)/\sqrt{6}\\(\sqrt{2}+2)/\sqrt{6}\\(\sqrt{2}-2\sqrt{3}-1)/\sqrt{6}\end{bmatrix}$ and $\vec{y} = \begin{bmatrix}(\sqrt{18}-\sqrt{2}+1)/\sqrt{6}\\(\sqrt{18}-2)/\sqrt{6}\\(\sqrt{18}+\sqrt{12}+1)/\sqrt{6}\end{bmatrix}$. With a little effort, we find that $\|\vec{x}\|^2 = \sqrt{\vec{x} \cdot \vec{x}} = 6$ and $\vec{x} \cdot \vec{y} = -2$.

4. The result is easily verified.

Section 7.2

1. We want to find all $\vec{v} = \begin{bmatrix}v_1\\v_2\\v_3\\v_4\end{bmatrix} \in \mathbb{R}^4$ such that $\vec{v} \cdot \begin{bmatrix}1\\1\\1\\0\end{bmatrix} = 0$, $\vec{v} \cdot \begin{bmatrix}-1\\0\\1\\1\end{bmatrix} = 0$, and $\vec{v} \cdot \begin{bmatrix}0\\1\\-1\\1\end{bmatrix} = 0$. This gives the homogeneous system of equations

$$\begin{aligned} v_1 + v_2 + v_3 &= 0 \\ -v_1 + v_3 + v_4 &= 0 \\ v_2 - v_3 + v_4 &= 0 \end{aligned}$$

which has solution space Span $\left\{\begin{bmatrix}1\\-1\\0\\1\end{bmatrix}\right\}$. Hence, $\mathbb{S}^\perp = \text{Span}\left\{\begin{bmatrix}1\\-1\\0\\1\end{bmatrix}\right\}$.

2. We have

$$\text{proj}_{\mathbb{S}}\,\vec{x} = \frac{\vec{x} \cdot \vec{v}_1}{\|\vec{v}_1\|^2}\vec{v}_1 + \frac{\vec{x} \cdot \vec{v}_2}{\|\vec{v}_2\|^2}\vec{v}_2 = \frac{7}{6}\begin{bmatrix}1\\2\\1\end{bmatrix} + \frac{-4}{3}\begin{bmatrix}-1\\1\\-1\end{bmatrix} = \begin{bmatrix}5/2\\1\\5/2\end{bmatrix}$$

$$\text{perp}_{\mathbb{S}}\,\vec{x} = \vec{x} - \text{proj}_{\mathbb{S}}\,\vec{x} = \begin{bmatrix}2\\1\\3\end{bmatrix} - \begin{bmatrix}5/2\\1\\5/2\end{bmatrix} = \begin{bmatrix}-1/2\\0\\1/2\end{bmatrix}$$

Section 7.4

1. We verify that $\langle\ ,\ \rangle$ satisfies the three properties of an inner product. Let $A = \begin{bmatrix} a_1 & a_2 \\ a_3 & a_4 \end{bmatrix}$, $B = \begin{bmatrix} b_1 & b_2 \\ b_3 & b_4 \end{bmatrix}$, and $C = \begin{bmatrix} c_1 & c_2 \\ c_3 & c_4 \end{bmatrix}$.

   (a) $\langle A, A\rangle = \text{tr}(A^TA) = a_1^2 + a_3^2 + a_2^2 + a_4^2 \geq 0$ and $\langle A, A\rangle = 0$ if and only if $a_1 = a_2 = a_3 = a_4 = 0$, so $A = O_{2,2}$. Thus, it is positive definite.

   (b) $\langle A, B\rangle = \text{tr}(B^TA) = b_1a_1 + b_3a_3 + b_2a_2 + b_4a_4 = a_1b_1 + a_3b_3 + a_2b_2 + a_4b_4 = \langle A, B\rangle$, so it is symmetric.

   (c) For any $A, B, C \in M(2, 2)$ and $s, t \in \mathbb{R}$,

$$\begin{aligned}\langle A, sB + tC\rangle &= \text{tr}\left((sB + tC)^TA\right) = \text{tr}\left((sB^T + tC^T)A\right) \\ &= \text{tr}(sB^TA + tC^TA) = s\,\text{tr}(B^TA) + t\,\text{tr}(C^TA) = s\langle A, B\rangle + t\langle A, C\rangle\end{aligned}$$

   So, $\langle\ ,\ \rangle$ is bilinear. Thus, $\langle\ ,\ \rangle$ is an inner product on $M(2, 2)$. We observe that $\langle A, B\rangle = a_1b_1 + a_2b_2 + a_3b_3 + a_4b_4$, so it matches the dot product on the isomorphic vectors in $\mathbb{R}^4$.

2. $\|1\| = \sqrt{\langle 1, 1\rangle} = \sqrt{1^2 + 1^2 + 1^2} = \sqrt{3}$. $\|x\| = \sqrt{\langle x, x\rangle} = \sqrt{0^2 + 1^2 + 2^2} = \sqrt{5}$

3. $\|x\| = \sqrt{\langle x, x\rangle} = \sqrt{(-1)^2 + 0^2 + 1^2} = \sqrt{2}$

CHAPTER 8

Section 8.1

1. We have $C(\lambda) = (\lambda - 8)(\lambda - 2)$. Thus, the eigenvalues are $\lambda_1 = 8$ and $\lambda_2 = 2$. For $\lambda_1 = 8$, we get

$$A - 8I = \begin{bmatrix} -3 & -3 \\ -3 & -3 \end{bmatrix} \sim \begin{bmatrix} 1 & 1 \\ 0 & 0 \end{bmatrix}$$

   Thus, a basis for its eigenspace is $\left\{\begin{bmatrix} -1 \\ 1 \end{bmatrix}\right\}$. For $\lambda_2 = 2$, we get

$$A - 8I = \begin{bmatrix} 3 & -3 \\ -3 & 3 \end{bmatrix} \sim \begin{bmatrix} 1 & -1 \\ 0 & 0 \end{bmatrix}$$

   Thus, a basis for its eigenspace is $\left\{\begin{bmatrix} 1 \\ 1 \end{bmatrix}\right\}$. We normalize the vectors and find that A is orthogonally diagonalized to $D = \begin{bmatrix} 8 & 0 \\ 0 & 2 \end{bmatrix}$ by $P = \begin{bmatrix} -1/\sqrt{2} & 1/\sqrt{2} \\ 1/\sqrt{2} & 1/\sqrt{2} \end{bmatrix}$.

2. We have $C(\lambda) = -\lambda(\lambda - 3)^2$. Thus, the eigenvalues are $\lambda_1 = 0$ and $\lambda_2 = 3$. For $\lambda_1 = 0$, we get

$$A - 0I = \begin{bmatrix} 2 & -1 & -1 \\ -1 & 2 & -1 \\ -1 & -1 & 2 \end{bmatrix} \sim \begin{bmatrix} 1 & 0 & -1 \\ 0 & 1 & -1 \\ 0 & 0 & 0 \end{bmatrix}$$

Thus, a basis for its eigenspace is $\left\{\begin{bmatrix}1\\1\\1\end{bmatrix}\right\}$. For $\lambda_2 = 3$, we get

$$A - 8I = \begin{bmatrix}-1 & -1 & -1\\-1 & -1 & -1\\-1 & -1 & -1\end{bmatrix} \sim \begin{bmatrix}1 & 1 & 1\\0 & 0 & 0\\0 & 0 & 0\end{bmatrix}$$

Thus, a basis for its eigenspace is $\left\{\begin{bmatrix}-1\\1\\0\end{bmatrix}, \begin{bmatrix}-1\\0\\1\end{bmatrix}\right\}$. But we need an orthonormal basis for each eigenspace, so we apply the Gram-Schmidt Procedure and normalize the vectors to get $\left\{\begin{bmatrix}-1/\sqrt{2}\\1/\sqrt{2}\\0\end{bmatrix}, \begin{bmatrix}-1/\sqrt{6}\\-1/\sqrt{6}\\2/\sqrt{6}\end{bmatrix}\right\}$.

Therefore, A is diagonalized to $D = \begin{bmatrix}0 & 0 & 0\\0 & 3 & 0\\0 & 0 & 3\end{bmatrix}$ by $P = \begin{bmatrix}1/\sqrt{3} & -1/\sqrt{2} & -1/\sqrt{6}\\1/\sqrt{3} & 1/\sqrt{2} & -1/\sqrt{6}\\1/\sqrt{3} & 0 & 2/\sqrt{6}\end{bmatrix}$.

Section 8.2

1. (a) $Q(\vec{x}) = 4x_1^2 + x_1x_2 + \sqrt{2}x_2^2$

(b) $Q(\vec{x}) = x_1^2 - 2x_1x_2 + 2x_2^2 + 6x_2x_3 - x_3^2$

2. (a) $\begin{bmatrix}1 & -1\\-1 & -3\end{bmatrix}$ (b) $\begin{bmatrix}2 & 3/2 & -1/2\\3/2 & 4 & 0\\-1/2 & 0 & 1\end{bmatrix}$ (c) $\begin{bmatrix}1 & 0 & 0 & 0\\0 & 2 & 0 & 0\\0 & 0 & 3 & 0\\0 & 0 & 0 & 4\end{bmatrix}$

3. The corresponding symmetric matrix is $A = \begin{bmatrix}0 & 2\\2 & -3\end{bmatrix}$. We have $C(\lambda) = (\lambda+4)(\lambda - 1)$. Thus, the eigenvalues are $\lambda_1 = -4$ and $\lambda_2 = 1$. For $\lambda_1 = -4$, we get

$$A + 4I = \begin{bmatrix}4 & 2\\2 & 1\end{bmatrix} \sim \begin{bmatrix}2 & 1\\0 & 0\end{bmatrix}$$

Thus, a basis for its eigenspace is $\left\{\begin{bmatrix}-1\\2\end{bmatrix}\right\}$. For $\lambda_2 = 1$, we get

$$A - I = \begin{bmatrix}-1 & 2\\2 & -4\end{bmatrix} \sim \begin{bmatrix}1 & -2\\0 & 0\end{bmatrix}$$

Thus, a basis for its eigenspace is $\left\{\begin{bmatrix}2\\1\end{bmatrix}\right\}$. Thus, taking $P = \begin{bmatrix}-1/\sqrt{5} & 2/\sqrt{5}\\2/\sqrt{5} & 1/\sqrt{5}\end{bmatrix}$ and $\vec{x} = P\vec{y}$, we get the diagonal form of $Q(\vec{x})$ is $Q(\vec{x}) = -4y_1^2 + y_2^2$.

4. (a) The corresponding symmetric matrix is $A = \begin{bmatrix}4 & 8\\8 & 4\end{bmatrix}$. We have $C(\lambda) = (\lambda-12)(\lambda + 4)$. Thus, the eigenvalues are $\lambda_1 = 12$ and $\lambda_2 = -4$. Thus, $Q(\vec{x})$ is indefinite.

(b) The corresponding symmetric matrix is $A = \begin{bmatrix}2 & -3 & -3\\-3 & 3 & 2\\-3 & 2 & 3\end{bmatrix}$. We have $C(\lambda) = -(\lambda-1)(\lambda+1)(\lambda-8)$. Thus, the eigenvalues are $\lambda_1 = 1$, $\lambda_2 = -1$, and $\lambda_3 = 8$. Since $Q(\vec{x})$ has both positive and negative eigenvalues, it is indefinite.

Section 8.3

1. The corresponding symmetric matrix is $A = \begin{bmatrix} 1 & 1 \\ 1 & 1 \end{bmatrix}$. We have $C(\lambda) = \lambda(\lambda - 2)$. Thus, the eigenvalues are $\lambda_1 = 0$ and $\lambda_2 = 2$. For $\lambda_1 = 0$, we get

$$A - 0I = \begin{bmatrix} 1 & 1 \\ 1 & 1 \end{bmatrix} \sim \begin{bmatrix} 1 & 1 \\ 0 & 0 \end{bmatrix}$$

Thus, a basis for its eigenspace is $\left\{ \begin{bmatrix} -1 \\ 1 \end{bmatrix} \right\}$. For $\lambda_2 = 2$, we get

$$A - 2I = \begin{bmatrix} -1 & 1 \\ 1 & -1 \end{bmatrix} \sim \begin{bmatrix} 1 & -1 \\ 0 & 0 \end{bmatrix}$$

Thus, a basis for its eigenspace is $\left\{ \begin{bmatrix} 1 \\ 1 \end{bmatrix} \right\}$. We take $\begin{bmatrix} -1 \\ 1 \end{bmatrix}$ to define the y_1-axis and $\begin{bmatrix} 1 \\ 1 \end{bmatrix}$ to define the y_2-axis. Then the original equation is equivalent to $0y_1^2 + 2y_2^2 = 2$. We get the pair of parallel lines depicted below.

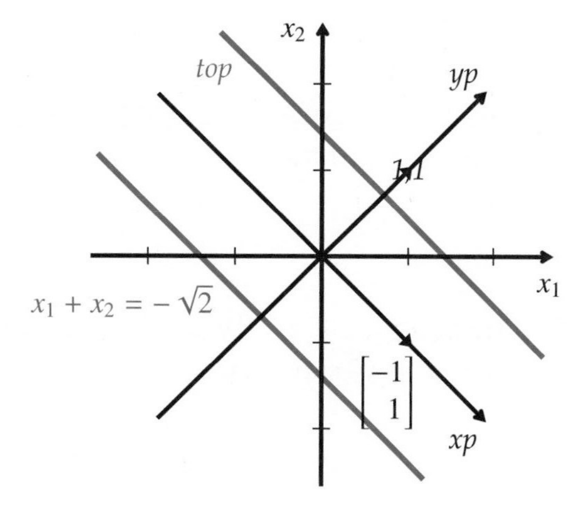

CHAPTER 9

Section 9.1

1. (a) $3 + i$ (b) $-2 + 2i$ (c) $5 - 5i$ (d) 13

2. (a) $\overline{\overline{z_1}} = \overline{x - yi} = x + yi = z_1$

(b) $x + iy = z_1 = \overline{z_1} = x - iy$ if and only if $y = 0$.

(c) Let $z_2 = a + ib$. Then

$$\overline{z_1 + z_2} = \overline{x + iy + a + ib} = \overline{x + a + i(y + b)} = x+a-i(y+b) = x-iy+a-ib = \overline{z_1} + \overline{z_2}$$

3. (a) $\dfrac{1+i}{1-i} = \dfrac{1+i}{1-i}\dfrac{1+i}{1+i} = \dfrac{2i}{2} = i$

(b) $\dfrac{2i}{1+i} = \dfrac{2i}{1+i}\dfrac{1-i}{1-i} = \dfrac{2+2i}{2} = 1+i$

(c) $\dfrac{4-i}{1+5i} = \dfrac{4-i}{1+5i}\dfrac{1-5i}{1-5i} = -\dfrac{1}{26} - \dfrac{21}{26}i$

4.

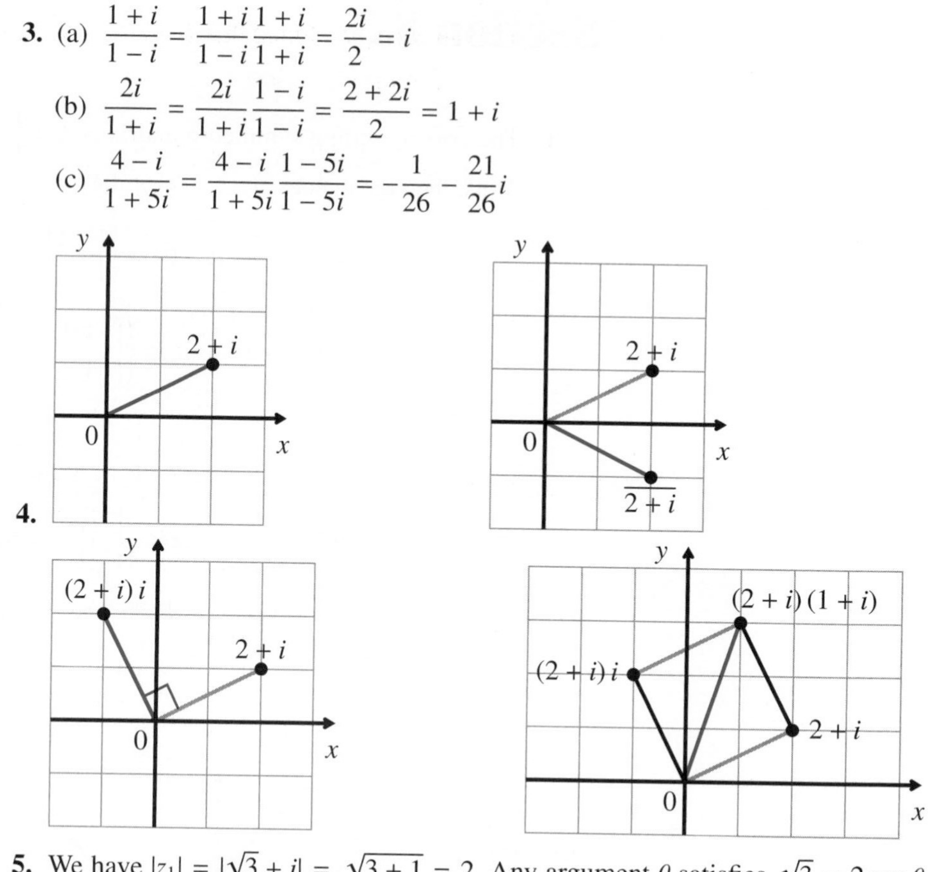

5. We have $|z_1| = |\sqrt{3}+i| = \sqrt{3+1} = 2$. Any argument θ satisfies $\sqrt{3} = 2\cos\theta$ and $1 = 2\sin\theta$. Thus, $\theta = \frac{\pi}{6} + 2\pi k$, $k \in \mathbb{Z}$. Hence,

$$z_1 = 2\left(\cos\frac{\pi}{6} + i\sin\frac{\pi}{6}\right)$$

We have $|z_2| = |-1-i| = \sqrt{1+1} = \sqrt{2}$. Any argument θ satisfies $-1 = \sqrt{2}\cos\theta$ and $-1 = \sqrt{2}\sin\theta$. Thus, $\theta = \frac{5\pi}{4} + 2\pi k$, $k \in \mathbb{Z}$. Hence,

$$z_2 = \sqrt{2}\left(\cos\frac{5\pi}{4} + i\sin\frac{5\pi}{4}\right)$$

6. We have

$$|\overline{z}| = |r\cos\theta - r\sin\theta| = \sqrt{r^2\cos^2\theta + r^2\sin^2\theta} = \sqrt{r^2} = |r| = |z|$$

Using the trigonometric identities $\cos\theta = \cos(-\theta)$ and $-\sin\theta = \sin(-\theta)$ gives

$$\overline{z} = r\cos\theta - r\sin\theta = r\cos(-\theta) + r\sin(-\theta) = r(\cos(-\theta) + \sin(-\theta))$$

Hence, an argument of $\overline{z}$ is $-\theta$.

7. Theorem 3 says the modulus of a quotient is the quotient of the moduli of the factors, while the argument of the quotient is the difference of the arguments. Taking $z_1 = 1 = 1(\cos 0 + i\sin 0)$ and $z_2 = z$ in Theorem 3 gives

$$\frac{1}{z_2} = \frac{1}{r}(\cos(0-\theta) + i\sin(0-\theta)) = \frac{1}{r}(\cos(-\theta) + i\sin(-\theta))$$

8. In Example 5, we found that $2 - 2i = 2\sqrt{2}\left(\cos\frac{-\pi}{4} + i\sin\frac{-\pi}{4}\right)$ and $-1 + \sqrt{3}i = 2\left(\cos\frac{2\pi}{3} + i\sin\frac{2\pi}{3}\right)$. Hence,

$$\begin{aligned}(2-2i)(-1+\sqrt{3}i) &= 4\sqrt{2}\left(\cos\left(-\frac{\pi}{4}+\frac{2\pi}{3}\right) + i\sin\left(-\frac{\pi}{4}+\frac{2\pi}{3}\right)\right)\\ &= 4\sqrt{2}\left(\cos\frac{5\pi}{12} + i\sin\frac{5\pi}{12}\right) = 1.464 + i(5.464)\\ \frac{2-2i}{-1+\sqrt{3}i} &= \frac{2\sqrt{2}}{2}\left(\cos\left(-\frac{\pi}{4}-\frac{2\pi}{3}\right) + i\sin\left(-\frac{\pi}{4}-\frac{2\pi}{3}\right)\right)\\ &= \sqrt{2}\left(\cos\frac{-11\pi}{12} + i\sin\frac{-11\pi}{12}\right) = -1.366 - i(0.366)\end{aligned}$$

9. We have $(1-i) = \sqrt{2}\left(\cos\frac{-\pi}{4} + i\sin\frac{-\pi}{4}\right)$, so

$$(1-i)^5 = 2^{5/2}\left(\cos\frac{-5\pi}{4} + i\sin\frac{-5\pi}{4}\right) = -4 + 4i$$

We have $(-1 - \sqrt{3}i) = 2\left(\cos\frac{4\pi}{3} + i\sin\frac{4\pi}{3}\right)$, so

$$(-1-\sqrt{3}i)^5 = 2^5\left(\cos\frac{20\pi}{3} + i\sin\frac{20\pi}{3}\right) = -16 + 16\sqrt{3}i$$

Section 9.2

1. We row reduce the corresponding augmented matrix to get

$$\left[\begin{array}{ccc|c} i & 1 & 3 & -1-2i \\ i & 1 & 1+2i & 2+i \\ 2 & 1+i & 2 & 5-i \end{array}\right] \sim \left[\begin{array}{ccc|c} 1 & 0 & 0 & \frac{5}{2}+i \\ 0 & 1 & 0 & 0 \\ 0 & 0 & 1 & -\frac{3}{2}i \end{array}\right]$$

Thus, the only solution is $z_1 = \frac{5}{2} + i$, $z_2 = 0$, and $z_3 = -\frac{3}{2}i$.

Section 9.3

1. All 10 vector spaces axioms are easily verified.

2. The reduced row echelon form of A is $\begin{bmatrix} 1 & 0 & 1 \\ 0 & 1 & -i \\ 0 & 0 & 0 \\ 0 & 0 & 0 \end{bmatrix}$. Hence, a basis for Row(A) is $\left\{\begin{bmatrix} 1 \\ 0 \\ 1 \end{bmatrix}, \begin{bmatrix} 0 \\ 1 \\ -i \end{bmatrix}\right\}$.

A basis for Col(A) is $\left\{\begin{bmatrix} 1 \\ 1 \\ 1+i \\ -i \end{bmatrix}, \begin{bmatrix} -1 \\ 1 \\ 1 \\ 1-2i \end{bmatrix}\right\}$ A basis for Null(A) is $\left\{\begin{bmatrix} -1 \\ i \\ 1 \end{bmatrix}\right\}$.

Section 9.4

1. For $\lambda_1 = 2 + 3i$, we have $A - \lambda_1 I = \begin{bmatrix} 3-3i & -6 \\ 3 & -3-3i \end{bmatrix} \sim \begin{bmatrix} 1-i & -2 \\ 0 & 0 \end{bmatrix}$.

An eigenvector corresponding to λ is $\begin{bmatrix} 2 \\ 1-i \end{bmatrix} = \begin{bmatrix} 2 \\ 1 \end{bmatrix} + i\begin{bmatrix} 0 \\ -1 \end{bmatrix}$. So, $P = \begin{bmatrix} 2 & 0 \\ 1 & -1 \end{bmatrix}$.

2. $\det(A - \lambda I) = \lambda^2 - 6\lambda + 11$; So, $\lambda_1 = 3 + \sqrt{2}i$. Hence, a real canonical form of A is $C = \begin{bmatrix} 3 & \sqrt{2} \\ -\sqrt{2} & 3 \end{bmatrix}$.

We have $A - \lambda_1 I = \begin{bmatrix} -\sqrt{2}i & 2 \\ -1 & -\sqrt{2}i \end{bmatrix} \sim \begin{bmatrix} 1 & \sqrt{2}i \\ 0 & 0 \end{bmatrix}$.

An eigenvector corresponding to λ is $\begin{bmatrix} -\sqrt{2}i \\ 1 \end{bmatrix} = \begin{bmatrix} 0 \\ 1 \end{bmatrix} + i\begin{bmatrix} -\sqrt{2} \\ 0 \end{bmatrix}$. So,

$P = \begin{bmatrix} 0 & -\sqrt{2} \\ 1 & 0 \end{bmatrix}$.

Section 9.5

1. We have

$$\begin{aligned} \langle \vec{u}, \vec{v} \rangle &= i(2-2i) + (1+2i)(1+3i) = -3 + 7i \\ \langle 2i\vec{u}, \vec{v} \rangle &= 2i\langle \vec{u}, \vec{v} \rangle = -14 - 6i \\ \langle \vec{u}, 2i\vec{v} \rangle &= \overline{2i}\langle \vec{u}, \vec{v} \rangle = -2i(-3+7i) = 14 + 6i \end{aligned}$$

2. We have

$$\langle \vec{z}, \vec{z} \rangle = \vec{z} \cdot \overline{\vec{z}} = z_1\overline{z_1} + \cdots + z_n\overline{z_n} = |z_1|^2 + \cdots + |z_n|^2$$

Thus, $\langle \vec{z}, \vec{z} \rangle \geq 0$ for all $\vec{z}$ and equal to 0 if and only if $\vec{z} = \vec{0}$:

$$\begin{aligned} \langle \vec{z}, \vec{w} \rangle &= \vec{z} \cdot \overline{\vec{w}} = \vec{z}^T\overline{\vec{w}} = (\vec{z}^T\overline{\vec{w}})^T = \overline{\vec{w}}^T\vec{z} = \overline{\vec{w}^T\overline{\vec{z}}} = \overline{\langle \vec{w}, \vec{z} \rangle} \\ \langle \vec{v} + \vec{z}, \vec{w} \rangle &= (\vec{v} + \vec{z})^T\overline{\vec{w}} = (\vec{v}^T + \vec{z}^T)\overline{\vec{w}} = \vec{v}^T\overline{\vec{w}} + \vec{z}^T\overline{\vec{w}} = \langle \vec{v}, \vec{w} \rangle + \langle \vec{z}, \vec{w} \rangle \\ \langle \vec{z}, \vec{w} + \vec{u} \rangle &= \vec{z}^T\overline{\vec{w} + \vec{u}} = \vec{z}^T(\overline{\vec{w}} + \overline{\vec{u}}) = \vec{z}^T\overline{\vec{w}} + \vec{z}^T\overline{\vec{u}} = \langle \vec{z}, \vec{w} \rangle + \langle \vec{z}, \vec{u} \rangle \\ \langle \alpha\vec{z}, \vec{w} \rangle &= (\alpha\vec{z})^T\overline{\vec{w}} = \alpha\vec{z}^T\overline{\vec{w}} = \alpha\langle \vec{z}, \vec{w} \rangle \\ \langle \vec{z}, \alpha\vec{w} \rangle &= \vec{z}^T\overline{\alpha\vec{w}} = \overline{\alpha}\vec{z}^T\overline{\vec{w}} = \overline{\alpha}\langle \vec{z}, \vec{w} \rangle \end{aligned}$$

3. The columns of A are not orthogonal under the standard complex inner product, so A is not unitary. We have $B^*B = I$, so B is unitary.

APPENDIX B

Answers to Practice Problems and Chapter Quizzes

CHAPTER 1

Section 1.1

A Problems

A1

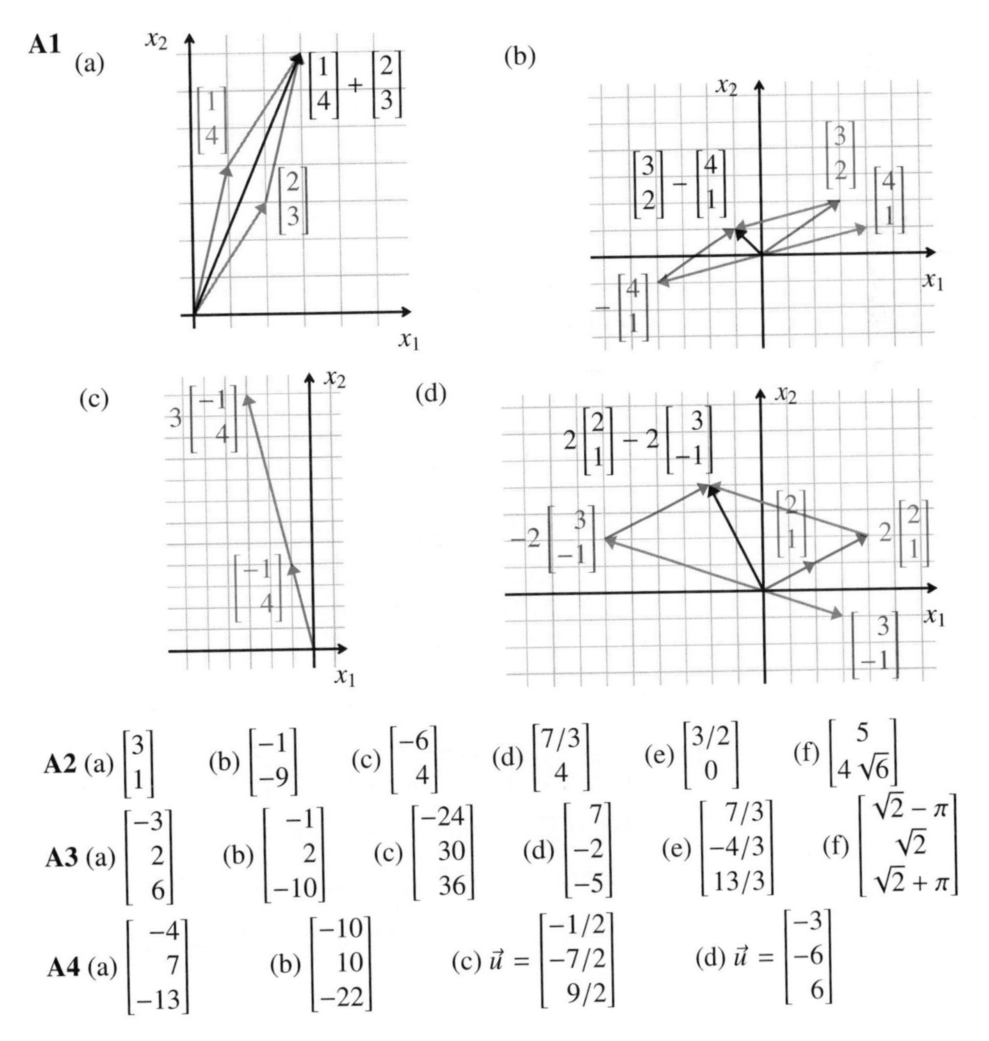

A2 (a) $\begin{bmatrix}3\\1\end{bmatrix}$ (b) $\begin{bmatrix}-1\\-9\end{bmatrix}$ (c) $\begin{bmatrix}-6\\4\end{bmatrix}$ (d) $\begin{bmatrix}7/3\\4\end{bmatrix}$ (e) $\begin{bmatrix}3/2\\0\end{bmatrix}$ (f) $\begin{bmatrix}5\\4\sqrt{6}\end{bmatrix}$

A3 (a) $\begin{bmatrix}-3\\2\\6\end{bmatrix}$ (b) $\begin{bmatrix}-1\\2\\-10\end{bmatrix}$ (c) $\begin{bmatrix}-24\\30\\36\end{bmatrix}$ (d) $\begin{bmatrix}7\\-2\\-5\end{bmatrix}$ (e) $\begin{bmatrix}7/3\\-4/3\\13/3\end{bmatrix}$ (f) $\begin{bmatrix}\sqrt{2}-\pi\\\sqrt{2}\\\sqrt{2}+\pi\end{bmatrix}$

A4 (a) $\begin{bmatrix}-4\\7\\-13\end{bmatrix}$ (b) $\begin{bmatrix}-10\\10\\-22\end{bmatrix}$ (c) $\vec{u}=\begin{bmatrix}-1/2\\-7/2\\9/2\end{bmatrix}$ (d) $\vec{u}=\begin{bmatrix}-3\\-6\\6\end{bmatrix}$

A5 (a) $\begin{bmatrix} 4 \\ 0 \\ -1/2 \end{bmatrix}$ (b) $\begin{bmatrix} 25 \\ -5 \\ -10 \end{bmatrix}$ (c) $\begin{bmatrix} -1 \\ -3 \\ -4 \end{bmatrix}$ (d) $\begin{bmatrix} 8 \\ -8/3 \\ -14/3 \end{bmatrix}$

A6

$$\vec{PQ} = \vec{OQ} - \vec{OP} = \begin{bmatrix} 3 \\ 1 \\ -2 \end{bmatrix} - \begin{bmatrix} 2 \\ 3 \\ 1 \end{bmatrix} = \begin{bmatrix} 1 \\ -2 \\ -3 \end{bmatrix}$$

$$\vec{PR} = \vec{OR} - \vec{OP} = \begin{bmatrix} 1 \\ 4 \\ 0 \end{bmatrix} - \begin{bmatrix} 2 \\ 3 \\ 1 \end{bmatrix} = \begin{bmatrix} -1 \\ 1 \\ -1 \end{bmatrix}$$

$$\vec{PS} = \vec{OS} - \vec{OP} = \begin{bmatrix} -5 \\ 1 \\ 5 \end{bmatrix} - \begin{bmatrix} 2 \\ 3 \\ 1 \end{bmatrix} = \begin{bmatrix} -7 \\ -2 \\ 4 \end{bmatrix}$$

$$\vec{QR} = \vec{OR} - \vec{OQ} = \begin{bmatrix} 1 \\ 4 \\ 0 \end{bmatrix} - \begin{bmatrix} 3 \\ 1 \\ -2 \end{bmatrix} = \begin{bmatrix} -2 \\ 3 \\ 2 \end{bmatrix}$$

$$\vec{SR} = \vec{OR} - \vec{OS} = \begin{bmatrix} 1 \\ 4 \\ 0 \end{bmatrix} - \begin{bmatrix} -5 \\ 1 \\ 5 \end{bmatrix} = \begin{bmatrix} 6 \\ 3 \\ -5 \end{bmatrix}$$

$$\vec{PQ} + \vec{QR} = \begin{bmatrix} 1 \\ -2 \\ -3 \end{bmatrix} + \begin{bmatrix} -2 \\ 3 \\ 2 \end{bmatrix} = \begin{bmatrix} -1 \\ 1 \\ -1 \end{bmatrix} = \begin{bmatrix} -7 \\ -2 \\ 4 \end{bmatrix} + \begin{bmatrix} 6 \\ 3 \\ -5 \end{bmatrix} = \vec{PS} + \vec{SR}$$

A7 (a) $\vec{x} = \begin{bmatrix} 3 \\ 4 \end{bmatrix} + t\begin{bmatrix} -5 \\ 1 \end{bmatrix}, \quad t \in \mathbb{R}$ (b) $\vec{x} = \begin{bmatrix} 2 \\ 3 \end{bmatrix} + t\begin{bmatrix} -4 \\ -6 \end{bmatrix}, \quad t \in \mathbb{R}$

(c) $\vec{x} = \begin{bmatrix} 2 \\ 0 \\ 5 \end{bmatrix} + t\begin{bmatrix} 4 \\ -2 \\ -11 \end{bmatrix}, \quad t \in \mathbb{R}$ (d) $\vec{x} = \begin{bmatrix} 4 \\ 1 \\ 5 \end{bmatrix} + t\begin{bmatrix} -2 \\ 1 \\ 2 \end{bmatrix}, \quad t \in \mathbb{R}$

A8 Note that alternative correct answers are possible.

(a) $\vec{x} = \begin{bmatrix} -1 \\ 2 \end{bmatrix} + t\begin{bmatrix} 3 \\ -5 \end{bmatrix}, \quad t \in \mathbb{R}$ (b) $\vec{x} = \begin{bmatrix} 4 \\ 1 \end{bmatrix} + t\begin{bmatrix} -6 \\ -2 \end{bmatrix}, \quad t \in \mathbb{R}$

(c) $\vec{x} = \begin{bmatrix} 1 \\ 3 \\ -5 \end{bmatrix} + t\begin{bmatrix} -3 \\ -2 \\ 5 \end{bmatrix}, \quad t \in \mathbb{R}$ (d) $\vec{x} = \begin{bmatrix} -2 \\ 1 \\ 1 \end{bmatrix} + t\begin{bmatrix} 6 \\ 1 \\ 1 \end{bmatrix}, \quad t \in \mathbb{R}$

(e) $\vec{x} = \begin{bmatrix} 1/2 \\ 1/4 \\ 1 \end{bmatrix} + t\begin{bmatrix} -3/2 \\ 3/4 \\ -2/3 \end{bmatrix}, \quad t \in \mathbb{R}$

A9 (a) $x_1 = -1 + t,\ x_2 = -1 + 3t, \quad t \in \mathbb{R};\ \vec{x} = \begin{bmatrix} -1 \\ -1 \end{bmatrix} + t\begin{bmatrix} 1 \\ 3 \end{bmatrix}, \quad t \in \mathbb{R}$

(b) $x_1 = 1 + 3t,\ x_2 = 1 - 2t, \quad t \in \mathbb{R};\ \vec{x} = \begin{bmatrix} 1 \\ 1 \end{bmatrix} + t\begin{bmatrix} 3 \\ -2 \end{bmatrix}, \quad t \in \mathbb{R}$

A10 (a) Three points P, Q, and R are collinear if $\vec{PQ} = t\vec{PR}$ for some $t \in \mathbb{R}$.

(b) Since $-2\vec{PQ} = \begin{bmatrix} -6 \\ 2 \end{bmatrix} = \vec{PR}$, the points P, Q, and R must be collinear.

(c) The points S, T, and U are not collinear because $\vec{SU} \neq t\vec{ST}$ for any t.

Section 1.2

A Problems

A1 (a) $\begin{bmatrix} 5 \\ 9 \\ 0 \\ 1 \end{bmatrix}$ (b) $\begin{bmatrix} 10 \\ -7 \\ 10 \\ -5 \end{bmatrix}$ (c) $\begin{bmatrix} 0 \\ 0 \\ 1 \\ 1 \\ 3 \end{bmatrix}$

A2 (a) The set is not a subspace of $\mathbb{R}^3$.

(b) The set is a subspace of $\mathbb{R}^3$.

(c) The set is a subspace of $\mathbb{R}^2$.

(d) The set is not a subspace of $\mathbb{R}^3$.

(e) The set is a subspace of $\mathbb{R}^3$.

(f) The set is a subspace of $\mathbb{R}^4$.

A3 (a) The set is a subspace of $\mathbb{R}^4$.

(b) The set is not a subspace of $\mathbb{R}^4$.

(c) The set is not a subspace of $\mathbb{R}^4$.

(d) The set is not a subspace of $\mathbb{R}^4$.

(e) The set is a subspace of $\mathbb{R}^4$.

(f) The set is not a subspace of $\mathbb{R}^4$.

A4 Alternative correct answers are possible.

(a) $1\begin{bmatrix} 0 \\ 0 \\ 0 \end{bmatrix} + 0\begin{bmatrix} 1 \\ 0 \\ 1 \end{bmatrix} + 0\begin{bmatrix} 2 \\ 1 \\ -1 \end{bmatrix} = \begin{bmatrix} 0 \\ 0 \\ 0 \end{bmatrix}$

(b) $0\begin{bmatrix} 2 \\ -1 \\ 3 \end{bmatrix} - 2\begin{bmatrix} 0 \\ 2 \\ 1 \end{bmatrix} + 1\begin{bmatrix} 0 \\ 4 \\ 2 \end{bmatrix} = \begin{bmatrix} 0 \\ 0 \\ 0 \end{bmatrix}$

(c) $1\begin{bmatrix} 1 \\ 1 \\ 0 \end{bmatrix} + 1\begin{bmatrix} 1 \\ 1 \\ 1 \end{bmatrix} - 1\begin{bmatrix} 2 \\ 2 \\ 1 \end{bmatrix} = \begin{bmatrix} 0 \\ 0 \\ 0 \end{bmatrix}$

(d) $1\begin{bmatrix} 1 \\ 1 \end{bmatrix} - 2\begin{bmatrix} 1 \\ 2 \end{bmatrix} + 1\begin{bmatrix} 1 \\ 3 \end{bmatrix} = \begin{bmatrix} 0 \\ 0 \end{bmatrix}$

A5 Alternative correct answers are possible.

(a) The plane in $\mathbb{R}^4$ with basis $\left\{\begin{bmatrix} 1 \\ 0 \\ 1 \\ 1 \end{bmatrix}, \begin{bmatrix} 1 \\ 2 \\ 1 \\ 3 \end{bmatrix}\right\}$

(b) The hyperplane in $\mathbb{R}^4$ with basis $\left\{\begin{bmatrix} 1 \\ 0 \\ 0 \\ 0 \end{bmatrix}, \begin{bmatrix} 0 \\ 1 \\ 0 \\ 0 \end{bmatrix}, \begin{bmatrix} 0 \\ 0 \\ 0 \\ 1 \end{bmatrix}\right\}$

(c) The line in $\mathbb{R}^4$ with basis $\left\{\begin{bmatrix}3\\1\\-1\\0\end{bmatrix}\right\}$

(d) The plane in $\mathbb{R}^4$ with basis $\left\{\begin{bmatrix}1\\1\\0\\2\end{bmatrix}, \begin{bmatrix}1\\0\\0\\-1\end{bmatrix}\right\}$

A6 If $\vec{x} = \vec{p} + t\vec{d}$ is a subspace of $\mathbb{R}^n$, then it contains the zero vector. Hence, there exists t_1 such that $\vec{0} = \vec{p} + t_1\vec{d}$. Thus, $\vec{p} = -t_1\vec{d}$ and so $\vec{p}$ is a scalar multiple of $\vec{d}$. On the other hand, if $\vec{p}$ is a scalar multiple of $\vec{d}$, say $\vec{p} = t_1\vec{d}$, then we have $\vec{x} = \vec{p} + t\vec{d} = t_1\vec{d} + t\vec{d} = (t_1 + t)\vec{d}$. Hence, the set is Span$\{\vec{d}\}$ and thus is a subspace.

A7 Assume that there is a non-empty subset $\mathcal{B}_1 = \{\vec{v}_1, \dots, \vec{v}_\ell\}$ of $\mathcal{B}$ that is linearly dependent. Then there exists c_i not all zero such that

$$\vec{0} = c_1\vec{v}_1 + \cdots + c_\ell\vec{v}_\ell = c_1\vec{v}_1 + \cdots + c_\ell\vec{v}_\ell + 0\vec{v}_{\ell+1} + \cdots + 0\vec{v}_n$$

This contradicts the fact that $\mathcal{B}$ is linearly independent. Hence, $\mathcal{B}_1$ must be linearly independent.

Section 1.3

A Problems

A1 (a) $\sqrt{29}$ (b) 1 (c) $\sqrt{2}$ (d) $\sqrt{17}$
(e) $\sqrt{251}/5$ (f) 1 (g) $\sqrt{6}$ (h) 1

A2 (a) $\begin{bmatrix}3/5\\-4/5\end{bmatrix}$ (b) $\begin{bmatrix}1/\sqrt{2}\\1/\sqrt{2}\end{bmatrix}$ (c) $\begin{bmatrix}-1/\sqrt{5}\\0\\2/\sqrt{5}\end{bmatrix}$

(d) $\begin{bmatrix}0\\-1\\0\end{bmatrix}$ (e) $\begin{bmatrix}-2/3\\-2/3\\1/3\\0\end{bmatrix}$ (f) $\begin{bmatrix}1/\sqrt{2}\\0\\0\\-1/\sqrt{2}\end{bmatrix}$

A3 (a) $2\sqrt{10}$ (b) 5 (c) $\sqrt{170}$ (d) $3\sqrt{6}$

A4 (a) $\|\vec{x}\| = \sqrt{26}$; $\|\vec{y}\| = \sqrt{30}$; $\|\vec{x} + \vec{y}\| = 2\sqrt{22}$; $|\vec{x} \cdot \vec{y}| = 16$; the triangle inequality: $2\sqrt{22} \approx 9.38 \leq \sqrt{26} + \sqrt{30} \approx 10.58$; the Cauchy-Schwarz inequality: $16 \leq \sqrt{26(30)} \approx 27.93$.

(b) $\|\vec{x}\| = \sqrt{6}$; $\|\vec{y}\| = \sqrt{29}$; $\|\vec{x} + \vec{y}\| = \sqrt{41}$; $|\vec{x} \cdot \vec{y}| = 3$; the triangle inequality: $\sqrt{41} \approx 6.40 \leq \sqrt{6} + \sqrt{29} \approx 7.83$; the Cauchy-Schwarz inequality: $3 \leq \sqrt{6(29)} \approx 13.19$.

A5 (a) $\begin{bmatrix}1\\3\\2\end{bmatrix} \cdot \begin{bmatrix}2\\-2\\2\end{bmatrix} = 0$; these vectors are orthogonal.

(b) $\begin{bmatrix}-3\\1\\7\end{bmatrix} \cdot \begin{bmatrix}2\\-1\\1\end{bmatrix} = 0$; these vectors are orthogonal.

(c) $\begin{bmatrix}2\\1\\1\end{bmatrix} \cdot \begin{bmatrix}-1\\4\\2\end{bmatrix} = 4 \neq 0$; these vectors are not orthogonal.

(d) $\begin{bmatrix}4\\1\\0\\-2\end{bmatrix} \cdot \begin{bmatrix}-1\\4\\3\\0\end{bmatrix} = 0$; these vectors are orthogonal.

(e) $\begin{bmatrix}0\\0\\0\\0\end{bmatrix} \cdot \begin{bmatrix}x_1\\x_2\\x_3\\x_4\end{bmatrix} = 0$; these vectors are orthogonal.

(f) $\begin{bmatrix}1/3\\2/3\\-1/3\\3\end{bmatrix} \cdot \begin{bmatrix}3/2\\0\\-3/2\\1\end{bmatrix} = 4$; these vectors are not orthogonal.

A6 (a) $k = 6$ (b) $k = 0$ or $k = 3$ (c) $k = -3$ (d) any $k \in \mathbb{R}$

A7 (a) $2x_1 + 4x_2 - x_3 = 9$ (b) $3x_1 + 5x_3 = 26$
(c) $3x_1 - 4x_2 + x_3 = 8$ (d) $-4x_1 - 2x_2 - 2x_3 = -12$

A8 (a) $3x_1 + x_2 + 4x_3 = 0$ (b) $x_2 + 3x_3 + 3x_4 = 1$
(c) $x_1 - 4x_2 + 5x_3 - 2x_4 = 0$ (d) $x_2 + 2x_3 - x_4 + x_5 = 1$

A9 (a) $\vec{n} = \begin{bmatrix}2\\1\end{bmatrix}$ (b) $\vec{n} = \begin{bmatrix}3\\-2\\3\end{bmatrix}$ (c) $\vec{n} = \begin{bmatrix}-4\\3\\-5\end{bmatrix}$

(d) $\vec{n} = \begin{bmatrix}1\\-1\\2\\-3\end{bmatrix}$ (e) $\vec{n} = \begin{bmatrix}1\\1\\-1\\2\\-1\end{bmatrix}$

A10 (a) $2x_1 - 3x_2 + 5x_3 = 6$

(b) $x_2 = -2$

(c) $x_1 - x_2 + 3x_3 = 2$

A11 (a) False. One possible counterexample is $\begin{bmatrix}1\\0\end{bmatrix} \cdot \begin{bmatrix}2\\2\end{bmatrix} = 2 = \begin{bmatrix}1\\0\end{bmatrix} \cdot \begin{bmatrix}2\\-97\end{bmatrix}$.

(b) Our counterexample in part (a) has $\vec{u} \neq \vec{0}$, so the result does not change.

Section 1.4

A Problems

A1 (a) $\text{proj}_{\vec{v}}\,\vec{u} = \begin{bmatrix}0\\-5\end{bmatrix}$, $\text{perp}_{\vec{v}}\,\vec{u} = \begin{bmatrix}3\\0\end{bmatrix}$

(b) $\text{proj}_{\vec{v}}\,\vec{u} = \begin{bmatrix}36/25\\48/25\end{bmatrix}$, $\text{perp}_{\vec{v}}\,\vec{u} = \begin{bmatrix}-136/25\\102/25\end{bmatrix}$

(c) $\text{proj}_{\vec{v}}\,\vec{u} = \begin{bmatrix} 0 \\ 5 \\ 0 \end{bmatrix}$, $\text{perp}_{\vec{v}}\,\vec{u} = \begin{bmatrix} -3 \\ 0 \\ 2 \end{bmatrix}$

(d) $\text{proj}_{\vec{v}}\,\vec{u} = \begin{bmatrix} -4/9 \\ 8/9 \\ -8/9 \end{bmatrix}$, $\text{perp}_{\vec{v}}\,\vec{u} = \begin{bmatrix} 40/9 \\ 1/9 \\ -19/9 \end{bmatrix}$

(e) $\text{proj}_{\vec{v}}\,\vec{u} = \begin{bmatrix} 0 \\ 0 \\ 0 \\ 0 \end{bmatrix}$, $\text{perp}_{\vec{v}}\,\vec{u} = \begin{bmatrix} -1 \\ -1 \\ 2 \\ -1 \end{bmatrix}$

(f) $\text{proj}_{\vec{v}}\,\vec{u} = \begin{bmatrix} -1/2 \\ 0 \\ 0 \\ -1/2 \end{bmatrix}$, $\text{perp}_{\vec{v}}\,\vec{u} = \begin{bmatrix} 5/2 \\ 3 \\ 2 \\ -5/2 \end{bmatrix}$

A2 (a) $\text{proj}_{\vec{v}}\,\vec{u} = \begin{bmatrix} 0 \\ 0 \end{bmatrix}$, $\text{perp}_{\vec{v}}\,\vec{u} = \begin{bmatrix} 3 \\ -3 \end{bmatrix}$

(b) $\text{proj}_{\vec{v}}\,\vec{u} = \begin{bmatrix} -2/17 \\ -3/17 \\ 2/17 \end{bmatrix}$, $\text{perp}_{\vec{v}}\,\vec{u} = \begin{bmatrix} 70/17 \\ -14/17 \\ 49/17 \end{bmatrix}$

(c) $\text{proj}_{\vec{v}}\,\vec{u} = \begin{bmatrix} 14/3 \\ -7/3 \\ 7/3 \end{bmatrix}$, $\text{perp}_{\vec{v}}\,\vec{u} = \begin{bmatrix} 1/3 \\ 4/3 \\ 2/3 \end{bmatrix}$

(d) $\text{proj}_{\vec{v}}\,\vec{u} = \begin{bmatrix} 3/2 \\ 3/2 \\ -3 \end{bmatrix}$, $\text{perp}_{\vec{v}}\,\vec{u} = \begin{bmatrix} 5/2 \\ -1/2 \\ 1 \end{bmatrix}$

(e) $\text{proj}_{\vec{v}}\,\vec{u} = \begin{bmatrix} 1/3 \\ -2/3 \\ -1/3 \\ 1 \end{bmatrix}$, $\text{perp}_{\vec{v}}\,\vec{u} = \begin{bmatrix} 5/3 \\ -1/3 \\ 7/3 \\ 0 \end{bmatrix}$

A3 (a) $\hat{u} = \frac{\vec{u}}{\|\vec{u}\|} = \begin{bmatrix} 2/7 \\ 6/7 \\ 3/7 \end{bmatrix}$

(b) $\text{proj}_{\vec{u}}\,\vec{F} = \begin{bmatrix} 220/49 \\ 660/49 \\ 330/49 \end{bmatrix}$

(c) $\text{perp}_{\vec{u}}\,\vec{F} = \begin{bmatrix} 270/49 \\ 222/49 \\ -624/49 \end{bmatrix}$

A4 (a) $\hat{u} = \frac{\vec{u}}{\|\vec{u}\|} = \begin{bmatrix} 3/\sqrt{14} \\ 1/\sqrt{14} \\ -2/\sqrt{14} \end{bmatrix}$

(b) $\text{proj}_{\vec{u}}\,\vec{F} = \begin{bmatrix} 24/7 \\ 8/7 \\ -16/7 \end{bmatrix}$

(c) $\text{perp}_{\vec{u}}\,\vec{F} = \begin{bmatrix} -3/7 \\ 69/7 \\ 30/7 \end{bmatrix}$

A5 (a) $R(5/2, 5/2)$, $5/\sqrt{2}$

(b) $R(58/17, 91/17)$, $6/\sqrt{17}$

(c) $R(17/6, 1/3, -1/6)$, $\sqrt{29/6}$

(d) $R(5/3, 11/3, -1/3)$, $\sqrt{6}$

A6 (a) $2/\sqrt{26}$

(b) $13/\sqrt{38}$

(c) $4/\sqrt{5}$

(d) $\sqrt{6}$

A7 (a) $R(1/7, 3/7, -3/7, 4/7)$

(b) $R(15/14, 13/7, 17/14, 3)$

(c) $R(0, 14/3, 1/3, 10/3)$

(d) $R(-12/7, 11/7, 9/7, -9/7)$

Section 1.5

A Problems

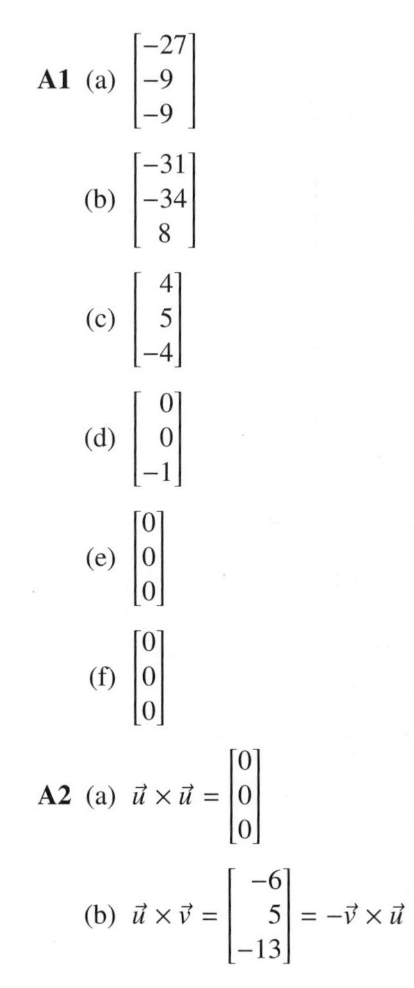

A1 (a) $\begin{bmatrix} -27 \\ -9 \\ -9 \end{bmatrix}$

(b) $\begin{bmatrix} -31 \\ -34 \\ 8 \end{bmatrix}$

(c) $\begin{bmatrix} 4 \\ 5 \\ -4 \end{bmatrix}$

(d) $\begin{bmatrix} 0 \\ 0 \\ -1 \end{bmatrix}$

(e) $\begin{bmatrix} 0 \\ 0 \\ 0 \end{bmatrix}$

(f) $\begin{bmatrix} 0 \\ 0 \\ 0 \end{bmatrix}$

A2 (a) $\vec{u} \times \vec{u} = \begin{bmatrix} 0 \\ 0 \\ 0 \end{bmatrix}$

(b) $\vec{u} \times \vec{v} = \begin{bmatrix} -6 \\ 5 \\ -13 \end{bmatrix} = -\vec{v} \times \vec{u}$

(c) $\vec{u} \times 3\vec{w} = \begin{bmatrix} 6 \\ 9 \\ -15 \end{bmatrix} = 3(\vec{u} \times \vec{w})$

(d) $\vec{u} \times (\vec{v} + \vec{w}) = \begin{bmatrix} -4 \\ 8 \\ -18 \end{bmatrix} = \vec{u} \times \vec{v} + \vec{u} \times \vec{w}$

(e) $\vec{u} \cdot (\vec{v} \times \vec{w}) = -14 = \vec{w} \cdot (\vec{u} \times \vec{v})$

(f) $\vec{u} \cdot (\vec{v} \times \vec{w}) = -14 = -\vec{v} \cdot (\vec{u} \times \vec{w})$

A3 (a) $\sqrt{35}$

(b) $\sqrt{11}$

(c) 9

(d) 13

A4 (a) $x_1 - 4x_2 - 10x_3 = -85$

(b) $2x_1 - 2x_2 + 3x_3 = -5$

(c) $-5x_1 - 2x_2 + 6x_3 = 15$

(d) $-17x_1 - x_2 + 10x_3 = 0$

A5 (a) $39x_1 + 12x_2 + 10x_3 = 140$

(b) $11x_1 - 21x_2 - 17x_3 = -56$

(c) $-12x_1 + 3x_2 - 19x_3 = -14$

(d) $x_2 = 0$

A6 (a) $\vec{x} = \begin{bmatrix} 46/11 \\ 3/11 \\ 0 \end{bmatrix} + t\begin{bmatrix} -2 \\ -3 \\ -11 \end{bmatrix}, \quad t \in \mathbb{R}$

(b) $\vec{x} = \begin{bmatrix} 7/2 \\ 4 \\ 0 \end{bmatrix} + t\begin{bmatrix} 3 \\ -4 \\ 2 \end{bmatrix}, \quad t \in \mathbb{R}$

A7 (a) 1

(b) 126

(c) 5

(d) 35

(e) 16

A8 $\vec{u} \cdot (\vec{v} \times \vec{w}) = 0$ means that $\vec{u}$ is orthogonal to $\vec{v} \times \vec{w}$. Therefore, $\vec{u}$ lies in the plane through the origin that contains $\vec{v}$ and $\vec{w}$. We can also see this by observing that $\vec{u} \cdot (\vec{v} \times \vec{w}) = 0$ means that the parallelepiped determined by $\vec{u}$, $\vec{v}$, and $\vec{w}$ has volume zero; this can happen only if the three vectors lie in a common plane.

A9

$$\begin{aligned}(\vec{u} - \vec{v}) \times (\vec{u} + \vec{v}) &= \vec{u} \times (\vec{u} + \vec{v}) - \vec{v} \times (\vec{u} + \vec{v}) \\ &= \vec{u} \times \vec{u} + \vec{u} \times \vec{v} - \vec{v} \times \vec{u} - \vec{v} \times \vec{v} \\ &= \vec{0} + \vec{u} \times \vec{v} + \vec{u} \times \vec{v} - \vec{0} \\ &= 2(\vec{u} \times \vec{v})\end{aligned}$$

Chapter 1 Quiz

E Problems

E1 $\vec{x} = \begin{bmatrix} -2 \\ 1 \\ -4 \end{bmatrix} + t \begin{bmatrix} 7 \\ -3 \\ 5 \end{bmatrix}, \quad t \in \mathbb{R}$

E2 $8x_1 - x_2 + 7x_3 = 9$

E3 To show that $\left\{ \begin{bmatrix} 1 \\ 2 \end{bmatrix}, \begin{bmatrix} -1 \\ 2 \end{bmatrix} \right\}$ is a basis, we need to show that it spans $\mathbb{R}^2$ and that it is linearly independent.
Consider

$$\begin{bmatrix} x_1 \\ x_2 \end{bmatrix} = t_1 \begin{bmatrix} 1 \\ 2 \end{bmatrix} + t_2 \begin{bmatrix} -1 \\ 2 \end{bmatrix} = \begin{bmatrix} t_1 - t_2 \\ 2t_1 + 2t_2 \end{bmatrix}$$

This gives $x_1 = t_1 - t_2$ and $x_2 = 2t_1 + 2t_2$. Solving using substitution and elimination, we get $t_1 = \frac{1}{4}(2x_1 + x_2)$ and $t_2 = \frac{1}{4}(-2x_1 + x_2)$. Hence, every vector $\begin{bmatrix} x_1 \\ x_2 \end{bmatrix}$ can be written as

$$\begin{bmatrix} x_1 \\ x_2 \end{bmatrix} = \frac{1}{4}(2x_1 + x_2) \begin{bmatrix} 1 \\ 2 \end{bmatrix} + \frac{1}{4}(-2x_1 + x_2) \begin{bmatrix} -1 \\ 2 \end{bmatrix}$$

So, it spans $\mathbb{R}^2$. Moreover, if $x_1 = x_2 = 0$, then our calculations above show that $t_1 = t_2 = 0$, so the set is also linearly independent. Therefore, it is a basis for $\mathbb{R}^2$.

E4 If $d \neq 0$, then $a_1(0) + a_2(0) + a_3(0) = 0 \neq d$, so $\vec{0} \notin S$. Thus, S is not a subspace of $\mathbb{R}^3$.
On the other hand, assume $d = 0$. Observe that, by definition, S is a subset of $\mathbb{R}^3$ and that $\vec{0} = \begin{bmatrix} 0 \\ 0 \\ 0 \end{bmatrix} \in S$ since taking $x_1 = 0$, $x_2 = 0$, and $x_3 = 0$ satisfies $a_1x_1 + a_2x_2 + a_3x_3 = 0$.
Let $\vec{x} = \begin{bmatrix} x_1 \\ x_2 \\ x_3 \end{bmatrix}, \vec{y} = \begin{bmatrix} y_1 \\ y_2 \\ y_3 \end{bmatrix} \in S$. Then they must satisfy the condition of the set, so $a_1x_1 + a_2x_2 + a_3x_3 = 0$ and $a_1y_1 + a_2y_2 + a_3y_3 = 0$.
To show that S is closed under addition, we must show that $\vec{x} + \vec{y}$ satisfies the condition of S. We have $\vec{x} + \vec{y} = \begin{bmatrix} x_1 + y_1 \\ x_2 + y_2 \\ x_3 + y_3 \end{bmatrix}$ and

$$a_1(x_1 + y_1) + a_2(x_2 + y_2) + a_3(x_3 + y_3) = a_1x_1 + a_2x_2 + a_3x_3 + a_1y_1 + a_2y_2 + a_3y_3 = 0 + 0 = 0$$

Hence, $\vec{x} + \vec{y} \in S$. Similarly, for any $t \in \mathbb{R}$, we have $t\vec{x} = \begin{bmatrix} tx_1 \\ tx_2 \\ tx_3 \end{bmatrix}$ and

$$a_1(tx_1) + a_2(tx_2) + a_3(tx_3) = t(a_1x_1 + a_2x_2 + a_3x_3) = t(0) = 0$$

So, S is closed under scalar multiplication. Therefore, S is a subspace of $\mathbb{R}^3$.

E5 The coordinate axes have direction vectors given by the standard basis vectors. The cosine of the angle between $\vec{v}$ and $\vec{e}_1$ is

$$\cos\alpha = \frac{\vec{v}\cdot\vec{e}_1}{\|\vec{v}\|\|\vec{e}_1\|} = \frac{2}{\sqrt{14}}$$

The cosine of the angle between $\vec{v}$ and $\vec{e}_2$ is

$$\cos\beta = \frac{\vec{v}\cdot\vec{e}_2}{\|\vec{v}\|\|\vec{e}_2\|} = \frac{-3}{\sqrt{14}}$$

The cosine of the angle between $\vec{v}$ and $\vec{e}_3$ is

$$\cos\gamma = \frac{\vec{v}\cdot\vec{e}_3}{\|\vec{v}\|\|\vec{e}_3\|} = \frac{1}{\sqrt{14}}$$

E6 Since the origin $O(0,0,0)$ is on the line, we get that the point Q on the line closest to P is given by $\vec{OQ} = \text{proj}_{\vec{d}}\vec{OP}$, where $\vec{d} = \begin{bmatrix} 3 \\ -2 \\ 3 \end{bmatrix}$ is a direction vector of the line.

Hence,

$$\vec{OQ} = \frac{\vec{OP}\cdot\vec{d}}{\|\vec{d}\|^2}\vec{d} = \begin{bmatrix} 18/11 \\ -12/11 \\ 18/11 \end{bmatrix}$$

and the closest point is $Q(18/11, -12/11, 18/11)$.

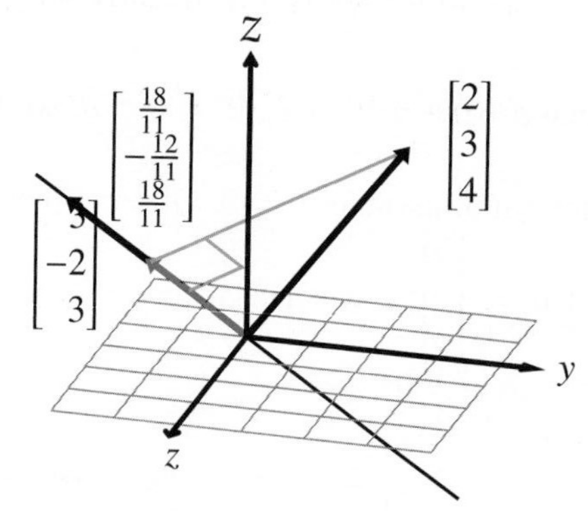

E7 Let $Q(0,0,0,1)$ be a point in the hyperplane. A normal vector to the plane is $\vec{n} = \begin{bmatrix} 1 \\ 1 \\ 1 \\ 1 \end{bmatrix}$. Then, the point R in the hyperplane closest to P satisfies $\vec{PR} = \text{proj}_{\vec{n}}\vec{PQ}$.

Hence,

$$\vec{OR} = \vec{OP} + \text{proj}_{\vec{n}}\vec{PQ} = \begin{bmatrix} 3 \\ -2 \\ 0 \\ 2 \end{bmatrix} - \frac{2}{4}\begin{bmatrix} 1 \\ 1 \\ 1 \\ 1 \end{bmatrix} = \begin{bmatrix} 5/2 \\ -5/2 \\ -1/2 \\ 3/2 \end{bmatrix}$$

Then the distance from the point to the line is the length of $\vec{PR}$:

$$\|\vec{PR}\| = \left\|\begin{bmatrix} -1/2 \\ -1/2 \\ -1/2 \\ -1/2 \end{bmatrix}\right\| = 1$$

E8 A vector orthogonal to both vectors is $\begin{bmatrix}1\\2\\0\end{bmatrix} \times \begin{bmatrix}-3\\1\\1\end{bmatrix} = \begin{bmatrix}2\\-1\\7\end{bmatrix}$.

E9 The volume of the parallelepiped determined by $\vec{u} + k\vec{v}$, $\vec{v}$, and $\vec{w}$ is

$$\begin{aligned}|(\vec{u} + k\vec{v}) \cdot (\vec{v} \times \vec{w})| &= |\vec{u} \cdot (\vec{v} \times \vec{w}) + k(\vec{v} \cdot (\vec{v} \times \vec{w}))| \\ &= |\vec{u} \cdot (\vec{v} \times \vec{w}) + k(0)|\end{aligned}$$

This equals the volume of the parallelepiped determined by $\vec{u}$, $\vec{v}$, and $\vec{w}$.

E10 (i) False. The points $P(0,0,0)$, $Q(0,0,1)$, and $R(0,0,2)$ lie in every plane of the form $t_1x_1 + t_2x_2 = 0$ with t_1 and t_2 not both zero.

(ii) True. This is the definition of a line, reworded in terms of a spanning set.

(iii) True. The set contains the zero vector and hence is linearly dependent.

(iv) False. The dot product of the zero vector with itself is 0.

(v) False. Let $\vec{x} = \begin{bmatrix}1\\0\end{bmatrix}$ and $\vec{y} = \begin{bmatrix}1\\1\end{bmatrix}$. Then, $\text{proj}_{\vec{x}}\, \vec{y} = \begin{bmatrix}1\\0\end{bmatrix}$, while $\text{proj}_{\vec{y}}\, \vec{x} = \begin{bmatrix}1/\sqrt{2}\\1/\sqrt{2}\end{bmatrix}$.

(vi) False. If $\vec{y} = \vec{0}$, then $\text{proj}_{\vec{x}}\, \vec{y} = \vec{0}$. Thus, $\{\text{proj}_{\vec{x}}\, \vec{y}, \text{perp}_{\vec{x}}\, \vec{y}\}$ contains the zero vector, so it is linearly dependent.

(vii) True. We have

$$||\vec{u} \times (\vec{v} + 3\vec{u})|| = ||\vec{u} \times \vec{v} + 3(\vec{u} \times \vec{u})|| = ||\vec{u} \times \vec{v} + \vec{0}|| = ||\vec{u} \times \vec{v}||$$

so the parallelograms have the same area.

CHAPTER 2

Section 2.1

A Problems

A1 (a) $\vec{x} = \begin{bmatrix}17\\4\end{bmatrix}$

(b) $\vec{x} = \begin{bmatrix}13\\0\\6\end{bmatrix} + t\begin{bmatrix}-2\\1\\0\end{bmatrix}, \quad t \in \mathbb{R}$

(c) $\vec{x} = \begin{bmatrix}32\\-8\\2\end{bmatrix}$

(d) $\vec{x} = \begin{bmatrix}-1\\-3\\2\\0\end{bmatrix} + t\begin{bmatrix}-1\\1\\-1\\1\end{bmatrix}, \quad t \in \mathbb{R}$

A2 (a) A is in row echelon form.

(b) B is in row echelon form.

(c) C is not in row echelon form because the leading 1 in the third row is not further to the right than the leading 1 in the second row.

(d) D is not in row echelon form because the leading 1 in the third row is to left of the leading 1 in the second row.

A3 Alternate correct answers are possible.

(a) $\begin{bmatrix} 1 & -3 & 2 \\ 0 & 13 & -7 \end{bmatrix}$

(b) $\begin{bmatrix} 1 & -1 & 2 & 3 \\ 0 & 0 & 1 & 2 \\ 0 & 0 & 0 & 1 \end{bmatrix}$

(c) $\begin{bmatrix} 5 & 0 & 0 \\ 0 & -1 & -2 \\ 0 & 0 & 1 \\ 0 & 0 & 0 \end{bmatrix}$

(d) $\begin{bmatrix} 2 & 0 & 2 & 0 \\ 0 & 2 & 2 & 4 \\ 0 & 0 & 4 & 8 \\ 0 & 0 & 0 & 0 \end{bmatrix}$

(e) $\begin{bmatrix} 1 & 2 & 1 & 1 \\ 0 & 1 & 2 & 1 \\ 0 & 0 & 7 & 5 \\ 0 & 0 & 0 & 2 \end{bmatrix}$

(f) $\begin{bmatrix} 1 & 0 & 3 & 0 & 1 \\ 0 & 1 & -1 & 2 & 1 \\ 0 & 0 & 24 & 1 & 11 \\ 0 & 0 & 0 & 0 & 1 \end{bmatrix}$

A4 (a) Inconsistent.

(b) Consistent. The solution is $\vec{x} = \begin{bmatrix} 2 \\ 0 \\ 3 \end{bmatrix} + t\begin{bmatrix} 0 \\ 1 \\ 0 \end{bmatrix}, \quad t \in \mathbb{R}.$

(c) Consistent. The solution is $\vec{x} = \begin{bmatrix} 1 \\ -1 \\ 0 \\ 3 \end{bmatrix} + t\begin{bmatrix} -1 \\ -1 \\ 1 \\ 0 \end{bmatrix}, \quad t \in \mathbb{R}$

(d) Consistent. The solution is $\vec{x} = \begin{bmatrix} 19/2 \\ 0 \\ 5/2 \\ -2 \end{bmatrix} + t\begin{bmatrix} -1 \\ 1 \\ 0 \\ 0 \end{bmatrix}, \quad t \in \mathbb{R}.$

(e) Consistent. The solution is $\vec{x} = s\begin{bmatrix} -1 \\ 0 \\ 1 \\ 0 \end{bmatrix} + t\begin{bmatrix} 1 \\ 0 \\ 0 \\ 1 \end{bmatrix}, \quad s, t \in \mathbb{R}.$

A5 (a) $\left[\begin{array}{cc|c} 3 & -5 & 2 \\ 1 & 2 & 4 \end{array}\right] \sim \left[\begin{array}{cc|c} 1 & 2 & 4 \\ 0 & -11 & -10 \end{array}\right]$. Consistent with solution $\vec{x} = \begin{bmatrix} 24/11 \\ 10/11 \end{bmatrix}$.

(b) $\left[\begin{array}{ccc|c} 1 & 2 & 1 & 5 \\ 2 & -3 & 2 & 6 \end{array}\right] \sim \left[\begin{array}{ccc|c} 1 & 2 & 1 & 5 \\ 0 & -7 & 0 & -4 \end{array}\right]$. Consistent with solution $\vec{x} = \begin{bmatrix} 27/7 \\ 4/7 \\ 0 \end{bmatrix} + t\begin{bmatrix} -1 \\ 0 \\ 1 \end{bmatrix}$, $t \in \mathbb{R}$.

(c) $\left[\begin{array}{ccc|c} 1 & 2 & -3 & 8 \\ 1 & 3 & -5 & 11 \\ 2 & 5 & -8 & 19 \end{array}\right] \sim \left[\begin{array}{ccc|c} 1 & 2 & -3 & 8 \\ 0 & 1 & -2 & 3 \\ 0 & 0 & 0 & 0 \end{array}\right]$. Consistent with solution $\vec{x} = \begin{bmatrix} 2 \\ 3 \\ 0 \end{bmatrix} + t\begin{bmatrix} -1 \\ 2 \\ 1 \end{bmatrix}$, $t \in \mathbb{R}$.

(d) $\left[\begin{array}{ccc|c} -3 & 6 & 16 & 36 \\ 1 & -2 & -5 & -11 \\ 2 & -3 & -8 & -17 \end{array}\right] \sim \left[\begin{array}{ccc|c} 1 & -2 & -5 & -11 \\ 0 & 1 & 2 & 5 \\ 0 & 0 & 1 & 3 \end{array}\right]$. Consistent with solution $\vec{x} = \begin{bmatrix} 2 \\ -1 \\ 3 \end{bmatrix}$.

(e) $\left[\begin{array}{ccc|c} 1 & 2 & -1 & 4 \\ 2 & 5 & 1 & 10 \\ 4 & 9 & -1 & 19 \end{array}\right] \sim \left[\begin{array}{ccc|c} 1 & 2 & -1 & 4 \\ 0 & 1 & 3 & 2 \\ 0 & 0 & 0 & 1 \end{array}\right]$. The system is inconsistent.

(f) $\left[\begin{array}{cccc|c} 1 & 2 & -3 & 0 & -5 \\ 2 & 4 & -6 & 1 & -8 \\ 6 & 13 & -17 & 4 & -21 \end{array}\right] \sim \left[\begin{array}{cccc|c} 1 & 2 & -3 & 0 & -5 \\ 0 & 1 & 1 & 4 & 9 \\ 0 & 0 & 0 & 1 & 2 \end{array}\right]$. Consistent with solution $\vec{x} = \begin{bmatrix} -7 \\ 1 \\ 0 \\ 2 \end{bmatrix} + t\begin{bmatrix} 5 \\ -1 \\ 1 \\ 0 \end{bmatrix}$, $t \in \mathbb{R}$.

(g) $\left[\begin{array}{ccccc|c} 0 & 2 & -2 & 0 & 1 & 2 \\ 1 & 2 & -3 & 1 & 4 & 1 \\ 2 & 4 & -5 & 3 & 8 & 3 \\ 2 & 5 & -7 & 3 & 10 & 5 \end{array}\right] \sim \left[\begin{array}{ccccc|c} 1 & 2 & -3 & 1 & 4 & 1 \\ 0 & 2 & -2 & 0 & 1 & 2 \\ 0 & 0 & 1 & 1 & 0 & 1 \\ 0 & 0 & 0 & 1 & 3/2 & 2 \end{array}\right]$. The system is consistent with solution $\vec{x} = \begin{bmatrix} -4 \\ 0 \\ -1 \\ 2 \\ 0 \end{bmatrix} + t\begin{bmatrix} 0 \\ 1 \\ 3/2 \\ -3/2 \\ 1 \end{bmatrix}$, $t \in \mathbb{R}$.

A6 (a) If $a \neq 0$, $b \neq 0$, this system is consistent, the solution is unique. If $a = 0$, $b \neq 0$, the system is consistent, but the solution is not unique. If $a \neq 0$, $b = 0$, the system is inconsistent. If $a = 0$, $b = 0$, this system is consistent, but the solution is not unique.

(b) If $c \neq 0$, $d \neq 0$, the system is consistent, and the solution is unique. If $d = 0$, the system is consistent only if $c = 0$. If $c = 0$, the system is consistent for all values of d, but the solution is not unique.

A7 600 apples, 400 bananas, and 500 oranges.

A8 75% in algebra, 90% in calculus, and 84% in physics.

Section 2.2

A Problems

A1 (a) $\begin{bmatrix} 1 & 0 \\ 0 & 1 \\ 0 & 0 \end{bmatrix}$; the rank is 2.

(b) $\begin{bmatrix} 1 & 0 & 0 \\ 0 & 1 & 0 \\ 0 & 0 & 1 \end{bmatrix}$; the rank is 3.

(c) $\begin{bmatrix} 1 & 0 & 0 \\ 0 & 1 & 0 \\ 0 & 0 & 1 \end{bmatrix}$; the rank is 3.

(d) $\begin{bmatrix} 1 & 0 & 0 \\ 0 & 1 & 0 \\ 0 & 0 & 1 \end{bmatrix}$; the rank is 3.

(e) $\begin{bmatrix} 1 & 2 & 0 \\ 0 & 0 & 1 \\ 0 & 0 & 0 \\ 0 & 0 & 0 \end{bmatrix}$; the rank is 2.

(f) $\begin{bmatrix} 1 & 1 & 0 & 0 \\ 0 & 0 & 1 & 0 \\ 0 & 0 & 0 & 1 \end{bmatrix}$; the rank is 3.

(g) $\begin{bmatrix} 1 & 0 & 0 & 0 \\ 0 & 1 & 0 & -2 \\ 0 & 0 & 1 & 3 \end{bmatrix}$; the rank is 3.

(h) $\begin{bmatrix} 1 & 0 & 0 & -1/2 \\ 0 & 1 & 0 & 3/2 \\ 0 & 0 & 1 & 1/2 \\ 0 & 0 & 0 & 0 \end{bmatrix}$; the rank is 3.

(i) $\begin{bmatrix} 1 & 0 & 0 & 0 & -56 \\ 0 & 1 & 0 & 0 & 17 \\ 0 & 0 & 1 & 0 & 23 \\ 0 & 0 & 0 & 1 & -6 \end{bmatrix}$; the rank is 4.

A2 (a) There is one parameter. The general solution is $\vec{x} = t\begin{bmatrix} -2 \\ 1 \\ 1 \\ 0 \end{bmatrix}, \quad t \in \mathbb{R}$.

(b) There are two parameters. The general solution is $\vec{x} = s\begin{bmatrix} 1 \\ 0 \\ 0 \\ 0 \end{bmatrix} + t\begin{bmatrix} 0 \\ -2 \\ 1 \\ 0 \end{bmatrix}, \quad s, t \in \mathbb{R}$

(c) There are two parameters. The general solution is $\vec{x} = s\begin{bmatrix} 3 \\ 1 \\ 0 \\ 0 \end{bmatrix} + t\begin{bmatrix} -2 \\ 0 \\ 1 \\ 0 \end{bmatrix}, \quad s, t \in \mathbb{R}$

(d) There are two parameters. The general solution is $\vec{x} = s\begin{bmatrix}-2\\1\\1\\0\\0\end{bmatrix} + t\begin{bmatrix}0\\2\\0\\-1\\1\end{bmatrix}$, $s, t \in \mathbb{R}$.

(e) There are two parameters. The general solution is $\vec{x} = s\begin{bmatrix}0\\1\\0\\0\\0\end{bmatrix} + t\begin{bmatrix}-4\\0\\5\\1\\0\end{bmatrix}$, $s, t \in \mathbb{R}$

(f) There is one parameter. The general solution is $\vec{x} = t\begin{bmatrix}0\\-1\\1\\0\\0\end{bmatrix}$, $t \in \mathbb{R}$.

A3 (a) $\begin{bmatrix}0 & 2 & -5\\1 & 2 & 3\\1 & 4 & -3\end{bmatrix} \sim \begin{bmatrix}1 & 0 & 0\\0 & 1 & 0\\0 & 0 & 1\end{bmatrix}$; the rank is 3; there are zero parameters. The only solution is $\vec{x} = \vec{0}$.

(b) $\begin{bmatrix}3 & 1 & -9\\1 & 1 & -5\\2 & 1 & -7\end{bmatrix} \sim \begin{bmatrix}1 & 0 & -2\\0 & 1 & -3\\0 & 0 & 0\end{bmatrix}$; the rank is 2; there is one parameter. The general solution is $\vec{x} = t\begin{bmatrix}2\\3\\1\end{bmatrix}$, $t \in \mathbb{R}$.

(c) $\begin{bmatrix}1 & -1 & 2 & -3\\3 & -3 & 8 & -5\\2 & -2 & 5 & -4\\3 & -3 & 7 & -7\end{bmatrix} \sim \begin{bmatrix}1 & -1 & 0 & -7\\0 & 0 & 1 & 2\\0 & 0 & 0 & 0\\0 & 0 & 0 & 0\end{bmatrix}$; the rank is 2; there are two parameters. The general solution is $\vec{x} = s\begin{bmatrix}1\\1\\0\\0\end{bmatrix} + t\begin{bmatrix}7\\0\\-2\\1\end{bmatrix}$, $s, t \in \mathbb{R}$.

(d) $\begin{bmatrix}0 & 1 & 2 & 2 & 0\\1 & 2 & 5 & 3 & -1\\2 & 1 & 5 & 1 & -3\\1 & 1 & 4 & 2 & -2\end{bmatrix} \sim \begin{bmatrix}1 & 0 & 0 & -2 & 0\\0 & 1 & 0 & 0 & 2\\0 & 0 & 1 & 1 & -1\\0 & 0 & 0 & 0 & 0\end{bmatrix}$; the rank is 3; there are two parameters. The general solution is $\vec{x} = s\begin{bmatrix}2\\0\\-1\\1\\0\end{bmatrix} + t\begin{bmatrix}0\\-2\\1\\0\\1\end{bmatrix}$, $s, t \in \mathbb{R}$.

A4 (a) $\left[\begin{array}{cc|c}3 & -5 & 2\\1 & 2 & 4\end{array}\right] \sim \left[\begin{array}{cc|c}1 & 0 & 24/11\\0 & 1 & 10/11\end{array}\right]$. Consistent with solution $\vec{x} = \begin{bmatrix}24/11\\10/11\end{bmatrix}$.

(b) $\left[\begin{array}{ccc|c} 1 & 2 & 1 & 5 \\ 2 & -3 & 2 & 6 \end{array}\right] \sim \left[\begin{array}{ccc|c} 1 & 0 & 1 & 27/7 \\ 0 & 1 & 0 & 4/7 \end{array}\right]$. Consistent with solution $\vec{x} = \begin{bmatrix} 27/7 \\ 4/7 \\ 0 \end{bmatrix} + t\begin{bmatrix} -1 \\ 0 \\ 1 \end{bmatrix}$, $t \in \mathbb{R}$.

(c) $\left[\begin{array}{ccc|c} 1 & 2 & -3 & 8 \\ 1 & 3 & -5 & 11 \\ 2 & 5 & -8 & 19 \end{array}\right] \sim \left[\begin{array}{ccc|c} 1 & 0 & 1 & 2 \\ 0 & 1 & -2 & 3 \\ 0 & 0 & 0 & 0 \end{array}\right]$. Consistent with solution $\vec{x} = \begin{bmatrix} 2 \\ 3 \\ 0 \end{bmatrix} + t\begin{bmatrix} -1 \\ 2 \\ 1 \end{bmatrix}$, $t \in \mathbb{R}$.

(d) $\left[\begin{array}{ccc|c} -3 & 6 & 16 & 36 \\ 1 & -2 & -5 & -11 \\ 2 & -3 & -8 & -17 \end{array}\right] \sim \left[\begin{array}{ccc|c} 1 & 0 & 0 & 2 \\ 0 & 1 & 0 & -1 \\ 0 & 0 & 1 & 3 \end{array}\right]$. Consistent with solution $\vec{x} = \begin{bmatrix} 2 \\ -1 \\ 3 \end{bmatrix}$.

(e) $\left[\begin{array}{ccc|c} 1 & 2 & -1 & 4 \\ 2 & 5 & 1 & 10 \\ 4 & 9 & -1 & 19 \end{array}\right] \sim \left[\begin{array}{ccc|c} 1 & 0 & -7 & 0 \\ 0 & 1 & 3 & 0 \\ 0 & 0 & 0 & 1 \end{array}\right]$. The system is inconsistent.

(f) $\left[\begin{array}{cccc|c} 1 & 2 & -3 & 0 & -5 \\ 2 & 4 & -6 & 1 & -8 \\ 6 & 13 & -17 & 4 & -21 \end{array}\right] \sim \left[\begin{array}{cccc|c} 1 & 0 & -5 & 0 & -7 \\ 0 & 1 & 1 & 0 & 1 \\ 0 & 0 & 0 & 1 & 2 \end{array}\right]$. Consistent with solution $\vec{x} = \begin{bmatrix} -7 \\ 1 \\ 0 \\ 2 \end{bmatrix} + t\begin{bmatrix} 5 \\ -1 \\ 1 \\ 0 \end{bmatrix}$, $t \in \mathbb{R}$.

(g) $\left[\begin{array}{ccccc|c} 0 & 2 & -2 & 0 & 1 & 2 \\ 1 & 2 & -3 & 1 & 4 & 1 \\ 2 & 4 & -5 & 3 & 8 & 3 \\ 2 & 5 & -7 & 3 & 10 & 5 \end{array}\right] \sim \left[\begin{array}{ccccc|c} 1 & 0 & 0 & 0 & 0 & -4 \\ 0 & 1 & 0 & 0 & -1 & 0 \\ 0 & 0 & 1 & 0 & -3/2 & -1 \\ 0 & 0 & 0 & 1 & 3/2 & 2 \end{array}\right]$. The system is consistent with solution $\vec{x} = \begin{bmatrix} -4 \\ 0 \\ -1 \\ 2 \\ 0 \end{bmatrix} + t\begin{bmatrix} 0 \\ 1 \\ 3/2 \\ -3/2 \\ 1 \end{bmatrix}$, $t \in \mathbb{R}$.

A5 (a) $\left[\begin{array}{ccc|c} 2 & -1 & 4 & 1 \\ 1 & 3 & 0 & 0 \\ 1 & 1 & 2 & 2 \end{array}\right] \sim \left[\begin{array}{ccc|c} 1 & 0 & 0 & -3 \\ 0 & 1 & 0 & 1 \\ 0 & 0 & 1 & 2 \end{array}\right]$. The solution to $\left[A \mid \vec{b}\right]$ is $\vec{x} = \begin{bmatrix} -3 \\ 1 \\ 2 \end{bmatrix}$. The solution to the homogeneous system is $\vec{x} = \begin{bmatrix} 0 \\ 0 \\ 0 \end{bmatrix}$.

(b) $\left[\begin{array}{ccc|c} 1 & 7 & 5 & 5 \\ 1 & 0 & 5 & -2 \\ -1 & 2 & -5 & 4 \end{array}\right] \sim \left[\begin{array}{ccc|c} 1 & 0 & 5 & -2 \\ 0 & 1 & 0 & 1 \\ 0 & 0 & 0 & 0 \end{array}\right]$. The solution to $\left[A \mid \vec{b}\right]$ is $\vec{x} = \begin{bmatrix} -2 \\ 1 \\ 0 \end{bmatrix} + t\begin{bmatrix} -5 \\ 0 \\ 1 \end{bmatrix}$, $t \in \mathbb{R}$. The solution to the homogeneous system is $\vec{x} = t\begin{bmatrix} -5 \\ 0 \\ 1 \end{bmatrix}$, $t \in \mathbb{R}$.

(c) $\left[\begin{array}{cccc|c} 0 & -1 & 5 & -2 & -1 \\ -1 & -1 & -4 & -1 & 4 \end{array}\right] \sim \left[\begin{array}{cccc|c} 1 & 0 & 9 & -1 & -5 \\ 0 & 1 & -5 & 2 & 1 \end{array}\right]$. The solution to $\left[A \mid \vec{b}\,\right]$ is $\vec{x} = \begin{bmatrix} -5 \\ 1 \\ 0 \\ 0 \end{bmatrix} + s \begin{bmatrix} -9 \\ 5 \\ 1 \\ 0 \end{bmatrix} + t \begin{bmatrix} 1 \\ -2 \\ 0 \\ 1 \end{bmatrix}$, $s, t \in \mathbb{R}$. The solution to the homogeneous system is $\vec{x} = s \begin{bmatrix} -9 \\ 5 \\ 1 \\ 0 \end{bmatrix} + t \begin{bmatrix} 1 \\ -2 \\ 0 \\ 1 \end{bmatrix}$, $s, t \in \mathbb{R}$.

(d) $\left[\begin{array}{cccc|c} 1 & 0 & -1 & -1 & 3 \\ 4 & 3 & 2 & -4 & 3 \\ -1 & -4 & -3 & 5 & 5 \end{array}\right] \sim \left[\begin{array}{cccc|c} 1 & 0 & 0 & 0 & 2 \\ 0 & 1 & 0 & -2 & -1 \\ 0 & 0 & 1 & 1 & -1 \end{array}\right]$. The solution to $\left[A \mid \vec{b}\,\right]$ is $\vec{x} = \begin{bmatrix} 2 \\ -1 \\ -1 \\ 0 \end{bmatrix} + t \begin{bmatrix} 0 \\ 2 \\ -1 \\ 1 \end{bmatrix}$, $t \in \mathbb{R}$. The solution to the homogeneous system is $\vec{x} = t \begin{bmatrix} 0 \\ 2 \\ -1 \\ 1 \end{bmatrix}$, $t \in \mathbb{R}$.

Section 2.3

A Problems

A1 (a) $2\begin{bmatrix} 1 \\ 0 \\ 1 \\ 1 \end{bmatrix} + (-1)\begin{bmatrix} 2 \\ 1 \\ 0 \\ 1 \end{bmatrix} + 3\begin{bmatrix} -1 \\ 1 \\ 2 \\ 1 \end{bmatrix} = \begin{bmatrix} -3 \\ 2 \\ 8 \\ 4 \end{bmatrix}$

(b) $\begin{bmatrix} 5 \\ 4 \\ 6 \\ 7 \end{bmatrix}$ is not in the span.

(c) $3\begin{bmatrix} 1 \\ 0 \\ 1 \\ 1 \end{bmatrix} + (-1)\begin{bmatrix} 2 \\ 1 \\ 0 \\ 1 \end{bmatrix} + (-1)\begin{bmatrix} -1 \\ 1 \\ 2 \\ 1 \end{bmatrix} = \begin{bmatrix} 2 \\ -2 \\ 1 \\ 1 \end{bmatrix}$

A2 (a) $\begin{bmatrix} 3 \\ 2 \\ -1 \\ -1 \end{bmatrix}$ is not in the span.

(b) $(-2)\begin{bmatrix} 1 \\ -1 \\ 1 \\ 0 \end{bmatrix} + 3\begin{bmatrix} -1 \\ 1 \\ 0 \\ 2 \end{bmatrix} + (-2)\begin{bmatrix} 1 \\ 1 \\ -1 \\ -1 \end{bmatrix} = \begin{bmatrix} -7 \\ 3 \\ 0 \\ 8 \end{bmatrix}$

(c) $\begin{bmatrix} 1 \\ 1 \\ 1 \\ 1 \end{bmatrix}$ is not in the span.

A3 (a) $x_3 = 0$

(b) $x_1 - 2x_2 = 0$, $x_3 = 0$

(c) $x_1 + 3x_2 - 2x_3 = 0$

(d) $x_1 + 3x_2 + 5x_3 = 0$

(e) $-x_1 - x_2 + x_3 = 0$, $x_2 + x_4 = 0$

(f) $-4x_1 + 5x_2 + x_3 + 4x_4 = 0$

A4 (a) It is a basis for the plane.

(b) It is not a basis for the plane.

(c) It is a basis for the hyperplane.

A5 (a) Linearly independent.

(b) Linearly dependent. $-3t\begin{bmatrix} 1 \\ 0 \\ 1 \\ 0 \end{bmatrix} - 2t\begin{bmatrix} 0 \\ 1 \\ 1 \\ 1 \end{bmatrix} - t\begin{bmatrix} 0 \\ 0 \\ 1 \\ 1 \end{bmatrix} + t\begin{bmatrix} 3 \\ 2 \\ 6 \\ 3 \end{bmatrix} = \begin{bmatrix} 0 \\ 0 \\ 0 \\ 0 \end{bmatrix}$, $t \in \mathbb{R}$

(c) Linearly dependent. $2t\begin{bmatrix} 1 \\ 1 \\ 0 \\ 1 \\ 1 \end{bmatrix} - t\begin{bmatrix} 2 \\ 3 \\ 1 \\ 3 \\ 3 \end{bmatrix} + t\begin{bmatrix} 0 \\ 1 \\ 1 \\ 1 \\ 1 \end{bmatrix} = \begin{bmatrix} 0 \\ 0 \\ 0 \\ 0 \\ 0 \end{bmatrix}$, $t \in \mathbb{R}$

A6 (a) Linearly independent for all $k \neq -3$.

(b) Linearly independent for all $k \neq -5/2$.

A7 (a) It is a basis.

(b) Only two vectors, so it cannot span $\mathbb{R}^3$. Therefore, it is not a basis.

(c) It has four vectors in $\mathbb{R}^3$, so it is linearly dependent. Therefore, it is not a basis.

(d) It is linearly dependent, so it is not a basis.

Section 2.4

A Problems

A1 To simplify writing, let $\alpha = \frac{1}{\sqrt{2}}$.
Total horizontal force: $R_1 + R_2 = 0$.
Total vertical force: $R_V - F_V = 0$.
Total moment about A: $R_1 s + F_V(2s) = 0$.
The horizontal and vertical equations at the joints A, B, C, D, and E are
$\alpha N_2 + R_2 = 0$ and $N_1 + \alpha N_2 + R_V = 0$;
$N_3 + \alpha N_4 + R_1 = 0$ and $-N_1 + \alpha N_4 = 0$

$-N_3 + \alpha N_6 = 0$ and $-\alpha N_2 + N_5 + \alpha N_6 = 0$
$-\alpha N_4 + N_7 = 0$ and $-\alpha N_4 - N_5 = 0$
$-N_7 - \alpha N_6 = 0$ and $-\alpha N_6 - F_V = 0$

A2 $\left[\begin{array}{ccccc|c} R_1+R_2 & -R_2 & 0 & 0 & 0 & E_1 \\ -R_2 & R_2+R_3 & -R_3 & 0 & 0 & 0 \\ 0 & -R_3 & R_3+R_4+R_8 & 0 & -R_8 & 0 \\ 0 & 0 & 0 & R_5+R_6 & -R_6 & 0 \\ 0 & 0 & -R_8 & -R_6 & R_6+R_7+R_8 & E_2 \end{array}\right]$

A3

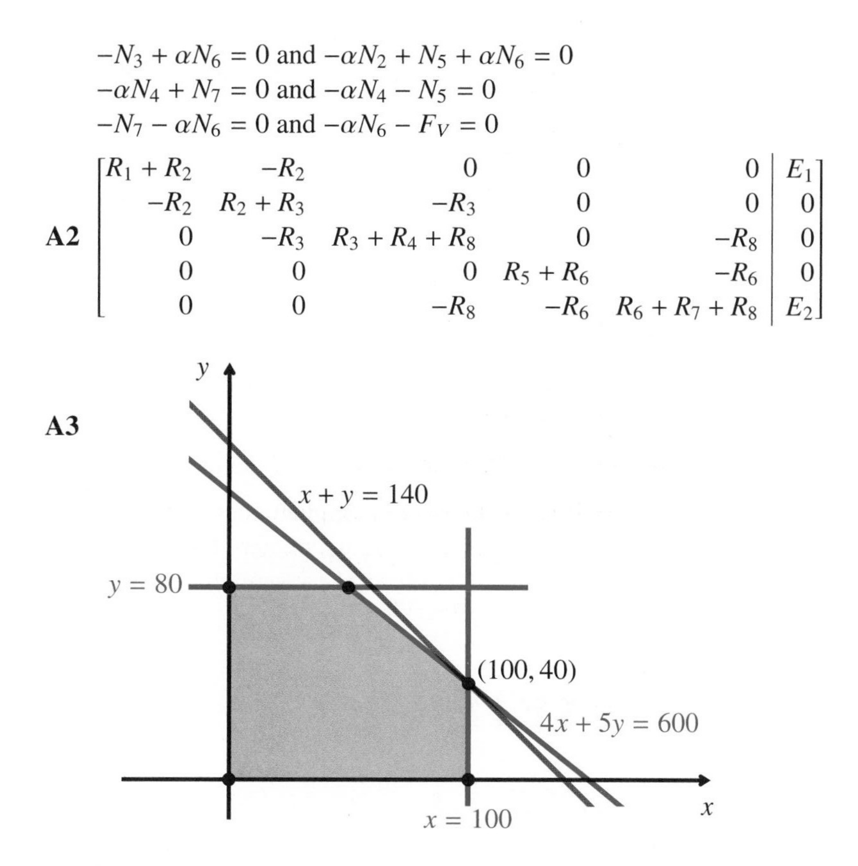

Chapter 2 Quiz

E Problems

E1 $\left[\begin{array}{cccc|c} 1 & -1 & 1 & 0 & 2 \\ 0 & 1 & 0 & 0 & 7 \\ 0 & 0 & -2 & 1 & -5 \\ 0 & 0 & 0 & 0 & 1 \end{array}\right]$. Inconsistent.

E2 $\begin{bmatrix} 1 & 0 & 0 & 0 & 1 \\ 0 & 1 & 0 & 0 & 0 \\ 0 & 0 & 1 & 0 & -1/3 \\ 0 & 0 & 0 & 1 & 1/3 \end{bmatrix}$

E3 (a) The system is inconsistent for all (a, b, c) of the form $(a, b, 1)$ or $(a, -2, c)$ and is consistent for all (a, b, c) where $b \neq -2$ and $c \neq 1$.

(b) The system has a unique solution if and only if $b \neq -2$, $c \neq 1$ and, $c \neq -1$.

E4 (a) $s\begin{bmatrix} -2 \\ -1 \\ 1 \\ 0 \\ 0 \end{bmatrix} + t\begin{bmatrix} 11/4 \\ -11/2 \\ 0 \\ -5/4 \\ 1 \end{bmatrix}, \quad s, t \in \mathbb{R}$

(b) $\vec{x} \cdot \vec{u} = 0$, $\vec{x} \cdot \vec{v} = 0$, and $\vec{x} \cdot \vec{w} = 0$, yields a homogeneous system of three linear equations with five variables. Hence, the rank of the matrix is at most three and thus there are at least # of variables - rank = 5 − 3 = 2 parameters. So, there are non-trivial solutions.

E5 Consider $\vec{x} = t_1 \begin{bmatrix} 3 \\ 1 \\ 2 \end{bmatrix} + t_2 \begin{bmatrix} 1 \\ 1 \\ 6 \end{bmatrix} + t_3 \begin{bmatrix} 4 \\ 1 \\ 5 \end{bmatrix}$. Row reducing the corresponding coefficient matrix gives

$$\begin{bmatrix} 3 & 1 & 4 \\ 1 & 1 & 1 \\ 2 & 6 & 5 \end{bmatrix} \sim \begin{bmatrix} 1 & 0 & 0 \\ 0 & 1 & 0 \\ 0 & 0 & 1 \end{bmatrix}$$

Thus, $\mathcal{B}$ is linearly independent and spans $\mathbb{R}^3$. Hence, it is a basis for $\mathbb{R}^3$.

E6 (a) False. The system may have infinitely many solutions.

(b) False. The system $x_1 = 1$, $2x_1 = 2$ has more equations than variables but is consistent.

(c) True. The system $x_1 = 0$ has a unique solution.

(d) True. If there are more variables than equations and the system is consistent, then there must be parameters and hence the system cannot have a unique solution. Of course, the system may be inconsistent.

CHAPTER 3

Section 3.1

A Problems

A1 (a) $\begin{bmatrix} -1 & -6 & 4 \\ 6 & -4 & 2 \end{bmatrix}$

(b) $\begin{bmatrix} -3 & 6 \\ -6 & -3 \\ -12 & 6 \end{bmatrix}$

(c) $\begin{bmatrix} -1 & 9 \\ -11 & -17 \end{bmatrix}$

A2 (a) $\begin{bmatrix} -11 & -1 & 12 \\ 8 & 27 & -1 \end{bmatrix}$

(b) $\begin{bmatrix} 12 & 11 \\ 9 & 4 \\ 3 & 15 \end{bmatrix}$

(c) $\begin{bmatrix} -4 & 6 & 13 & -4 \\ 0 & -3 & -5 & 1 \\ 8 & 15 & 19 & -1 \end{bmatrix}$

(d) The product is not defined since the number of columns of the first matrix does not equal the number of rows of the second matrix.

A3 (a) $A(B + C) = \begin{bmatrix} -13 & 10 \\ 14 & 7 \end{bmatrix} = AB + AC \qquad A(3B) = \begin{bmatrix} -9 & 12 \\ 30 & -18 \end{bmatrix} = 3(AB)$

(b) $A(B + C) = \begin{bmatrix} 6 & -16 \\ 4 & 14 \end{bmatrix} = AB + AC \qquad A(3B) = \begin{bmatrix} 3 & -18 \\ -27 & 51 \end{bmatrix} = 3(AB)$

A4 (a) $A+B$ is defined. AB is not defined. $(A+B)^T = \begin{bmatrix} -3 & 2 & 1 \\ -1 & 2 & 3 \end{bmatrix} = A^T + B^T$.

(b) $A+B$ is not defined. AB is defined. $(AB)^T = \begin{bmatrix} -21 & -10 \\ 15 & -27 \end{bmatrix} = B^T A^T$.

A5 (a) $AB = \begin{bmatrix} 13 & 31 & 2 \\ 10 & 12 & 10 \end{bmatrix}$.

(b) Does not exist because the matrices are not of the correct size for this product to be defined.

(c) Does not exist because the matrices are not of the correct size for this product to be defined.

(d) Does not exist because the matrices are not of the correct size for this product to be defined.

(e) Does not exist because the matrices are not of the correct size for this product to be defined.

(f) $\begin{bmatrix} 11 & 7 & 3 & 15 \\ 7 & 9 & 11 & 1 \end{bmatrix}$

(g) $\begin{bmatrix} 52 & 139 \\ 62 & 46 \end{bmatrix}$

(h) $\begin{bmatrix} 13 & 10 \\ 31 & 12 \\ 2 & 10 \end{bmatrix}$

(i) $D^T C = (C^T D)^T = \begin{bmatrix} 11 & 7 \\ 7 & 9 \\ 3 & 11 \\ 15 & 1 \end{bmatrix}$

A6 (a) $A\vec{x} = \begin{bmatrix} 12 \\ 17 \\ 3 \end{bmatrix}, A\vec{y} = \begin{bmatrix} 8 \\ 4 \\ -4 \end{bmatrix}, A\vec{z} = \begin{bmatrix} -2 \\ 5 \\ 1 \end{bmatrix}$

(b) $\begin{bmatrix} 12 & 8 & -2 \\ 17 & 4 & 5 \\ 3 & -4 & 1 \end{bmatrix}$

A7 (a) $\begin{bmatrix} 1 \\ 0 \\ 0 \end{bmatrix}$

(b) $\begin{bmatrix} 2 \\ 0 \\ 0 \end{bmatrix}$

(c) $\begin{bmatrix} 0 \\ 0 \\ 3 \end{bmatrix}$

(d) $\begin{bmatrix} 1 \\ 1 \\ 0 \end{bmatrix}$

A8 (a) $\begin{bmatrix} -6 & -18 \\ 2 & 6 \end{bmatrix}$

(b) $[0]$

(c) $\begin{bmatrix} 10 & 8 & -6 \\ -5 & -4 & 3 \\ 15 & 12 & -9 \end{bmatrix}$

(d) $[-3]$

A9 Both sides give $\begin{bmatrix} -13 & 16 \\ -27 & 0 \end{bmatrix}$.

A10 It is in the span since $3\begin{bmatrix} 1 & 2 \\ 1 & 0 \end{bmatrix} + (-2)\begin{bmatrix} 0 & 1 \\ -1 & 2 \end{bmatrix} + (-1)\begin{bmatrix} 1 & 1 \\ 3 & -1 \end{bmatrix} = \begin{bmatrix} 2 & 3 \\ 2 & -3 \end{bmatrix}$.

A11 It is linearly independent.

A12 Using the second view of matrix-vector multiplication and the fact that the i-th component of $\vec{e}_i$ is 1 and all other components are 0, we get

$$A\vec{e}_i = 0\vec{a}_1 + \cdots + 0\vec{a}_{i-1} + 1\vec{a}_i + 0\vec{a}_{i+1} + \cdots + 0\vec{a}_n = \vec{a}_i$$

Section 3.2

A Problems

A1 (a) Domain $\mathbb{R}^2$, codomain $\mathbb{R}^4$

(b) $f_A(2,-5) = \begin{bmatrix} -19 \\ 6 \\ -23 \\ 38 \end{bmatrix}$, $f_A(-3,4) = \begin{bmatrix} 18 \\ -9 \\ 17 \\ -36 \end{bmatrix}$

(c) $f_A(1,0) = \begin{bmatrix} -2 \\ 3 \\ 1 \\ 4 \end{bmatrix}$, $f_A(0,1) = \begin{bmatrix} 3 \\ 0 \\ 5 \\ -6 \end{bmatrix}$

(d) $f_A(\vec{x}) = \begin{bmatrix} -2x_1 + 3x_2 \\ 3x_1 + 0x_2 \\ x_1 + 5x_2 \\ 4x_1 - 6x_2 \end{bmatrix}$

(e) The standard matrix of f_A is

$$[f_A] = \begin{bmatrix} f_A(1,0) & f_A(0,1) \end{bmatrix} = \begin{bmatrix} -2 & 3 \\ 3 & 0 \\ 1 & 5 \\ 4 & -6 \end{bmatrix}$$

A2 (a) Domain $\mathbb{R}^4$, codomain $\mathbb{R}^3$

(b) $f_A(2,-2,3,1) = \begin{bmatrix} -11 \\ 9 \\ 7 \end{bmatrix}$, $f_A(-3,1,4,2) = \begin{bmatrix} -13 \\ -1 \\ 3 \end{bmatrix}$

(c) $f_A(\vec{e}_1) = \begin{bmatrix} 1 \\ 2 \\ 1 \end{bmatrix}$, $f_A(\vec{e}_2) = \begin{bmatrix} 2 \\ -1 \\ 0 \end{bmatrix}$, $f_A(\vec{e}_3) = \begin{bmatrix} -3 \\ 0 \\ 2 \end{bmatrix}$, $f_A(\vec{e}_4) = \begin{bmatrix} 0 \\ 3 \\ -1 \end{bmatrix}$

(d) $f_A(\vec{x}) = \begin{bmatrix} x_1 + 2x_2 - 3x_3 \\ 2x_1 - x_2 + 3x_4 \\ x_1 + 2x_3 - x_4 \end{bmatrix}$

(e) The standard matrix of f_A is

$$[f_A] = \begin{bmatrix} f_A(\vec{e}_1) & f_A(\vec{e}_2) & f_A(\vec{e}_3) & f_A(\vec{e}_4) \end{bmatrix} = \begin{bmatrix} 1 & 2 & -3 & 0 \\ 2 & -1 & 0 & 3 \\ 1 & 0 & 2 & -1 \end{bmatrix}$$

A3 (a) f is not linear.

(b) g is linear.

(c) h is not linear.

(d) k is linear.

(e) ℓ is not linear.

(f) m is not linear.

A4 (a) Domain $\mathbb{R}^3$, codomain $\mathbb{R}^2$, $[L] = \begin{bmatrix} 2 & -3 & 1 \\ 0 & 1 & -5 \end{bmatrix}$

(b) Domain $\mathbb{R}^4$, codomain $\mathbb{R}^2$, $[K] = \begin{bmatrix} 5 & 0 & 3 & -1 \\ 0 & 1 & -7 & 3 \end{bmatrix}$

(c) Domain $\mathbb{R}^4$, codomain $\mathbb{R}^4$, $[M] = \begin{bmatrix} 1 & 0 & -1 & 1 \\ 1 & 2 & -1 & -3 \\ 0 & 1 & 1 & 0 \\ 1 & -1 & 1 & -1 \end{bmatrix}$

A5 (a) The domain is $\mathbb{R}^3$ and codomain is $\mathbb{R}^2$ for both mappings.

(b) $[S+T] = \begin{bmatrix} 3 & 3 & 2 \\ 1 & 2 & 5 \end{bmatrix}$, $[2S - 3T] = \begin{bmatrix} 1 & -4 & 9 \\ -8 & -6 & -5 \end{bmatrix}$

A6 (a) The domain of S is $\mathbb{R}^4$, and the codomain of S is $\mathbb{R}^2$. The domain of T is $\mathbb{R}^2$, and the codomain of T is $\mathbb{R}^4$.

(b) $[S \circ T] = \begin{bmatrix} 6 & -19 \\ 10 & -10 \end{bmatrix}$, $[T \circ S] = \begin{bmatrix} -3 & 5 & 16 & 9 \\ 6 & 8 & 4 & 0 \\ -6 & -8 & -4 & 0 \\ -9 & -17 & -16 & -5 \end{bmatrix}$

A7 (a) The domain and codomain of $L \circ M$ are both $\mathbb{R}^3$.

(b) The domain and codomain of $M \circ L$ are both $\mathbb{R}^2$.

(c) Not defined

(d) Not defined

(e) Not defined

(f) The domain of $N \circ M$ is $\mathbb{R}^3$, and the codomain is $\mathbb{R}^4$.

A8 $[\text{proj}_{\vec{v}}] = \frac{1}{5}\begin{bmatrix} 4 & -2 \\ -2 & 1 \end{bmatrix}$

A9 $[\text{perp}_{\vec{v}}] = \frac{1}{17}\begin{bmatrix} 16 & -4 \\ -4 & 1 \end{bmatrix}$

A10 $[\text{proj}_{\vec{v}}] = \frac{1}{9}\begin{bmatrix} 4 & 4 & -2 \\ 4 & 4 & -2 \\ -2 & -2 & 1 \end{bmatrix}$

Section 3.3

A Problems

A1 (a) $\begin{bmatrix} 0 & -1 \\ 1 & 0 \end{bmatrix}$ (b) $\begin{bmatrix} -1 & 0 \\ 0 & -1 \end{bmatrix}$ (c) $\frac{1}{\sqrt{2}}\begin{bmatrix} 1 & 1 \\ -1 & 1 \end{bmatrix}$ (d) $\begin{bmatrix} 0.309 & -0.951 \\ 0.951 & 0.309 \end{bmatrix}$

A2 (a) $[S] = \begin{bmatrix} 1 & 0 \\ 0 & 5 \end{bmatrix}$ (b) $[R_\theta \circ S] = \begin{bmatrix} \cos\theta & -5\sin\theta \\ \sin\theta & 5\cos\theta \end{bmatrix}$

(c) $[S \circ R_\theta] = \begin{bmatrix} \cos\theta & -\sin\theta \\ 5\sin\theta & 5\cos\theta \end{bmatrix}$

A3 (a) $[R] = \begin{bmatrix} 4/5 & -3/5 \\ -3/5 & -4/5 \end{bmatrix}$ (b) $[S] = \begin{bmatrix} -3/5 & 4/5 \\ 4/5 & 3/5 \end{bmatrix}$

A4 (a) $\frac{1}{3}\begin{bmatrix} 1 & -2 & -2 \\ -2 & 1 & -2 \\ -2 & -2 & 1 \end{bmatrix}$ (b) $\frac{1}{9}\begin{bmatrix} 1 & 8 & 4 \\ 8 & 1 & -4 \\ 4 & -4 & 7 \end{bmatrix}$

A5 (a) $\begin{bmatrix} 5 & 0 & 0 \\ 0 & 5 & 0 \\ 0 & 0 & 0 \\ 0 & 0 & 5 \end{bmatrix}$ (b) $\begin{bmatrix} 0 & 1 & 0 \\ 2 & 0 & 1 \end{bmatrix}$

(c) There is no shear T such that $T \circ P = P \circ S$. (d) $\begin{bmatrix} 1 & 0 & 0 \\ 0 & 1 & 0 \end{bmatrix}$

Section 3.4

A Problems

A1 (a) $\begin{bmatrix} 3 \\ 1 \\ 6 \\ 1 \end{bmatrix}$ is not in the range of L. (b) $L(1, 1, -2) = \begin{bmatrix} 3 \\ -5 \\ 1 \\ 5 \end{bmatrix}$

A2 (a) A basis for Range(L) is $\left\{\begin{bmatrix} 1 \\ 0 \end{bmatrix}, \begin{bmatrix} 0 \\ 1 \end{bmatrix}\right\}$. A basis for Null($L$) is $\left\{\begin{bmatrix} 0 \\ 2 \\ 1 \end{bmatrix}\right\}$.

(b) A basis for Range(M) is $\left\{\begin{bmatrix} 0 \\ 0 \\ 0 \\ 0 \\ 1 \end{bmatrix}, \begin{bmatrix} 0 \\ 0 \\ 0 \\ 1 \\ 1 \end{bmatrix}, \begin{bmatrix} 0 \\ 1 \\ 0 \\ 0 \\ -1 \end{bmatrix}, \begin{bmatrix} 1 \\ 0 \\ 0 \\ 0 \\ 0 \end{bmatrix}\right\}$. A basis for Null($M$) is the empty set since Null(M) = $\{\vec{0}\}$.

A3 The matrix of L is any multiple of $\begin{bmatrix} 1 & -1 \\ 2 & -2 \\ 3 & -3 \end{bmatrix}$.

A4 The matrix of L is any multiple of $\begin{bmatrix} 2 & 1 \\ 2 & 1 \\ 2 & 1 \end{bmatrix}$.

A5 (a) The number of variables is 4. The rank of A is 2. The dimension of the solution space is 2.

(b) The number of variables is 5. The rank of A is 3. The dimension of the solution space is 2.

(c) The number of variables is 5. The rank of A is 2. The dimension of the solution space is 3.

(d) The number of variables is 6. The rank of A is 3. The dimension of the solution space is 3.

A6 (a) A basis for the rowspace is $\left\{\begin{bmatrix}1\\0\\0\end{bmatrix}, \begin{bmatrix}0\\1\\0\end{bmatrix}, \begin{bmatrix}0\\0\\1\end{bmatrix}\right\}$. A basis for the columnspace is $\left\{\begin{bmatrix}1\\1\\1\end{bmatrix}, \begin{bmatrix}2\\1\\0\end{bmatrix}, \begin{bmatrix}8\\5\\-2\end{bmatrix}\right\}$. A basis for the nullspace is the empty set.
Then, rank(A)+nullity$(A) = 3+0 = 3$, the number of columns of A as predicted by the Rank Theorem.

(b) A basis for the rowspace is $\left\{\begin{bmatrix}1\\0\\-1\\0\end{bmatrix}, \begin{bmatrix}0\\1\\-2\\0\end{bmatrix}, \begin{bmatrix}0\\0\\0\\1\end{bmatrix}\right\}$. A basis for the columnspace is $\left\{\begin{bmatrix}1\\2\\0\end{bmatrix}, \begin{bmatrix}1\\3\\1\end{bmatrix}, \begin{bmatrix}1\\4\\3\end{bmatrix}\right\}$. A basis for the nullspace is $\left\{\begin{bmatrix}1\\2\\1\\0\end{bmatrix}\right\}$.
Then, rank(A)+nullity$(A) = 3+1 = 4$, the number of columns of A as predicted by the Rank Theorem.

(c) A basis for the rowspace is $\left\{\begin{bmatrix}1\\2\\0\\3\\0\end{bmatrix}, \begin{bmatrix}0\\0\\1\\4\\0\end{bmatrix}, \begin{bmatrix}0\\0\\0\\0\\1\end{bmatrix}\right\}$. A basis for the columnspace is $\left\{\begin{bmatrix}1\\1\\2\\3\end{bmatrix}, \begin{bmatrix}0\\1\\0\\1\end{bmatrix}, \begin{bmatrix}0\\1\\1\\2\end{bmatrix}\right\}$. A basis for the nullspace is $\left\{\begin{bmatrix}-2\\1\\0\\0\\0\end{bmatrix}, \begin{bmatrix}-3\\0\\-4\\1\\0\end{bmatrix}\right\}$.
Then, rank(A)+nullity$(A) = 3+2 = 5$, the number of columns of A as predicted by the Rank Theorem.

A7 (a) A basis for the nullspace is $\left\{\begin{bmatrix}2\\1\\0\end{bmatrix}, \begin{bmatrix}2\\0\\-1\end{bmatrix}\right\}$ (or any pair of linearly independent vectors orthogonal to $\begin{bmatrix}1\\-2\\2\end{bmatrix}$); a basis for the range is $\left\{\begin{bmatrix}1\\-2\\2\end{bmatrix}\right\}$.

(b) For the nullspace, a basis is $\left\{\begin{bmatrix}3\\1\\2\end{bmatrix}\right\}$; for the range, $\left\{\begin{bmatrix}1\\-3\\0\end{bmatrix}, \begin{bmatrix}0\\-2\\1\end{bmatrix}\right\}$.

(c) A basis for the nullspace is the empty set; the range is $\mathbb{R}^3$, so take any basis for $\mathbb{R}^3$.

A8 (a) $n = 5$ (b) $\left\{\begin{bmatrix}1\\0\\2\\0\\3\end{bmatrix}, \begin{bmatrix}0\\1\\-1\\0\\1\end{bmatrix}, \begin{bmatrix}0\\0\\0\\1\\1\end{bmatrix}\right\}$ (c) $m = 4$ (d) $\left\{\begin{bmatrix}1\\2\\1\\1\end{bmatrix}, \begin{bmatrix}1\\3\\1\\2\end{bmatrix}, \begin{bmatrix}1\\2\\3\\-1\end{bmatrix}\right\}$

(e) $\vec{x} = s\begin{bmatrix}-2\\1\\1\\0\\0\end{bmatrix} + t\begin{bmatrix}-3\\-1\\0\\-1\\1\end{bmatrix}$, $s, t \in \mathbb{R}$. So, a spanning set is $\left\{\begin{bmatrix}-2\\1\\1\\0\\0\end{bmatrix}, \begin{bmatrix}-3\\-1\\0\\-1\\1\end{bmatrix}\right\}$.

(f) $\left\{\begin{bmatrix}-2\\1\\1\\0\\0\end{bmatrix}, \begin{bmatrix}-3\\-1\\0\\-1\\1\end{bmatrix}\right\}$ is also linearly independent, so it a basis.

(g) The rank of A is 3 and a basis for the solution space has two vectors in it, so the dimension of the solution space is 2. We have $3 + 2 = 5$ is the number of variables in the system.

Section 3.5

A Problems

A1 (a) $\frac{1}{23}\begin{bmatrix}5 & 4\\-2 & 3\end{bmatrix}$

(b) $\begin{bmatrix}2 & 0 & -1\\-1 & 1 & -1\\-1 & 0 & 1\end{bmatrix}$

(c) It is not invertible.

(d) $\begin{bmatrix}0 & -1 & 1\\-1 & 1 & 0\\1 & 0 & 0\end{bmatrix}$

(e) $\begin{bmatrix}6 & 10 & -5/2 & -7/2\\1 & 2 & -1/2 & -1/2\\-2 & -3 & 1 & 1\\0 & -3 & 0 & 1\end{bmatrix}$

(f) $\begin{bmatrix}1 & 0 & -1 & 1 & -2\\0 & 1 & 0 & -1 & 2\\0 & 0 & 1 & -1 & 1\\0 & 0 & 0 & 1 & -2\\0 & 0 & 0 & 0 & 1\end{bmatrix}$

A2 (a) $\begin{bmatrix}1\\-1\\0\end{bmatrix}$

(b) $\begin{bmatrix}-3\\0\\2\end{bmatrix}$

(c) $\begin{bmatrix}-2\\-1\\2\end{bmatrix}$

A3 (a) $A^{-1} = \begin{bmatrix} 2 & -1 \\ -3 & 2 \end{bmatrix}$, $B^{-1} = \begin{bmatrix} -5 & 2 \\ 3 & -1 \end{bmatrix}$

(b) $(AB) = \begin{bmatrix} 5 & 9 \\ 9 & 16 \end{bmatrix}$, $(AB)^{-1} = \begin{bmatrix} -16 & 9 \\ 9 & -5 \end{bmatrix}$

(c) $(3A)^{-1} = \begin{bmatrix} 2/3 & -1/3 \\ -1 & 2/3 \end{bmatrix}$

(d) $(A^T)^{-1} = \begin{bmatrix} 2 & -3 \\ -1 & 2 \end{bmatrix}$

A4 (a) $[R_{\pi/6}]^{-1} = [R_{-\pi/6}] = \begin{bmatrix} \sqrt{3}/2 & 1/2 \\ -1/2 & \sqrt{3}/2 \end{bmatrix}$

(b) $\begin{bmatrix} 1 & 3 \\ 0 & 1 \end{bmatrix}$

(c) $\begin{bmatrix} 1/5 & 0 \\ 0 & 1 \end{bmatrix}$

(d) $\begin{bmatrix} 1 & 0 & 0 \\ 0 & -1 & 0 \\ 0 & 0 & 1 \end{bmatrix}$

A5 (a) $[S] = \begin{bmatrix} 1 & 0 \\ 2 & 1 \end{bmatrix}$, $[S^{-1}] = \begin{bmatrix} 1 & 0 \\ -2 & 1 \end{bmatrix}$

(b) $[R] = \begin{bmatrix} 0 & 1 \\ 1 & 0 \end{bmatrix} = [R^{-1}]$

(c) $[(R \circ S)^{-1}] = \begin{bmatrix} 0 & 1 \\ 1 & -2 \end{bmatrix}$, $[(S \circ R)^{-1}] = \begin{bmatrix} -2 & 1 \\ 1 & 0 \end{bmatrix}$

A6 Let $\vec{v}, \vec{y} \in \mathbb{R}^n$ and $t \in \mathbb{R}$. Then there exists $\vec{u}, \vec{x} \in \mathbb{R}^n$ such that $\vec{x} = M(\vec{y})$ and $\vec{u} = M(\vec{v})$. Then $L(\vec{x}) = \vec{y}$ and $L(\vec{u}) = \vec{v}$. Since L is linear $L(t\vec{x} + \vec{u}) = tL(\vec{x}) + L(\vec{u}) = t\vec{y} + \vec{v}$. It follows that

$$M(t\vec{y} + \vec{v}) = t\vec{x} + \vec{u} = tM(\vec{y}) + M(\vec{v})$$

So, M is linear.

Section 3.6

A Problems

A1 (a) $E = \begin{bmatrix} 1 & -5 & 0 \\ 0 & 1 & 0 \\ 0 & 0 & 1 \end{bmatrix}$, $EA = \begin{bmatrix} 6 & -13 & -17 \\ -1 & 3 & 4 \\ 4 & 2 & 0 \end{bmatrix}$

(b) $E = \begin{bmatrix} 1 & 0 & 0 \\ 0 & 0 & 1 \\ 0 & 1 & 0 \end{bmatrix}$, $EA = \begin{bmatrix} 1 & 2 & 3 \\ 4 & 2 & 0 \\ -1 & 3 & 4 \end{bmatrix}$

(c) $E = \begin{bmatrix} 1 & 0 & 0 \\ 0 & 1 & 0 \\ 0 & 0 & -1 \end{bmatrix}$, $EA = \begin{bmatrix} 1 & 2 & 3 \\ -1 & 3 & 4 \\ -4 & -2 & 0 \end{bmatrix}$

(d) $E = \begin{bmatrix} 1 & 0 & 0 \\ 0 & 6 & 0 \\ 0 & 0 & 1 \end{bmatrix}$, $EA = \begin{bmatrix} 1 & 2 & 3 \\ -6 & 18 & 24 \\ 4 & 2 & 0 \end{bmatrix}$

(e) $E = \begin{bmatrix} 1 & 0 & 0 \\ 0 & 1 & 0 \\ 4 & 0 & 1 \end{bmatrix}$, $EA = \begin{bmatrix} 1 & 2 & 3 \\ -1 & 3 & 4 \\ 8 & 10 & 12 \end{bmatrix}$

A2 (a) $\begin{bmatrix} 1 & 0 & 0 & 0 \\ 0 & 1 & 0 & 0 \\ 0 & 0 & 1 & 0 \\ 0 & 0 & -3 & 1 \end{bmatrix}$

(b) $\begin{bmatrix} 1 & 0 & 0 & 0 \\ 0 & 0 & 0 & 1 \\ 0 & 0 & 1 & 0 \\ 0 & 1 & 0 & 0 \end{bmatrix}$

(c) $\begin{bmatrix} 1 & 0 & 0 & 0 \\ 0 & 1 & 0 & 0 \\ 0 & 0 & -3 & 0 \\ 0 & 0 & 0 & 1 \end{bmatrix}$

(d) $\begin{bmatrix} 1 & 0 & 0 & 0 \\ 0 & 1 & 0 & 0 \\ 2 & 0 & 1 & 0 \\ 0 & 0 & 0 & 1 \end{bmatrix}$

(e) $\begin{bmatrix} 3 & 0 & 0 & 0 \\ 0 & 1 & 0 & 0 \\ 0 & 0 & 1 & 0 \\ 0 & 0 & 0 & 1 \end{bmatrix}$

(f) $\begin{bmatrix} 0 & 0 & 1 & 0 \\ 0 & 1 & 0 & 0 \\ 1 & 0 & 0 & 0 \\ 0 & 0 & 0 & 1 \end{bmatrix}$

A3 (a) It is elementary. The corresponding elementary row operation is $R_3 + (-4)R_2$.

(b) It is not elementary. Both row 1 and row 3 have been multiplied by -1.

(c) It is not elementary. We have multiplied row 1 by 3 and then added row 3 to row 1.

(d) It is elementary. The corresponding elementary row operation is $R_1 \leftrightarrow R_3$.

(e) It is not elementary. All three rows have been swapped.

(f) It is elementary. A corresponding elementary row operation is $(1)R_1$.

A4 (a) $E_1 = \begin{bmatrix} 1 & 0 & 0 \\ 0 & 0 & 1 \\ 0 & 1 & 0 \end{bmatrix}$, $E_2 = \begin{bmatrix} 1 & 0 & 0 \\ 0 & 1 & 0 \\ 0 & 0 & 1/2 \end{bmatrix}$, $E_3 = \begin{bmatrix} 1 & 0 & -4 \\ 0 & 1 & 0 \\ 0 & 0 & 1 \end{bmatrix}$, $E_4 = \begin{bmatrix} 1 & -3 & 0 \\ 0 & 1 & 0 \\ 0 & 0 & 1 \end{bmatrix}$

$$A^{-1} = \begin{bmatrix} 1 & -2 & -3 \\ 0 & 0 & 1 \\ 0 & 1/2 & 0 \end{bmatrix}$$

$$A = \begin{bmatrix} 1 & 0 & 0 \\ 0 & 0 & 1 \\ 0 & 1 & 0 \end{bmatrix}\begin{bmatrix} 1 & 0 & 0 \\ 0 & 1 & 0 \\ 0 & 0 & 2 \end{bmatrix}\begin{bmatrix} 1 & 0 & 4 \\ 0 & 1 & 0 \\ 0 & 0 & 1 \end{bmatrix}\begin{bmatrix} 1 & 3 & 0 \\ 0 & 1 & 0 \\ 0 & 0 & 1 \end{bmatrix}$$

(b) $E_1 = \begin{bmatrix} 1 & 0 & 0 \\ 0 & 1 & 0 \\ -2 & 0 & 1 \end{bmatrix}, E_2 = \begin{bmatrix} 1 & 0 & 0 \\ 0 & 1 & -3 \\ 0 & 0 & 1 \end{bmatrix}, E_3 = \begin{bmatrix} 1 & 0 & -2 \\ 0 & 1 & 0 \\ 0 & 0 & 1 \end{bmatrix}, E_4 = \begin{bmatrix} 1 & -2 & 0 \\ 0 & 1 & 0 \\ 0 & 0 & 1 \end{bmatrix}$

$$A^{-1} = \begin{bmatrix} -7 & -2 & 4 \\ 6 & 1 & -3 \\ -2 & 0 & 1 \end{bmatrix}$$

$$A = \begin{bmatrix} 1 & 0 & 0 \\ 0 & 1 & 0 \\ 2 & 0 & 1 \end{bmatrix}\begin{bmatrix} 1 & 0 & 0 \\ 0 & 1 & 3 \\ 0 & 0 & 1 \end{bmatrix}\begin{bmatrix} 1 & 0 & 2 \\ 0 & 1 & 0 \\ 0 & 0 & 1 \end{bmatrix}\begin{bmatrix} 1 & 2 & 0 \\ 0 & 1 & 0 \\ 0 & 0 & 1 \end{bmatrix}$$

(c) $E_1 = \begin{bmatrix} 1 & 0 & 0 \\ 2 & 1 & 0 \\ 0 & 0 & 1 \end{bmatrix}, E_2 = \begin{bmatrix} 1 & 0 & 0 \\ 0 & 1 & 0 \\ 4 & 0 & 1 \end{bmatrix}, E_3 = \begin{bmatrix} 1 & 0 & 0 \\ 0 & -1/4 & 0 \\ 0 & 0 & 1 \end{bmatrix},$

$E_4 = \begin{bmatrix} 1 & 0 & 0 \\ 0 & 0 & 1 \\ 0 & 1 & 0 \end{bmatrix}, E_5 = \begin{bmatrix} 1 & 0 & 1 \\ 0 & 1 & 0 \\ 0 & 0 & 1 \end{bmatrix}$

$$A^{-1} = \begin{bmatrix} 1/2 & -1/4 & 0 \\ 4 & 0 & 1 \\ -1/2 & -1/4 & 0 \end{bmatrix}$$

$$A = \begin{bmatrix} 1 & 0 & 0 \\ -2 & 1 & 0 \\ 0 & 0 & 1 \end{bmatrix}\begin{bmatrix} 1 & 0 & 0 \\ 0 & 1 & 0 \\ -4 & 0 & 1 \end{bmatrix}\begin{bmatrix} 1 & 0 & 0 \\ 0 & -4 & 0 \\ 0 & 0 & 1 \end{bmatrix}\begin{bmatrix} 1 & 0 & 0 \\ 0 & 0 & 1 \\ 0 & 1 & 0 \end{bmatrix}\begin{bmatrix} 1 & 0 & -1 \\ 0 & 1 & 0 \\ 0 & 0 & 1 \end{bmatrix}$$

(d) $E_1 = \begin{bmatrix} 1 & 0 & 0 & 0 \\ 1 & 1 & 0 & 0 \\ 0 & 0 & 1 & 0 \\ 0 & 0 & 0 & 1 \end{bmatrix}, E_2 = \begin{bmatrix} 1 & 0 & 0 & 0 \\ 0 & 1 & 0 & 0 \\ 0 & 0 & 1 & 0 \\ 2 & 0 & 0 & 1 \end{bmatrix}, E_3 = \begin{bmatrix} 1 & 2 & 0 & 0 \\ 0 & 1 & 0 & 0 \\ 0 & 0 & 1 & 0 \\ 0 & 0 & 0 & 1 \end{bmatrix},$

$E_4 = \begin{bmatrix} 1 & 0 & 0 & 0 \\ 0 & 1 & 0 & 0 \\ 0 & -1 & 1 & 0 \\ 0 & 0 & 0 & 1 \end{bmatrix}, E_5 = \begin{bmatrix} 1 & 0 & 0 & 0 \\ 0 & 1 & 0 & 0 \\ 0 & 0 & 1/2 & 0 \\ 0 & 0 & 0 & 1 \end{bmatrix}, E_6 = \begin{bmatrix} 1 & 0 & -4 & 0 \\ 0 & 1 & 0 & 0 \\ 0 & 0 & 1 & 0 \\ 0 & 0 & 0 & 1 \end{bmatrix},$

$E_7 = \begin{bmatrix} 1 & 0 & 0 & -1 \\ 0 & 1 & 0 & 0 \\ 0 & 0 & 1 & 0 \\ 0 & 0 & 0 & 1 \end{bmatrix}$

$$A^{-1} = \begin{bmatrix} 3 & 4 & -2 & -1 \\ 1 & 1 & 0 & 0 \\ -1/2 & -1/2 & 1/2 & 0 \\ 2 & 0 & 0 & 1 \end{bmatrix}$$

$$A = \begin{bmatrix} 1 & 0 & 0 & 0 \\ -1 & 1 & 0 & 0 \\ 0 & 0 & 1 & 0 \\ 0 & 0 & 0 & 1 \end{bmatrix}\begin{bmatrix} 1 & 0 & 0 & 0 \\ 0 & 1 & 0 & 0 \\ 0 & 0 & 1 & 0 \\ -2 & 0 & 0 & 1 \end{bmatrix}\begin{bmatrix} 1 & -2 & 0 & 0 \\ 0 & 1 & 0 & 0 \\ 0 & 0 & 1 & 0 \\ 0 & 0 & 0 & 1 \end{bmatrix}\begin{bmatrix} 1 & 0 & 0 & 0 \\ 0 & 1 & 0 & 0 \\ 0 & 1 & 1 & 0 \\ 0 & 0 & 0 & 1 \end{bmatrix}$$

$$\begin{bmatrix} 1 & 0 & 0 & 0 \\ 0 & 1 & 0 & 0 \\ 0 & 0 & 2 & 0 \\ 0 & 0 & 0 & 1 \end{bmatrix}\begin{bmatrix} 1 & 0 & 4 & 0 \\ 0 & 1 & 0 & 0 \\ 0 & 0 & 1 & 0 \\ 0 & 0 & 0 & 1 \end{bmatrix}\begin{bmatrix} 1 & 0 & 0 & 1 \\ 0 & 1 & 0 & 0 \\ 0 & 0 & 1 & 0 \\ 0 & 0 & 0 & 1 \end{bmatrix}$$

Section 3.7

A Problems

A1 (a) $\begin{bmatrix} 1 & 0 & 0 \\ 2 & 1 & 0 \\ -1 & 0 & 1 \end{bmatrix}\begin{bmatrix} -2 & -1 & 5 \\ 0 & 2 & -12 \\ 0 & 0 & 8 \end{bmatrix}$

(b) $\begin{bmatrix} 1 & 0 & 0 \\ 3 & 1 & 0 \\ 2 & 3/2 & 1 \end{bmatrix}, \begin{bmatrix} 1 & -2 & 4 \\ 0 & 4 & -8 \\ 0 & 0 & -1 \end{bmatrix}$

(c) $\begin{bmatrix} 1 & 0 & 0 \\ 1 & 1 & 0 \\ 1 & 1/3 & 1 \end{bmatrix}\begin{bmatrix} 2 & -4 & 5 \\ 0 & 9 & -3 \\ 0 & 0 & 1 \end{bmatrix}$

(d) $\begin{bmatrix} 1 & 0 & 0 & 0 \\ -2 & 1 & 0 & 0 \\ 0 & 1/2 & 1 & 0 \\ 0 & 0 & 0 & 1 \end{bmatrix}\begin{bmatrix} 1 & 5 & 3 & 4 \\ 0 & 4 & 5 & 11 \\ 0 & 0 & -7/2 & -13/2 \\ 0 & 0 & 0 & 0 \end{bmatrix}$

(e) $\begin{bmatrix} 1 & 0 & 0 & 0 \\ 0 & 1 & 0 & 0 \\ 3 & -1 & 1 & 0 \\ 0 & -4/3 & 17/9 & 1 \end{bmatrix}\begin{bmatrix} 1 & -2 & 1 & 1 \\ 0 & -3 & -2 & 1 \\ 0 & 0 & -3 & -3 \\ 0 & 0 & 0 & 7 \end{bmatrix}$

(f) $\begin{bmatrix} 1 & 0 & 0 & 0 \\ -2 & 1 & 0 & 0 \\ -3/2 & 3/2 & 1 & 0 \\ -1 & -2 & 2 & 1 \end{bmatrix}\begin{bmatrix} -2 & -1 & 2 & 0 \\ 0 & 1 & 2 & 2 \\ 0 & 0 & 4 & 0 \\ 0 & 0 & 0 & 0 \end{bmatrix}$

A2 (a) $LU = \begin{bmatrix} 1 & 0 & 0 \\ -2 & 1 & 0 \\ -1 & 4 & 1 \end{bmatrix}\begin{bmatrix} 1 & 0 & 3 \\ 0 & 1 & 3 \\ 0 & 0 & -4 \end{bmatrix}; \vec{x}_1 = \begin{bmatrix} -3 \\ -4 \\ 2 \end{bmatrix}, \vec{x}_2 = \begin{bmatrix} 5 \\ 2 \\ -1 \end{bmatrix}$

(b) $LU = \begin{bmatrix} 1 & 0 & 0 \\ -1 & 1 & 0 \\ 3 & 1 & 1 \end{bmatrix}\begin{bmatrix} 1 & 0 & -2 \\ 0 & -4 & 2 \\ 0 & 0 & 3 \end{bmatrix}; \vec{x}_1 = \begin{bmatrix} 3 \\ 3 \\ 2 \end{bmatrix}, \vec{x}_2 = \begin{bmatrix} -4 \\ -2 \\ -3 \end{bmatrix}$

(c) $LU = \begin{bmatrix} 1 & 0 & 0 \\ -3 & 1 & 0 \\ -3 & 2 & 1 \end{bmatrix}\begin{bmatrix} 1 & 0 & 1 \\ 0 & 2 & 2 \\ 0 & 0 & 1 \end{bmatrix}; \vec{x}_1 = \begin{bmatrix} 3 \\ 2 \\ 0 \end{bmatrix}, \vec{x}_2 = \begin{bmatrix} -3 \\ -3 \\ -1 \end{bmatrix}$

(d) $LU = \begin{bmatrix} 1 & 0 & 0 & 0 \\ 0 & 1 & 0 & 0 \\ -3 & 2 & 1 & 0 \\ -1 & 0 & 0 & 1 \end{bmatrix}\begin{bmatrix} -1 & 2 & -3 & 0 \\ 0 & -1 & 3 & 1 \\ 0 & 0 & -12 & 0 \\ 0 & 0 & 0 & 1 \end{bmatrix}; \vec{x}_1 = \begin{bmatrix} -1 \\ 1 \\ 3 \\ -1 \end{bmatrix}, \vec{x}_2 = \begin{bmatrix} -5 \\ -3 \\ -2 \\ 0 \end{bmatrix}$

Chapter 3 Quiz

E Problems

E1 (a) $\begin{bmatrix} -14 & 1 & -17 \\ -1 & 10 & -39 \end{bmatrix}$

(b) Not defined

(c) $\begin{bmatrix} -3 & -38 \\ 0 & -23 \\ -8 & -42 \end{bmatrix}$

E2 (a) $f_A(\vec{u}) = \begin{bmatrix} -11 \\ 0 \end{bmatrix}$, $f_A(\vec{v}) = \begin{bmatrix} -16 \\ 17 \end{bmatrix}$

(b) $\begin{bmatrix} -16 & -11 \\ 17 & 0 \end{bmatrix}$

E3 (a) $[R] = [R_{\pi/3}] = \begin{bmatrix} 1/2 & -\sqrt{3}/2 & 0 \\ \sqrt{3}/2 & 1/2 & 0 \\ 0 & 0 & 1 \end{bmatrix}$

(b) $[M] = [\text{refl}_{(-1,-1,2)}] = \dfrac{1}{3}\begin{bmatrix} 2 & -1 & 2 \\ -1 & 2 & 2 \\ 2 & 2 & -1 \end{bmatrix}$

(c) $[R \circ M] = \dfrac{1}{6}\begin{bmatrix} 2+\sqrt{3} & -1-2\sqrt{3} & 2-2\sqrt{3} \\ 2\sqrt{3}-1 & -\sqrt{3}+2 & 2\sqrt{3}+2 \\ 4 & 4 & -2 \end{bmatrix}$

E4 The solution space of $A\vec{x} = \vec{0}$ is $\vec{x} = s\begin{bmatrix} -2 \\ 1 \\ 1 \\ 0 \\ 0 \end{bmatrix} + t\begin{bmatrix} -1 \\ 0 \\ 0 \\ 1 \\ 0 \end{bmatrix}$, $s, t \in \mathbb{R}$. The solution set of $A\vec{x} = \vec{b}$ is $\vec{x} = \begin{bmatrix} 5 \\ 6 \\ 0 \\ 0 \\ 7 \end{bmatrix} + s\begin{bmatrix} -2 \\ 1 \\ 1 \\ 0 \\ 0 \end{bmatrix} + t\begin{bmatrix} -1 \\ 0 \\ 0 \\ 1 \\ 0 \end{bmatrix}$, $s, t \in \mathbb{R}$. In particular, the solution set is obtained from the solution space of $A\vec{x} = \vec{0}$ by translating by the vector $\begin{bmatrix} 5 \\ 6 \\ 0 \\ 0 \\ 7 \end{bmatrix}$.

E5 (a) $\vec{u}$ is not in the columnspace of B. $\vec{v}$ is in the columnspace of B.

(b) $\vec{x} = \begin{bmatrix} -1 \\ -2 \\ -3 \end{bmatrix}$

(c) $\vec{y} = \begin{bmatrix} 0 \\ 1 \\ 0 \end{bmatrix}$

E6 A basis for the rowspace of A is $\left\{ \begin{bmatrix} 1 \\ 0 \\ 1 \\ 0 \\ -1 \end{bmatrix}, \begin{bmatrix} 0 \\ 1 \\ -1 \\ 0 \\ 3 \end{bmatrix}, \begin{bmatrix} 0 \\ 0 \\ 0 \\ 1 \\ 2 \end{bmatrix} \right\}$. A basis for the columnspace of A is $\left\{ \begin{bmatrix} 1 \\ 2 \\ 0 \\ 3 \end{bmatrix}, \begin{bmatrix} 0 \\ 1 \\ 2 \\ 3 \end{bmatrix}, \begin{bmatrix} 1 \\ 2 \\ 1 \\ 4 \end{bmatrix} \right\}$. A basis for the nullspace is $\left\{ \begin{bmatrix} -1 \\ 1 \\ 1 \\ 0 \\ 0 \end{bmatrix}, \begin{bmatrix} 1 \\ -3 \\ 0 \\ -2 \\ 1 \end{bmatrix} \right\}$.

E7 $A^{-1} = \begin{bmatrix} 2/3 & 0 & 0 & 1/3 \\ 1/6 & 0 & 1/2 & -1/6 \\ 0 & 1 & 0 & 0 \\ -1/3 & 0 & 0 & 1/3 \end{bmatrix}$

E8 The matrix is invertible only for $p \neq 1$. The inverse is $\frac{1}{1-p}\begin{bmatrix} 1 & p & -p \\ -1 & 1-2p & p \\ -1 & -1 & 1 \end{bmatrix}$.

E9 By definition, the range of L is a subset of $\mathbb{R}^m$. We have $L(\vec{0}) = \vec{0}$, so $\vec{0} \in \text{Range}(L)$. If $\vec{x}, \vec{y} \in \text{Range}(L)$, then there exists $\vec{u}, \vec{v} \in \mathbb{R}^n$ such that $L(\vec{u}) = \vec{x}$ and $L(\vec{v}) = \vec{y}$. Hence, $L(\vec{u} + \vec{v}) = L(\vec{u}) + L(\vec{v}) = \vec{x} + \vec{y}$, so $\vec{x} + \vec{y} \in \text{Range}(L)$. Similarly, $L(t\vec{u}) = tL(\vec{u}) = t\vec{x}$, so $t\vec{x} \in \text{Range}(L)$. Thus, L is a subspace of $\mathbb{R}^m$.

E10 Consider $c_1L(\vec{v}_1) + \cdots + c_kL(\vec{v}_k) = \vec{0}$. Since L is linear, we get $L(c_1\vec{v}_1 + \cdots + c_k\vec{v}_k) = \vec{0}$. Thus, $c_1\vec{v}_1 + \cdots + c_k\vec{v}_k \in \text{Null}(L)$ and so $c_1\vec{v}_1 + \cdots + c_k\vec{v}_k = \vec{0}$. This implies that $c_1 = \cdots = c_k = 0$ since $\{\vec{v}_1, \dots, \vec{v}_k\}$ is linearly independent. Therefore, $\{L(\vec{v}_1), \dots, L(\vec{v}_k)\}$ is linearly independent.

E11 (a) $E_1 = \begin{bmatrix} 1 & 0 & 0 \\ 0 & 1/2 & 0 \\ 0 & 0 & 1 \end{bmatrix}$, $E_2 = \begin{bmatrix} 1 & 0 & 0 \\ 0 & 1 & 0 \\ 0 & 0 & 1/4 \end{bmatrix}$, $E_3 = \begin{bmatrix} 1 & 0 & 2 \\ 0 & 1 & 0 \\ 0 & 0 & 1 \end{bmatrix}$, $E_4 = \begin{bmatrix} 1 & 0 & 0 \\ 0 & 1 & 3/2 \\ 0 & 0 & 1 \end{bmatrix}$

(b) $A = \begin{bmatrix} 1 & 0 & 0 \\ 0 & 2 & 0 \\ 0 & 0 & 1 \end{bmatrix}\begin{bmatrix} 1 & 0 & 0 \\ 0 & 1 & 0 \\ 0 & 0 & 4 \end{bmatrix}\begin{bmatrix} 1 & 0 & -2 \\ 0 & 1 & 0 \\ 0 & 0 & 1 \end{bmatrix}\begin{bmatrix} 1 & 0 & 0 \\ 0 & 1 & -3/2 \\ 0 & 0 & 1 \end{bmatrix}$

E12 (a) $K = I_3$

(b) There is no matrix K.

(c) The range cannot be spanned by $\begin{bmatrix} 1 \\ 1 \end{bmatrix}$ because this vector is not in $\mathbb{R}^3$.

(d) The matrix of L is any multiple of $\begin{bmatrix} 1 & -2/3 \\ 1 & -2/3 \\ 2 & -4/3 \end{bmatrix}$

(e) This contradicts the Rank Theorem, so there can be no such mapping L.

(f) This contradicts Theorem 3.5.2, so there can be no such matrix.

CHAPTER 4

Section 4.1

A Problems

A1 (a) $-1 - 6x + 4x^2 + 6x^3$

(b) $-3 + 6x - 6x^2 - 3x^3 - 12x^4$

(c) $-1 + 9x - 11x^2 - 17x^3$

(d) $-3 + 2x + 6x^2$

(e) $7 - 2x - 5x^2$

(f) $\frac{7}{3} - \frac{4}{3}x + \frac{13}{3}x^2$

(g) $\sqrt{2} - \pi + \sqrt{2}x + (\sqrt{2} + \pi)x^2$

A2 (a) $0 = 0(1 + x^2 + x^3) + 0(2 + x + x^3) + 0(-1 + x + 2x^2 + x^3)$

(b) $2 + 4x + 3x^2 + 4x^3$ is not in the span.

(c) $-x + 2x^2 + x^3 = 2(1 + x^2 + x^3) + (-1)(2 + x + x^3) + 0(-1 + x + 2x^2 + x^3)$

(d) $-4 - x + 3x^2 = 1(1 + x^2 + x^3) + (-2)(2 + x + x^3) + 1(-1 + x + 2x^2 + x^3)$

(e) $-1 - 7x + 5x^2 + 4x^3 = (-3)(1 + x^2 + x^3) + 3(2 + x + x^3) + 4(-1 + x + 2x^2 + x^3)$

(f) $2 + x + 5x^3$ is not in the span.

A3 (a) The set is linearly independent.

(b) The set is linearly dependent. We have

$$0 = (-3t)(1 + x + x^2) + tx + t(x^2 + x^3) + t(3 + 2x + 2x^2 - x^3), \quad t \in \mathbb{R}$$

(c) The set is linearly independent.

(d) The set is linearly dependent. We have

$$0 = (-2t)(1+x+x^3+x^4)+t(2+x-x^2+x^3+x^4)+t(x+x^2+x^3+x^4), \quad t \in \mathbb{R}$$

A4 Consider

$$a_1 + a_2x + a_3x^2 = t_1 1 + t_2(x-1) + t_3(x-1)^2 = (t_1 - t_2 + t_3) + (t_2 - 2t_3)x + t_3x^2$$

The corresponding augmented matrix is $\left[\begin{array}{ccc|c} 1 & -1 & 1 & a_1 \\ 0 & 1 & -2 & a_2 \\ 0 & 0 & 1 & a_3 \end{array}\right]$.

Since there is a leading 1 in each row, the system is consistent for all polynomials $a_1 + a_2x + a_3x^2$. Thus, $\mathcal{B}$ spans P_2. Moreover, since there is a leading 1 in each column, there is a unique solution and so $\mathcal{B}$ is also linearly independent. Therefore, it is a basis for P_2.

Section 4.2

A Problems

A1 (a) It is a subspace.

(b) It is a subspace.

(c) It is a subspace.

(d) It is not a subspace.

(e) It is not a subspace.

(f) It is a subspace.

A2 (a) It is a subspace.

(b) It is not a subspace.

(c) It is a subspace.

(d) It is a subspace.

A3 (a) It is a subspace.

(b) It is a subspace.

(c) It is a subspace.

(d) It is not a subspace.

(e) It is a subspace.

A4 (a) It is a subspace.

(b) It is not a subspace.

(c) It is a subspace.

(d) It is not a subspace.

A5 Let the set be $\{\mathbf{v}_1, \dots, \mathbf{v}_k\}$ and assume that $\mathbf{v}_i$ is the zero vector. Then we have

$$\mathbf{0} = 0\mathbf{v}_1 + \cdots + 0\mathbf{v}_{i-1} + \mathbf{v}_i + 0\mathbf{v}_{i+1} + \cdots + 0\mathbf{v}_k$$

Hence, by definition, $\{\mathbf{v}_1, \dots, \mathbf{v}_k\}$ is linearly dependent.

Section 4.3

A Problems

A1 (a) It is a basis.

(b) Since it only has two vectors in $\mathbb{R}^3$, it cannot span $\mathbb{R}^3$ and hence cannot be a basis.

(c) Since it has four vectors in $\mathbb{R}^3$, it is linearly dependent and hence cannot be a basis.

(d) It is not a basis.

(e) It is a basis.

A2 Show that it is a linearly independent spanning set.

A3 (a) One possible basis is $\left\{ \begin{bmatrix} 1 \\ -2 \\ 1 \end{bmatrix}, \begin{bmatrix} 0 \\ 1 \\ 2 \end{bmatrix} \right\}$. Hence, the dimension is 2.

(b) One possible basis is $\left\{ \begin{bmatrix} 1 \\ 3 \\ 2 \end{bmatrix}, \begin{bmatrix} -1 \\ -1 \\ 2 \end{bmatrix}, \begin{bmatrix} 0 \\ 1 \\ 1 \end{bmatrix} \right\}$. Hence, the dimension is 3.

A4 (a) One possible basis is $\left\{ \begin{bmatrix} 1 & 1 \\ -1 & 1 \end{bmatrix}, \begin{bmatrix} 0 & 1 \\ 3 & -1 \end{bmatrix}, \begin{bmatrix} 1 & -1 \\ 2 & -3 \end{bmatrix} \right\}$. Hence, the dimension is 3.

(b) One possible basis is $\left\{ \begin{bmatrix} 1 & -1 \\ -1 & -1 \end{bmatrix}, \begin{bmatrix} 1 & 1 \\ 1 & -1 \end{bmatrix}, \begin{bmatrix} 0 & 1 \\ 2 & 1 \end{bmatrix}, \begin{bmatrix} 0 & 1 \\ 0 & 1 \end{bmatrix} \right\}$. Hence, the dimension is 4.

(c) $\mathcal{B}$ is a basis. Thus, the dimension is 4.

A5 (a) The dimension is 3.

(b) The dimension is 3.

(c) The dimension is 4.

A6 Alternate correct answers are possible.

(a) $\left\{\begin{bmatrix}1\\2\\0\end{bmatrix}, \begin{bmatrix}1\\0\\2\end{bmatrix}\right\}$

(b) $\left\{\begin{bmatrix}1\\2\\0\end{bmatrix}, \begin{bmatrix}1\\0\\2\end{bmatrix}, \begin{bmatrix}2\\-1\\-1\end{bmatrix}\right\}$

A7 Alternate correct answers are possible.

(a) $\left\{\begin{bmatrix}1\\0\\0\\1\end{bmatrix}, \begin{bmatrix}1\\0\\-1\\0\end{bmatrix}, \begin{bmatrix}1\\1\\0\\0\end{bmatrix}\right\}$

(b) $\left\{\begin{bmatrix}1\\0\\0\\1\end{bmatrix}, \begin{bmatrix}1\\0\\-1\\0\end{bmatrix}, \begin{bmatrix}1\\1\\0\\0\end{bmatrix}, \begin{bmatrix}1\\0\\0\\0\end{bmatrix}\right\}$

A8 (a) One possible basis is $\{x, 1-x^2\}$. Hence, the dimension is 2.

(b) One possible basis is $\left\{\begin{bmatrix}1 & 0\\0 & 0\end{bmatrix}, \begin{bmatrix}0 & 1\\0 & 0\end{bmatrix}, \begin{bmatrix}0 & 0\\0 & 1\end{bmatrix}\right\}$. Hence, the dimension is 3.

(c) One possible basis is $\left\{\begin{bmatrix}0\\1\\-1\end{bmatrix}, \begin{bmatrix}1\\0\\-1\end{bmatrix}\right\}$. Hence, the dimension is 2.

(d) One possible basis is $\{x-2, x^2-4\}$. Hence, the dimension is 2.

(e) One possible basis is $\left\{\begin{bmatrix}1 & 0\\-1 & 0\end{bmatrix}, \begin{bmatrix}0 & 1\\0 & 0\end{bmatrix}, \begin{bmatrix}0 & 0\\0 & 1\end{bmatrix}\right\}$. Hence, the dimension is 3.

Section 4.4

A Problems

A1 (a) $[\mathbf{x}]_{\mathcal{B}} = \begin{bmatrix}3\\-2\end{bmatrix}$, $[\mathbf{y}]_{\mathcal{B}} = \begin{bmatrix}2\\3\end{bmatrix}$

(b) $[\mathbf{x}]_{\mathcal{B}} = \begin{bmatrix}5\\1\\-2\end{bmatrix}$, $[\mathbf{y}]_{\mathcal{B}} = \begin{bmatrix}2\\2\\-2\end{bmatrix}$

(c) $[\mathbf{x}]_{\mathcal{B}} = \begin{bmatrix}-2\\3\\1\end{bmatrix}$, $[\mathbf{y}]_{\mathcal{B}} = \begin{bmatrix}-2\\3\\-1\end{bmatrix}$

(d) $[\mathbf{x}]_{\mathcal{B}} = \begin{bmatrix}1\\1\end{bmatrix}$, $[\mathbf{y}]_{\mathcal{B}} = \begin{bmatrix}3\\-2\end{bmatrix}$

(e) $[\mathbf{x}]_{\mathcal{B}} = \begin{bmatrix}5\\-3\\4\end{bmatrix}$, $[\mathbf{y}]_{\mathcal{B}} = \begin{bmatrix}-1\\2\\-1\end{bmatrix}$

A2 (a) Show that it is linearly independent and spans the plane.

(b) $\begin{bmatrix}3\\2\\1\end{bmatrix}$ and $\begin{bmatrix}5\\2\\3\end{bmatrix}$ are not in the plane. We have $\begin{bmatrix}3\\2\\2\end{bmatrix}_{\mathcal{B}} = \begin{bmatrix}3\\-2\end{bmatrix}$.

A3 (a) Show that it is linearly independent and spans P_2.

(b) $[p(x)]_{\mathcal{B}} = \begin{bmatrix}-1\\1\\-1\end{bmatrix}$, $[q(x)]_{\mathcal{B}} = \begin{bmatrix}4\\1\\1\end{bmatrix}$, $[r(x)]_{\mathcal{B}} = \begin{bmatrix}2\\-1\\3\end{bmatrix}$

(c) $[2 - 4x + 10x^2]_{\mathcal{B}} = \begin{bmatrix}6\\0\\4\end{bmatrix}$. We have

$$\begin{aligned}[4 - 2x + 7x^2]_{\mathcal{B}} + [-2 - 2x + 3x^2]_{\mathcal{B}} &= \begin{bmatrix}4\\1\\1\end{bmatrix} + \begin{bmatrix}2\\-1\\3\end{bmatrix} = \begin{bmatrix}6\\0\\4\end{bmatrix}\\ &= [2 - 4x + 10x^2]_{\mathcal{B}} = [(4-2) + (-2-2)x + (7+3)x^2]_{\mathcal{B}}\end{aligned}$$

A4 (a) i. A is in the span of $\mathcal{B}$.

ii. $\mathcal{B}$ is linearly independent, so it forms a basis for Span $\mathcal{B}$.

iii. $[A]_{\mathcal{B}} = \begin{bmatrix}-1/2\\1/2\\3/4\end{bmatrix}$

(b) i. A is not in the span of $\mathcal{B}$.

ii. $\mathcal{B}$ is linearly independent, so it forms a basis for Span $\mathcal{B}$.

A5 (a) The change of coordinates matrix Q from $\mathcal{B}$-coordinates to $\mathcal{S}$-coordinates is $\begin{bmatrix}3 & 0 & -2\\4 & 1 & -3\\1 & 0 & 3\end{bmatrix}$. The change of coordinates matrix P from $\mathcal{S}$-coordinates to $\mathcal{B}$-coordinates is $P = \begin{bmatrix}3/11 & 0 & 2/11\\-15/11 & 1 & 1/11\\-1/11 & 0 & 3/11\end{bmatrix}$.

(b) The change of coordinates matrix Q from $\mathcal{B}$-coordinates to $\mathcal{S}$-coordinates is $\begin{bmatrix}1 & 1 & 0\\-2 & 0 & 1\\5 & -2 & 1\end{bmatrix}$. The change of coordinates matrix P from $\mathcal{S}$-coordinates to $\mathcal{B}$-coordinates is $P = \begin{bmatrix}2/9 & -1/9 & 1/9\\7/9 & 1/9 & -1/9\\4/9 & 7/9 & 2/9\end{bmatrix}$.

(c) The change of coordinates matrix Q from $\mathcal{B}$-coordinates to $\mathcal{S}$-coordinates is $\begin{bmatrix}1 & 0 & 2\\-1 & -4 & 1\\-1 & -1 & 1\end{bmatrix}$. The change of coordinates matrix P from $\mathcal{S}$-coordinates to $\mathcal{B}$-coordinates is $P = \begin{bmatrix}1/3 & 2/9 & -8/9\\0 & -1/3 & 1/3\\1/3 & -1/9 & 4/9\end{bmatrix}$.

Section 4.5

A Problems

A1 Show that each mapping preserves addition and scalar multiplication.

A2 (a) det is not linear. (b) L is linear.
(c) T is not linear. (d) M is linear.

A3 (a) $\mathbf{y}$ is in the range of L. We have $L\left(\begin{bmatrix}1\\2\\1\end{bmatrix}\right) = \mathbf{y}$.

(b) $\mathbf{y}$ is in the range of L. We have $L(1 + 2x + x^2) = \mathbf{y}$.

(c) $\mathbf{y}$ is not in the range of L.

(d) $\mathbf{y}$ is not in the range of L.

A4 Alternate correct answers are possible.

(a) A basis for Range(L) is $\left\{\begin{bmatrix}1\\1\end{bmatrix}, \begin{bmatrix}0\\1\end{bmatrix}\right\}$. A basis for Null($L$) is $\left\{\begin{bmatrix}1\\-1\\0\end{bmatrix}\right\}$. We have $\text{rank}(L) + \text{Null}(L) = 2 + 1 = \dim \mathbb{R}^3$.

(b) A basis for Range(L) is $\{1 + x, x\}$. A basis for Null(L) is $\left\{\begin{bmatrix}1\\-1\\0\end{bmatrix}\right\}$. We have $\text{rank}(L) + \text{Null}(L) = 2 + 1 = \dim \mathbb{R}^3$.

(c) A basis for Range(L) is $\{1\}$. A basis for Null(L) is $\left\{\begin{bmatrix}1 & 0\\0 & -1\end{bmatrix}, \begin{bmatrix}0 & 1\\0 & 0\end{bmatrix}, \begin{bmatrix}0 & 0\\1 & 0\end{bmatrix}\right\}$. We have $\text{rank}(L) + \text{Null}(L) = 1 + 3 = \dim M(2, 2)$.

(d) A basis for Range(L) is $\left\{\begin{bmatrix}1 & 0\\0 & 0\end{bmatrix}, \begin{bmatrix}0 & 1\\0 & 0\end{bmatrix}, \begin{bmatrix}0 & 0\\1 & 0\end{bmatrix}, \begin{bmatrix}0 & 0\\0 & 1\end{bmatrix}\right\}$. A basis for Null($L$) is the empty set. We have $\text{rank}(L) + \text{Null}(L) = 4 + 0 = \dim P_3$.

Section 4.6

A Problems

A1 (a) $[L]_\mathcal{B} = \begin{bmatrix}0 & 2\\1 & -1\end{bmatrix}$, $[L(\mathbf{x})]_\mathcal{B} = \begin{bmatrix}6\\1\end{bmatrix}$

(b) $[L]_\mathcal{B} = \begin{bmatrix}2 & 2 & 0\\0 & 0 & 4\\-1 & -1 & 5\end{bmatrix}$, $[L(\mathbf{x})]_\mathcal{B} = \begin{bmatrix}12\\-4\\-11\end{bmatrix}$

A2 (a) $[L]_\mathcal{B} = \begin{bmatrix}-3 & 0\\0 & 4\end{bmatrix}$

(b) $[L]_\mathcal{B} = \begin{bmatrix}0 & 2\\1 & 0\end{bmatrix}$

A3 (a) $[L(\mathbf{v}_1)]_{\mathcal{B}} = \begin{bmatrix} 0 \\ 1 \\ 0 \end{bmatrix}, [L(\mathbf{v}_2)]_{\mathcal{B}} = \begin{bmatrix} 1 \\ 0 \\ 0 \end{bmatrix}, [L(\mathbf{v}_3)]_{\mathcal{B}} = \begin{bmatrix} 5 \\ 0 \\ 0 \end{bmatrix}, [L]_{\mathcal{B}} = \begin{bmatrix} 0 & 1 & 5 \\ 1 & 0 & 0 \\ 0 & 0 & 0 \end{bmatrix}$

(b) $[L(\mathbf{v}_1)]_{\mathcal{B}} = \begin{bmatrix} 0 \\ 0 \\ 1 \end{bmatrix}, [L(\mathbf{v}_2)]_{\mathcal{B}} = \begin{bmatrix} 1 \\ 0 \\ 0 \end{bmatrix}, [L(\mathbf{v}_3)]_{\mathcal{B}} = \begin{bmatrix} 0 \\ 1 \\ 0 \end{bmatrix}, [L]_{\mathcal{B}} = \begin{bmatrix} 0 & 1 & 0 \\ 0 & 0 & 1 \\ 1 & 0 & 0 \end{bmatrix}$

(c) $[L(\mathbf{v}_1)]_{\mathcal{B}} = \begin{bmatrix} 1 \\ 0 \\ 0 \end{bmatrix}, [L(\mathbf{v}_2)]_{\mathcal{B}} = \begin{bmatrix} 1 \\ 1 \\ 1 \end{bmatrix}, [L(\mathbf{v}_3)]_{\mathcal{B}} = \begin{bmatrix} 3 \\ -2 \\ -1 \end{bmatrix}, [L]_{\mathcal{B}} = \begin{bmatrix} 1 & 1 & 3 \\ 0 & 1 & -2 \\ 0 & 1 & -1 \end{bmatrix}$

A4 (a) $\mathcal{B} = \left\{ \begin{bmatrix} 1 \\ -2 \end{bmatrix}, \begin{bmatrix} 2 \\ 1 \end{bmatrix} \right\}, [\text{refl}_{(1,-2)}]_{\mathcal{B}} = \begin{bmatrix} -1 & 0 \\ 0 & 1 \end{bmatrix}$

(b) $\mathcal{B} = \left\{ \begin{bmatrix} 2 \\ 1 \\ -1 \end{bmatrix}, \begin{bmatrix} 1 \\ 0 \\ 2 \end{bmatrix}, \begin{bmatrix} 0 \\ 1 \\ 1 \end{bmatrix} \right\}, [\text{proj}_{(2,1,-1)}]_{\mathcal{B}} = \begin{bmatrix} 1 & 0 & 0 \\ 0 & 0 & 0 \\ 0 & 0 & 0 \end{bmatrix}$

(c) $\mathcal{B} = \left\{ \begin{bmatrix} -1 \\ -1 \\ 1 \end{bmatrix}, \begin{bmatrix} 1 \\ 0 \\ 1 \end{bmatrix}, \begin{bmatrix} 0 \\ 1 \\ 1 \end{bmatrix} \right\}, [\text{refl}_{(-1,-1,1)}]_{\mathcal{B}} = \begin{bmatrix} -1 & 0 & 0 \\ 0 & 1 & 0 \\ 0 & 0 & 1 \end{bmatrix}$

A5 (a) $\begin{bmatrix} 1 \\ 2 \\ 4 \end{bmatrix}_{\mathcal{B}} = \begin{bmatrix} 2 \\ -1 \\ 1 \end{bmatrix}$

(b) $[L]_{\mathcal{B}} = \begin{bmatrix} 2 & 0 & 0 \\ -1 & 0 & 2 \\ 1 & 1 & 0 \end{bmatrix}$

(c) $L(1,2,4) = \begin{bmatrix} 4 \\ 1 \\ 6 \end{bmatrix}$

A6 (a) $\begin{bmatrix} 5 \\ 3 \\ -5 \end{bmatrix}_{\mathcal{B}} = \begin{bmatrix} 2 \\ 3 \\ -3 \end{bmatrix}$

(b) $[L]_{\mathcal{B}} = \begin{bmatrix} 0 & -2 & 2 \\ 0 & 0 & 3 \\ 1 & 0 & -3 \end{bmatrix}$

(c) $L(5,3,-5) = \begin{bmatrix} -21 \\ -7 \\ 23 \end{bmatrix}$

A7 (a) $[L]_{\mathcal{B}} = \begin{bmatrix} 11 & 16 \\ -4 & -3 \end{bmatrix}$

(b) $[L]_{\mathcal{B}} = \begin{bmatrix} 5 & 0 \\ 0 & -5 \end{bmatrix}$

(c) $[L]_{\mathcal{B}} = \begin{bmatrix} -40 & -118 \\ 18 & 52 \end{bmatrix}$

(d) $[L]_{\mathcal{B}} = \begin{bmatrix} 4 & 0 \\ 0 & 6 \end{bmatrix}$

(e) $[L]_{\mathcal{B}} = \begin{bmatrix} 4 & 2 & 0 \\ 0 & 4 & 2 \\ 0 & 0 & 4 \end{bmatrix}$

(f) $[L]_{\mathcal{B}} = \begin{bmatrix} 2 & 0 & 0 \\ 0 & -2 & 0 \\ 0 & 0 & 3 \end{bmatrix}$

A8 (a) $[L]_{\mathcal{B}} = \begin{bmatrix} 2 & 1 & 0 \\ 0 & 1 & 1 \\ -2 & -3 & -2 \end{bmatrix}$

(b) $[L]_{\mathcal{B}} = \begin{bmatrix} 1 & -1 & 1 \\ 0 & 2 & -1 \\ 0 & 2 & -1 \end{bmatrix}$

(c) $[D]_{\mathcal{B}} = \begin{bmatrix} 0 & 1 & 0 \\ 0 & 0 & 2 \\ 0 & 0 & 0 \end{bmatrix}$

(d) $[T]_{\mathcal{B}} = \begin{bmatrix} -1 & -1 & -2 \\ 0 & 0 & -1 \\ 2 & 2 & 4 \end{bmatrix}$

Section 4.7

A Problems

A1 In each case, verify that the given mapping is linear, one-to-one, and onto.

(a) Define $L(a + bx + cx^2 + dx^3) = \begin{bmatrix} a \\ b \\ c \\ d \end{bmatrix}$.

(b) Define $L\left(\begin{bmatrix} a & b \\ c & d \end{bmatrix}\right) = \begin{bmatrix} a \\ b \\ c \\ d \end{bmatrix}$.

(c) Define $L(a + bx + cx^2 + dx^3) = \begin{bmatrix} a & b \\ c & d \end{bmatrix}$.

(d) Define $L(a_1(x - 2) + a_2(x^2 - 2x)) = \begin{bmatrix} a_1 & 0 \\ 0 & a_2 \end{bmatrix}$.

Chapter 4 Quiz

E Problems

E1 (a) The given set is a subset of $M(4, 3)$ and is non-empty since it clearly contains the zero matrix. Let A and B be any two vectors in the set. Then $a_{11}+a_{12}+a_{13} = 0$ and $b_{11} + b_{12} + b_{13} = 0$. Then the first row of $A + B$ satisfies

$$a_{11}+b_{11}+a_{12}+b_{12}+a_{13}+b_{13} = a_{11}+a_{12}+a_{13}+b_{11}+b_{12}+b_{13} = 0+0 = 0$$

so the subset is closed under addition. Similarly, for any $t \in \mathbb{R}$, the first row of tA satisfies

$$ta_{11} + ta_{12} + ta_{13} = t(a_{11} + a_{12} + a_{13}) = 0$$

so the subset is also closed under scalar multiplication. Thus, it is a subspace of $M(4,3)$ and hence a vector space.

(b) The given set is a subset of the vector space of all polynomials, and it clearly contains the zero polynomial, so it is non-empty. Let $p(x)$ and $q(x)$ be in the set. Then $p(1) = 0$, $p(2) = 0$, $q(1) = 0$, and $q(2) = 0$. Hence, $p + q$ satisfies

$$(p+q)(1) = p(1) + q(1) = 0 \qquad \text{and} \qquad (p+q)(2) = p(2) + q(2) = 0$$

so the subset is closed under addition. Similarly, for any $t \in \mathbb{R}$, the first row of tp satisfies

$$(tp)(1) = tp(1) = 0 \qquad \text{and} \qquad (tp)(2) = tp(2) = 0$$

so the subset is also closed under scalar multiplication. Thus, it is a subspace and hence a vector space.

(c) The set is not a vector space since it is not closed under scalar multiplication. For example, $\frac{1}{2}\begin{bmatrix} 1 & 1 \\ 1 & 1 \end{bmatrix}$ is not in the set since it contains rational entries.

(d) The given set is a subset of $\mathbb{R}^3$ and is non-empty since it clearly contains $\vec{0}$. Let $\vec{x}$ and $\vec{y}$ be in the set. Then $x_1 + x_2 + x_3 = 0$ and $y_1 + y_2 + y_3 = 0$. Then $\vec{x} + \vec{y}$ satisfies

$$x_1 + y_1 + x_2 + y_2 + x_3 + y_3 = x_1 + x_2 + x_3 + y_1 + y_2 + y_3 = 0 + 0 = 0$$

so the subset is closed under addition. Similarly, for any $t \in \mathbb{R}$, $t\vec{x}$ satisfies

$$tx_1 + tx_2 + tx_3 = t(x_1 + x_2 + x_3) = 0$$

so the subset is also closed under scalar multiplication. Thus, it is a subspace of $\mathbb{R}^3$ and hence a vector space.

E2 (a) A set of five vectors in $M(2,2)$ is linearly dependent, so the set cannot be a basis.

(b) Consider

$$t_1 \begin{bmatrix} 1 & 1 \\ 2 & 1 \end{bmatrix} + t_2 \begin{bmatrix} 0 & 1 \\ 1 & -1 \end{bmatrix} + t_3 \begin{bmatrix} 0 & 1 \\ 1 & 3 \end{bmatrix} + t_4 \begin{bmatrix} 2 & 2 \\ 4 & -2 \end{bmatrix} = \begin{bmatrix} 0 & 0 \\ 0 & 0 \end{bmatrix}$$

Row reducing the coefficient matrix of the corresponding system gives

$$\begin{bmatrix} 1 & 0 & 0 & 2 \\ 1 & 1 & 1 & 2 \\ 2 & 1 & 1 & 4 \\ 1 & -1 & 3 & -2 \end{bmatrix} \sim \begin{bmatrix} 1 & 0 & 0 & 2 \\ 0 & 1 & 0 & 1 \\ 0 & 0 & 1 & -1 \\ 0 & 0 & 0 & 0 \end{bmatrix}$$

Thus, the system has infinitely many solutions, so the set is linearly dependent, and hence it is not a basis.

(c) A set of three vectors in $M(2,2)$ cannot span $M(2,2)$, so the set cannot be a basis.

E3 (a) Consider

$$t_1\begin{bmatrix}1\\0\\1\\1\\3\end{bmatrix}+t_2\begin{bmatrix}1\\1\\0\\1\\1\end{bmatrix}+t_3\begin{bmatrix}3\\3\\1\\0\\2\end{bmatrix}+t_4\begin{bmatrix}1\\1\\1\\-2\\0\end{bmatrix}=\begin{bmatrix}0\\0\\0\\0\\0\end{bmatrix}$$

Row reducing the coefficient matrix of the corresponding system gives

$$\begin{bmatrix}1&1&3&1\\0&1&3&1\\1&0&1&1\\1&1&0&-2\\3&1&2&0\end{bmatrix}\sim\begin{bmatrix}1&0&0&0\\0&1&0&-2\\0&0&1&1\\0&0&0&0\\0&0&0&0\end{bmatrix}$$

Thus, $\mathcal{B}=\{\vec{v}_1,\vec{v}_2,\vec{v}_3\}$ is a linearly independent set. Moreover, $\vec{v}_4$ can be written as a linear combination of the $\vec{v}_1,\vec{v}_2$, and $\vec{v}_3$, so $\mathcal{B}$ also spans $\mathbb{S}$. Hence, it is a basis for $\mathbb{S}$ and so $\dim\mathbb{S}=3$.

(b) We need to find constants $t_1, t_2,$ and t_3 such that

$$t_1\begin{bmatrix}1\\0\\1\\1\\3\end{bmatrix}+t_2\begin{bmatrix}1\\1\\0\\1\\1\end{bmatrix}+t_3\begin{bmatrix}3\\3\\1\\0\\2\end{bmatrix}=\begin{bmatrix}0\\2\\-1\\-3\\-5\end{bmatrix}$$

$$\begin{bmatrix}1&1&3&0\\1&1&3&2\\1&0&1&-1\\1&1&0&-3\\3&1&2&-5\end{bmatrix}\sim\begin{bmatrix}1&0&0&-2\\0&1&0&-1\\0&0&1&1\\0&0&0&0\\0&0&0&0\end{bmatrix}$$

Thus, $[\vec{x}]_{\mathcal{B}}=\begin{bmatrix}-2\\-1\\1\end{bmatrix}$.

E4 (a) Let $\vec{v}_1=\begin{bmatrix}0\\1\\0\end{bmatrix}$ and $\vec{v}_2=\begin{bmatrix}1\\0\\1\end{bmatrix}$. Then $\{\vec{v}_1,\vec{v}_2\}$ is a basis for the plane since it is a set of two linearly independent vectors in the plane.

(b) Since $\vec{v}_3=\begin{bmatrix}1\\0\\-1\end{bmatrix}$ does not lie in the plane, the set $\mathcal{B}=\{\vec{v}_1,\vec{v}_2,\vec{v}_3\}$ is linearly independent and hence a basis for $\mathbb{R}^3$.

(c) We have $L(\vec{v}_1)=\vec{v}_1$, $L(\vec{v}_2)=\vec{v}_2$, and $L(\vec{v}_3)=-\vec{v}_3$, so

$$[L]_{\mathcal{B}}=\begin{bmatrix}1&0&0\\0&1&0\\0&0&-1\end{bmatrix}$$

(d) The change of coordinates matrix from $\mathcal{B}$-coordinates to $\mathcal{S}$-coordinates (standard coordinates) is

$$P=\begin{bmatrix}0&1&1\\1&0&0\\0&1&-1\end{bmatrix}$$

It follows that

$$[L]_S = P[L]_{\mathcal{B}}P^{-1} = \begin{bmatrix} 0 & 0 & 1 \\ 0 & 1 & 0 \\ 1 & 0 & 0 \end{bmatrix}$$

E5 The change of coordinates matrix from $\mathcal{S}$ to $\mathcal{B}$ is

$$P = \begin{bmatrix} 1 & 0 & 1 \\ 1 & 1 & -1 \\ 0 & 1 & 1 \end{bmatrix}$$

Hence,

$$[L]_{\mathcal{B}} = P^{-1} \begin{bmatrix} 1 & -1 & 2 \\ -1 & 0 & 1 \\ -2 & 1 & 0 \end{bmatrix} P = \begin{bmatrix} 0 & 2/3 & 11/3 \\ -1 & 2/3 & -10/3 \\ 0 & 1/3 & 1/3 \end{bmatrix}$$

E6 If $t_1L(\mathbf{v}_1) + \cdots + t_kL(\mathbf{v}_k) = \mathbf{0}$, then

$$\mathbf{0} = L(t_1\mathbf{v}_1 + \cdots + t_k\mathbf{v}_k)$$

and hence $t_1\mathbf{v}_1 + \cdots + t_k\mathbf{v}_k \in \text{Null}(L)$. Thus,

$$t_1\mathbf{v}_1 + \cdots + t_k\mathbf{v}_k = \mathbf{0}$$

and hence $t_1 = \cdots = t_k = 0$ since $\{\mathbf{v}_1, \dots, \mathbf{v}_k\}$ is linearly independent. Thus, $\{L(\mathbf{v}_1), \dots, L(\mathbf{v}_k)\}$ is linearly independent.

E7 (a) False. $\mathbb{R}^n$ is an n-dimensional subspace of $\mathbb{R}^n$.

(b) True. The dimension of P_2 is 3, so a set of four polynomials in P_2 must be linearly dependent.

(c) False. The number of components in a coordinate vector is the number of vectors in the basis. So, if $\mathcal{B}$ is a basis for a 4 dimensional subspace, the $\mathcal{B}$-coordinate vector would have only four components.

(d) True. Both ranks are equal to the dimension of the range of L.

(e) False. Consider a linear mapping $L : P_2 \to P_2$. Then, the range of L is a subspace of P_2, while for any basis $\mathcal{B}$, the columnspace of $[L]_{\mathcal{B}}$ is a subspace of $\mathbb{R}^2$. Hence, Range(L) cannot equal the columnspace of $[L]_{\mathcal{B}}$.

(f) False. The mapping $L : \mathbb{R} \to \mathbb{R}^2$ given by $L(x_1) = (x_1, 0)$ is one-to-one, but $\dim \mathbb{R} \neq \dim \mathbb{R}^2$.

CHAPTER 5

Section 5.1

A Problems

A1 (a) 38 (b) −5 (c) 0
(d) 0 (e) 0 (f) 48

A2 (a) 3 (b) 0 (c) 196 (d) −136

A3 (a) 0 (b) 20 (c) 18
(d) −90 (e) 76 (f) 420

A4 (a) −26 (b) 98

A5 (a) −1 (b) 1 (c) −3

Section 5.2

A Problems

A1 (a) $\det A = 30$, so A is invertible.

(b) $\det A = 1$, so A is invertible.

(c) $\det A = 8$, so A is invertible.

(d) $\det A = 0$, so A is not invertible.

(e) $\det A = -1120$, so A is invertible.

A2 (a) 14 (b) −12 (c) −5 (d) 716

A3 (a) $\det A = 3p - 14$, so A is invertible for all $p \neq \frac{14}{3}$.

(b) $\det A = -5p - 20$, so A is invertible for all $p \neq -4$.

(c) $\det A = 2p - 116$, so A is invertible for all $p \neq 58$.

A4 (a) $\det A = 13$, $\det B = 14$, $\det AB = \det\begin{bmatrix} 7 & 0 \\ 4 & 26 \end{bmatrix} = 182$

(b) $\det A = -2$, $\det B = 56$, $\det AB = \det\begin{bmatrix} 2 & 7 & 25 \\ -7 & -4 & 7 \\ 11 & 11 & 8 \end{bmatrix} = -112$

A5 (a) Since rA is the matrix where each of the n rows of A must been multiplied by r, we can use Theorem 5.2.1 n times to get $\det(rA) = r^n \det A$.

(b) We have AA^{-1} is I, so

$$1 = \det I = \det AA^{-1} = (\det A)(\det A^{-1})$$

by Theorem 5.2.7. Since $\det A \neq 0$, we get $\det A^{-1} = \frac{1}{\det A}$.

(c) By Theorem 5.2.7, we have $1 = \det I = \det A^3 = (\det A)^3$. Taking cube roots of both sides gives $\det A = 1$.

Section 5.3

A Problems

A1 (a) $\frac{1}{-2}\begin{bmatrix} 10 & -3 \\ -4 & 1 \end{bmatrix}$ (b) $\frac{1}{7}\begin{bmatrix} -1 & 5 \\ -2 & 3 \end{bmatrix}$

(c) $\frac{1}{8}\begin{bmatrix} -3 & 21 & 11 \\ -1 & 7 & 5 \\ 3 & -13 & -7 \end{bmatrix}$ (d) $\frac{1}{4}\begin{bmatrix} -1 & -8 & -4 \\ -2 & -12 & -4 \\ -2 & -8 & -4 \end{bmatrix}$

A2 (a) $\text{cof}\, A = \begin{bmatrix} -1 & 5 & -2 \\ 3+t & 2-3t & -11 \\ -3 & -2-2t & -6 \end{bmatrix}$

(b) $A(\text{cof}\, A)^T = \begin{bmatrix} -2t-17 & 0 & 0 \\ 0 & -2t-17 & 0 \\ 0 & 0 & -2t-17 \end{bmatrix}$. So $\det A = -2t - 17$ and $A^{-1} = \frac{1}{-2t-17}\begin{bmatrix} -1 & 3+t & -3 \\ 5 & 2-3t & -2-2t \\ -2 & -11 & -6 \end{bmatrix}$, provided $-2t - 17 \neq 0$.

A3 (a) $\begin{bmatrix} 51/19 \\ -4/19 \end{bmatrix}$ (b) $\begin{bmatrix} 7/5 \\ -11/15 \end{bmatrix}$ (c) $\begin{bmatrix} 21/11 \\ -26/11 \\ 2 \end{bmatrix}$ (d) $\begin{bmatrix} 3/5 \\ -12/5 \\ -8/5 \end{bmatrix}$

Section 5.4

A Problems

A1 (a) 11

(b) $A\vec{u} = \begin{bmatrix} 18 \\ 19 \end{bmatrix}, A\vec{v} = \begin{bmatrix} 34 \\ 20 \end{bmatrix}$

(c) -26

(d) $\left|\det \begin{bmatrix} 18 & 34 \\ 19 & 20 \end{bmatrix}\right| = 286 = |(-26)|(11)$

A2 $A\vec{u} = \begin{bmatrix} 3 \\ 1 \end{bmatrix}, A\vec{v} = \begin{bmatrix} 2 \\ -2 \end{bmatrix}$. Area $= \left|\det \begin{bmatrix} 3 & 2 \\ 1 & -2 \end{bmatrix}\right| = |-8| = 8$.

A3 (a) 63

(b) 42

(c) 2646

A4 (a) 41

(b) 78

(c) 3198

A5 (a) 5

(b) 245

A6 The n-volume of the parallelotope induced by $\vec{v}_1, \ldots, \vec{v}_n$ is $\left|\det \begin{bmatrix} \vec{v}_1 & \cdots & \vec{v}_n \end{bmatrix}\right|$. Since adding a multiple of one column to another does not change the determinant (see Problem 5.2.D8), we get that

$$\left|\det \begin{bmatrix} \vec{v}_1 & \cdots & \vec{v}_n \end{bmatrix}\right| = \left|\det \begin{bmatrix} \vec{v}_1 & \cdots & \vec{v}_n + t\vec{v}_1 \end{bmatrix}\right|$$

This is the volume of the parallelotope induced by $\vec{v}_1, \ldots, \vec{v}_{n-1}, \vec{v}_n + t\vec{v}_1$.

Chapter 5 Quiz

E Problems

E1 $\begin{vmatrix} -2 & 4 & 0 & 0 \\ 1 & -2 & 2 & 9 \\ -3 & 6 & 0 & 3 \\ 1 & -1 & 0 & 0 \end{vmatrix} = 2(-1)^{2+3} \begin{vmatrix} -2 & 4 & 0 \\ -3 & 6 & 3 \\ 1 & -1 & 0 \end{vmatrix} = (-2)(3)(-1)^{2+3} \begin{vmatrix} -2 & 4 \\ 1 & -1 \end{vmatrix} = -12$

E2 $\begin{vmatrix} 3 & 2 & 7 & -8 \\ -6 & -1 & -9 & 20 \\ 3 & 8 & 21 & -17 \\ 3 & 5 & 12 & 1 \end{vmatrix} = \begin{vmatrix} 3 & 2 & 7 & -8 \\ 0 & 3 & 5 & 4 \\ 0 & 0 & 4 & -17 \\ 0 & 0 & 0 & 5 \end{vmatrix} = 180$

E3 $\begin{vmatrix} 0 & 2 & 0 & 0 & 0 \\ 0 & 0 & 0 & 3 & 0 \\ 0 & 0 & 0 & 0 & 1 \\ 0 & 0 & 4 & 0 & 0 \\ 5 & 0 & 0 & 0 & 6 \end{vmatrix} = 5(2)(4)(3)(1) = 120$

E4 The matrix is invertible for all $k \neq -\frac{7}{8} \pm \frac{\sqrt{145}}{8}$.

E5 (a) $3(7) = 21$

(b) $(-1)^4(7) = 7$

(c) $2^5(7) = 224$

(d) $\frac{1}{\det A} = \frac{1}{7}$

(e) $7(7) = 49$

E6 $(A^{-1})_{31} = -\frac{1}{3}$

E7 $x_2 = \frac{1}{\det A} \begin{vmatrix} 2 & 1 & 1 \\ 1 & -1 & -1 \\ -2 & 1 & 2 \end{vmatrix} = -\frac{1}{2}$

E8 (a) $\left| \det \begin{bmatrix} 1 & 2 & 0 \\ 1 & -1 & 3 \\ -2 & 3 & 4 \end{bmatrix} \right| = 33$

(b) $|-24|(33) = 792$

CHAPTER 6

Section 6.1

A Problems

A1 $\begin{bmatrix} 1 \\ 0 \\ 1 \end{bmatrix}$ and $\begin{bmatrix} 1 \\ -1 \\ 1 \end{bmatrix}$ are not eigenvectors of A. $\begin{bmatrix} 1 \\ 0 \\ 2 \end{bmatrix}$ is an eigenvector with eigenvalue $\lambda = 0$; $\begin{bmatrix} 1 \\ 1 \\ -1 \end{bmatrix}$ is an eigenvector with eigenvalue $\lambda = -6$; and $\begin{bmatrix} 1 \\ 1 \\ 1 \end{bmatrix}$ is an eigenvector with eigenvalue $\lambda = 4$.

A2 (a) The eigenvalues are $\lambda_1 = 2$ and $\lambda_2 = 3$. The eigenspace of λ_1 is Span$\left\{\begin{bmatrix}1\\2\end{bmatrix}\right\}$. The eigenspace of λ_2 is Span$\left\{\begin{bmatrix}1\\3\end{bmatrix}\right\}$.

(b) The only eigenvalue is $\lambda = 1$. The eigenspace of λ is Span$\left\{\begin{bmatrix}1\\0\end{bmatrix}\right\}$.

(c) The eigenvalues are $\lambda_1 = 2$ and $\lambda_2 = 3$. The eigenspace of λ_1 is Span$\left\{\begin{bmatrix}1\\0\end{bmatrix}\right\}$. The eigenspace of λ_2 is Span$\left\{\begin{bmatrix}0\\1\end{bmatrix}\right\}$.

(d) The eigenvalues are $\lambda_1 = -1$ and $\lambda_2 = 4$. The eigenspace of λ_1 is Span$\left\{\begin{bmatrix}5\\2\end{bmatrix}\right\}$. The eigenspace of λ_2 is Span$\left\{\begin{bmatrix}1\\3\end{bmatrix}\right\}$.

(e) The eigenvalues are $\lambda_1 = 5$ and $\lambda_2 = -2$. The eigenspace of λ_1 is Span$\left\{\begin{bmatrix}3\\4\end{bmatrix}\right\}$. The eigenspace of λ_2 is Span$\left\{\begin{bmatrix}-1\\1\end{bmatrix}\right\}$.

(f) The eigenvalues are $\lambda_1 = 0$ and $\lambda_2 = -3$. The eigenspace of λ_1 is Span$\left\{\begin{bmatrix}1\\1\end{bmatrix}\right\}$. The eigenspace of λ_2 is Span$\left\{\begin{bmatrix}1\\2\end{bmatrix}\right\}$.

A3 (a) $\lambda_1 = 2$ has algebraic multiplicity 1. A basis for its eigenspace is $\left\{\begin{bmatrix}1\\0\end{bmatrix}\right\}$, so it has geometric multiplicity 1. $\lambda_2 = 3$ has algebraic multiplicity 1. A basis for its eigenspace is $\left\{\begin{bmatrix}2\\-1\end{bmatrix}\right\}$, so it has geometric multiplicity 1.

(b) $\lambda_1 = 2$ has algebraic multiplicity 2; a basis for its eigenspace is $\left\{\begin{bmatrix}1\\0\end{bmatrix}\right\}$, so it has geometric multiplicity 1.

(c) $\lambda_1 = 2$ has algebraic multiplicity 2; a basis for its eigenspace is $\left\{\begin{bmatrix}1\\1\end{bmatrix}\right\}$, so it has geometric multiplicity 1.

(d) $\lambda_1 = 2$ has algebraic multiplicity 1; a basis for its eigenspace is $\left\{\begin{bmatrix}-1\\1\\1\end{bmatrix}\right\}$, so it has geometric multiplicity 1. $\lambda_2 = 1$ has algebraic multiplicity 1; a basis for its eigenspace is $\left\{\begin{bmatrix}3\\0\\1\end{bmatrix}\right\}$, so it has geometric multiplicity 1. $\lambda_3 = -2$ has algebraic multiplicity 1; a basis for its eigenspace is $\left\{\begin{bmatrix}1\\1\\1\end{bmatrix}\right\}$, so it has geometric multiplicity 1.

(e) $\lambda_1 = 0$ has algebraic multiplicity 2; a basis for its eigenspace is $\left\{\begin{bmatrix}-1\\1\\0\end{bmatrix}, \begin{bmatrix}-1\\0\\1\end{bmatrix}\right\}$, so it has geometric multiplicity 2. $\lambda_2 = 6$ has algebraic multiplicity 1; a basis for its eigenspace is $\left\{\begin{bmatrix}1\\1\\1\end{bmatrix}\right\}$, so it has geometric multiplicity 1.

(f) $\lambda_1 = 2$ has algebraic multiplicity 2; a basis for its eigenspace is $\left\{\begin{bmatrix}-1\\1\\0\end{bmatrix}, \begin{bmatrix}-1\\0\\1\end{bmatrix}\right\}$, so it has geometric multiplicity 2. $\lambda_2 = 5$ has algebraic multiplicity 1; a basis for its eigenspace is $\left\{\begin{bmatrix}1\\1\\1\end{bmatrix}\right\}$, so it has geometric multiplicity 1.

Section 6.2

A Problems

A1 (a) $\begin{bmatrix}2\\1\end{bmatrix}$ is an eigenvector of A with eigenvalue 14, and $\begin{bmatrix}-1\\3\end{bmatrix}$ is an eigenvector of A with eigenvalue -7. $P^{-1} = \frac{1}{7}\begin{bmatrix}3 & 1\\-1 & 2\end{bmatrix}$, $P^{-1}AP = \begin{bmatrix}14 & 0\\0 & -7\end{bmatrix}$.

(b) P does not diagonalize A.

(c) $\begin{bmatrix}2\\1\end{bmatrix}$ is an eigenvector of A with eigenvalue 1, and $\begin{bmatrix}1\\1\end{bmatrix}$ is an eigenvector of A with eigenvalue -3. $P^{-1} = \begin{bmatrix}1 & -1\\-1 & 2\end{bmatrix}$, $P^{-1}AP = \begin{bmatrix}1 & 0\\0 & -3\end{bmatrix}$.

(d) $\begin{bmatrix}-1\\1\\0\end{bmatrix}$ and $\begin{bmatrix}-1\\0\\1\end{bmatrix}$ are both eigenvectors of A with eigenvalue -2, and $\begin{bmatrix}1\\1\\1\end{bmatrix}$ is an eigenvector of A with eigenvalue 10. $P^{-1} = \frac{1}{3}\begin{bmatrix}-1 & 2 & -1\\-1 & -1 & 2\\1 & 1 & 1\end{bmatrix}$, $P^{-1}AP = \begin{bmatrix}-2 & 0 & 0\\0 & -2 & 0\\0 & 0 & 10\end{bmatrix}$

A2 Alternate correct answers are possible.

(a) The eigenvalues are $\lambda_1 = 1$ and $\lambda_2 = 8$, each with algebraic multiplicity 1. A basis for the eigenspace of λ_1 is $\left\{\begin{bmatrix}-1\\1\end{bmatrix}\right\}$, so the geometric multiplicity is 1. A basis for the eigenspace of λ_2 is $\left\{\begin{bmatrix}2\\5\end{bmatrix}\right\}$, so the geometric multiplicity is 1. Therefore, by Corollary 6.2.3, A is diagonalizable with $P = \begin{bmatrix}-1 & 2\\1 & 5\end{bmatrix}$ and $D = \begin{bmatrix}1 & 0\\0 & 8\end{bmatrix}$.

(b) The eigenvalues are $\lambda_1 = -6$ and $\lambda_2 = 1$, each with algebraic multiplicity 1. A basis for the eigenspace of λ_1 is $\left\{\begin{bmatrix}-3\\4\end{bmatrix}\right\}$, so the geometric multiplicity is 1. A basis for the eigenspace of λ_2 is $\left\{\begin{bmatrix}1\\1\end{bmatrix}\right\}$, so the geometric multiplicity is 1. Therefore, by Corollary 6.2.3, A is diagonalizable with $P = \begin{bmatrix}-3 & 1\\4 & 1\end{bmatrix}$ and $D = \begin{bmatrix}-6 & 0\\0 & 1\end{bmatrix}$.

(c) The eigenvalues of A are $\pm\sqrt{12}i$. Hence, A is not diagonalizable over $\mathbb{R}$.

(d) The eigenvalues are $\lambda_1 = -1$, $\lambda_2 = 2$, and $\lambda_3 = 0$, each with algebraic multiplicity 1. A basis for the eigenspace of λ_1 is $\left\{\begin{bmatrix}-1\\1\\0\end{bmatrix}\right\}$, so the geometric multiplicity is 1. A basis for the eigenspace of λ_2 is $\left\{\begin{bmatrix}1\\2\\3\end{bmatrix}\right\}$, so the geometric multiplicity is 1. A basis for the eigenspace of λ_3 is $\left\{\begin{bmatrix}-1\\0\\1\end{bmatrix}\right\}$, so the geometric multiplicity is 1. Therefore, by Corollary 6.2.3, A is diagonalizable with $P = \begin{bmatrix}-1 & 1 & -1\\1 & 2 & 0\\0 & 3 & 1\end{bmatrix}$ and $D = \begin{bmatrix}-1 & 0 & 0\\0 & 2 & 0\\0 & 0 & 0\end{bmatrix}$.

(e) The eigenvalues are $\lambda_1 = 1$ with algebraic multiplicity 2 and $\lambda_2 = 5$ with algebraic multiplicity 1. A basis for the eigenspace of λ_1 is $\left\{\begin{bmatrix}1\\0\\1\end{bmatrix}\right\}$, so the geometric multiplicity is 1. A basis for the eigenspace of λ_2 is $\left\{\begin{bmatrix}1\\-1\\2\end{bmatrix}\right\}$, so the geometric multiplicity is 1. Therefore, by Corollary 6.2.3, A is not diagonalizable since the geometric multiplicity of λ_1 does not equal its algebraic multiplicity.

(f) The eigenvalues of A are 1, i, and $-i$. Hence, A is not diagonalizable over $\mathbb{R}$.

(g) The eigenvalues are $\lambda_1 = 2$ with algebraic multiplicity 2 and $\lambda_2 = -1$ with algebraic multiplicity 1. A basis for the eigenspace of λ_1 is $\left\{\begin{bmatrix}2\\1\\0\end{bmatrix}, \begin{bmatrix}1\\0\\1\end{bmatrix}\right\}$, so the geometric multiplicity is 2. A basis for the eigenspace of λ_2 is $\left\{\begin{bmatrix}1\\-1\\2\end{bmatrix}\right\}$, so the geometric multiplicity is 1. Therefore, by Corollary 6.2.3, A is diagonalizable with $P = \begin{bmatrix}2 & 1 & 1\\1 & 0 & -1\\0 & 1 & 2\end{bmatrix}$ and $D = \begin{bmatrix}2 & 0 & 0\\0 & 2 & 0\\0 & 0 & -1\end{bmatrix}$.

A3 Alternate correct answers are possible.

(a) The only eigenvalue is $\lambda_1 = 3$ with algebraic multiplicity 2. A basis for the eigenspace of λ_1 is $\left\{\begin{bmatrix} 0 \\ 1 \end{bmatrix}\right\}$, so the geometric multiplicity is 1. Therefore, by Corollary 6.2.3, A is not diagonalizable since the geometric multiplicity of λ_1 does not equal its algebraic multiplicity.

(b) The eigenvalues are $\lambda_1 = 0$ and $\lambda_2 = 8$, each with algebraic multiplicity 1. A basis for the eigenspace of λ_1 is $\left\{\begin{bmatrix} -1 \\ 1 \end{bmatrix}\right\}$, so the geometric multiplicity is 1. A basis for the eigenspace of λ_2 is $\left\{\begin{bmatrix} 1 \\ 1 \end{bmatrix}\right\}$, so the geometric multiplicity is 1. Therefore, by Corollary 6.2.3, A is diagonalizable with $P = \begin{bmatrix} -1 & 1 \\ 1 & 1 \end{bmatrix}$ and $D = \begin{bmatrix} 0 & 0 \\ 0 & 8 \end{bmatrix}$.

(c) The eigenvalues are $\lambda_1 = 3$ and $\lambda_2 = -7$, each with algebraic multiplicity 1. A basis for the eigenspace of λ_1 is $\left\{\begin{bmatrix} 1 \\ 1 \end{bmatrix}\right\}$, so the geometric multiplicity is 1. A basis for the eigenspace of λ_2 is $\left\{\begin{bmatrix} -1 \\ 1 \end{bmatrix}\right\}$, so the geometric multiplicity is 1. Therefore, by Corollary 6.2.3, A is diagonalizable with $P = \begin{bmatrix} 1 & -1 \\ 1 & 1 \end{bmatrix}$ and $D = \begin{bmatrix} 3 & 0 \\ 0 & -7 \end{bmatrix}$.

(d) The eigenvalues are $\lambda_1 = 2$ with algebraic multiplicity 2 and $\lambda_2 = -2$ with algebraic multiplicity 1. A basis for the eigenspace of λ_1 is $\left\{\begin{bmatrix} -1 \\ 1 \\ 1 \end{bmatrix}\right\}$, so the geometric multiplicity is 1. A basis for the eigenspace of λ_2 is $\left\{\begin{bmatrix} 1 \\ 1 \\ 1 \end{bmatrix}\right\}$, so the geometric multiplicity is 1. Therefore, by Corollary 6.2.3, A is not diagonalizable since the geometric multiplicity of λ_1 does not equal its algebraic multiplicity.

(e) The eigenvalues are $\lambda_1 = -2$ with algebraic multiplicity 2 and $\lambda_2 = 4$ with algebraic multiplicity 1. A basis for the eigenspace of λ_1 is $\left\{\begin{bmatrix} -1 \\ 0 \\ 1 \end{bmatrix}, \begin{bmatrix} -1 \\ 1 \\ 0 \end{bmatrix}\right\}$, so the geometric multiplicity is 2. A basis for the eigenspace of λ_2 is $\left\{\begin{bmatrix} 1 \\ 1 \\ 1 \end{bmatrix}\right\}$, so the geometric multiplicity is 1. Therefore, by Corollary 6.2.3, A is diagonalizable with $P = \begin{bmatrix} -1 & -1 & 1 \\ 0 & 1 & 1 \\ 1 & 0 & 1 \end{bmatrix}$ and $D = \begin{bmatrix} -2 & 0 & 0 \\ 0 & -2 & 0 \\ 0 & 0 & 4 \end{bmatrix}$.

(f) The eigenvalues are $\lambda_1 = 2$ with algebraic multiplicity 2 and $\lambda_2 = 1$ with algebraic multiplicity 1. A basis for the eigenspace of λ_1 is $\left\{ \begin{bmatrix} -2 \\ 1 \\ 0 \end{bmatrix}, \begin{bmatrix} 1 \\ 0 \\ 1 \end{bmatrix} \right\}$, so the geometric multiplicity is 2. A basis for the eigenspace of λ_2 is $\left\{ \begin{bmatrix} 0 \\ 1 \\ 1 \end{bmatrix} \right\}$, so the geometric multiplicity is 1. Therefore, by Corollary 6.2.3, A is diagonalizable with $P = \begin{bmatrix} -2 & 1 & 0 \\ 1 & 0 & 1 \\ 0 & 1 & 1 \end{bmatrix}$ and $D = \begin{bmatrix} 2 & 0 & 0 \\ 0 & 2 & 0 \\ 0 & 0 & 1 \end{bmatrix}$.

(g) The eigenvalues of are 2, $2 + i$ and $2 - i$. Hence, A is not diagonalizable over $\mathbb{R}$.

Section 6.3

A Problems

A1 (a) A is not a Markov matrix.

(b) B is a Markov matrix. The invariant state is $\begin{bmatrix} 6/13 \\ 7/13 \end{bmatrix}$.

(c) C is not a Markov matrix.

(d) D is a Markov matrix. The invariant state is $\begin{bmatrix} 1/3 \\ 1/3 \\ 1/3 \end{bmatrix}$.

A2 (a) In the long run, 25% of the population will be rural dwellers and 75% will be urban dwellers.

(b) After five decades, approximately 33% of the population will be rural dwellers and 67% will be urban dwellers.

A3 $T = \frac{1}{10}\begin{bmatrix} 8 & 3 & 3 \\ 1 & 6 & 1 \\ 1 & 1 & 6 \end{bmatrix}$. In the long run, 60% of the cars will be at the airport, 20% of the cars will be at the train station, and 20% of the cars will be at the city centre.

A4 (a) The dominant eigenvalue is $\lambda = 5$.

(b) The dominant eigenvalue is $\lambda = 6$.

Section 6.4

A Problems

A1 (a) $ae^{-5t}\begin{bmatrix} -1 \\ 4 \end{bmatrix} + be^{4t}\begin{bmatrix} 2 \\ 1 \end{bmatrix}, \quad a, b \in \mathbb{R}$.

(b) $ae^{-0.5t}\begin{bmatrix} -1 \\ 1 \end{bmatrix} + be^{0.3t}\begin{bmatrix} 7 \\ 1 \end{bmatrix}, \quad a, b \in \mathbb{R}$.

Chapter 6 Quiz

E Problems

E1 (a) $\begin{bmatrix}3\\1\\0\end{bmatrix}$ is not an eigenvector of A.

(b) $\begin{bmatrix}1\\0\\1\end{bmatrix}$ is an eigenvector with eigenvalue 1.

(c) $\begin{bmatrix}4\\1\\0\end{bmatrix}$ is an eigenvector with eigenvalue 1.

(d) $\begin{bmatrix}2\\1\\-1\end{bmatrix}$ is an eigenvector with eigenvalue -1.

E2 The matrix is diagonalizable. $P = \begin{bmatrix}1 & 1 & 1\\3 & -1 & 1\\-1 & 2 & 1\end{bmatrix}$ and $D = \begin{bmatrix}0 & 0 & 0\\0 & -4 & 0\\0 & 0 & -2\end{bmatrix}$

E3 $\lambda_1 = 2$ has algebraic and geometric multiplicity 2; $\lambda_2 = 4$ has algebraic and geometric multiplicity 1. Thus, A is diagonalizable.

E4 Since A is invertible, 0 is not an eigenvalue of A (see Problem 6.2. D8). Then, if $A\vec{x} = \lambda\vec{x}$, we get $\vec{x} = \lambda A^{-1}\vec{x}$, so $A^{-1}\vec{x} = \frac{1}{\lambda}\vec{x}$.

E5 (a) One-dimensional

(b) Zero-dimensional

(c) Rank$(A) = 2$

E6 The invariant state is $\vec{x} = \begin{bmatrix}1/4\\1/4\\1/2\end{bmatrix}$.

E7 $ae^{-0.1t}\begin{bmatrix}-1\\1\end{bmatrix} + be^{0.4t}\begin{bmatrix}2\\3\end{bmatrix}$

CHAPTER 7

Section 7.1

A Problems

A1 (a) The set is orthogonal. $P = \begin{bmatrix}1/\sqrt{5} & 2/\sqrt{5}\\2/\sqrt{5} & -1/\sqrt{5}\end{bmatrix}$.

(b) The set is not orthogonal.

(c) The set is orthogonal. $P = \begin{bmatrix} 1/\sqrt{11} & 3/\sqrt{10} & 1/\sqrt{110} \\ 1/\sqrt{11} & 0 & -10/\sqrt{110} \\ 3/\sqrt{11} & -1/\sqrt{10} & 3/\sqrt{110} \end{bmatrix}$.

(d) The set is not orthogonal.

A2 (a) $[\vec{w}]_{\mathcal{B}} = \begin{bmatrix} 8/3 \\ 19/3 \\ 5/3 \end{bmatrix}$ (b) $[\vec{x}]_{\mathcal{B}} = \begin{bmatrix} 7 \\ -2 \\ -3 \end{bmatrix}$ (c) $[\vec{y}]_{\mathcal{B}} = \begin{bmatrix} 22/3 \\ 2/3 \\ 4/3 \end{bmatrix}$ (d) $[\vec{z}]_{\mathcal{B}} = \begin{bmatrix} 0 \\ 9 \\ 0 \end{bmatrix}$

A3 (a) $[\vec{x}]_{\mathcal{B}} = \begin{bmatrix} 4 \\ -5 \\ -5/\sqrt{2} \\ -1/\sqrt{2} \end{bmatrix}$ (b) $[\vec{y}]_{\mathcal{B}} = \begin{bmatrix} -5/2 \\ 3/2 \\ 7/\sqrt{2} \\ 6/\sqrt{2} \end{bmatrix}$

(c) $[\vec{w}]_{\mathcal{B}} = \begin{bmatrix} 5/2 \\ 1/2 \\ -3/\sqrt{2} \\ 0 \end{bmatrix}$ (d) $[\vec{z}]_{\mathcal{B}} = \begin{bmatrix} 1/2 \\ 1/2 \\ 3/\sqrt{2} \\ 6/\sqrt{2} \end{bmatrix}$

A4 (a) It is orthogonal.

(b) It is not orthogonal. The columns of the matrix are not orthogonal.

(c) It is not orthogonal. The columns are not unit vectors.

(d) It is not orthogonal. The third column is not orthogonal to the first or second column.

(e) It is orthogonal.

A5 (a) $\vec{g}_3 \cdot \vec{g}_1 = 0$. $\vec{g}_2 = \begin{bmatrix} 6 \\ -3 \\ 6 \end{bmatrix}$

(b) $P = \frac{1}{3}\begin{bmatrix} -1 & 2 & 2 \\ 2 & -1 & 2 \\ 2 & 2 & -1 \end{bmatrix}$

(c) $[L]_{\mathcal{B}} = \frac{1}{\sqrt{2}}\begin{bmatrix} 1 & -1 & 0 \\ 1 & 1 & 0 \\ 0 & 0 & \sqrt{2} \end{bmatrix}$

(d) $[L]_S = \frac{1}{9\sqrt{2}}\begin{bmatrix} 5+4\sqrt{2} & -1+4\sqrt{2} & 8-2\sqrt{2} \\ -7+4\sqrt{2} & 5+4\sqrt{2} & -4-2\sqrt{2} \\ -4-2\sqrt{2} & 8-2\sqrt{2} & 8+\sqrt{2} \end{bmatrix}$

A6 Since $\mathcal{B}$ is orthonormal, the matrix $P = \begin{bmatrix} 1/\sqrt{6} & 1/\sqrt{3} & 1/\sqrt{2} \\ -2/\sqrt{6} & 1/\sqrt{3} & 0 \\ 1/\sqrt{6} & 1/\sqrt{3} & -1/\sqrt{2} \end{bmatrix}$ is orthogonal. Hence, the rows of P are orthonormal. Thus, we can pick another orthonormal basis be $\left\{ \begin{bmatrix} 1/\sqrt{6} \\ 1/\sqrt{3} \\ 1/\sqrt{2} \end{bmatrix}, \begin{bmatrix} -2/\sqrt{6} \\ 1/\sqrt{3} \\ 0 \end{bmatrix}, \begin{bmatrix} 1/\sqrt{6} \\ 1/\sqrt{3} \\ -1/\sqrt{2} \end{bmatrix} \right\}$.

Section 7.2

A Problems

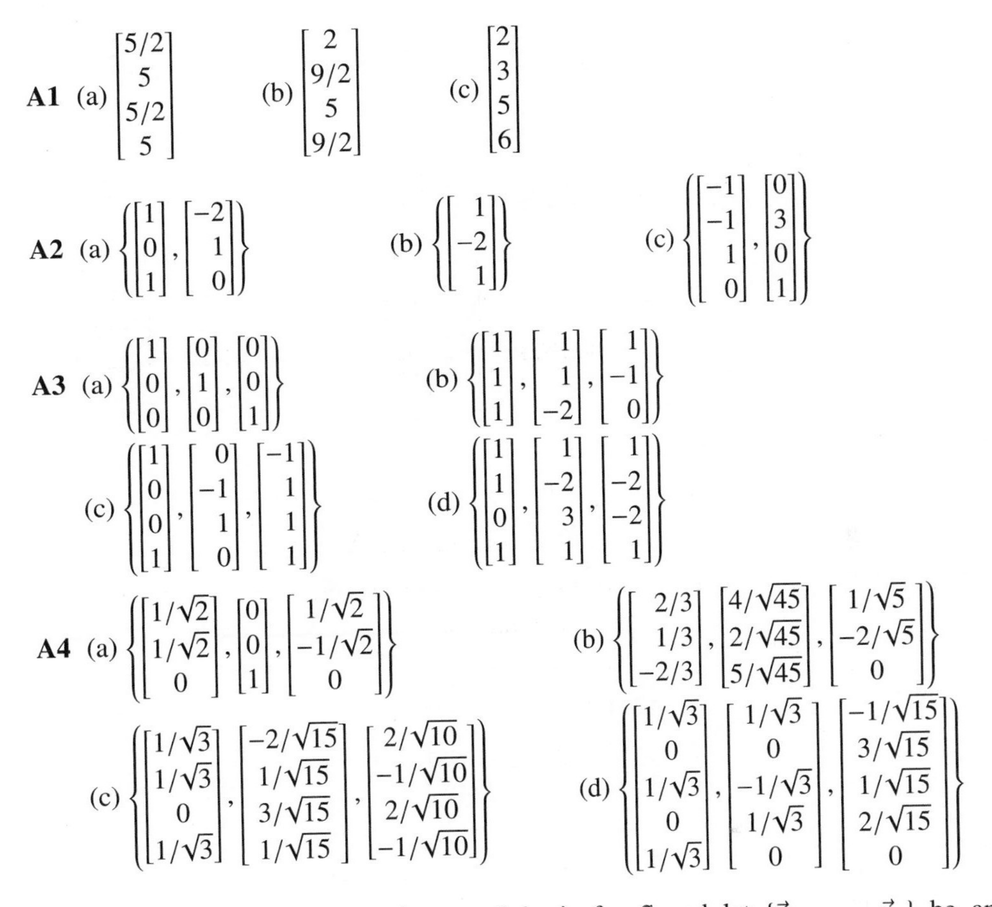

A1 (a) $\begin{bmatrix} 5/2 \\ 5 \\ 5/2 \\ 5 \end{bmatrix}$ (b) $\begin{bmatrix} 2 \\ 9/2 \\ 5 \\ 9/2 \end{bmatrix}$ (c) $\begin{bmatrix} 2 \\ 3 \\ 5 \\ 6 \end{bmatrix}$

A2 (a) $\left\{ \begin{bmatrix} 1 \\ 0 \\ 1 \end{bmatrix}, \begin{bmatrix} -2 \\ 1 \\ 0 \end{bmatrix} \right\}$ (b) $\left\{ \begin{bmatrix} 1 \\ -2 \\ 1 \end{bmatrix} \right\}$ (c) $\left\{ \begin{bmatrix} -1 \\ -1 \\ 1 \\ 0 \end{bmatrix}, \begin{bmatrix} 0 \\ 3 \\ 0 \\ 1 \end{bmatrix} \right\}$

A3 (a) $\left\{ \begin{bmatrix} 1 \\ 0 \\ 0 \end{bmatrix}, \begin{bmatrix} 0 \\ 1 \\ 0 \end{bmatrix}, \begin{bmatrix} 0 \\ 0 \\ 1 \end{bmatrix} \right\}$ (b) $\left\{ \begin{bmatrix} 1 \\ 1 \\ 1 \end{bmatrix}, \begin{bmatrix} 1 \\ 1 \\ -2 \end{bmatrix}, \begin{bmatrix} 1 \\ -1 \\ 0 \end{bmatrix} \right\}$

(c) $\left\{ \begin{bmatrix} 1 \\ 0 \\ 0 \\ 1 \end{bmatrix}, \begin{bmatrix} 0 \\ -1 \\ 1 \\ 0 \end{bmatrix}, \begin{bmatrix} -1 \\ 1 \\ 1 \\ 1 \end{bmatrix} \right\}$ (d) $\left\{ \begin{bmatrix} 1 \\ 1 \\ 0 \\ 1 \end{bmatrix}, \begin{bmatrix} 1 \\ -2 \\ 3 \\ 1 \end{bmatrix}, \begin{bmatrix} 1 \\ -2 \\ -2 \\ 1 \end{bmatrix} \right\}$

A4 (a) $\left\{ \begin{bmatrix} 1/\sqrt{2} \\ 1/\sqrt{2} \\ 0 \end{bmatrix}, \begin{bmatrix} 0 \\ 0 \\ 1 \end{bmatrix}, \begin{bmatrix} 1/\sqrt{2} \\ -1/\sqrt{2} \\ 0 \end{bmatrix} \right\}$ (b) $\left\{ \begin{bmatrix} 2/3 \\ 1/3 \\ -2/3 \end{bmatrix}, \begin{bmatrix} 4/\sqrt{45} \\ 2/\sqrt{45} \\ 5/\sqrt{45} \end{bmatrix}, \begin{bmatrix} 1/\sqrt{5} \\ -2/\sqrt{5} \\ 0 \end{bmatrix} \right\}$

(c) $\left\{ \begin{bmatrix} 1/\sqrt{3} \\ 1/\sqrt{3} \\ 0 \\ 1/\sqrt{3} \end{bmatrix}, \begin{bmatrix} -2/\sqrt{15} \\ 1/\sqrt{15} \\ 3/\sqrt{15} \\ 1/\sqrt{15} \end{bmatrix}, \begin{bmatrix} 2/\sqrt{10} \\ -1/\sqrt{10} \\ 2/\sqrt{10} \\ -1/\sqrt{10} \end{bmatrix} \right\}$ (d) $\left\{ \begin{bmatrix} 1/\sqrt{3} \\ 0 \\ 1/\sqrt{3} \\ 0 \\ 1/\sqrt{3} \end{bmatrix}, \begin{bmatrix} 1/\sqrt{3} \\ 0 \\ -1/\sqrt{3} \\ 1/\sqrt{3} \\ 0 \end{bmatrix}, \begin{bmatrix} -1/\sqrt{15} \\ 3/\sqrt{15} \\ 1/\sqrt{15} \\ 2/\sqrt{15} \\ 0 \end{bmatrix} \right\}$

A5 Let $\{\vec{v}_1, \dots, \vec{v}_k\}$ be an orthonormal basis for $\mathbb{S}$ and let $\{\vec{v}_{k+1}, \dots, \vec{v}_n\}$ be an orthonormal basis for $\mathbb{S}^\perp$. Then $\{\vec{v}_1, \dots, \vec{v}_n\}$ is an orthonormal basis for $\mathbb{R}^n$. Thus, any $\vec{x} \in \mathbb{R}^n$ can be written

$$\vec{x} = (\vec{x} \cdot \vec{v}_1)\vec{v}_1 + \cdots + (\vec{x} \cdot \vec{v}_n)\vec{v}_n$$

Then

$$\begin{aligned} \operatorname{perp}_{\mathbb{S}}(\vec{x}) &= \vec{x} - \operatorname{proj}_{\mathbb{S}} \vec{x} \\ &= \vec{x} - [(\vec{x} \cdot \vec{v}_1)\vec{v}_1 + \cdots + (\vec{x} \cdot \vec{v}_k)\vec{v}_k] \\ &= (\vec{x} \cdot \vec{v}_{k+1})\vec{v}_{k+1} + \cdots + (\vec{x} \cdot \vec{v}_n)\vec{v}_n \\ &= \operatorname{proj}_{\mathbb{S}^\perp} \vec{x} \end{aligned}$$

Section 7.3

A Problems

A1 (a) $y = 10.5 - 1.9t$ (b) $y = 3.4 + 0.8t$

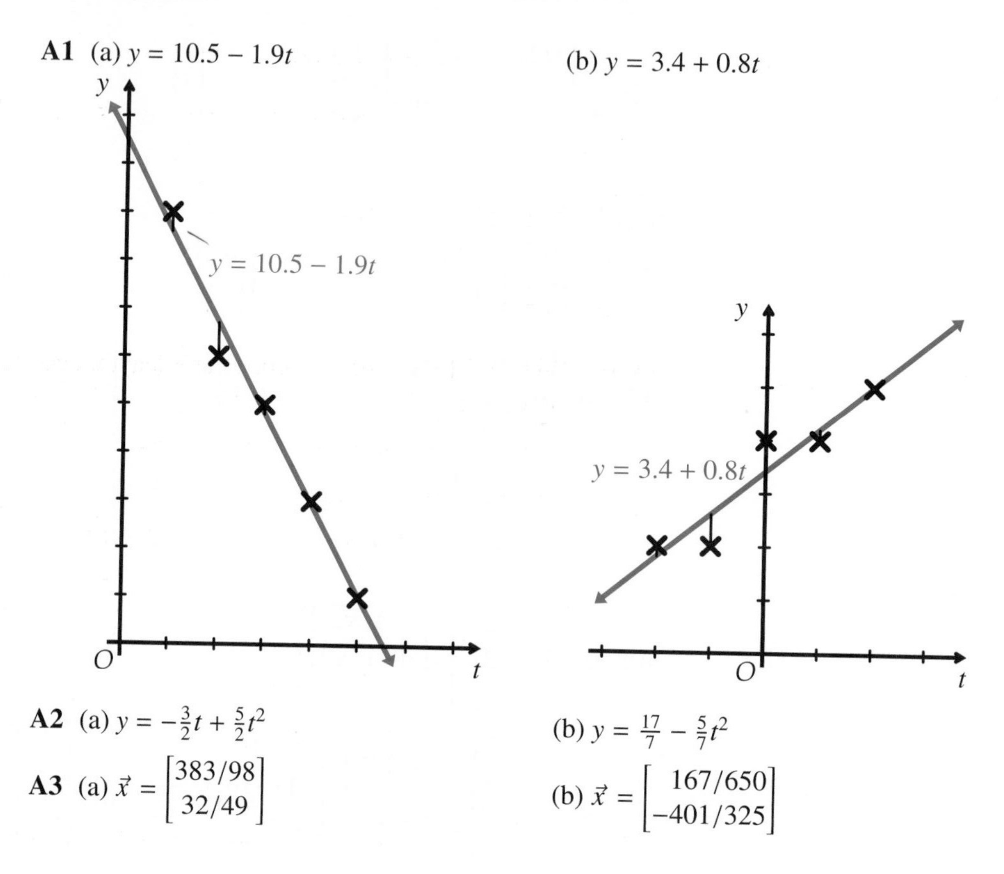

A2 (a) $y = -\frac{3}{2}t + \frac{5}{2}t^2$ (b) $y = \frac{17}{7} - \frac{5}{7}t^2$

A3 (a) $\vec{x} = \begin{bmatrix} 383/98 \\ 32/49 \end{bmatrix}$ (b) $\vec{x} = \begin{bmatrix} 167/650 \\ -401/325 \end{bmatrix}$

Section 7.4

A Problems

A1 (a) $\langle x - 2x^2, 1 + 3x \rangle = -46$ (b) $\langle 2 - x + 3x^2, 4 - 3x^2 \rangle = -84$
(c) $\|3 - 2x + x^2\| = \sqrt{22}$ (d) $\|9 + 9x + 9x^2\| = 9\sqrt{59}$

A2 (a) It does not define an inner product since $\langle x - x^2, x - x^2 \rangle = 0$.

(b) It does not define an inner product since $\langle -p, q \rangle \neq -\langle p, q \rangle$.

(c) It does define an inner product.

(d) It does not define an inner product since $\langle x, x \rangle = -2$.

A3 (a) $\left\{ \frac{1}{\sqrt{3}} \begin{bmatrix} 1 & 0 \\ -1 & 1 \end{bmatrix}, \frac{1}{\sqrt{15}} \begin{bmatrix} 1 & 3 \\ 2 & 1 \end{bmatrix}, \frac{1}{\sqrt{15}} \begin{bmatrix} 3 & -1 \\ 1 & -2 \end{bmatrix} \right\}$, $\text{proj}_{\mathbb{S}} \begin{bmatrix} 4 & 3 \\ -2 & 1 \end{bmatrix} = \begin{bmatrix} 4 & 5/3 \\ -2/3 & 7/3 \end{bmatrix}$

(b) $\left\{ \frac{1}{\sqrt{3}} \begin{bmatrix} 1 & 1 \\ 0 & 1 \end{bmatrix}, \frac{1}{\sqrt{3}} \begin{bmatrix} 0 & -1 \\ 1 & 1 \end{bmatrix}, \frac{1}{\sqrt{3}} \begin{bmatrix} 1 & -1 \\ -1 & 0 \end{bmatrix} \right\}$, $\text{proj}_{\mathbb{S}} \begin{bmatrix} 4 & 3 \\ -2 & 1 \end{bmatrix} = \begin{bmatrix} 11/3 & 3 \\ -7/3 & 4/3 \end{bmatrix}$

A4 (a) $\mathcal{B} = \left\{ \begin{bmatrix} 1 \\ 1 \\ 0 \end{bmatrix}, \begin{bmatrix} -1 \\ 2 \\ 0 \end{bmatrix}, \begin{bmatrix} 0 \\ 0 \\ 1 \end{bmatrix} \right\}$ (b) $[\vec{x}]_{\mathcal{B}} = \begin{bmatrix} 2/\sqrt{3} \\ -2\sqrt{6} \\ 0 \end{bmatrix}$.

A5 We have

$$\begin{aligned}\langle \mathbf{v}_1 + \cdots + \mathbf{v}_k, \mathbf{v}_1 + \cdots + \mathbf{v}_k \rangle =& \langle \mathbf{v}_1, \mathbf{v}_1 \rangle + \cdots + \langle \mathbf{v}_1, \mathbf{v}_k \rangle + \langle \mathbf{v}_2, \mathbf{v}_1 \rangle + \langle \mathbf{v}_2, \mathbf{v}_2 \rangle + \\ &+ \cdots + \langle \mathbf{v}_2, \mathbf{v}_k \rangle + \cdots + \langle \mathbf{v}_k, \mathbf{v}_1 \rangle + \cdots + \langle \mathbf{v}_k, \mathbf{v}_k \rangle\end{aligned}$$

Since $\{\mathbf{v}_1, \dots, \mathbf{v}_k\}$ is orthogonal, we have $\langle \mathbf{v}_i, \mathbf{v}_j \rangle = 0$ if $i \neq j$. Hence,

$$\|\mathbf{v}_1 + \cdots + \mathbf{v}_k\|^2 = \langle \mathbf{v}_1, \mathbf{v}_1 \rangle + \cdots + \langle \mathbf{v}_k, \mathbf{v}_k \rangle = \|\mathbf{v}_1\|^2 + \cdots + \|\mathbf{v}_k\|^2$$

Chapter 7 Quiz

E Problems

E1 (a) Neither. The vectors are not of unit length, and the first and third vectors are not orthogonal.

(b) Neither. The vectors are not orthogonal.

(c) The set is orthonormal.

E2 Let $\vec{v}_1, \vec{v}_2,$ and $\vec{v}_3$ denote the vectors in $\mathcal{B}$. We have

$$\vec{v}_1 \cdot \vec{x} = 2, \qquad \vec{v}_2 \cdot \vec{x} = \frac{9}{\sqrt{3}}, \qquad \vec{v}_3 \cdot \vec{x} = \frac{6}{\sqrt{3}}$$

Hence, $[\vec{x}]_{\mathcal{B}} = \begin{bmatrix} 2 \\ 9/\sqrt{3} \\ 6/\sqrt{3} \end{bmatrix}$.

E3 (a) We have

$$1 = \det I = \det(P^T P) = (\det P^T)(\det P) = (\det P)^2$$

Thus, $\det P = \pm 1$.

(b) We have $P^T P = I$ and $R^T R = I$. Hence,

$$(PR)^T (PR) = R^T P^T P R = R^T I R = R^T R = I$$

Thus, PR is orthogonal.

E4 (a) Denote the vectors in the spanning set for $\mathbb{S}$ by $\vec{z}_1, \vec{z}_2,$ and $\vec{z}_3$. Let $\vec{w}_1 = \vec{z}_1$. Then

$$\vec{w}_2 = \operatorname{perp}_{\mathbb{S}_1} \vec{z}_2 = \begin{bmatrix} 1 \\ 0 \\ 1 \\ 0 \end{bmatrix}$$

$$\vec{w}_3 = \operatorname{perp}_{\mathbb{S}_2} \vec{z}_3 = \begin{bmatrix} -1 \\ 1 \\ 1 \\ -1 \end{bmatrix}$$

Normalizing $\{\vec{w}_1, \vec{w}_2, \vec{w}_3\}$ gives us the orthonormal basis $\left\{ \begin{bmatrix} 1/\sqrt{2} \\ 0 \\ 1/\sqrt{2} \\ 0 \end{bmatrix}, \begin{bmatrix} 0 \\ 1/\sqrt{2} \\ 0 \\ 1/\sqrt{2} \end{bmatrix}, \begin{bmatrix} -1/2 \\ 1/2 \\ 1/2 \\ -1/2 \end{bmatrix} \right\}$.

(b) The closest point in $\mathbb{S}$ to $\vec{x}$ is $\text{proj}_{\mathbb{S}}\,\vec{x}$. We find that

$$\text{proj}_{\mathbb{S}}\,\vec{x} = \vec{x}$$

Hence, $\vec{x}$ is already in $\mathbb{S}$.

E5 (a) Let $A = \begin{bmatrix} 1 & 1 \\ 2 & 2 \end{bmatrix}$. Since $\det A = 0$, it follows that

$$\langle A, A\rangle = \det(AA) = (\det A)(\det A) = 0$$

Hence, it is not an inner product.

(b) We verify that $\langle\,,\rangle$ satisfies the three properties of the inner product:

$$\langle A, A\rangle = a_{11}^2 + 2a_{12}^2 + 2a_{21}^2 + a_{22}^2 \geq 0$$

and equals zero if and only if $A = O_{2,2}$:

$$\begin{aligned}
\langle A, B\rangle &= a_{11}b_{11} + 2a_{12}b_{12} + 2a_{21}b_{21} + a_{22}b_{22} \\
&= b_{11}a_{11} + 2b_{12}a_{12} + 2b_{21}a_{21} + b_{22}a_{22} = \langle B, A\rangle \\
\langle A, sB + tC\rangle &= a_{11}(sb_{11} + tc_{11}) + 2a_{12}(sb_{12} + tc_{12}) + \\
&\quad + 2a_{21}(sb_{21} + tc_{21}) + a_{22}(sb_{22} + tc_{22}) \\
&= s(a_{11}b_{11} + 2a_{12}b_{12} + 2a_{21}b_{21} + a_{22}b_{22}) + \\
&\quad + t(a_{11}c_{11} + 2a_{12}c_{12} + 2a_{21}c_{21} + a_{22}c_{22}) \\
&= s\langle A, B\rangle + t\langle A, C\rangle
\end{aligned}$$

Thus, it is an inner product.

CHAPTER 8

Section 8.1

A Problems

A1 (a) A is symmetric.
(b) B is symmetric.
(c) C is not symmetric since $c_{12} \neq c_{21}$.
(d) D is symmetric.

A2 Alternate correct answers are possible.

(a) $P = \begin{bmatrix} -1/\sqrt{2} & 1/\sqrt{2} \\ 1/\sqrt{2} & 1/\sqrt{2} \end{bmatrix}$, $D = \begin{bmatrix} 3 & 0 \\ 0 & -1 \end{bmatrix}$

(b) $P = \begin{bmatrix} -1/\sqrt{10} & 3/\sqrt{10} \\ 3/\sqrt{10} & 1/\sqrt{10} \end{bmatrix}$, $D = \begin{bmatrix} -4 & 0 \\ 0 & 6 \end{bmatrix}$

(c) $P = \begin{bmatrix} 1/\sqrt{3} & -1/\sqrt{2} & -1/\sqrt{6} \\ 1/\sqrt{3} & 1/\sqrt{2} & -1/\sqrt{6} \\ 1/\sqrt{3} & 0 & 2/\sqrt{6} \end{bmatrix}$, $D = \begin{bmatrix} 2 & 0 & 0 \\ 0 & -1 & 0 \\ 0 & 0 & -1 \end{bmatrix}$

(d) $P = \begin{bmatrix} 2/3 & 2/3 & 1/3 \\ -2/3 & 1/3 & 2/3 \\ 1/3 & -2/3 & 2/3 \end{bmatrix}$, $D = \begin{bmatrix} 0 & 0 & 0 \\ 0 & 3 & 0 \\ 0 & 0 & -3 \end{bmatrix}$

(e) $P = \begin{bmatrix} 1/\sqrt{5} & 4/\sqrt{45} & -2/3 \\ 0 & 5/\sqrt{45} & 2/3 \\ 2/\sqrt{5} & -2/\sqrt{45} & 1/3 \end{bmatrix}$, $D = \begin{bmatrix} 9 & 0 & 0 \\ 0 & 9 & 0 \\ 0 & 0 & -9 \end{bmatrix}$

Section 8.2

A Problems

A1 (a) $Q(x_1, x_2) = x_1^2 + 6x_1x_2 - x_2^2$

(b) $Q(x_1, x_2, x_3) = x_1^2 - 2x_2^2 + 6x_2x_3 - x_3^2$

(c) $Q(x_1, x_2, x_3) = -2x_1^2 + 2x_1x_2 + 2x_1x_3 + x_2^2 - 2x_2x_3$

A2 (a) $A = \begin{bmatrix} 1 & -3/2 \\ -3/2 & 1 \end{bmatrix}$; $Q(\vec{x}) = \frac{5}{2}y_1^2 - \frac{1}{2}y_2^2$, $P = \begin{bmatrix} -1/\sqrt{2} & 1/\sqrt{2} \\ 1/\sqrt{2} & 1/\sqrt{2} \end{bmatrix}$; $Q(\vec{x})$ is indefinite.

(b) $A = \begin{bmatrix} 5 & -2 \\ -2 & 2 \end{bmatrix}$; $Q(\vec{x}) = y_1^2 + 6y_2^2$, $P = \begin{bmatrix} 1/\sqrt{5} & -2/\sqrt{5} \\ 2/\sqrt{5} & 1/\sqrt{5} \end{bmatrix}$; $Q(\vec{x})$ is positive definite.

(c) $A = \begin{bmatrix} -2 & 6 \\ 6 & 7 \end{bmatrix}$; $Q(\vec{x}) = 10y_1^2 - 5y_2^2$, $P = \begin{bmatrix} 1/\sqrt{5} & -2/\sqrt{5} \\ 2/\sqrt{5} & 1/\sqrt{5} \end{bmatrix}$; $Q(\vec{x})$ is indefinite.

(d) $A = \begin{bmatrix} 1 & -1 & 3 \\ -1 & 1 & 3 \\ 3 & 3 & -3 \end{bmatrix}$; $Q(\vec{x}) = 2y_1^2 + 3y_2^2 - 6y_3^2$, $P = \begin{bmatrix} -1/\sqrt{2} & 1/\sqrt{3} & 1/\sqrt{6} \\ 1/\sqrt{2} & 1/\sqrt{3} & 1/\sqrt{6} \\ 0 & 1/\sqrt{3} & -2/\sqrt{6} \end{bmatrix}$; $Q(\vec{x})$ is indefinite.

(e) $A = \begin{bmatrix} -4 & 1 & 0 \\ 1 & -5 & -1 \\ 0 & -1 & -4 \end{bmatrix}$, $Q(\vec{x}) = -3y_1^2 - 6y_2^2 - 4y_3^2$, $P = \begin{bmatrix} -1/\sqrt{3} & -1/\sqrt{6} & 1/\sqrt{2} \\ -1/\sqrt{3} & 2/\sqrt{6} & 0 \\ 1/\sqrt{3} & 1/\sqrt{6} & 1/\sqrt{2} \end{bmatrix}$; $Q(\vec{x})$ is negative definite.

A3 (a) Positive definite.
(b) Positive definite.
(c) Indefinite.
(d) Negative definite.
(e) Positive definite.
(f) Indefinite.

Section 8.3

A Problems

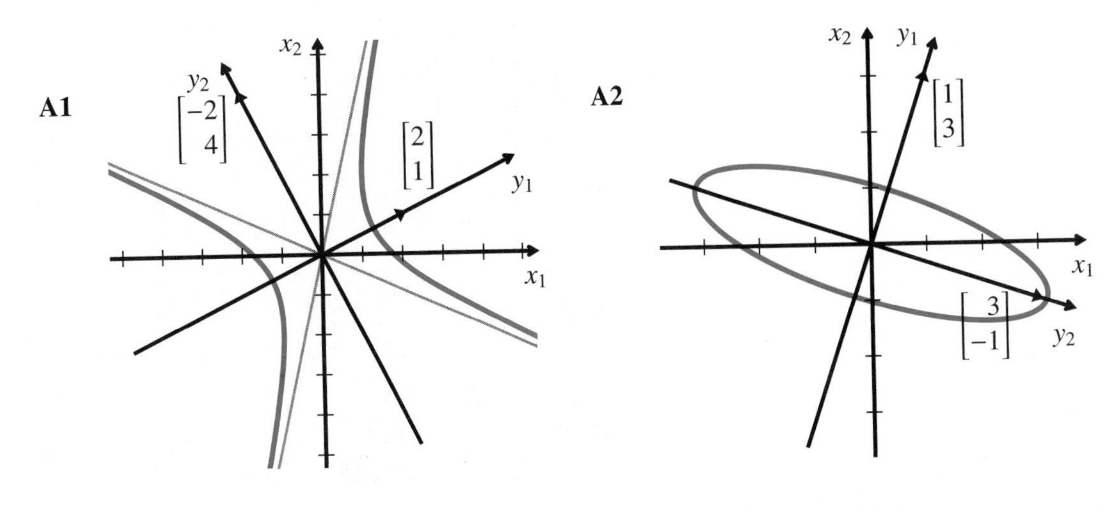

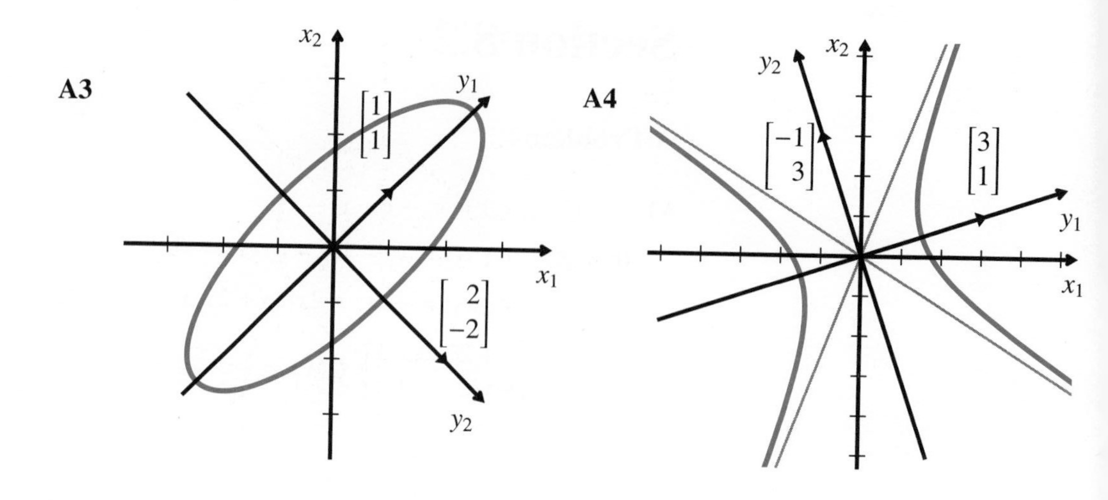

A5 (a) The graph of $\vec{x}^T A\vec{x} = 1$ is a set of two parallel lines. The graph of $\vec{x}^T A\vec{x} = -1$ is the empty set.

(b) The graph of $\vec{x}^T A\vec{x} = 1$ is a hyperbola. The graph of $\vec{x}^T A\vec{x} = -1$ is a hyperbola.

(c) The graph of $\vec{x}^T A\vec{x} = 1$ is a hyperboloid of two sheets. The graph of $\vec{x}^T A\vec{x} = -1$ is a hyperboloid of one sheet.

(d) The graph of $\vec{x}^T A\vec{x} = 1$ is a hyperbolic cylinder. The graph of $\vec{x}^T A\vec{x} = -1$ is a hyperbolic cylinder.

(e) The graph of $\vec{x}^T A\vec{x} = 1$ is a hyperboloid of one sheet. The graph of $\vec{x}^T A\vec{x} = -1$ is a hyperboloid of two sheets.

Chapter 8 Quiz

E Problems

E1 $P = \begin{bmatrix} 1/\sqrt{6} & 1/\sqrt{3} & 1/\sqrt{2} \\ -2/\sqrt{6} & 1/\sqrt{3} & 0 \\ -1/\sqrt{6} & -1/\sqrt{3} & 1/\sqrt{2} \end{bmatrix}$ and $D = \begin{bmatrix} 6 & 0 & 0 \\ 0 & -3 & 0 \\ 0 & 0 & 4 \end{bmatrix}$.

E2 (a) $A = \begin{bmatrix} 5 & 2 \\ 2 & 5 \end{bmatrix}$, $Q(\vec{x}) = 7y_1^2 + 3y_2^2$, and $P = \begin{bmatrix} 1/\sqrt{2} & -1/\sqrt{2} \\ 1/\sqrt{2} & 1/\sqrt{2} \end{bmatrix}$. $Q(\vec{x})$ is positive definite. $Q(\vec{x}) = 1$ is an ellipse, and $Q(\vec{x}) = 0$ is the origin.

(b) $A = \begin{bmatrix} 2 & -3 & -3 \\ -3 & -3 & 2 \\ -3 & 2 & -3 \end{bmatrix}$, $Q(\vec{x}) = -5y_1^2+5y_2^2-4y_3^2$, and $P = \begin{bmatrix} 0 & -2/\sqrt{6} & 1/\sqrt{3} \\ -1/\sqrt{2} & 1/\sqrt{6} & 1/\sqrt{3} \\ 1/\sqrt{2} & 1/\sqrt{6} & 1/\sqrt{3} \end{bmatrix}$.
$Q(\vec{x})$ is indefinite. $Q(\vec{x}) = 1$ is a hyperboloid of two sheets, and $Q(\vec{x}) = 0$ is a cone.

E3

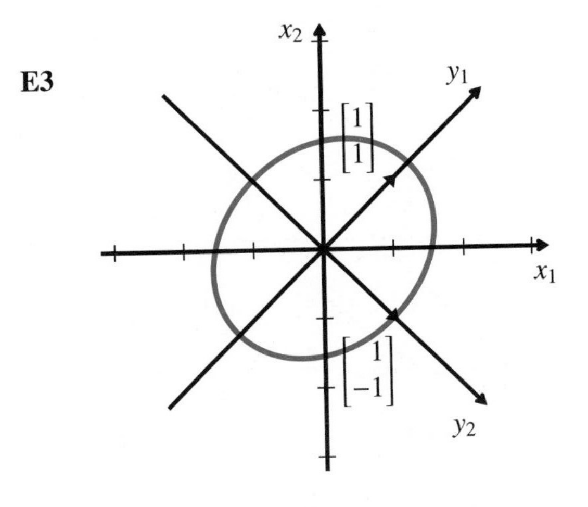

E4 Since A is positive definite, we have

$$\langle \vec{x}, \vec{x} \rangle = \vec{x}^T A \vec{x} \geq 0$$

and $\langle \vec{x}, \vec{x} \rangle = 0$ if and only if $\vec{x} = \vec{0}$.
Since A is symmetric, we have

$$\langle \vec{x}, \vec{y} \rangle = \vec{x}^T A \vec{y} = (\vec{x}^T A \vec{y})^T = \vec{y}^T A^T \vec{x} = \vec{y}^T A \vec{x} = \langle \vec{y}, \vec{x} \rangle$$

For any $\vec{x}, \vec{y}, \vec{z} \in \mathbb{R}^n$ and $s, t \in \mathbb{R}$, we have

$$\begin{aligned}\langle \vec{x}, s\vec{y} + t\vec{z} \rangle &= \vec{x}^T A(s\vec{y} + t\vec{z}) = \vec{x}^T A(s\vec{y}) + \vec{x}^T A(t\vec{z}) \\ &= s\vec{x}^T A\vec{y} + t\vec{x}^T A\vec{z} = s\langle \vec{x}, \vec{y} \rangle + t\langle \vec{x}, \vec{z} \rangle\end{aligned}$$

Thus, $\langle \vec{x}, \vec{y} \rangle$ is an inner product on $\mathbb{R}^n$.

E5 Since A is a 4×4 symmetric matrix, there exists an orthogonal matrix P that diagonalizes A. Since the only eigenvalue of A is 3, we must have $P^T AP = 3I$. Then we multiply on the left by P and on the right by P^T, and we get

$$A = P(3I)P^T = 3PP^T = 3I$$

CHAPTER 9

Section 9.1

A Problems

A1 (a) $5 + 7i$ (b) $-3 - 4i$ (c) $-7 + 2i$ (d) $-14 + 5i$

A2 (a) $9 + 7i$ (b) $-10 - 10i$ (c) $2 + 25i$ (d) -2

A3 (a) $3 + 5i$ (b) $2 - 7i$ (c) 3 (d) $4i$

A4 (a) $\text{Re}(z) = 3$, $\text{Im}(z) = -6$ (b) $\text{Re}(z) = 17$, $\text{Im}(z) = -1$
(c) $\text{Re}(z) = 24/37$, $\text{Im}(z) = 4/37$ (d) $\text{Re}(z) = 0$, $\text{Im}(z) = 1$

A5 (a) $\frac{2}{13} - \frac{3}{13}i$ (b) $\frac{6}{53} + \frac{21}{53}i$ (c) $-\frac{4}{13} - \frac{19}{13}i$ (d) $-\frac{2}{17} + \frac{25}{17}i$

A6 (a) $z_1 z_2 = 2\sqrt{2}\left(\cos\frac{7\pi}{12} + i\sin\frac{7\pi}{12}\right)$, $\frac{z_1}{z_2} = \frac{2}{\sqrt{2}}\left(\cos\frac{-\pi}{12} + i\sin\frac{-\pi}{12}\right)$

(b) $z_1z_2 = 2\sqrt{2}\left(\cos\frac{11\pi}{12} + i\sin\frac{11\pi}{12}\right)$, $\frac{z_1}{z_2} = \frac{2}{\sqrt{2}}\left(\cos\frac{17\pi}{12} + i\sin\frac{17\pi}{12}\right)$

(c) $z_1z_2 = 4 - 7i$, $\frac{z_1}{z_2} = -\frac{8}{13} - \frac{1}{13}i$ (This answer can be checked using Cartesian form.)

(d) $z_1z_2 = -17 + 9i$, $\frac{z_1}{z_2} = -\frac{9}{37} + \frac{3}{37}i$ (This answer can be checked using Cartesian form.)

A7 (a) -4 (b) $-54 - 54i$ (c) $-8 - 8\sqrt{3}i$ (d) $512(\sqrt{3} + i)$

A8 (a) The roots are $\cos\left(\frac{\pi+2k\pi}{5}\right) + i\sin\left(\frac{\pi+2k\pi}{5}\right)$, $0 \leq k \leq 4$.

(b) The roots are $2\left[\cos\left(\frac{-\frac{\pi}{2}+2k\pi}{4}\right) + i\sin\left(\frac{-\frac{\pi}{2}+2k\pi}{4}\right)\right]$, $0 \leq k \leq 3$.

(c) The roots are $2^{1/3}\left[\cos\left(\frac{-\frac{5\pi}{6}+2k\pi}{3}\right) + i\sin\left(\frac{-\frac{5\pi}{6}+2k\pi}{3}\right)\right]$, $0 \leq k \leq 2$.

(d) The roots are $17^{1/6}\left[\cos\left(\frac{\theta+2k\pi}{3}\right) + i\sin\left(\frac{\theta+2k\pi}{3}\right)\right]$, $0 \leq k \leq 2$, where $\theta = \arctan 4$.

Section 9.2

A Problems

A1 (a) The general solution is $\vec{z} = \begin{bmatrix} 1 - i \\ \frac{4}{5} + \frac{2}{5}i \\ -\frac{1}{5} - \frac{3}{5}i \end{bmatrix}$.

(b) The general solution is $\vec{z} = \begin{bmatrix} 5 + i \\ -2 + 2i \\ -i \\ 0 \end{bmatrix} + t\begin{bmatrix} -1 + 2i \\ 0 \\ -i \\ 1 \end{bmatrix}$, $t \in \mathbb{C}$.

Section 9.3

A Problems

A1 (a) $\begin{bmatrix} -5 - 3i \\ i \end{bmatrix}$ (b) $\begin{bmatrix} 5 - 3i \\ 7 + 8i \\ -1 - 9i \end{bmatrix}$ (c) $\begin{bmatrix} -10 + 4i \\ 4 + 6i \end{bmatrix}$ (d) $\begin{bmatrix} -4 - 3i \\ -1 - 7i \\ -12 + i \end{bmatrix}$

A2 (a) $[L] = \begin{bmatrix} 1 + 2i & 3 + i \\ 1 & 1 - i \end{bmatrix}$

(b) $L(2 + 3i, 1 - 4i) = \begin{bmatrix} 3 - 4i \\ -1 - 2i \end{bmatrix}$

(c) A basis for Range(L) is $\left\{\begin{bmatrix} 1 + 2i \\ 1 \end{bmatrix}\right\}$. A basis for Null($L$) is $\left\{\begin{bmatrix} -1 + i \\ 1 \end{bmatrix}\right\}$

A3 (a) A basis for Row(A) is $\left\{\begin{bmatrix} 1 \\ 0 \\ i \end{bmatrix}, \begin{bmatrix} 0 \\ 1 \\ 1 \end{bmatrix}\right\}$. A basis for Col($A$) is $\left\{\begin{bmatrix} 1 + i \\ -2i \end{bmatrix}, \begin{bmatrix} 1 \\ 2i \end{bmatrix}\right\}$, and a basis for Null($A$) is $\left\{\begin{bmatrix} -i \\ -1 \\ 1 \end{bmatrix}\right\}$.

(b) A basis for Row(B) is $\left\{\begin{bmatrix}1\\0\end{bmatrix}, \begin{bmatrix}0\\1\end{bmatrix}\right\}$. A basis for Col($B$) is $\left\{\begin{bmatrix}1\\1+i\\-1\end{bmatrix}, \begin{bmatrix}i\\-1+i\\i\end{bmatrix}\right\}$, and a basis for Null($B$) is the empty set.

(c) A basis for Row(C) is $\left\{\begin{bmatrix}1\\0\\i\\-1\end{bmatrix}, \begin{bmatrix}0\\1\\i\\0\end{bmatrix}\right\}$. A basis for Col($C$) is $\left\{\begin{bmatrix}1\\2\\1+i\end{bmatrix}, \begin{bmatrix}i\\1+2i\\i\end{bmatrix}\right\}$,

and a basis for Null(C) is $\left\{\begin{bmatrix}-i\\-i\\1\\0\end{bmatrix}, \begin{bmatrix}1\\0\\0\\1\end{bmatrix}\right\}$.

Section 9.4

A Problems

A1 (a) $D = \begin{bmatrix}2i & 0\\0 & -2i\end{bmatrix}$, $P = \begin{bmatrix}0 & -2\\1 & 0\end{bmatrix}$, $P^{-1}AP = \begin{bmatrix}0 & 2\\-2 & 0\end{bmatrix}$

(b) $D = \begin{bmatrix}-2+i & 0\\0 & -2-i\end{bmatrix}$, $P = \begin{bmatrix}-1 & -1\\1 & 0\end{bmatrix}$, $P^{-1}AP = \begin{bmatrix}-2 & 1\\-1 & -2\end{bmatrix}$

(c) $D = \begin{bmatrix}0 & 0 & 0\\0 & 1+2i & 0\\0 & 0 & 1-2i\end{bmatrix}$, $P = \begin{bmatrix}1 & 1 & 0\\0 & 0 & 1\\2 & 1 & 0\end{bmatrix}$, $P^{-1}AP = \begin{bmatrix}0 & 0 & 0\\0 & 1 & 2\\0 & -2 & 1\end{bmatrix}$

(d) $D = \begin{bmatrix}1 & 0 & 0\\0 & 2+i & 0\\0 & 0 & 2-i\end{bmatrix}$, $P = \begin{bmatrix}0 & 1 & 2\\1 & 3 & 1\\1 & 5 & 0\end{bmatrix}$, $P^{-1}AP = \begin{bmatrix}1 & 0 & 0\\0 & 2 & 1\\0 & -1 & 2\end{bmatrix}$

Section 9.5

A Problems

A1 (a) $\langle \vec{u}, \vec{v}\rangle = 2 - 5i$, $\langle \vec{v}, \vec{u}\rangle = 2 + 5i$, $\|\vec{u}\| = \sqrt{18}$, $\|\vec{v}\| = \sqrt{33}$

(b) $\langle \vec{u}, \vec{v}\rangle = 6i$, $\langle \vec{v}, \vec{u}\rangle = -6i$, $\|\vec{u}\| = \sqrt{22}$, $\|\vec{v}\| = \sqrt{20}$

(c) $\langle \vec{u}, \vec{v}\rangle = 3 + i$, $\langle \vec{v}, \vec{u}\rangle = 3 - i$, $\|\vec{u}\| = \sqrt{11}$, $\|\vec{v}\| = 2$

(d) $\langle \vec{u}, \vec{v}\rangle = 4 - i$, $\langle \vec{v}, \vec{u}\rangle = 4 + i$, $\|\vec{u}\| = \sqrt{15}$, $\|\vec{v}\| = \sqrt{5}$

A2 (a) A is not unitary. (b) B is unitary. (c) C is unitary. (d) D is unitary.

A3 (a) $\langle \vec{u}, \vec{v}\rangle = 0$ (b) $\begin{bmatrix}\frac{2}{3}+i\\2+\frac{1}{3}i\\3+\frac{4}{3}i\end{bmatrix}$

A4 (a) We have $1 = \det I = \det(U^*U) = \det(U^*)\det U = \overline{\det U}\det U = |\det U|^2$.

(b) The matrix $U = \begin{bmatrix}i & 0\\0 & 1\end{bmatrix}$ is unitary and $\det U = i$.

Section 9.6

A Problems

A1 (a) A is Hermitian. $U = \begin{bmatrix} (\sqrt{2}+i)/\sqrt{12} & (\sqrt{2}+i)/2 \\ -3\sqrt{12} & 1/2 \end{bmatrix}$, $D = \begin{bmatrix} 1 & 0 \\ 0 & 5 \end{bmatrix}$

(b) B is not Hermitian.

(c) C is Hermitian. $U = \begin{bmatrix} (\sqrt{3}-i)/\sqrt{5} & (\sqrt{3}-i)/\sqrt{20} \\ 1/\sqrt{5} & -4/\sqrt{20} \end{bmatrix}$, $D = \begin{bmatrix} 7 & 0 \\ 0 & 2 \end{bmatrix}$

(d) F is Hermitian. $D = \begin{bmatrix} 0 & 0 & 0 \\ 0 & \sqrt{5} & 0 \\ 0 & 0 & -\sqrt{5} \end{bmatrix}$,

$$U = \begin{bmatrix} -2/\sqrt{10} & (3+\sqrt{5})/a & (3-\sqrt{5})/b \\ (1-i)/\sqrt{10} & (1-i)(1+\sqrt{5})/a & (1-i)(1-\sqrt{5})/b \\ 2/\sqrt{10} & 2/a & 2/b \end{bmatrix}$$

where $a \approx 52.361$ and $b \approx 7.639$.

Chapter 9 Quiz

E Problems

E1 $z_1 z_2 = 4\sqrt{2}e^{-i\pi/12}$, $\frac{z_1}{z_2} = \frac{1}{\sqrt{2}}e^{-7\pi/12}$

E2 The square roots of i are $e^{i\pi/4}$ and $e^{i5\pi/4}$.

E3 (a) $\begin{bmatrix} 7-i \\ 3+5i \\ 9+3i \end{bmatrix}$ (b) $\begin{bmatrix} 3+i \\ -i \\ 2 \end{bmatrix}$

(c) $11 + 4i$ (d) $11 - 4i$

(e) $\sqrt{27}$ (f) $\frac{1}{15}\begin{bmatrix} 29-23i \\ 4+11i \\ 22-8i \end{bmatrix}$

E4 (a) $P = \begin{bmatrix} 2-3i & 2+3i \\ 1 & 1 \end{bmatrix}$ and $P^{-1}AP = D = \begin{bmatrix} 2+3i & 0 \\ 0 & 2-3i \end{bmatrix}$

(b) $P = \begin{bmatrix} 2 & -3 \\ 1 & 0 \end{bmatrix}$ and $C = \begin{bmatrix} 2 & 3 \\ -3 & 2 \end{bmatrix}$

E5
$$UU^* = \frac{1}{\sqrt{3}}\begin{bmatrix} 1-i & -i \\ 1 & -1+i \end{bmatrix}\frac{1}{\sqrt{3}}\begin{bmatrix} 1+i & 1 \\ i & -1-i \end{bmatrix} = \begin{bmatrix} 1 & 0 \\ 0 & 1 \end{bmatrix}$$

E6 (a) A is Hermitian if $A^* = A$. Thus, we must have $3 + ki = 3 - i$ and $3 - ki = 3 + i$, which is only true when $k = -1$. Thus, A is Hermitian if and only if $k = -1$.

(b) If $k = -1$, then $A = \begin{bmatrix} 0 & 3-i \\ 3+i & 3 \end{bmatrix}$ and

$$\det(A - \lambda I) = \begin{vmatrix} 0-\lambda & 3-i \\ 3+i & 3-\lambda \end{vmatrix} = (\lambda + 2)(\lambda - 5)$$

Thus the eigenvalues are $\lambda_1 = -2$ and $\lambda_2 = 5$. For $\lambda_1 = -2$,

$$A - \lambda_1 I = \begin{bmatrix} 2 & 3-i \\ 3+i & 5 \end{bmatrix} \sim \begin{bmatrix} 2 & 3-i \\ 0 & 0 \end{bmatrix}$$

Thus, a basis for the eigenspace of λ_1 is $\left\{ \begin{bmatrix} 3-i \\ -2 \end{bmatrix} \right\}$. For $\lambda_2 = 5$,

$$A - \lambda_2 I = \begin{bmatrix} -5 & 3-i \\ 3+i & -2 \end{bmatrix} \sim \begin{bmatrix} -5 & 3-i \\ 0 & 0 \end{bmatrix}$$

Thus, a basis for the eigenspace of λ_2 is $\left\{ \begin{bmatrix} 3-i \\ 5 \end{bmatrix} \right\}$.

We can easily verify that $\left\langle \begin{bmatrix} 3-i \\ -2 \end{bmatrix}, \begin{bmatrix} 3-i \\ 5 \end{bmatrix} \right\rangle = 0$.

Index

M

N

O